Sclerotinia Stem Rot of Crucifers

Volume 1: Biology and Ecology

Professor (Dr) G.S. Saharan
Former Professor and Head
Department of Plant Pathology,
CCS Haryana Agricultural University,
Hisar, 125 004, Haryana, India

Professor (Dr) Naresh Kumar Mehta
Former Associate Dean and Professor,
Department of Plant Pathology,
CCS Haryana Agricultural University,
Hisar-125 004, Haryana, India

Dr Prabhu Dayal Meena
Principal Scientist and In-Charge, Crop Protection Unit
ICAR-Indian Institute of Rapeseed-Mustard Research,
Sewar, Bharatpur-321303, Rajasthan, India

Sclerotinia Stem Rot of Crucifers

Volume 1: Biology and Ecology

Govind Singh Saharan

Naresh K. Mehta

and

Prabhu Dayal Meena

CABI

CABI is a trading name of CAB International

CABI
Nosworthy Way
Wallingford
Oxfordshire OX10 8DE
UK

Tel: +44 (0)1491 832111
E-mail: info@cabi.org
Website: www.cabi.org

CABI
200 Portland Street
Boston
MA 02114
USA

T: +1 (617)682-9015
E-mail: cabi-nao@cabi.org

A catalogue record for this book is available from the British Library, London, UK.

ISBN-13: 9781800629578 (hardback)
9781800629585 (ePDF)
9781800629592 (ePub)

DOI: 10.1079/9781800629592.0000

Commissioning Editor: Rebecca Stubbs
Editorial Assistant: Emma McCann
Production Editor: Rosie Hayden

Typeset by Exeter Premedia Services Pvt Ltd, Chennai, India
Printed and bound by CPI Group (UK) Ltd, Croydon, CR0 4YY

Contents

About the Authors

Professor (Dr) Govind Singh Saharan

Dr Govind Singh Saharan, former Professor and Head, Department of Plant Pathology, CCS HAU, Hisar, India, has conducted research in diverse fields of plant pathology and has published 250 articles in national and international journals. He has been editor/author of several books, monographs, and Crop Production Compendium (CABI). He is on the panel of experts of SAU, ICAR, IARI, CSIR, UGC, and Department of Biotechnology. He has been a visiting professor at University of Alberta, Edmonton, Canada; Agriculture and Agri-Food Canada, Saskatoon, Canada, and Rothamsted Research, Harpenden, UK. He has been President (NZ) of the Indian Phytopathological Society, editor-in-chief of *Journal of Mycology and Plant Pathology*, *Journal of Oilseed Brassica*, and President of the Indian Society of Mycology and Plant Pathology. He has been awarded with Dr Y.L. Nene Outstanding Plant Pathology Teacher Award by ISMPP, Udaipur, and Life Time Achievement Award by the Society for Rapeseed–Mustard Research, Bharatpur, India.

Professor (Dr) Naresh Kumar Mehta

Dr Naresh Kumar Mehta, former Associate Dean, Professor, and consultant faculty, Department of Plant Pathology, CCS HAU, Hisar. He has been completely associated with teaching to undergraduate and postgraduate students. He has conducted research in diverse fields of plant pathology, with a particular focus on rapeseed–mustard crops, and has guided several MSc and PhD students. He has published about 200 research articles and is the editor/author of several books, book chapters, review articles, and teaching manuals. He has been president of INSOPP, Ludhiana, and editor-in-chief of ISMPP in addition to serving as member of editorial boards of various phytopathological societies of India. He has been awarded Dr Y. L. Nene, Outstanding Plant Pathology Teacher Award by ISMPP, Udaipur; Dr T.S. Thind Distinguished Plant Pathologist Award by INSOPP, Ludhiana; and Ms. Manju Utereja Memorial Gold Medal, HAU, Hisar. He is on the panel of experts of SAU, ICAR, UGC, IARI, and member of various national and international committees on plant pathology. He has been a visiting scientist to the University of Alberta, Edmonton, Canada.

Dr Prabhu Dayal Meena

Dr Prabhu Dayal Meena, Principal Scientist (Plant Pathology) and In-Charge, Crop Protection Unit at ICAR–Indian Institute of Rapeseed–Mustard Research, Bharatpur, Rajasthan, India, has conducted research on various aspects of rapeseed–mustard diseases, such as host resistance, management, epidemiology, biology of crucifer pathogens, and also reported new pathogens, that is, *Nigrospora oryzae* (Berk. & Broome) Petch, *Alternaria citri* Ellis & Pears, and *Fusarium equiseti* (Corda) Sacc. He reported for the first time Rhizoctonia leaf spot of groundnut caused by *Rhizoctonia solani* in Rajasthan, India. He has analyzed yield losses and developed various screening techniques for different *Brassica* pathogens. He has been associated with the development of nine high-yielding varieties and registered seven trait-specific germplasms of rapeseed–mustard. He has published 120 research papers, reviews, book chapters, and 16 books. He has been honored with Best Worker Award by ICAR-DRMR, Fellow of ISMPP, PPAI, ISOR, SRMR, IPS, Dr P.R. Kumar Outstanding *Brassica* Scientist Award, Gold Medal Award and Life Time Achievement Award of Society for Rapeseed–Mustard Research (SRMR), JF Dastur Memorial Lecture Award 2022 by IPS, PR Verma Memorial Lecture Award 2022 by ISMPP. He is founder secretary of SRMR, managing editor and chief editor of *Journal of Oilseed Brassica*, and currently a member of the editorial board of *Indian Phytopathology*, and *Journal of Mycology and Plant Pathology*. He has organized five National *Brassica* Conferences, in India, and has been visiting scientist at Rothamsted Research, Harpenden, UK under Indo-UK Collaborative Research on Oilseed *Brassica* crops. He has also guided PhD and ten MSc students.

Foreword

The hemibiotrophic pathogen *Sclerotinia sclerotiorum* has always been a fascinating and interesting subject of research for mycologists, pathologists, and, more recently, molecular biologists since genomic sequences are available for both *Brassica* hosts and the pathogen. More than 900 plant species in 98 families of crucifers have been reported as hosts of *Sclerotinia* all over the world that cause heavy economic losses. It was a challenge to discover the weak links in the disease cycle to manage *Sclerotinia* diseases of a large number of crops. For researchers and students, it has been a matter of concern how to access voluminous literature on *Sclerotinia*, scattered across different articles and reviews in journals, proceedings of symposia and workshops, books, abstracts, and so on, to get a comprehensive picture of this pathogen with a wide host range. The authors of this book have made an excellent attempt to compile all the available information on various aspects of the pathogen infecting crucifers. The information generated so far has been presented in two volumes, each containing ten chapters. Volume 1 contains, after introducing the subject with historical research developments, various aspects, *viz.*, the diseases; symptomatology; disease assessment; its distribution; economic importance; the pathogen; its taxonomy; nomenclature; reproduction; reproductive structures with fine details; infection and pathogenesis; and biochemical, molecular, and physiological aspects of host–pathogen interaction; and seed infection. Volume 2 contains, after introducing the subject with historical developments, various aspects such as disease cycle; epidemiology and forecasting; pathogenic variability; host resistance with sources of resistance; molecular mechanism of resistance; and management strategies including biological control. The inclusion of numerous laboratory and field techniques is an additional quality of the book for researchers, teachers, and students. The use of *Sclerotinia* as a mycoherbicide, phytotoxin, phytoalexins, hypovirulence, resistance to fungicides, volatile compounds of *Sclerotinia*, sporigermin from sclerotia, and *Sclerotinia* as a health hazard problem give futuristic insight to the book. The outlining of future research priorities on biology, ecology, and disease management strategies speaks to the knowledge and experience of the authors.

I congratulate Dr Govind Singh Saharan, former Professor and Head, Department of Plant Pathology; Dr Naresh Kumar Mehta, former Associate Dean and Professor of Plant Pathology, CCS Haryana Agricultural University, Hisar, Haryana; and Dr Prabhu Dayal Meena, Principal Scientist and In-Charge, Crop Protection Unit, ICAR–Indian Institute of Rapeseed–Mustard Research, Bharatpur, Rajasthan, for their stupendous, incredible, and splendid task of bringing out a comprehensive treatise on Sclerotinia, which will provide the fraternity of agriculture with a bounty of knowledge at one edifice. I am sure this book will be of immense help to scientists, teachers, students, extension specialists, and all those who are interested in protecting plant health from Sclerotinia diseases.

Professor (Dr) Chittaranjan Kole
Chairman, Professor Chittaranjan Kole Foundation for Science and Society
President, International Climate Resilient Crop Genomics Consortium
President, International Phytomedomics & Nutriomics Consortium
President, Genome India International

Preface

Sclerotinia is ranked number three among the most devastating and cosmopolitan plant pathogen on crucifers at the global level. More than 60 names have been used to refer to this disease caused by the hemibiotrophic fungal pathogen. The fungus infects more than 900 species of plants worldwide, including important field crops, fruit crops, ornamental plants, trees, shrubs, and numerous weeds. Annual yield losses due to *Sclerotinia* disease exceed several hundred million dollars worldwide. Extensive crop damage, lack of high levels of host resistance, and the general difficulty of managing diseases caused by *Sclerotinia* have been the impetus for sustainable research on this pathogen. Contrary to the earlier notion of the necrotrophic nature of *Sclerotinia*, application of multi-omics technology has revealed a two-phase model, involving an initial brief biotrophic phase followed by a necrotrophic phase, indicating a hemibiotrophic life cycle of this pathogen.

The genus *Sclerotinia*—belonging to phylum *Ascomycota*, class Discomycetes, order Helotiales, family Sclerotiniaceae—has been redefined to include only those species that produce tuberous sclerotia without incorporating host tissue within the sclerotial medulla, develop an apothecial ectal excipulum composed of globose cells, and lack a disseminative conidial state. The taxonomy and nomenclature of 259 epithets previously referred to *Sclerotinia* have been reviewed, with 21 placed in synonymy under the three accepted species and 25 included as imperfectly known. Two hundred and ten epithets have been excluded and either assigned or accepted to other genera. With the increased information available through omics technology on molecular biology, genetics, variability, and epidemiology of these species—as well as with re-examination in the light of microanatomical and cultural characters employed—only three species, that is, *S. sclerotiorum*, *S. minor*, and *S. trifoliorum* have been retained in this genus.

S. sclerotiorum, which causes stem rot in crucifers, has the broadest host range, including both herbaceous and succulent plants and even some woody ornamentals and monocots. Recently, a new species of *Sclerotinia*, known as *Sclerotinia subarctica*, has been observed to be associated with crucifers. Recent advances in genomics and the sequences available on *Brassica* species and *S. sclerotiorum* have provided unique information on the molecular mechanisms of the host–pathogen interaction, revealing pathogen biology, pathogenesis, effector genes, and molecular mechanisms of host resistance in *Brassica* crops.

The authors congratulate and are thankful to all the *Brassica* scientists for their excellent contribution on all aspects of stem rot of crucifers. The compilation of voluminous literature has compelled us to divide the information into two volumes, considering the easy handling of the book by the readers. Volume 1, titled *Sclerotinia Stem Rot of Crucifers: Biology, and Ecology*, with ten chapters, deals with various aspects of *Sclerotinia* such as taxonomy, nomenclature, geographical distribution, economic importance, host range, diseases, symptomatology, disease assessment, reproduction, ultrastructures, infection and pathogenesis, biochemical, molecular and physiological aspects of host–pathogen interaction, and seed infection. Volume 2, titled *Sclerotinia Stem Rot of Crucifers: Epidemiology and Disease Management*, also contains ten chapters, on disease cycle, epidemiology and disease forecasting, pathogenic variability, host resistance including molecular and biometabolomics, and disease management strategies, including biological control. In addition, the latest molecular laboratory and field techniques developed so far for *Sclerotinia* study have been included. Some newly emerging areas of *Sclerotinia* research that are likely to have a bearing on its management— *Sclerotinia* as a mycoherbicide, phytotoxin, phytoalexin elicitor, hypovirulence, volatile compound imitator, sporigermin from sclerotia, resistance to fungicides, and *Sclerotinia* diseases as a health hazard problem—have been discussed.

The subject matter is vividly illustrated with photographs (macroscopic, microscopic, electron micrographs, scanning electron micrographs), drawings, figures, histograms, graphs, tables, and flow charts of techniques to make it more interesting, stimulating, effective, and easy to understand for readers and researchers. Each chapter is organized chronologically, using numbered headings and subheadings, to ensure continuity of the subject matter. The book has been provided with important references and websites that will be helpful in original consultations by *Sclerotinia* researchers, teachers and students, as well as by policy makers.

We are sure that this comprehensive treatise on *S. sclerotiorum* will be of immense use to scientists, teachers, students, and all others in the diagnosis and management of *Sclerotinia* disease of crucifers worldwide.

Govind Singh Saharan
Naresh Kumar Mehta
Prabhu Dayal Meena

Acknowledgments

The authors are grateful to all the scientists, publishers, societies, journals, institutes, websites, and all others whose valuable materials, such as photographs (macroscopic, microscopic, electron micrographs, scanning electron micrographs), drawings, figures, histograms, graphs, tables, and flowcharts, have been reproduced in the present document. The address of the author(s)/source(s) from where material was adopted/produced—such as institutes, websites, societies, and so on—can be obtained through the references, which have been cited in the References section placed at the end of each chapter. We, the authors, hereby acknowledge and formally recognize all contributing authors for their outstanding and valuable publications and articles. Their contributions are duly noted in this manuscript by citing their names alongside each table, figure, plate, and dendrogram, among others. We have sought the necessary permission to reproduce their material in the present work, and are thankful to all the scientists, publishers, journals, institutes, societies, and websites for granting permission. We apologize if we have inadvertently failed to acknowledge their name. We are also grateful to and sincerely acknowledge the help and information provided by the ICAR–Indian Institute of Rapeseed–Mustard Research, Bharatpur, Rajasthan, in updating the present work. The prompt assistance in supplying the required permissions, information, photographs, figures, and consents for reproduction in the current manuscript has been duly recognized, and the authors express their gratitude and sincere thanks to all involved. There might be some errors, mistakes, and shortcomings in this manuscript. We would greatly appreciate suggestions for the further improvement of this publication in the future.

Govind Singh Saharan
Naresh Kumar Mehta
Prabhu Dayal Meena

Abbreviations

aa	amino acid
AA	antifungal activity
ABA	abscisic acid
ABC	ATP-binding cassette
ACC	1-aminocyclopropane-1-carboxylate
ACIAR	Australian Centre for International Agricultural Research
AFLP-RGA	amplified fragment length polymorphism based on resistance gene analogs
AFSs	antifungal substances
AGS	aliphatic glucosinolates
AL	amorphous layer
AMOVA	analysis of molecular variance
AMP	antimicrobial peptide
ANOVA	analysis of variance
AP	apetalous
APX	ascorbate peroxidase
AS	abiotic stress
AS	alternate splicing
ASA	acetyl salicylic acid
ASD	anaerobic soil disinfestation
ASNA	asparagine synthetase enzymes
ASO	allele-specific oligonucleotide
At	*Arabidopsis thaliana*
ATMT	*Agrobacterium tumefaciens*-mediated transformation
ATP	Adenosine triphosphate
AUC	area under the receiver operating characteristics curve
AUDPC	area under disease progress curve
BABA	β-amino butyric acid
BAP	benzylaminopurine
BCA	biological control agent
BGCs	biosynthetic gene clusters
BIK1	*Botrytis*-induced kinase1
BLAST	basic local alignment search tool

BLUP	best linear unbiased prediction
BnMPK4	*B. napus* mitogen-activated protein kinase
BP	biological process
BPT	blue plate test
BRAD	*Brassica* database
CABI	Centre for Agriculture and Bioscience International
cAMP	cyclic adenosine monophosphate
CaMs	calmodulins
CAMTA	calmodulin-binding transcription activators
CARS	competitive adaptive reweighted sampling
CAT	carnitine acetyl transferase
CAT	catalase
CAZymes	carbohydrate-active enzymes
CC	cellular component
CCB	Coomassie Brilliant Blue
CCR	cinnamoyl-CoA reductase
CDA	candidate defense-associated
CDI	cotyledon disease index
CDIPs	cell death-inducing proteins
CDPKs	calcium-dependent protein kinases
CDS	coding sequence
CE	carbohydrate esterase
CF	culture filter
CF	culture filtrate
CFH	culture filtrates treated with heat
CFs	culture filtrates
CG	carpogenic germination
CHS	chalcone synthase
CHS	chalcone synthase
CMC	carboxyl methyl cellulose
CMI	Commonwealth Mycological Institute
CMS	cytoplasmic male sterile
CNGCs	cyclic nucleotide gated ion channels
COI1	coronatine-insensitive protein 1
COL	colonies per dish
CP	cerato-platanin
CPC	coding potential calculator
CPs	Cerato-platanin proteins
CRK	CDPK-related kinase
CTAB	cetyltrimethyl ammonium bromide
CVNH	Cyanovirin-N homology
Cvs	cultivars
Cvt	cytoplasm-to-vacuole targeting
CWDEs	cell wall degradation enzymes
CWDEs	enzymes that break down cell walls
CWS	cell walls
CWT	cell wall thickness
Cx	cellulase
DAI	days after inoculation
DAMP	damage-associated molecular pattern
DAPC	discriminant analysis of principal components
DAS	days after sowing

DCFs	dicarboximide fungicides
DCT	diazinon 6%, captan 18%, and thiophanate-methyl 4%
DEGs	differentially expressed genes
DEP	dielectrophoresis
DEPs	differently expressed proteins
DETC	diethyl-dithiocarbamate
DGE	differential gene expression
DGE	differential gene expression
DH	doubled haploid
DHN	dihydroxynaphthalene
DI	disease incidence
DI	disease index
DIECA	diethyldithiocarbamate
DLD	dihydrolipoyl dehydrogenase
DMATS	dimethylallyl tryptophan synthases
DMI	demethylation inhibitor
DMIs	demethylation inhibitors
DNA	deoxyribonucleic acid
dNTP	deoxynucleotide triphosphate
DOM	dissolved organic matter
DPI	days postinfection
DRMR	Directorate of Rapeseed–Mustard Research
DS	direct standardization
DSI	disease severity index
dsRNA	double-stranded ribonucleic acids
dsRNAs	double-stranded ribonucleic acids
DTF	days to flowering
DW	days to wilt
EC	emulsifiable concentrate
EF	elongation factor
EF-TU	elongation factor Tu
EIN3	ethylene-insensitive 3
ELISA	enzyme-linked immunosorbent assay
EMSA	electrophoretic mobility shift assay
e-PCR	electronic PCR
EPS	exopolysaccharides
EPS	extracellular polysaccharide
EPTC	ethyl dipropylthiocarbamate
ER	endoplasmic reticulum
ERK	extracellular signal-regulated kinase
ERS	ethylene response sensor
ES	exon skipping
ESTs	expressed sequence tags
ET	ethylene
ET/JA	ethylene/jasmonate
ETH	ethephon
ETI	effector-triggered immunity
EV	empty vector
EW	emulsion-in-water
FDR	false discovery rate
FeEDTA	ferric-potassium salt of EDTA
Fig	figure

FISH	fluorescence in situ hybridization
FKH	fork head
FOX	fork head box
FP	fully petalled
FT	flowering time
FWIoU	Frequency Weighted Intersection over Union
GAM	genetic advance mean
GAPDH	glyceraldehyde-3-phosphate-dehydrogenase
GBS	genotyping by sequencing
GBS	glucobrassicin
GBS	glucosinolate glucobrassicin
GC	guanine and cytosine
GCA	general combining ability
GCV	genotypic coefficient of variability
GEBVs	genomic estimated breeding values
GFP	green fluorescent protein
GH	glycoside hydrolase
GIB	glucoiberin
GISH	genomic in situ hybridization
GIV	glucoiberverin
GLM	generalized linear model
GM	genetically modified
GO	gene ontology
GP	Genomic prediction
GR	genetic resistance
GR	glutathione reductase
GS	genomic selection
GS	growth stage
GSH	glutathione
GSLs	glucosinolates
GST	glutathione S-transferase
GTR	glycosyl-transferase
GUS	β-glucuronidase
GWAM	genome-wide association mapping
GWAS	genome-wide association study
HAU	Haryana Agricultural University
HCN	hydrogen cyanide
HEK	human embryonic kidney
Hel	helicase
HGIB	high glucoiberin
HIGS	host-induced gene silencing
HMG	igh mobility group
HPIS	host–pathogen interaction system
HPLC	high-performance liquid chromatography
HR	highly resistant
HR	hypersensitive response
HR	hypersensitive response/reaction/resistance
HRE	heat-released elicitor
HS	highly susceptible
HSD	honest significant difference
HSIN	high sinigrin
HSPs	heat shock proteins

Hsps	heat-shock proteins
HZAU	Huazhong Agricultural University
IA	index of association
IAA	indole-3-acetic acid
IAEA	International Atomic Energy Agency
IAPT	International Association of Plant Taxonomists
ICL1	isocitrate lyase
IDA	inflorescence deficient in abscission
IDM	integrated disease management
IG	inhibition of growth
IGS	indole glucosinolates
IGS	intergenic spacer
IGV	Integrative Genomics Viewer
ILs	introgression lines
Inh	infection hours
IPI	infected petal inoculation
IR	induce resistance
IS	intercellular space
ISI	Inoculation Severity Index
ISR	induced systemic resistance
ITCs	isothiocyanates
ITS	internal transcribed spacer
JA	jasmonic acid
KASP	Kompetitive Allele Specific PCR
KEGG	Kyoto Encyclopedia of Genes and Genomes
KNN	k-nearest neighbors
KOH	potassium hydroxide
L	Linnaeus
LAP	linearazol (in)e-containing peptides
LC-QTOF	liquid chromatography-quadrupole time-of-flight
LC-QTOF	liquid chromatography-quadrupole time-of-flight
LD	linkage disequilibrium
LDIA	leaf disc inoculation assay
LDT	latency time
LFCs	log-fold changes
LG	linkage group
LGIB	low glucoiberin
LL	lesion length
LMM	linear mixed model
LR	leaf resistance
LRR	leucine-rich repeat
LSD	least significant difference
LSIN	low sinigrin
LS-SVM	least squares-support vector machine
LSTM	long short-term memory
LSU	large subunit
LTP	lipid transfer proteins
LTR	long terminal repeats
LU	spring oilseed rape
LW	lesion width
MAF	minor allele frequency
MAM-IL	methylthioalkylmalate isomerase large subunit

MAMPs	microbe-associated molecular patterns
MAPEG	membrane-associated protein in eicosanoid and glutathione metabolism
MAPK	mitogen-activated protein kinase
MAPKK	MAPK kinase
MAPKKK	MAPK kinase kinase
MAS	molecular-assisted selection
Mat	mating-type
MB	methyl bromide
MBC	benzimidazole carbamate
MBC	methyl-2-benzimidazole carbamate
MCD	modified Czapek–Dox broth
MCG	mycelial compatibility group
MCGs	mycelial compatibility groups
MCMC	Markov Chain Monte Carlo model
MDA	malondialdehyde
MDH	malate dehydrogenase
MDHAR	monodehydroascorbate reductase
MDHAR	mono-dehydroascorbate reductase
MDI	mean decrease in impurity
MES	metham sodium
MF	molecular function
MFS	major facilitator superfamily
M-GLM	multivariate generalized linear model
MIC	minimum inhibitory concentration
MKK3	mitogen-activated protein kinase kinase 3
MLHs	multilocus haplotypes
MLM	mixed linear model
MLO2	Mildew Locus O2
MLOs	mycoplasma-like organisms
MLP	major latex-related protein
MPI	mycelial plug inoculation
MPI	mycelial plug inoculation
MPK4	MAP kinase 4 ()
MPKs	mitogen-activated protein kinases
MPPI	Model Predictive Path Integral
MR	moderately resistant
mRNAs	messenger RNA
MS medium	Murashige and Skoog medium
MSL1	maleate synthetase
MT	melatonin
MTA	marker-trait association
mtDNA	mitochondrial DNA
MTI	MAMP-triggered immunity
MTI	mycelial toothpick inoculation
MTIS	melatonin induce resistance
MTR	methyltransferase
NAA	naphthaleneacetic acid
NAD	nicotinamide adenine dinucleotide
NADPH	nicotinamide adenine dinucleotide phosphate
NASC	Nottingham *Arabidopsis* Stock Centre
NbHIR2	*Nicotiana benthamiana* hypersensitive-induced response protein 2
NBS	nucleotide-binding site

NBS-LRR	nucleotide-binding site leucine-rich repeat
NC	nucleotidyl cyclase
NCBI	National Centre for Biotechnology Information
NEB	New England Biolabs
NEP	necrosis and ethylene-inducing protein
nF-EIS	nonfaradaic electrochemical impedance spectroscopy
nF-EIS	non-faradaic electrochemical impedance spectroscopy
NGB	Nordic Gene Bank
NGSEP-GBS	Next-generation Sequencing Eclipse Plug-in for Genotyping-by-Sequencing
NIBIO	Norwegian Institute of Bioeconomy Research
NILs	near isogenic lines
NIR	near-infrared region
NIRS	near-infrared reflectance spectroscopy
NLPs	necrosis- and ethylene-inducing protein 1 (Nep1)-like proteins (NLPs)
NLRs	nucleotide-binding site leucine-rich repeats
NO	nitric oxide
NOX	nicotinamide adenine dinucleotide phosphate (NADPH) oxidases
NPR1	nonexpressor of pathogenesis-related genes 1
NRPS	nonribosomal peptide synthetase
nsLTPs	nonspecific lipid transfer proteins
OA	oxalic acid
OAH	oxaloacetate acetyl hydrolase
OCRI	Oil Crops Research Institute, Chinese Academy of Agricultural Sciences
OE	overexpression
OER	oxalate exchanger-related
OFR	oxygen free radicals
OGs	oligogalacturonides
OMV3a	*Ophiostoma mitovirus* 3a
ORF	open reading flame
OSR	oilseed rape
OxDC	oxalate decarboxylase
P	calculated probability
PA	phosphatidic acid
PAD3	phytoalexin-deficient 3
PAL	phenylalanine ammonia lyase
PAMP	pathogen-associated molecular pattern
PAS	periodic acid–Schiff reaction
PAU	Punjab Agricultural University
PBD	phloem-based defense
PC	principal components
PCA	principal component analysis
PCD	programmed cell death
PCNB	pentachloronitrobenzene
PCR	polymerase chain reaction
PCR	polymerase chain reaction
PDA	potato dextrose agar
PDB	potato dextrose broth
PDB	potato dextrose broth
PDI	percent disease index
PDI	protein disulfide isomerase
PEIs	pectinesterase inhibitors
pFDR	positive false discovery rate

PFGE	pulsed field gel electrophoresis
PFGE	pulsed-field gel electrophoresis
PG	polygalacturonase
PG	polygalacturonase
PGIP	polygalacturonase-inhibiting protein
PGP	plant growth promotion
PGTE	polygalacturonase transeliminase
PI	petal infestations
PKA	protein kinase A
PKS	polyketide synthase
PLS-DA	partial least-squares discriminant analysis
PM	plasma membrane
PME	pectin transeliminase
PMEIs	pectin methylesterase inhibitors
PMG	polymethyl-galacturonase
PO	peroxidase
POD	peroxidase
POQR	prolyl oligopeptidase associated with quantitative disease resistance
PPI	percent petal infection
PPI	percentage petal infestation
PPO	polyphenol oxidase
PR	pathogenesis related
PRRs	pattern recognition receptors
PTGS	posttranscriptional gene silencing
PTI	PAMPs-triggered immunity
PTI	pathogen-associated molecular pattern-triggered immunity
PTI	pattern-triggered immunity
QDR	quantitative disease resistance
qPCR	quantitative polymerase chain reaction
qRT-PCR	quantitative real-time polymerase chain reaction
QTL	quantitative trait loci
R	resistance
RA	resistance assessment
RACE	rapid amplification of cDNA ends
RALFs	rapid alkalinization factors
RAPD	random amplification of polymorphic DNA
RCP	rape crop phase
RCPS	representative concentration pathway scenario
rDNA	ribosomal deoxyribonucleic acid
RdRp	RNA-dependent RNA polymerase
REML	residual maximum likelihood
RF	random frog
RFLP	restricted fragment length polymorphism
RGAs	resistance gene analogs
RH	relative humidity
RILs	recombinant inbred lines
RIP	ribonucleoprotein immunoprecipitation
RLK	receptor-like kinase 902
RLKs	receptor-like kinases
RLPs	receptor-like proteins
RNA	ribonucleic acid
RNAi	RNA interference

RNAP II	RNA polymerase II
RNASeq	ribonucleic acid sequencing
ROS	reactive oxygen species
ROS	reactive oxygen species
rrBLUP	ridge regression best linear unbiased prediction
RR-BLUP	ridge regression BLUP
rRNA	ribosomal ribonucleic acid
RT-PCR	reverse transcription-polymerase chain reaction
RT-qPCR	real-time quantitative PCR
S	susceptible
SA	salicylic acid
SAMs	shoot apical meristem
SAR	systemic acquired resistance
SBI	sterol biosynthesis inhibitors
SCA	specific combining ability
SCFE1	*Sclerotinia* culture filtrate elicitor 1
SD	standard deviation
SD	stem diameter
SDH	succinate dehydrogenase
SDHIs	succinate dehydrogenase inhibitors
SDS	sodium dodecyl sulfate
SE	standard error
SEM	scanning electron microscopy
SIGS	spray-induced gene silencing
SIMCA	soft independent modeling of class analogies
SIN	sinigrin
SLL	stem lesion length
SMW	standard meteorological week
SNN	spiking neural network
SNP	single nucleotide polymorphism
SOD	superoxide dismutase
SPA	successive projections algorithm
SPSMR	stem-physical-strength-mediated-resistance
SPSS	statistical package for the social sciences
SR	stem resistance
SRAP	sequence-related amplified polymorphism
SRC	*Sclerotinia* rot of carrot
SsDRV	*S. sclerotiorum* debilitation-associated RNA virus
SsHADV-1	*S. sclerotiorum* hypovirulence-associated DNA virus-1
SSITL	*Sclerotinia sclerotiorum integrin-like*
SSM	semiselective medium
SSR	*Sclerotinia* stem rot
SSR	simple sequence repeat
SsRV-L	*S. sclerotiorum* RNA virus L
ST	seed treatment
SVF1	survival factor 1
SVM	support vector machine
SWIR	shortwave infrared
TCA	tricarboxylic acid
T-DNA	transfer DNA
TE	transposable elements
TEM	transmission electron microscopy

TFs	transcription factors
TLPs	thaumatin-like proteins
TMM	trimmed mean of means
TMT	Tandem Mass Tags
TMV	tobacco mosaic virus
Tor	target of rapamycin
TR	*Trichoderma roseum*
TuMV	Turnip mosaic virus
UAV	unmanned aerial vehicle
UK	United Kingdom
UPGMA	unweighted pair group mean analysis
UPGMA	unweighted pair group method with arithmetic mean
USA	United States of America
USDA	United States Department of Agriculture
UTR	untranslated region
UV	ultraviolet
VCF	variant call format
VIGS	virus-induced gene silencing
VOC's	volatile organic compounds
WAK1	wall-associated kinase 1
WapA	wall-associated protein AWEGOWeb Gene Ontology
WG	
WGCNA	weighted gene co-expression network analysis
WGD	whole-genome duplications
WOR	winter oilseed rape
WP	wettable powder
WT	wild type
XLGs	extra-large G proteins
YD	tyrosine–aspartate
ZFP	zinc finger protein
ZIM	zinc finger inflorescence meristem
Symbols	
@	at
β	beta
%	percentage
≈	is approximately equal to
/	per
:	ratio
~	approximately
+	plus
<	less than
>	more than
=	is equal to
±	plus or minus sign
×	multiply
≤	less than or equal to
≥	more than or equal to
μg	microgram
$\mu g\ ml^{-1}$	microgram per milliliter
μl	microliter
μm	micrometer
μmol	micromole

a.m.	ante meridiem
a.i. ha^{-1}	active ingredient per hectare
Al	aluminum
Al^{3+}	aluminum cation
B	boron
bp	base pair
Ca	calcium
Ca^{2+}	calcium ion
Cl^-	chlorine
CO2	carbon dioxide
cfu	colony-forming unit
cm	centimeter
cm^2	square centimeter
cm^{-2}	per square centimeter
cm^{-3}	per cubic centimeter
CuZn	zinc–copper couple
Cu	copper
cvs.	cultivars
dia	diameter
dpi	days postinoculation
EC50	median effective concentration
e.g.	for example
et al.	"et alia" meaning "and others"
etc.	etcetera; and so on
Fe	iron
ft	feet
ft^2	square feet
g	gram
g^{-1}	per gram
g m^{-2}	grams per square meter
g kg^{-1}	grams per kilogram
g l a.i.$^{-1}$	grams per liter active ingredient
h	hours
HCO3	bicarbonate
H2O2	hydrogen peroxide
hp	hairpin
hpi	hours postinoculation
i.e.	id est, meaning "that is"
K	potassium
kb	kilobase
kg ha^{-1}	kilograms per hectare
kg	kilogram
kHz	kilohertz
kPa	kilopascals
l	liter
l^{-1}	per liter
LD50	median lethal dose
m	meter
M	molar
m^2	square meter
m^{-3}	per cubic meter
MB	methyl bromide

Mb	megabases
Mg	magnesium
mg	milligram
MgCl2	magnesium chloride
Mha	million hectares
min	minute
ml	milliliter
ml m^{-2}	milliliter per square meter
ml^{-1}	per milliliter
mm	millimeter
mM	millimolar
mmole	millimoles
Mn	manganese
Mo	molybdenum
mol m^{-2} s^{-1}	mol per square meter per second
mol wt	molecular weight
MPa	megapascals
N	nitrogen
Na^+	sodium ion
NaCl	sodium chloride
nm	nanometer
nmol/g	nanomole per gram
°C	degree Celsius
pH	potential of hydrogen
p.m.	post meridiem
ppm	parts per million
RH	relative humidity
R^2	coefficient of determination
S	sulfur
SO4	sulphate ion
S	susceptible
s	second
s^{-1}	per second
Subsp.	subspecies
t/ha	tonne per hectare
v/v	volume by volume
var	variety
viz.	videlicet, meaning “namely” or “which is”
w/v	weight by volume
μM	micromolar
Σ	sigma, meaning “sum”
Zn	zinc

1 Historical Research Developments in the Biology and Ecology of *Sclerotinia*

Abstract

Sclerotinia sclerotiorum (Lib.) de Bary was established by Whetzel in 1945 and later recognized as a highly adaptable and popular fungus that infected various plants. In 1870, Fuckel described the genus *Sclerotinia*, and in 1837, Libert referred to it as *Peziza sclerotiorum*. It has a wide host range, and a complex life cycle, employing consecutive hemibiotrophic infection strategies to invade and colonize its host plants. Historical research developments have established the status of *Sclerotinia* species, and development of asexual and sexual structures like sclerotia, apothecia, and ascospores in its life cycle. The analysis of *S. sclerotiorum* secondary metabolites and use of multi-omics technology have aided in understanding the *Brassica–Sclerotinia* pathosystem in more holistic, targeted, and accurate ways. Over the last 187 years (1837–2024), the research progress of *Sclerotinia* has been assessed through several parameters, including the key research areas, resources devoted by more than 142 research organizations, and disease management strategies adopted throughout the world.

1.1 Introduction

In the global agricultural economy, crucifers are a major group of oilseeds and vegetable crops contributing significantly to edible quality oil and vegetable production along with fodder for animals with high-quality nutrition. These groups of crops encounter various biotic stresses due to fungi, bacteria, viruses, viroids, Mycoplasma Like Organism (MLOs), and nematodes, along with different abiotic stresses that are responsible for the reduction in crop yield and seed quality. The pathogen *Sclerotinia* has a very wide host range, is a notable plant pathogen (Schwartz *et al.*, 1978; Purdy, 1979; Lumsden, 1979; Adams and Ayers, 1979; Adams and Wong, 1991; Boland and Hall, 1994; Saharan and Mehta, 2008; Derbyshire *et al.*, 2022; Mehta *et al.*, 2023), and poses a threat to the effective management of the disease it causes. *Sclerotinia*, a fascinating hemibioroph fungal pathogen, is classified based on its existence and interface with the plant species it infects, as it utilizes a sequential method of biotrophic and necrotrophic infection tactics to infiltrate and establish colonies within cruciferous host plants. In the life cycle of *Sclerotinia*, the shift from the biotrophic stage to the destructive necrotrophic stage is referred to as the biotrophy–necrotrophy transition (Walker, 1969; Lumsden, 1979; Steadman, 1979; Saharan and Mehta, 2008; Saharan *et al.*, 2021a, b; Hossain *et al.*, 2023). This particular fungus is commonly found in areas that tend to have temperate and humid conditions (Purdy, 1979). However, there have been reports of its occurrence in certain semiarid regions, where the conditions would typically be considered unfavorable for disease development (Blad *et al.*, 1978; Purdy, 1979). In the arid regions, disease development and outbreaks take place in irrigated fields during the summer months. This is because irrigation creates microclimatic conditions, which are conducive to the development of disease, even if the overall macroclimate is unfavorable (Rotem and Palti, 1969; Blad *et al.*, 1978). Furthermore, the ailment is also prevalent throughout the cooler and humid winter season. Seed yield losses in susceptible crops can vary, reaching as high as 100% in some cases (Purdy, 1979). The *Sclerotinia* rot disease is typically considered a simple interest disease, with ascospores and microconidia serving as the primary inoculum. However, the specific role of these spores in the pathogen's life cycle remains unknown, as indicated by Walker (1969) and Willetts and Wong (1980). Furthermore, the disease can also spread through the growth of mycelium

 Sclerotinia Stem Rot of Crucifers Volume 1: Biology and Ecology (G.S. Saharan *et al.*)
DOI: 10.1079/9781800629592.0001

near the bases of the stems or through contact between infected plants and adjacent healthy plants during the later stages of the crop's growth period (Huang and Hoes, 1980; Morrall and Dueck, 1982). The manipulation of plant sphingolipid pathways by the *Sclerotinia sclerotiorum*, regulated by the *LAZ 5* (leucine-rich repeat) gene, is responsible for triggering programmed cell death and causing disease development (Barbacci *et al.*, 2020). *S. sclerotiorum* possesses several noteworthy attributes, such as the ability to infect a diverse array of hosts, particularly dicotyledonous plants. Additionally, it exhibits a relatively gradual rate of evolution and is known for generating protein effectors that display activity across multiple host species. The hosts' resistance against *S. sclerotiorum* is highly intricate and typically involves numerous polymorphisms with minimal effects, posing a significant challenge to the development of disease-resistant crop varieties (Derbyshire *et al.*, 2022; Chen *et al.*, 2023).

1.2 *Sclerotinia* Species as Plant Pathogens

Several species of *Sclerotinia* that are pathogenic to plants have been identified. Nevertheless, the very broad host range, distribution, and substantial produce losses caused in economically important crops have led to the recognition of three species of *Sclerotinia* as having significant consequences. These species include *S. sclerotiorum*, *Sclerotinia minor*, and *Sclerotinia trifoliorum*. Based on Kohn's (1979) monographic modification of *Sclerotinia*, it is proposed that *Sclerotinia homoeocarpa*, the pathogen responsible for dollar spot in turf grasses, could potentially be more appropriately categorized under *Lanzia* sp. or *Moellerodiscus* sp. Among the three species, *S. sclerotiorum* and *S. trifoliorum* have large sclerotia that grow terminally through the frequent branching of prime hyphae. In spring, *S. sclerotiorum* forms apothecia on the soil surface and has a wide range of food sources, while *S. trifoliorum* does the same in autumn and is restricted to alfalfa and clover species. In contrast, *S. minor* exhibits smaller sclerotia that emerge sideways through the repeated division of short aerial hyphae. Initially, it demonstrates less frequent coalescence compared to the aforementioned two species. The differentiation between these taxa is additionally reinforced by observations indicating unrestricted expansion in the contact zone when nonspecific isolates encounter hyphal contact. Conversely, incompatibility reactions have been documented between *S. sclerotiorum*, *S. minor*, and *S. trifoliorum*.

These three *Sclerotinia* species could be distinguished from one another by the electrophoretic dissolution of their proteins, which also revealed slight variation within each species. Furthermore, these *Sclerotinia* species exhibited differences in the electrophoretic profiles of soluble proteins and enzymes, such as aryl esterase, acid phosphatase, tetrazolium oxidase, glucose-6-phosphate dehydrogenase, and NADPH (nicotinamide adenine dinucleotide phosphate) phosphate dehydrogenase. Furthermore, different Canadian isolates of *S. sclerotiorum* showed substantial divergence in morphology, pathogenesis, and physiology, while cultures from different hosts in the UK showed a more progressive spectrum of diversity. Lower temperatures further encouraged the formation of unsaturated lipids in sclerotia formed from diseased plant parts, which had a higher proportion of unsaturated lipids than those created in a controlled environment. Usually, oleic acid made up the majority of the sclerotia, with smaller levels of palmitic and linoleic acids.

1.3 *Sclerotinia sclerotiorum* as a Pathogen of Crucifers

The *S. sclerotiorum* is an extensively polyphagous plant pathogen. There is a wealth of phytopathological information pertaining to this organism that is accessible globally through various mediums such as research articles, books, reviews, leaflets, bulletins, websites, and various databases. This pathogen is distributed in temperate zones across the globe and infects numerous host species. However, it primarily affects agriculturally important plants like beans, potatoes, lettuce, sunflowers, rapeseed, safflower, soybeans, peanuts, carrots, linseed, eggplants, crucifers (cabbages, cauliflowers), tomatoes, celery, chickpeas, peas, lentils, buckwheat, capsicum, opium poppies, and various other vegetables. Further, it has the ability to infect a substantial number of decorative plants, undesirable vegetation, and therapeutic flora. The *Sclerotinia* produces resting structures called sclerotia, as well as mushroom-like reproductive structures known as apothecia. During its life cycle, it produces ascospores. The combination of dry and wet sieving methods allows for the quantitative collection of sclerotia from infested soils. To determine their viability, freshly cut carrot disks can be used as a testing medium. However, it is uncommon to find reports of mycelial isolations from soil. The fruiting structure, or ascospore, is considered the chief source of inoculum that infects aboveground plant parts, specifically entering the leaf axis through the spores present on the petals. These ascospores act as a primary energy source, disintegrating within a period of 72 hour, which facilitates the pathogen's further spread. They have a remarkable ability to persist for up to 7 months under minimal humidity conditions, especially in well-maintained environments, and can germinate even at osmotic pressure potentials as low as −56 bars. However, they are unable to germinate at −91 bars. Sclerotia, on the contrary, can be produced *in vitro* conditions at osmotic pressure potentials ranging from −1 to −64 bars. These sclerotia primarily enter the soil through decayed plant debris, and in clay soil, they can produce daughter sclerotia. The pathogen *S. sclerotiorum* has been found to be present in the seeds of minimum 27 different host genera, indicating its seed-borne nature. Additionally, the sclerotia of this fungus can be spread through a variety of seeds, with a particular affinity for cruciferous species. Normally, the sclerotia can perpetuate and survive in the soil for approximately 5–7 years, or even up to 10 years, particularly under desiccated situations. The survival of sclerotia is adversely affected by higher soil temperatures and moisture levels; however, it is more influenced by microbial activities rather than temperature and humidity. Sclerotia that are buried at depths of 5–20 cm in the soil can remain viable and produce fruiting bodies known as apothecia even after 3–5 years, while sclerotia closer to the soil surface decompose fast. Nevertheless, apothecia are not formed by sclerotia when the soil depth exceeds 5 cm, and the quantity of apothecia derived from a sclerotium diminishes with increasing depth. A previously named species of *Sclerotinia*, known as *Sclerotinia subarctica* nom. prev. (*Sclerotinia* species 1), has been discovered in Norway, causing *Sclerotinia* rot on *Brassica rapa* subsp. *oleifera* (Brodal *et al.*, 2017), and in the UK (Clarkson *et al.*, 2010), affecting *Ranunculus acris* (meadow buttercup). Nevertheless, there exists a dearth of exhaustive data pertaining to its pathogenesis, epidemiology, and effector genes.

1.4 Growth of Fruiting Bodies of *Sclerotinia sclerotiorum*

Apothecia, which are fruiting bodies, can be formed from sclerotia. The sclerotia are located underground in the soil, at a maximum depth of 5 cm, whether they are outdoors or in a controlled environment. In order to generate apothecia, fully developed sclerotia are subjected to low temperatures (ranging from 0 to 3°C) for a period of two to several months. Subsequently, they are placed on a damp substrate with limited nutrients and exposed to temperatures ranging between 10 and 15°C, under illuminated conditions. It has been observed that maintaining a constant temperature of 12–21°C for a duration of 60–90 days during the maturation of the sclerotium leads to the most efficient reduction in germination. On the other hand, temperatures ranging from 15 to 20°C lead to 10% abnormal fruiting bodies. Light plays a vital role not only in the differentiation of disks but also in promoting the growth of stipe initiation. Apothecia are exclusively formed in soils that are either saturated or nearly saturated with moisture. In heavy clay, a minimum water potential of −7.5 bars at 15°C is required for this process. Numerous enzymes, including glucose-6-phosphate dehydrogenases, laccase, esterases, tyrosinase, succinate, and alkaline and acid phosphatases, are found in small amounts in elderly sclerotia but in markedly increased amounts in developing apothecial initials, hypothecium, and stipes. With a rise in 1, 3-β-D-glucanase activity, germination uses the sclerotial 1, 3-β-D-glucanase. Carpogenic sclerotium germination is selectively restricted by various factors such as EDTA at 1×10^{-3} molar; glucose and sugars at a concentration of 0.25%; Cd^{2+} at 5×10^{-6}, NaCl at 5×10^{-3}, Hg^{2+} at 5×10^{-5}, CO^{2+} at 10^{-4}, Ni^{2+} at 5×10^{-5}, and Tris buffer at 2.5×10^{-3} molar concentrations; as well as by fungicides like benomyl, dicloran, and some other chemicals. A device used to collect quantitative ascospores is known as a Millipore filter. It works by using a funnel connected to a vacuum pump, positioned towards sporulating apothecia, to gather ascospores under pressure. To trigger discharge, a brief air flush is administered every 3 hour. High relative humidity is required for optimal ascospore germination, particularly when temperatures range between 5 and 10°C. In the absence of light, there is a significant decrease in the formation of sclerotia, with only about half as many being formed compared to normal light intensity ranging from 40 to 100%. The formation of sclerotia is influenced by temperature, with a negative correlation between the size and number of sclerotia and at different temperatures. On the other hand, the rate of growth and creation of sclerotia are positively correlated with the diameter of the growth area. In large plates, sclerotia are formed in circular rings that are spaced 0.5 cm apart. Acid-staling compounds can induce sclerotium formation *in vitro*. Only aggressively growing hyphae from the edge of the culture in plates exhibit matching patterns of sclerotium formation in cultures under laboratory conditions. Sclerotia can form within the osmotic pressure potential range of −1 to −64 bars, but it is absent at −73 bars. Mycelial growth is strongly increased as the osmotic pressure potential decreases from −1 to −14 bars, but it declines below the previous value. The zone between −37 and −47 bars is where the ideal growth rate is reached, where the growth rate is half of the maximum rate. There have been reports indicating that sclerotium formation can be supported by a variety of carbon sources, including lactose, maltose, D-galactose, sucrose, D-mannose, D-glucose, D-fructose, raffinose, lactose, and L-arabinose. On the other hand, certain sugars, along with specific inorganic ions and organic compounds, have the ability to be discharged from developing or desiccated sclerotia. Throughout the sclerotium formation process, a notable amount of CO_2 is emitted and a substantial amount of glucose is assimilated. The ideal temperature range for mycelial growth varies between 15 and 25°C, depending upon the particular isolate. Similarly, the ideal pH for growth falls between 4 and 5.5. Further, apart from the above mentioned carbon sources like inulin, araban, and starch, growth is also accelerated, thereby promoting the formation of sclerotium. The accumulation of metabolites does not impede the growth of mycelium over long distances in agar tubes. The breakdown of pectin involves the identification of endo-polygalacturonase, along with some pectin methylesterase, exo-polygalacturonase, and pectin transeliminase. The growth on sterilized bean hypocotyls shows the formation of phospholipase A2, which functions best at a pH of 4.0, and cellulases, which work best at a pH of 3.0. Other identified enzymes include arabanase, arbutine-splitting glucosidase, trehalase bound to mycelium, D-mannitol: NAD (nicotinamide adenine dinucleotide) oxidoreductase, D-mannitol-1-phosphate: NAD oxidoreductase, pentitol oxidoreductase (reportedly detected in cell-free extracts of mycelia and sclerotia grown on medium containing D-xylose, L-arabinose, or D-ribose), oxalacetate acetylhydrolase, citrate synthetase, malate dehydrogenase, aconitate hydratase, fumarate hydratase, and acid phosphatase. The following substances have been recognized and documented as appropriate sources of nitrogen (N): glycine, aspartic and glutamic acids, tyrosine, peptone, sodium and potassium nitrate, alanine, serine, and glycine. But substances like urea, cysteine, lysine, and methionine are not good for growth or apothecia production. In liquid culture, growth is hindered by deficiencies in phosphate, magnesium, and trace elements. Additionally, substantial progress does not occur once there are deficiencies in nitrogen and potassium. As the *Sclerotinia* develops, it abundantly produces oxalic acid, fumaric acid, succinic acid, and glycolic acid. Mannitol and phenols have been identified as metabolites that inhibit the growth of various organisms like *Bacillus subtilis*, *Fusarium oxysporum*, and pathogen *S. sclerotiorum*. Additionally, the mycotoxins 8-methoxypsoralen and 4,5,8-trimethyl-psoralen, both of which can cause skin problems like dermatitis in humans, are also present. It is worth noting that *Sclerotinia* possesses antibiotic properties against the pathogen *Staphylococcus aureus*. In higher plants, it has been discovered that certain compounds, such as the isocoumarin derivative sclerin, as well as sclerotinin A and B, sclerone, and isosclerone, have growth-promoting qualities at low concentrations and inhibitory effects at greater concentrations. Sclerin synthesis is directly linked to the development and coloration of sclerotium, and it is this substance that appears to drive the actions of peroxidase and phenoloxidase. Remarkably, even at a low pH of 3.7, the fungus *S. sclerotiorum* has a great tolerance to chemical Al^{3+}. On sunflower, however, the phytotoxic concentrations of the chemical Al3 +can increase its virulence (Domsch *et al.*, 1980; Livingstone *et al.*, 2005; Saharan and Mehta, 2008).

1.5 Terminology of *Sclerotinia* Diseases

Sclerotinia causes a variety of plant diseases, such as watery soft rot, leaf mold, white mold, cottony rot, stem rot, crown rot, wilt, white blight, blossom blight, twig blight, stem canker, stalk break, damping off, drop, and head rot. Three different species of *Sclerotinia* (*S. sclerotiorum* (Lib.) de Bary, *S. minor* Jagger, *S. trifoliorum* Erikss.) that affect numerous crops worldwide are responsible for these diseases. However, the specific *Sclerotinia* disease that affects crucifers is generally known as *Sclerotinia* rot of crucifers or *Sclerotinia* stem rot of crucifers. The disease is able to cause significant and random yield losses, both in the field and during storage as well. The disease incidence varies from minimal levels to 100% across different regions. The variations in incidence are attributed to various factors, such as agronomic practices, inoculum density, soil type and texture, drainage, environmental conditions, and disease management practices being used in different areas (Purdy, 1979). Unfortunately, there are no simple methods or techniques which can completely control *Sclerotinia* diseases. However, a well-implemented eco-friendly integrated approach can effectively control the disease and minimize losses in seed yield (Saharan and Mehta, 2008).

1.6 Major Symptoms of *Sclerotinia* Diseases

The symptoms resulting from *Sclerotinia* infection can differ to some degree depending on the affected plant or plant part, as well as the predominant ecological conditions. The initial and most noticeable sign of *Sclerotinia* diseases is the emergence of a white, fluffy mycelial growth, which is rapidly followed by the formation of big, solid inactive structures or sclerotia on the diseased plant. Initially, the stems of diseased succulent and herbaceous hosts may display dark or light/pale brown wounds near their base. The fungal mycelium frequently conceals these lesions with white cottony patches. During the initial phases of stem infection, the foliage might not display noticeable indications of infection, leading to the possibility of overlooking the existence of infected plants until the fungus completely infiltrates stem and initiates decaying. Leaves situated above the affected area wither and perish swiftly. In specific instances, the contamination might initiate on leaf and subsequently extend to the stem across the leaf. The pathogen's sclerotia can either form internally within the stem's pith, giving no visible indication of their presence, or they can form externally on the stem, where they are easily noticeable. The succulent foliage and stalks abruptly wilt and perish when the pathogen infiltrates the stem's foundation, and the lower leaves may also shrivel. The pathogen infiltrates and disseminates across the stem, ultimately leading to the demise and collapse of the entire plant. Consequently, each leaf hangs downward until it rests on the leaf beneath it. The decayed sections of the plant exhibit fluffy, white mycelium and sclerotia. Infected cruciferous plants can be identified in the field by their dried appearance, with few or no shriveled pods and foliage (Saharan and Mehta, 2008).

1.7 Establishment of *Sclerotinia* as Pathogen

Extensive research has been conducted on the disease caused by *Sclerotinia*. The causes, characteristics, spread, and control methods of significant *Sclerotinia* diseases have been thoroughly investigated and documented in various sources such as periodicals, conferences, books, and investigation publications (Abawi and Grogan, 1979; Adams and Ayers, 1979; Kohn, 1979; Le Tourneau, 1979; Lumsden, 1979; Purdy, 1979; Phillips, 1986; Rhodes *et al.*, 1992; Saharan and Mehta, 2008; Hossain *et al.*,

2023). To enhance understanding of the biology of the fungus *S. sclerotiorum*, reporter genes like the green fluorescent protein (*GFP*) gene have been utilized in both laboratory environments and plants (Lorang *et al.*, 2001; Guimarães and Stotz, 2004; de Silva *et al.*, 2005). The fungus overwinters as vegetative structures like hyphae, mycelium in either diseased debris or healthy living plants, or as sclerotia on or within diseased areas, or as resting structures known as sclerotia that have dropped onto the ground. Sclerotia found in the soil have the ability to sprout and give rise to slender stalks, ultimately forming small, cup-shaped apothecia measuring 5–15 mm in diameter. These apothecia produce asci and ascospores. The initial infections are triggered by ascospores, which require moisture and external sources of energy from fallen flower petals and necrotic tissues to germinate (Steadman, 1979). Leaves contaminated with ascospores that come into contact with moist soil, as well as leaves that come into contact with sclerotia on the soil surface, serve as other means of primary infection (Tu, 1989). Secondary infections occur when healthy plant parts come into contact with diseased ones. The epidemiology of *Sclerotinia* diseases relies on a number of factors, like inoculum availability in soil, soil and environmental conditions like temperature, moisture/humidity, host susceptibility, plant density, soil type, soil moisture, rainfall, and cultural practices followed in different agroclimatic conditions prevalent in an area.

1.8 Genome Organization

The Sanger technology was utilized to sequence the genome of *S. sclerotiorum* strain 1980. Archne (Jaffe *et al.*, 2003) was employed to assemble the sequences. The estimated size of the *S. sclerotiorum* optical map is 39.6 Mb, with approximately 38.0 Mb enclosed by Scaffolds. Consequently, there is a 1.6 Mb sequence absent from the *S. sclerotiorum* assembly (Fraissinet-Tachet *et al.*, 1996). The *S. sclerotiorum* genome, ranging from 38 to 39 Mb, encompasses 11,860–14,270 expected genes (Amselem *et al.*, 2011). The genome architecture of *S. sclerotiorum* has been elucidated through valuable insights gained from its complete genome sequence (Derbyshire *et al.*, 2017). Numerous researchers have described the detailed analysis of *S. sclerotiorum* transcriptome and the biosynthesis of secondary metabolites, elucidating key mechanisms of pathogenesis (Chittem *et al.*, 2020; Graham-Taylor *et al.*, 2020; Hossain *et al.*, 2023).

1.9 Application of Omics Technologies

Advancements in omics technologies, including genomics, proteomics, transcriptomics, and metabolomics, have significantly contributed to our comprehension of the *Brassica–Sclerotinia* host pathosystem during the second decade of this century. These technologies have enabled us to delve deeper into the interaction between these two species. By integrating multi-omics data and employing systems biology, it becomes feasible to develop and cultivate *Sclerotinia*-resistant *Brassica* crops in a comprehensive, targeted, and accurate manner (Wu *et al.*, 2016; Atri *et al.*, 2019; Rana *et al.*, 2019; Neik *et al.*, 2020; Saharan *et al.*, 2021a, b, 2023; Chen *et al.*, 2023).

1.10 Research Progress Assessment

Over the course of the past 187 years (1837–2024), the research progress of *Sclerotinia* has been thoroughly evaluated through a vast number of scientific publications, exceeding 4600 in total. These publications have undergone comprehensive analyses, categorically examining every aspect of research. To identify the primary research areas and leading institutions, the publications have been organized according to centuries and decades. Notable contributions in this field have been made by various researchers, including Saharan and Mehta, 2008; Sharma *et al.*, 2015a; Saharan *et al.*, 2021a, b, 2023; Derbyshire *et al.*, 2022; Chen *et al.*, 2023; Hossain *et al.*, 2023. The books authored by Saharan *et al.* (2021a, b, 2023) provide extensive coverage of research advancements and achievements regarding the mechanisms of host resistance, molecular aspects, and genomics of host–pathogen interaction.

1.10.1 Century-wise research progress

During the 19th century, only 1% of publications focused on stem rot. However, the 20th century witnessed a significant increase in research on *Sclerotinia* rot, with 52% of papers published during this period. During the initial two decades of the 21st century, nearly half of all publications focused on *Sclerotinia* rot. The remarkable progress in *Sclerotinia* rot research during the 20th and 21st centuries can be attributed to various factors. These include heightened awareness of the disease, its economic impact resulting in substantial yield losses, the establishment of additional research institutes, the recruitment of more research personnel, improved research facilities, advancements in biological sciences such as molecular biology, molecular plant pathology, and molecular breeding. Furthermore, the availability of increased funding, academic interest, collaboration among scientists, and the emergence of new fields within biological sciences have all contributed to the rapid advancement of *Sclerotinia* rot research.

1.10.2 Decade-wise research development

The research articles on *Sclerotinia* stem rot published in various decades across three centuries demonstrate a sluggish advancement in the field during the final decade of the 19th century. However, significant strides were made in *Sclerotinia* stem rot research from the sixth to the tenth decade of the 20th century. Moving into the 21st century, the first decade alone witnessed a remarkable surge, with more than 600 publications on the subject.

1.10.3 Major research areas identified and emphasized

The initial documentation of *Sclerotinia* dates back to 1837. Since then, more than 37 significant research domains pertaining to *Sclerotinia* stem rot have been identified and highlighted by researchers. This span of investigation ranges from 1837 to 2024, as documented in Table 1.1. A comprehensive analysis of these areas has been presented in a *Sclerotinia* stem rot review (Sharma *et al.*, 2015a). Additionally, Saharan and Mehta (2008) published books on the subject, while (Saharan *et al.*, 2021a, b, 2023) contributed to the literature with publications in addition to other review articles on this subject by various researchers throughout the world (Neik *et al.*, 2020; Chen *et al.*, 2023; Han *et al.*, 2023; Hossain *et al.*, 2023).

1.10.4 Publications of historical significance

Every year, numerous publications are released on *Sclerotinia* stem rot. These publications encompass a wide range of formats, including research publications, book chapters, review articles, popular scientific articles, conference proceedings, abstracts of lead lectures of conferences, group meetings, workshops, ad-hoc research scheme reports, key note addresses at symposia, and circulars released for farmers. However, Table 1.2 provides a comprehensive list of the significant publications, especially reviews, etc., that hold historical importance.

1.10.5 Research resources

The global significance of the *Sclerotinia* stem rot of crucifers is evident since it ranks as the third most significant pathogen of crucifers. This ranking is based on its wide distribution, the losses it causes, and the extensive research efforts dedicated to it by more than 145 research organizations worldwide (Table 1.3). Remarkable advancements have been achieved in managing this disease, encompassing various aspects such as the development of resistant genotypes.

Table 1.1. Major research areas identified for *Sclerotinia sclerotiorum*. (Adapted and updated from publication of Sharma *et al*. (2015a) with permission.)

Sr. no.	Major research areas
1.	Disease symptoms, host range, damages, dispersal, occurrence, existence, ecology, and new reports of *Sclerotinia sclerotiorum*
2.	Apothecia and ascospore of *Sclerotinia sclerotiorum* and their role in disease cycle
3.	Sclerotia formation, survival, and role in epidemiology of *Sclerotinia sclerotiorum*
4.	Enzymes, toxins, and biochemistry of *Sclerotinia sclerotiorum*
5.	Infection, pathogenesis, and inoculation procedure of *Sclerotinia sclerotiorum*
6.	Mycelium, inoculum, histopathology, and morphology of *Sclerotinia sclerotiorum*
7.	Epidemiology and disease forecasting of *Sclerotinia sclerotiorum*
8.	Pathological and morphological variability, genetic diversity, and Mycelial Compatibility Grouping (MCG) of *Sclerotinia sclerotiorum*
9.	Screening and identification of resistant sources of *Sclerotinia sclerotiorum*
10.	Genetical and molecular aspects of *Sclerotinia sclerotiorum*
11.	Genomics of *Sclerotinia*
12.	Proteomics of *Sclerotinia*
13.	Genome sequencing of *Sclerotinia*
14.	Identification of effector genes of *Sclerotinia*
15.	Identification of genes governing biochemical pathways
16.	Mapping and cloning of *R*-genes in *Brassica* against *Sclerotinia*
17.	Mechanisms of host resistance against *Sclerotinia*
18.	Mechanisms of variability in *Sclerotinia*
19.	Application of omics technology in *Sclerotinia*–host pathosystem
20.	Genomics of *Sclerotinia*–host pathosystem
21.	High-quality genome assembles of *Sclerotinia*
22.	Transcriptomics of *Sclerotinia* virulence-related genes
23.	Secretomics of *Sclerotinia*
24.	Interactomics of *Sclerotinia*
25.	Metabolomics of *Sclerotinia*
26.	Effector omics of *Sclerotinia*
27.	Bioinformatics of *Sclerotinia*
28.	Molecular mechanisms of *Sclerotinia*–host pathosystem
29.	Virulence gene detection of *Sclerotinia*
30.	Detection of *Sclerotinia* inoculum
31.	Spatial dispersal of *Sclerotinia*
32.	Detection, identification, and diagnosis techniques
33.	Management of *Sclerotinia sclerotiorum*
34.	Chemical and integrated management of *Sclerotinia sclerotiorum*
35.	Mycoparasitism and biocontrol of *Sclerotinia sclerotiorum*
36.	*Sclerotinia sclerotiorum* as myco-herbicide
37.	Mycoviruses, hypovirulence, and biocontrol of *Sclerotinia*

Table 1.2. Historical publications on *Sclerotinia* during 1837–2024. (Adapted and updated from publication of Sharma *et al*. (2015a) with permission.)

Authors	Year	Title of the article	Name of journal/Institution	Volume and page number
de Bary, A., de Bary, H.A. and Deutschland, B.	1884	Vergleichend Morphologie und Biologie der Pilze, Mycetozoen und Bacterien	Leipzig: Wilhelm Engelmann	525 pp.
de Bary, A.	1886	Ueber einige Sclerotinien und Sclerotien krankheiten	*Botanische Zeitimg*	44, 374–474
de Bary, A.	1887	Comparative morphology and biology of Fungi (Translated by H. E. F. Garnsey, revised by I. B. Balfour)	Clarendon, The Clarendon Press, Oxford	525 pp.
Whetzel, 1945	1945	Synopsis of the genera and species of *Sclerotiniaceae*, a family of somatic inoperculate discomycetes	*Mycologia*	37, 648–714
Ingold, C.T.	1960	Dispersal in fungi	Clarendon, Oxford	206 pp.
Saito, I.	1969	Effect of some nutritional conditions on the formation and germinability of sclerotia of *Sclerotinia sclerotiorum* (Lib.) de Bary	*Bulletin of the Hokkaido Prefectural Agricultural Experiment Station, Japan*	19, 1–7
Ingold, C.T.	1971	Fungal Spores: their liberation and dispersal	Oxford University Press, London	302 pp.
Dumont, K.P. and Korf, R.P.	1971	*Sclerotiniaceae* I. Genetic nomenclature	*Mycologia*	63, 157–168
Dumont, K.P.	1971	*Sclerotiniaceae* II. Lambertelia Mem.	*New York Botanical Garden*	22, 1–178
Coley-Smith, J.R. and Cooke, R.C.	1971	Survival and germination of fungal sclerotia	*Annual Review of Phytopathology*	9, 65–92
Saito, I.	1973	Initiation and development of apothecial stipe and primordia of *Sclerotinia sclerotiorum*	*Transactions of Mycological Society of Japan*	14, 343–351
Alabouvette, C. and Louvet, J.	1973	*Sclerotinia sclerotiorum*. Bibliographie Sclerotionnee (Dec. 1971)	Information Technology Centre for Technology Interpretation of Oleag Metrop	31, 7–26
Saito, I.	1974a	Utilization of β-glucans in germinating sclerotia of *Sclerotinia sclerotiorum* (Lib.) de Bary	*Annals of Phytopathological Society of Japan*	40, 372–374
Saito, I.	1974b	Ultra-structural aspects of the maturation of sclerotia of *Sclerotinia sclerotiorum* (Lib.) de Bary	*Transactions of Mycological Society of Japan*	15, 384–400
Chet, I. and Henis, Y.	1975	Sclerotial morphogenesis in fungi	*Annual Review of Phytopathology*	13, 169–192
Mordue, J.E.M. and Holliday, P.	1976	*Sclerotinia sclerotiorum (sclerotial state)*	CMI Descriptions of Pathogenic Fungi and Bacteria PP 52, Sheet No. 513. CMI, Kew, Surrey, UK	Sheet No. 513

Continued

Table 1.2. Continued

Authors	Year	Title of the article	Name of journal/Institution	Volume and page number
Saito, I.	1977	*Sclerotinia sclerotiorum* (Lib.) de Bary. Studies on the maturation and germination of sclerotia of *Sclerotinia sclerotiorum* (Lib.) de Bary, a causal fungus of bean stem rot	Report Hokkaido Prefectural Agricultural Experimentation Station No. 26, March, 1977, Hokkaido Central Agricultural Experiment Station, Japan	069–13, 106 p.
Kohn, L.M.	1979	A monographic revision of the genus *Sclerotinia*	*Mycotaxon*	9, 365–444
Willetts, H.J. and Wong, J.A.L.	1980	The biology of *Sclerotinia sclerotiorum*, *S. trifoliorum*, and *S. minor* with emphasis on specific nomenclature	*Botanical Review*	46, 100–165
Bell, A.A. and Wheeler, M.H.	1986	Biosynthesis and functions of fungal melanins	*Annual Review of Phytopathology*	24, 411–451
Phillips, A.J.L.	1987	Carpogenic germination of sclerotia of *Sclerotinia sclerotiorum*: a review	*Phytophylactica*	19, 279–283
Nordin, K.	1988	*Sclerotinia sclerotiorum* (Lib.) de Bary—A literature review with emphasis on epidemiology and development of forecasting methods for Sclerotinia stem rot	*Växtskydds rapporter Jordbruk*	5:77
Gladders, P., Davies, J.M.L. and Hardwick, N.V.	1991	Review of *Sclerotinia* epidemic in winter oilseed rape in England and Wales	*Bull OILB/SROP*	16, 1–8
Krüger, W.	1991	A review on assessment of diseases in oilseed rape—Comparison of various methods	*Bulletin SROP*	14, 91–111
Kohn, L.M.	1992	Developing new characters for fungal systematic: An experimental approach for determining the rank of *Sclerotinia*	*Mycologia*	84, 139–153
Willetts, H.J. and Bullock, S.	1992	Developmental biology of sclerotia	*Mycological Research*	96, 801–816
Gupta, S.K. and Dohroo, N.P.	1996	Chemical control of *Sclerotinia sclerotiorum* (Lib) de Bary—A review	*Agricultural Revolution*	17, 75–80
Mordue, J. E. M. and Holliday, P.	1998	*Sclerotinia sclerotiorum*	IMI Descriptions of Fungi and Bacteria. (52) Sheet No. 513, CABI Bioscience, Bakeham Lane, e.g.ham, Surrey, TW20 9TY, UK	Sheet No. 513
Henson, J.M., Butler, M.J. and Day, A.W.	1999	The dark side of mycelium: melanins of phytopathogenic fungi	*Annual Review of Phytopathology*	37, 447–471
Glass, N.L., Jacobson, D.J. and Shiu, P.K.	2000	The genetics of hyphal fusion and vegetative incompatibility in filamentous ascomycete fungi	*Annual Review of Genetics*	34, 165–186

Continued

Table 1.2. Continued

Authors	Year	Title of the article	Name of journal/Institution	Volume and page number
Thaning, C.	2000	Ways of managing *Sclerotinia sclerotiorum* inoculum	*Acta Universitatis Agriculturae Sueciae—Agraria*	124 pp.
Hegedus, D.D. and Rimmer, S.R.	2005	Mini-Review—*Sclerotinia sclerotiorum*: When 'to be or not to be' a pathogen?	*FEMS Microbiology Letters*	251, 177–184
Bolton, M.D., Thomma, B.P.H.T. and Nelson, B.D.	2006	Pathogen profile; *Sclerotinia sclerotiorum* (Lib.) de Bary: Biology and molecular traits of a cosmopolitan pathogen	*Molecular Plant Pathology*	79(1), 1–16
Erental, A., Dickman, M.B. and Yarden, O.	2008	Sclerotial development in *Sclerotinia sclerotiorum*: awakening molecular analysis of a 'Dormant' structure	*Fungal Biology Review*	22, 6–16
Gladdens, P., Ginsburg, D. and Smith, J.A.	2008	*Sclerotinia* in oilseed rape—A review of the 2007 epidemic in England	*HGCA Project Reporter*	433:44
Singh, R., Singh, D., Li, H., Sivasithampara, M. S., Yadav, N.R., Salisbury, P. and Barbetti, M.J.	2008	Management of Sclerotinia rot of oilseed *Brassicas*—a focus on India	*Journal of Oilseed Research*	10, 1–27
Saharan, G.S. and Mehta, N.	2008	Sclerotinia diseases of crop plants: Biology, ecology and disease management	Springer Science, The Netherlands	485 p.
Sharma, P., Verma, P. R., Meena, P.D., Kumar, V. and Singh, D.	2015a	Research progress analysis of sclerotinia rot (*Sclerotinia sclerotiorum*) of oilseed Brassicas through bibliography	*Journal of Oilseed Brassica*	6 (Special), 45–125
Sharma *et al.*, 2015b, P., Meena, P.D., Verma, P.R., Saharan, G.S., Mehta, N., Singh, D. and Kumar, A.	2015b	*Sclerotinia sclerotiorum* (Lib) de Bary causing Sclerotinia rot in oilseed Brassicas: A review	*Journal of Oilseed Brassica*	6(S), 1–44
Derbyshire and Denton-Giles, 2016	2016	Review-The control of sclerotinia stem rot on oilseed rape (*Brassica napus*): Current practices and future opportunities	*Plant Pathology*	65, 859–877 doi:10.1111/ppa.12517
Sharma, J., Rathore, G.S., Godhika, S., Meena, S. and Lal, C.	2017	*Sclerotinia sclerotiorum*—A threat and challenges for crops	*International Journal of Agriculture and Life Sciences* (I J A L S)	3 (2), 188–194
Smolińska, U. and Kowalska, B.	2018	Biological control of the soil borne fungal pathogen *Sclerotinia sclerotiorum*—A review	*Journal of Plant Pathology*	100, 1–12 doi:10.1007/s42161-018-0023-0

Continued

Table 1.2. Continued

Authors	Year	Title of the article	Name of journal/Institution	Volume and page number
Xu, L., Li, G., Jiang, D. and Chen, W.	2018	*Sclerotinia sclerotiorum*: An Evaluation of Virulence Theories-Annual Review	*Annual Review of Phytopathology*	56, 311–338 doi:10.1146/annurev-phyto-080417–050052
Rothmann, L.A. and McLaren, N.W.	2018	*Sclerotinia sclerotiorum* disease prediction: A review and potential applications in South Africa	*South African Journal of Science*	114(3/4), Art. #2017–0155, 9 doi:10.17159/sajs.2018/20170155
Neik, T. X., Amas, J., Barbetti, M., Edward, D. and Batley, J.	2020	Review—Understanding host–pathogen interaction in *Brassica napus* in the Omics era	*Plants*	9, 1336, 1–29 doi:10.3390/plants 9101336
O'Sullivan *et al*., 2021, C.A., Belt, K. and Thatcher, L.F.	2021	Tackling Control of a cosmopolitan phytopathogen: *Sclerotinia*	*Frontiers in Plant Science*	12:707509 doi:10.3389/fpls.2021.707509
Saharan, G.S., Mehta, N., and Meena, P.D.	2021a	Genomics of crucifer's host-resistance	Springer Nature, The Netherlands	773 p.
Saharan, G.S., Mehta, N., and Meena, P.D.	2021b	Molecular mechanism of crucifer's host resistance	Springer Nature, The Netherlands	doi:10.1007/978-981-16-0862-9, 790 p.
Derbyshire, M.C., Newman, T.E., Khentry, Y. and Taiwo, A.O.	2022	Pathogen profile: The evolutionary and molecular features of the broad host range plant pathogen *Sclerotinia sclerotiorum*	*Molecular Plant Pathology*	23, 1075–1090 doi:10.1111/mpp.13221, 1–16
Otun, O.S., Achilonu, I. and Ntushelo, K.	2022	Proteomics analysis of the phytopathogenic fungus *Sclerotinia sclerotiorum*: a narrative review	*Journal of Bio-X Research*	5, 112–117 doi:10.1097/JBR.0000000000000130
Saharan, G.S., Mehta, N., and Meena, P.D.	2023	Genomics of crucifer's host-pathosystem	Springer Nature, The Netherlands	doi:10.1007/978-981-19-3812-2, 930 p.
Hossain, M.M., Sultana, F., Li, W., Phan Tran, Lam-S. and Mostofa, M.G.	2023	Review-*Sclerotinia sclerotiorum* (Lib.) de Bary: Insights into the pathogenomic features of a global pathogen	*Cells*	12, 1063 doi:10.3390/cells 12071063
Mehta, N., Meena, P.D. and Saharan, G.S.	2023	Geographical distribution and host range of *sclerotinia sclerotiorum* causing stem rot of crucifers	*Plant Disease Research*	38(1), 11–46
Han, V.C., Michael, P.J., Swift, B. and Bennet, S.J.	2023	Biological control of *Sclerotinia sclerotiorum*: Modes of action of biocontrol agents, soil organic amendments, and soil microbiome manipulation	*Biological Control*	186, 105346 doi:10.1016/j.biocontrol.2023.105346
Chen, R.S., Wang, J.Y., Sarwar, R. and Tan, X.L.	2023	Review Article—Genetic breakthroughs in the *Brassica napus–Sclerotinia sclerotiorum* interactions	*Frontiers in Plant Sciences*	14:1,276055 doi:10.3389/fpls.2023.1276055

Table 1.3. Leading research institutes working on *Sclerotinia* in the world. (Adapted and updated from publication of Sharma *et al.* (2015a) with permission.)

Sr. no.	Country	Name of Institute
1.	Argentina	Cátedra de Microbiología Agrícola/INBA (CONICET/UBA), University de Buenos Aires, Buenos Aires
2.	Argentina	Universidad Nacional del Sur, 8000 Bahía Blanca
3.	Australia	University of New England, Armidale
4.	Australia	The University of Western Australia, 35 Stirling Highway, Crawley, WA 6009
5.	Australia	Charles Sturt University, Wagga Wagga, NSW 2678
6.	Belgium	ISK Biosciences Europe N.V., Diegem
7.	Brazil	Universidade Federal de Viçosa—36570–000, Viçosa, MG
8.	Brazil	Universidade Federal da Grande Dourados, UFGD, Dourados, MS
9.	Brazil	Universidade Federal de Goiás-UFG, Goiânia, GO
10.	Brazil	Universidade Estadual de Ponta Grossa
11.	Canada	Agriculture and Agri-Food Canada, 107 Science Place, Saskatoon, SK S7N 0X2
12.	Canada	Agriculture and Agri-Food Canada, Lethbridge Research Centre, 5403-1st Ave. S, PO Box 3000, Lethbridge, Alberta-T1J4B1
13.	Canada	University of Alberta, Edmonton, AB T6G 2P5
14.	Canada	University of Saskatchewan, 110 Science Place, Saskatoon, SK S7N 5C9
15.	Canada	University of Manitoba, Winnipeg, MB R3T 2N2
16.	Canada	Agriculture and Agri-Food Canada, P.O. Box 1240, Melfort, Saskatchewan, S0E 1A0
17.	Canada	University of Guelph, Guelph, ON N1G 2W1
18.	China	Nanjing Agricultural University, Nanjing—210,095
19.	China	Southwest University, Chongqing—400,716
20.	China	Huazhong Agricultural University, Wuhan—430,070
21.	China	Nanjing Agricultural University, Nanjing—210,095
22.	China	The Oil Crops Institute/National Oil Crops Improvement Center, Changsha, Hunan—410,128
23.	China	Oil Crops Research Institute, Chinese Academy of Agricultural Sciences, Wuhan—430,062
24.	China	Guizhou Institute of Oil Crops, Guiyang, Guizhou—550,006
25.	China	Anhui Agricultural University, Hefei—230,036
26.	China	South China Normal University, Guangzhou—510,631
27.	China	Northwest A & F University, Yangling, Shaanxi—712,100
28.	China	Longyan University, Longyan—364,000
29.	China	Jiangsu University, Zhenjiang—212,013
30.	China	Yangzhou University, Yangzhou—225,009
31.	China	Dalian Institute of Chemical Physics, Chinese Academy of Sciences, Dalian—116,023
32.	China	Yangtze University, Jingzhou, Hubei Province—434,025

Continued

Table 1.3. Continued

Sr. no.	Country	Name of Institute
33.	China	Zhenjiang Institute of Agricultural Science in Hilly Area of Jiangsu Province, Jurong—212,400
34.	China	Longyan University, Longyan, Fujian—364,000
35.	China	Zhejiang University, Hangzhou—310,058
36.	China	Shanghai Normal University, Shanghai—200,234
37.	China	Crops Research Institute, Jiangxi Academy of Agricultural Sciences, Nanchang—330,200
38.	China	Zhejiang University of Technology, 18# Chaowang Road, Hangzhou 310,032
39.	China	National Research Center of Rapeseed Engineering and Technology, Huazhong Agricultural University, Wuhan—430,070
40.	China	Zhengzhou University, Zhengzhou—450,001
41.	China	Industrial Crops Research Institute, Henan Academy of Agricultural Sciences, Zhengzhou—450,002
42.	China	Hunan Agricultural University, Changsha—410,128
43.	China	Agricultural Sciences Research Institute of Yichun City, Jiangxi Province,Yichun-336000
44.	China	Institute of Crop Sciences, Chinese Academy of Agricultural Sciences, Beijing—100,081
45.	China	Sichuan University, Chengdu—610,064
46.	Croatia	Poljoprivredni fakultet u Osijeku, Kralja Petra Svaèic´a 1d-31000, Osijek
47.	Croatia	Department of Phytopathology, Svetošimunska 25, 10,000 Zagreb
48.	Czech Republic	Academy of Sciences of the Czech Republic, Rozvojová 313, 165 02 Prague
49.	Czech Republic	Agrotest fyto, Ltd. s.r.o., Havlíèkova 2787/121, 767 01 Kromeríz
50.	Czech Republic	Svaz pestitelu a zpracovatelu olejnin, Na Fabiánce 146, 182 00 Praha 8
51.	Denmark	Afd. for Plantepatologi Planteværnscentret, Lottenborgvej 2, 2800 Lyngby
52.	France	CETIOM, Direction Technique—Pathologie végétale, Campus de Grignon, Av. L. Brétignières, FR-78850 Thiverval-Grignon
53.	France	DuPont Solutions, Défense Plaza, 23/25 rue Delarivière Lefoullon, F-92800 Puteaux
54.	France	DRAAF-SRAL Midi-Pyrénées, Bd Armand Duportal-31074 Toulouse Cedex
55.	France	Unité de Phytotechnologie, Sanofi Recherche, Centre de Labège, F-31676 Labège Cedex
56.	France	BASF Agro SAS, 21 Chemin de la Sauvegarde, F-69134 Ecully Cedex
57.	France	INRA, UMR 211 INRA Agro Paris Tech, 78,850 Thiverval-Grignon
58.	Germany	Georg-August-University of Göttingen, Grisebachstr 6, D-37077 Göttingen
59.	Germany	Dienstleistungszentrum Ländlicher Raum, Westerwald-Osteifel, Montabaur
60.	Germany	Prophyta GmbH, Inselstrasse 12, 23,999 Malchow
61.	Germany	Feinchemie Schwebda GmbH, Eschwege
62.	Germany	Du Pont de Nemours (Deutschland) GmbH, Neu-Isenburg

Continued

Table 1.3. Continued

Sr. no.	Country	Name of Institute
63.	Germany	Sächsisches Landesamt für Umwelt, Landwirtschaft und Geologie, Freiberg
64.	Germany	Christian-Albrechts-Universität of Kiel, Hermann, Rodewald Str. 9, D-24118 Kiel
65.	Germany	Julius Kühn-Institut, Braunschweig
66.	Germany	Inst. Pflanzenbau Pflanzenzuchtung, Giessen
67.	Germany	Universität-GH-Paderborn, Labor für Biotechnologie und Qualitätssicherung, Lübecker Ring 2, D-59494 Soest
68.	Hungary	National Institute for Agricultural Quality Control, Budapest
69.	India	Aligarh Muslim University, Aligarh—202001, Uttar Pradesh
70.	India	Assam Agricultural University, Shillongani—782 001, Assam
71.	India	Banaras Hindu University, Varanasi—221 005, Uttar Pradesh
72.	India	CCS Haryana Agricultural University, Hisar—125 004, Haryana
73.	India	CSA University of Agriculture and Technology, Kanpur—208 002, Uttar Pradesh
74.	India	CSK Himachal Pradesh Krishi Vishwavidyalaya, Kangra—176 001, Palampur
75.	India	Dr. Rajendra Prasad Central Agricultural University, Samastipur, Pusa—848125, Bihar
76.	India	GB Pant University of Agriculture and Technology, Pantnagar—263145, Uttarakhand
77.	India	Gurukul Kangri University, Haridwar—249 404, Uttarakhand
78.	India	ICAR-Indian Institute of Rapeseed-Mustard Research, Sewar, Bharatpur—321 303, Rajasthan
79.	India	ICAR-Indian Agricultural Research Institute, Pusa, New Delhi—110 012
80.	India	ICAR-National Centre for Integrated Pest Management, Pusa Campus, New Delhi—110 012
81.	India	Indira Gandhi Agricultural University, Jabalpur—494 005, Madhya Pradesh
82.	India	ND University of Agri. & Tech., Kumar Ganj, Faizabad—224 229, Uttar Pradesh
83.	India	Pulses and Oilseeds Research Station, Berhampur—742 101, West Bengal
84.	India	Punjab Agricultural University, Ludhiana—141 004, Punjab
85.	India	Rani Lakshmi Bai Central Agricultural University, Jhansi—484003, Uttar Pradesh
86.	India	RVRS Agricultural University, Morena—476 001, Madhya Pradesh
87.	India	Sardar Vallabhbhai Patel University of Agriculture and Technology, Modi Puram, Meerut—250 110, Uttar Pradesh
88.	India	SKN Agriculture University, Jobner, Jaipur—302 029, Rajasthan
89.	India	Swami Keshwanand Rajasthan Agricultural University, Bichhwal, Bikaner—334006, Rajasthan
90.	India	University of Delhi, South Campus, New Delhi—110 021
91.	India	Uttar Banga Krishi Vishwavidyalaya Majhian, Patiram, Dakshin Dinajpur—733 133

Continued

Table 1.3. Continued

Sr. no.	Country	Name of Institute
92.	Iran	Urmia University, Urmia
93.	Iran	Islamic Azad University, Tehran
94.	Iran	University of Tehran, Karaj, Tehran
95.	Iran	Agricultural and Natural Resources University of Sari, Mazandaran
96.	Iran	University of Zanjan, Zanjan
97.	Iran	Gorgan Univ. of Agricultural Sciences and Natural Resources, Gorgan
98.	Iran	Iranian Research Institute of Plant Protection, Tehran
99.	Iran	National Institute of Genetic Engineering and Biotechnology, P.O. Box 14965/161, Tehran
100.	Iran	Tarbiat Modares University, Tehran
101.	Iran	University of Mazandaran, Babolsar—47416–95447
102.	Irish Republic	University College, Cork
103.	Israel	Hebrew University, Jerusalem, Rehovot—76,100
104.	Italy	CRA CAT, Unità di ricerca per le Colture Alternative al Tabacco, via Vitiello 108, Scafati (SA)
105.	Italy	ENEA, Divisione Biotecnologie e Agricoltura, Centro Ricerche Trisaia, Rotondella (MT)
106.	Japan	Hiroaki University
107.	Japan	Tottori Daisen Agricultural Extension Service Station, 541–8 Daisen-cho, Saihaku-gun, Tottori—689–3303
108.	Japan	Tokyo Metropolitan Agriculture and Forestry Research Center, Chichi Jima, Ogasawara-mura, Tokyo—100–2101
109.	Latvia	Latvia University of Agriculture, Liela Street 2, Jelgava, LV 3001
110.	Lithuania	Lithuanian University of Agriculture, Studentu 9, LT-53361 Akademija, Kaunas District
111.	New Zealand	Lincoln University, Lincoln 7647
112.	Poland	Uniwersytet Przyrodniczy we Wroc³awiu, Katedra Ochrony Roœlin, Pl. Grunwaldzki 24A, 50–363 Wroc³aw
113.	Poland	Instytut Genetyki Roœlin Polskiej Akademii Nauk w Poznaniu, Poznan
114.	Poland	DuPont Poland Sp. z o.o., Postepu 17b, 02–676 Warszawa
115.	Poland	Uniwersytet Warmińsko-Mazurski w Olsztynie 1 Katedra Fitopatologii i.e.ntomologii Prawocheńskiego 17, 10–720 Olsztyn
116.	Poland	National Research Institute, Research Division in Poznań, ul. Strzeszyńska 36, 60–479 Poznań
117.	Poland	Stacja Doœwiadczalna Oceny Odmian, Kreczunowicza 13, 62–811 Koœcielna Wieœ
118.	Poland	Instytut Ochrony Roœlin, Państwowy Instytut Badawczy, Wladyslawa Wegorka 20, 60–318 Poznań
119.	Poland	Technical-Agricultural University, Bydgoszcz
120.	Romania	University of Bucharest, 1–3 Portocalelor Alley, RO-060101 Bucharest 35
121.	Romania	University of Agronomic Sciences and Veterinary Medicine of Bucharest, 59 Maras'ti Blvd., District 1, 011464 Bucharest

Continued

Table 1.3. Continued

Sr. no.	Country	Name of Institute
122.	Romania	University of Agricultural Sciences and Veterinary Medicine of Iași, Iași
123.	Romania	Banat's University of Agricultural Sciences and Veterinary Medicine, Timisoara, Aradului Street, No. 119, RO-300645
124.	Slovakia	Slovak Academy of Sciences, Nádrazná 52, SK-90028 Ivanka pri Dunaji
125.	South Africa	North-West University, Potchefstroom Campus, Private Bag X6001, Potchefstroom 2520
126.	South Africa	Stellenbosch University, Stellenbosch
127.	South Africa	Department of Molecular and Cell Biology, University of Witwatersrand, South Africa
128.	South Africa	Department of Agriculture and Animal Health, Science Campus, University of South Africa, Johannesburg, South Africa
129.	Sweden	Department of Plant Biology, SLU, Box 7080, SE-750 07 Uppsala
130.	Switzerland	Agroscope RAC Changins, case postale 254, CH-1260 Nyon 1
131.	Turkey	Mustafa Kemal University, 31,034 Antakya-Hatay
132.	UK	Central Science Laboratory, Sand Hutton, York—Y041 1LZ
133.	UK	ADAS Boxworth, Boxworth, Cambridge, CB23 4NN
134.	UK	NIAB TAG, Winchester, Hampshire S021 3NE
135.	UK	Rothamsted Research, Harpenden, Herts AL5 2JQ
136.	UK	ADAS Drayton, Defra Drayton, Alcester Rd., Stratford on Avon, Warwickshire CV37 9RQ
137.	USA	Washington State University, Pullman, WA 99164
138.	USA	University of Florida, Quincy, Florida
139.	USA	Monsanto Corporation, Spencer, IA 51301
140.	USA	North Dakota State University, Fargo, ND 58108
141.	USA	Texas A and M University, College Station, TX 77843
142.	USA	University of Georgia, Griffin Campus, Griffin, GA 30223
143.	USA	University of Alaska Fairbanks, Palmer, AK 99645
144.	USA	A DuPont Company, 7300 NW 62nd Avenue, P.O. Box 1004, Johnston, IA 50131
145.	USA	New York State Agricultural Experiment Station, 630 W. North St., Geneva, NY—14,456
146.	USA	Langdon Research Extension Center, North Dakota State University, Langdon, ND—58249, USA
147.	Venezuela	FONAIAP Experimental Station El Cuji, Barquisimeto

References

Abawi, G.S. and Grogan, R.G. (1979) Epidemiology of diseases caused by *Sclerotinia* species. *Phytopathology* 69, 899–904.

Adams, P.B. and Ayers, W.A. (1979) Ecology of *Sclerotinia* species. *Phytopathology* 69, 896–899.

Adams, P.B. and Wong, A.L. (1991) The effect of chemical pesticides on the infection of sclerotia of *Sclerotinia minor* by the biocontrol agent *Sporidesmium sclerotivorum*. *Phytopathology* 81(10), 1340. DOI: 10.1094/Phyto-81-1340.

Amselem, J., Cuomo, C.A., Kan, J.A.L., Viaud, M., Benito, E.P. *et al.* (2011) Genomic analysis of the necrotrophic fungal pathogens *Sclerotinia sclerotiorum* and *Botrytis cinerea*. *PLoS Genetics* 7(8), e1002230. DOI: 10.1371/journal.pgen.1002230.

Atri, C., Akhatar, J., Gupta, M., Gupta, N., Goyal, A. *et al.* (2019) Molecular and genetic analysis of defensive responses of *Brassica juncea—B. fruticulosa* introgression lines to *Sclerotinia* infection. *Scientific Reports* 9(1), 17089. DOI: 10.1038/s41598-019-53444-3.

Barbacci, A., Navaud, O., Mbengue, M., Barascud, M., Godiard, L. *et al.* (2020) Rapid identification of an *Arabidopsis NLR* gene as a candidate conferring susceptibility to *Sclerotinia sclerotiorum* using time-resolved automated phenotyping. *The Plant Journal* 103(2), 903–917. DOI: 10.1111/tpj.14747.

Blad, B.L., Steadman, J.R. and Weiss, A. (1978) Canopy structure and irrigation influence white mold disease and microclimate of dry edible beans. *Phytopathology* 68(10), 1431. DOI: 10.1094/Phyto-68-1431.

Boland, G.J. and Hall, R. (1994) Index of plant hosts of *Sclerotinia sclerotiorum*. *Canadian Journal of Plant Pathology* 16(2), 93–108. DOI: 10.1080/07060669409500766.

Bolton, M.D., Thomma, B.P.H.J. and Nelson, B.D. (2006) *Sclerotinia sclerotiorum* (Lib.) de Bary: Biology and molecular traits of a cosmopolitan pathogen. *Molecular Plant Pathology* 7(1), 1–16. DOI: 10.1111/j.1364-3703.2005.00316.x.

Brodal, G., Warmington, R., Grieu, C., Ficke, A. and Clarkson, J.P. (2017) First report of *Sclerotinia subarctica* nom. prov. (*Sclerotinia* sp. 1) causing stem rot on turnip rape (*Brassica rapa* subsp. *Oleifera*) in Norway. *Plant Disease* 102, 386. DOI: 10.1094/PDIS-06-16-0785-PDN.

Chen, R.S., Wang, J.Y., Sarwar, R. and Tan, X.L. (2023) Genetic breakthroughs in the *Brassica napus–Sclerotinia sclerotiorum* interactions. *Frontiers in Plant Science* 14, 1276055. DOI: 10.3389/fpls.2023.1276055.

Chittem, K., Yajima, W.R., Goswami, R.S. and Del Río Mendoza, L.E. (2020) Transcriptome analysis of the plant pathogen *Sclerotinia sclerotiorum* interaction with resistant and susceptible canola (*Brassica napus*) lines. *PLoS ONE* 15(3), e0229844. DOI: 10.1371/journal.pone.0229844.

Clarkson, J.P., Carter, H.E. and Coventry, E. (2010) First report of *Sclerotinia subarctica* nom. prov. (*Sclerotinia* species 1) in the UK on *Ranunculus acris*. *Plant Pathology* 59, 1173. DOI: 10.1111/j.1365-3059.2010.02271.x.

Derbyshire, M., Denton-Giles, M., Hegedus, D., Seifbarghy, S., Rollins, J. *et al.* (2017) The complete genome sequence of the phytopathogenic fungus *Sclerotinia sclerotiorum* reveals insights into the genome architecture of broad host range pathogens. *Genome Biology and Evolution* 9(3), 593–618. DOI: 10.1093/gbe/evx030.

Derbyshire, M.C. and Denton-Giles, M. (2016) The control of Sclerotinia stem rot on oilseed rape (*Brassica napus*): Current practices and future opportunities. *Plant Pathology* 65(6), 859–877. DOI: 10.1111/ppa.12517.

Derbyshire, M.C., Newman, T.E., Khentry, Y. and Owolabi Taiwo, A. (2022) The evolutionary and molecular features of the broad-host-range plant pathogen *Sclerotinia sclerotiorum*. *Molecular Plant Pathology* 23(8), 1075–1090. DOI: 10.1111/mpp.13221.

Domsch, K.H., Gams, W. and Anderson, T. (1980) *Compendium of Soil Fungi*, Vol. 1. Academic Press, New York, pp. 712–716.

Fraissinet-Tachet, L., Reymond-Cotton, P. and Fèvre, M. (1996) Molecular karyotype of the phytopathogenic fungus *Sclerotinia sclerotiorum*. *Current Genetics* 29(5), 496–501. DOI: 10.1007/BF02221520.

Fuckel, L. (1870) Symbolae Mycologicae. Beiträge Zur Kenntniss Der Rheinischen Pilze. *Jahrbücher Des Nassauischen Vereins Für Naturkunde* 23, 1–459. DOI: 10.5962/bhl.title.47117.

Graham-Taylor, C., Kamphuis, L.G. and Derbyshire, M.C. (2020) A detailed *in silico* analysis of secondary metabolite biosynthesis clusters in the genome of the broad host range plant pathogenic fungus *sclerotinia sclerotiorum*. *BMC Genomics* 21(1), 7. DOI: 10.1186/s12864-019-6424-4.

Guimarães, R.L. and Stotz, H.U. (2004) Oxalate production by *Sclerotinia sclerotiorum* deregulates guard cells during infection. *Plant Physiology* 136(3), 3703–3711. DOI: 10.1104/pp.104.049650.

Han, V.C., Michael, P.J., Swift, B. and Bennet, S.J. (2023) Biological control of *Sclerotinia sclerotiorum*: Modes of action of biocontrol agents, soil organic amendments, and soil microbiome manipulation. *Biological Control* 186, 105346. DOI: 10.1016/j.biocontrol.2023.105346.

Hegedus, D.D. and Rimmer, S.R. (2005) Mini review-*Sclerotinia sclerotiorum*: When "to be or not to be" a pathogen? *FEMS Microbiology Letters* 251, 177–184. DOI: 10.1016/j.femsle.2005.07.040.

Hossain, M.M., Sultana, F., Li, W., Tran, L.S.P. and Mostofa, M.G. (2023) *Sclerotinia sclerotiorum* (Lib.) de Bary: Insights into the pathogenomic features of a global pathogen. *Cells* 12(7), 1063. DOI: 10.3390/cells12071063.

Huang, H.C. and Hoes, J.A. (1980) Importance of plant spacing and sclerotial position to development of Sclerotinia wilt of sunflower. *Plant Disease* 64(1), 81. DOI: 10.1094/PD-64-81.

Jaffe, D.B., Butler, J., Gnerre, S., Mauceli, E., Lindblad-Toh, K. *et al*. (2003) Whole-genome sequence assembly for mammalian genomes: Arachne 2. *Genome Research* 13(1), 91–96. DOI: 10.1101/gr.828403.

Kohn, L.M. (1979) A monographic revision of the genus *Sclerotinia*. *Mycotaxon* 9, 365–444.

Le Tourneau, D.L. (1979) Morphology, cytology, and physiology of *Sclerotinia* species in culture. *Phytopathology* 69(8), 887. DOI: 10.1094/Phyto-69-887.

Libert, M.A. (1837) Plante *crytogamicae arduennae*. No. 326.

Livingstone, D.M., Hampton, J.L., Phipps, P.M. and Grabau, E.A. (2005) Enhancing resistance to *Sclerotinia* minor in peanut by expressing a barley oxalate oxidase gene. *Plant Physiology* 137(4), 1354–1362. DOI: 10.1104/pp.104.057232.

Lorang, J.M., Tuori, R.P., Martinez, J.P., Sawyer, T.L., Redman, R.S. *et al*. (2001) Green fluorescent protein is lighting up fungal biology. *Applied and Environmental Microbiology* 67(5), 1987–1994. DOI: 10.1128/AEM.67.5.1987-1994.2001.

Lumsden, R.D. (1979) Histology and physiology of pathogenesis in plant diseases caused by *Sclerotinia* species. *Phytopathology* 69(8), 890. DOI: 10.1094/Phyto-69-890.

Mehta, N., Meena, P.D. and Saharan, G.S. (2023) Geographical distribution and host range of *Sclerotinia sclerotiorum* causing stem rot of crucifers. *Plant Disease Research* 38(1), 11–46. DOI: 10.5958/2249-8788.2023.00002.1.

Morrall, R.A.A. and Dueck, J. (1982) Epidemiology of Sclerotinia stem rot of rapeseed in Saskatchewan. *Canadian Journal of Plant Pathology* 4(2), 161–168. DOI: 10.1080/07060668209501319.

Neik, T.X., Amas, J., Barbetti, M., Edward, D. and Batley, J. (2020) Review—Understanding host-pathogen interaction in *Brassica napus* in the omics era. *Plants* 9(10), 1336. DOI: 10.3390/plants%209101336.

O'Sullivan, C.A., Belt, K. and Thatcher, L.F. (2021) Tackling control of a cosmopolitan phytopathogen. Sclerotinia. *Frontiers in Plant Science* 12, 707509. DOI: 10.3389/fpls.2021.707509.

Otun, O.S., Achilonu, I. and Ntushelo, K. (2022) Proteomics analysis of the phytopathogenic fungus *Sclerotinia sclerotiorum*: A narrative review. *Journal of Bio-X Research* 5, 112–117. DOI: 10.1097/JBR.0000000000000130.

Phillips, A.J.L. (1986) Factors affecting the parasitic activity of *Gliocladium virens* on sclerotia of *Sclerotinia sclerotiorum* and a note on its host range. *Journal of Phytopathology* 116(3), 212–220. DOI: 10.1111/j.1439-0434.1986.tb00913.x.

Purdy, L.H. (1979) *Sclerotinia sclerotiorum*: History, diseases and symptomatology, host range, geographic distribution, and impact. *Phytopathology* 69(8), 875. DOI: 10.1094/Phyto-69-875.

Rana, K., Atri, C., Akhatar, J., Kaur, R., Goyal, A. *et al*. (2019) Detection of first marker trait associations for resistance against *Sclerotinia sclerotiorum* in *Brassica juncea–Erucastrum cardaminoides* introgression lines. *Frontiers in Plant Science* 10, 1015–1031. DOI: 10.3389/fpls.2019.01015.

Rhodes, L.H., Myers, D.K. and Kouren, R.W. (1992) A biological disease forecasting system for fungicidal control of Sclerotinia crown and stem rot. In: *Proceedings on the 33rd North American alfalfa improvement conference*, Beltsville, Maryland, June 14–18.

Rotem, J. and Palti, J. (1969) Irrigation and plant diseases. *Annual Review of Phytopathology* 7(1), 267–288. DOI: 10.1146/annurev.py.07.090169.001411.

Rothmann, L.A. and McLaren, N.W. (2018) *Sclerotinia sclerotiorum* disease prediction: A review and potential applications in South Africa. *South African Journal of Science* 114(3/4), 9. DOI: 10.17159/sajs.2018/20170155.

Saharan, G.S. and Mehta, N. (2008) *Sclerotinia Diseases of Crop Plants: Biology, Ecology and Disease Management*. Springer Science, Dordrecht, The Netherlands, p. 485.

Saharan, G.S., Mehta, N. and Meena, P.D. (2021a) *Genomics of Crucifer's Host-Resistance*. Springer Nature, Singapore, p. 773.

Saharan, G.S., Mehta, N. and Meena, P.D. (2021b) *Molecular Mechanism of Crucifer's Host-Resistance*. Springer Nature, Singapore, p. 766. DOI: 10.1007/978-981-16-1974-8.

Saharan, G.S., Mehta, N. and Meena, P.D. (2023) *Genomics of Crucifer's Host-Pathosystem*. Springer Nature, Singapore, p. 936.

Schwartz, H.F., Steadman, J.R. and Coyne, D.R.P. (1978) Influence of *Phaseolus vulgaris* blossoming characteristics and canopy structure upon reaction to *Sclerotinia sclerotiorum*. *Phytopathology* 68(3), 465. DOI: 10.1094/Phyto-68-465.

Sharma, J., Rathore, G.S., Godhika, S., Meena, S. and Lal, C. (2017) *Sclerotinia sclerotiorum*—A threat and challenges for crops. *International Journal of Agriculture and Life Sciences* 3, 188–194.

Sharma, P., Verma, P.R., Meena, P.D., Kumar, V. and Singh, D. (2015a) Research progress analysis of Sclerotinia rot (*Sclerotinia sclerotiorum*) of oilseed *Brassicas* through bibliography. *Journal of Oilseed Brassica* 6(Special), 45–125.

Sharma, P., Meena, P.D., Verma, P.R., Saharan, G.S., Mehta, N. *et al*. (2015b) *Sclerotinia sclerotiorum* (Lib) de Bary causing Sclerotinia rot in oilseed brassicas: A review. *Journal of Oilseed Brassica* 6(S), 1–4.

Silva, A., de Bolton, M.D. and Nelson, B.D. (2005) Transformation of *Sclerotinia sclerotiorum* with the green fluorescent protein gene and expression of fluorescence of host tissues. *Phytopathology* 95, 523.

Smolińska, U. and Kowalska, B. (2018) Biological control of the soil-borne fungal pathogen *Sclerotinia sclerotiorum*—A review. *Journal of Plant Pathology* 100(1), 1–12. DOI: 10.1007/s42161-018-0023-0.

Steadman, J.R. (1979) Control of plant diseases caused by *Sclerotinia* species. *Phytopathology* 69, 904–907. DOI: 10.1094/Phyto-69-904.

Tu, J.C. (1989) Modes of primary infection caused by *Sclerotinia sclerotiorum* in Navy bean. *Microbios* 57, 85–91.

Walker, J.C. (1969) *Plant Pathology*, 3rd edn. McGraw-Hill, New York, p. 819.

Whetzel, H.H. (1945) A synopsis of the genera and species of the Sclerotiniaceae, a family of stromatic inoperculate discomycetes. *Mycologia* 37(6), 648–714. DOI: 10.1080/00275514.1945.12024025.

Willetts, H.J. and Wong, J.A.L. (1980) The biology of *Sclerotinia sclerotiorum*, *S. trifoliorum* and *S. minor* with emphasis on specific nomenclature. *The Botanical Review* 46(2), 101–165. DOI: 10.1007/BF02860868.

Wu, J., Zhao, Q., Liu, S., Shahid, M., Lan, L. *et al*. (2016) Genome-wide association study identifies new loci for resistance to *Sclerotinia* stem rot in *Brassica napus*. *Frontiers in Plant Science* 7, 1418. DOI: 10.3389/fpls.2016.01418.

Xu, L., Li, G., Jiang, D. and Chen, W. (2018) *Sclerotinia sclerotiorum*: An evaluation of virulence theories. *Annual Review of Phytopathology* 56, 311–338. DOI: 10.1146/annurev-phyto-080417-050052.

2 Geographical Distribution and Host Range

Abstract

Sclerotinia sclerotiorum (Lib.) de Bary, originally designated as *Peziza sclerotiorum* (Lib.), is a soil-borne hemibiotrophic pathogen responsible for sizeable yield losses in crucifers across the globe. According to CABI distribution map of plant diseases, it is found across all continents, covering more than 220 countries and states. The pathogen has an extraordinarily wide host range, infecting over 900 host plant species from more than 98 families. According to USDA database, it is associated with 2048 host species and varieties. Most hosts are herbaceous plants belonging to dicots but several also belong to monocots. The molecular factors that determine the first choice of *S. sclerotiorum* for dicotyledonous plants have been uncovered. The pathogen causes necrosis and generates proteins similar to ethylene-inducing peptide 1 (NLPs), which trigger cell death in a majority of dicots but only in a limited number of monocots. The necrosis caused by NLPs is a consequence of their attachment to a particular form of membrane lipid glycosylinositol phosphorylceramide, which is prevalent in dicots but not found in monocots.

2.1 Introduction

The interaction between the pathogen *Sclerotinia* and various host plants has captivated researchers in the fields of mycology, plant pathology, breeding, and molecular biology. This fascination stems from the pathogen's ability to switch between biotrophic and necrotrophic lifestyles, its wide distribution and host range, the complexity of host resistance mechanisms, as well as its evolutionary and molecular characteristics. The study of *Sclerotinia* as a species—including its role as a causative agent, its classification within the *Sclerotiniaceae* family, the distribution of different species, with *Sclerotinia sclerotiorum* being the most prominent, and the development of management practices—has undergone numerous phases of research conducted worldwide.

2.1.1 Early phase of *Sclerotinia* species establishment

For a considerable duration, the recognition of *Sclerotinia* as a plant disease-causing organism has been extensively acknowledged. In 1979, Purdy provided a comprehensive historical account of *Sclerotinia*. In 1837, Libert described an earlier form of *Sclerotinia* as *Peziza sclerotiorum*. Later, in 1870, Fuckel established and described the genus *Sclerotinia*. Fuckel used a newly created binomial to rename *Peziza sclerotiorum* as *Sclerotinia libertiana* in honor of Libert's contributions. Wakefield's studies in 1924 showed that Fuckel had decided to create a new genus rather than endorse the combination of *S. sclerotiorum*. Wakefield subsequently showed, however, that Fuckel's use of *S. libertiana* did not conform to the International Rules of Botanical Nomenclature. As Massee had regularly used that binomial up until 1985, Wakefield concluded that Massee was the proper authority for *Sclerotinia sclerotiorum* (Lib.) Massee. It is important to remember that de Bary used it in his 1884 contribution earlier. The pathogen that causes clover stem rot is officially known as *Sclerotinia sclerotiorum* (Lib.) de Bary. It was identified as *Sclerotinia trifoliorum* Erikss by Ericksson in 1880. Wolf and Cromwell postulated in 1919 that clover (trifoliate) stem rot might have existed in the 1857 season close

DOI: 10.1079/9781800629592.0002

to Berber Beck in Hess, Germany. Additionally, it was noted that the disease might be closely linked to clover sickness, an ailment first discovered in England in the early 1800s. In 1849, researchers at Rothamsted Experiment Station began investigating this disease. But *S. trifoliorum* was acknowledged as the accurate name for the infection in 1897. According to Wolf and Cromwell (1919), Delaware published the first report of *S. trifoliorum* in the USA in 1890. Later in 1913, Jagger described a little *Sclerotinia* variety that was discovered in various parts of New York in crops such as celery and lettuce. He also talked about another kind, which he called *Sclerotinia minor*, that was discovered on lettuce in Sanford, Florida. According to a 1933 proposal by Valleau *et al.* (1933), *S. minor* and *S. trifoliorum* were essentially the same species because *S. minor* was detected on lettuce, a plant that was not frequently recognized as a host for *S. trifoliorum*. As a result, the basis for distinguishing between several *Sclerotinia* isolates was the associated plant and sclerotial size.

As a result, several species were discovered, such as *Sclerotinia sativa* Drayton and Groves, *S. trifoliorum* Erikss. var. *fabae* Keay, *Sclerotinia intermedia* Ramsy, and *Sclerotinia serica* Keay. Dennis did not include *S. intermedia* or *S. sativa* in his 1956 publication but did include *S. sclerotiorum*, *S. trifoliorum* (which includes other *Fabae* family host types), *S. minor*, *S. serica*, and *S. tuberosa*. He suggested that the latter two species were not documented and were only present in some other areas of the New World Notably, it indicates that although *S. intermedia*, *S. sativa*, *S. serica*, and *S. trifoliorum* var. *fabae* appear to be missing from the literature, seemingly neglected or disregarded, other researchers have adopted fundamental principles or elements of ideas proposed by Dennis.

2.1.2 Establishment of *Sclerotinia* as disease-causing agent

Certain diseases caused by *S. sclerotiorum* may serve as historical markers (Purdy, 1979). In one of the well-known cases, Smith (1900) clearly demonstrated that *S. sclerotiorum* (also known as *S. libertiana*) was the causative agent of the disease known as "lettuce drop," thereby separating it from *Botrytis cinerea* as a distinct fungus. While the diseases caused by *B. cinerea* and *S. sclerotiorum* were similar in lettuce crop, the true drop disease caused by *S. sclerotiorum* was more common in glasshouse lettuce in Massachusetts in the late 1890s. Additional cases of lettuce decline were reported by Stevens and Hall (1911) to have occurred in Massachusetts in 1890, Florida in 1896, North Carolina in 1897, and Wisconsin in 1904. Burger (1913), however, provided an explanation for the lettuce decline disease that occurred in North Carolina in 1897 and close to Gainesville in 1896. But it wasn't until 1900 that *S. libertiana* (also known as *S. sclerotiorum*) was linked to this well-known illness in Massachusetts. According to Jagger (1913), the fungus Smith (1900) reported was the same as the one he had seen in lettuce in different parts of New York. Jagger proposed that this fungus belonged to a *Sclerotinia* species that was "undescribed." As a result, Jagger identified the small sclerotia type known as *Sclerotinia minor* in 1920. According to Held and Haenseler (1953), *S. trifoliorum* may be the cause of severe episodes of lettuce drop in fields that are sown for the first time after clover or lucerne (alfalfa). Gilbert and Bennett (1917) and Wolf and Cromwell (1919) reported that Rhem had determined that *Peziza ciborioides* Fries (1872) was the pathogen causing clover stem rot. Ericksson (1880) disregarded this identification and determined that *S. trifoliorum* was the pathogen. In the USA, clover stem rot was originally documented by Wolf and Cromwell in 1890. According to Jones (1923), *S. libertiana* (*S. sclerotiorum*) is the causative agent of sunflower stem rot. Although the *Sclerotinia* blight of peanuts was reported in 1935 in China and in 1950 in Argentina, it was not until 1974 that it was first reported in the USA (Beute *et al.*, 1975). Smith, in 1929 and 1931, recognized cottony rot of lemons. The life cycle of *S. sclerotiorum*, in connection to the apricot green fruit rot disease, was also described by the mycologist Smith. The fig limb blight was first reported by Taubenhaus and Ezekiel (1931). Dickson (1930) detected the greenhouse tomato wilt disease and identified the fungus *S. libertiana* (*S. sclerotiorum*) as the cause of this disease. White mold on beans was extensively investigated, and the pathogen was thoroughly characterized by Harter and Zaumeyer (1944).

2.1.3 Establishment of family *Sclerotiniaceae*

The *Sclerotinia* is a fungal family that can be found in the *Helotiales* order within the *Ascomycota* phylum. Within this family, there are species that produce inoperculate asci that arise from brownish stipitate apothecia. These apothecia emerge from a *Sclerotinia* stroma that is either within or on related host plant (Whetzel, 1945). The growth of a *Sclerotinia* stroma, a melanized hyphal aggregate, is a common characteristic exhibited by all members of the *Sclerotinia* family. While the teleomorphic (perfect stage) characters remain consistent within this family, there are significant variations in the anamorphic state. This variation has led to the classification of different genera within the *Sclerotinia*, like *Monilinia* for *Monilia* and *Botryotinia* for *Botrytis* in anamorphic (imperfect stage) state. Kohn (1979b) further supported this separation. Over time, additional taxonomic criteria have been utilized to define the *Sclerotinia* beyond Whetzel's initial delimitation in 1945. The criteria encompass the following aspects: histochemistry; the ultra-structure of *Sclerotinia* as described by Backhouse and Willetts (1984); biochemical characteristics as outlined by Carbone and Kohn (1993); gene sequences as reported by Holst-Jensen *et al.* (1997 a, b); and the characteristics of sterile tissues derived from apothecia and *Sclerotinia* (Korf and Dumont, 1972; Kohn, 1979a, b). As of right now, this family has 33 genera that have been identified (Willetts and Wong, 1980; Willetts, 1997).

2.1.4 Distribution of *Sclerotinia* species

The classification of *Sclerotinia* species within the genus has undergone several revisions. *Sclerotinia* species have been distinguished using a variety of techniques, including size analysis (Jagger, 1920), host specificity (Kreitlow, 1949), size of ascus and ascospores (Ramsey, 1924), and nuclear and mitochondrial Restriction Fragment Length Polymorphism (RFLP) analysis (Kohn *et al.*, 1988). *Sclerotinia* currently has three officially recognized and identified species: *Sclerotiorum* (Lib.) de Bary, which is widely accepted worldwide, *S. minor* Jagger, and *S. trifoliorum* Eriks (Kohn *et al.*, 1988). Furthermore, *Sclerotinia nivalis* Saito (Saito, 1997; Li *et al.*, 2000) and *Sclerotinia asari* Wu and Wang (Wu and Wang, 1983) have been identified as separate *Sclerotinia* members through research. Though it hasn't been reclassified yet, the species *Sclerotinia homoeocarpa* F.T. Bannett isn't regarded as lawful (Kohn, 1979b; Rossman *et al.*, 1987).

2.1.5 Establishment of *Sclerotinia sclerotiorum* as a species

Libert officially named *P. sclerotiorum* and published the first description of *S. sclerotiorum* in 1837. Up until Fuckel moved the species to the new genera *Sclerotinia* in 1870, this classification remained in place. Then, in 1979, Purdy updated it as *S. libertiana* Fuckel in honor of Libert. *P. sclerotiorum* Lib. and *S. sclerotiorum* Fuckel were regarded as interchangeable during this time (Wakefield, 1924).

Mycologists and plant pathologists used the term *S. libertiana* extensively until Wakefield (1924) pointed out that it was in violation of the International Code of Botanical Nomenclature. The law states that a species should keep its original specific name when it is moved to a new genus, with the exception of the resultant, already used combination. But the name was not in use in the instance of *S. sclerotiorum*. Wakefield (1924) cited *S. sclerotiorum* (Lib.) Massee because it was wrongly believed that G.E. Massee used the combination *S. sclerotiorum* for the first time in 1895. Purdy (1979) subsequently ascertained that de Bary had, in fact, used the name in 1884. *S. sclerotiorum* (Lib.) de Bary is hence the pathogen's officially recognized correct name and authority.

In addition to the confusion over the exact nomenclature of the disease, there has also been uncertainty over the particular specimen that most closely resembles it (Korf and Dumont, 1972). In 1973, Buchwald and Neergaard suggested that the type species of *S. sclerotiorum* (Lib.) de Bary genus be conserved in order to address this problem (Kohn, 1979b). In 1981, this suggestion was acknowledged as a preserved name.

2.2 Geographical Distribution

The *Sclerotinia* is a highly destructive and widespread plant pathogen, causing diseases that have been given more than 60 different names. This fungal pathogen is capable of infecting over 900 plant species across the globe (O'Sullivan *et al.*, 2021; Mehta *et al.*, 2023). The presence of *S. sclerotiorum* has been documented in numerous countries spanning all continents. Davies *et al.* (1999) have provided a concise overview of the existence and distribution of all the three species of *Sclerotinia*, such as *S. sclerotiorum*, *S. trifoliorum*, and *S. minor*, on various crops in the UK. It is highly likely that this fungus can be found in almost every country or region on Earth, causing diseases in multiple plant species (Purdy, 1979; Boland and Hall, 1994; Saharan and Mehta, 2008; Mehta *et al.*, 2023).

2.2.1 *Sclerotinia sclerotiorum* distribution map

A distribution map of the fungus/pathogen *S. sclerotiorum* (Lib.) de Bary, which is a member of the *Helotiales* group of *Ascomycota*, is available. This fungus is known to infect a wide range of host plants, making it plurivorous. The pathogen has been documented across various regions of the world, infecting more than 900 different host plants, and up-to-date information has been provided in Table 2.1. The countries in all the continents where *Sclerotinia* has been reported are listed in the following section:

Africa: Nigeria, South Africa (St. Helena), Tanzania, Zimbabwe, Algeria, Congo, Egypt, Ethiopia, Kenya, Libya, Malawi, Mauritius, Morocco, North Africa, and Zimbabwe.

Asia: Azerbaijan, China (Anhui, Fujian, Guangdong, Guangxi, Guizhou, Hainan, Hebei, Heilongjiang, and Henan), Hong Kong (Hubei, Hunan, Jiangsu, Jiangxi, Jifin, Liaoning, Nei Menggu, Ningxia, Shaanxi, Shandong, Shanxi, Sichuan, Xinjiang, Xizhang, Yunnan, and Zhejiang), Republic of Georgia, India (Arunachal Pradesh, Assam, Bihar, Delhi, Haryana, Himachal Pradesh, Jammu and Kashmir, Madhya Pradesh, Nagaland, Punjab, Rajasthan, Tamil Nadu, Uttar Pradesh, and West Bengal), Iran, Israel, Japan (Hokkaido, Honshu, Kyushu, Ryukyu Archipelago, and Shikoku), Jordan, Korea Republic, Lebanon, Nepal, Pakistan, Singapore, Syria, Taiwan, Tajikistan, Thailand, Turkey, and Uzbekistan.

Australia: States including New South Wales, Queensland, South Australia, Tasmania, Victoria, Western Australia; Fiji, New Zealand, and Samoa.

Central America and Caribbean (Bermudas): Costa Rica, El Salvador, Guatemala, Nicaragua, and Panama.

Europe: Albania, Austria, Belarus, Belgium, Bulgaria, Croatia, Cyprus, Czech Republic, Denmark, Finland, France, Germany, Greece, Hungary, Iceland, Ireland, Italy, Latvia, Lithuania, Malta, Moldova, Netherlands, Norway, Poland, Portugal, Romania, Central Russia, Russian Far East, Southern Russia, Serbia and Montenegro, Slovakia, Spain, Sweden, Switzerland, UK, and Ukraine.

North America: The provinces of Canada include Alberta, British Columbia, Manitoba, New Brunswick, Newfoundland, Nova Scotia, Ontario, Labrador, Prince Edward Island, Quebec, Saskatchewan), USA includes Alaska, Alabama, Arizona, Arkansas, California, Colorado, Connecticut, Delaware, Florida, Georgia, Hawaii, Idaho, Illinois, Indiana, Owa, Kentucky, Louisiana, Maine, Maryland, Massachusetts, Michigan, Minnesota, Mississippi, Missouri, Montana, Nebraska, New Hampshire, New Jersey, New York, North Carolina, North Dakota, Ohio, Oklahoma, Oregon, Pennsylvania, Rhode Island, South Carolina, Texas, Vermont, Virginia, Washington, West Virginia, Wisconsin, Wyoming (according to DM PD, 2005).

South America: Argentina, Bolivia, Brazil (Mato Grosso, Minas Gerais, Parana, Rio Grande Do Sul, and São Paulo), Chile, Ecuador, Peru, Uruguay, Venezuela, Oceania, American Samoa, and Samoa (DMPD, 2005).

To obtain the distribution map of *S. minor*, *S. trifoliorum*, *Sclerotinia fructigena*, *Sclerotinia laxa*, *Sclerotinia fructicola*, *Sclerotinia*

Table 2.1. Distribution of *Sclerotinia sclerotiorum* on different hosts[a] (Boland and Hall, 1994; Saharan and Mehta, 2008; Mehta *et al*., 2023 updated 2024)

Sr. no.	Host name	English name	Country	Year	References
1.	*Abelmoschus esculentus* (L.) Moench.	Okra	Bangladesh	2017	Prova *et al*. (2017)
			USA	1960, 1984	Anon (1960); Alfieri Jr *et al*. (1984)
2.	*Abutilon theophrasti* Medik	Velvet leaf	New York	1990	Dillard *et al*. (1991)
			Croatia	1997	Jurkovic and Culek (1997)
			North Carolina	1985	Grand (1985)
3.	*Acalypha australis*	Asian copperleaf	China	1979	Tai (1979)
			Taiwan	1979	Anon (1979)
4.	*Acca sellowiana* (as *Feijoa sellowiana*)	Pineapple guava/guava Steen	Italy	1994	Aloj *et al*. (1994)
5.	*Acer saccharinum*	Silver maple	Florida	1991	Miller (1991a,b)
6.	*Aciphylla dieffenbachii*	Dumb cane or leopard lily	New Zealand	1999	McKenzie and Johnston (1999)
7.	*Aconitum carmichaelii* Debeaux	Azure monkshood	Colorado	1960	Anon (1960)
8.	*Acroptilon repens* (as *Centaurea repens* L.)	Russian knapweed	Canada	1989	Mortensen and Molloy (1989)
9.	*Actinidia deliciosa* Planch	Chinese gooseberry/kiwifruit	New Zealand	1989	Pennycook (1989)
			Korea	2014	Lee *et al*. (2015)
10.	*Actinidia chinensis*	Kiwifruit	Italy	2023	Aci *et al*. (2024)
11.	*Adenophora remotiflora*	Scattered lady bell	Korea	2003, 2004	Chang and Kim (2003); Cho and Shin (2004)
12.	*Aegopodium* spp.	Goutweed	USA	1974	Schwartz (1977)
13.	*Aeschynomene americana*	Shyleaf	Australia	1990	Lenne (1990)
14.	*Aesculus hippocastanum*	Horse chestnut	Canada	1972, 1986	Gourley and Delbridge (1972); Ginns (1986)
15.	*Ageratum conyzoides* L.	Billy goat weed	India	1986	Singh and Singh (1986)
16.	*Ajuga genevensis*	Blue bugle	Australia	1982	Sampson and Walker (1982)

Continued

Table 2.1. Continued

Sr. no.	Host name	English name	Country	Year	References
17.	*Alcea ficifolia* (L.) Cav.	Antwerp hollyhock	Canada	1970	Duczek and Morrall (1971)
			USA	1974	Schwartz (1977)
18.	*Alcea rosea* L. (Syn. *Althaea rosea*) (Crown rot)	Hollyhock	Brazil	1954	Neto (1954)
			California	1989	French (1989)
			Canada	1938, 1964, 1965, 1967, 1986	Bisby (1938); Creelman (1964, 1965); Conners (1967); Ginns (1986)
			Montana, New Jersey	1960, 1973	Anon (1960); Shaw (1973)
19.	*Alcea* sp.	Hollyhocks	British Columbia, Canada	1998	Joshi and Elmhirst (1998)
20.	*Althaea officinalis*	Marshmallow	Korea	2016	Kim *et al*. (2016)
21.	*Allium cepa* L.	Onion	Bulgaria	2009	Bobev (2009)
			California, Florida, Idaho, Michigan, Ohio	1960	Anon (1960)
			Greece	2000	Holevas *et al*. (2000)
			Hawaii	1981	Raabe *et al*. (1981)
			New Zealand	1989	Pennycook (1989)
			Washington	1973	Shaw (1973)
			West Virginia	1982	Williams and Hayne (1982)
22.	*Allium fistulosum*	Welse onion	China	1979	Tai (1979)
			Taiwan	1979	Anon (1979)
23.	*Allium sativum* L.	Garlic	Brazil	1998	Mendes *et al*. (1998)
			California	1989	French (1989)
			New Zealand	1989	Pennycook (1989)
			Taiwan	1979	Anon (1979)

24.	*Allium tuberosum*	Garlic chives	South Korea	2017	Choi *et al.* (2017a,b)
25.	*Alternanthera sessilis*	Stalkless Joy weed	China	1979	Tai (1979)
			Taiwan	1979	Anon (1979)
26.	*Amaranthus deflexus*	Perennial pigweed	South Africa	1992	Phillips (1992)
27.	*Amaranthus retroflexus* L.	Redroot pigweed	Colorado	1957	Burke *et al.* (1957)
			Montana	1960, 1973	Anon (1960); Shaw (1973)
28.	*Amaranthus* sp.	Undetermined	Brazil	1998	Mendes *et al.* (1998)
			Canada	1967	Conners (1967)
			France	2018	Leyronas *et al.* (2018)
			South Africa	2000	Crous *et al.* (2000)
29.	*Amaranthus viridis*	Slender amaranth/Green amaranth	China	1979	Tai (1979)
			Taiwan	1979	Anon (1979)
			India	2004	Ghasolia *et al.* (2004)
30.	*Ambrosia artemisiifolia* L.	Ragweed	USA	1960, 1984, 1986	Anon (1960); Alfieri Jr *et al.* (1984); Dillard and Hunter (1986)
			Europe	1998	Bohár and Kiss (1999)
31.	*Ambrosia hispida* Pursh (Stem rot)	Undetermined	Florida	1984	Alfieri Jr *et al.* (1984)
32.	*Ammi majus*	Bishop's weed	India	1994	Upadhyaya (1994); Kanno and Ohkubo (1999)
33.	*Anagallis arvensis* L.	Scarlet pimpernel	China	1979	Tai (1979)
			India	1986	Singh and Singh (1986)
			Taiwan	1979	Anon (1979)
34.	*Anchusa azurea* Mill.	Alkanet (Buglossum)	Canada	1967, 1986	Conners (1967); Ginns (1986)
			British Columbia	1964	Toms (1964)
35.	*Anchusa capensis* Thunb.	Alkanet, Bugloss	India	1973	Roy (1973)

Continued

Table 2.1. Continued

Sr. no.	Host name	English name	Country	Year	References
36.	*Andrographis paniculata*	Creat or green chiretta	China	2016	Shi *et al*. (2016)
37.	*Anemone coronaria* (L.) Wilt.	Poppy anemone	Bermuda	1941	Russell (1942)
			Florida	1984	Alfieri Jr *et al*. (1984)
			Italy	1981	Garibaldi (1981)
			Korea	2013	Han *et al*. (2013)
			New Zealand	1989	Pennycook (1989)
38.	*Anemone* sp.	Wind flowers	Australia	1989	Shivas (1989)
			California	1989	French (1989)
39.	*Anethum graveolens* L.	Dill	California	1989	French (1989)
			Canada	1986	Ginns (1986)
			Texas	1960	Anon (1960)
40.	*Anethum sowa*	Sowa/soya/dill/gardendill	India	1991	Sinha and Singh (1991)
41.	*Angelica archangelica* L.	Angelica	Korea	2003, 2004	Chang and Kim (2003); Cho and Shin (2004)
42.	*Angelica keiskei* Koidz.	Ashitaba	Japan	1996	Takeuchi and Horie (1996)
43.	*Anoectochilus lanceolatus*	Jewel orchid	Japan	1996	Takeuchi and Horie (1996)
44.	*Annona squamosa* L.	Sugar apple	Israel	1963	Palti (1963)
45.	*Anthemis tinctoria*	Golden chamomile/golden marguerite	Australia	1982	Sampson and Walker (1982)
46.	*Anthriscus cerefolium*	Garden chervil	California	1999	Koike (1999)
47.	*Anthriscus sylvestris*	Cow parsley	Korea	2003	Chang and Kim (2003)
48.	*Antirrhinum majus* L.	Common snapdragon	California	1989	French (1989)
			Canada	1967	Conners (1967)
			Chile	1967	Mujica and Oehrens (1967)
			Florida	1984	Alfieri Jr *et al*. (1984)
			Illinois	1964	Boew (1964)

			India	1964	Shukla and Pavgi (1964)
			Korea	2004	Cho and Shin (2004)
			Mauritius	1968	Orieux and Felix (1968)
			New Zealand	1932, 1989	Brien (1932); Pennycook (1989)
			Ohio	1957	Ellett (1957)
			Scotland	1961	Foister (1961)
			Texas	1960	Anon (1960)
			UK	1926	Dowson (1926)
49.	*Antirrhinum* sp.	Dragon flowers	Cyprus	1957	Georghiou and Papadopoulos (1957)
50.	*Apios americana* Medic.	America-Hodoimo/American *groundnut*/groundnut	Louisiana	1990	Holcomb (1990)
			USA	1965	Anon (1965)
			Virginia	1982	Phipps and Porter (1982)
51.	*Apium graveolens*	Celery	Australia	1982, 1989	Sampson and Walker (1982); Shivas (1989); Cook and Dubé (1989)
			California	1952, 2006	Middleton *et al.* (1952); Koike *et al.* (2006)
			Canada	1967, 1986	Conners (1967); Ginns (1986)
			China	1979	Tai (1979)
			Idaho	1973	Shaw (1973)
			Korea	2004	Cho and Shin (2004)
			Mauritius	1968	Orieux and Felix (1968)
			Montana	1973	Shaw (1973)
			New Zealand	1989	Pennycook (1989)
			New York	1944	Newhall (1944)
			Oregon	1973	Shaw (1973)

Continued

Table 2.1. Continued

Sr. no.	Host name	English name	Country	Year	References
52.	*Apium graveolens* var. *dulce* (Mill.) Pers	Celery	California	1989	French (1989)
			Florida	1984	Alfieri Jr *et al*. (1984)
			Hawaii	1981	Raabe *et al*. (1981)
			Range of host	1960	Anon (1960)
			Taiwan	1979	Anon (1979)
53.	*Apium graveolens* L. var. *rapaceum* (Mill.) Gaud Beaup.	Celeriac	California	1960	Anon (1960)
54.	*Aquilegia flabellata*	Dwarf columbine	Italy	2011	Garibaldi *et al*. (2011)
55.	*Aquilegia* spp.	Columbine	Canada	1967	Conners (1967)
56.	*Aquilegia vulgaris* L.	European crowfoot	Delaware, Florida, Ohio, Pennsylvania, Texas	1960, 1984	Anon (1960); Alfieri Jr *et al*. (1984)
57.	*Arabidopsis thaliana* (L.) Heynh.	Mouse-ear cress/Thale cress	China	2008	Wang *et al*. (2008)
			Maryland, USA	1971	Morgan (1971)
58.	*Arabis caucasica*	Garden arabis	New Zealand	1996	McKenzie and Dingley (1996)
59.	*Arachis hypogaea* L.	Peanut	Australia	1966, 1989	Simmonds (1966); Shivas (1989)
			Brazil	1998	Mendes *et al*. (1998)
			China	2014, 2022	Yan *et al*. (2014); Zhang *et al*. (2022)
			Georgia	2006	Woodward *et al*. (2006)
			Libya	1981	El-Buni and Rattan (1981)
			Mauritius	1968	Orieux and Felix (1968)
			North Carolina	1974, 1975	Porter and Beute (1974); Beute *et al*. (1975)
			New Zealand	1989	Pennycook (1989)
			New Mexico	2007	Sanogo and Puppala (2007)
			Oklahoma	1979	Wadsworth (1979)

			Texas	2008	Woodward *et al.* (2008)
			Virginia	1974, 1975	Porter and Beute (1974); Beute *et al.* (1975)
60.	*Aralia cordata* Thunb.	Spikenard	Maryland	1960	Anon (1960)
			USA	1923	Weimer (1923)
61.	*Arctium lappa* L.	Greater burdock	China	1979	Tai (1979)
			Japan	1991	Shiraishi and Sunaga (1991)
			Taiwan	1979	Anon (1979)
62.	*Arctium minus* (Hill) Bernh	Common burdock	Canada	1986	Ginns (1986)
63.	*Arctotheca calendula*	Capeweed/cape dandelion	Australia	1982, 1989	Sampson and Walker (1982); Cook and Dubé (1989)
64.	*Arctotis stoechadifolia* Bergius	African daisy	India	1977	Kaur and Chohan (1977)
65.	*Arctotis grandiflora*	*Silver arctotis*	Israel	1963	Palti (1963)
66.	*Ardisia crenata* Sims	Coral berry	Florida	1984	Alfieri Jr *et al.* (1984)
67.	*Ardisia crispa* (Thunb.) A. D.C.	Ardisia	Florida	1984	Alfieri Jr *et al.* (1984)
68.	*Aristolocha durior* Hill	Dutchman's pipe	Canada	1967	Conners (1967)
69.	*Argemone* sp.	Argemone	India	1924	Joshi (1924)
70.	*Argyranthemum frutescens*	Paris daisy	Italy	2008	Garibaldi *et al.* (2008d)
71.	*Arisaema dracontium*	Dragon root/green dragon	Canada	1967	Conners (1967)
72.	*Armoracia lapathifolia*	Horseradish	Korea	2003, 2004	Chang and Kim (2003); Cho and Shin (2004)
73.	*Armoracia rusticana* Gaertn., Mey & Scherb	Horseradish	New Zealand	1989	Pennycook (1989)
			USA	1940	Kadow and Anderson (1940)
74.	*Artemisia vulgaria/A. absinthium*	Mug wort/*Worm* wood	India	2004	Ghasolia *et al.* (2004)
75.	*Artemisia capillaris*	Medicinal Plant	China	2023	Liu *et al.* (2023)
76.	*Artocarpus heterophyllus*	Jackfruit	Bangladesh	2015	Rahman *et al.* (2015)
77.	*Asarum canadense* L.	Wild ginger	New York	1960	Anon (1960)

Continued

Table 2.1. Continued

Sr. no.	Host name	English name	Country	Year	References
78.	*Asclepias* sp.	Milkweed	USA	1974	Schwartz (1977)
79.	*Asimina* sp. Adans	Pawpaw	Bermuda	1946	Kinghorn (1947)
80.	*Asparagus officinalis*	Sparrow grass	Canada	1986	Ginns (1986)
			New Zealand	1989	Pennycook (1989)
			Southern states	1960	Anon (1960)
81.	*Asparagus officinalis* L. var. *altilis* L.	Asparagus	USA	1960	Anon (1960)
			Canada	1976	Morrall *et al*. (1976)
82.	*Asphodelus tenuifolius* Cav.	Wild onion	India	1993	Rathore *et al*. (1993)
83.	*Aster pilosus*	Hairy white old field aster/frost aster	Japan	1999	Takeuchi and Horie (1999)
84.	*Aster* sp.	Aster	California	1989	French (1989)
			Connecticut	1960	Anon (1960)
			Florida	1984	Alfieri Jr *et al*. (1984)
			Scotland	1961	Foister (1961)
85.	*Aster tataricus*	Atari now's aster	Korea	2003, 2004	Chang and Kim (2003); Cho and Shin (2004)
86.	*Asteromoea indica*	Indian aster or Indian kalimeris	China	1979	Tai (1979)
87.	*Astragalus sinicus*	Milk vetch	China	1979	Tai (1979)
			USA	1974	Schwartz (1977)
			Taiwan	1972	Peregrine and Siddiqi (1972)
88.	*Atractylodes lancea*	Chinese medicinal herb	China	2022	Han *et al*. (2022)
89.	*Avena* sp.	Oats	–	1994	Boland and Hall (1994); Saharan and Mehta (2008)
90.	*Azadirachta indica*	Neem	India	2004	Ghasolia and Shivpuri (2004a)
91.	*Azolla pinnata* R.Br.	Mosquito fern	India	1985	Balou and Iboton (1985)
92.	*Barbarea vulgaris* R Br.	Yellow rocket	USA	1971	Morgan (1971)

93.	*Barleria cristata*	Philippine violet	India	2015	Mondal *et al*. (2015)
94.	*Basella rubra* (*B. alba*)	Malabar nightshade	Japan	1999	Kanno and Ohkubo (1999)
95.	*Begonia tuberhybrida* Voss	Begonia	California	1960, 1989	Anon (1960); French (1989)
96.	*Bellis perennis* L.	English daisy	New Jersey	1960	Anon (1960)
			India	2014	Hansda *et al*. (2014)
97.	*Berberis* sp.	Barberry	USA	1974	Schwartz (1977)
98.	*Beta vulgaris* L.	Beet/sugar beet	Australia	1982	Sampson and Walker (1982)
			Bulgaria	2009	Bobev (2009)
			California	1989	French (1989)
			China	1979, 1990	Tai (1979); Zhang and Huang (1990)
			Connecticut	1960	Anon (1960)
			Florida	1984	Alfieri Jr *et al*. (1984)
			Illinois	1960	Anon (1960)
			Japan	1986	Naito and Sugimoto (1986)
			Minnesota	2021	Khan *et al*. (2021)
			Montana	1973	Shaw (1973)
			North Dakota	2020	Khan *et al*. (2020)
			New Zealand	1989	Pennycook (1989)
			South Africa	2000	Crous *et al*. (2000)
			Taiwan	1979	Anon (1979)
99.	*Beta vulgaris* var. *cicla*	Chard	Korea	2003, 2004	Chang and Kim (2003); Cho and Shin (2004)
100.	*Beta vulgaris* var. *saccharifera*	Sugar beet	Korea	2004	Cho and Shin (2004)
101.	*Bidens biternata* (Lour.) Merr. & Scherff.	Blackjack	Korea	1987	Shin and Lee (1987)
102.	*Bidens cosmoides*	Hitchhikers/cosmos flower	South Africa	1992, 2000	Phillips (1992); Crous *et al*. (2000)

Continued

Table 2.1. Continued

Sr. no.	Host name	English name	Country	Year	References
103.	*Bidens pilosa*	Beggar's tick/common blackjack	Brazil	1998	Mendes *et al*. (1998)
			China	1979	Tai (1979)
			South Africa	1992	Phillips (1992)
104.	*Blumea sericans*	Blumea	China	1979	Tai (1979)
			Taiwan	1979	Anon (1979)
105.	*Boehmeria frutescens* (Syn. *Leucosyke puya*)	Chinese silk plant	Taiwan	1979	Anon (1979)
106.	*Boehmeria nivea*	True ramie/China grass	China	1979	Tai (1979)
107.	*Borago officinalis*	Borage	Saskatchewan, Canada	2000	Hilton (2000)
			Italy	2008	Garibaldi *et al*. (2008c)
			North Dakota	2005	Bradley *et al*. (2005)
			Tanzania	1960	Riley (1960)
108.	*Borreria alata* (syn. *Spermacoce alata*)	Winged false buttonweed	Brazil	1998	Mendes *et al*. (1998)
109.	*Bothriospermum tenellum* (Hornem.)	Leaf between flower	China	1979	Tai (1979)
			Taiwan	1979	Anon (1979)
110.	*Bouvardia* sp.	Firecracker bush	Japan	2002	Horie and Hoshi (2002)
111.	*Brachiaria decumbens*	Signal grass	Brazil	1998	Mendes *et al*. (1998)
112.	*Brachycombe iberidifolia* Benth.	Swan river daisy	India	1964	Shukla and Pavgi (1964)
113.	*Brassica arvensis*	Charlock mustard/field mustard/ wild mustard	India	2004	Ghasolia *et al*. (2004)
114.	*Brassica campestris* L.	Bird rape	California	1960	Anon (1960); French (1989)
			Canada	1967, 1980, 1983	Conners (1967); Williams and Stelfox (1980); Dueck and Frederick (1983)
			Chile	1967	Mujica and Oehrens (1967)

			China	1979, 1990	Tai (1979); Zhang and Huang (1990)
			India	1987	Saxena and Rai (1987)
			New Zealand	1932	Brien (1932)
			Bermuda	1941	Russell (1942)
			Taiwan	1979	Anon (1979)
115.	*Brassica rapa* subsp. *chinensis* (Syn. *Brassica campestris* subsp. *Chinensis* (L.; Makino)	Pak soy or Chinese cabbage	Taiwan	1979	Anon (1979)
			Japan	1996	Takeuchi and Horie (1996)
116.	*Brassica campestris* subsp. *napus*	Rape, field mustard	Korea	2004	Cho and Shin (2004)
117.	*Brassica campestris* subsp. *Pekinensis* (Syn. *Brassica rapa* subsp. *pekinensis*)	Napa cabbage	Korea	2004	Cho and Shin (2004)
118.	*Brassica campestris* var. *marinosa*	Rape/field mustard	Korea	2003, 2004	Cho and Shin (2004); Chang and Kim (2003)
119.	*Brassica campestris* var. *napobrassica* (L.) D. C.	Rutabaga (Swede)	Maine	1960	Anon (1960)
			Mississippi	1959	Parris (1959)
			Washington	1960, 1973	Anon (1960); Shaw (1973)
			New Zealand	1932	Brien (1932)
			Canada	1967	Conners (1967)
			British Columbia	1964	Toms (1964)
120.	*Brassica campestris* var. *oleifera*	Rape/field mustard/bird's rape	China	1979	Tai (1979)
121.	*Brassica campestris* var. *purpuraria*	Common *turnip*/field mustard	China	1979	Tai (1979)
122.	*Brassica campestris* var. *rapa*	Bird's rape/field mustard	Australia	1989	Shivas (1989)
123.	*Brassica carinata*	Abyssinian mustard/Ethiopian rapeseed	Florida	2000, 2012	Corato and Baviello (2000); Young *et al*. (2012); Kelly *et al*. (2012)
			Australia	2016	Uloth *et al*. (2016)
124.	*Brassica oleracea var. caulorapa*	Stem turnip	Washington	1973	Shaw (1973)

Continued

Table 2.1. Continued

Sr. no.	Host name	English name	Country	Year	References
125.	*Brassica cernua*	Kurashina	China	1979	Tai (1979)
126.	*Brassica chinensis* L.	Pak choi	China	1979	Tai (1979)
			Mauritius	1948	Wiehe (1948)
			Washington	1960	Anon (1960)
127.	*Brassica cretica* subsp. *botrytis*	Common cauliflower	Greece	1973, 2000	Pantidou (1973); Holevas *et al.* (2000)
128.	*Brassica hirta* Moench.	White mustard	Canada	1965, 1972, 1986	Petrie and Vanterpool (1965); Morrall *et al.* (1972); Ginns (1986)
			Washington	1960	Anon (1960)
129.	*Brassica juncea* (L.) Coss.	Leaf mustard	Canada	1972, 1986	Morrall *et al.* (1972); Ginns (1986)
			China	1979	Tai (1979)
			Brazil	1954	Neto (1954)
			India	1973, 1980, 1987	Roy (1973); Dhawan (1980); Saxena and Rai (1987)
			Taiwan	1979	Anon (1979)
			Tennessee	2018	Shrestha *et al.* (2018)
			USA	2018	Shrestha *et al.* (2018)
130.	*Brassica juncea* (L.) Coss var. *crispifolia* Bailey	Curled mustard	Washington	1960	Anon (1960)
131.	*Brassica juncea* var. *tumida*	Tumorous stem mustard	China	1979	Tai (1979)
132.	*Brassica kaber* (D.C.) Wheeler	Charlock	Canada	1986	Ginns (1986)
			Montana	1960	Anon (1960)
			Manitoba	1975	Hoes and Huang (1976)
133.	*Brassica napobrassica*	Rutabaga	Montana	1973	Shaw (1973)
			Washington	1973	Shaw (1973)
134.	*Brassica napus* L.	Rape	Argentina	2005	Gaetan and Madia (2005)

			Australia	1989, 2016	Shivas (1989); Derbyshire and Denton-Giles (2016); van de Wouw *et al*. (2016)
			Brazil	1998	Mendes *et al*. (1998)
			Bulgaria	2009	Bobev (2009)
			California	1960, 1989	Anon (1960); French (1989)
			Canada	1980, 1983, 1986, 2016	Williams and Stelfox (1980); Thomson and Stelfox (1983); Dueck and Frederick (1983); Ginns (1986); Derbyshire and Denton-Giles (2016)
			Manitoba, Saskatchewan, Canada	2000	Hilton (2000)
			Alberta	1982	Berkenkamp (1982)
			China	2020	Yu *et al*. (2020)
			Florida	1984	Alfieri Jr *et al*. (1984)
			France	2018	Leyronas *et al*. (2018)
			USA	1939	Raabe (1939)
			Georgia	1991	Brenneman and Sumner (1991)
			Greece	2008	Tziros *et al*. (2008)
			Idaho	1960, 1973	Anon (1960); Shaw (1973)
			Mississippi	1959, 1960	Parris (1959); Anon (1960)
			New Zealand	1932, 1989	Brien (1932); Pennycook (1989)
			Norway	2018	Ficke *et al*. (2018)
			Texas	2010	Isakeit *et al*. (2010)
			USA	2016	Derbyshire and Denton-Giles (2016)
135.	*Brassica napus* subsp. *napus*	Oilseed rape	China	2014	Zhou *et al*. (2014)
136.	*Brassica napus* var. *napobrassica*	Rutabaga	Australia	1982	Sampson and Walker (1982)

Continued

Table 2.1. Continued

Sr. no.	Host name	English name	Country	Year	References
			Canada	1986	Ginns (1986)
			New Zealand	1989	Pennycook (1989)
137.	*Brassica napus* var. *napus*	Rape	Australia	1989	Cook and Dubé (1989)
			Florida	1984	Alfieri Jr *et al*. (1984)
138.	*Brassica napus* var. *oleifera*	Turnip	South Africa	2000	Crous *et al*. (2000)
139.	*Brassica nigra* (L.) Koch	Black mustard	California	1960, 1989	Anon (1960); French (1989)
			New Zealand	1932, 1989	Brien (1932); Pennycook (1989)
			Texas	1960	Anon (1960)
140.	*Brassica oleracea*	Cabbage	Australia	1966	Simmonds (1966)
			California	1971, 1989	Farmer *et al*. (1971); French (1989)
			China	1933, 1979	Tai and Wei (1933); Tai (1979)
			Mauritius	1968	Orieux and Felix (1968)
			Oregon	1973	Shaw (1973)
			Panama	2006	Piepenbring (2006)
			Sri Lanka	2017, 2020	Mahalingam *et al*. (2017); Adikaram and Yakandawala (2020)
			Taiwan	1979	Anon (1979)
			Tennessee	2018	Shrestha *et al*. (2018)
			USA	2018	Shrestha *et al*. (2018)
			Washington	1973	Shaw (1973)
141.	*Brassica oleracea* L. var. *acephala* D.C.	Kale	Brazil	1998	Mendes *et al*. (1998)
			Korea	2004	Cho and Shin (2004)
			New Zealand	1989	Pennycook (1989)
			USA	1974	Schwartz (1977)
			North Carolina	1985	Grand (1985)

			Georgia	1963	Hanlin (1963)
142.	*Brassica oleracea* L. var. *botrytis* L.	Broccoli/cauliflower	Australia	1982, 1989	Sampson and Walker (1982); Shivas (1989); Cook and Dubé (1989)
			Brazil	1998	Mendes *et al*. (1998)
			British Columbia	1964	Toms (1964)
			California	1960	Anon (1960)
			Canada	1967, 1986	Conners (1967); Ginns (1986)
			China	1979	Tai (1979)
			Korea	2004	Cho and Shin (2004)
			North Carolina	1985	Grand (1985)
			New Zealand	1989	Pennycook (1989)
			Range of host	1960	Anon (1960)
			South Africa	1977, 2000	Gorter (1977); Crous *et al*. (2000)
			Taiwan	1979	Anon (1979)
			Texas	1960	Anon (1960)
			India	1973, 1986, 1988, 2014	Roy (1973); Kapoor *et al*. (1986); Gupta and Agarwala (1988); Hansda *et al*. (2014)
143.	*Brassica oleracea* L. var. *capitata* L.	Cabbage	Australia	1982, 1989	Sampson and Walker (1982); Shivas (1989); Cook and Dubé (1989)
			Brazil	1954, 1998	Neto (1954); Mendes *et al*. (1998)
			British Columbia,	1964	Toms (1964)
			Canada	1938, 1967, 1986	Bisby (1938); Conners (1967); Ginns (1986)
			China	1979	Tai (1979)
			Costa Rica	1967	McGuire Jr and Crandall (1967)

Continued

Table 2.1. Continued

Sr. no.	Host name	English name	Country	Year	References
			Florida	1984	Alfieri Jr *et al.* (1984)
			Georgia	1963	Hanlin (1963)
			Greece	1973, 2000	Pantidou (1973); Holevas *et al.* (2000)
			Hawaii	1981	Raabe *et al.* (1981)
			India	1973	Roy (1973)
			Korea	2004	Cho and Shin (2004)
			North Carolina	1985	Grand (1985)
			New Zealand	1989	Pennycook (1989)
			New Mexico	2015	Sanogo *et al.* (2015)
			New York	1986	Dillard and Hunter (1986)
			Pakistan	1988	Mirza and Ahmad (1988)
			Panama	1967	McGuire Jr and Crandall (1967)
			Range of host	1960	Anon (1960)
			South Africa	1977, 2000	Gorter (1977); Crous *et al.* (2000)
			Taiwan	1979	Anon (1979)
144.	*Brassica oleracea* L. var. *caulorapa* D.C.	Knol khol	India	1989	Singh and Saha (1989)
145.	*Brassica oleracea* L. var. *gemmifera* Zenk.	Brussels sprouts	Australia	1982, 1989	Sampson and Walker (1982); Shivas (1989); Cook and Dubé (1989)
			Brazil	1954	Neto (1954)
			California; Louisiana	1960	Anon (1960)
			India	1973	Roy (1973)
			North Carolina	1985	Grand (1985)
			New Zealand	1989	Pennycook (1989)

			South Africa	1977, 2000	Gorter (1977); Crous *et al*. (2000)
			Texas	1960	Anon (1960)
			Mexico	1923	Ayvar-Serna *et al*. (2023)
146.	*Brassica oleracea* L. var. *gongylodes* L.	Kohlrabi	Indiana	1960	Anon (1960)
			India	1973	Roy (1973)
			Korea	2014	Kim *et al*. (2014)
147.	*Brassica oleracea* var. *italica*	Broccoli	Brazil	1998	Mendes *et al*. (1998)
			New Zealand	1989	Pennycook (1989)
			South Africa	1977, 2000	Gorter (1977); Crous *et al*. (2000)
148.	*Brassica oleracea* L. var. *ramosa* Alef.	Thousand-head kale	Australia	1982	Sampson and Walker (1982)
			USA	1974	Schwartz (1977)
			New Zealand	1932, 1989	Brien (1932); Pennycook (1989)
149.	*Brassica oleracea* L. var. *viridis*	Kale	Georgia	1960, 1963	Anon (1960); Hanlin (1963)
			USA	1960, 1985	Anon (1960); Grand (1985)
150.	*Brassica pekinensis* (Lour.) Rupr.	Chinese cabbage	Arizona	1960	Anon (1960)
			China	1979, 1990	Tai (1979); Zhang and Huang (1990)
			Hawaii	1966, 1981	Raabe (1966); Raabe *et al*. (1981)
			USA	1960	Anon (1960)
			Taiwan	1979	Anon (1979)
151.	*Brassica petsai*	Petsai	Mauritius	1968	Orieux and Felix (1968)
152.	*Brassica rapa* L.	Turnip	Australia	1982	Sampson and Walker (1982)
			Canada	1967, 1986	Conners (1967); Ginns (1986)
			China	1979	Tai (1979)
			Connecticut	1960	Anon (1960)
			USA	1939	Raabe (1939)
			Idaho	1973	Shaw (1973)

Continued

Table 2.1. Continued

Sr. no.	Host name	English name	Country	Year	References
			Maryland	1951, 1960	Jeffers *et al*. (1951); Anon (1960)
			Mississippi	1959, 1960	Parris (1959); Anon (1960)
			New Zealand	1989	Pennycook (1989)
			Scotland	1961	Foister (1961)
			Taiwan	1979	Anon (1979)
			Texas	1960	Anon (1960)
			Washington	1960, 1973	Anon (1960); Shaw (1973)
153.	*Brassica rapa* subsp. *sylvestris*	Wild turnip	Australia	1982	Sampson and Walker (1982)
154.	*Brassica rugosa*	Ornamental rai	India	1973	Roy (1973)
155.	*Brassica* sp.	Mustard, crucifers	Brazil	1998	Mendes *et al*. (1998)
			Bulgaria	2009	Bobev (2009)
			Canada	1982	Mederick and Piening (1982)
			Alberta, Canada	1990	Richardson (1990)
			Manitoba, Canada	1990	Richardson (1990)
			Saskatchewan, Canada	1990, 1997	Richardson (1990); Gernandt *et al*. (1997)
			Pakistan	1997	Ahmad *et al*. (1997)
			UK	1990	Richardson (1990)
156.	*Broussonetia kazinoki* (Syn. *Morus papyrifera* L.)	Paper mulberry	Japan	2007	Kobayashi (2007)
157.	*Bupleurum falcatum*	Sickle-leaved hare's-ear	Korea	2004	Cho and Shin (2004)
158.	*Bupleurum rotundifolium*	Bupleurum	Japan	2000	Togawa (2000)
159.	*Cajanus cajan*	Pigeon pea	China	1979, 2002	Tai (1979); Chen (2002)
			India	2015	Gupta *et al*. (2015)

160.	*Calceolaria crenatiflora* Cav.	Slipper wort	New York	1960	Anon (1960)
161.	*Calceolaria integrifolia*	Bush slipperwort	Italy	2008	Garibaldi *et al*. (2008a)
162.	*Calceolaria* sp.	Slipperwort	New York	1960	Anon (1960)
			Washington	1960, 1973	Anon (1960); Shaw (1973)
163.	*Calendula arvensis*	Field marigold	China	1979	Tai (1979)
164.	*Calendula officinalis* L.	Pot marigold	Australia	1966, 1982, 1989	Simmonds (1966); Sampson and Walker (1982); Shivas (1989)
			Bulgaria	2009	Bobev (2009)
			California	1960, 1989	Anon (1960); French (1989)
			Canada	1967	Conners (1967)
			Florida	1960, 1984	Anon (1960); Alfieri Jr *et al*. (1984)
			Italy	2001	Garibaldi *et al*. (2001b)
			Louisiana	1960	Anon (1960)
			Missouri	1960	Anon (1960)
			Ohio	1960	Anon (1960)
			Texas	1960	Anon (1960)
165.	*Calendula officinalis* var. *subspathulata*	Pot marigold	Taiwan	1979	Anon (1979)
166.	*Calendula* sp.	Mary's gold/Scotch marigold	California	1989	French (1989)
167.	*Calibrachoa hybrida*	Callie pink morn	Brazil	2020	Borrelli *et al*. (2020)
168.	*Callistephus chinensis* (L.) Hees	China aster	Australia	1982, 1989	Sampson and Walker (1982); Cook and Dubé (1989)
			California	1989	French (1989)
			Canada	1967	Conners (1967)
			Hong Kong	2000	Lu *et al*. (2000)
			New Zealand	1989	Pennycook (1989)

Continued

Table 2.1. Continued

Sr. no.	Host name	English name	Country	Year	References
169.	*Camellia japonica* L.	Camellia	Chile	1945	Mujica and Vergara (1945)
			Carolina	1954	Winstead and Haasis (1954)
			USA	1954	Price (1954)
170.	*Camellia* sp. (On flowers)	Camellia	Georgia	1950	Watson (1950)
			North Carolina	1950	Watson (1950)
171.	*Camelina sativa*	Gold-of-pleasure/false flax	Romania	2016	Cristea and Jurcoane (2016)
172.	*Campanula acuminata*	American bellflower	China	1979	Tai (1979)
173.	*Campanula carpatica*	Tussock bellflower	Italy	2002	Garibaldi *et al*. (2002b)
174.	*Campanula medium* L.	Canterbury bells	UK	1926	Dowson (1926)
			California	1989	French (1989)
			Canada	1964, 1967, 1986	Toms (1964); Conners (1967); Ginns (1986)
			Washington	1960	Anon (1960)
175.	*Campanula persicifolia* L. (Stem rot)	Willow bellflower	Maryland	1960	Anon (1960)
			New Zealand	1989	Pennycook (1989)
			UK	1951	Beaumont (1951)
176.	*Campanula rapunculoides* L.	Creeping bellflower	Canada	1976, 1986	Morrall *et al*. (1976); Ginns (1986)
177.	*Campanula pyramidalis* L.	Chimney bellflower	UK	1951	Beaumont (1951)
178.	*Campanula* sp.	Bellflowers	Canada	1967	Conners (1967)
179.	*Canavalia gladiata*	Sword bean	South Korea	2020	Han *et al*. (2020)
180.	*Cannabis sativa* (Stem rot)	Marijuana, hemp	New South Wales, Australia	1941	Anon (1941)
			USA	1939	Raabe (1939)
			British Columbia, Canada	1967, 2021	Conners (1967); Punja and Ni (2021)

			China	1979	Tai (1979)
			France	2018	Leyronas *et al.* (2018)
			Montana	1960, 1973	Anon (1960); Shaw (1973)
			Oregon	2021	Garfinkel (2021)
			Canada	2000	Bains *et al.* (2000)
181.	*Cannabis sativa* subsp. *sativa*	Hemp	Manitoba, Canada	2000	Hilton (2000)
			Prince Edward Island, Canada	2000	Hilton (2000)
182.	*Capsella bursa-pastoris* (L.) Medic	Shepherd's purse	China	1979	Tai (1979)
			Taiwan	1979	Anon (1979)
			USA	1971	Morgan (1971)
183.	*Capsicum annuum* L.	Pepper	Australia	1989	Cook and Dubé (1989)
			Brazil	1998	Mendes *et al.* (1998)
			Bulgaria	2009	Bobev (2009)
			China	1979	Tai (1979)
			Cook Islands	1981	Dingley *et al.* (1981)
			Florida	1984	Alfieri Jr *et al.* (1984)
			Greece	2000	Holevas *et al.* (2000)
			India	1987	Srivastava and Diwakar (1987)
			Korea	1987, 2004	Shin and Lee (1987); Cho and Shin (2004)
			New Zealand	1989	Pennycook (1989)
			Spain	2016	Liu *et al.* (2016)
			India	2014	Hansda *et al.* (2014)
184.	*Capsicum annuum* var. *acuminatum*	Cayenne pepper	Taiwan	1979	Anon (1979)

Continued

Table 2.1. Continued

Sr. no.	Host name	English name	Country	Year	References
185.	*Capsicum annuum* var. *annuum*	Cayenne pepper	Canada	1986	Ginns (1986)
			Florida	1984	Alfieri Jr *et al*. (1984)
186.	*Capsicum frutescens* L.	Tabasco	Australia	1966	Simmonds (1966)
			California	1960, 1989	Anon (1960); French (1989)
			Canada	1964, 1967	Toms (1964); Conners (1967)
			Connecticut	1960	Anon (1960)
			Florida	1960, 1984	Anon (1960); Alfieri Jr *et al*. (1984)
			Hawaii	1981	Raabe *et al*. (1981)
			India	1973	Roy (1973)
			Massachusetts	1960	Anon (1960)
			Taiwan	1979	Anon (1979)
			Texas	1960	Anon (1960)
187.	*Capsicum frutescens* L. cv. *grossum*	Sweet pepper	India	1973	Roy (1973)
188.	*Capsicum longum*	Long peppers	China	1979	Tai (1979)
189.	*Capsicum* sp.	Pepper	Brazil	1998	Mendes *et al*. (1998)
190.	*Cardamine debilis*	Roadside bitter-cress	New Zealand	1989	Pennycook (1989)
191.	*Cardamine flexuosa debilis*	Wavy bittercress/wood bitter-cress	China	1979	Tai (1979)
192.	*Cardamine angustata* (formerly *Cardamine heterophylla*) (Forst.) Schutz.	Bitter-cress	USA	1974	Schwartz (1977)
			New Zealand	1932	Brien (1932)
193.	*Cardamine hirsuta*	Hairy bitter-cress	China	1979	Tai (1979)
194.	*Cardamine regeliana*	Japanese bitter-cress	Taiwan	1979	Anon (1979)
195.	*Carduus* sp.	Plumeless thistle	England	1978	Dennis (1978)
196.	*Carduus tenuiflorus*	Sheep thistle	New Zealand	1989	Pennycook (1989)
197.	*Carica papaya*	Papaya	Australia	1966	Simmonds (1966)

198.	*Carpinus betulus*	European/common horn beam	Poland	2006	Chlebicki and Chmiel (2006)
199.	*Carthamus* sp.	Distaff thistles	Mexico	1976	Alvarez (1976)
200.	*Carthamus tinctorius* L.	Safflower	Australia	1966	Simmonds (1966)
			California	1963, 1989	Zimmer *et al*. (1963); French (1989)
			Canada	1967, 1986, 1990	Conners (1967); Ginns (1986); Richardson (1990)
			Indiana	1960	Anon (1960)
			Mexico	1976	Alvarez (1976)
			North Dakota	1960	Anon (1960)
			New Zealand	1989	Pennycook (1989)
			Virginia	1960	Anon (1960)
			Washington	1960	Anon (1960)
			India	1924	Joshi (1924)
201.	*Carum carvi* L. (Stem rot)	Caraway	California	1989	French (1989)
			Canada	1967	Conners (1967)
			Netherlands	1990	Richardson (1990)
			Washington	1960	Anon (1960)
202.	*Carum petroselinum*	Parsley	China	1979	Tai (1979)
			Taiwan	1979	Anon (1979)
203.	*Casimiroa edulis*	White sapote	Israel	1963	Palti (1963)
204.	*Catharanthus roseus* (L.) Don (as *Vinca rosea* L.)	Madagascar periwinkle	Canada, Ontario	2016	Sun and Hsiang (2016)
			Bermuda	1946	Kinghorn (1947)
			China	1979	Tai (1979)
			Taiwan	1979	Anon (1979)
205.	*Celosia* sp.	Cock's comb	Mauritius	1968	Orieux and Felix (1968)
206.	*Centaurea cyanus* L.	Bachelor's button/cornflower	California	1989	French (1989)

Continued

Table 2.1. Continued

Sr. no.	Host name	English name	Country	Year	References
			China	1979	Tai (1979)
			Florida	1984	Alfieri Jr *et al*. (1984)
			Illinois	1964	Boew (1964)
			Missouri	1960	Anon (1960)
			Mississippi	1960, 1959	Anon (1960); Parris (1959)
			Taiwan	1979	Anon (1979)
			India	1991	Sinha and Singh (1991)
207.	*Centaurea dealbata* Willd.	Persian cornflower	India	1977	Kaur and Chohan (1977)
208.	*Centaurea diffusa* Lam.	Diffuse knapweed	Canada	1974, 1986	Watson *et al*. (1974); Ginns (1986)
209.	*Centaurea montana* L.	Mountain bluet	Indiana	1960	Anon (1960)
210.	*Centaurea moschata* L. (as *C. Oderata* Hort.)	Sweet sultan	USA	1974	Schwartz (1977)
211.	*Centaurea solstitialis* L.	Yellow star thistle	California	1986	Klisiewicz (1986)
212.	*Centaurea* sp.	Knapweed	California	1960	Anon (1960)
			Mississippi	1960, 1959	Anon (1960); Parris (1959)
			Texas	1960	Anon (1960)
213.	*Centella asiatica*	Gotu kola	China	1979	Tai (1979)
			Taiwan	1979	Anon (1979)
214.	*Centipeda minima*	Sneeze weed	China	1979	Tai (1979)
			Taiwan	1979	Anon (1979)
215.	*Chamaecyparis lawsoniana* (Murr.) Parl.	Lawson white cedar	New Zealand	1989	Pennycook (1989)
216.	*Chamaecytisus palmensis*	Tagasaste/tree lucerne	New Zealand	1989	Pennycook (1989)
217.	*Chamaedaphne calyculata*	Leather leaf/cassandra	Poland	2015	Kowalik *et al*. (2015)
218.	*Cheiranthus cheiri* L.	Wallflower	Canada	1967	Conners (1967)
			New Zealand	1932, 1989	Brien (1932); Pennycook (1989)

			Taiwan	1961	Foister (1961)
			India	1987	Saxena and Rai (1987)
219.	*Chenopodium album* L. (Root rot)	Lamb's quarters	China	1979	Tai (1979)
			Montana	1960, 1973	Anon (1960); Shaw (1973)
			New Zealand	1989	Pennycook (1989)
220.	*Chenopodium album* L. var. *centrorubrum* Makino	Lamb's quarters	Korea	1987	Shin and Lee (1987)
221.	*Chenopodium ficifolium*	Fig-leaved goosefoot	China	1979	Tai (1979)
222.	*Chenopodium serotinum*	Fig-leaved goosefoot	Taiwan	1979	Anon (1979)
223.	*Chrysanthemum cinerariifolium* (Trever) Viz.	Dalmatian pyrethrum	China	1979	Tai (1979)
			Kenya	1961	Nattrass (1961)
			Taiwan	1979	Anon (1979)
			Tanzania	1960	Riley (1960)
224.	*Chrysanthemum coccineum* Willd. (Stem rot)	Common pyrethrum	Virginia	1960	Anon (1960)
225.	*Chrysanthemum coronarium* L. var. *spatiosum* (Syn. *Glebionis coronaria*)	Crown daisy	China	1979	Tai (1979)
			Korea	2004	Cho and Shin (2004)
			Taiwan	1979	Anon (1979)
226.	*Chrysanthemum indicum*	Indian chrysanthemum	China	1979	Tai (1979)
			New Zealand	1969	Dingley (1969)
227.	*Chrysanthemum leucanthemum* L. var. *pinnatifidum*	Ox-eye daisy	Texas	1960	Anon (1960)
228.	*Chrysanthemum maximum* Ramond	Daisy	Montana	1960, 1973	Anon (1960); Shaw (1973)
			Washington	1960, 1973	Anon (1960); Shaw (1973)
229.	*Chrysanthemum monilifera* sp. ssp. *rotundata*	Bitou bush	Australia	2000	Cother (2000)

Continued

Table 2.1. Continued

Sr. no.	Host name	English name	Country	Year	References
230.	*Chrysanthemum morifolium* Ramat.	Florist's daisy/hardy garden mum/florists' chrysanthemum	Arizona	1960	Anon (1960)
			California	1989	French (1989)
			Canada	1967, 1986	Conners (1967); Ginns (1986)
			Florida	1984	Alfieri Jr *et al.* (1984)
			Hawaii	1981	Raabe *et al.* (1981)
			Korea	2004	Cho and Shin (2004)
			Louisiana	1960	Anon (1960)
			Michigan	1960	Anon (1960)
			North Carolina	1985	Grand (1985)
			New Zealand	1989	Pennycook (1989)
			Scotland	1961	Foister (1961)
			Virginia	1960	Anon (1960)
			Argentina	2003	Wright and Palmucci (2003)
231.	*Chrysanthemum* sp.	Chrysanthemum	Australia	1982	Sampson and Walker (1982)
			Brazil	1998	Mendes *et al.* (1998)
			California	1989	French (1989)
			Washington	1973	Shaw (1973)
232.	*Chrysogonum virginianum*	Golden-star/Golden-knee	Tennessee	2018	Trigiano *et al.* (2018)
233.	*Cicer arietinum* L.	Chick pea	Arizona	2000	Matheron and Porchas (2000)
			Australia	1987	Bretag and Mebalds (1987)
			Hungary	1989	Simay (1989)
			Saskatchewan, Alberta, Canada	2000	Hilton (2000)
			Greece	1973	Pantidou (1973)

			India	1963, 1979, 1998	Bedi (1963); Singh *et al*. (1979); Pande and Rao (1998)
			North Dakota, Washington, DC	2006	Chen *et al*. (2006)
			Syria	1988	Haware (1988)
234.	*Cichorium endivia* L.	Endive	Arizona	1969	Gilbertson and McHenry (1969)
			Brazil	1998	Mendes *et al*. (1998)
			California	1989	French (1989)
			Canada	1967	Conners (1967)
			Florida	1984	Alfieri Jr *et al*. (1984)
			France	2018	Leyronas *et al*. (2018)
			Korea	2004	Cho and Shin (2004)
			Montana	1973	Shaw (1973)
235.	*Cichorium intybus* L.	Chicory	Australia	1982, 1989	Sampson and Walker (1982); Shivas (1989)
			Canada	1986	Ginns (1986)
			Korea	2003, 2004	Chang and Kim (2003); Cho and Shin (2004)
			Libya	1981	El-Buni and Rattan (1981)
			New Zealand	1989	Pennycook (1989)
			Texas	1960	Anon (1960)
			Washington	1960, 1973	Anon (1960); Shaw (1973)
236.	*Cirsium arvense* (L.) Scop.	Canada thistle	Canada	1967	Conners (1967)
			Montana	1973, 1986	Shaw (1973); Brosten and Sanda (1986)
			New Zealand	1989	Pennycook (1989)
			Washington	1973	Shaw (1973)

Continued

Table 2.1. Continued

Sr. no.	Host name	English name	Country	Year	References
237.	*Cirsium japonicum*	Korean thistle	China	1967	Conners (1967)
238.	*Cirsium japonicum* var. *australe*	Formosan thistle	Taiwan	1979	Anon (1979)
239.	*Cirsium vulgare* (Savi.) Ten. (as *C. lanceolatum* (L.) Juk.	Bull thistle	New Zealand	1989	Pennycook (1989)
			USA	1974	Schwartz (1977)
240.	*Citrullus edulis*	Bitter cucumber	Bulgaria	2009	Bobev (2009)
241.	*Citrullus lanatus* (Thunb) Matsum & Nakal (as *C. vulgaris* Schrad.)	Watermelon	Canada	1967	Conners (1967)
			Korea	1987, 2004	Shin and Lee (1987); Cho and Shin (2004)
			USA	1960	Anon (1960)
			New Zealand	1932, 1989	Brien (1932); Pennycook (1989)
242.	*Citrullus vulgaris* var. *citroides* (Bailey) Mansf.	Citron	Canada	1967	Conners (1967)
			New Jersey	1960	Anon (1960)
			Texas	1960	Anon (1960)
243.	*Citrullus vulgaris* var. *fistulosus*	Round gourd	India	1980	Agrawal *et al*. (1980)
244.	*Citrus aurantifolia* Christm.	Lime	California	1987, 1989	French (1987, 1989)
245.	*Citrus aurantifolia* Christm. var. *dulcis*	Lime	USA	1974	Schwartz (1977)
246.	*Citrus aurantium* L.	Seville orange	Chile	1945	Mujica and Vergara (1945)
			Cyprus	1957	Georghiou and Papadopoulos (1957)
			New Zealand	1932	Brien (1932)
			Turkey	2012	Baysal-Gurel *et al*. (2012)
			USA	1960, 1974	Anon (1960); Schwartz (1977)
247.	*Citrus grandis*	Pomelo	California	1987	French (1987)
248.	*Citrus junos*	Yuzu	Japan	2007	Kobayashi (2007)
249.	*Citrus latifolia* Tanaka	Persian lime	California	1987, 1989	French (1987, 1989)
250.	*Citrus limon* Burn. (as *C. medica* L. var. *limonum*)	Citron	Australia	1989	Shivas (1989); Cook and Dubé (1989)

			California	1960, 1987, 1989	Anon (1960); French (1987, 1989)
			Chile	2014	Ferrada *et al*. (2014)
			Greece	1973	Pantidou (1973)
			Sicily, Italy	1991	Greuter *et al*. (1991)
			New Zealand	1932, 1989	Pennycook (1989); Brien (1932)
			Texas, USA	1960, 1974	Anon (1960); Schwartz (1977)
251.	*Citrus limonia*	Rangpur	Chile	1945	Mujica and Vergara (1945)
			South Africa	1977, 2000	Gorter (1977); Crous (2000)
252.	*Citrus limonium*	Lemon	Cyprus	1957	Georghiou and Papadopoulos (1957)
253.	*Citrus maxima* (Burm.) Merril (as. *C. grandis* (L.) Osbeck)	Pummelo	California	1974, 1989	French (1989); Schwartz (1977)
254.	*Citrus medica* L. (as *C. medica* L. var. *ethrog* Engl.)	Citron	Israel	1963	Palti (1963)
			USA	1974	Schwartz (1977)
255.	*Citrus meyerii*	Meyer lemon	California	1987	French (1987)
256.	*Citrus reticulata* Blanco	Mandrin orange tree/Mandrin orange	California	1987, 1989	French (1987, 1989)
			New Zealand	1989	Pennycook (1989)
257.	*Citrus sinensis* Osbeck.	Sweet orange	Australia	1989	Shivas (1989); Cook and Dubé (1989)
			California	1960, 1987, 1989	Anon (1960), French (1987, 1989)
			China	1979	Tai (1979)
			Cyprus	1957	Georghiou and Papadopoulos (1957)
			Sicily, Italy	1991	Greuter *et al*. (1991)
			New Zealand	1989	Pennycook (1989)

Continued

Table 2.1. Continued

Sr. no.	Host name	English name	Country	Year	References
			South Africa	1977, 2000	Gorter (1977); Crous *et al.* (2000)
			Spain	1988	Tuset and Marti (1988)
258.	*Citrus* sp.	Citrus fruits	California	1960, 1987, 1989	Anon (1960); French (1987, 1989)
			Japan	2007	Kobayashi (2007)
			Taiwan	1979	Anon (1979)
			Texas	1960	Anon (1960)
259.	*Citrus unshiu*	Miyagawa mandarin	Japan	2007	Kobayashi (2007)
			Korea	2004	Cho and Shin (2004)
			Cheju, Korea	1999	Song and Koh (1999)
260.	*Citrus volkameriana*	Vol Kamer Lemon	Italy	2011	Polizzi *et al.* (2011)
261.	*Citrus paradisi* Macfady	Grapefruit	California	1960, 1987, 1989	Anon (1960); French (1987, 1989)
			Pakistan	2016	Hanif *et al.* (2016)
			South Africa	1987, 1989	French (1987, 1989)
262.	*Citrus tangelo*	Tangelo	California	1987, 1989	French (1987, 1989)
263.	*Cleome spinosa*	Spiny spider flower	China	2002	Liu *et al.* (2002)
264.	*Cleome viscosa* L.	Tick weed	Bermuda	1941	Russell (1942)
265.	*Cnicus arvensis* Hoffm. (as *Carduus arvensis* (L.) Robs)	Thistle	USA	1974	Schwartz (1977)
266.	*Cnicus benedictus L.*	Blessed thistle	Germany	1985	Cristea *et al.* (1985)
			USA	1936	Brown and Butler (1936)
267.	*Cnicus* spp.	Blessed thistle	–	1994	Boland and Hall (1994); Saharan and Mehta (2008); Ficker (2019)
268.	*Codonopsis lanceolata*	Deo deok or Lance Asia bell	Korea	2020	Choi *et al.* (2020)
269.	*Coleus blumei* (Cottony blight)	Coleus	Florida	1984	Alfieri Jr *et al.* (1984)

270.	*Commelina nudiflora*	Dayflowers	China	1979	Tai (1979)
			Taiwan	1979	Anon (1979)
271.	*Conium maculatum* L.	Poison hemlock	UK	1936	Ogilvie (1936)
272.	*Consolida ambigua*	Larkspur	New Zealand	1989	Pennycook (1989)
273.	*Consolida orientalis* (Gray) Schrod. (as *Delphinium ajacis* L.)	Rocket larkspur	New Zealand	1932	Brien (1932)
			USA	1960	Anon (1960)
274.	*Convolvulus arvensis* L.	Field bindweed	India	1986	Singh and Singh (1986)
275.	*Convolvulus* sp.	Bindweed	UK	1936	Ogilvie (1936)
276.	*Corchorus capsularis* L.	Jute	Taiwan	1979	Anon (1979)
			New South Wales	1941	Anon (1941)
277.	*Coreopsis drummondii*	Tickseed	India	2014	Khatua *et al*. (2014)
278.	*Coreopsis grandiflora* Hagg. Ex. sweet	Tickseed	Ohio, Washington	1969	Gilbertson and McHenry (1969)
			USA	1960	Anon (1960)
279.	*Coreopsis* sp.	Tickseed	Washington	1973, 1974	Shaw (1973); Schwartz (1977)
280.	*Coreopsis stillmanii* (Gray)Blake	Tickseed	India	1959, 1971	Rai and Agnihotri (1970); Sarbhoy *et al*. (1971)
281.	*Coreopsis tinctoria* Nutt.	Calliopsis	Ohio,	1960	Anon (1960)
			Washington	1960	Anon (1960)
282.	*Coriandrum sativum* L.	Coriander	Germany	1985	Cristea *et al*. (1985)
			India	1963	Gupta (1963)
			Argentina,		Gaeten *et al*. (1997)
283.	*Coronilla varia* L.	Crown vetch	USA	1974	Schwartz (1977)
284.	*Coronopus didymus*	Lesser swine-cress	New Zealand	1996	McKenzie and Dingley (1996)
285.	*Corydalis bulbosa*	Fumewort/bird-in-a-bush	China	1979	Tai (1979)

Continued

Table 2.1. Continued

Sr. no.	Host name	English name	Country	Year	References
286.	*Cosmos bipinnatus* Cay.	Cosmos	Canada	1967, 1986	Conners (1967); Ginns (1986)
			India	1985	Roy (1985)
			USA	1973	Shaw (1973)
			British Columbia, Canada	1964	Toms (1964)
287.	*Cosmos sulphureus*	Garden cosmos/Mexican aster	India	2015	Mondal *et al*. (2015)
288.	*Crambe abyssinica* Hochst. ex. R E. Fr. (Root rot)	Kale	Louisiana	1970	Holcomb and Newman (1970)
289.	*Craspedia globosa*	Golf beauty/Billy buttons	Argentina	2005	Wolcan and Grego (2005)
290.	*Crepis japonica* (L.) Benth. (as *Youngia japonica* (L.) D.C.)	Hawk's beard	Washington	1960	Anon (1960)
			China	1979	Tai (1979)
			Florida	1984	Alfieri Jr *et al*. (1984)
			Taiwan	1979	Anon (1979)
			New Zealand	1980	Cole (1980)
			Germany	1985	Cristea *et al*. (1985)
			Italy	1989	Granata and Tirro (1989)
			Vietnam	1985	Marras *et al*. (1985)
			USA	1973	Shaw (1973)
			California	1989	French (1989)
291.	*Crotalaria juncea*	Brown hemp/Indian hemp	China	1979	Tai (1979)
			Taiwan	1979	Anon (1979)
			USA	1974	Schwartz (1977)
292.	*Crotalaria retusa*	Devil-bean/Rattle weed	India	1998	Pande and Rao (1998)
293.	*Crotalaria spectabilis*	Rattlebox/showy rattle pod	Brazil	2015	Oliveira *et al*. (2015)

294.	*Crotalaria striata*	Giant striata	China	1979	Tai (1979)
			Taiwan	1979	Anon (1979)
295.	*Cryptomeria japonica* (L. f.) Don	Japanese cedar	Japan	2007	Kobayashi (2007)
296.	*Cryptotaenia canadensis*	Canadian honewort	Taiwan	1979	Anon (1979)
297.	*Cryptotaenia japonica*	Honewort	China	1979	Tai (1979)
298.	*Cucumis anguria* L.	Gherkin	USA	1960	Anon (1960)
299.	*Cucumis melo* L.	Melon	Bulgaria	1979	Tai (1979)
			California	1987, 1989	French (1987, 1989)
			Canada	1967	Conners (1967)
			France	2018	Leyronas *et al*. (2018)
			Greece	2000	Holevas *et al*. (2000)
			Korea	2004	Cho and Shin (2004)
			New Zealand	1989	Pennycook (1989)
300.	*Cucumis melo* L. var. *cantalupensis* Naud.	Melon	USA	1974	Schwartz (1977)
			New Zealand	1932	Brien (1932)
301.	*Cucumis melo* L. var. *makuwa* Makino.	Oriental melon	Korea	1987	Shin and Lee (1987)
302.	*Cucumis melo* var. *reticulatus* Naud.	Muskmelon	Arkansas	1960	Anon (1960)
			Massachusetts	1960	Anon (1960)
			Texas	1960	Anon (1960)
			Japan	1984	Kawakubo and Takeshita (1984)
303.	*Cucumis melo* var. *utilissmus*	Snake cucumber	India	1980	Agrawal *et al*. (1980)
304.	*Cucumis sativus* L.	Cucumber	Australia	1989	Cook and Dubé (1989)
			Bulgaria	2009	Bobev (2009)
			Bermuda	1939	Waterston (1939)
			California	1960	Anon (1960); French (1989)

Continued

Table 2.1. Continued

Sr. no.	Host name	English name	Country	Year	References
			Canada	1938, 1964, 1967, 1986	Bisby (1938); Toms (1964); Conners (1967); Ginns (1986)
			Chile	1967	Mujica and Oehrens (1967)
			Colorado	1957	Burke *et al*. (1957)
			Eastern states	1960	Anon (1960)
			Florida	1960, 1984	Alfieri Jr *et al*. (1984); Anon (1960)
			Germany	1975	Apelt and Bochow (1975)
			Greece	1973, 2000	Pantidou (1973); Holevas *et al*. (2000)
			Illinois	1960	Anon (1960)
			India	2014	Hansda *et al*. (2014)
			Korea	1987, 2004	Shin and Lee (1987); Cho and Shin (2004)
			Minnesota	1960	Anon (1960)
			North Carolina	1985	Grand (1985)
			New Zealand	1932, 1989	Brien (1932); Pennycook (1989)
			Northwestern states, USA	1960	Anon (1960)
			Oregon	1973	Shaw (1973)
			Scotland	1961	Foister (1961)
			Texas	1960	Anon (1960)
			Washington	1973	Shaw (1973)
305.	*Cucumis sativus* var. *sativus*	Common cucumber	Korea	1991	Lee *et al*. (1991)
306.	*Cucurbita maxima* Duchesne	Winter squash	Canada	1967, 1971, 1986	Conners (1967); Duczek and Morrall (1971); Ginns (1986)
			Egypt	1979	El-Tobshy and Zayed (1979)

			Florida	1984	Alfieri Jr *et al.* (1984)
			Idaho	1960	Anon (1960)
			Massachusetts	1960	Anon (1960)
			Maine	1960	Anon (1960)
			Montana	1960, 1973	Shaw (1973); Anon (1960)
			New Zealand	1989	Pennycook (1989)
			New Hampshire	1960	Anon (1960)
			New York	1960	Anon (1960)
			Pennsylvania	1960	Anon (1960)
			Washington	1960, 1973	Shaw (1973); Anon (1960)
			USA	1936	Young (1936)
307.	*Cucurbita moschata* Duchesne	Butternut squash	China	1979	Tai (1979)
			Korea	1987	Shin and Lee (1987)
308.	*Cucurbita pepo* L. (Fruit rot)	Pumpkin	Australia	1966, 1982	Sampson and Walker (1982); Simmonds (1966)
			Canada	1964, 1967, 1971, 1986	Creelman (1965); Conners (1967); Duczek and Morrall (1971); Ginns (1986)
			Idaho	1960	Anon (1960)
			Montana	1960, 1973	Shaw (1973); Anon (1960)
			New Zealand	1932, 1989	Brien (1932); Pennycook (1989)
			USA	1936, 1987	Young (1936); Chrominski *et al.* (1987)
309.	*Cucurbita pepo* var. *melopepo* (L.) Alef. (Fruit rot)	Bush pumpkin	New York	1960	Anon (1960)
310.	*Cucurbita pepo* var. *ovifera* (L.) Alef.	Yellow-flowered gourd	USA	1974	Schwartz (1977)

Continued

Table 2.1. Continued

Sr. no.	Host name	English name	Country	Year	References
311.	*Cucurbita* sp.	Squash/pumpkin	Bulgaria	2009	Bobev (2009)
			California	1989	French (1989)
			Canada	1990	Richardson (1990)
			Greece	2000	Holevas *et al.* (2000)
			Korea	2004	Cho and Shin (2004)
312.	*Cuminum cyminum*	Cumin	India	2017	Prasad *et al.* (2017)
313.	*Cuphea* sp.	Hawaiian Heather/Cigarette Plant	Minnesota	2006	Gulya *et al.* (2006)
			North Dakota	2006	Gulya *et al.* (2006)
314.	*Cynara scolymus* L.	Globe Artichoke	New York	1960	Anon (1960)
			New Zealand	1980	Cole (1980)
			Germany	1985	Cristea *et al.* (1985)
			Vietnam	1985	Marras *et al.* (1985)
			Italy	1989	Granata and Tirro (1989)
			Brazil	1998	Mendes *et al.* (1998)
			Greece	1973, 2000	Pantidou (1973); Holevas *et al.* (2000)
			Oregon	1973	Shaw (1973)
			Washington	1990	Richardson (1990)
315.	*Cyperus rotundus* L.	Nut grass	India	1986	Singh and Singh (1986)
316.	*Cynoglossum amabile* Staff. & Drum.	Chinese forget-me-not	Washington	1960, 1973	Shaw (1973); Anon (1960)
317.	*Cynoglossum officinale*	Hounds tongue	Canada	2005	Huang *et al.* (2005)
318.	*Cyphomandra betaceae* Sendt.	Tree tomato	New Zealand	1989	Pennycook (1989)
			Ecuador	1991	Velastegui and Ball (1991)
			USA	1974	Schwartz (1977)

319.	*Dahlia pinnata* Cav. (as *D. variabilis* (Willd.) Desf.) (Stem rot)	Dahlia	California	1989	French (1989)
			China	1979	Tai (1979)
			Washington	1960, 1973	Shaw (1973); Anon (1960)
			Florida	1984	Alfieri *et al*. (1984)
			Taiwan	1979	Anon (1979)
320.	*Dahlia pluvialis* (L.) Moench.	Lo-Blush	Canada	1967	Conners (1967)
321.	*Dahlia* sp.	Dahlia	Canada	1938	Bisby (1938)
			India	1973	Roy (1973)
			Washington	1973	Shaw (1973)
			Mondul, Cambodia	1944	Wallace (1944)
			Australia	1982, 1989	Sampson and Walker (1982); Shivas (1989)
			Bulgaria	2009	Bobev (2009)
			California	1989	French (1989)
			New Zealand	1969, 1989	Dingley (1969); Pennycook (1989)
			Rhode Island	2010	Goos (2010)
			Scotland	1961	Foister (1961)
			Tanzania	1960	Riley (1960)
322.	*Dahlia variabilis*	Fireworks mixed	California	1960	Anon (1960)
			Canada	1967	Conners (1967)
			Maine	1960	Anon (1960)
			New York	1960	Anon (1960)
			Scotland	1990	Richardson (1990)
323.	*Daphne odora*	Winter daphne	Japan	2007	Kobayashi (2007)
			New Zealand	1989	Pennycook (1989)

Continued

Table 2.1. Continued

Sr. no.	Host name	English name	Country	Year	References
324.	*Datura alba*	Indian thorn apple	China	1979	Tai (1979)
			Taiwan	1979	Anon (1979)
325.	*Datura innoxa* Mill.	Angel's trumpet	Germany	1985	Cristea *et al*. (1985)
326.	*Datura stramonium* L.	Datura/Jimson weed	Korea	1987	Shin and Lee (1987)
			South Africa	2000	Crous *et al*. (2000)
			India	2004	Ghasolia *et al*. (2004)
327.	*Daucus carota* L.	Carrot/Queen Anne's lace/wild carrot	Australia	1966, 1982, 1989	Simmonds (1966); Sampson and Walker (1982); Shivas (1989); Cook and Dubé (1989)
			Brazil	1998	Mendes *et al*. (1998)
			Canada	1986	Ginns (1986)
			Chile	1967	Mujica and Oehrens (1967)
			China	1979	Tai (1979)
			Fiji	1972, 1981	Firman (1972); Dingley *et al*. (1981)
			France	2018	Leyronas *et al*. (2018)
			South Africa	2000	Crous *et al*. (2000)
			Taiwan	1979	Anon (1979)
			California	1989	French (1989)
			India	2015	Mondal *et al*. (2015)
328.	*Daucus carota* subsp. *sativus*	Carrot	Canada	1967, 1986, 1991	Conners (1967); Ginns (1986); Arcelin and Kushalappa (1991)
			Florida	1984	Alfieri Jr *et al*. (1984)
			Georgia	1963	Hanlin (1963)
			Korea	2004	Cho and Shin (2004)
			New Zealand	1989	Pennycook (1989)
			Range of host	160	Anon (1960)
			Scotland	1961	Foister (1961)

329.	*Daucus sativus*	Carrot	Bulgaria	2009	Bobev (2009)
			USSR	1990	Jovaisiene and Strukeinskas (1990)
330.	*Daucus* sp.	Carrot	England	1978	Dennis (1978)
			UK	1986	Dennis (1986)
331.	*Delphinium ajacis*	Rocket larkspur	Missouri	1960	Anon (1960)
			Texas	1960	Anon (1960)
332.	*Delphinium cheilanthum Voss*	Garden larkspur	USA	1960	Anon (1960)
333.	*Delphinium cultorum* Voss	Larkspur/rosy future	USA	1960	Anon (1960)
			Delaware	1960	Anon (1960)
			Illinois	1960	Anon (1960)
			Massachusetts	1960	Anon (1960)
			Michigan	1960	Anon (1960)
			Ohio	1960	Anon (1960)
			Washington	1960	Anon (1960)
334.	*Delphinium elatum* L.	Candle larkspur	Delaware	1960	Anon (1960)
			Illinois	1960	Anon (1960)
			Massachusetts	1960	Anon (1960)
			Michigan	1960	Anon (1960)
			Ohio	1960	Anon (1960)
			Washington	1960	Anon (1960)
335.	*Delphinium grandiflorum* L.	Bouquet larkspur	Delaware	1960	Anon (1960)
			Illinois	1960	Anon (1960)
			Massachusetts	1960	Anon (1960)
			Michigan	1960	Anon (1960)
			Ohio	1960	Anon (1960)
			Washington	1960	Anon (1960)

Continued

Table 2.1. Continued

Sr. no.	Host name	English name	Country	Year	References
336.	*Delphinium* sp.	Larkspur	Australia	1982	Sampson and Walker (1982)
			California	1989	French (1989)
			Canada	1967	Conners (1967)
			Florida	1984	Alfieri Jr *et al*. (1984)
			India	1964, 1977	Shukla and Pavgi (1964); Kaur and Chohan (1977)
			North Carolina	1985	Grand (1985)
			Texas	1960	Anon (1960)
			Washington	1973	Shaw (1973)
337.	*Dendranthema grandiflorum*	Florist's chrysanthemum	Argentina	2003	Wright and Palmucci (2003)
338.	*Descurainia sophia* (L.) Webb.	Tansy mustard	Canada	1967	Conners (1967)
339.	*Desmodium triflorum* L. D C	Creeping tick trefoil	India	1986, 1990, 1998	Singh and Singh (1986); Lenne (1990); Pande and Rao (1998)
340.	*Dianthus caryophyllus* L. D.C.	Carnation/clove pink	California	1989	French (1989)
			Canada	1967	Conners (1967)
			Germany	1980	Jacob (1980)
			Greece	1973	Pantidou (1973)
			India	2015	Vinod Kumar *et al*. (2015)
			Korea	2004	Cho and Shin (2004)
			New Zealand	1989	Pennycook (1989)
341.	*Dianthus chinensis*	Rainbow pink/china pink	China	1979	Tai (1979)
342.	*Dianthus* sp.	Pinks	California	1989	French (1989)
343.	*Dicentra spectabilis* (L.) Lem.	Bleeding heart	Minnesota	1960	Anon (1960)
344.	*Dichondra* sp.	Silky kidney weed	California	1989	French (1989)
345.	*Dichrocephala bicolor*	Bicolor buttonweed	Taiwan	1979	Anon (1979)

346.	*Dichrocephala latifolia*	Bicolor buttonweed	China	1979	Tai (1979)
347.	*Didiscus caeruleus*	Blue lace flower	Japan	1999	Kanno and Ohkubo (1999)
348.	*Digitaria sanguinalis* var. *ciliaris* (Retz.) Parl.	Large crab grass	Japan	1957	Kobayashi (1957)
349.	*Digitalis purpurea* L. (Stem rot)	Common foxglove	New York	1960	Anon (1960)
350.	*Dimorphotheca aurantiaca* D.C.	Cape marigold	India	1959, 1964, 1971	Rai and Agnihotri (1970); Shukla and Pavgi (1964); Sarbhoy *et al.* (1971)
351.	*Dimorphotheca pluvialis*	White African daisy	New Zealand	1996	McKenzie and Dingley (1996)
352.	*Dimorphotheca sinuata*	Cape marigold	California	1989	French (1989)
353.	*Dimorphotheca* sp. Moench.	Cape marigold	California	1965, 1989	Leppik (1965); French (1989)
354.	*Diplotaxis tenuifolia*	Wild rocket/perennial wall-rocket	Italy	2005, 2013	Garibaldi *et al.* (2005a); Minuto *et al.* (2005a); Gilardi *et al.* (2013)
355.	*Dipsacus fullonum* L. (as *D. sylvestris* Huds.)	Common teasel	New Zealand	1989	Pennycook (1989)
			Texas	1960	Anon (1960)
			Russia	1938	Masalab (1938)
356.	*Dipsacus sylvestris* (stem rot)	Wild teasel/Fuller's teasel	Texas	1960	Anon (1960)
357.	*Dolichos biflorus* L.	Horse gram	India	1980	Agrawal *et al.* (1980)
358.	*Dolichos lablab* L.	Egyptian bean	India	1973, 1980, 1998	Roy (1973); Agrawal *et al.* (1980); Pande and Rao (1998)
359.	*Dolichos* sp.	Dolichos plant	China	1979	Tai (1979)
			Taiwan	1979	Anon (1979)
360.	*Dorotheanthus bellidiformis* (Burm.f.) N.E.Br.	Livingstone daisy	India	1985	Roy (1985)
361.	*Draba* sp.	Dill	New Jersey, USA	1960	Cappellini (1960)

Continued

Table 2.1. Continued

Sr. no.	Host name	English name	Country	Year	References
362.	*Echeveria gigantea*	Blue atoll	Mexico	2023	Terrones-Salgado *et al*. (2023)
363.	*Echinacea purpurea*	Cone flower	Alberta, Canada	1997	Chang *et al*. (1997b)
364.	*Echium plantagineum*	Purple viper's-bugloss/ Patterson's curse	Australia	1989	Cook and Dubé (1989)
365.	*Echium vulgare*	Echium	North Dakota	2005	del Río *et al*. (2005)
366.	*Eclipta alba*	False daisy	China	1979	Tai (1979)
			Taiwan	1979	Anon (1979)
367.	*Eclipta prostrata*	False daisy	USA	1992	Melouk *et al*. (1992)
368.	*Edgeworthia papyrifera*	Chinese paper bush	Japan	2007	Kobayashi (2007)
369.	*Eleusine indica*	Indian goosegrass/Yard-grass	China	1979	Tai (1979)
			Taiwan	1979	Anon (1979)
370.	*Emilia sonchifolia*	Lilac tassel flower/cupid's shaving brush	China	1979	Tai (1979)
			Taiwan	1979	Anon (1979)
371.	*Enhydra fluctuans*	Buffalo spinach	India	2015	Mondal *et al*. (2015)
372.	*Engelhardtia spicata* Blume	Undetermined	India	1989	Mehrotra (1989)
373.	*Epipremnum aureum* (Linden & Andre) Bunt. (Cottony blight)	Pathos	Florida	1984	Alfieri Jr *et al*. (1984)
374.	*Erechtites hieracifolia* (L.) Raf. (Stem rot)	Pilewort	Florida	1960, 1984	Anon (1960); Alfieri Jr *et al*. (1984)
375.	*Eremophila maculata*	Potted emu bush or spotted fuchsia-bush	Australia	1989	Cook and Dubé (1989)
376.	*Erigeron annuus* (L.) Pers.	Sweet scabious	Ohio	1957	Ellett (1957)
377.	*Erigeron crispus* (Syn. *Erigeron bonariensis*)	Wavy-leaf fleabane	China	1979	Tai (1979)
378.	*Erigeron canadensis* L.	Hog weed	England	1936	Ogilvie (1936)

379.	*Erigeron linifolius*	Asthmaweed	Taiwan	1979	Anon (1979)
380.	*Erodium moschatum*	Musk stork's-bi	Australia	1982	Sampson and Walker (1982)
381.	*Eruca vesicaria* (L.) Cav. (as *E. sativa* Mill.)	Rocket salad	India	1963	Gupta (1963)
382.	*Eryngium maritimum*	Sea holly/sea eryngo	Australia	1982	Sampson and Walker (1982)
383.	*Erysimum asperum* (Nutt.) D.C.	Western wallflower	Canada	1979	Anon (1979)
384.	*Erysimum hieraciifolium* (as *Cheiranthus allanii* (D.C.) Kuntze)	Wallflower	UK	1926	Dowson (1926)
385.	*Erysimum perofskianum*	Wallflower/Afghan blister cress	New Zealand	1996	McKenzie and Dingley (1996)
386.	*Escallonia rubra*	Red claws/Escallonia	New Zealand	1989	Pennycook (1989)
387.	*Eschscholzia californica* Cham.	California poppy	Australia	1982	Sampson and Walker (1982)
			USA	1974	Schwartz (1977)
388.	*Eschscholzia* sp.	California poppy	Tanzania	1960	Riley (1960)
			Cambodia	1944	Wallace (1944)
389.	*Eucalyptus* sp.	Eucalyptus	California	1989	French (1989)
			Israel	1963	Palti (1963)
			USA	1974	Schwartz (1977)
390.	*Euonymus alatus* (Thunb.) Siebold var. *campacta*	Winged spindle tree	USA	1974, 1975	Jackson and Campanini (1974, 1975)
391.	*Eupatorium formosanum*	Boneset	China	1979	Tai (1979)
			Taiwan	1979	Anon (1979)
392.	*Euphorbia dentata* Michx	Toothed spurge	Colorado, USA	1957	Burke *et al*. (1957)
			USA	1974	Schwartz (1977)
393.	*Euphorbia heterophylla*	Painted euphorbia	Brazil	1998	Mendes *et al*. (1998)
394.	*Euphorbia pulcherrima* Willd. ex. Klotzsch	Point settia	California	1989	French (1989)
			Florida	1984	Alfieri Jr *et al*. (1984)

Continued

Table 2.1. Continued

Sr. no.	Host name	English name	Country	Year	References
			North Carolina	1985	Grand (1985)
			Viet Nam	2012	Trinh *et al*. (2012)
			Washington	1973	Shaw (1973); Anon (1960)
395.	*Euphorbia serpyllifolia* Pers. (Stem rot)	Thyme leaved spurge	Michigan	1960	Anon (1960)
396.	*Euphorbia* spp. (Stem rot)	Spurge	California	1989	French (1989)
			Florida	1984	Alfieri Jr *et al*. (1984)
397.	*Euphorbia wulfenii*	Mediterranean spurge	New Zealand	1989	Pennycook (1989)
398.	*Eureka limon*	Lemon	-	1999	Fogliata *et al*. (1999)
399.	*Euryops tenuis*	Grey-leaved euryops	New Zealand	1989	Pennycook (1989)
400.	*Eustoma grandiflorum*	Lisianthus	Argentina	1996	Wolcan *et al*. (1996)
401.	*Eustoma russellianum*	Texas blue bells	Taiwan	2012	Shen *et al*. (2012)
402.	*Fagopyrum esculentum* Moench. (Stem rot)	Buckwheat	Canada	1972, 1975, 1986	Morrall *et al*. (1972, 1976); Ginns (1986)
			Delaware	1960	Anon (1960)
			India	2002	Mondal *et al*. (2002, 2003)
			Italy	2007	Lahoz *et al*. (2007)
			Korea	2004	Cho and Shin (2004)
403.	*Fagopyrum tataricum* Gaertn.	Buckwheat	India	2002	Mondal *et al*. (2002)
404.	*Felicia amelloides*	Blue marguerite	Argentina	2005	Wright *et al*. (2005)
			Italy	2004	Garibaldi *et al*. (2004a); Minuto *et al*. (2004a)
405.	*Ferula communis*	Giant fennel	Cyprus	1957	Georghiou and Papadopoulos (1957)
406.	*Ficus carica* L. (canker, stem canker, flush blast, branch disease)	Fig	California	1960, 1987, 1989	Anon (1960); French (1987); French (1989)
			Central America	1977	Wellman (1977)

			Chile	1945, 1977	Mujica and Vergara (1945); Wellman (1977)
			Greece	1973	Pantidou (1973)
			Japan	2007	Kobayashi (2007)
			Texas	1960	Anon (1960)
			West Indies	1977	Wellman (1977)
407.	*Ficus magnifolia* Muell	Magnolia leaf fig	USA	1931	Taubenhaus and Ezekiel (1931)
408.	*Foeniculum officinale*	Fennel	China	1979	Tai (1979)
409.	*Foeniculum vulgare* Mill.	Common fennel	India	1971	Sarbhoy *et al.* (1971); Sehgal and Agrawal (1971)
			Italy	1975, 1978, 2005	Noviello *et al.* (1975); Aloj and Garibaldi (1978); Balmas *et al.* (2005)
			Germany	1985	Cristea *et al.* (1985)
			Korea	2016	Choi *et al.* (2016)
			New Zealand	1989	Pennycook (1989)
			Taiwan	1979	Anon (1979)
			India	2015	Mondal *et al.* (2015)
410.	*Foeniculum vulgare* Mill. var. *dulce* Batt. & Trab. (Stem rot)	Fennel	Illinois	1960	Anon (1960)
			New Jersey	1960	Anon (1960)
			Texas	1960	Anon (1960)
411.	*Forsythia* sp. (Twig blight.)	Golden bells	California	1989	French (1989)
			North Carolina	1960	Anon (1960)
			New Zealand	1989	Pennycook (1989)
			Michigan	1983	Gourley (1982)
			Romania	2023	Florea *et al.* (2023)

Continued

Table 2.1. Continued

Sr. no.	Host name	English name	Country	Year	References
412.	*Forsythia suspensa* (Thunb.) Vahl	Golden bells	Japan	2007	Kobayashi (2007)
413.	*Forsythia viridissima* Lindl.	Golden bells	USA	1947	Wolf (1947)
			Scotland	1961	Foister (1961)
			USA	1960	Anon (1960)
414.	*Fragaria ananassa* Duchesne (as *F. chiloensis* (L.) Duchesne var. *ananassa* Bailey)	Strawberry	Australia	1966	Simmonds (1966)
			California	1960, 1987, 1989	Anon (1960); French (1987, 1989)
			Egypt	2007	Embaby (2007)
			Florida	1984, 2020	Alfieri Jr *et al*. (1984); Marin and Peres (2020)
			Iowa	1960	Anon (1960)
			Illinois	1960	Anon (1960)
			Korea	2004	Cho and Shin (2004)
			Maryland	2022	Hellman *et al*. (2022)
			Minnesota	1960	Anon (1960)
			North Carolina	1985	Grand (1985)
			New Zealand	1989	Pennycook (1989)
			South-eastern states	1960	Anon (1960)
			Texas	1960	Anon (1960)
415.	*Fragaria* sp. (on fruits)	Strawberry	Washington	1957	McLean (1957)
416.	*Freesia hybrida*	Freesia	Bulgaria	2009	Bobev (2009)
417.	*Freesia* spp. Eckl. Ex. Klatt	Freesia	Canada	1967	Conners (1967)
418.	*Fuchsia* sp.	Humming bird fuchsia	Moscow	1963	Shumilenko (1963)
419.	*Gaillardia aristata*	Gaillardia	New Zealand	1989	Pennycook (1989)
			Canada	1967	Conners (1967)

420.	*Gaillardia pulchella* Foug.	Firewheel	China	1979	Tai (1979)
			Taiwan	1979	Anon (1979)
			India	1959, 1977	Rai and Agnihotri (1970); Kaur and Chohan (1977)
421.	*Gaillardia grandiflora*	Blanket flower	California	1997	Koike (1997)
			Italy	2015	Garibaldi *et al*. (2015)
422.	*Gaillardia* sp.	Gaillardia	Israel	1963	Palti (1963)
			USA	1974	Schwartz (1977)
423.	*Galinsoga parviflora* Cav.	Small-flowered galinsoga	Australia	1989	Shivas (1989)
			USA	1974	Adams *et al*. (1974)
			Brazil	1998	Mendes *et al*. (1998)
424.	*Galium aparine*	Cleavers	China	1979	Tai (1979)
425.	*Gazania* hybrid	Gazania/treasure Flower	Italy	2002	Garibaldi *et al*. (2002a)
426.	*Gazania rigens* (L.) Gaertn.	Treasure flower	USA	1974	Schwartz (1977)
			Louisiana	2006	Holcomb (2006)
			New Zealand	1989	Pennycook (1989)
427.	*Gazania* spp.	Gazania	Australia	1989	Shivas (1989)
			Italy	2001	Perez *et al*. (2001)
			Italy	2001, 2004	Garibaldi *et al*. (2001a); Wolcan (2004)
428.	*Gentiana lutea* L.	Yellow gentian	Belgrade, Serbia	1953	Kljajic (1953)
429.	*Gerbera jamesonii* Bolus (crown rot/rot of gerberas)	African daisy/Transvaal daisy	Arizona	1994	Matheron and Matejka (1994)
			Bermuda	1935, 1939	Russell (1935); Waterson (1939)
			Korea	2004	Cho and Shin (2004)
			Maryland, New York	1960	Anon (1960)

Continued

Table 2.1. Continued

Sr. no.	Host name	English name	Country	Year	References
			New Zealand	1989	Pennycook (1989)
			Poland	1976	Orlikowski (1976)
			India	1973	Roy (1973)
			Argentina	2004	Wolcan (2004)
430.	*Gerbera* spp.	Gerbera	Sicily, Italy	1991	Greuter *et al*. (1991)
			USA	1974	Schwartz (1977)
			New Zealand	1946	Brien (1946)
			Hungary	1985	Folk and Tusnadin (1985)
			Italy	1976	Garibaldi and Gullino (1976)
431.	*Glaucium flavum* Cranz.	Yellow-horned poppy	Germany	1985	Cristea *et al*. (1985)
432.	*Gladiolus* spp.	Gladiolus	USA	1974	Schwartz (1977)
433.	*Gloxinia* sp.	Sinningia hybrid	Bulgaria	2009	Bobev (2009)
434.	*Glycine max* (L.) Merrill (stem rot)	Soybean	Australia	1982, 1966	Sampson and Walker (1982); Simmonds (1966)
			Brazil	1998	Mendes *et al*. (1998)
			Bulgaria	2009	Bobev (2009)
			Canada	1967, 1986	Conners (1967); Ginns (1986)
			China	1979, 1981	Tai (1979); Li and Fu (1981)
			Florida	1984	Alfieri Jr *et al*. (1984)
			Ethiopia	2019	Pawlowski *et al*. (2019)
			Greece	2000	Holevas *et al*. (2000)
			Iowa	1949, 1960	Anon (1960); Gilman (1949)
			Illinois, New Hampshire	1951, 1997	Chamberlain (1951); Zhang *et al*. (1997)
			India	1998	Pande and Rao (1998)
			Korea	1987, 2004	Shin and Lee (1987); Cho and Shin (2004)

			Maryland, New York	1960	Anon (1960)
			Maine	1998	Johnson *et al*. (1998)
			Mexico	1976	Alvarez (1976)
			Minnesota	2011	Bienapfl *et al*. (2011)
			New Zealand	1989	Pennycook (1989)
			Nigeria	1992	Akem and Dashiell (1992)
			Portugal	1978	de Sousa and Lucas (1978)
			Romania	1990	Richardson (1990)
			South Africa	2000	Crous *et al*. (2000)
			Taiwan	1979	Anon (1979)
			Virginia	1960, 1982	Anon (1960); Phipps and Porter (1982)
435.	*Glycine soja*	Wild soybean	China	1979	Tai (1979)
436.	*Gnaphalium formosanum* (Syn. *Pseudognaphalium adnatum*)	*Cudweeds*	Taiwan	1979	Anon (1979)
437.	*Gnaphalium indicum*	Cudweeds	China	1979	Tai (1979)
			Taiwan	1979	Anon (1979)
438.	*Gnaphalium luteoalbum* (Syn. *Helichrysum luteoalbum* (L.) Rchb.)	Jersey cudweed	China	1979	Tai (1979)
439.	*Gnaphalium purpureum* var. *spathulatum* (crown rot)	Purple cudweed	Florida	1984	Alfieri Jr *et al*. (1984)
			India	1973	Roy (1973)
440.	*Goodyera schlechtendaliana*	Jewel orchid	India	2006	Bag (2006)
441.	*Gossypium hirsutum* L.	Upland cotton	Arizona	1974, 2018	Schwartz (1977); Hu *et al*. (2018)
442.	*Gossypium* sp.	Cotton	Israel	1960	Plati (1960)
			Brazil	1999	Charchar *et al*. (1999)

Continued

Table 2.1. Continued

Sr. no.	Host name	English name	Country	Year	References
443.	*Guizotia abyssinica*	Niger	India	2009	Nagaraja and Krishnappa (2009)
			North Dakota	2003	Bradley *et al*. (2003)
444.	*Gynandropsis gynandra*	African spider flower	China	1979	Tai (1979)
445.	*Gynura* sp.	Longevity spinach	Florida	1984	Alfieri Jr *et al*. (1984)
446.	*Gypsophila elegans*	Showy baby's breath	Kenya	1939	Keay (1939)
447.	*Gypsophila paniculata* L.	Baby's breath	New Zealand	1957, 1989	Brien and Dingley (1957); Pennycook (1989)
448.	*Hibiscus rosa-sinensis*	Shoe black plant	China	2023	Lin *et al*. (2024)
449.	*Hebe* sp.	Hebe plant	New Zealand	1989	Pennycook (1989)
450.	*Hedera helix* L. (Stem rot)	English ivy	Florida	1984	Alfieri Jr *et al*. (1984)
451.	*Helianthus annuus* L.	Sunflower	Australia	1966, 1989	Shivas (1989); Simmonds (1966)
			Brazil	1954, 1998	Neto (1954); Mendes *et al*. (1998)
			Bulgaria	2009	Bobev (2009)
			California	1976, 1989, 1991	Klisiewicz and Beard (1976); French (1989); Gulya *et al*. (1991)
			Manitoba, Ontario Saskatchewan, Manitoba, Canada	1938, 1967, 1972, 1976, 1982, 1986, 1990	Bisby (1938); Conners (1967); Morrall *et al*. (1972); Hoes and Huang (1976); Mederick and Piening (1982); Ginns (1986); Richardson (1990)
			British Columbia, Canada	1964, 2000	Toms (1964); Hilton (2000)
			Chile	1945, 1967	Mujica and Oehrens (1967); Mujica and Vergara (1945)
			China	1979, 1988	Tai (1979); Yang *et al*. (1988); Liu and Li (1988)

Cyprus	1957	Georghiou and Papadopoulos (1957)
Greece	2000	Holevas *et al*. (2000)
Hawaii	1981, 1985	Raabe *et al*. (1981); Gulya *et al*. (1985)
Iran	1974	Ale-Agha (1974)
Israel	1961	Chorin and Rotem (1961)
India	1973	Roy (1973)
Kenya	1961	Nattrass (1961)
Korea	2004	Cho and Shin (2004)
Mauritius	1968	Orieux and Felix (1968)
Mexico	1976	Alvarez (1976)
Montana	1973	Shaw (1973)
North Dakota	1986	Holley and Nelson (1986)
New Zealand	1989	Pennycook (1989)
North central states	1960	Anon (1960)
Oregon	1973	Shaw (1973)
Romania	1990	Richardson (1990)
Russia	1990	Richardson (1990)
South Africa	2000	Crous *et al*. (2000)
Taiwan	1979	Anon (1979)
Tanzania	1960	Riley (1960)
Turkey	2017	Ozer *et al*. (2017)
USSR	1990	Richardson (1990)
Washington	1973	Shaw (1973)
Western states	1960	Anon (1960)

Continued

Table 2.1. Continued

Sr. no.	Host name	English name	Country	Year	References
			Yugoslavia	1990	Richardson (1990)
452.	*Helianthus* sp.	Common sunflower	England	1978	Dennis (1978)
			North Carolina	1985	Grand (1985)
			Romania	1990	Richardson (1990)
			UK	1986	Dennis (1986)
453.	*Helianthus tuberosus* L. (Syn. *Helianthemum tuberosus*)	Jerusalem artichoke/prickly free rock roses	California	1989	French (1989)
			Chile	1967	Mujica and Oehrens (1967)
			Massachusetts, Minnesota	1960	Anon (1960)
			Washington, Oregon	1973	Shaw (1973)
			New Zealand	1932, 1989	Brien (1932); Pennycook (1989)
			Turkey	2017	Ozer *et al*. (2017)
			Ontario, Canada	1967	Conners (1967)
			Scotland	1961	Foister (1961)
454.	*Helichrysum bracteatum* (Venten.) Andr.	Golden everlasting/Strawflower	Brazil	2009	Duarte and Barreto (2009)
			China	1979	Tai (1979)
			Italy	2020	Garibaldi *et al*. (2020)
			Taiwan	1979	Anon (1979)
			India	1973, 1977	Mookerjee (1973); Kaur and Chohan (1977)
			Japan	1999	Takeuchi and Horie (1999)
455.	*Helichrysum* sp.	Straw flower	Bermunda	1927	Ogilvie (1927)
456.	*Heliotropium ellipticum*	Indian heliotrope	India	2004	Ghasolia *et al*. (2004)
457.	*Helipterum almcans* (Cunn.) D. C. (as *Acroclium album* Gray)	Everlasting	India	1964, 1977	Shukla and Pavgi (1964); Kaur and Chohan (1977)

458.	*Helipterum roseum* (Hook.) Benth. (as *Acroclium roseum* Hook.)	Paper flower	India	1964	Shukla and Pavgi (1964)
459.	*Hemigraphis alternata* (Bum.) Anderson	Red Ivy	Florida	1984	Alfieri Jr *et al*. (1984)
460.	*Hemisteptia carthamoides (Candolle) Kuntze* [Syn. *Saussurea lyrata*]	Saw-wort	Taiwan	1979	Anon (1979)
461.	*Hemisteptia lyrata* [Syn. *Saussurea lyrata*]	Saw-wort/snow lotus	China	1979	Tai (1979)
462.	*Heracleum mantegazzianum*	Giant hogweed	Scotland	1961	Foister (1961)
			Netherlands	2003	Erneberg *et al*. (2003)
			Denmark	2003	de Voogd *et al*. (2003)
463.	*Hibiscus cannabinus* L.	Indian hemp/Deccan hemp/ Java jute	Florida	1984	Alfieri Jr *et al*. (1984)
			India	2015	Tripathi *et al*. (2015)
			Kenya	1961	Nattrass (1961)
			Italy	1996	Corato (1996)
464.	*Hibiscus esculentus*	Okra	China	1979	Tai (1979)
			Florida	1984	Alfieri Jr *et al*. (1984)
			Massachusetts	1960	Anon (1960)
			Taiwan	1979	Anon (1979)
465.	*Hibiscus rosa-sinensis* L.	Chinese hibiscus	Australia	1989	Shivas (1989)
			China	1979	Tai (1979)
			Taiwan	1979	Anon (1979)
			USA	1974	Schwartz (1977)
466.	*Hibiscus sabdariffa* L.	Jamaica sorrel	China	1979	Tai (1979)
			India	1963	Gupta (1963)
			Taiwan	1979	Anon (1979)

Continued

Table 2.1. Continued

Sr. no.	Host name	English name	Country	Year	References
467.	*Hibiscus schizopetalus*	Spider hibiscus	China	1979	Tai (1979)
			Taiwan	1979	Anon (1979)
468.	*Hibiscus* sp.	Rose mallow, hemp	Alberta, Canada	2000	Bains *et al*. (2000)
469.	*Hibiscus trionum*	Flower-of-an-hour	New York	2009	Strauss and Dillard (2009)
470.	*Holoptelea integrifolia*	Indian elm/jungle cork tree	India	2004	Ghasolia *et al*. (2004)
471.	*Hordeum vulgare* L.	Barley	Czechoslovakia	1986	Hysek *et al*. (1986); Miclaus *et al*. (1988)
472.	*Humulus lupulus* L.	Common hop	Australia	1982	Sampson and Walker (1982)
			Kenya	1939	Keay (1939)
			New Zealand	1989	Pennycook (1989)
			UK	1936	Ogilvie (1936)
			USA	1974, 2012	Schwartz (1977); Kropf *et al*. (2012)
473.	*Humulus* sp.	Hop	–	1994	Boland and Hall (1994); Saharan and Mehta (2008); Ficker (2019)
474.	*Hutchinsia alpina*	Chamois cress	Italy	2004	Garibaldi *et al*. (2004b); Minuto *et al*. (2004a)
475.	*Hydrangea* sp.	Hydrangea	Canada	1964, 1967, 1986	Toms (1964); Conners (1967); Ginns (1986)
476.	*Hydrocotyle rotundifolia*	*Lawn marsh penny wort*	Taiwan	1979	Anon (1979)
477.	*Hydrocotyle sibthorpioides*	Lawn penny wort	China	1979	Tai (1979)
478.	*Hyoscyamus niger* L.	Henbane	New Zealand	1946, 1989	Brien (1946); Pennycook (1989)
479.	*Iberis amara* L.	Rocket candytuff	India	1987	Saxena and Rai (1987)
480.	*Iberis sempervirens*	Evergreen candytuft	Italy	2007	Garibaldi *et al*. (2007a)
481.	*Iberis* sp.	Candytuft	California	1989	French (1989)
482.	*Iberis umbellata* L.	Globe candytuff	India	1964, 1977	Shukla and Pavgi (1964); Kaur and Chohan (1977)

483.	*Iliamna rivularis* (Dougl.) Greene	Streambank wild hollyhock	Montana	1973	Shaw (1973)
484.	*Indigofera arrecta*	Natal indigo	China	1979	Tai (1979)
485.	*Indigofera natalensis*	Forest indigo	Japan	2007	Kobayashi (2007)
486.	*Ipomoea aquatica*	Water spinach	India	2014	Hansda *et al*. (2014)
487.	*Ipomoea batatas* (L.) Lam.	Sweet potato	China	1979, 1990	Tai (1979); Zhang and Huang (1990)
			Korea	2004	Cho and Shin (2004)
			New Zealand	1989	Pennycook (1989)
			South Africa	2000	Crous *et al*. (2000)
			Taiwan	1979	Anon (1979)
488.	*Ipomoea reptans*	Water spinach	China	1979	Tai (1979)
			Taiwan	1979	Anon (1979)
489.	*Ipomoea* sp.	Morning glories	Brazil	1998	Mendes *et al*. (1998)
490.	*Iris* sp.	Iris	Bulgaria	2009	Bobev (2009)
			Canada	1986	Ginns (1986)
			Washington	1973	Shaw (1973)
491.	*Iris tingitana*	Bulbous Spanish iris	Australia	1989	Shivas (1989)
492.	*Iris xiphioides* Ehrh.	English iris	USA	1974	Schwartz (1977)
493.	*Iva sp.*	Marsh-elder	Montana	1973	Shaw (1973)
494.	*Iva xanthifolia* Nutt.	False ragweed	Canada	1938	Bisby (1938)
495.	*Ixeridium dentatum*	Toothed ixeridium	South Korea	2017	Park *et al*. (2017)
496.	*Ixeris debilis*	Uncertain	China	1979	Tai (1979)
497.	*Ixeris denticulata*	Gutweed	China	1979	Tai (1979)
498.	*Ixeris dentata* (Thunb) Nakai (Syn. *Ixeridium dentatum*)	Shiro-bana-niga-na (Japanese name)	South Korea	1987	Shin and Lee (1987)
499.	*Ixeris oldhamii* (Syn. *Ixeridium laevigatum* (Blume) Pak & Kawano)	*Giant timber bamboo*	Taiwan	1979	Anon (1979)

Continued

Table 2.1. Continued

Sr. no.	Host name	English name	Country	Year	References
500.	*Lablab purpureus* (Stem and branch rot)	Hyacinth bean	Australia	1990	Lenne (1990)
			Bangladesh	2014, 2018	Prova *et al*. (2014, 2018)
501.	*Lactuca debilis*	*Prickly lettuce*	Taiwan	1979	Anon (1979)
502.	*Lactuca formosana*	*Maxim*	China	1979	Tai (1979)
			Taiwan	1979	Anon (1979)
503.	*Lactuca gracilifolia*	Grassleaf lettuce	China	1979	Tai (1979)
504.	*Lactuca indica*	Indian lettuce	China	1979	Tai (1979)
			Taiwan	1979	Anon (1979)
505.	*Lactuca indica* var. *dracoglossa*	*Indian lettuce*	Korea	2003, 2004	Cho and Shin (2004); Chang and Kim (2003)
506.	*Lactuca laciniata*	Lettuce	Taiwan	1979	Anon (1979)
507.	*Lactuca oldhamii*	*Prickly lettuce/milk thistle compass plant*	China	1979	Tai (1979)
508.	*Lactuca sativa* L.	Garden lettuce	Alabama	1931	Blain (1931)
			Australia	1982, 1989	Sampson and Walker (1982); Shivas (1989); Simmonds (1966); Cook and Dubé (1989)
			Arizona	1979	Kohn (1979a,b)
			Brazil	1954, 1990, 1998	Neto (1954); Richardson (1990); Mendes *et al*. (1998)
			California	1985, 1989	Patterson and Grogan (1985); French (1989)
			Canada	1967, 1986	Conners (1967); Ginns (1986)
			Chile	1967	Mujica and Oehrens (1967)
			China	1979	Tai (1979)
			British Columbia	1964	Toms (1964)

Costa Rica	1967	McGuire Jr and Crandall (1967)
Cyprus	1957	Georghiou and Papadopoulos (1957)
Fiji	1972, 1981	Firman (1972); Dingley *et al*. (1981)
Florida	1984	Alfieri Jr *et al*. (1984)
France	2018	Leyronas *et al*. (2018)
Georgia	1963	Hanlin (1963)
Greece	1973, 2000	Pantidou (1973); Holevas *et al*. (2000)
Guatemala, Honduras	1967	McGuire Jr and Crandall (1967)
Hong Kong	2000	Lu *et al*. (2000)
Idaho	1973	Shaw (1973)
Jorhat, India	1973	Roy (1973)
Kenya	1961	Nattrass (1961)
Korea	2003, 2004	Chang and Kim (2003); Cho and Shin (2004)
Kentucky	1958	Johnson and Chapman (1958)
Libya	1981	El-Buni and Rattan (1981)
Mauritius	1948, 1968	Wiehe (1948); Orieux and Felix (1968)
Mexico	1976	Alvarez (1976)
Mississippi	1959	Parris (1959)
Montana	1973	Shaw (1973)
North Carolina	1985	Grand (1985)
New Jersey	1922	Poole (1922)
New Zealand	1932, 1989	Brien (1932); Pennycook (1989)

Continued

Table 2.1. Continued

Sr. no.	Host name	English name	Country	Year	References
			New York	1986	Wymore and Lorbeer (1986)
			Oklahoma	1945	Preston (1945)
			Oregon	1973	Shaw (1973)
			USA	1936, 1960	Young (1936); Anon (1960)
			South Africa	1977, 2000	Gorter (1977); Crous *et al*. (2000)
			Taiwan	1979	Anon (1979)
			Tanzania	1960	Riley (1960)
			Tennessee	2018	Shrestha *et al*. (2018)
			USA	2018	Shrestha *et al*. (2018)
			Washington	1973	Shaw (1973)
509.	*Lactuca sativa* var. *capitata* L.	Head lettuce	Australia	1978	Merriman *et al*. (1978)
			China	1979	Tai (1979)
			Hawaii	1981	Raabe *et al*. (1981)
510.	*Lactuca sativa* var. *crispa* L.	Leaf lettuce	USA	1960	Anon (1960)
511.	*Lactuca sativa* var. *longifolia* Lam.	Romaine lettuce	USA	1960	Anon (1960)
512.	*Lactuca serriola* L. (Stem rot)	Prickly lettuce	California	1960, 1989	Anon (1960); French (1989)
513.	*Lactuca* sp. (Stem rot)	Lettuce	Bulgaria	2009	Bobev (2009)
			Bermuda	1938	Waterston (1939)
			Ontario, Canada	1997	Holst-Jensen *et al*. (1997a,b)
			Maryland, Mississippi	1960	Anon (1960)
			USA	1998	Holst-Jensen *et al*. (1998)
514.	*Lagenaria siceraria* (Mol.) Standl.	Bottle gourd	India	1973, 2002	Roy (1973); Pandey and Pandey (2002)

515.	*Lamium amplexicaule* L.	Henbit	North Carolina	1977	Welty (1977)
516.	*Lamium* spp.	Deadnettle	Arizona	1936	Brown and Butler (1936)
517.	*Lanaria canadensis* (L.) Dum	Blue toadflax	–	1994	Boland and Hall (1994); Saharan and Mehta (2008)
518.	*Lanaria* spp.	Toadflax	–	1994	Boland and Hall (1994); Saharan and Mehta (2008)
519.	*Lanaria vulgaris* Mill.	Butter-and-eggs	–	1994	Boland and Hall (1994); Saharan and Mehta (2008)
520.	*Lantana camara*	Lantana or shrub verbena	Italy	2008	Garibaldi *et al*. (2008e)
521.	*Lapsana apogonoides*	Japanese nipplewort	Japan	1999	Kanno and Ohkubo (1999)
522.	*Larix kaempferi* (lamb.) Carriere	Japanese larch	–	1994	Boland and Hall (1994); Saharan and Mehta (2008)
523.	*Lathyrus esculenta* Moench (as *Lens culinaris* Medic.)	Lentil	USA	1974, 1992	Schwartz (1977); Kaiser (1992)
524.	*Lathyrus odoratus*	Sweet pea	New Zealand	1989	Pennycook (1989)
			USA	1920, 1960	Taubenhaus (1920); Anon (1960)
525.	*Lathyrus sativus*	Grass pea	India	1998	Pande and Rao (1998)
			Canada	1990	Zimmer and Campbell (1990)
526.	*Lathyrus* sp.	Pea vines	Greece	1973	Pantidou (1973)
527.	*Lavandula angustifolia*	English lavender	Europe	2007	Orlikowski and Valjustkaite (2007)
528.	*Lavatera arborea* L. (Stem rot)	Tree mallow	Florida	1974, 1984	Schwartz (1977); Alfieri Jr *et al*. (1984)
			New Zealand	1932, 1989	Brien (1932), Pennycook (1989)
529.	*Lens culinaris* Medic. (isolated from roots, stems, and leaves of seedlings)	Lentil	Bangladesh	2015	Ahmed and Akhond (2015)
			Canada	1986	Ginns (1986)
			Alberta, Canada	1994	Hwang *et al*. (1994)

Continued

Table 2.1. Continued

Sr. no.	Host name	English name	Country	Year	References
			Manitoba, Canada	2000	Hilton (2000)
			Saskatchewan, Canada	1970, 1972, 2000	Duczek and Morrall (1971); Morrall *et al*. (1972); Hilton (2000)
			India	1991	Khare (1991)
			Iran	2017	Mahdikhani and Aghaalikhani (2017)
			North-western states, USA	1965, 1973	Wilson and Brandsberg (1965); Shaw (1973)
530.	*Lens esculentum*	Lentil	India	1998	Pande and Rao (1998)
531.	*Leonurus sibiricus*	Siberian mother wort	China	1979	Tai (1979)
			Korea	2004	Cho and Shin (2004)
			Taiwan	1979	Anon (1979)
			Korea	2002	Kwon and Park (2002)
532.	*Lepidagathis formosensis*	Orange daylily	China	1979	Tai (1979)
			Taiwan	1979	Anon (1979)
533.	*Lepidium sativum* L. (Crown rot)	Garden cress	Massachusetts	1960	Anon (1960)
534.	*Lepidium virginicum* L.	Poor man's pepper cress	USA	1971	Morgan (1971)
535.	*Liatris* sp. (Stem rot)	Gay feather	New York	1960	Anon (1960)
			Bermuda	1938	Waterston (1939)
536.	*Liatris spicata*	Dense blazing star or prairie feather	Florida	1991	Miller (1991a,b)
537.	*Ligusticum acutilobum*	Licorice root	Korea	2004	Cho and Shin (2004)
538.	*Lilium candidum* L. (Stem rot)	Madonna lily	Connecticut	1960	Anon (1960)
			Mauritius	1968	Orieux and Felix (1968)
539.	*Lilium longiflorum* Thurb (Stem rot)	Trumpet lily	Connecticut	1960	Anon (1960)
			Florida	1960, 1984	Anon (1960); Alfieri Jr *et al*. (1984)

540.	*Lilium* sp. (Stem rot)	Lily	California	1989	French (1989)
			Texas	1960	Anon (1960)
			Washington	1960	Anon (1960)
541.	*Linaria canadensis* (Stem rot)	Blue toadflax	Florida	1984	Alfieri Jr *et al*. (1984)
542.	*Linaria* sp. (Stem rot)	Golden flax	Australia	1966	Simmonds (1966)
			Arizona	1960	Anon (1960)
			Canada	1967	Conners (1967)
543.	*Linaria vulgaris* Mill.	Butter-and-eggs	India	1977	Kaur and Chohan (1977)
544.	*Linum flavum* (Stem rot)	Yellow flax	California	1960	Anon (1960)
			Moscow	1981	Karpunin (1981)
545.	*Linum usitatissimum* L. (Stem rot)	Common flax	Arizona	1960	Anon (1960)
			Alberta	1982	Mederick and Piening (1982)
			Bulgaria	2009	Bobev (2009)
			California	1960	Anon (1960); French (1989)
			Manitoba, Canada	2000	Hilton (2000)
			Saskatchewan, Canada	2000	Hilton (2000)
			China	1979	Tai (1979)
			France	2018	Leyronas *et al*. (2018)
			Italy	1990	Richardson (1990)
			New Zealand	1996	McKenzie and Dingley (1996)
			Oregon	1960	Anon (1960)
			Taiwan	1959, 1979	Sawada (1959); Anon (1979)
			Texas	1960	Anon (1960)

Continued

Table 2.1. Continued

Sr. no.	Host name	English name	Country	Year	References
			UK	1986, 1990, 1991	Mitchell *et al*. (1986); Richardson (1990); Pope and Sweet (1991); Rashid (2001)
546.	*Lobelia chinensis*	Asian lobelia	China	1979	Tai (1979)
547.	*Lobelia erinus* L.	Edging lobelia	Italy	1922	Voglino (1922)
			New Zealand	1989	Pennycook (1989)
548.	*Lobelia radicans*	Asian lobelia	Taiwan	1979	Anon (1979)
549.	*Lobelia* sp.	Cardinal flower	California	1989	French (1989)
550.	*Lobularia maritima* (L.) Deso. (crown rot)	Sweet alyssum	Australia	1966	Simmonds (1966)
			Canada	1967	Conners (1967)
			India	1987	Saxena and Rai (1987)
			New Zealand	1996	McKenzie and Dingley (1996)
			Virginia	1960	Anon (1960)
551.	*Lotus corniculatus* L.	Bird's foot trefoils	New York	1960	Anon (1960)
			Washington	1973	Shaw (1973)
552.	*Lotus purshianus*	American bird's-foot Trefoil	California	1989	French (1989)
553.	*Lotus* sp.	Bird's foot trefoils	USA	1974	Schwartz (1977)
554.	*Luffa cylindrical* Mill. (as *L. Leucantha* (Dutch.) Rasby	Loofah	Washington	1973	Shaw (1973)
555.	*Lupinus albus*	White lupin/field lupine	India	1998	Pande and Rao (1998)
556.	*Lupinus angustifolius* L. (Stem rot)	European blue lupine/narrow leaf lupin/blue lupin	Australia	1966, 1982, 1989	Simmonds (1966); Sampson and Walker (1982); Shivas (1989)
			Brazil	1998	Mendes *et al*. (1998)
			Florida	1960	Anon (1960)
			Louisiana	1960	Anon (1960)
			Mississippi	1959	Parris (1959)
			New Zealand	1932, 1989	Brien (1932); Pennycook (1989)

			South Africa	2000	Crous *et al*. (2000)
557.	*Lupinus cosentinii*	Sandplain lupin	Australia	1989	Shivas (1989)
558.	*Lupinus hirsutus*	Blue lupin	India	1998	Pande and Rao (1998)
559.	*Lupinus nootkatensis* Donn. (*L. regalis* Berg.)	Lupine	Canada	1967	Conners (1967)
560.	*Lupinus perennis* L. (Stem rot)	Sundial lupine	New York	1960	Anon (1960)
			Texas	1960	Anon (1960)
561.	*Lupinus polyphyllus* Lindl. (Stem rot)	Washington lupine	UK	1954	Beaumont (1951)
562.	*Lupinus* sp.	Lupine	Brazil	1998	Mendes *et al*. (1998)
			Florida	1984	Alfieri Jr *et al*. (1984)
			Louisiana, New York	1960	Anon (1960)
			Scotland	1961	Foister (1961)
			South Africa	1977	Gorter (1977)
			USA	1974	Schwartz (1977)
563.	*Lupinus texensis*	Texas lupine	USA	1993	Woodard and Newman (1993)
564.	*Lycopersicon esculentum* Mill.	Tomato	Australia	1966, 1982, 1989	Simmonds (1966); Sampson and Walker (1982); Shivas (1989); Cook and Dubé (1989)
			Bermuda	1939	Waterston (1939)
			Brazil	1998	Mendes *et al*. (1998)
			Bulgaria	2009	Bobev (2009)
			California	1951, 1989	Bardin (1951); French (1989)
			Canada	1964, 1971	Toms (1964); Duczek and Morrall (1971)
			China	1979	Tai (1979)

Continued

Table 2.1. Continued

Sr. no.	Host name	English name	Country	Year	References
			El Salvador	1967	McGuire Jr and Crandall (1967)
			Florida	1984	Alfieri Jr *et al*. (1984)
			Greece	1973, 2000	Pantidou (1973); Holevas *et al*. (2000)
			Hawaii	1972, 1981	Goos and Anderson (1972); Raabe *et al*. (1981)
			India	1973	Roy (1973)
			Korea	2004	Cho and Shin (2004)
			Louisiana	1957	Plakidas (1957)
			Mauritius	1968	Orieux and Felix (1968)
			North Carolina	1985	Grand (1985)
			New Zealand	1932, 1989	Brien (1932); Pennycook (1989)
			Scotland	1961	Foister (1961)
			South Africa	1977, 2000	Gorter (1977); Crous *et al*. (2000)
			Taiwan	1979	Anon (1979)
			Tanzania	1960	Riley (1960)
			USA	1960	Anon (1960)
			Washington	1973	Shaw (1973)
565.	*Lycopersicon lycopersicum*	Tomato	Canada	1986	Ginns (1986)
566.	*Lycopersicon* sp.	Tomato	Rhode Island	2010	Goos (2010)
567.	*Macleaya cordata*	Five-seeded plume-poppy	China	2019	Zhou *et al*. (2019)
568.	*Malus sylestris* Mill. (as *Malus domestica* Borkh.)	Apple	Nova Scotia, Canada	1959, 1967	Hockey (1959); Conners (1967)
			Italy	1982	Marciano *et al*. (1982)
			Japan	1986	Batra and Harada (1986)
			New Zealand	1989	Pennycook (1989)

			Washington	1985	Covey (1985)
569.	*Malus* sp. (as *Malus pumila* Mill.)	Apple	Canada	1986	Ginns (1986)
			USA	1974	Schwartz (1977)
570.	*Malva parviflora*	Cheese weed	California	1989	French (1989)
571.	*Malva sylvestris*	Common mallow	China	1979	Tai (1979)
572.	*Malva verticillata*	Chinese mallow or cluster mallow	China	1979	Tai (1979)
			Korea	2004	Cho and Shin (2004)
			Taiwan	1979	Anon (1979)
573.	*Malvaviscus arboreus* var. *arboreus*	Lanka jaba	India	2014	Hansda *et al.* (2014)
			Texas	1960	Anon (1960)
574.	*Malvaviscus* sp. (twig blight)	Sleepy mallow	Texas	1960	Anon (1960)
575.	*Mangifera indica*	Mango	Japan	2018	Ajitomi *et al.* (2018)
576.	*Manihot esculenta*	Cassava manioc/Yuca	China	1979	Tai (1979)
577.	*Manihot utilissima*	Cassava	Taiwan	1979	Anon (1979)
578.	*Matthiola incana* (L.) Br. (crown rot)	Common stock	Australia	1982	Sampson and Walker (1982)
			California	1960, 1989	Anon (1960); French (1989)
			Greece	2000	Holevas *et al.* (2000)
			Korea	2004, 2017	Cho and Shin (2004); Choi *et al.* (2017a)
			Michigan	1960	Anon (1960)
			New Zealand	1989	Pennycook (1989)
			Pennsylvania	1960	Anon (1960)
			Scotland	1961	Foister (1961)
			South Africa	1977, 2000	Gorter (1977); Crous *et al.* (2000)

Continued

Table 2.1. Continued

Sr. no.	Host name	English name	Country	Year	References
579.	*Matthiola* sp.	Stock	California	1989	French (1989)
			Mauritius	1968	Orieux and Felix (1968)
			Washington	1973	Shaw (1973)
580.	*Mazus japonicus*	Japanese mazus	China	1979	Tai (1979)
581.	*Mazus stachydifolius*	Baby jump-up	China	1979	Tai (1979)
582.	*Mazus stolonifer* (Syn. *Mazus miquelii* Makino]	Creeping mazus	Taiwan	1979	Anon (1979)
583.	*Meconopsis* sp.	Welsh poppy	Scotland	1961	Foister (1961)
584.	*Medicago hispida* Gaertn. (Collar rot)	Bur clover	California	1960, 1989	Anon (1960); French (1989)
585.	*Medicago lupulina* L.	Black medick	India	1998	Pande and Rao (1998)
			USA	1974	Schwartz (1977)
586.	*Medicago sativa* L. (collar rot)	Alfalfa	Australia	1989	Shivas (1989)
			California	1960, 1989	Anon (1960); French (1989)
			Alberta, Canada	2021	Gossen and Howard (2021)
			Manitoba, Canada	2000, 2021	Hilton (2000); Gossen and Howard (2021)
			Saskatchewan, Canada	1938, 2000, 2021	Bisby (1938); Hilton (2000); Gossen and Howard (2021)
			Chile	1967	Mujica and Oehrens (1967)
			China	1979	Tai (1979)
			Delaware, NewYork	1960	Anon (1960)
			Idaho	1960, 1973	Anon (1960); Shaw (1973)
			Korea	1987	Shin and Lee (1987)

			Mississippi	1991	Pratt and Rowe (1991)
			New Zealand	1932	Brien (1932)
			Oregon	1973	Shaw (1973)
			Taiwan	1979	Anon (1979)
			Washington	1960, 1973, 1991	Anon (1960); Shaw (1973); Gilbert (1991)
587.	*Melilotus alba* Desr.	White sweet clover	California	1960, 1989	Anon (1960); French (1989)
			Chile	1967	Mujica and Oehrens (1967)
			Montana	1960, 1973	Shaw (1973); Anon (1960)
			USA	1936	Young (1936)
588.	*Melilotus indica* (L.) All. (Stem rot)	Sweet clover	California	1960, 1989	Anon (1960); French (1989)
589.	*Melilotus officinalis* (L.) Pall. (Stem rot)	Yellow sweet clover	Montana	1973	Anon (1960); Shaw (1973)
			Washington	1936, 1973	Young (1936); Shaw (1973)
590.	*Melilotus* sp.	Sweet clover	Canada	1967, 1982	Conners (1967); Mederick and Piening (1982)
591.	*Melinis minutiflora*	Molasses grass	Brazil	1998	Mendes *et al.* (1998)
592.	*Mentha arvensis* var. *piperascens*	Japanese mint	China	1979	Tai (1979)
593.	*Mentha arvensis* var. *vulgaris*	Corn mint	Taiwan	1979	Anon (1979)
594.	*Mentha canadensis* var. *piperascens*	American wild mint	Taiwan	1979	Anon (1979)
595.	*Mentha spicata*	Spear mint	Italy	2013	Garibaldi *et al.* (2013)
596.	*Mentha piperita* L.	Peppermint	Washington	1957, 1973	Skotland and Menzies (1957); Shaw (1973)
597.	*Mertensia lanceolata* (Pursh.) D.C. (Stem rot)	Blue bell	Colorado	1960	Anon (1960)
598.	*Mimosa pudica*	Shame plant	India	2018	Borah *et al.* (2018)

Continued

Table 2.1. Continued

Sr. no.	Host name	English name	Country	Year	References
599.	*Mollucella laevis* L. (Stem rot)	Bells of Ireland	Florida	1984	Alfieri Jr *et al*. (1984)
600.	*Morus alba* L.	White mulberry	Bulgaria	1961	Spaulding (1961)
			China	1979, 1990	Tai (1979); Zhang and Huang (1990)
			Japan	2007	Kobayashi (2007)
			Korea	2004	Cho and Shin (2004)
			Taiwan	1979	Anon (1979)
			USA	1960	Anon (1960)
601.	*Morus* spp.	Mulberry	Bulgaria	2009	Bobev (2009)
			Iran	2013	Arzanlou and Dokhanchi (2013)
			Japan	2007	Kobayashi (2007)
			USA	1974	Schwartz (1977)
602.	*Musa acuminata* (Fruit rot)	Banana	Australia	1966	Simmonds (1966)
603.	*Musa paradisiaca* L.	Edible plantain	Brazil	1998	Mendes *et al*. (1998)
604.	*Musa* spp.	Banana	Israel	1961	Chorin and Rotem (1961)
605.	*Myosotis arvensis* (L.) Hill	Forget-me-not	Brazil	1954	Neto (1954)
606.	*Myosotis scorpioides* L. (crown rot)	Forget-me-not	Illinois,	1960	Anon (1960)
			Washington	1960	Anon (1960)
607.	*Myosotis* spp.	Forget-me-nots/scorpion grasses	Washington	1973	Shaw (1973)
608.	*Myosotis sylvatica* Hoffm. (as *M. oblongata* Link) (Crown rot)	Garden forget-me-not	Illinois, Washington	1960	Anon (1960)
609.	*Nasturtium officinale* Br. (leaf and stem blight)	Watercress	Florida	1984	Alfieri Jr *et al*. (1984)
			Hungary	2022	Csüllög *et al*. (2022)
			Italy	2019	Garibaldi *et al*. (2019)

			USA	1974, 1984	Schwartz (1977); Alfieri Jr *et al.* (1984)
610.	*Nasturtium sublyratum* (Syn. *rorippa dubia* (Pers.) H. Hara)	Garden nasturtium/Capuchina	Taiwan	1979	Anon (1979)
611.	*Nemesia* sp.	Nemesia flower	Canada	1967	Conners (1967)
612.	*Neonotonia wightii*	Perennial soybean	Australia	1990	Lenne (1990)
613.	*Nicandra physalodes* (L.) Gaertn. (Stem Rot)	Apple of Peru	Hawaii	1981	Raabe *et al.* (1981)
614.	*Nicotiana rustica* L.	Wild tobacco	India	1988	Monga (1988)
			Russia	1929	Popova (1929)
615.	*Nicotiana tabacina*	Tobacco	Taiwan	1979	Anon (1979)
616.	*Nicotiana tabacum* (Seedling blight)	Tobacco	Brazil	1998	Mendes *et al.* (1998)
			Bulgaria	2009	Bobev (2009)
			Canada	1986	Ginns (1986)
			China	1979, 1990	Tai (1979); Zhang and Huang (1990); Lu *et al.* (2020a)
			Connecticut	1960	Anon (1960)
			Edward islands, Canada	1987	Martin and Arsenault (1987)
			Greece	1973	Pantidou (1973)
			India	1963	Gupta (1963)
			Japan	1984	Akazawa and Ono (1984)
			Kentucky	1960	Anon (1960)
			North Carolina	1960, 2011	Anon (1960); Blanco-Meneses and Ristaino (2011)
			New Zealand	1932, 1989	Brien (1932); Pennycook (1989)
			Venezuela	1970	Dennis (1970)

Continued

Table 2.1. Continued

Sr. no.	Host name	English name	Country	Year	References
617.	*Ocimum basilicum* L.	Basil	California	2000	Koike (2000)
			China	1979	Tai (1979)
			Germany	1985	Cristea *et al*. (1985)
			Louisiana	1994	Holcomb and Reed (1994)
			Taiwan	1959, 1979	Sawada (1959); Anon (1979)
			Turkey	2008	Tok (2008)
			Canada	1997	Paulitz (1997)
618.	*Oenanthe javanica*	Water celery/water dropwort	China	1979	Tai (1979)
			Korea	2004	Cho and Shin (2004)
619.	*Oenanthe stolonifera*	Water celery/water dropwort	Taiwan	1979	Anon (1979)
620.	*Oenothera* sp.	Evening primrose	USA	1974	Schwartz (1977)
621.	*Olea europaea*	Olive tree	Italy	2017	Ruano-Rosa *et al*. (2017)
622.	*Onobrychis vicifolia* Scop.	Sainfoin	USA	1974	Schwartz (1977)
623.	*Onopordum acanthium* L. (as *Cnicus lanceolatus* L.)	Scotch thistle	New Zealand	1946	Brien (1946)
624.	*Origanum vulgare*	Wild marjoram	Italy	2007	Garibaldi *et al*. (2007b)
625.	*Orobanche cernua* Loefl.	Broomrape	India	1963	Gupta (1963)
626.	*Orobanche cumana*	Sunflower, broomrape	China	2010	Ding *et al*. (2012)
627.	*Orobanche* spp.	Broomrape	USA	1974	Schwartz (1977)
628.	*Osteospermum ecklonis* (D.C.) Norl.	African daisy	California	1975	Gill (1975)
629.	*Osteospermum fruticosum* (L.) Norl.	African daisy	California	1975, 1989	Gill (1975); French (1989)
630.	*Osteospermum* sp.	African daisy	Argentina	2005	Wright *et al*. (2005)
			Italy	2008	Garibaldi *et al*. (2008b)
			Louisiana	2005	Holcomb (2005)
631.	*Oxalis corniculata*	Creeping wood sorrel	China	1979	Tai (1979)
632.	*Oxalis repens*	Creeping wood sorrel	Taiwan	1979	Anon (1979)

633.	*Paeonia lactiflora* Pall. (Stem rot)	Garden peony	Canada	1967, 1986	Conners (1967); Ginns (1986)
			Illinois	1960	Anon (1960)
			Maryland	1960	Anon (1960)
			Maine	1960	Anon (1960)
			Minnesota	1960	Anon (1960)
			New Jersey	1960	Anon (1960)
			New York	1960	Anon (1960)
			Ohio	1960	Anon (1960)
			Oklahoma	1960	Anon (1960)
634.	*Paeonia officinalis* L. (Stem rot, wilt)	Peony	North Carolina	1985	Grand (1985)
			New Jarsey	1960	Cappellini (1960)
635.	*Paeonia* sp.	Peony	Bulgaria	2009	Bobev (2009)
			California	1989	French (1989)
636.	*Paeonia suffruticosa*	Moutan peony	Korea	2004	Cho and Shin (2004)
637.	*Panax pseudoginseng* Wallich	Ginseng	Russia	1981	Latysheve and Sevryuk (1981)
638.	*Panax quinquefolius* L.	American ginseng	China	2022	Guan *et al.* (2022)
			Michigan	1960	Anon (1960)
			New York	1960	Anon (1960)
			Ohio	1960	Anon (1960)
			Pennsylvania	1960	Anon (1960)
			Wisconsin	1960	Anon (1960)
639.	*Panicum repens*	Torpedo grass/creeping panic	China	1979	Tai (1979)
			Taiwan	1979	Anon (1979)
640.	*Papaver nudicaule*	Arctic poppy	Australia	1966	Simmonds (1966)
641.	*Papaver somniferum* Crantz.	Opium poppy	Australia	1982	Sampson and Walker (1982)
			Germany	1980, 1983	Leiteritz and Roder (1980); Laughlin and Munro (1983)

Continued

Table 2.1. Continued

Sr. no.	Host name	English name	Country	Year	References
			India	2003	Singh *et al*. (2003)
			Oregon	1973	Shaw (1973)
			Sweden	1943	Reinmuth (1943)
642.	*Papaver* sp.	Opium poppy or bread seed poppy	Bulgaria	2009	Bobev (2009)
			California	1989	French (1989)
643.	*Parthenium argentatum* Gray (Stem rot)	Guayule	California	1960, 1989	Anon (1960); French (1989)
644.	*Parthenium hysterophorus*	Santa Maria feverfew	India	2004	Ghasolia and Shivpuri (2004b)
645.	*Passiflora caerulea*	Blue crown passion flower	China	1979	Tai (1979)
			Taiwan	1979	Anon (1979)
646.	*Passiflora edulis* Sims.	Purple granadilla/passion fruit	Australia	1982, 1989	Sampson and Walker (1982); Shivas (1989)
			Japan	1957, 1990	Kobayashi (1957); Kagiwata (1990)
			New Zealand	1932, 1989	Brien (1932); Pennycook (1989)
647.	*Passiflora* sp.	Passion fruit	-	1994	Boland and Hall (1994); Saharan and Mehta (2008); Ficker (2019)
648.	*Pastinaca sativa* L.	Parsnip	Australia	1966, 1982, 1989	Sampson and Walker (1982); Shivas (1989); Simmonds (1966)
			California	1960, 1989	Anon (1960); French (1989)
			Canada	1986	Ginns (1986)
			Idaho	1960	Anon (1960)
			Indiana	1960	Anon (1960)
			Louisiana	1960	Anon (1960)
			Massachusetts	1960	Anon (1960)
			New Zealand	1989	Pennycook (1989)
			New York	1960	Anon (1960)

			Oregon	1960	Anon (1960)
			Scotland	1960	Anon (1960)
			Texas	1960	Anon (1960)
649.	*Paulownia* sp.	Empress tree	Taiwan	1983	Hsieh (1983)
650.	*Pedicellaria viscida*	Sea urchins	Taiwan	1977	Anon (1979)
651.	*Pelargonium hortorum* Bailey	Zonal geranium	New Zealand	1989	Pennycook (1989)
			USA	1974	Schwartz (1977)
652.	*Pelargonium peltatum*	Ivy geraniums	New Zealand	1996	McKenzie and Dingley (1996)
653.	*Pelargonium* sp.	Omeo stork's bill	Italy	1935	Ferraris (1935)
654.	*Pelargonium zonale*	Geranium	New Zealand	1996	McKenzie and Dingley (1996)
655.	*Pennantia baylisiana*	Three Kings kaikōmako	New Zealand	1996	McKenzie and Dingley (1996)
656.	*Pennisetum americanum* Schum.	Pearl millet	India	1983	Singh and Singh (1983)
657.	*Penstemon gloxinioides* Hort.	Foxglove beard-tongue	Japan	1996	Takeuchi and Horie (1996)
658.	*Penstemon* sp.	Beardtongue	–	1994	Boland and Hall (1994); Saharan and Mehta (2008)
659.	*Pentas* sp.	Egyptian star cluster	Florida	1991	Miller (1991a,b)
660.	*Perilla frutescens* (L.) Britt. var. *japonica* (Hassk.) Hara	Perilla	Korea	1987	Shin and Lee (1987)
661.	*Persea americana* Mill. (collar rot)	Avocado	California	1960, 1987, 1989	Anon (1960); French (1987, 1989)
			Florida	1984	Alfieri Jr *et al*. (1984)
662.	*Persea borbonia* (L.) Spreng (collar rot)	Laurel tree	California	1960	Anon (1960)
663.	*Petasites* sp.	Butterburs	England	1978	Dennis (1978)
664.	*Petroselinum crispum* (Mill.) Nym. Ex Hill	Parsley	California	1989	French (1989)
			Connecticut	1960	Anon (1960)
			Florida	1997	Miller (1997)

Continued

Table 2.1. Continued

Sr. no.	Host name	English name	Country	Year	References
			Georgia	1960	Anon (1960)
			Greece	1973	Pantidou (1973)
			Hawaii	1981	Raabe *et al*. (1981)
			Korea	2003	Chang and Kim (2003)
			Louisiana	1960	Anon (1960)
			New York	1960	Anon (1960)
			Pennsylvania	1960	Anon (1960)
			Turkey	2017	Kurt *et al*. (2017a)
			Texas	1960	Anon (1960)
			Virginia	1960	Anon (1960)
665.	*Petroselinum sativum*	Parsley or coriander	Korea	2004	Cho and Shin (2004)
666.	*Petunia axillaris* (Lam.) BSP	White moon petunia	Washington	1960	Anon (1960)
667.	*Petunia hybrida* Hort.	Garden petunia	Australia	1966, 1982	Simmonds (1966); Sampson and Walker (1982)
			California	1989	French (1989)
			Canada	1966, 1967	Creelman (1966); Conners (1967)
			Florida	1984	Alfieri *et al*. (1984)
			India	1973	Roy (1973)
			Italy	2009	Garibaldi *et al*. (2009)
			Louisiana	2001	Holcomb (2001)
			New Zealand	1989	Pennycook (1989)
			Ohio	1957	Ellett (1957)
			Saskatchewan	1976	Morrall *et al*. (1976)
			Texas	1960	Anon (1960)
668.	*Petunia* spp.	Petunia	Canada	1967	Conners (1967)
			California	1989	French (1989)

			Washington	1973	Shaw (1973)
669.	*Petunia violacea*	Violet petunia	China	1979	Tai (1979)
			Taiwan	1979	Anon (1979)
			Tanzania	1960	Riley (1960)
670.	*Phacelia tanacetifolia*	Lacy phacelia	Romania	1993	Ivancia (1993)
671.	*Phaseolus aureus*	Mung bean	Texas	1960	Anon (1960)
			Virginia	1960	Anon (1960)
672.	*Phaseolus coccineus* L.	Scarlet runner bean	Canada	1966, 1986	Creelman (1966); Ginns (1986)
			Scotland	1961	Foister (1961)
			New York	1981	Hunter *et al.* (1981)
673.	*Phaseolus limensis* Macfady	Lima bean	California	1960	Anon (1960)
			Indiana	1960	Anon (1960)
			North Carolina	1985	Grand (1985)
			New York	1960	Anon (1960)
			Texas	1960	Anon (1960)
674.	*Phaseolus lunatus* L.	Civet bean	Brazil	1998	Mendes *et al.* (1998)
			California	1989	French (1989)
			Mexico	1976	Alvarez (1976)
			New York	1960	Anon (1960)
675.	*Phaseolus mungo*	Black gram	Virginia	1960	Anon (1960)
676.	*Phaseolus radiatus* L.	Green bean	China	1979	Tai (1979)
			India	1973, 1998	Roy (1973); Pande and Rao (1998)
			New York	1960	Anon (1960)
677.	*Phaseolus* sp.	Common bean	Australia	1990	Richardson (1990)
			Canada	1986	Ginns (1986)
			England	1978	Dennis (1978)
			Idaho	1956	Wilson (1956)

Continued

Table 2.1. Continued

Sr. no.	Host name	English name	Country	Year	References
			Nebraska	1972	Cook *et al*. (1972)
			Poland	1990	Richardson (1990)
678.	*Phaseolus trinervius* (Syn. *Vigna radiata*]	Green gram	China	1979	Tai (1979)
			Taiwan	1979	Anon (1979)
679.	*Phaseolus vulgaris* (field blight and nestiness or watery soft rot in transit, storage rot)	Kidney bean/common bean	Alaska	1960	Logsdon and Strobel (1960)
			Argentina	2018	Abán *et al*. (2018)
			Australia	1966, 1982, 1989	Simmonds (1966); Sampson and Walker (1982); Shivas (1989); Cook and Dubé (1989)
			Brazil	1954, 1998	Neto (1954); Mendes *et al*. (1998)
			Bulgaria	2009	Bobev (2009)
			Burundi	1995	Allen (1995)
			California	1989	French (1989)
			Canada	1967, 1986	Conners (1967); Ginns (1986)
			Chile	1945, 1967	Mujica and Vergara (1945); Mujica and Oehrens (1967)
			China	1979	Tai (1979)
			Colorado	1957	Burke *et al*. (1957)
			Costa Rica	1967	McGuire Jr and Crandall (1967)
			Cuba	2013	Martinez de la Parte *et al*. (2013)
			Cyprus	1957	Georghiou and Papadopoulos (1957)
			El Salvador	1967	McGuire Jr and Crandall (1967)
			Ethiopia	1995	Allen (1995)
			Fiji	1972, 1981	Firman (1972); Dingley *et al*. (1981)

Florida	1951, 1984	Stoner (1951); Alfieri *et al*. (1984)
France	2018	Leyronas *et al*. (2018)
Greece	1973, 2000	Pantidou (1973); Holevas *et al*. (2000)
Guatemala	1967	McGuire Jr and Crandall (1967)
Hawaii	1981	Raabe *et al*. (1981)
Idaho	1950, 1973	Meiners (1950); Shaw (1973)
India	1973, 2014	Roy (1973); Hansda *et al*. (2014)
Japan	1981	Akai (1981)
Kenya	1995	Allen (1995)
Korea	2004	Cho and Shin (2004)
Mauritius	1948, 1995	Wiehe (1948); Allen (1995)
Mexico	1967, 1976	McGuire Jr and Crandall (1967); Alvarez (1976)
Montana	1973	Shaw (1973)
North Carolina	1985	Grand (1985)
Nebraska	2011, 2018	Walker *et al*. (2011); Salgado-Salazar *et al*. (2018)
New Zealand	1932, 1989	Brien (1932); Pennycook (1989)
New York	1981	Hunter *et al*. (1981)
Nicaragua	1967	McGuire Jr and Crandall (1967)
Oregon	1973	Shaw (1973)
Panama	1967, 2006	McGuire Jr and Crandall (1967); Piepenbring (2006)
Range of host	1960	Anon (1960)
Rwanda	1995	Allen (1995)

Continued

Table 2.1. Continued

Sr. no.	Host name	English name	Country	Year	References
			South Africa	1977, 1995, 2000	Gorter (1977); Allen (1995); Crous *et al*. (2000)
			Taiwan	1979	Anon (1979)
			Tanzania	1995	Allen (1995)
			Tonga	1981	Dingley *et al*. (1981)
			UK	1926	Dowson (1926)
			USA	1936	Young (1936)
			Viet Nam	2010	Luong *et al*. (2010)
			Washington	1973	Shaw (1973)
			Wyoming	1958	Davison (1958)
			Zimbabwe	1995	Allen (1995)
680.	*Philodendron scandens* Koch & Sello (Cottony blight)	Heart-leaf philodendron	Florida	1984	Alfieri Jr *et al*. (1984)
681.	*Philodendron selloum* C. Koch (Cottony blight)	Philodendron	Florida	1984	Alfieri Jr *et al*. (1984)
682.	*Phlox drummondii* Hook	Annual phlox	Australia	1982	Sampson and Walker (1982)
			India	1964, 1973, 1977	Shukla and Pavgi (1964); Roy (1973); Kaur and Chohan (1977)
			USA	1974	Schwartz (1977)
683.	*Phlox* sp.	Phlox	–	1994	Boland and Hall (1994); Saharan and Mehta (2008); Ficker (2019)
684.	*Phyllanthus fraternus*	Gulf leaf flower	India	2004	Ghasolia *et al*. (2004)
685.	*Phyllanthus niruri*	Gale of the wind	China	1979	Tai (1979)
			Taiwan	1979	Anon (1979)
686.	*Physalis alkekengi*	Chinese lantern	New Zealand	1996	McKenzie and Dingley (1996)
687.	*Physalis angulata* L.	Undetermined	Korea	1987	Shin and Lee (1987)

688.	*Physalis peruviana* L.	Cape gooseberry	New Zealand	1989	Pennycook (1989)
			USA	1974	Schwartz (1977)
689.	*Physostegia virginiana* (L.) Benth. (Stem rot)	Obedience	Korea	2004	Cho and Shin (2004)
			Maine	1960	Anon (1960)
690.	*Phyteuma japonicum*	*Rampion*	Korea	2003	Chang and Kim (2003
691.	*Pimpinella anisum* L. (Stem rot)	Aniseed	California	1960, 1989	Anon (1960); French (1989)
			Texas	1960	Anon (1960)
692.	*Pimpinella brachycarpa*	Chamnamul	Korea	2004	Cho and Shin (2004)
693.	*Pimpinella* sp.	Undetermined	-	1960	Anon (1960)
694.	*Pinus densiflora* Slebd. & Zucc	Japanese red pine	Japan	1957	Kobayashi (1957)
695.	*Pinus palustris*	Longleaf pine	Florida	1990	Miller (1991a,b)
696.	*Pistacia vera*	Pistachio	California	1994	Michailides and Morgan (1994)
697.	*Pisum sativum* L. (Stem and pod rot)	Pea	Australia	1966	Simmonds (1966)
			Bangladesh	2020	Islam *et al*. (2020)
			Brazil	1954, 1998	Neto (1954); Mendes *et al*. (1998)
			California	1960, 1989	Anon (1960); French (1989)
			Alberta, Manitoba, Saskatchewan, Canada	1967, 1986, 2000	Conners (1967); Ginns (1986); Hilton (2000)
			Chile	1967	Mujica and Oehrens (1967)
			China	1979	Tai (1979)
			Delaware	1960	Anon (1960)
			Florida	1960, 1984	Anon (1960); Alfieri Jr *et al*. (1984)
			France	2018	Leyronas *et al*. (2018)
			Idaho	1960, 1973	Anon (1960); Shaw (1973)

Continued

Table 2.1. Continued

Sr. no.	Host name	English name	Country	Year	References
			India	1973, 1998	Roy (1973); Pande and Rao (1998)
			Mauritius	1948, 1968	Wiehe (1948); Orieux and Felix (1968)
			Mexico	1976	Alvarez (1976)
			Michigan	1957	Lockwood *et al.* (1957)
			Montana	1960, 1973	Anon (1960); Shaw (1973)
			New Zealand	1932, 1989	Brien (1932); Pennycook (1989)
			New Jersey	1960	Anon (1960)
			Oregon	1973	Shaw (1973)
			Pakistan	1998	Iqbal *et al.* (1998)
			Pennsylvania	1960	Anon (1960)
			Scotland	1961, 1990	Foister (1961); Richardson (1990)
			South Africa	2000	Crous *et al.* (2000)
			Taiwan	1979	Anon (1979)
			Texas	1960	Anon (1960)
			UK	1990	Richardson (1990)
			USA	1936	Young (1936)
			Virginia	1960	Anon (1960)
			Washington	1960, 1973	Anon (1960); Shaw (1973)
698.	*Pisum sativum* var. *arvense* (L.) Poir.	Field pea	Canada, Manitoba	1971, 2000	Morrall *et al.* (1972); Hilton (2000)
			Georgia	1960	Anon (1960)
			Idaho	1960	Anon (1960)
			Poland	1997	Marcinkowska (1997)
699.	*Pisum* sp.	Common pea	Australia	1982	Sampson and Walker (1982)

700.	*Plantago asiatica*	Chinese plantain	China	1979	Tai (1979)
701.	*Plantago formosana*	Plantains/fleaworts	Taiwan	1979	Anon (1979)
702.	*Plantago lanceolata* L.	Buckhorn	Australia	1982	Sampson and Walker (1982)
			USA	1974	Schwartz (1977)
703.	*Platycodon grandiflorus*	Balloon flower	Korea	2004	Cho and Shin (2004)
704.	*Polygonum aviculare*	Common knotgrass	China	1979	Tai (1979)
705.	*Polygonum lapathifolium*	Pale persicaria	China	1979	Tai (1979)
			Taiwan	1979	Anon (1979)
706.	*Poncirus trifoliata*	Hardy orange/trifoliate orange	Sicily, Italy	1991	Greuter *et al*. (1991)
707.	*Populus canadensis*	Canadian poplar	China	2002	Chen (2002)
708.	*Portulaca oleracea* L.	Common purslane	Australia	1989	Shivas (1989)
			China	1979	Tai (1979)
			Japan	1999	Kanno and Ohkubo (1999)
			Taiwan	1979	Anon (1979)
			USA	1974	Schwartz (1977)
709.	*Primula polyantha*	False oxlip	New Zealand	1989	Pennycook (1989)
710.	*Proboscidea louisianica* (Mill.) Thell (Stem rot)	Probiscus flower	Massachusetts	1960	Anon (1960)
711.	*Prunella vulgaris*	Common self-heal	Australia	1982	Sampson and Walker (1982)
712.	*Prunus americana* March	American plum	USA	1974, 1987	Schwartz (1977); French (1987)
713.	*Prunus amygdalus* Batsch	Almond	California	1960	Anon (1960)
714.	*Prunus armeniaca*	Apricot	Australia	1966, 1982, 1989	Sampson and Walker (1982); Shivas (1989); Simmonds (1966)
			California	1960, 1979, 1987, 1989	Anon (1960); El-Goorani and Sommer (1979); French (1987, 1989)

Continued

Table 2.1. Continued

Sr. no.	Host name	English name	Country	Year	References
			Chile	2014	Ferrada *et al*. (2014)
			North Carilona	1929	Smith (1929)
			Mexico	1976	Alvarez (1976)
			New Zealand	1989	Pennycook (1989)
			South Africa	1977, 2000	Gorter (1977); Crous *et al*. (2000)
715.	*Prunus avium*	Cherry wild	California	1987, 1989	French (1987, 1989)
			Chile	2014	Ferrada *et al*. (2014)
			Japan	2007	Kobayashi (2007)
			Oregon	2007	Serdani and Spotts (2007)
716.	*Prunus domestica* L.	Garden plum	Australia	1989	Shivas (1989)
			USA	1974	Schwartz (1977)
717.	*Prunus dulcis*	Almond	California	1987, 1989	French (1987, 1989)
718.	*Prunus mume*	Chinese plum	Japan	2007	Kobayashi (2007)
719.	*Prunus persica* (L.) Batsch	Peach	Australia	1989	Shivas (1989)
			California	1960, 1987, 1989	Anon (1960); French (1987, 1989)
			Fujian, China	1981, 1990	Gao (1981); Zhang and Huang (1990)
			Mexico	1976	Alvarez (1976)
720.	*Prunus persica* var.*nectarina*	Nectarines	Chile	2014	Ferrada *et al*. (2014)
721.	*Prunus persica* var.*vulgaris*	Peach	Japan	2007	Kobayashi (2007)
722.	*Prunus salicina*	Japanese plum	Chile	2014	Ferrada *et al*. (2014)
723.	*Prunus* sp.	Undetermined	California	1960	Anon (1960)
724.	*Psoralea corylifolia*	Babchi	China	1979	Tai (1979)
			India	2014	Hansda *et al*. (2014)
725.	*Pterocarya stenoptera*	Chinese wingnut	China	2002	Chen (2002)

726.	*Pyrus communis* L. (Fruit rot)	Pear	California	1960, 1987, 1989	Anon (1960); French (1987, 1989)
			Canada	1986	Ginns (1986)
			New Zealand	1932, 1989	Brien (1932); Pennycook (1989)
			Washington	1960, 1973	Anon (1960); Shaw (1973)
727.	*Pyrus serotina* var. *culta*	Asian pear	Japan	2007	Kobayashi (2007)
728.	*Quercus* sp.	Oak	Czechoslovakia	1957, 1990	Urosevic (1957); Richardson (1990)
729.	*Ranunculus asiaticus* L.	Persian buttercup	Argentina	2005	Wright *et al*. (2005)
			Italy	2003	Garibaldi *et al*. (2003)
			Korea	2015	Han *et al*. (2015)
			New Zealand	1989	Pennycook (1989)
			USA	1960, 1974	Anon (1960); Schwartz (1977)
730.	*Ranunculus cantoniensis*	Chinese Buttercup	China	1979	Tai (1979)
731.	*Ranunculus ficaria*	Lesser celandine or pilewort	Norway	1997, 1997, 1998, 2002, 2004	Holst-Jensen *et al*. (1997a,b, 1998); Nielsen *et al*. (2002); Holst-Jensen *et al*. (2004)
732.	*Ranunculus glacialis*	Glacier buttercup	Switzerland	1995	Graf and Schumachar (1995)
733.	*Ranunculus* sp. (Stem rot)	Buttercup	Arizona	1960	Anon (1960)
			California	1960, 1989	Anon (1960); French (1989)
734.	*Raphanus raphanistrum* L. (Crown rot)	Wild radish	California	1960, 1989	Anon (1960); French (1989)
735.	*Raphanus sativus* L.	Garden radish	Brazil	1998	Mendes *et al*. (1998)
			California, Indiana, New Jersey, Texas	1960, 1989	Anon (1960); French (1989)
			Chile	1967	Mujica and Oehrens (1967)
			China	1979	Tai (1979)
			Canada	1967	Conners (1967)

Continued

Table 2.1. Continued

Sr. no.	Host name	English name	Country	Year	References
			India	1980, 1987	Agrawal *et al*. (1980); Saxena and Rai (1987)
			Minnesota	1955, 1960	Preston and Dosdall (1955); Anon (1960)
			New Zealand	1932, 1989	Brien (1932); Pennycook (1989)
			Taiwan	1979	Anon (1979)
736.	*Raphanus sativus* L. var. *hortensis* Makino	Undetermined	Korea	1987	Shin and Lee (1987)
737.	*Raphanus sativus* var. *longipinnatus*	Daikon or radish	South Africa	2000	Crous *et al*. (2000)
738.	*Raphanus sativus* var. *oleiferus*	Fodder Radish	China	1979	Tai (1979)
739.	*Rheum rhaponticum* L. (as *R. officinale* Baill.)	Rhubarb	Canada	1986	Ginns (1986)
			USA	1974	Schwartz (1977)
740.	*Ricinus communis* L. (stem rot)	Caster bean	China	1979	Tai (1979)
			Florida	1955, 1984	Alfieri Jr *et al*. (1984); Cook (1955)
			Taiwan	1979	Anon (1979)
741.	*Romneya coulteri*	California tree poppy	Canada	1986	Ginns (1986)
742.	*Romneya* sp.	Matilija poppy	British Columbia, Canada	1964	Toms (1964)
			Ontario, Canada	1967	Conners (1967)
743.	*Rorippa montana*	Variable leaf/yellow cress	China	1979	Tai (1979)
744.	*Rorippa* sp. (crown rot)	Yellow cress	Florida	1984	Alfieri Jr *et al*. (1984)
745.	*Rorippa sylvestris* (L.) Besser (crown rot)	Creeping yellow cress	Florida	1984	Alfieri Jr *et al*. (1984)
746.	*Rosa* sp.	Rose	Canada	1967	Conners (1967)

747.	*Rosmarinus officinalis*	Rosemary	India	1994, 2005	Mohan (1994); Garibaldi *et al.* (2005b)
			Italy	2005, 2017	Minuto *et al.* (2005b), Garibaldi *et al.* (2017)
			Oregon	2004	Putnam (2004)
			USA	2017	Garibaldi *et al.* (2017)
748.	*Rubus* sp.	Raspberry	USA	1960	Anon (1960)
749.	*Rudbeckia laciniata* (L.) Per. (Stem rot)	Coneflower	Connecticut	1960	Anon (1960)
750.	*Rudbeckia laciniata* (L.) var. *hortensis* Bailey	Golden glow	Connecticut	1960	Anon (1960)
			Canada	1938	Bisby (1938)
751.	*Rumex crispus* L.	Yellow dock	New Zealand	1989	Pennycook (1989)
			USA	1974	Schwartz (1977)
752.	*Rumex dentatus*	Toothed dock/Aegean dock	China	1979	Tai (1979)
753.	*Rumohra adiantiformis* (G. Forst.) Ching (Root and stem rot)	Leather leaf fern	Florida	1984	Alfieri Jr *et al.* (1984)
754.	*Salpiglossis* sp.	Bitter little stick	California	1989	French (1989)
755.	*Salvia coccinea*	Scarlet sage	India	1999	Rajappan *et al.* (1999)
756.	*Salvia miltiorrhiza*	Salvia root/rhizome	Taiwan	2020	Lu *et al.* (2020b)
757.	*Salvia officinalis*	Common saga	Canada	1986	Ginns (1986)
			Italy	2004	Garibaldi *et al.* (2004a)
			New Zealand	1989, 2004	Pennycook (1989); Minuto *et al.* (2004b)
758.	*Salvia reflexa*	Lance leaf sage	North Dakota	1996	Venette *et al.* (1996); Jons *et al.* (1996)
759.	*Salvia* sp.	Common sage/garden sage	Australia	1982	Sampson and Walker (1982)
			Canada	1967	Conners (1967)

Continued

Table 2.1. Continued

Sr. no.	Host name	English name	Country	Year	References
760.	*Salvia splendens*	Scarlet sage	Australia	1966	Simmonds (1966)
			Bangladesh	2019	Islam *et al*. (2019)
			Portugal	1980	de Sousa and Lucas (1980)
			Taiwan	1979	Anon (1979)
761.	*Satsuma mandrin*	Mandrin	Korea	1996	Song and Young (1999)
762.	*Scabiosa* sp. (stem rot)	Sweet scabious	New York	1960	Anon (1960)
			Canada	1967	Conners (1967)
763.	*Schefflera arboricola* Hayata	Undetermined	Florida	1984	Alfieri Jr *et al*. (1984)
			Mississippi	1959, 1960	Parris (1959); Anon (1960)
764.	*Schizanthus pinnatus*	Butterfly flower	Mississippi	1960	Anon (1960)
765.	*Schizanthus retusus* Hook.	Butterfly flower	UK	1926	Dowson (1926)
			USA	1974	Schwartz (1977)
766.	*Schizanthus* sp. (stem rot)	Butterfly flower	Missouri	1960	Anon (1960)
			Mississippi	1959, 1960	Parris (1959); Anon (1960)
767.	*Schizanthus wisetonensis*	Butterfly flower	Italy	2002	Bag (2003); Garibaldi *et al*. (2002b)
			UK	1926	Dowson (1926)
768.	*Scorzonera hispanica* Hort.	Black salsify	Germany	1990, 1964	Richardson (1990); Siedel and Daebeler (1964)
			France	1974	Champion *et al*. (1974)
769.	*Scorzonera* sp.	Salsify	Canada	1967	Conners (1967)
770.	*Sculus hippocastanum* L.	Horse chestnut	–	1994	Boland and Hall (1994); Saharan and Mehta (2008)
771.	*Secale cereal* L.	Rye	New Jersey	1922	Poole (1922)
772.	*Senecio cruentus* (Masson)D.C.	Florists' cineraria	USA	1960, 1973	Anon (1960); Shaw (1973)
773.	*Senecio jacobaea*	Common ragwort	New Zealand	1996	McKenzie and Dingley (1996)

774.	*Senecio* spp. (stem rot)	Groundsel	Florida	1984	Alfieri Jr *et al*. (1984)
			Louisiana	1960	Anon (1960)
			Bermuda	1926	Ogilvie (1927)
775.	*Senecio vulgaris* L.	Groundsel	Canada	1976, 1986	Morrall *et al*. (1976); Ginns (1986)
776.	*Sesamum indicum* L.	Sesame	Bulgaria	2009	Bobev (2009)
			China	1990	Zhang and Huang (1990)
			Korea	2004	Cho and Shin (2004)
777.	*Sesbania* sp.	River hemp	Israel	1963	Palti (1963)
778.	*Setaria viridis* (L.) Beauv.	Green foxtail	Colorado	1957	Burke *et al*. (1957)
779.	*Sida rhombifolia*	Arrowleaf sida	Brazil	1998	Mendes *et al*. (1998)
780.	*Sigesbeckia orientalis*	Common St. Paul's wort	China	1979	Tai (1979)
			Taiwan	1979	Anon (1979)
781.	*Silene gallica*	Common catchfly	South Africa	1981, 2000	Gorter (1981); Crous *et al*. (2000)
782.	*Silene vulgaris*	Bladder campion	Japan	1996	Takeuchi and Horie (1996)
783.	*Silybum marianum* (L) Graertn.	Milk thistle	New Zealand	1932, 1989	Brien (1932); Pennycook (1989)
784.	*Sinapsis alba*	White mustard	California	1989	French (1989)
			Canada	1976	Morrall *et al*. (1976)
			Texas	1960	Anon (1960)
			USA	1974	Schwartz (1977)
785.	*Sinapis arvensis*	Wild mustard	California	1989	French (1989)
786.	*Sinningia* sp.	Gloxinia	California	1989	French (1989)
787.	*Sinningia speciosa* (Lodd.) Hiern.	Gloxinia	California	1960, 1966, 1989	Anon (1960); Raabe and Hendrix Jr (1966); French (1989)
			Florida	1984	Alfieri Jr *et al*. (1984)
			North Carolina	1985	Grand (1985)
			Washington	1973	Shaw (1973)

Continued

Table 2.1. Continued

Sr. no.	Host name	English name	Country	Year	References
788.	*Sisymbrium officinale* (L.) Scop	Hedge mustard	USA	1974	Schwartz (1977)
789.	*Solanum chacoense* Bitter (Stem blight)	Undetermined	Hawaii	1966	Raabe (1966)
790.	*Solanum citrullifolium* Br.	Melon leaf nightshade	Arizona	1936	Brown and Butler (1936)
791.	*Solanum elaeagnifolium* Cav. (Stem rot)	Silverleaf nightshade	Texas	1960	Anon (1960)
792.	*Solanum gilo*	Scarlet eggplant	Brazil	1998	Mendes *et al*. (1998)
793.	*Solanum lycopersicum*	Tomato	France	2018	Leyronas *et al*. (2018)
			Iceland	1963	Jorstad (1963)
			India	2014	Hansda *et al*. (2014)
			New Jersey	2018	Salgado-Salazar *et al*. (2018)
794.	*Solanum melongena* L. (stem blight)	Eggplant	Australia	1966	Simmonds (1966)
			Brazil	1998	Mendes *et al*. (1998)
			Bulgaria	2009	Bobev (2009)
			California	1989	French (1989)
			China	1979	Tai (1979)
			Greece	1973, 2000	Pantidou (1973); Holevas *et al*. (2000)
			Hawaii	1981	Raabe *et al*. (1981)
			Hong Kong	2000	Lu *et al*. (2000)
			India	1973, 1980	Roy (1973); Agrawal *et al*. (1980)
			Korea	1987, 2004	Shin and Lee (1987); Cho and Shin (2004)
			Nepal	1992	Dahal *et al*. (1992)
			Pakistan	2019	Kamran *et al*. (2019)
			Spain	1991	Tores and Moreno (1991)
			Taiwan	1979	Anon (1979)
			USA	1960	Anon (1960)

795.	*Solanum melongena* var. *esculentum* Nees. (stem rot)	Eggplant	California	1960	Anon (1960)
			Florida	1960, 1984	Anon (1960); Alfieri Jr *et al*. (1984)
			India	1988	Kapoor (1988)
			Indiana	1960	Anon (1960)
			Louisiana	1960	Anon (1960)
			Oregon	1960	Anon (1960)
			Texas	1960	Anon (1960)
796.	*Solanum nigrum* L. (stem rot)	Black nightshade	Florida	1984	Alfieri *et al*. (1984)
			Korea	1987	Shin and Lee (1987)
797.	*Solanum quitoense* Lam	Pepino/melon shrub/lulo	New Zealand	1989	Pennycook (1989)
			Toronto, Canada	1993	Abawi (1993)
798.	*Solanum torvum* Swartz. (stem rot)	Devil's fig	Florida	1984	Alfieri Jr *et al*. (1984)
799.	*Solanum tuberosum* L. (stem rot)	Potato	Australia	1966, 1982, 1989	Sampson and Walker (1982); Shivas (1989); Simmonds (1966); Cook and Dubé (1989)
			Bolivia	1963	Farr and Stevenson (1963)
			Brazil	1998	Mendes *et al*. (1998)
			Bulgaria	2009	Bobev (2009)
			Canada	1964, 1967, 1986	Toms (1964); Conners (1967); Ginns (1986)
			Manitoba, Canada	2000	Hilton (2000)
			China	1979, 2020	Tai (1979); Ojaghian (2020)
			Colorado	1957	Burke *et al*. (1957)
			Cyprus	1957	Georghiou and Papadopoulos (1957)

Continued

Table 2.1. Continued

Sr. no.	Host name	English name	Country	Year	References
			England	1978	Dennis (1978)
			Florida	1960, 1984	Anon (1960); Alfieri Jr *et al*. (1984)
			Greece	1973, 2000	Pantidou (1973); Holevas *et al*. (2000)
			Idaho	2020	Woodhall *et al*. (2020)
			India	2009	Dutta *et al*. (2009)
			Iran	2009	Ojaghian (2009)
			Kenya	1989	Maykuss (1989)
			Korea	2004	Cho and Shin (2004)
			Massachusetts	1960	Anon (1960)
			Montana	1960, 1973	Anon (1960); Shaw (1973)
			New Zealand	1932, 1989	Brien (1932); Pennycook (1989)
			New York	1960	Anon (1960)
			Oregon	1960	Anon (1960)
			Pakistan	2021	Alam *et al*. (2021)
			Scotland	1961	Foister (1961)
			South Africa	2000	Crous *et al*. (2000)
			Taiwan	1979	Anon (1979)
			Tanzania	1960	Riley (1960)
			Turkey	2017	Kurt *et al*. (2017b)
			Texas	1960	Anon (1960)
			UK	1986	Dennis (1986)
			Venezuela	1986	Urtiaga (1986)
			Washington	1960, 1973	Anon (1960); Shaw (1973)
			USA	1936	Young (1936)

800.	*Solenostemon scutellarioides* (L.) Codd	Coleus	Florida	1984	Alfieri Jr *et al*. (1984)
801.	*Solidago canadensis* L. var. *salebrosa* (Piper) Jones	Golden rod	Canada	1970	Duczek and Morrall (1971)
802.	*Soliva anthemifolia*	Button burr weed	China	1979	Tai (1979)
			Taiwan	1979	Anon (1979)
803.	*Sonchus arvensis* L. (stem rot)	Field sow thistle	China	1979	Tai (1979)
			Taiwan	1979	Anon (1979)
			Texas	1960	Anon (1960)
			Canada	1967, 1970	Conners (1967); Duczek and Morrall (1971)
804.	*Sonchus asper* (L.) Hill. (stem rot)	Spiny sow thistle	Texas	1960	Anon (1960)
805.	*Sonchus oleraceus* L. (stem rot)	Common sow thistle	Canada	1976, 1986	Hoes and Haung (1976); Ginns (1986)
			China	1979	Tai (1979)
			Florida, Texas	1960	Anon (1960)
			New Zealand	1932, 1989	Brien (1932); Pennycook (1989)
			South Africa	1981, 2000	Gorter (1981); Crous *et al*. (2000)
			Taiwan	1979	Anon (1979)
806.	*Sonchus* sp. (Stem rot)	Sow thistle	Canada	1965, 1986	Petrie and Vanterpool (1965); Ginns (1986)
			Florida	1984	Alfieri *et al*. (1984)
			Texas	1960	Anon (1960)
807.	*Sophora flavescens*	Shrubby sophora	China	1961, 1979, 2002	Spaulding (1961); Tai (1979); Chen (2002)
808.	*Sorghum bicolor* (L.) Moench	Broomcorn	California	1989	French (1989)
809.	*Sorghum vulgare* Pers. (Culm rot)	Sorghum	California	1960	Anon (1960)

Continued

Table 2.1. Continued

Sr. no.	Host name	English name	Country	Year	References
810.	*Spinacia oleracea* L. (Crown rot)	Spinach	Brazil	1998	Mendes *et al*. (1998)
			China	1979	Tai (1979)
			Georgia	1960	Anon (1960)
			Korea	1987	Shin and Lee (1987)
			New Zealand	1989	Pennycook (1989)
			New York	1960	Anon (1960)
			Taiwan	1979	Anon (1979)
811.	*Stachys floridana* Shuttleworth ex. Benth (Cottony blight)	Florida hedgenettle	Florida	1984	Alfieri Jr *et al*. (1984)
812.	*Stellaria media* (L.) Cyr (Stem rot)	Chickweed	California	1960, 1989	Anon (1960); French (1989)
			China	1979	Tai (1979)
			South Carolina	1960	Anon (1960)
			UK	1936	Ogilvie (1936)
813.	*Stephanotis floribunda* Brongn.	Madagascar jasmine	Hawaii	1971	Raabe (1971)
814.	*Stephania hernandifolia*	*Tape vine*	China	1979	Tai (1979)
815.	*Stephania japonica*	Snake vine	Taiwan	1979	Anon (1979)
816.	*Stevia rebaudiana*	Stevia	Alberta, Canada	1997	Chang *et al*. (1997a)
			North Carolina	2014	Koehler and Shew (2014)
817.	*Stevia* sp.	Candy leaf/sweet leaf	Korea	2004	Cho and Shin (2004)
818.	*Stokesia laevis* (Hill.) Greene (Basal stem rot.)	Stokes aster	Florida	1984	Alfieri Jr *et al*. (1984)
819.	*Stylosanthes guianensis* (Aubl.) Sw.	Brazilian stylo	India	1960, 1982	Maiti (1983); Lenne (1990)
			Zimbabwe	1960	Lenne (1990)
820.	*Stylosanthes hamata* (L.) Taub.	Caribbean stylo	India	1960, 1982	Maiti (1983); Lenne (1990)

821.	*Stylosanthes humilis* Kunth.	Townsville stylo	India	1960, 1982	Maiti (1983); Lenne (1990)
822.	*Symphyotrichum dumosum*	Rice button aster and bushy aster	Poland	2018	Pieczul (2018)
823.	*Syringa vulgaris* (Shoot blight)	Lilac	California	1989	French (1989)
			Canada	1967	Conners (1967)
			Japan	2007	Kobayashi (2007)
			Washington	1960	Anon (1960)
824.	*Tagetes erecta* L.	African marigold	Australia	1982	Sampson and Walker (1982)
			Canada	1975, 1986	Morrall *et al*. (1976); Ginns (1986)
			India	2014	Hansda *et al*. (2014)
			New Zealand	1996	McKenzie and Dingley (1996)
825.	*Tagetes minuta*	Huacatay or Wacatay/tall khaki weed	South Africa	1992, 2000	Phillips (1992); Crous *et al*. (2000)
826.	*Tagetes patula* L. (as *Tanmpteises patula* L.)	African marigold	China	1979	Tai (1979)
			India	1973	Roy (1973); Kumar *et al*. (2022)
			Taiwan	1979	Anon (1979); Minuto *et al*. (2001)
827.	*Tagetes* sp.	Marigold	Florida	1984	Alfieri Jr *et al*. (1984)
			New York	1960	Anon (1960)
			Canada	1967	Conners (1967)
828.	*Tagetes tenuifolia* Cav.	Signet marigold	Germany	1985	Cristea *et al*. (1985)
829.	*Taraxacum kok-saghyz* Rodin.	Russian dandelion	Canada	1967	Conners (1967)
			USSR	1951	Cheremisinov (1951)
830.	*Taraxacum officinale* Wiggers	Common dandelion	Florida	1984	Alfieri Jr *et al*. (1984)
			Italy	2007	Garibaldi *et al*. (2007b)
			Pennsylvania	1960	Anon (1960)
			Washington	1951, 1973	McLean (1951); Shaw (1973)
			Canada	1991	Riddle *et al*. (1991)

Continued

Table 2.1. Continued

Sr. no.	Host name	English name	Country	Year	References
831.	*Tetragonia expansa*	New Zealand spinach	China	1979	Tai (1979)
			Taiwan	1979	Anon (1979)
832.	*Tetragonia tetragonioides* (Pall.) Kuntze (as *T. expansa* Murr.)	New Zealand spinach	USA	1977	Schwartz (1977)
833.	*Thlaspi arvense* L. (Crown rot)	Penny cress	Canada	1965, 1986	Petrie and Vanterpool (1965); Ginns (1986)
			USA	1960, 1973	Anon (1960); Shaw (1973)
834.	*Thymus citriodorus*	Thyme	Italy	2004	Garibaldi *et al*. (2004a); Minuto *et al*. (2004b)
835.	*Tithonia diversifolia*	Marigold, Mexican tournesol	Kenya	1961	Nattrass (1961)
836.	*Torenia* sp.	Bluewings/wishbone flower	California	1989	French (1989)
837.	*Trachellium caeruleum*	Blue throatwort	Argentina	2005	Wolcan and Grego (2005)
838.	*Trachyspermum ammi*	Bishop weed	India	2001	Singh and Singh (2001)
839.	*Tradescantia fluminensis*	Wandering jew	New Zealand	2006	Waipara (2006)
840.	*Tragopogon porrifolius* L. (crown rot)	Salsify	Illinois	1960	Anon (1960)
			Canada	1967	Conners (1967)
841.	*Trianthema portulacastrum*	Desert horse purslane/black pigweed/horse purslane	India	2004	Ghasolia *et al*. (2004)
842.	*Tridax procumbens*	Coat buttons/tridax daisy	India	2004	Ghasolia *et al*. (2004)
843.	*Trifolium alexandrinum* L.	Egyptian clover	Pakistan	2017	Saira *et al*. (2017)
			India	1962, 1998	Parasar and Chahal (1962); Pande and Rao (1998)
844.	*Trifolium hybridum* L.	Alsike clover	USA	1979	Kohn (1979a,b)
			Canada	1961	Creelman (1962)
845.	*Trifolium incarnatum* L.	Crimson clover	Oregon	1973	Shaw (1973)
			USA	1974	Schwartz (1977)

846.	*Trifolium pratense* L.	Red clover	Canada	1938, 1976	Bisby (1938); Morrall *et al*. (1976)
			USA	1979	Kohn (1979a,b)
847.	*Trifolium pratense* L. var. *praecox*	Red spring clover	USA	1974	Schwartz (1977)
848.	*Trifolium pratense* L. var. *seratinum*	Undetermined	USA	1974	Schwartz (1977)
849.	*Trifolium procumbens* L. (as *Trifolium dubium* Sibth)	Least hop clover	USA	1974	Schwartz (1977)
850.	*Trifolium repens* L.	White clover	Australia	1982	Sampson and Walker (1982)
			Korea	1987	Shin and Lee (1987)
			Scotland	1961	Foister (1961)
			USA	1974	Schwartz (1977)
			New Jersey, USA	1960	Cappellini (1960)
851.	*Trifolium* sp.	Clover	Canada	1967	Conners (1967)
852.	*Trifolium subterraneum* L.	Subterranean clover	USA	1974	Schwartz (1977)
853.	*Trifolium wormskjoldii* Lehm. (as *T. Fimbriatum* Lindl.)	Sierra clover	USA	1974	Schwartz (1977)
854.	*Trigonella foenum-graecum* L.	Fenugreek	Tunisia	2017	Gargouri *et al*. (2017)
			USA	1974	Schwartz (1977)
855.	*Trigonella* sp.	Fenugreek	UK	1934	Petri (1934)
856.	*Trillium decipiens*	Chattahoochee river wake robin	Florida	1990	Miller (1991a,b)
857.	*Trillium foetidissimum* Freeman	Red lily/stinking trillium	Louisiana	1990, 1991	Holcomb (1990, 1991)
858.	*Triticum aestivum* L.	Wheat	Australia	1966	Simmonds (1966)
859.	*Trollius* sp.	Globeflower	Moskow	1967	Shavrova (1967)
860.	*Tropaeolum majus L.*	Garden nasturtium	Cambodia	1944	Wallace (1944)
861.	*Tropaeolum* sp.	Nasturtium	USA	1974	Schwartz (1977)
862.	*Tulipa gesneriana* L.	Tulip	Washington	1973	Shaw (1973)

Continued

Table 2.1. Continued

Sr. no.	Host name	English name	Country	Year	References
863.	*Tulipa* sp.	Tulip	New Zealand	1989	Pennycook (1989)
864.	*Urtica dioica* L. ssp. *gracilis* (Ait) C.J. Gustafsson var. *gracilis*	Stinging nettle	Saskatchewan, Canada	1976	Morrall *et al*. (1976)
865.	*Urtica gracilis* Ait	Nettle	Saskatchewan, Canada	1938, 1976	Bisby (1938); Morrall *et al*. (1976)
866.	*Urtica holosericea*	Stinging nettle	California	1989	French (1989)
867.	*Urtica* sp. (Stem rot)	Nettle	California	1960	Anon (1960)
			Florida	1984	Alfieri Jr *et al*. (1984)
868.	*Urtica urens*	Small nettle	New Zealand	1996	McKenzie and Dingley (1996)
869.	*Vaccinium ashei*	Rabbiteye blueberry	Japan	2007	Umemoto *et al*. (2007)
870.	*Vaccinium corymbosum*	Highbush blueberry	Argentina	2011	Perez *et al*. (2011)
			Europe	2015	Bustos Lopez *et al*. (2015)
			Japan	2007	Umemoto *et al*. (2007)
			New Zealand	1989	Pennycook (1989)
871.	*Valeriana officinalis* L.	Common valerian	Bulgaria	2009	Bobev (2009)
			Russia	1938	Masalab (1938)
872.	*Venidium decurrens* Less. (Syn. *Arctotis arctotoides*)	Venidium	Canada	1967	Conners (1967)
873.	*Venidium fastuosum* (Jacq). Stapf.	Cape daisy	India	1959, 1971	Rai and Agnihotri (1970); Sarbhoy *et al*. (1971)
874.	*Verbascum blattaria* L.	Moth mullein	USA	1974	Schwartz (1977)
875.	*Verbesina* spp.	Crown beard	USA	1984	Alfieri Jr *et al*. (1984)
876.	*Verbena bonariensis*	Purple top vervain	Poland	2019	Pieczul (2019)
877.	*Verbena officinalis* L.	Common vervain/common verbena	Japan	1996	Takeuchi and Horie (1996)
878.	*Vernonia cinerea*	Common vernonia/little ironweed	China	1979	Tai (1979)
			India	2004	Ghasolia *et al*. (2004)

			Taiwan	1979	Anon (1979)
879.	*Vicia cracca*	Bird vetch	China	1979	Tai (1979)
880.	*Vicia faba* L. (Stem rot)	Broad bean	Australia	1982, 1989	Sampson and Walker (1982); Shivas (1989)
			Bulgaria	2009	Bobev (2009)
			California	1960, 1965, 1989	Anon (1960); Anon (1965); French (1989)
			China	1979	Tai (1979)
			Canada	1967, 1973	Conners (1967); McKenzie and Morrall (1975)
			Italy	1982	Caruso and D'Anna (1982)
			Cyprus	1957	Georghiou and Papadopoulos (1957)
			Greece	1973, 1989	Pantidou (1973); Lithourgidis *et al.* (1989)
			India	1998	Pande and Rao (1998)
			Italy, Sicily	1991	Greuter *et al.* (1991)
			Mauritius	1968	Orieux and Felix (1968)
			Mexico	1976	Alvarez (1976)
			North Dakota	2018	Chapara *et al.* (2018)
			New Zealand	1932, 1989	Brien (1932); Pennycook (1989)
			Taiwan	1979	Anon (1979)
			UK	1990	Jellis *et al.* (1990)
881.	*Vicia faba* var. *equina*	Horse bean	China	1979	Tai (1979)
882.	*Vicia sativa* L.	Spring vetch	China	1979	Tai (1979)
			Greece	1973	Pantidou (1973)
			India	1986, 1998	Singh and Singh (1986); Pande and Rao (1998)

Continued

Table 2.1. Continued

Sr. no.	Host name	English name	Country	Year	References
			New Zealand	1989	Pennycook (1989)
883.	*Vicia* sp.	Vetch	Oregon	1973	Shaw (1973)
884.	*Vicia villosa* Roth.	Hairy vetch	USA	1974	Schwartz (1977)
885.	*Vigna angularis* (Willd.) Ohwi & Ohashi (as *Phaseolus angularis* (Willd.) Wright)	Azuki bean	Japan	1959, 1981	Sugimoto (1959); Akai (1981)
886.	*Vigna mungo* (L.) Hepper (as *Phseolus mungo* L.)	Black gram	Texas	1960	Anon (1960)
887.	*Vigna radiata* (L.) Wilcz. (as *Phseolus mungo* L.)	Mung bean	Canada	1985	Huang (1985)
			Texas	1960	Anon (1960)
888.	*Vigna sinensis* (Stem rot, runner blight)	Black-eyed pea	Australia	1966	Simmonds (1966)
			Mexico	1976	Alvarez (1976)
			Texas	1960	Anon (1960)
889.	*Vigna unguiculata* (L.) Walp. (as *V. sesquipedalis* (L.) Frow.) (Stem rot)	Cowpea/yard-long bean	Brazil	1998	Mendes *et al*. (1998)
			Florida	1984	Alfieri Jr *et al*. (1984)
			USA	1974	Schwartz (1977)
890.	*Vigna unguiculata* (L.) Walp. subsp. *sesquipedalis* (as *V. sinensis* (Torner) Savi.)	Cowpea	California	1989	French (1989)
			Texas	1960	Anon (1960)
891.	*Vinca minor* L.	Common periwinkle	Bermuda	1946	Kinghorn (1947)
892.	*Viola odorata* L.	Sweet violet	USA	1920	Taubenhaus (1920)
893.	*Viola* sp.	Pansy	California	1989	French (1989)
			Holland	1950	Houtman (1950)
894.	*Vitis* sp.	Wild grape	Australia	2018	Jayawardena *et al*. (2018)
			Chile	2018	Jayawardena *et al*. (2018)
			Greece	2018	Jayawardena *et al*. (2018)
			Japan	2018	Jayawardena *et al*. (2018)

			Mexico	2018	Jayawardena *et al*. (2018)
			New Zealand	2018	Jayawardena *et al*. (2018)
			Spain	2018	Jayawardena *et al*. (2018)
			Switzerland	2018	Jayawardena *et al*. (2018)
			USA	2018	Jayawardena *et al*. (2018)
			Chile	2001	Latorre and Guerrero (2001)
895.	*Vitis vinifera* (Shoot blight)	Grape	Australia	1989	Shivas (1989)
			California	1960, 1987, 1989	Anon (1960); French (1987, 1989)
			Chile	2001, 2014	Ferrada *et al*. (2014); Latorre and Guerrero (2001)
			Greece	2000	Holevas *et al*. (2000)
			Japan	2007	Kobayashi (2007)
			Mexico	1976	Alvarez (1976)
			New Zealand	1989	Pennycook (1989)
			Spain	2011	González and Tello (2011)
			Switzerland	2009	Casieri *et al*. (2009)
			South Australia	2002	Hall *et al*. (2002)
896.	*Vitis vinifera* ssp. *roptundata*	European wine grape	India	2004	Ghasolia *et al*. (2004)
897.	*Vitis vinifera* var. *vinifera*	Grapevine	Greece	1973	Pantidou (1973)
898.	*Withania somnifera*	Winter cherry	India	2004	Ghasolia *et al*. (2004)
899.	*Xanthium pensylvanicum* Wallr.	Cocklebur	USA	1983	Adams *et al*. (1983)
900.	*Xanthium spinosum*	Spiny cocklebur/prickly burweed	Australia	1990	Nikandrow *et al*. (1990)
901.	*Xanthium strumarium*	Rough cocklebur	China	1979	Tai (1979)
			South Africa	2000	Crous *et al*. (2000)

Continued

Table 2.1. Continued

Sr. no.	Host name	English name	Country	Year	References
902.	*Xanthorrhoea australis*	Grass tree/Austral grasstree/ black boy	Australia	1987	Spooner (1987)
903.	*Zea* sp. (*Zea mays*)	Maize	Brazil	1998	Mendes *et al*. (1998)
			China	1979	Tai (1979)
			Russia	1990	Richardson (1990)
904.	*Zephyranthes candida (Zephyranthes grandiflora.)*	Pink rain lily	China	2023	Yin *et al*. (2024)
905.	*Zinnia elegans* Jacq. (*Z. elegans* L.) (Stem rot)	Zinnia	California	1960, 1989	Anon (1960); French (1989)
			Colorado	1960	Anon (1960)
			Massachusetts	1960	Anon (1960)
			Montana	1960, 1973	Shaw (1973); Anon (1960)
			New Zealand	1932, 1989	Brien (1932); Pennycook (1989)
			Oregon, Pennsylvania, Washington, USA	1960	Anon (1960)
			Poland	1979	Lacicowa *et al*. (1979)
			USA	1936	Young (1936)
906.	*Zinnia* spp.	Zinnia	Australia	1982	Sampson and Walker (1982)
			California	1989	French (1989)
			Chile	1967	Mujica and Oehrens (1967)
			Oregon	1973	Shaw (1973)
			Washington	1973	Shaw (1973)
			Canada	1961	Creelman (1962)

Source: Systematic Botany and Mycology Laboratory: https://nt.arsgrin.gov/fungal data bases/new_all View. cfm?
[a]Name of the hosts as reported by the authors.

squamosa, *Sclerotinia narcissicola*, *Sclerotinia borealis*, and *Sclerotinia fuckeliana*, please refer to Chapter 2 of the book "Sclerotinia Diseases of Crop Plants: Biology, Ecology and Disease Management" authored by Saharan and Mehta (2008).

2.2.2 Distribution map of *Sclerotinia* in India

The *Sclerotinia* stem rot is considered as the most damaging plant disease, causing huge losses in rapeseed–mustard crop in India since 1999. The disease reached its peak during 2012, 2013, and 2014 and has remained as a major concern thereafter. The disease significantly decreased later on, which may be due to climatic effects and technological advances for its management. However, the disease has been identified and considered as major problem since 1999, particularly in Rajasthan, Haryana, Madhya Pradesh, Uttar Pradesh, Bihar, and Punjab in India. The intensity of *Sclerotinia* stem rot was recorded maximum on *Brassica rapa* ssp. Yellow Sarson (51.2%), followed by *B. rapa* ssp. Toria (50.2%), *Eruca sativa* (49%), *Brassica napus* (38.5%), and the least on *Brassica carinata* (33.4%).

S. sclerotiorum (Lib.) de Bary, the causal organism of *Sclerotinia* stem rot of *Brassica*, is distributed worldwide and has a wide host range of more than 900 plants (Saharan and Mehta, 2008; Sharma *et al.*, 2015; Mehta *et al.*, 2023). In India, the disease appeared to be serious since 1999 and resulted in yield reduction of up to 40%, with an average disease incidence of 67%. Currently, this disease ranks third among the major problems affecting oilseed *Brassica* crops in India. However, *Sclerotinia* stem rot is a menace to the cultivation of oilseed *Brassicas* all over the world. The pathogen initial infection occurs on leaves, then on flowers, followed by stems and pods at different developmental stages of plant growth, causing seed yield losses of up to 80%, as well as significant reductions in oil content and quality of the seed. The distribution of *Sclerotinia* stem rot disease in different states of India indicated that, on an average, more than 20% incidence has been recorded in different states, such as Uttarakhand (40.2%), Haryana (35.7%), Rajasthan (33.2%), Madhya Pradesh (28.3%), Uttar Pradesh (27.2%), Punjab (25.3%), and Bihar (24.8%). Other states where *Sclerotinia* stem rot incidence was observed to be less than 20% were Delhi (19.1%), West Bengal (10.0%), Himachal Pradesh (6.3%), and Assam (4.7%). The other mustard-growing states of the country have recorded low disease incidence (in trace amounts), which are Chhattisgarh, Jharkhand, Jammu and Kashmir, Manipur, Gujarat, Maharashtra, and Tamil Nadu (Fig. 2.1). The geographical distribution map of *Sclerotinia* stem rot disease on rapeseed–mustard indicated the wide spread of the disease in different parts of India (Fig. 2.2).

Among the different *Brassica* species observed for infection by *Sclerotinia*, it was observed that the most susceptible were *B. rapa* ssp. Brown Sarson cv. BSH-1 (55.8%), *E. sativa* cultivar RTM 314 (44.7%), *B. rapa* ssp. Toria cv. PT 303 (36.12%), *Brassica juncea* cv. Rohini (34.9%), and *B. rapa* ssp. Yellow Sarson cv. YSPB 9 (29.3%). However, low incidence was observed on *B. napus* cv. GSL 1 (15.3%) and *B. carinata* cv. DLSC 1 (20.6%), indicating a level of tolerance (Fig. 2.3). The *Sclerotinia* stem rot incidence was about 9% on Indian mustard during 2000–2001, which continuously increased up to 81.8% within a period of 14 years (2001–2014), with variable changes (81.8–7.5%) in its intensity between 2003 and 2020. However, the severity of the *Sclerotinia* stem rot decreased continuously (from 57 to 17%) during the 2015/19 crop seasons (Fig. 2.4; Meena *at al.*, 2022).

2.3 Host Range

2.3.1 *Sclerotinia sclerotiorum* (Lib.) de Bary

S. sclerotiorum is considered to be one of the most versatile, all-encompassing, effective, and successful plant pathogens. Its ability to infect a wide range of hosts is crucial in managing diseases in agricultural plants, as it limits the inclusion of non-host crops in various crop rotations. This helps in reducing the quantity of sclerotia or primary inoculum in infested

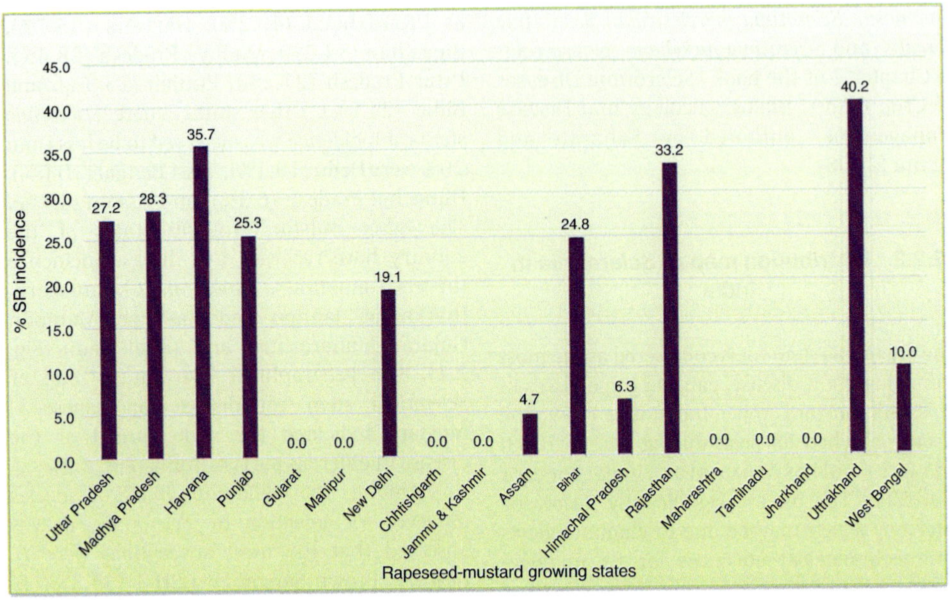

Fig. 2.1. Geographical distribution of *Sclerotinia* stem rot (%) on rapeseed–mustard in India. (Adapted from the publication of Meena *et al*., 2022 with permission.)

soils. However, the wide host range of *Sclerotinia* hinders its use as a mycoherbicide, as it has the potential to spread the pathogen to unintended hosts. Assessing the risk of this disease in non-target plants is challenging due to the lack of an up-to-date, easily accessible, and wide-ranging host index for *S. sclerotiorum*. Information on vulnerable hosts of *Sclerotinia* is scattered across both published and unpublished scientific literature.

Out of 37 plant families, Partyka and Mai (1962) found roughly 172 species with 118 genera as vulnerable hosts, and Farr *et al.* (1989) named 148 genera that *S. sclerotiorum* attacked. A host range of 374 plant species, comprising 237 genera in 65 families, was noted by Schwartz in 1977. Purdy (1979) cited a P.B. Adams collection that included 361 species in 225 genera across 64 groups. Following Boland and Hall (1994), who reported that it included 42 subspecies or variants, more than 408 species, with 278 genera and 75 families of plants, the most recent host index for *S. sclerotiorum* (Table 2.1) was created. According to Derbyshire *et al.* (2022), the host list of *S. sclerotiorum* has documented 425 species in 74 families of plants. The study carried out by Mehta *et al.* (2023) indicates that the host range of *S. sclerotiorum* has surpassed 900 plant species, encompassing approximately 427 genera (Table 2.1). It is possible that there are numerous hosts yet to be documented and included in the list. According to research from the US Department of Agriculture database, *S. sclerotiorum* is linked to 2048 distinct host species and variations. The information provided in reports on new hosts of *S. sclerotiorum* varies greatly in terms of clarity. The changes in the nomenclature of the fungus and the hosts contribute to the observed variability. Compiling this information into a comprehensive index of *S. sclerotiorum* hosts will greatly facilitate the investigation of the risk associated with using biocontrol products that contain *Sclerotinia*. It is crucial to determine the host range of *Sclerotinia* in order to recognize prospective sources of disease resistance, which may enhance understanding of its host specificity and contribute to knowledge of fungal evolution. With the exception of one species of Pteridophyta, all hosts of *S. sclerotiorum* belong to the Gymnospermae and Angiospermae classes from Spermatophyta division. Majority of the hosts are herbaceous

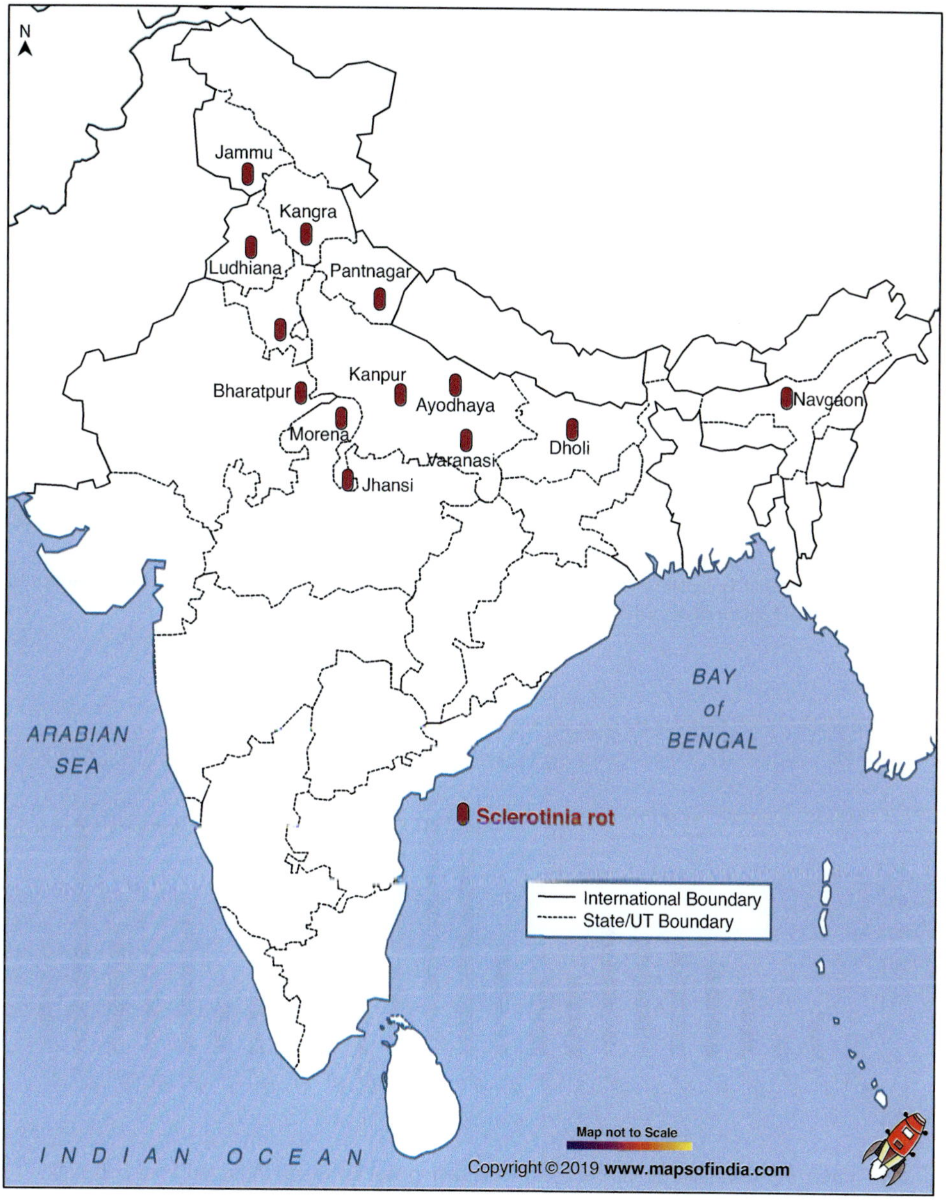

Fig. 2.2. Geographical distribution map of Sclerotinia stem rot on rapeseed-mustard in India (Adapted from publication of Meena *et al*,. 2022 with permission.)

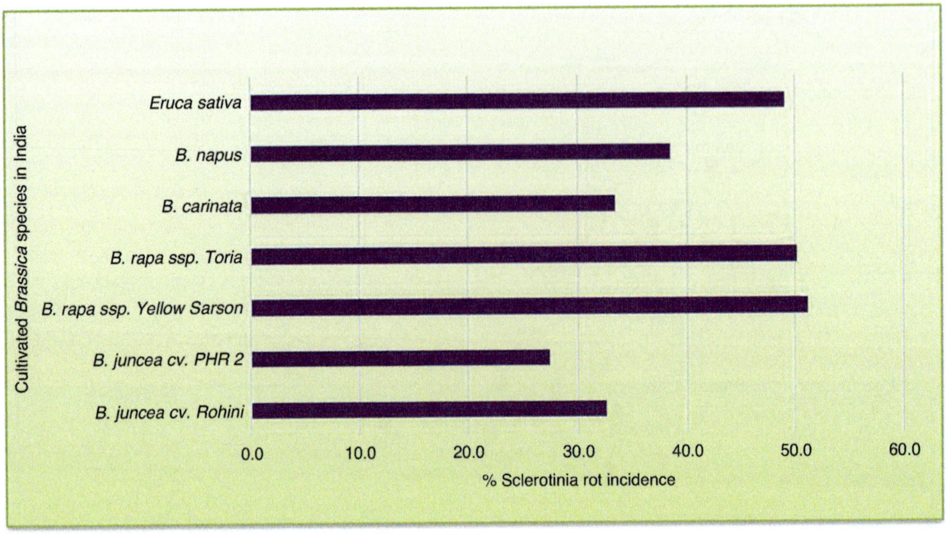

Fig. 2.3. Incidence of Sclerotinia stem rot on different cultivated *Brassica* species in India. (Adapted from publication of Meena *et al.*, 2022 with permission.)

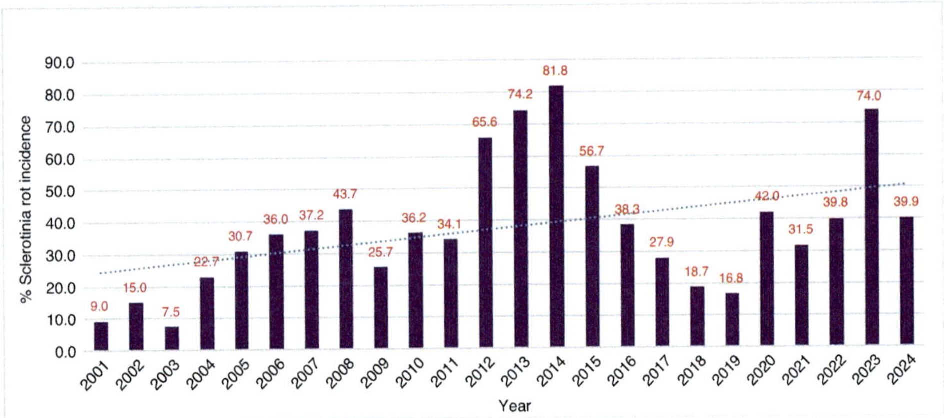

Fig. 2.4. Trends of Sclerotinia rot incidence in India during last 20 years. (Adapted from publication of Meena *et al.*, 2022 with permission.)

plants from the Dicotyledoneae subclass of the Angiospermae; there are a number of hosts in the subclass Monocotyledoneae. For information on the host range of *S. minor* Jagger and *S. trifoliorum* Erikss, please refer to the book "*Sclerotinia Diseases of Crop Plants: Biology, Ecology and Disease Management*" by Saharan and Mehta (2008).

2.4 The Molecular Evolutionary Features of *Sclerotinia* as Host Generalist

Sclerotinia is a group of fungus that causes disease in a variety of plants, including *sclerotiorum*. The clade comprising other well-known

Sclerotinia species most likely evolved around 15 million years ago, although *Sclerotinia* is polyphyletic (Navaud *et al.*, 2018). *S. trifoliorum*, *S. minor*, and *S. nivalis* are the *Sclerotinia* species in this group that have been investigated the most, apart from *Sclerotinia*. The geographic distributions and host ranges of all three species are extensive. Several species of the *Fabaceae* family are infected by *S. trifoliorum*, including fenugreek (*Trigonella foenumgraecum*) in Tunisia (Gargouri *et al.*, 2017), clover (*Trifolium* spp.), and chickpea (*Cicer arietinum*) in the USA (Njambere *et al.*, 2014; Baturo-Ciesniewska *et al.*, 2017). *S. minor* infects members of the *Asteraceae* family, including peanut (*Arachis hypogaea*) and (Crutcher *et al.*, 2018) lettuce (*Lactuca sativa*) in the USA (Subbarao, 1998) and safflower (*Carthamus tinctorius*) in Turkey (Erper *et al.*, 2020). The least researched of the three species, *S. nivalis*, may infect plants from at least ten distinct families and has a broad host range (Wang *et al.*, 2021). There are substantially fewer published and identified research articles on *S. trifoliorum*, *S. minor*, and *S. nivalis* than there are on *S. sclerotiorum* (4297 publications). The numbers are 120, 371, and 13, respectively. Majority of the studies on these species concentrate on useful control strategies, with molecular research receiving less attention. Each species has certain host range constraints even if it infects different host species. *S. trifoliorum*, for example, only feeds on members of the *Fabaceae* family (Kusch *et al.*, 2022). Comparing the molecular pathways of *S. sclerotiorum* with those involved in its pathogenesis and infection would be interesting.

Studies have revealed that distinct *Sclerotinia* species may be dominant in different parts of the same geographic region. In the San Joaquin Valley of California, USA, Wu and Subbarao (2006) discovered that *S. sclerotiorum* was the most usually found species in lettuce fields, while *S. minor* was the most regularly found species in the Salinas Valley of USA. This disparity may be attributed to the smaller sclerotia of *S. minor*, which are less able to withstand the harsh circumstances of the San Joaquin Valley, and the absence of conditions in the Salinas Valley that are conducive to the carpogenic germination of *S. sclerotiorum*. *Sclerotinia sclerotiorum* and *S. minor* are commonly found together in the same location and settings, despite the preponderance of distinct *Sclerotinia* species in different field sites (Marinelli *et al.*, 1998; Deng *et al.*, 2021). Similar levels of *S. sclerotiorum* and *S. minor* prevalence were also noted by Wu and Subbarao (2006) in the Santa Maria Valley, suggesting that the environment was equally favorable for the development of both pathogen species However, there is limited knowledge regarding the geographical, ecological, and host niche relationships among various *Sclerotinia* species, as well as how these factors influence their dispersal prior to their links to global agriculture. The method of sequencing and comparing the whole genomes of these species has only been applied in one study. This strategy seeks to identify the genes that are common to them and those that are unique, as well as how they relate to host preferences or other aspects of lifestyle. According to Kusch *et al.* (2022), synteny was observed between 67.2% of the genomes of *S. trifoliorum* and *S. sclerotiorum*. Furthermore, *S. sclerotiorum* homologs were found for 93% of the 10,626 genes that were found and predicted in the genome of *S. trifoliorum*. A more thorough study on *Botrytis* (Valero-Jiménez *et al.*, 2019) has offered a wider view on the variations in gene content among *Sclerotinia*, but the overall result of these changes is not the major focus.

This work may help with future investigations into these species' evolutionary histories.

With the exception of *S. sclerotiorum*, *B. cinerea* is the most extensively researched species in the *Sclerotinia* family. It is a highly versatile pathogen that may infect a variety of plant species, much like *S. sclerotiorum* (Hua *et al.*, 2018). According to Navaud *et al.* (2018), *B. cinerea* and *S. sclerotiorum* are thought to have diverged approximately 40 million years ago, indicating that they have seen substantial evolutionary changes. Researchers have found several discrepancies between the genomes of *S. sclerotiorum* and *B. cinerea* (Amselem *et al.*, 2011). Protein sequences were determined to be 83% similar on average, showing a significant degree of amino acid variation. Furthermore, a greater number and diversity of secondary metabolite genes are present in *B. cinerea*, though it is unclear how important these differences are for the species' biological niche. There are common virulence mechanisms between *B. cinerea* and *S. sclerotiorum* despite their divergent evolutionary histories. Oxalic acid is used by both of these species to increase their pathogenicity (Rascle *et al.*, 2018). This could be connected to the requirement for pH regulation in hosts whose tissues are inherently alkaline, as seen in the case of *S. sclerotiorum* (Müller *et al.*, 2018). Research on *S. minor* also points to a common virulence mechanism within the *Sclerotinia* family and a function for oxalic acid in the pathogenicity of the organism (Livingstone *et al.*, 2005). One feature of niche generalism is its capacity to infect a broad variety of host species. Despite the paucity of research on the development of position generalism in fungi, investigations on *S. sclerotiorum* and similar species shed important light on the general evolutionary dynamics of these species.

According to its taxonomic classification, *S. sclerotiorum* belongs to the *Sclerotiniaceae* family, which is made up of a variety of species that infect a broad variety of host plants. The family *Fabaceae* has been suggested as the most likely ancestral host family for the *Sclerotinia* based on a thorough phylogenetic study of the *Sclerotinia* and 263 of their corresponding host plants (Navaud *et al.*, 2018). It was discovered that almost half of the *Sclerotinia* species investigated in this study infected plant species that diverged before the Fabaceae emerged, indicating multiple host jumps. Within the *Ranunculales*, a noteworthy host hop took place approximately 20 million years ago, at the same time as the *Sclerotinia* genus originated. *S. sclerotiorum* has since amassed hundreds of identified hosts. Navaud *et al.* (2018) also looked at *Sclerotinia*'s evolutionary history and found that different clades within the *Sclerotinia* underwent variable rates of diversification throughout their evolutionary histories. The different genera of *Monilinia*, *Encoelia*, and *Ciboria* are among the early distinct clades that showed comparatively low rates of change. But the rates of variation in the more divergent genera *Sclerotinia* and *Myriosclerotinia* were intermediate. *Botrytis*, the most diverse genus, showed tremendous rates of diversification. There was a considerable correlation between host generalism and the intermediate diversification rate seen in the *Sclerotinia* and *Myriosclerotinia* lineages. The accumulation of new hosts, as opposed to other mechanisms like continuous connection with a diversified host lineage, was the main driver of the growth of host generalism (Saharan *et al.*, 2021a,b, 2023).

Few studies have been conducted on pathogenic fungi, although many have examined the increase of host range observed in *Sclerotinia* in herbivorous insects (Peccoud *et al.*, 2008). Further research is necessary to understand the methods and causes for the notable increase of *S. sclerotiorum* and its host range, which are now mostly unknown. Factors such as preadaptation to the new host, successful establishment, and subsequent fungal selection can all lead to the extension of the host range. The retention of eight crucial virulence factors suggests that several molecular traits necessary for the colonization of contemporary host plants were shared by the progenitor of *Sclerotinia*. The presence of positive selection in five of these genes suggests an evolutionary arms race with hosts, which aligns with the occurrence of collection following host transitions and range expansions. The synthesis of oxalic acid requires the expression of oxaloacetate acetylhydrolase, a gene that is expressed 10–300 times more in host generalists than in specialists. This coincided with greater oxalic acid accumulation during infection. According to Andrew *et al.* (2012), this gene may thus be a preadaptation factor, whose expression has now grown to support host generalism.

The genomes of living organisms are still not well recognized in terms of how host generalism shapes them. Host generalist fungi, in particular, tend to have greater produced proteins compared to host specialists. This is only due to host generalists' need to secrete a wide range of carbohydrate-based active enzymes in order to metabolize various host substrates. However, the process of translation, which is energetically costly, poses a strong selective pressure on translation efficiency in host generalists. This is because the cell loses the energy required to produce the secreted proteins. The basic hypothesis put forth by Badet *et al.* (2017) is consistent with the strong selective restriction on polymorphisms in codons matching the most prevalent in wild populations of *S. sclerotiorum*. Moreover, Prat *et al.* (2009) suggested that gene age increases both gene length and translation efficiency. This suggests that, in comparison to specialists, host generalists have a slower pace of gene diversification. As opposed to the qualitative resistance mediated by resistance genes or susceptibility loci that are either present or absent, this observation is compatible with the lower selective pressure exerted by hosts, such as quantitative disease resistance (Saharan *et al.*, 2021a,b; Derbyshire *et al.*, 2022; Chen *et al.*, 2023).

2.5 Molecular Determinants of Host Range of *Sclerotinia sclerotiorum*

The majority of the documented hosts for *S. sclerotiorum* consist of dicotyledonous plants. However, there have been no direct studies conducted to determine the molecular factors that limit the host range of *S. sclerotiorum*. Nonetheless, several speculations have been proposed by Boland and Hall (1994), Saharan and Mehta (2008), Derbyshire *et al.* (2022), and Chen *et al.* (2023). Necrosis and ethylene inducing peptide 1-like protein (NLPs), are generated by a variety of organisms, including *S. sclerotiorum* (Bashi *et al.*, 2010). According to Gijzen and Nurnberger (2006), these proteins cause cell death in the majority of dicots but exert very mild effects on monocots. The ability of NLPs to induce necrosis is ascribed to their attachment to glycosylinositol phosphorylceramide, a particular type of membrane lipid that is frequently present in dicots but uncommon in monocots. The monocot *Phaenopsis amabilisi* is an exception, though, as it has NLP homologs that cause necrosis in *Nicotiana benthamiana*. According to Bashi *et al.* (2010), these homologs are significantly expressed during infection, and inhibiting their expression results in a decrease in *B. napus* pathogenicity. Nevertheless, a subsequent study failed to replicate the necrosis-inducing activity of these homologs (Seifbarghi *et al.*, 2020).

It is important to note that Albert *et al.* (2015) proposed that NLPs play a role in disease by inducing an inherent invulnerable response in *Arabidopsis* through the presence of a sustained peptide sequence. The leucine-rich repeat (LRR) receptor-like protein RLP23 mediates this immunological response, and plants with higher RLP23 expression levels show increased resistance to *S. sclerotiorum*. The conservation of NLPs throughout a wide range of microbes, including *biotrophs*, suggests that they may have purposes other than causing necrosis. Thus, they might function as broad-spectrum basal immunity elicitors (Raaymakers and Van den Ackerveken, 2016). However, without investigating the two *S. sclerotiorum* NLPs, it is difficult to determine whether their principal roles are physiological or just connected with their aptitude to induce necrosis in dicot plants. Oxalic acid, a key factor in *S. sclerotiorum's* pathogenicity across various hosts, is another element that could explain its affinity for dicot plants. Furthermore, many monocot species secrete proteins called germins in reaction to pathogen infection (Davidson *et al.*, 2009). When expressed in different dicot plants, a subset of these organisms exhibits the enzymatic activity of oxalate oxidase, which can boost resistance against *S. sclerotiorum* by breaking down oxalic acid (Yang *et al.*, 2019).

The capacity of most monocot plants to metabolize one of *S. sclerotiorum's* primary virulence factors may be one of the reasons it is unable to infect them. The investigation of oxalate oxidases in monocot plants after exposure to *S. sclerotiorum* has been suggested by Williams *et al.* (2011). *S. sclerotiorum* does not seem to be a parasite of most monocots, but a new study indicates that it may be able to live as an endophyte inside of a few species (Tian *et al.*, 2020). *S. sclerotiorum* was able to grow both intracellularly and intercellularly, without exhibiting any symptoms, when it was added to the roots of wheat seedlings.

Gene expression alterations associated with *S. sclerotiorum* endophytic development in wheat mirrored the plant's defense mechanism against biotrophic infections. Salicylic acid-related genes were upregulated, but those linked to jasmonate and abscisic acid were downregulated. It has been discovered that the endophytic growth of *S. sclerotiorum* within monocots improves resistance against a number of diseases that are important to agriculture, including rice blast and wheat stripe rust. Until now, only a few *Colletotrichum* species were known to lead this distinctive lifestyle (Redman *et al.*, 2001). Though this phenomenon has only recently come to light, it's possible that other plant pathogenic fungi are also affected by it. The fact that *S. sclerotiorum* was able to enter monocots' cell walls without causing necrosis suggests that, rather than any potential physical barriers, what separates the *S. sclerotiorum* hosts from nonhosts is most likely their responses to cell death. Distinctive responses to *S. sclerotiorum* triggers of programmed cell death are clearly present in dicots, but are largely missing in monocots (Derbyshire *et al.*, 2022; Chen *et al.*, 2023).

References

Abán, C.L., Taboada, G., Spedaletti, Y., Aparicio, M., Curti, R.N. *et al.* (2018) Molecular, morphological and pathogenic diversity of *Sclerotinia sclerotiorum* isolates from common bean (*Phaseolus vulgaris*) fields in Argentina. *Plant Pathology* 67(8), 1740–1748. DOI: 10.1111/ppa.12880.

Abawi, G.S. (1993) Lulu (*Solanum guitoense)* as a host for *S. sclerotiorum*. In: *Proceeding 8th International Sclerotinia Workshop*, Poster, University of Toronto, Toronto, Ontario, 8–12 August 1993.

Aci, M.M., Moscal, S., Zangari, A., Malacrino, L. and Schena, L. (2024) First report of *Sclerotinia sclerotiorum* causing rots at the shoot base of *Actinidia chinensis*. *Plant Disease Reports*. DOI: 10.1002/ndr2.12281.

Adams, P.B., Lumsden, R.D. and Tate, C.J. (1974) *Galinsoga parviflora*: A new host for *Whetzelinia sclerotiorum*. *Plant Disease Reporter* 58, 700–701.

Adams, P.B., Marose, B.H. and Dutky, E.M. (1983) Cocklebur: A new host for several *Sclerotinia* species. *Plant Disease* 67, 484–485. DOI: 10.1094/PD-67-484.

Adikaram, N.K.B. and Yakandawala, D.M.D. (2020) A checklist of plant pathogenic fungi and Oomycota in Sri Lanka. *Ceylon Journal of Science* 49(1), 93. DOI: 10.4038/cjs.v49i1.7709.

Agrawal, J.M., Chhipa, H.P. and Sobti, A.K. (1980) A note on two post-harvest diseases of vegetables for Rajasthan. *Indian Journal of Mycology and Plant Pathology* 10(1), 97–98.

Ahmad, S., Iqba, S.H. and Khalid, A.N. (1997) *Fungi of Pakistan*. Sultan Ahmad Mycological Society of Pakistan, Lahore, 248 pp.

Ahmed, A.U. and Akhond, M.A.Y. (2015) First report of Sclerotinia rot caused by *Sclerotinia sclerotiorum* on *Lens culinaris* in Bangladesh. *New Disease Reports* 31:23. DOI: 10.5197/j.2044-0588.2015.031.023.

Ajitomi, A., Takushi, T., Ooshiro, A., Yamashiro, M. and Taba, S. (2018) First report of Sclerotinia rot of mango caused by *Sclerotinia sclerotiorum* in Japan. *Journal of General Plant Pathology* 84(1), 70–72. DOI: 10.1007/s10327-017-0758-z.

Akai, J. (1981) Studies on the epidemiology and control of Sclerotinia disease of beans caused by *Sclerotinia sclerotiorum* (lib.) de Bary. *Report, Hokkaido National Agricultural Experiment Station* 36:83.

Akazawa, Y. and Ono, K. (1984) Factors affecting the occurrence of tobacco stem rot caused by *Sclerotinia sclerotiorum* (lib.) de Bary. *Okayama Tabako Shiken Jo Hokoku* 43, 47–58.

Akem, C.N. and Dashiell, K.E. (1992) Sclerotinia stem rot caused by *Sclerotinia sclerotiorum* on soybeans in Nigeria. *Plant Disease* 76(1): 101. DOI: 10.1094/PD-76-0101C.

Alam, M.W., Rehman, A., Malik, A., Mehboob, S., Sarwar, M. *et al.* (2021) First report of white mould of potato caused by *Sclerotinia sclerotiorum* in Pakistan. *Journal of Plant Pathology* 103(2), 669–669. DOI: 10.1007/s42161-021-00756-x.

Albert, I., Böhm, H., Albert, M., Feiler, C.E., Imkampe, J. *et al.* (2015) An RLP23-SOBIR1-BAK1 complex mediates NLP-triggered immunity. *Nature Plants* 1, 15140. DOI: 10.1038/nplants.2015.140.

Ale-Agha, N. (1974) Etude du *Sclerotinia sclerotiorum* (lib) de Bary. *En Iran. Annals of Phytopathology* 6(4), 385–393.

Alfieri, S.A., Jr., Langdon, K.R., Wehlburg, C. and Kimbrough, J.W. (1984) Index of plant diseases in Florida (Revised). *Florida, Department of Agriculture and Consumer Services, Division of Plant Indexing Bulletin* 11, 1–389.

Allen, D.J. (1995) An annotated list of diseases, pathogens and associated fungi of the common bean (*Phaseolus vulgaris*) in eastern and southern Africa. *Phytopathology Papers* 34, 1–42.

Aloj, B. and Garibaldi, A. (1978) *Control Trials in Campania against Sclerotinia*. Annals of Faculty of Science, Agricultural University, Studi Napoli Portici, Naples, Italy, 22, 213–217.

Aloj, B., Scalcione, M., Nanni, B. and Pugliano, G. (1994) Observations on some rots of fruits of Feyoa. 11. *Annali Dalla Facolta Di Geieuze Agrarie Della Universita Degli Gtudi Di Napoli Partrei,* 28, 36–47.

Alvarez, M.G. (1976) Primer catalogo de enfermedades de plantas Mexicanas. *Fitofilo* 71, 1–169.

Amselem, J., Cuomo, C.A., van Kan, J.A.L., Viaud, M., Benito, E.P. *et al.* (2011) Genomic analysis of the necrotrophic fungal pathogens *Sclerotinia sclerotiorum* and *Botrytis cinerea*. *PLoS Genetics* 7(8): e1002230. DOI: 10.1371/journal.pgen.1002230.

Andrew, M., Barua, R., Short, S.M. and Kohn, L.M. (2012) Evidence for a common toolbox based on necrotrophy in a fungal lineage spanning necrotrophs, biotrophs, endophytes, host generalists and specialists. *PLoS ONE* 7(1): e29943. DOI: 10.1371/journal.pone.0029943.

Anon (1941) Plant diseases. Notes contributed by the Biological Branch. *The Agricultural Gazette of New South Wales* 2(5), 286.

Anon (1960) Index of plant diseases in the United States. *USDA Agricultural Handbook* 165, 1–531.

Anon (1965) Annual report of the department of agriculture for the year 1964. *Department of Agriculture Plant Pathology* 29, 63–67.

Anon (1979) *List of plant diseases in Taiwan*. Plant Protection Society, Republic of China, Taichung, Taiwan, 404 pp.

Apelt, G. and Bochow, H. (1975) For the prophylactic control of *Sclerotinia sclerotiorum* (lib.) de Bary in cucumbers. *Archive for Phytopathology and Plant Protection* 11, 101–110. DOI: 10.1080/03235407509431161.

Arcelin, R. and Kushalappa, A.C. (1991) A survey of carrot diseases on muck soils in the south western part of Quebec. *Canadian Plant Disease Survey* 71, 147–153.

Arzanlou, M. and Dokhanchi, H. (2013) Morphological and molecular characterization of *Diplodia seriata*, the causal agent of canker and twig dieback disease on mulberry in Iran. *Archives of Phytopathology and Plant Protection* 46(6), 682–694. DOI: 10.1080/03235408.2012.749703.

Ayvar-Serna, S., Díaz-Nájera, J.F., Tejeda-Reyes, M.A., Mora-Romero, G.A., Pérez-Mora, J.L. *et al.* (2023) First report of white mold caused by *Sclerotinia sclerotiorum* on brussels sprouts in Mexico. *Plant Disease* 107(10), 3300. DOI: 10.1094/PDIS-05-23-0849-PDN.

Backhouse, D. and Willetts, H.J. (1984) A histochemical study of *Botrytis cinerea* and *Botrytis fabae*. *Canadian Journal of Microbiology* 30, 171–178.

Badet, T., Peyraud, R., Mbengue, M., Navaud, O., Derbyshire, M. *et al.* (2017) Codon optimization underpins generalist parasitism in fungi. *eLife* 6:e22472. DOI: 10.7554/eLife.22472.

Bag, T.K. (2003) *Sclerotinia sclerotiorum* on butterfly flower—A new record. *Indian Phytopathology* 56:121.

Bag, T.K. (2006) A new report on *Sclerotinia* white rot of two jewel orchids from Darjeeling, India. *Journal of Mycology and Plant Pathology* 36:304.

Bains, P.S., Bennypaul, H.S., Blade, S.F. and Weeks, C. (2000) First report of hemp canker caused by *Sclerotinia sclerotiorum* in Alberta. *Canadian Plant Disease* 84, 372–373. DOI: 10.1094/PDIS.2000.84.3.372B.

Balmas, V., Scherm, B., Ghignone, S., Salem, A.O.M., Cacciola, S.O. *et al.* (2005) Characterisation of *Phoma tracheiphila* by RAPD-PCR, microsatellite-primed PCR and ITS rDNA sequencing and development of specific primers for in planta PCR detection. *European Journal of Plant Pathology* 111(3), 235–247. DOI. 10.1007/s10658-004-4173-x.

Balou, S. and Iboton, N. (1985) Azolla diseases in Manipur, India. *International Rice Research Newsletter* 10:28.

Bardin, R. (1951) Unusual occurrence of ascosporic infection of tomatoes by *Sclerotinia sclerotiorum*. *Plant Disease Reporter* 35:246.

Bashi, Z.D., Hegedus, D.D., Buchwaldt, L., Rimmer, S.R. and Borhan, M.H. (2010) Expression and regulation of *Sclerotinia sclerotiorum* necrosis and ethylene-inducing peptides (NEPs). *Molecular Plant Pathology* 11(1), 43–53. DOI: 10.1111/j.1364-3703.2009.00571.x.

Batra, L.R. and Harada, Y. (1986) A field record of apothecia of *Monilinia fructigena* in Japan and its significance. *Mycologia* 78(6), 913–917. DOI: 10.1080/00275514.1986.12025349.

Baturo-Ciesniewska, A., Groves, C.L., Albrecht, K.A., Grau, C.R., Willis, D.K. *et al.* (2017) Molecular identification of *Sclerotinia trifoliorum* and *Sclerotinia sclerotiorum* isolates from the United States and Poland. *Plant Disease* 101(1), 192–199. DOI: 10.1094/PDIS-06-16-0896-RE.

Baysal-Gurel, F., Bozan, O., Onelge, N. and Cinar, A. (2012) First report of Sclerotinia stem and twig blight caused by *Sclerotinia sclerotiorum* on sour orange rootstock in Turkey. *New Disease Reports* 26(1), 24–24. DOI: 10.5197/j.2044-0588.2012.026.024.

Beaumont, A. (1951) Campanula diseases. *Gardener's Chronicle* 129, 157.

Bedi, K.S. (1963) The age and the size of *sclerotia* of *Sclerotinia sclerotiorum* (lib.) de Bary in relation to the formation of apothecia. *Indian Botanical Society* 42, 204–207.

Berkenkamp, B. (1982) Diseases of rapeseed in central and northern Alberta in 1982. *Plant Disease Reporter* 72:62.

Beute, M.K., Porter, D.M. and Hadley, B. A. (1975) Sclerotinia blight of peanut in North Carolina and Virginia and its chemical control. *Plant Disease Reporter* 59, 697–700.

Bienapfl, J.C., Malvick, D.K. and Percich, J.A. (2011) Specific molecular detection of *Phytophthora sojae* using conventional and real-time PCR. *Fungal Biology* 115, 733–740.

Bisby, G.R. (1938) *The fungi of Manitoba and Saskatchewan*. National Research Council of Canada, Ottawa, Ontario, 200 pp.

Blain, W.L. (1931) A list of diseases of economic plants in Alabama. *Mycologia* 23, 300–304.

Blanco-Meneses, M. and Ristaino, J.B. (2011) Detection and quantification of *Peronospora tabacina* using a real-time polymerase chain reaction assay. *Plant Disease* 95(6), 673–682. DOI: 10.1094/PDIS-05-10-0333.

Bobev, S. (2009) *Reference Guide for the Diseases of Cultivated Plants (Russian)*. Makros Publication, St. Petersburg, Russia, p. 466.

Boew, G.H. (1964) Some plant diseases new to Illinois. *Plant Disease Reporter* 48, 866–870.

Bohár, G. and Kiss, L. (1999) First report of *Sclerotinia sclerotiorum* on common ragweed (*Ambrosia artemisiifolia*) in Europe. *Plant Disease* 83(3), 302. DOI: 10.1094/PDIS.1999.83.3.302C.

Boland, G.J. and Hall, R. (1994) Index of plant hosts of *Sclerotinia sclerotiorum*. *Canadian Journal of Plant Pathology* 16(2), 93–108. DOI: 10.1080/07060669409500766.

Borah, T.R., Dutta, S. and Barman, A.R. (2018) First report of *Sclerotinia sclerotiorum* on *Mimosa pudica* in India. *New Disease Reports* 38, 14.

Borrelli, N.P., Papone, M.L., Moreno, M.V., Stenglein, S., Stancanelli, S. *et al.* (2020) First report of basal rot caused by *Sclerotinia sclerotiorum* on *Calibrachoa hybrida*. *Plant Disease* 104(12), 3254.

Bradley, C.A., del Rio, L.E. and Johnson, B.L. (2003) First report of *Sclerotinia sclerotiorum* on Niger (*Guizotia abyssinica*). *Plant Disease* 87, 602.

Bradley, C.A., del Rio, L.E., Chesrown, C.D. and Johnson, B.L. (2005) First report of soft rot caused by *Sclerotinia sclerotiorum* on Borage in North Dakota. *Plant Disease* 89, 208.

Brenneman, T.B. and Sumner, D.R. (1991) *Sclerotinia sclerotiorum* on canola in Georgia and its potential as a pathogen on peanut. *Plant Disease* 75, 319.

Bretag, T.W. and Mebalds, M.I. (1987) Pathogenicity of fungi isolated from *Cicer arietinum* (chickpea) grown in northwestern Victoria. *Australian Journal of Experimental Agriculture* 27(1), 141. DOI: 10.1071/EA9870141.

Brien, R.M. (1932) Host range of *Sclerotinia sclerotiorum* in New Zealand. *New Zealand Journal of Agriculture* 44–45, 127–129.

Brien, R.M. (1946) Second supplement to "A list of plant diseases recorded in New Zealand". *New Zealand Journal of Science and Technology, Section A* 28, 221–224.

Brien, R.M. and Dingley, J.M. (1957) Third supplement to "A revised list of plant diseases recorded in New Zealand": 1955–1957. *New Zealand Journal of Science and Technology, Section A* 38, 771–781.

Brosten, B.S. and Sanda, D.C. (1986) Field trials of *Sclerotinia sclerotiorum* to control Canada thistle *(Cirsium arvense)*. *Weed Science* 34, 377–380.

Brown, J.G. and Butler, K.D. (1936) *Sclerotinia* of lettuce in Arizona. *Arizona Agriculture Experiment Station Bulletin* 63, 475–506.

Burger, O.F. (1913) Lettuce drop. *Florida Agriculture Experiment Station Bulletin* 116, 8.

Burke, D.W., Gomes, J.C. and Foeppel, W.G. (1957) Observations on Sclerotinia wilt of beans in north-eastern Colorado. *Plant Disease Reporter* 41, 72–73.

Bustos Lopez, M.P., Spadaro, D. and Gullino, M.L. (2015) First report of *Sclerotinia sclerotiorum* causing postharvest Sclerotinia rot on highbush blueberry in Europe. *Plant Disease* 99(11), 1648–1648. DOI: 10.1094/PDIS-01-15-0026-PDN.

Cappellini, R.A. (1960) Field inoculations of forage legumes and temperature studies with isolates of *Sclerotinia trifoliorum* and *Sclerotinia sclerotiorum*. *Plant Disease Reporter* 44, 862–864.

Carbone, I. and Kohn, L.M. (1993) Ribosomal DNA sequence divergence within internal transcribed spacer 1 of the *Sclerotinia*. *Mycologia* 85(3), 415–427. DOI: 10.1080/00275514.1993.12026293.

Caruso, P. and D'Anna, F. (1982) Experiments on the seed yield of the broad bean (*Faba vulgaris* Moench var. *major* Harz.). *Review of Ortic Floriculture, University of Palermo* 66, 401–418.

Casieri, L., Hofstetter, V., Viret, O. and Gindro, K. (2009) Fungal communities living in the wood of different cultivars of young *Vitis vinifera* plants. *Phytopathologia Mediterranea* 48, 73–83.

Chamberlain, D.W. (1951) Sclerotinia stem rot of soybeans. *Plant Disease Reporter* 35, 490–491.

Champion, R., Brunet, D. and Bourdin, J. (1974) The state of health of salsify and scorzonera seed and disinfection trials. *Defense Des Vegetaux* 28(167), 148–153.

Chang, S.W. and Kim, S.K. (2003) First report of Sclerotinia rot caused by *Sclerotinia sclerotiorum* on some vegetable crops in Korea. *The Plant Pathology Journal* 19(2), 79–84. DOI: 10.5423/PPJ.2003.19.2.079.

Chang, K.F., Howard, R.J., Gaudiel, R.G. and Hwang, H.F. (1997a) First report of stevia as a host of *Sclerotinia sclerotiorum*. *Plant Disease* 81(3), 311. DOI: 10.1094/PDIS.1997.81.3.311B.

Chang, K.F., Howard, R.J., Gaudiel, R.G. and Hwang, R.P. (1997b) First report of *Sclerotinia sclerotiorum* on coneflower. *Plant Disease* 81(9), 1093. DOI: 10.1094/PDIS.1997.81.9.1093A.

Chapara, V., Chittem, K. and Mendoza, L.E.R. (2018) First report of white mold caused by *Sclerotinia sclerotiorum* on faba beans in North Dakota. *Plant Disease* 102(8), 1669–1669. DOI: 10.1094/PDIS-12-17-1954-PDN.

Charchar, M.J.D., Anjos, J.R.N. and Ossipi, E. (1999) Occurrence of a new disease of irrigated cotton, in Brazil caused by *Sclerotinia sclerotiorum* A. *Pesquisa Agropecuaria Brasileira* 34, 1101–1106. DOI: 10.1590/S0100-204X1999000600024.

Chen, M.M. (2002) *Forest fungi phytogeography: Forest fungi phytogeography of China, North America, and Siberia and international quarantine of tree pathogens*. Pacific Mushroom Research and Education Center, Sacramento, California, 469 pp.

Chen, W., Schatz, B., Henson, B., McPhee, K.E. and Muehlbauer, F.J. (2006) First report of Sclerotinia stem rot of chickpea caused by *Sclerotinia sclerotiorum* in North Dakota and Washington. *Plant Disease* 90(1), 114. DOI: 10.1094/PD-90-0114A.

Chen, R.S., Wang, J.Y., Sarwar, R. and Tan, X.L. (2023) Genetic breakthroughs in the *Brassica napus–Sclerotinia sclerotiorum* interactions. *Frontiers in Plant Science* 14, 1276055. DOI: 10.3389/fpls.2023.1276055.

Cheremisinov, N.A. (1951) Michurin teaching-a basis for control measures for the diseases of Kok-Saghyz. *Nature* 6, 30–38.

Chlebicki, A. and Chmiel, M.A. (2006) Microfungi of *Carpinus betulus* from Poland. I. Annotated list of micro fungi. *Acta Mycologica* 41(2), 253–278. DOI: 10.5586/am.2006.027.

Cho, W.D. and Shin, H.D. (2004) *List of plant diseases in Korea*, 4th edn. Korean Society of Plant Pathology, Suwon, Korea, p. 779.

Choi, I.Y., Kim, J.H., Kim, B.S., Park, M.J. and Shin, H.D. (2016) First report of Sclerotinia stem rot of fennel caused by *Sclerotinia sclerotiorum* in Korea. *Plant Disease* 100(1), 223–223. DOI: 10.1094/PDIS-05-15-0512-PDN.

Choi, I.Y., Kim, J., Lee, W.H., Cho, S.E. and Shin, H.D. (2017a) First report of Sclerotinia stem rot caused by *Sclerotinia sclerotiorum* on Chinese chives in Korea. *Plant Disease* 101(11), 1953–1953. DOI: 10.1094/PDIS-04-17-0587-PDN.

Choi, I.-Y., Kim, J.-H., Kim, J., Han, K.-S., Galea, V. *et al.* (2017b) Confirmation of *Sclerotinia sclerotiorum* as the causal agent of stem rot of stock in Korea. *Australasian Plant Disease Notes* 12(1), 22. DOI: 10.1007/s13314-017-0247-4.

Choi, H.W., Hong, S.K., Kim, J.S., Lee, Y.K., Yang, D.C. *et al.* (2020) First report of *Sclerotinia sclerotiorum* causing *Sclerotinia* rot on *Codonopsis lanceolata* in Korea. *Plant Disease* 104(6), 1863. DOI: 10.1094/PDIS-08-19-1574-PDN.

Chorin, M. and Rotem, J. (1961) Fruit diseases of the Cavendish banana in Israel. *Israel Journal of Agricultural Research* 11, 43–50.

Chrominski, A., Abia, J.A. and Smith, B.N. (1987) Calcium deficiency and gibberellic acid enhance susceptibility of pumpkin and sunflower seedlings to *Sclerotinia sclerotiorum* infection. *Journal of Plant Nutrition* 10(17), 2181–2193. DOI: 10.1080/01904168709363771.

Cole, A.L.J. (1980) A blind-seed disease of globe artichoke. *Transactions of the British Mycological Society* 74(1): 199. DOI: 10.1016/S0007-1536(80)80028-3.

Conners, I.L. (1967) An annotated index of plant diseases in Canada and fungi recorded on plants in Alaska, Canada and Greenland. *Research Branch Canada, Department Agriculture* 1251, 1–381.

Cook, A.A. (1955) *Sclerotinia sclerotiorum* on castor bean in Florida. *Plant Disease Reporter* 39, 585–586.

Cook, R.P. and Dubé, A.J. (1989) *Host-Pathogen Index of Plant Diseases in South Australia*. South Australian Department of Agriculture, Adelaide, 1–142.

Cook, G.E., Steadman, J.R. and Kerr, E.D. (eds) (1972) Epidemiology of white mold in western Nebraska. *Phytopathology* 62:1107.

Corato, U. de (1996) Research on stem rot of kenaf caused by *Sclerotinia sclerotiorum* in Basilicata. *Informatore Fitopatologico* 46, 30–32.
Corato, U. de and Baviello, G. (2000) Occurrence of *Sclerotinia sclerotiorum* on *Brassica carinata* in Southern Italy (Basilicata). *Informatore Fitopatologico* 50, 61–63.
Cother, E.J. (2000) Pathogenicity of *Sclerotinia sclerotiorum* to *Chrysanthemoides monilifera* ssp. *rotundata* (Bitoubush) and selected species of the coastal flora in eastern Australia. *Biological Control* 18(1), 10–17. DOI: 10.1006/bcon.1999.0810.
Covey, R.P. (1985) Calyx end rot (*Sclerotinia sclerotiorum*) of apples in Washington. *Plant Disease* 69(8): 726. DOI: 10.1094/PD-69-726a.
Creelman, D.W. (1962) Summary of the prevalence of plant diseases in 1961. *Canadian Plant Disease Survey* 42, 23–102.
Creelman, D.W. (1964) Summary of the prevalence of plant diseases in Canada in 1963. *Canadian Plant Disease Survey* 44, 1–81.
Creelman, D.W. (1965) Summary of the prevalence of plant diseases in Canada in 1964, a compilation. *Canadian Plant Disease Survey* 45, 42–78.
Creelman, D.W. (1966) Summary of the prevalence of plant diseases in Canada in 1965, a compilation. *Canadian Plant Disease Survey* 46, 37–76.
Cristea, S. and Jurcoane, S. (2016) *Sclerotinia sclerotiorum* (Lib) De bary–first report on cultivated *Camelina sativa* in Romania. *Romanian Biotechnological Letters* 21(3), 11682–11683.
Cristea, G., Hiescu, H., Operea, M. and Teodorescu, T. (1985) *Sclerotinia sclerotiorum* (Lib) de Bary, pathogen of sunflower white rot, parasitic on some medicinal and aromatic species. *Bulletin Protectia Plantelor* 1–2, 25–33.
Crous, P.W., Phillips, A.J.L. and Baxter, A.P. (2000) *Phytopathogenic Fungi from South Africa*. University of Stellenbosch, Department of Plant Pathology Press, Stellenbosch, South Africa, p. 358.
Crutcher, F.K., Henry-Gregory, M.A., Wilkinson, H.H., Duke, S.E., Wheeler, T. *et al.* (2018) Characterization of *Sclerotinia minor* populations in Texas peanut fields. *Plant Pathology* 67(4), 839–847. DOI: 10.1111/ppa.12792.
Csüllög, K., Tóth, B., Lelesz, É.J., Fehér, M., Virág, C.I. *et al.* (2022) First report of *Sclerotinia sclerotiorum* on watercress (*Nasturtium officinale*) in an aquaponic system in Hungary. *Plant Disease* 106(2), 767. DOI: 10.1094/PDIS-07-21-1472-PDN.
Dahal, G., Amatya, P. and Manandhar, H. (1992) Plant diseases in Nepal. *Review of Plant Pathology* 71, 797–807.
Davidson, R.M., Reeves, P.A., Manosalva, P.M. and Leach, J.E. (2009) Germins: A diverse protein family important for crop improvement. *Plant Science* 177(6), 499–510. DOI: 10.1016/j.plantsci.2009.08.012.
Davies, J.M.L., Gladders, P., Young, C., Dyer, C., Hiron, L. *et al.* (1999) Petal culturing to forecast *Sclerotinia* stem rot in winter oilseed rape: 1993–1998. *Aspects of Applied Biology* 56, 129–134.
Davison, A.D. (1958) Plant diseases of economic crops occurring in Wyoming during 1958. *Plant Disease Reporter* 42, 1409–1410.
de Sousa, D.M.M. and Lucas, M.T. (1978) Fungi lusitaniae. *Agron Lusitaniae* 38, 285–295.
de Sousa, D.M.M. and Lucas, M.T. (1980) Fungi of Madeira and Selvagem Islands. *Boletim Da Sociedade Broteriana* 53, 469–476.
De Voogd, W.B., de Jong, M.D. and Nielsen, C. (2003) Use of *Sclerotinia sclerotiorum* as a mycoherbicide to control the spread of giant hogweed (*Heracleum mantegazzianum*) in the Netherlands and Denmark. *Biological Electronic News* 318, 1–3.
del Río, L.E., Bradley, C.A. and Johnson, B.L. (2005) First report of white mold caused by *Sclerotinia sclerotiorum* on echium (*Echium vulgare*). *Plant Disease* 89(6): 684. DOI: 10.1094/PD-89-0684C.
Deng, D., Sun, S., Du, C., Xiang, C., Long, J. *et al.* (2021) Three *Sclerotinia* species as the cause of white mold on pea in Chongqing and Sichuan of China. *Journal of Integrative Agriculture* 20(11), 2957–2965. DOI: 10.1016/S2095-3119(21)63629-8.
Dennis, R.W.G. (1956) A revision of the British Helotiaceae in the herbarium of the Royal Botanic Garden, Kew, with notes of related European species. *Mycology Paper* 62, 1–206.
Dennis, R.W.G. (1970) *Kew Bulletin Additional Series III. Fungus Flora of Venezuela and Adjacent Countries*. Verlag von J Cramer, Royal Botanic Gardens, Kew, UK, p. 531.
Dennis, R.W.G. (1978) *British Ascomycetes*. Vaduz, Germany, p. 585.
Dennis, R.W.G. (1986) *Fungi of the Hebrides*. Royal Botanical Gardens, Kew, UK, p. 383.

Derbyshire, M.C. and Denton-Giles, M. (2016) The control of Sclerotinia stem rot on oilseed rape (*Brassica napus*): Current practices and future opportunities. *Plant Pathology* 65(6), 859–877. DOI: 10.1111/ppa.12517.

Derbyshire, M.C., Newman, T.E., Khentry, Y. and Owolabi Taiwo, A. (2022) Pathogen profile: The evolutionary and molecular features of the broad-host-range plant pathogen *Sclerotinia sclerotiorum*. *Molecular Plant Pathology* 23(8), 1075–1090. DOI: 10.1111/mpp.13221.

Dhawan, S. (1980) Production of protease by *Sclerotinia sclerotiorum* cause white rot of *Brassica juncea*. *Current Science* 49(9), 361–362.

Dickson, J.G. (1930) Studies on *Sclerotinia sclerotiorum* (Lib) de Bary. PhD. thesis, Cornell University, Ithaca, New York.

Dillard, H.R. and Cobb, A.C. (1991) Velvetleaf, a new host for *Sclerotinia sclerotiorum*. *Plant Disease* 75(8), 863. DOI: 10.1094/PD-75-0863B.

Dillard, H.R. and Hunter, J.E. (1986) Association of common ragweed with Sclerotinia rot of cabbage in New York State. *Plant Disease* 70, 26–28. DOI: 10.1094/PD-70-26.

Ding, L.L., Zhao, S.F., Zhang, X.K., Yao, Z.Q. and Zhang, J. (2012) Sclerotinia rot of broomrape (*Orobanche cumana*) caused by *Sclerotinia sclerotiorum* in China. *Plant Disease* 96(6), 916. DOI: 10.1094/PDIS-09-11-0762-PDN.

Dingley, J.M. (1969) Host plants, pathogens, and physiological diseases. *New Zealand Department Science and Industrial Research Bulletin* 192, 10–87.

Dingley, J.M., Fullerton, R.A. and McKenzie, E.H.C. (1981) Survey of agricultural pests and diseases [Technical Report]. In: *Records of Fungi, Bacteria, Algae, and Angiosperms Pathogenic on Plants in Cook Islands*. FAO, Suva, p. 485.

Distribution Maps of Plant Diseases (DMPD). (2005a) *Distribution Maps of Plant Diseases*, CABI International, Sclerotinia sclerotiorum. [Distribution Map 971]. DOI: 10.1079/DMPD/20066500971.

Dowson, W.J. (1926) A blossom wilt and stem rot of cultivated antirrhinums and schizanthus due to *Sclerotinia sclerotiorum* (Lib.) Massee. *Journal of the Royal Horticultural Society* 51, 252–265.

Duarte, L.L. and Barreto, R.W. (2009) First report of stem rot of *Helichrysum bracteatum* by *Sclerotinia sclerotiorum* in Brazil. *Australasian Plant Disease Notes* 4(1), 100–101. DOI: 10.1071/DN0 9042.

Duczek, L.J. and Morrall, R.A.A. (1971) *Sclerotinia* in Saskatchewan in 1970. *Canadian Plant Disease Survey* 51, 116–221.

Dueck, J. and Frederick, S.S. (1983) Distribution of *Sclerotinia sclerotiorum* in western Canada as indicated by *Sclerotinia* levels in rapeseed unloaded in Vancouver, 1973–1981. *Canadian Plant Disease Survey* 63, 27–29.

Dutta, S., Ghosh, P.P. and Kuiry, S.P. (2009) Stem rot, a new disease of potato in West Bengal, India. *Australasian Plant Disease Notes* 4(1), 80–81. DOI: 10.1071/DN09034.

El-Buni, A.M. and Rattan, S. (1981) *Check List of Libyan Fungi*. Al Faateh University, Faculty of Sciences, Dept Botany, Tripoli, 169 pp.

El-Goorani, M.A. and Sommer, N.F. (1979) Suppression of postharvest plant pathogenic fungi by carbon monoxide. *Phytopathology* 69(834), 839. DOI: 10.1094/Phyto-69-834.

Ellett, C.W. (1957) Diseases not previously reported in Ohio. *Plant Disease Reporter* 41, 369–371.

El-Tobshy, Z.M. and Zayed, M.A. (1979) Post harvest Sclerotinia soft rot of squash and carrot. *Agriculture Research Review* 57, 47–56.

Embaby, E.M. (2007) *Pestalotia* fruit rot on strawberry plants in Egypt. *Egyptian Journal of Phytopathology* 35(2), 99–110.

Ericksson, J. (1880) Om klofverrotan med sarskildt afseende pa dess upptradante I vart land under aren 1878–1879. *Kgl Landtbruks Akad Handl Tidskr* 19, 28–42.

Erneberg, D., Strandberg, B. and Jensen, B.D. (2003) Susceptibility of a plant invader to a pathogenic fungus: An experimental study of *Heracleum mantegazzianum* (Giant Hogweed) and *Sclerotinia sclerotiorum*. In: Child, L.E., Brock, J.H., Brundu, G., Prach, K., Pysek, P. *et al.* (eds) *Plant Invasions: Ecological Threats and Management Solutions*. Bachhuys Publishing, Leiden, the Netherlands, pp. 355–372.

Erper, I., Ozer, G., Yildirim, E. and Turkkan, M. (2020) First report of white mold caused by *Sclerotinia minor* on safflower in Turkey. *Journal of Plant Pathology* 102, 977–978.

Farmer, L.J., Firstman, A. and Betz, J. (1971) A firm head rot of broccoli. *Plant Disease Reporter* 55, 1136.

Farr, M.L. and Stevenson, J.A. (1963) Eine erganzungsliste bolivianischer pilze. *Sydowia* 17, 37–69.

Farr, D.F., Bills, G.F., Chamuris, G.P. and Rossman, A.Y. (1989) *Fungi on Plants and Plant Products in the United States*. APS, St. Paul, Minnesota, p. 1252.

Ferrada, E.E., Diaz, G.A., Zoffoli, J.P. and Latorre, B.A. (2014) First report of blossom blight caused by *Sclerotinia sclerotiorum* on Japanese plum, nectarine, and sweet cherry orchards in Chile. *Plant Disease* 98(5), 695.
Ferraris, T. (1935) Foot rot of pelargoniums. *Riv Agric Roma* 31, 356–357.
Ficke, A., Grieu, C., Brurberg, M.B. and Brodal, G. (2018) The role of precipitation, and petal and leaf infections in Sclerotinia *stem* rot of spring oilseed *Brassica* crops in Norway. *European Journal of Plant Pathology* 152(4), 885–900. DOI: 10.1007/s10658-018-1569-6.
Ficker, A.L. (2019) *Sclerotinia sclerotiorum* impacts on host crops. MSc. thesis, Iowa State University, Ames, 35 pp.
Firman, I.D. (1972) A list of fungi and plant parasitic bacteria, viruses and nematodes in Fiji. *Phytopathology Paper* 15, 1–36.
Florea, A.-M., Gafencu, A.-M., Lipşa, F.-D., Gabur, I. and Ulea, E. (2023) A first report of *Sclerotinia sclerotiorum* causing forsythia twig blight in Romania. *Plants* 12(20), 3516. DOI: 10.3390/plants12203516.
Fogliata, G.M., Ramallo, N.V. and de Ploper, L.D. (1999) Shoot wilt in lemons caused by *Sclerotinia sclerotiorum* (Sclerotinia twig blight). *Advance Agroindustrial* 20, 28–30.
Foister, C.E. (1961) The economic plant diseases of Scotland. *Technical Bulletin, Department of Agriculture and Fisheries, Scotland* 1, 1–210.
Folk, G. and Tusnadin, C.K. (1985) A new disease of Gerbera in Hungary *Sclerotinia* wilt. *Novenyvedelem* 21, 557–561.
French, A.M. (1987) *California Plant Disease Host Index. Part 1: Fruit and Nuts*. California Department of Food and Agriculture, Sacramento, p. 39.
French, A.M. (1989) *California Plant Disease Host Index*. California Department of Food and Agriculture, Sacramento, p. 394.
Fries, E. (1872) *Scleromycetes Sueciae*. DOI: 10.5962/bhl.title.139654.
Fuckel, L. (1870) Symbolic mycologicae. Beitrage zur Kenntniss der Rheinischen Pilze. I Jhrb. *Nassauischen Vercins Naturk* 23, 1–459.
Gaetan, S. and Madia, M. (2005) Occurrence of stem rot on canola caused by *Sclerotinia sclerotiorum* in Argentina. *Plant Disease* 89, 530.
Gaetan, S., Madia, M. and Reyna, S. (1997) Aromatic Umbelliferae, hosts of *Sclerotinia sclerotiorum* (Lib.) de Bary. *Boletin de Sanidad Vegetal, Plagas* 23, 377–383.
Gao, R. (1981) Occurrence of Sclerotinia blight of peach in Fujian. *Acta Phytopathologica Sinica* 11, 54–57.
Garfinkel, A.R. (2021) First report of *Sclerotinia sclerotiorum* causing stem canker on *Cannabis sativa* in Oregon. *Plant Disease* 105(8), 2245.
Gargouri, S., Berraies, S., Gharbi, M.S., Paulitz, T., Murray, T.D. *et al.* (2017) Occurrence of Sclerotinia stem rot of fenugreek caused by *Sclerotinia trifoliorum* and S. *sclerotiorum* in Tunisia. *European Journal Forest Pathology* 149(3), 587–597.
Garibaldi, A. (1981) Principal fungus diseases of Anemone. *Colture Protette* 10, 43–48.
Garibaldi, A. and Gullino, G. (1976) Diseases of flower and ornamental plants new or little known in Italy. *III Root and Vascular Diseases*. *Agriculture Italy (Pisa)* 105, 273–284.
Garibaldi, A., Minuto, A., Gilardi, G. and Gullino, M.L. (2001a) First report of *Sclerotinia sclerotiorum* on *Gazania* sp. hybrid in Italy. *Plant Disease* 85(11), 1207. DOI: 10.1094/PDIS.2001.85.11.1207D.
Garibaldi, A., Minuto, A. and Gullino, M.L. (2001b) First report of *Sclerotinia sclerotiorum* on calendula officinalis in Italy. *Plant Disease* 85(4), 446. DOI: 10.1094/PDIS.2001.85.4.446C.
Garibaldi, A., Minuto, A., Pensa, P., Bruzzone, C., Salvi, D. *et al.* (2002a) *Sclerotinia sclerotiorum* agent of basal rot of Gazania, Campanula and Schizantus. *Colture Protette* 31, 105–107.
Garibaldi, A., Minuto, A. and Gullino, M.L. (2002b) First report of *Sclerotinia sclerotiorum* on *Campanula carpatica* and *Schizanthus* and *Wisetonensis* in Italy. *Plant Disease* 86:71.
Garibaldi, A., Minuto, A. and Gullino, M.L. (2003) First report of white mold caused by *Sclerotinia sclerotiorum* on Persian buttercup (*Ranunculus asiaticus)* in Italy. *Plant Disease* 87(9), 1151. DOI: 10.1094/PDIS.2003.87.9.1151A.
Garibaldi, A., Minuto, A. and Gullino, M.L. (2004a) First report of white mold caused by *Sclerotinia sclerotiorum* on blue marguerite (*Felicia amelloides*) in Italy. *Plant Disease* 88(5), 575. DOI: 10.1094/PDIS.2004.88.5.575B.
Garibaldi, A., Minuto, A. and Gullino, M.L. (2004b) First report of white mold caused by *Sclerotinia sclerotiorum* on *Hutchinsia alpina* in Italy. *Plant Disease* 88, 1045.

Garibaldi, A., Minuto, A. and Gullino, M.L. (2005a) First report of Sclerotinia stem rot and watery soft rot caused by *Sclerotinia sclerotiorum* on sand rocket (*Diplotaxis tenuifolia*) in Italy. *Plant Disease* 89(11), 1241–1241. DOI: 10.1094/PD-89-1241C.

Garibaldi, A., Minuto, A. and Gullino, M.L. (2005b) First report of white mold caused by *Sclerotinia sclerotiorum* on *Rosmarinus officinalis* "Prostratus" in Italy. *Plant Disease* 89(9), 1016. DOI: 10.1094/PD-89-1016A.

Garibaldi, A., Minuto, A. and Gullino, M.L. (2007a) First report of white mold caused by *Sclerotinia sclerotiorum* on *Iberis sempervirens* in Italy. *Plant Disease* 91(4), 464. DOI: 10.1094/PDIS-91-4-0464C.

Garibaldi, A., Minuto, A. and Gullino, M.L. (2007b) First report of white mold caused by *Sclerotinia sclerotiorum* on *Oreganum vulgare* and *Taraxacum officinale* in Italy. *Plant Disease* 91, 1360.

Garibaldi, A., Minuto, A. and Gullino, M.L. (2008a) First report of *Sclerotinia sclerotiorum* on *Calceolaria integrifolia* in Italy. *Plant Disease* 92(7), 1133. DOI: 10.1094/PDIS-92-7-1133C.

Garibaldi, A., Minuto, A., Pensa, P. and Gullino, M.L. (2008b) First report of *Sclerotinia sclerotiorum* on African daisy (*Osteospermum* sp.) in Italy. *Plant Disease* 92, 982.

Garibaldi, A., Pensa, P., Bertetti, D. and Gullino, M.L. (2008c) First report of white mold caused by *Sclerotinia sclerotiorum* on *Borago officinalis* in Italy. *Plant Disease* 92(12), 1711. DOI: 10.1094/PDIS-92-12-1711A.

Garibaldi, A., Pensa, P. and Gullino, M.L. (2008d) First report of *Sclerotinia sclerotiorum* on *Argyranthemum frutescens* in Italy. *Plant Disease* 92(8), 1250. DOI: 10.1094/PDIS-92-8-1250C.

Garibaldi, A., Pensa, P., Minuto, A. and Gullino, M.L. (2008e) First report of *Sclerotinia sclerotiorum* on *Lantana camara* in Italy. *Plant Disease* 92, 1369.

Garibaldi, A., Bertetti, D., Pensa, P. and Gullino, M.L. (2009) First report of Sclerotinia blight caused by *Sclerotinia sclerotiorum* on *Petunia × hybrida* in Italy. *Plant Disease* 93(12), 1353. DOI: 10.1094/PDIS-93-12-1353A.

Garibaldi, A., Bertetti, D., Poli, A. and Gullino, M.L. (2011) First report of Sclerotinia blight caused by *Sclerotinia sclerotiorum* on fan columbine (*Aquilegia flabellata*) in Italy. *Plant Disease* 95, 1481. DOI: 10.1094/PDIS-07-11-0542.

Garibaldi, A., Pensa, P., Bertetti, D., Poli, A. and Gullino, M.L. (2013) First report of Sclerotinia blight caused by *Sclerotinia sclerotiorum* on spearmint in Northern Italy. *Plant Disease* 97(10), 1384. DOI: 10.1094/PDIS-04-13-0398-PDN.

Garibaldi, A., Bartetti, D., Ortu, G. and Gullino, M.L. (2015) First report of Sclerotinia blight caused by *Sclerotinia sclerotiorum* on *Gaillardia x grandiflora* in Northern Italy. *Plant Disease* 99, 729.

Garibaldi, A., Barbetti, D., Pensa, P., Matic, S. and Gullino, M.L. (2017) First report of white mould caused by *Sclerotinia sclerotiorum* on rosemary in Italy. *Journal of Plant Pathology* 99(2), 543.

Garibaldi, A., Bertetti, D., Pensa, P., Matic, S. and Gullino, M.L. (2019) First report of white mold caused by *Sclerotinia sclerotiorum* on watercress (*Nasturtium officinale*) in Italy. *Plant Disease* 103(1), 152.

Garibaldi, A., Bertetti, D., Matic, S., Pensa, P. and Gullino, M.L. (2020) First report of white mold caused by *Sclerotinia sclerotiorum* on everlasting flower in Italy. *Plant Disease* 104(11), 3078.

Georghiou, G.P. and Papadopoulos, C. (1957) *A Second List of Cyprus Fungi*. Government of Cyprus, Department of Agriculture, Nicosia, Cyprus, p. 38.

Gernandt, D.S., Camacho, F.J. and Stone, J.K. (1997) *Meria laricis*, an anamorph of *Rhabdocline*. *Mycologia* 89, 735–744.

Ghasolia, R.P. and Shivpuri, A. (2004a) Neem—A new host of *Sclerotinia sclerotiorum*. *Journal of Mycology and Plant Pathology* 34, 200–201.

Ghasolia, R.P. and Shivpuri, A. (2004b) *Parthenium hysterophorus* L—A new host record for *Sclerotinia sclerotiorum* (Lib.) de Bary. *Journal of Mycology and Plant Pathology* 34, 242–243.

Ghasolia, R.P., Shivpuri, A. and Bhargava, A.K. (2004) Some common weed species as hosts for *Sclerotinia sclerotiorum*. *Journal of Mycology and Plant Pathology* 34, 670–671.

Gijzen, M. and Nürnberger, T. (2006) Nep1-like proteins from plant pathogens: Recruitment and diversification of the NPP1 domain across taxa. *Phytochemistry* 67(16), 1800–1807. DOI: 10.1016/j.phytochem.2005.12.008.

Gilardi, G., Gullino, M.L. and Garibaldi, A. (2013) New diseases of wild and cultivated rocket in Italy. *Acta Horticulturae* 1005, 569–572. DOI: 10.17660/ActaHortic.2013.1005.70.

Gilbert, R.G. (1991) Burning to reduce *Sclerotia* of *sclerotinia sclerotiorum* in alfalfa seed fields of southeastern Washington. *Plant Disease* 75, 141–142. DOI: 10.1094/PD-75-0141.

Gilbert, A.H. and Bennett, C.W. (1917) *Sclerotinia trifoliorum*, the cause of stem rot of clovers and alfalfa. *Phytopathology* 7, 432–442.

Gilbertson, R.L. and McHenry, J. (1969) Check list and host index for Arizona rust fungi. *University of Arizona Agricultural and Experimental Station Technical Bulletin* 186, 1–40.
Gill, H.S. (1975) An unusual outbreak of Whetzelinia rot on African Daisies in California. *Plant Disease Reporter* 59, 82–83.
Gilman, J.C. (1949) Second supplementary list of parasitic fungi from Iowa. *Iowa State College Journal of Science* 23(3), 261–272.
Ginns, J.H. (1986) *Compendium of Plant Disease and Decay Fungi in Canada 1960–1980*. Vol. 1816. Research Branch, Department of Agriculture, Ottawa, Canada, p. 416.
González, V. and Tello, M.L. (2011) The endophytic mycota associated with *Vitis vinifera* in central Spain. *Fungal Diversity* 47(1), 29–42. DOI: 10.1007/s13225-010-0073-x.
Goos, R.D. (2010) *The Mycota of Rhode Island: A Check List of the Fungi Recorded in Rhode Island (Including Lichens and Myxomycetes)*. Vol. 4. Biota of Rhode Island, Kingston, Rhode Island Natural History Survey, Kingston, RI, p. 222.
Goos, R.D. and Anderson, J.H. (1972) The meliolaceae of Hawaii. *Sydowia* 26, 73–80.
Gorter, G.J.M.A. (1977) Index of plant pathogens and the diseases they cause in cultivated plants in South Africa. *Republic South Africa, Department of Agriculture, Technical Survey of Plant Protection Research Institute Science Bulletin* 392, 1–177.
Gorter, G.J.M.A. (1981) Index of plant pathogens (II) and the diseases they cause in wild growing plants in South Africa. *Republic South Africa Department of Agriculture and Fisheries Science Bulletin* 398, 1–84.
Gossen, B.D. and Howard, R.J. (2021) Symptoms and distribution of blossom blight in alfalfa seed production on the Canadian prairies. *Canadian Journal of Plant Pathology* 43(3), 421–430. DOI: 10.1080/07060661.2020.1825015.
Gourley, C.O. (1982) An annotated index of the fungi of Nova Scotia. *Proceedings of the Nova Scotian Institute of Science* 32, 75–293.
Gourley, C.O. and Delbridge, R.W. (1972) *Sclerotinia sclerotiorum* on horse chestnut trees. *Canadian Journal of Plant Disease Survey* 52, 97–98.
Graf, F. and Schumacher, T. (1995) *Sclerotinia glacialis* sp. nv., from the alpine zone of Switzerland. *Mycological Research* 99, 113–117. DOI: 10.1016/S0953-7562(09)80324-9.
Granata, G. and Tirro, A. (1989) Serious damage by *Sclerotinia sclerotiorum* (Lib.) Massee on globe artichoke (*Cynara scolymus* L). *Difesa Delle Piante* 12, 127–131.
Grand, L.F. (1985) North Carolina plant disease index. *North Carolina Agricultural Research Survey Technical Bulletin* 240, 1–157.
Greuter, W., Poelt, J. and Raimondo, F.M. (1991) A check list of Sicilians fungi. *Bocconea* 2, 222.
Guan, Y.M., Ma, Y.Y., Zhang, L.L., Pan, X.X., Liu, N. *et al.* (2022) Occurrence of *Sclerotinia sclerotiorum* causing Sclerotinia root rot on American ginseng in Northeastern China. *Plant Disease* 106(5), 1518. DOI: 10.1094/PDIS-05-21-1115-PDN.
Gulya, T.J., Ooka, J.J. and Mancl, M.K. (1985) Diseases of cultivated sunflower in Hawaii. *Plant Disease* 69, 542. DOI: 10.1094/PD-69-542e.
Gulya, T.J., Woods, D.M., Bell, R. and Mancl, M.K. (1991) Diseases of sunflower in California. *Plant Disease* 75, 572–574. DOI: 10.1094/PD-75-0572.
Gulya, T.J., Gesch, R.W., Bradley, C.A., del Rio, L.E. and Johnson, B.L. (2006) First report of *Sclerotinia sclerotiorum* infection on *Cuphea*. *Plant Disease* 90, 1554. DOI: 10.1094/PD-90-1554A.
Gupta, J.S. (1963) Effect of soil temperature on the foot rot disease of coriander (*Sclerotinia sclerotiorum* (Lib.) de Bary). *Indian Phytopathology* 16, 211–213.
Gupta, S.K. and Agarwala, R.K. (1988) Biological control of Sclerotinia stalk rot of cauliflower. *Indian Journal of Plant Pathology* 6, 71–74.
Gupta, A.K., Bashyal, B.M., Choudhary, R., Kumar, M. and Solanki, I.S. (2015) First report of Sclerotinia rot of pigeon pea caused by *Sclerotinia sclerotiorum* (Lib) de Bary in India. *Canadian Journal of Plant Pathology* 37(4), 514–518. DOI: 10.1080/07060661.2015.1115782.
Hall, B.H., McMahon, R.L. and Wicks, T.J. (2002) First report of *Sclerotinia sclerotiorum* on grape (*Vitis vinifera*) in South Australia. *Australasian Plant Pathology* 31(4), 417. DOI: 10.1071/AP02045.
Han, K.S., Kim, J.Y., Park, J.H. and Shin, H.D. (2013) First report of Sclerotinia stem rot of anemone caused by *Sclerotinia sclerotiorum* in Korea. *Plant Disease* 97, 997. DOI: 10.1094/PDIS-12-12-1192-PDN.
Han, K.S., Park, M.J., Park, J.H., Cho, S.E. and Shin, H.D. (2015) First report of Sclerotinia stem rot of *Ranunculus asiaticus* caused by *Sclerotinia sclerotiorum* in Korea. *Plant Disease* 99(11), 1653–1654.

Han, I., Park, K., Lee, H., Lee, S.M., Shin, J. *et al.* (2020) First report of Sclerotinia rot in sword bean caused by *Sclerotinia sclerotiorum* in South Korea. *Plant Disease* 104(3), 988–988. DOI: 10.1094/PDIS-09-19-1840-PDN.

Han, F., Huang, J., Lin, M. and Song, X. (2022) First report of *Sclerotinia sclerotiorum* on *Atractylodes lancea* in China. *Journal of Plant Pathology* 104, 1557. DOI: 10.1007/s42161-022-01190-3.

Hanif, S., Hafeez, R., Akram, W., Ashfaq, M. and Ali, A. (2016) First report of Sclerotinia fruit rot of *Citrus paradisi* caused by *Sclerotinia sclerotiorum* in Pakistan. *Plant Disease* 100(4), 863.

Hanlin, R.T. (1963) A revision of the Ascomycetes of Georgia. *Georgia Agricultural Experiment Station Athens, GA. Mimeograph Series* 175, 1–65.

Hansda, S., Ray, S.K., Dutta, S. and Khatua, D.C. (2014) Sclerotinia rot in West Bengal. *Journal of Mycopathological Research* 52(2), 273–278.

Harter, L.L. and Zaumeyer, W.J. (1944) A monographic study of bean diseases and methods for their control. *Technical Bulletin US Department of Agriculture* 868, 160.

Haware, M.P. (1988) Fusarium wilt and other important diseases of chickpea in the Mediterranean area. *Options Mediterraneennes Serie A: Seminaires Mediterraneens* 9, 61–64.

Held, V.M. and Haenseler, C.M. (1953) Cross inoculation with New Jersey isolates of *Sclerotinia sclerotiorum, S. Minor* and *S. Trifoliorum*. *Plant Disease Reporter* 37, 515–517.

Hellman, E.M., Koivunen, E.E. and Swett, C.L. (2022) First report of Sclerotinia fruit and crown rot caused by *Sclerotinia sclerotiorum* on short-day strawberry in Maryland. *Plant Disease* 106(2), P764. DOI: 10.1094/PDIS-07-21-1357-PDN.

Hilton, S. (2000) Disease highlights, Canadian plant disease survey, agriculture and agri-food Canada. *CPDS* 80, 1–151.

Hockey, J.F. (1959) Calyx-end rot of apples. *Plant Disease Reporter* 43, 804–805.

Hoes, J.A. and Huang, H.C. (1976) Importance of disease to sunflower in Manitoba in (1975). *Canadian Plant Disease Survey* 76, 75–76.

Holcomb, G.E. (1990) *Apios americana* and *Trillium foetidissimum* as new hosts of *Sclerotinia sclerotiorum*. *Plant Disease* 74, 720. DOI: 10.1094/PD-74-0720B.

Holcomb, G.E. (1991) Severe disease outbreak in red *Trillium wildflower* populations. *Proceedings of Louisiana Academic Science* 54, 1–4.

Holcomb, G.E. (2001) First report of occurrence of Sclerotinia blight on Petunia in Louisiana. *Plant Disease* 85(1), 95. DOI: 10.1094/PDIS.2001.85.1.95C.

Holcomb, G.E. (2005) First report of Sclerotinia stem rot and death of *Osteospermum* spp. hybrid cultivars caused by *Sclerotinia sclerotiorum* in Louisiana. *Plant Disease* 89(8), 911. DOI: 10.1094/PD-89-0911A.

Holcomb, G.E. (2006) First report of crown rot of *Gazania rigens* caused by *Sclerotinia sclerotiorum* in Louisiana. *Plant Disease* 90, 1114. DOI: 10.1094/PD-90-1114A.

Holcomb, G.E. and Newman, B.E. (1970) *Alternaria circinans* and other fungal pathogens on *Crambe abyssinica* in Louisiana. *Plant Disease Reporter* 54, 28.

Holcomb, G.E. and Reed, M.J. (1994) Stem rot of Basil caused by *Sclerotinia sclerotiorum*. *Plant Disease* 78, 924. DOI: 10.1094/PD-78-0924E.

Holevas, C.D., Chitzanidis, A., Pappas, A.C., Tzamos, E.C., Elena, K. *et al.* (2000) Disease agents of cultivated plants observed in Greece from 1981 to 1990. *Benaki Phytopathology Institute Kiphissia, Athens* 19, 1–96.

Holley, R.C. and Nelson, B.D. (1986) Effect of plant population and inoculum density on incidence of Sclerotinia wilt of sunflower. *Phytopathology* 76, 71–74. DOI: 10.1094/Phyto-76-71.

Holst-Jensen, A., Kohn, L.M., Jakobsen, K.S. and Schumacher, T. (1997a) Molecular phylogeny and evolution of *Monilinia* (*Sclerotiniaceae*) based on coding and noncoding rDNA sequences. *American Journal of Botany* 84, 686–701.

Holst-Jensen, A., Kohn, L.M. and Schumacher, T. (1997b) Nuclear rDNA phylogeny of the *Sclerotinia*. *Mycologia* 89, 885–899. DOI: 10.1080/00275514.1997.12026859.

Holst-Jensen, J.A., Vaage, M. and Schumacher, T. (1998) An approximation to the phylogeny of *Sclerotinia* and related genera. *Nordic Journal of Botany* 18(6), 705–719. DOI: 10.1111/j.1756-1051.1998.tb01553.x.

Holst-Jensen, A., Vrålstad, T. and Schumacher, T. (2004) *Kohninia linnaeicola*, a new genus and species of the *Sclerotinia* pathogenic to *Linnaea borealis*. *Mycologia* 96, 135–142.

Horie, H. and Hoshi, H. (2002) First report of Sclerotinia rot of hybrid bouvardia in Japan. *Annual Report Kanto Tosan Plant Protection Society* 49, 69–71.

Houtman, G. (1950) On a pathogenic fungus: *Sclerotinia sclerotiorum*. *Cult En Handel* 16, 370–372.

Hsieh, H. (1983) Study on paulownia diseases found in Taiwan. *Bulletin of Taiwan Forest Research Institute* 388, 14–24.

Hu, J., Handique, U. and Norton, E.R. (2018) First report of Sclerotinia boll rot and stem blight of cotton in Arizona. *Plant Disease* 102(8), 1663. DOI: 10.1094/PDIS-12-17-1897-PDN.

Hua, L., Yong, C., Zhanquan, Z., Boqiang, L., Guozheng, Q. *et al*. (2018) Pathogenic mechanisms and control strategies of *Botrytis cinerea* causing post-harvest decay in fruits and vegetables. *Food Quality and Safety* 2, 111–119. DOI: 10.1093/fqsafe/fyy016.

Huang, H.C. (1985) Factors affecting myceliogenic germination of sclerotia of *Sclerotinia sclerotiorum*. *Phytopathology* 75(4), 433. DOI: 10.1094/Phyto-75-433.

Huang, H.C., Erickson, R.S., Van Hezewijk, B. and Clerck, F.R. de (2005) White mold of houndstongue (*Cynoglossum officinale*) caused by *Sclerotinia sclerotiorum* in Canada. *Plant Disease* 89, 1013. DOI: 10.1094/PD-89-1013C.

Hunter, J.E., Dickson, M.H. and Cigna, J.A. (1981) Limited-term inoculation: A method to screen bean plants for partial resistance to white mold. *Plant Disease* 65, 414. DOI: 10.1094/PD-65-414.

Hwang, S.F., Howard, R.J., Chang, K.F., Park, B. and Burnett, P.A. (1994) Etiology and severity of Fusarium root rot of lentil in Alberta. *Canadian Journal of Plant Pathology* 16(4), 295–303. DOI: 10.1080/07060669409500734.

Hysek, J., Tempirova, Z., Vlk, J. and Curiova, A. (1986) Mycoflora on stored cereal seed. *Rostinna Vywheatroba* 32, 963–970.

Iqbal, S.M., Mirza, M.S., Bakhsh, A. and Mahmood, T. (1998) Sclerotinia rot of peas (*Pisum sativum* L.) in Pakistan. *Sarhad Journal of Agriculture* 14, 157–159.

Isakeit, T., Woodward, J.E., Niu, C. and Wright, R.J. (2010) First report of Sclerotinia stem rot of canola caused by *Sclerotinia sclerotiorum* in Texas. *Plant Disease* 94, 792. DOI: 10.1094/PDIS-94-6-0792B.

Islam, M.M., E-Elahi, F. and Khatun, F. (2019) First report of white mold caused by *Sclerotinia sclerotiorum* in red salvia (*Salvia splendens*) in Bangladesh. *Plant Disease* 103(11), 2955.

Islam, M.R., Prova, A., Akanda, A.M. and Hossain, M.M. (2020) First report of white mould caused by *Sclerotinia sclerotiorum* on pea in Bangladesh. *Journal of Plant Pathology* 102(3), 941–941. DOI: 10.1007/s42161-020-00518-1.

Ivancia, V. (1993) A nona planta gazda pentric cruperca *Sclerotinia sclerotiorum* (Lib) de bary, in Romania. *Probleme de Protectia Plantelar* 21, 35–36.

Jackson, N. and Campanini, J.T. (1974) *Whetzelinia sclerotinia* the cause of a dieback disease on *Euomymus alatus*. *Proceeding of American Phytopathological Society* 1, 137.

Jackson, N. and Campanini, J.T. (1975) A dieback of *Euomymus alatus* caused by *Whetzelinia sclerotiorum*. *Plant Disease Reporter* 59, 310–311.

Jacob, M. (1980) Phytosanitary measures required for high yields in carnation production. *Horticultural Science Abstracts* 51, 6310.

Jagger, I.C. (1913) The small lettuce *Sclerotinia* an undescribed species. *Phytopathology* 3, 74.

Jagger, I.C. (1920) *Sclerotinia minor* n. sp., the cause of a decay of lettuce, celery and other crops. *Journal of Agricultural Research* 20, 331–334.

Jayawardena, R.S., Purahong, W., Zhang, W., Wubet, T., Li, X. *et al*. (2018) Biodiversity of fungi on *Vitis vinifera* L. revealed by traditional and high-resolution culture-independent approaches. *Fungal Diversity* 90(1), 1–84. DOI: 10.1007/s13225-018-0398-4.

Jeffers, W.F., Cox, C.E., Jehle, R.A., Weaver, L.O., Morgan, O.D. *et al*. (1951) Notes on occurrence of some plant diseases in Maryland during 1950. *Plant Disease Reporter* 35, 385–386.

Jellis, G.J., Smith, D.B. and Scott, E.S. (1990) Identification of *Sclerotinia* spp. on *Vicia faba*. *Mycological Research* 94(3), 407–409. DOI: 10.1016/S0953-7562(09)80366-3.

Johnson, E.M. and Chapman, R.A. (1958) Unusual occurrence of certain plant diseases in Kentucky in 1958. *Plant Disease Reporter* 42, 1411–1413.

Johnson, S.B., Kleinhenz, M.D., Dwyer, J.D., Griffin, T., Olanya, O.M. *et al*. (1998) First report of white mold (*Sclerotinia sclerotiorum*) on soybean in Maine. *Plant Disease* 82, 832. DOI: 10.1094/PDIS.1998.82.7.832B.

Jones, E.S. (1923) Taxonomy of the *Sclerotinia* on *Helianthus annuus* L. *Phytopathology* 13, 496.

Jons, V.L., Venette, J.R., Lamey, H.A. and Mc Mullen, M.P. (1996) First report of *Sclerotinia sclerotiorum* on Salvia in the United States. *Plant Disease* 80, 1302. DOI: 10.1094/PD-80-1302C.

Jorstad, I. (1963) Icelandic parasitic fungi apart from Uredinales. *Skr Norske Vidensk-Akad Oslo, Mat-Naturvidensk Kl* 10, 1–72.

Joshi, S.D. (1924) The wilt disease of saffloweres. *Indian Botany Services, Department of Agriculture and Memories* 13, 39–46.

Joshi, V. and Elmhirst, J.F. (1998) Diseases diagnosed on commercial crops in British Columbia. *Canadian Plant Disease Survey* 78, 4–16.

Jovaisiene, Z. and Strukeinskas, M. (1990) Pathogenic micromycetes on umbelliferous plants (*Apiaceae lindl*) in Lithuania. *Ekologija* 4, 84–89.

Jurkovic, D. and Culek, M. (1997) Abutilon theophrasti Medik- a new host for *Sclerotinia sclerotiorum* (Lib) de Bary in Croatia. *Acta Phytopathologica et Entomologica Hungarica* 32, 307–312.

Kadow, K.J. and Anderson, H.W. (1940) A study of horseradish diseases and their control. *University Illinois Agricultural Experiment Station Bulletin* 469, 531–583.

Kagiwata, T. (1990) Fruit rot of passion fruit [*Passiflora edulis*] caused by *Botrytis cineraria* and *Sclerotinia sclerotiorum*. *Japanese Journal of Tropical Agriculture* 34, 35–39.

Kaiser, W.J. (1992) Fungi associated with the seeds of commercial lentils from the U.S. Pacific Northwest. *Plant Disease* 76(605), 610. DOI: 10.1094/PD-76-0605.

Kamran, M., Ehetisham Ul Haq, M., Ullah, I., Ali, S., Idrees, M. *et al.* (2019) First report of *Sclerotinia sclerotiorum* causing stem rot of eggplant (*Solanum melongena*) in Pakistan. *Plant Disease* 103(3), 589. DOI: 10.1094/PDIS-06-18-1009-PDN.

Kanno, H. and Ohkubo, H. (1999) *Sclerotinia* rot of Malabar nightshade (*Basella rubra* L.), *Lapsana apogonoides* Maxim., Bishop's weed (*Ammi majus* L.), blue lace flower (*Didiscus caeruleus* DC.) and *Portulaca oleracea* L caused by *Sclerotinia sclerotiorum*. *Annual Report of the Society of Plant Protection of North Japan, Miyagi Japan* 50, 115–119. DOI: 10.11455/kitanihon1966.1999.115.

Kapoor, K.S. (1988) *Sclerotinia sclerotiorum*—A threat to seed production of eggplant. *Capsicum Newsletter* 7:92.

Kapoor, K.S., Gill, H.S. and Sharma, S.R. (1986) A technique for artificial inoculation of curd stage of cauliflower with *Sclerotinia sclerotiorum*. *Cruciferae Newsletter* 11, 113.

Karpunin, F.M. (1981) Decrease in flax infection by diseases. *Zashch Rast (Mosc)* 2, 30.

Kaur, S. and Chohan, J.S. (1977) Additions to the host-range of *Sclerotinia sclerotiorum* (Lib.) de Bary. *Punjab Journal of Research* 15, 131–132.

Kawakubo, Y. and Takeshita, S. (1984) Sclerotinia rot of netted melon caused by *Sclerotinia sclerotiorum* (Libert) de Bary. *Proceeding of the Association for Plant Protection of Hokuriku* 32, 101–104.

Keay, M.A. (1939) A study of certain species of the genus *Sclerotinia*. *Annals of Applied Biology* 26, 227–246. DOI: 10.1111/j.1744-7348.1939.tb06968.x.

Kelly, H.M., Srivastava, P., Paret, M.L., Dankers, H., Wright, D.L. *et al.* (2012) First report of Sclerotinia stem rot caused by *Sclerotinia sclerotiorum* on *Brassica Carinata* in Florida. *Plant Disease* 96(10), 1581–1581. DOI: 10.1094/PDIS-06-12-0525-PDN.

Khan, M.F.R., Bhuiyan, M.Z.R., Chittem, K., Shahoveisi, F., Haque, M.E. *et al.* (2020) First report of *Sclerotinia sclerotiorum* causing leaf blight in sugar beet (*Beta vulgaris*) in North Dakota, USA. *Plant Disease* 104(4), 1258. DOI: 10.1094/PDIS-11-19-2304-PDN.

Khan, M.F.R., Bhuiyan, Z.R., Liu, Y., Lakshman, D. and Bloomquist, M. (2021) First report of leaf blight of sugar beet (*Beta vulgaris*) caused by *Sclerotinia sclerotiorum* in Minnesota, USA. *Plant Health Progress* 22(2), 149–150. DOI: 10.1094/PHP-02-21-0048-BR.

Khare, M.N. (1991) Lentil diseases with special reference to seed quality. *Indian Journal of Mycology Plant Pathology* 21, 1–13.

Khatua, D.C., Mondal, B. and Hansda, S. (2014) White rot of *Coreopsis drummondii*—A new record from West Bengal India. *Journal of Mycopathological Research* 52(1), 157–158.

Kim, J.Y., Aktaruzzaman, M.D., Afroz, T., Hahm, Y.I. and Kim, B.S. (2014) The first report of postharvest stem rot of kohlrabi caused by *Sclerotinia sclerotiorum* in Korea. *Mycobiology* 42(4), 409–411. DOI: 10.5941/MYCO.2014.42.4.409.

Kim, K.M., Won, G.Y. and Choi, I.Y. (2016) First report of Sclerotinia stem rot of marshmallow caused by *Sclerotinia sclerotiorum* in Korea. *Plant Disease* 100(4), 858. DOI: 10.1094/PDIS-07-15-0769-PDN.

Kinghorn, W.J. (1947) *Report of the Director of Agriculture, 1946*. Department of Agriculture and Fisheries, Bermuda, pp. 1–14.

Klisiewicz, J.M. (1986) Susceptibility of yellow starthistle to selected plant pathogens. *Plant Disease* 70(4), 295. DOI: 10.1094/PD-70-295.

Klisiewicz, J.M. and Beard, B.H. (1976) Diseases of sunflower in California. *Plant Disease Reporter* 60, 298–301.

Kljajic, R. (1953) Contribution to the study of the parasitic flora of medicinal plants during 1952. *Review of Research, Faculty of Agriculture, University of Belgrade* 1, 185–200.

Kobayashi, T. (1957) Another Sclerotinia disease of Japanese Cedar, *Cryptomeria japonica* D. Don. caused by *Sclerotinia libertiana* Funckel. *Bulletin of Government Forestry Experiment Station* 96, 1–16.

Kobayashi, T. (2007) Index of fungi inhabiting woody plants in Japan, Host, Distribution and Literature. Zenkoku-Noson-Kyoiku Kyokai Publishing, Tokyo, Japan, p. 1227.

Koehler, A. and Shew, H. (2014) First report of stem rot of stevia caused by *Sclerotinia sclerotiorum* in North Carolina. *Plant Disease* 98(10), 1433. DOI: 10.1094/PDIS-03-14-0307-PDN.

Kohn, L.M. (1979a) A monographic revision of the genus *Sclerotinia*. *Mycotaxon* 9, 365–444.

Kohn, L.M. (1979b) Delimitation of the economically important plant pathogenic *Sclerotinia* species. *Phytopathology* 69(8), 881. DOI: 10.1094/Phyto-69-881.

Kohn, L.M., Petsche, D.M., Bailey, S.R., Novak, L.A. and Anderson, J.B. (1988) Restriction fragment length polymorphisms in nuclear and mitochondrial DNA of *Sclerotinia* species. *Phytopathology* 78(8), 1047. DOI: 10.1094/Phyto-78-1047.

Koike, S.T. (1997) Occurrence of stem and crown rot of *Gaillardia grandiflora*, caused by *Sclerotinia sclerotiorum*, in California. *Plant Disease* 81, 1334. DOI: 10.1094/PDIS.1997.81.11.1334D.

Koike, S.T. (1999) Stem and crown rot of chervil, caused by *Sclerotinia sclerotiorum*, in California. *Plant Disease* 83, 1177. DOI: 10.1094/PDIS.1999.83.12.1177A.

Koike, S.T. (2000) Occurrence of stem rot of basil, caused by *Sclerotinia sclerotiorum*, in coastal California. *Plant Disease* 84(12), 1342. DOI: 10.1094/PDIS.2000.84.12.1342B.

Koike, S.T., Daugovish, O. and Downer, J.A. (2006) *Sclerotinia* petiole and crown rot of celery caused by *Sclerotinia minor* in California. *Plant Disease* 90, 829. DOI: 10.1094/PD-90-0829A.

Korf, R.P. and Dumont, K.P. (1972) Whetzelinia, a new generic name for *Sclerotinia sclerotiorum* and *Sclerotinia tuberosa*. *Mycologia* 64, 248–251. DOI: 10.1080/00275514.1972.12019259.

Kowalik, M., Bonio, J. and Duda-Franiak, K. (2015) Micromycetes on ericaceous plant leaves. *Acta Mycologica* 50(1), 1055. DOI: 10.5586/am.1055.

Kreitlow, K.W. (1949) *Sclerotinia trifoliorum,* a pathogen of ladino clover. *Phytopathology* 39, 158–166.

Kropf, S.M., Putnam, M.L., Serdani, M., Twomey, M.C., Woods, J.L. *et al.* (2012) Sclerotinia wilt of hop (*Humulus lupulus*) caused by *Sclerotinia sclerotiorum* in the Pacific Northwest United States. *Plant Disease* 96(4), 583. DOI: 10.1094/PDIS-12-11-1039.

Kumar, P., Sharma, S., Singh, R., Singh, P. and Kumar, A. (2022) First report of *Sclerotinia sclerotiorum* causing white rot of marigold in Punjab, India. *Journal of Plant Pathology* 104(1), 435. DOI: 10.1007/s42161-021-00996-x.

Kurt, S., Uysal, A., Kara, M., Soylu, S. and Soylu, E.M. (2017a) First report of stem rot disease of parsley caused by *Sclerotinia sclerotiorum* in Turkey. *Journal of Plant Pathology* 99(1), 301.

Kurt, Ş., Uysal, A., Kara, M., Soylu, S. and Soylu, E.M. (2017b) Natural infection of potato by *Sclerotinia sclerotiorum* causing stem rot disease in Turkey. *Australasian Plant Disease Notes* 12, 39. DOI: 10.1007/s13314-017-0266-1.

Kusch, S., Larrouy, J., Ibrahim, H.M.M., Mounichetty, S., Gasset, N. *et al.* (2022) Transcriptional response to host chemical cues underpins the expansion of host range in a fungal plant pathogen lineage. *The ISME Journal* 16, 138–148. DOI: 10.1038/s41396-021-01058-x.

Kwon, J.H. and Park, C.S. (2002) Occurrence of Sclerotinia rot of *Leonurus sibiricus* caused by *Sclerotinia sclerotiorum*. *Research in Plant Disease* 8, 113–116. DOI: 10.5423/RPD.2002.8.2.113.

Lacicowa, B., Filipowicz, A. and Wagner, A. (1979) Fungi pathogenic to *Zinnia elegans* L. *Acta Mycology* 15, 11–20.

Lahoz, E., Caiazzo, R., Carella, A. and Cozzolino, E. (2007) First report of *Sclerotinia sclerotiorum* on buckwheat (*Fagopyrum esculentum*) in Italy. *Plant Disease* 91, 1519. DOI: 10.1094/PDIS-91-11-1519C.

Latorre, B.A. and Guerrero, M.J. (2001) First report of shoot blight of grapevine caused by *Sclerotinia sclerotiorum* in Chile. *Plant Disease* 85(10), 1122. DOI: 10.1094/PDIS.2001.85.10.1122C.

Latysheve, T.A. and Sevryuk, N.I. (1981) Diseases of Ginseng. *Zashch. Rast. (Mosco)* 11, 23–24.

Laughlin, J.C. and Munro, D. (1983) The effect of Sclerotinia stem infection on morphine production and distribution in poppy (*Papaver somniferum* L.) plants. *The Journal of Agricultural Science* 100, 299–303.

Lee, Y.H., Cho, W.D., Kim, W.K., Jin, K.S. and Lee, E.J. (1991) Report on host-unrecorded diseases identified from economical crops in Korea. *Research Report on Rural Development Administration* 33, 15–19.

Lee, J.H., Kwon, Y.H., Kwack, Y.B. and Kwak, Y.S. (2015) Report of postharvest rot of kiwifruit in Korea caused by *Sclerotinia sclerotiorum*. *International Journal of Food Microbiology* 206, 81–83. DOI: 10.1016/j.ijfoodmicro.2015.05.001.

Leiteritz, R. and Roder, W. (1980) On important plant production methods in expanded cultivation of opium poppy in the Dresden district. *Nachrichtenbl. Pflanzenschutz DDR* 34(8), 166–169.

Lenne, J.M. (1990) World list of fungal diseases of tropical pasture species. *Phytopathology Paper* 31, 1–162.

Leppik, E.E. (1965) *Plasmopara halstedii* and other diseases on *Dimorphotheca*. *Plant Disease Reporter* 49, 940–942.

Leyronas, C., Troulet, C., Duffaud, M., Villeneuve, F., Benigni, M. *et al*. (2018) First report of *Sclerotinia subarctica* in France detected with a rapid PCR-based test. *Canadian Journal of Plant Pathology* 40(2), 248–253. DOI: 10.1080/07060661.2018.1438515.

Li, H. and Fu, C. (1981) Studies on the Sclerotinia rot of soybean. *Acta Phytopathological Sinica* 11, 19–24.

Li, G.Q., Wang, D.B., Jiang, D.H., Huang, H.C. and Laroche, A. (2000) First report of *Sclerotinia navalis* on lettuce in central China. *Mycological Research* 104, 232–237.

Libert, M.A. (1837) *Plante Crytogamicae Arduennae* (Exsiccate) no. 326.

Lin, X., Li, Y., Liao, Q., Lu, G., Yuan, G. *et al*. (2024) First report of *Sclerotinia sclerotiorum* causing *Hibiscus rosa-sinensis* stem rot in Guangxi, China. *Plant Disease* 108. DOI: 10.1094/PDIS-07-23-1466-PDN.

Lithourgidis, A.S., Tzavella-Klonari, K. and Roupakias, D.G. (1989) Methods of inoculation of faba bean plants with *Sclerotinia sclerotiorum*. *Journal of Phytopathology* 127, 123–128. DOI: 10.1111/j.1439-0434.1989.tb01120.x.

Liu, L. and Li, X. (1988) The geographical distribution of sunflower diseases in China. *Plant Disease Reporter* 34, 186–187. DOI: 10.1111/j.1365-3059.1988.tb02104.x.

Liu, X.M., Liu, S.Y. and Xi, J.W. (2002) Studies on the host range of *Sclerotinia sclerotiorum* from *Cleome spinosa*. *Journal of Jilin Agricultural University* 24, 91–93.

Liu, Q.L., Li, G.Q., Li, J.Q. and Chen, S.F. (2016) *Botrytis eucalypti,* a novel species isolated from diseased *Eucalyptus* seedlings in South China. *Mycological Progress* 15, 1057–1079. DOI: 10.1007/s11557-016-1229-1.

Liu, Y.M., Zhong, S., Wang, R., Wei, J.H., Qiao, X. *et al*. (2023) First report of Sclerotinia rot caused by *Sclerotinia sclerotiorum* on *Artemisia capillaris* in China. *Plant Disease*. DOI: 10.1094/PDIS-09-23-1748-PDN.

Livingstone, D.M., Hampton, J.L., Phipps, P.M. and Grabau, E.A. (2005) Enhancing resistance to *Sclerotinia minor* in peanut by expressing a barley oxalate oxidase gene. *Plant Physiology* 137, 1354. DOI: 10.1104/pp.104.057232.

Lockwood, J.L., Zeeuw, D.J., Andersen, A.L. and Hagedorn, D.J. (1957) Pea diseases in Michigan, 1955 and 1956. *Plant Disease Reporter* 41, 478–480.

Logsdon, C.E. and Strobel, G. (1960) Additional records of vegetable diseases in Alaska. *Plant Disease Reporter* 44, 92–93.

Lu, B., Hyde, K.D., Ho, W.H., Tsui, K.M., Taylor, J.E. *et al*. (2000) *Check list of Hong Kong Fungi*. Fungal Diversity Press, Hong Kong, p. 207.

Lu, C.H., Liu, J.Y., Lin, Z.L., Zhen, A.Z., Xia, Z.Y. *et al*. (2020a) First report of Sclerotinia blight caused by *Sclerotinia sclerotiorum* on oriental tobacco in the Yunnan Province of China. *Plant Disease* 104(6), 1867.

Lu, P.K., Lin, R.J., Hsieh, C.M., Chen, Y.H. and Hsieh, W.T. (2020b) First report of Sclerotinia blight caused by *Sclerotinia sclerotiorum* on *Salvia miltiorrhiza* in Taiwan. *Plant Disease* 104(9), 2523. DOI: 10.1094/PDIS-09-19-1878-PDN.

Luong, T.M., Huynh, L.M.T., Le, T.V., Burgess, L.W. and Phan, H.T. (2010) First report of Sclerotinia blight caused by *Sclerotinia sclerotiorum* in Quang Nam, Vietnam. *Australasian Plant Disease Notes* 5(42), 44. DOI: 10.1071/DN10016.

Mahalingam, T., Guruge, B.M.A., Somachandra, K.P., Rajapakse, C.S. and Attanayake, R.N. (2017) First report of white mold caused by *Sclerotinia sclerotiorum* on cabbage in Sri Lanka. *Plant Disease* 101(1), 249. DOI: 10.1094/PDIS-05-16-0693-PDN.

Mahdikhani, M. and Aghaalikhani, A. (2017) First report of *Sclerotinia sclerotiorum* causing Sclerotinia rot on *Lens culinaris* in west of Iran. *Journal of Plant Pathology* 99(1), 296.

Maiti, S. (1983) Sclerotinia foot rot of *Stylosanthes* in India. *Indian Journal of Mycology and Plant Pathology* 12, 224.

Marciano, P., Di, Lenna, P. and Magro, P. (1982) Polygalacturonase isoenzymes produced by *Sclerotinia sclerotiorum in vivo* and *in vitro*. *Physiological Plant Pathology* 20, 210–212.
Marcinkowska, J. (1997) Micromycetes on *Pisum sativum* var. *arvense*. *Acta Mycology* 32, 30–39.
Marin, M.V. and Peres, N.A. (2020) First report of *Sclerotinia sclerotiorum* causing strawberry fruit rot in Florida. *Plant Disease* 104(12), 3250.
Marinelli, A., March, G.J., Rago, A. and Giuggia, J. (1998) Assessment of crop loss in peanut caused by *Sclerotinia sclerotiorum S. minor* and *Sclerotinia rolfsii* in Argentina. *International Journal of Pest Management* 44, 251–254.
Marras, F., Foddai, A. and Corda, P. (1985) Protection of globe artichoke from cryptogamic diseases. *Informatore Fitopatologico* 20, 210–212.
Martin, R.A. and Arsenault, W.J. (1987) Prevalence and severity of Sclerotinia stalk rot of tobacco on Prince Edward Island 1985 and 1986. *Canadian Plant Disease Survey* 67, 41–43.
Martinez, de la Parte, E., Trujillo, M., Cantillo-Perez, T. and Garcia, D. (2013) First report of white mould of beans caused by *Sclerotinia sclerotiorum* in Cuba. *New Disease Reports* 27, 5.
Masalab, N.A. (1938) Diseases of medicinal and of some industrial plants by species of Sclerotinia. *Sov Botany* 6, 41–43.
Matheron, M.E. and Matejka, J.C. (1994) First report of *Sclerotinia sclerotiorum* on African Daisy in Arizona. *Plant Disease* 78, 924.
Matheron, M.E. and Porchas, M. (2000) First report of stem and crown rot of garbanzo caused by *Sclerotinia minor* in the United States and by *Sclerotinia sclerotiorum* in Arizona. *Plant Disease* 84, 1250.
Maykuss, F. (1989) Plant protection in irrigated potatoes. *Kartoffelbau* 40(5), 168–170.
McGuire, J.U., Jr. and Crandall, B.S. (1967) Survey of insect pests and plant diseases of selected food crops of Mexico, Central America and Panama. In: *International Agricultural Development Service*. Vol. 2. AID, ARS, USDA, Washington, DC, p. 157.
McKenzie, E.H.C. and Dingley, J.M. (1996) New plant disease records in New Zealand: Miscellaneous fungal pathogens III. *New Zealand Journal of Botany* 34, 263–272.
McKenzie, E.H.C. and Johnston, P.R. (1999) New records of phytopathogenic fungi in the Chatham Islands, New Zealand. *Australasian Plant Pathology* 28, 131–138.
McKenzie, D.L. and Morrall, R.A.A. (1975) Faba bean diseases in Saskatchewan in 1973. *Canadian Plant Disease Survey* 55, 1–7.
McLean, D.M. (1951) Sclerotinia stem rot and *Ramularia* leaf spot of dandelion grown for seed in Skagit County, Washington. *Plant Disease Reporter* 35, 162–163.
McLean, D.M. (1957) Sclerotinia on strawberry fruit in northwest Washington. *Plant Disease Reporter* 41, 1057.
Mederick, F.M. and Piening, L.J. (1982) *Sclerotinia sclerotiorum* on oil and fibre flax in Alberta. *Canadian Plant Disease Survey* 62, 11.
Meena, P.D., Rai, P.K. and Saharan, G.S. (2022) Geographical distribution and severity of rapeseed-mustard diseases under changing climatic scenario in India. ICAR-Directorate of Rapeseed-Mustard, Research, Bharatpur, India, p. 44.
Mehta, N., Meena, P.D. and Saharan, G.S. (2023). Geographical distribution and host range of *Sclerotinia sclerotiorum* causing stem rot of crucifers. *Plant Disease Research*, 38(1), 11-46. DOI: 10.5958/2249-8788.2023.00002.1.
Mehrotra, M.D. (1989) Leaf blight of some hardwood species of Assam and Meghalaya and its control in the nursery. *Indian Forest* 115, 378–384.
Meiners, J.P. (1950) Bean diseases in southern Idaho in 1949. *Plant Disease Reporter* 34:14.
Melouk, H.A., Damicone, J.P. and Jackson, K.E. (1992) *Eclipta prostrata*—A new weed host for *Sclerotinia minor*. *Plant Disease* 76, 101. DOI: 10.1094/PD-76-0101E.
Mendes, M.A.S., da Silva, V.L., Dianese, J.C., Ferreira, M.A.S.V., dos Santos, D. *et al.* (1998) *Fungos em Plants no Brasil*. Embrapa-SPI/Embrapa-Cenargen, Brasilia, p. 555.
Merriman, P.R., Pywell, M. and Harrison, G. (1978) Distribution of benomyl in lettuce. *Australasian Plant Pathology* 7, 30–31. DOI: 10.1071/APP9780030.
Michailides, T.J. and Morgan, D.P. (1994) Occurrence of Sclerotinia and Botrytis shoot blights on *Pistachio* in California. *Plant Disease* 78, 641. DOI: 10.1094/PD-78-0641D.
Miclaus, D., Sin, G., Damian, V. and Guran, M. (1988) Presence and spread of *Sclerotinia sclerotiorum* (Lib) de Bary on seeds of various cereal and industrial crops. *An Inst Cercet Cereale Plante Tehran* 56, 169–188.

Middleton, J.T., Kendrick, J.B., Jr. and Hall, B.J. (1952) Parasitic and non-parasitic diseases of celery in southern California. *Plant Disease Reporter* 36, 188–193.

Miller, J.W. (1991a) Bureau of plant pathology. *Triology Technical Report, Division of Plant Industry* 30(8), 1–2.

Miller, J.W. (1991b) Bureau of plant pathology. *Triology Technical Report, Division of Plant Industry* 30(2), 1–3.

Miller, J.W. (1997) Plant pathology. *Triology Technical Report, Division of Plant Industry* 36(1), 12–13.

Minuto, A., Pensa, P., Gilardi, G. and Garibaldi, A. (2001) *Sclerotinia sclerotiorum,* a new disease of pot marigold. *Colture Protette* 30, 95–96.

Minuto, A., Pensa, P., Gullino, M.L. and Garibaldi, A. (2004a) *Sclerotinia sclerotiorum* disease agent of collar rot of *Felicia amelloides* and *Hutchinsia alpina* in Italy. *Informatore Fitopatologico* 54, 40–42.

Minuto, A., Pensa, P., Minuto, G. and Garibaldi, A. (2004b) *Sclerotinia sclerotiorum* agent of stem rot on time (*Thymus citriodorus*) and common sage (*Salvia officinalis). Informatore Fitopatologico* 54, 36–39.

Minuto, A., Pensa, P., Rapa, B. and Garibaldi, A. (2005a) Basal and foliar rot of wild rocket [*Diplotaxis tenuifolia* (L) DC] caused by *Sclerotinia sclerotiorum* (Lib) de Bary in Northern i\Italy. *Informatore Fitopatologico* 55, 43–45.

Minuto, A., Pensa, P. and Garibaldi, A. (2005b) Attacks of *Sclerotinia sclerotiorum* on *Rosmarinus officinalis* L var. *prostratus* hort. *Informatore Fitopatologico* 55, 50–52.

Mirza, M.S. and Ahmad, Y. (1988) Cabbage: A new host of *Sclerotinia sclerotiorum* for Pakistan. *Revue d' Agronomie* 8(1), 85–87. DOI: 10.1051/agro:19880111.

Mitchell, S.J., Jellis, G.J. and Cox, T.W. (1986) *Sclerotinia sclerotiorum* on linseed. *Plant Pathology* 35, 403–405. DOI: 10.1111/j.1365-3059.1986.tb02034.x.

Mohan, L. (1994) Sclerotinia rot in rosemary. *Indian Phytopathology* 47, 443.

Mondal, K.K., Rana, S.S. and Sood, P. (2002) Sclerotinia root rot: A new threat to buckwheat seedlings in India. *Plant Disease* 86, 1404.

Mondal, K.K., Sood, P. and Rana, S.S. (2003) Occurrence of *Sclerotinia sclerotiorum* on buckwheat (*Fagopyrum esculentum*) seedlings in Himachal Pradesh. *Plant Disease Research* 18, 199.

Mondal, B., Khatua, D.C., Hansda, S. and Sharma, R. (2015) Addition to the host range of *Sclerotinia sclerotiorum* in West Bengal. *Sclerotinia Academic Journal of Biosciences* 3(4), 361–364.

Monga, D. (1988) Peduncle rot of Motihari tobacco in Cooch Behar district. *Indian Phytopathology* 41(3), 504.

Mookerjee, K.G. (1973) Additions to plant diseases of Delhi. *Angew Botany* 47, 205–213.

Morgan, O.D. (1971) A study of four weed hosts of *Sclerotinia* species in alfalfa fields. *Plant Disease Reporter* 55, 1087–1089.

Morrall, R.A.A., McKenzie, D.L., Dueck, L.J. and Verma, P.R. (1972) A qualitative survey of diseases of some specialty crops in Saskatchewan in 1970 and 1971, sunflower, safflower, buckwheat, lentil, mustards, and field pea. *Canadian Plant Disease Survey* 52, 143–148.

Morrall, R.A.A., Dueck, J., McKenzie, D.L. and McGee, D.C. (1976) Some aspects of *Sclerotinia sclerotiorum* in Saskatchewan 1970–75. *Canadian Plant Disease Survey* 56, 56–62.

Mortensen, K. and Molloy, M.M. (1989) Fungi detected on *Acroptilon repens* (Russian knapweed) during surveys from 1981 to 1988. *Canadian Plant Disease Survey* 69(2), 143–145.

Mujica, F. and Oehrens, B.E. (1967) *Segunda addenda* a flora fungosa Chilena. *Boletin Tecnico* 27, 1–78.

Mujica, F. and Vergara, C. (1945) *Flora fungosa Chilena. Indice preliminar de los huespedes de los hongos chilenos y sus referencias bibliograficas*. Imprenta Stanley, Stanley Santiago, Chile, p. 199.

Müller, N., Leroch, M., Schumacher, J., Zimmer, D., Könnel, A. *et al.* (2018) Investigations on VELVET regulatory mutants confirm the role of host tissue acidification and secretion of proteins in the pathogenesis of *Botrytis cinerea*. *New Phytologist* 219, 1062–1074.

Nagaraja, O. and Krishnappa, M. (2009) Seed-borne mycoflora of Niger (*Guizotia abyssinica* Cass.) and its effect on germination. *Indian Phytopathology* 62, 513–517.

Naito, S. and Sugimoto, T. (1986) Sclerotinia stalk rot of sugar beet. *Annals of the Phytopathological Society of Japan* 52, 217–224.

Nattrass, R.M. (1961) Host lists of Kenya fungi and bacteria. *Mycology Paper* 81, 1–46.

Navaud, O., Barbacci, A., Taylor, A., Clarkson, J.P. and Raffaele, S. (2018) Shifts in diversification rates and host jump frequencies shaped the diversity of host range among *Sclerotinia* fungal plant pathogens. *Molecular Ecology* 27, 1309–1323. DOI: 10.1111/mec.14523.

Neto, J.P. (1954) Occurrence and apothecial state of *Sclerotinia sclerotiorum* (Lib.) Mass. In: *The State of Rio Grande Do Sul*. Vol. 17. Rev Agron Porto Alegre, Brazil, pp. 109–112.

Newhall, A.G. (1944) A serious storage rot of celery caused by the fungus *Ansatospora macrospora* n. gen. *Phytopathology* 34, 92–105.

Nielsen, K., Yohalem, D.S. and Jensen, D.F. (2002) PCR detection and RFLP differentiation of *Botrytis* species associated with neck rot of onion. *Plant Disease* 86(6), 682–686. DOI: 10.1094/PDIS.2002.86.6.682.

Nikandrow, A., Weidemann, G.J. and Auld, B.A. (1990) Incidence and pathogenicity of *Colletotrichum orbiculare* and a *Phomopsis* sp. on *Xanthium* sp. *Plant Disease* 74(10), 796. DOI: 10.1094/PD-74-0796.

Njambere, E.N., Peever, T.L., Vandemark, G. and Chen, W. (2014) Genotypic variation and population structure of *Sclerotinia trifoliorum* infecting chickpea in California. *Plant Pathology* 63, 994–1004. DOI: 10.1111/ppa.12176.

Noviello, C., Marziano, F., Aloj, B. and Garibaldi, A. (1975) Observations over three years on fennel diseases in Campania. *Annals of Faculty of Science and Agricultural University, Studi Napoli Portici* 9–10, 259–271.

Ogilvie, L. (1927) *Report of the Plant Pathologists for the Year 1926*. Department of Agriculture and Fisheries, Bermuda, pp. 35–41.

Ogilvie, L. (1936) Sclerotinia wilt of the hop. In: *Report Agriculture and Horticulture Research Station*. University of Bristol, UK, pp. 107–109.

Ojaghian, M.R. (2009) First report of *Sclerotinia sclerotiorum* on potato plants in Iran. *Australasian Plant Disease Notes* 4, 39–41.

Ojaghian, S. (2020) A new record of *Sclerotinia sclerotiorum* infecting potato plants in China. *Journal of Plant Pathology* 102, 1273.

Oliveira, R.R., Braz, G.B.P., Jr., Oliveira, R.S., Hentges, M., Takano, H.K. *et al*. (2015) First report of Sclerotinia blight caused by *Sclerotinia sclerotiorum* on *Crotalaria spectabilis* in Brazil. *Plant Disease* 99:1037.

Orieux, L. and Felix, S. (1968) List of plant diseases in Mauritius. *Phytopathology, Papers* 7, 1–48.

Orlikowski, L. (1976) The pathogens of diseases of gerbera *Jamesonii bolus* in some glasshouses in Poland. *PR Institute Sadow Skierniewicach Serveces, B* 2, 197–201.

Orlikowski, L.B. and Valjuskate, A. (2007) New record of Phytophthora root and stem rot of *Lavandula angustifolia*. *Acta Mycologica* 42, 193–198. DOI: 10.5586/am.2007.021.

O'Sullivan, C.A., Belt, K. and Thatcher, L.F. (2021) Tackling control of a cosmopolitan phytopathogen: *Sclerotinia*. *Frontiers in Plant Science* 12, 707509. DOI: 10.3389/fpls.2021.707509.

Ozer, G., Sameeullah, M., Bayraktar, H. and Gore, M.E. (2017) Genetic diversity among phytopathogenic *Sclerotinia*, based on retrotransposon molecular markers. *Phytopathologia Mediterranea* 56(2), 251–258.

Palti, J. (1963) *Sclerotinia sclerotiorum* in Israel. *Phytopathologia Mediterranea* 2(2), 60–64.

Pande, A. and Rao, V.G. (1998) *A Compendium- Fungi on Legumes from India*. Scientific Publishers, Jodhpur, India, p. 188.

Pandey, K.K. and Pandey, P.K. (2002) A new report of *Sclerotinia sclerotiorum* on bottle gourd. *Indian Phytopathology* 55, 352–353.

Pantidou, M.E. (1973) *Fungus-host index for Greece*. Benaki Phytopathological Institute, Kiphissia, Athens, p. 382.

Parasar, C.S. and Chahal, D.S. (1962) Stem rot of berseem (*Trifolium alexandrinum* L.) caused by *Sclerotinia sclerotiorum* (Lib.) de Bary. *Indian Phytopathology* 15, 293–294.

Park, M.S., Kim, Y.G., Lee, S.W., Park, C.G., Kim, Y.I. *et al*. (2017) First report of *Sclerotinia sclerotiorum* causing Sclerotinia rot on *Ixeridium dentatum* in Korea. *Korean Journal of Mycology* 45(4), 381–385.

Parris, G.K. (1959) A Revised Host Index of Mississippi Plant Diseases. Vol. 1. Botany Department Miscellaneous Publication, Mississippi State University, pp. 1–146.

Partyka, R.E. and Mai, W.F. (1962) Effect of environment and some chemicals on *Sclerotinia sclerotiorum* in laboratory and potato field. *Phytopathology* 52, 766–770.

Patterson, C.L. and Grogan, R.G. (1985) Differences in epidemiology and control of lettuce drop caused by *Sclerotinia minor* and *S. Sclerotiorum*. *Plant Disease* 69, 766–770.

Paulitz, T.C. (1997) First report of *Sclerotinia sclerotiorum* on basil in Canada. *Plant Disease* 81(2), 229. DOI: 10.1094/PDIS.1997.81.2.229B.

Pawlowski, M.L., Murithi, H., Hailemariam, M., Tesfaye, A.A. and Hartman, G.L. (2019) First report of *Sclerotinia sclerotiorum* causing stem rot on soybean (*Glycine max*) in Ethiopia. *Plant Disease* 103(10), 2676.

Peccoud, J., Figueroa, C.C., Silva, A.X., Ramirez, C.C., Mieuzet, L. *et al*. (2008) Host range expansion of an introduced insect pest through multiple colonizations of specialized clones. *Molecular Ecology* 17(21), 4608–4618. DOI: 10.1111/j.1365-294X.2008.03949.x.

Peregrine, W.T.H. and Siddiqi, M.A. (1972) A revised and annotated list of plant diseases in Malawi. *Phytopathology Papers*, 16, 1–51.

Pennycook, S.R. (1989) Plant diseases recorded in New Zealand. *Plant Disease* 3, 1–51.

Perez, W.G., Gamboa, J.S., Falcon, Y.V., Coca, M., Raymundo, R.M. *et al*. (2001) Genetic structure of peruvian populations of *Phytophthora infestans*. *Phytopathology* 91(10), 956–965. DOI: 10.1094/PHYTO.2001.91.10.956.

Perez, B.A., Farinon, O.M. and Berretta, M.F. (2011) First report of Sclerotinia rot on blueberry caused by *Sclerotinia sclerotiorum* in Argentina. *Plant Disease* 95, 774. DOI: 10.1094/PDIS-01-11-0074.

Petri, L. (1934) Review of phytopathological records noted in 1933. In: *Review of Applied Mycology*. Vol. 13. Kew, Surrey, UK.

Petrie, R.G. and Vanterpool, T.C. (1965) Diseases of rape and cruciferous weeds in Saskatchewan in 1965. *Canadian Plant Disease Survey* 45, 111–112.

Phillips, A.J.L. (1992) Some common weed spp. as alternative host for *Sclerotinia sclerotiorum*. *Phytophylactica* 24, 207–210.

Phipps, P.M. and Porter, D.M. (1982) Sclerotinia blight of soybean caused by *Sclerotinia minor* and *Sclerotinia sclerotiorum*. *Plant Disease* 66, 163–165. DOI: 10.1094/PD-66-163.

Pieczul, K. (2018) First report of *Sclerotinia sclerotiorum* (Sclerotinia blight) on *Symphyotrichum dumosum* in Poland. *Plant Disease* 102(5), 1027–1028. DOI: 10.1094/PDIS-08-17-1148-PDN.

Pieczul, K. (2019) First report of *Sclerotinia sclerotiorum* on *Verbena bonariensis* in Poland. *Journal of Plant Pathology* 101(3), 761–761. DOI: 10.1007/s42161-018-00226-x.

Piepenbring, M. (2006) Checklist of fungi in Panama. *Preliminary Version. Puente Biology* 1, 1–190.

Plakidas, A.G. (1957) New or unusual plant diseases in Louisiana. *Plant Disease Reporter* 41, 643–645.

Plati, J. (1960) Sclerotinia diseases of irrigated crops. *Hassendeh* 40, 450–451.

Polizzi, G., Aiello, D., Scuderi, G., Cirvilleri, G. and Sofia, V.S. (2011) First report of Sclerotinia stem and twig blight caused by *Sclerotinia sclerotiorum* on *Citrus volkameriana* rootstock in Italy. *Plant Disease* 95, 1030. DOI: 10.1094/PDIS-05-11-0367.

Poole, R.F. (1922) The Sclerotinia rot of celery. *New Jersey Agricultural Experiment Stations Bulletin* 359, 1–27.

Pope, S.J. and Sweet, J.B. (1991) Sclerotinia stem rot resistance in linseed cultivars. *Aspects of Applied Biology* 28, 79–84.

Popova, A.A. (1929) Diseases of the tobacco *Nicotiana rustica* L 18, 45–53.

Porter, D.M. and Beute, M.K. (1974) Sclerotinia blight of peanuts. *Phytopathology* 64, 263–264.

Prasad, L., Kamil, D., Solanki, R.K. and Singh, B. (2017) First report of *Sclerotinia sclerotiorum* infecting cumin *(Cuminum cyminum)* in India. *Plant Disease* 101(5), 833.

Prat, Y., Fromer, M., Linial, N. and Linial, M. (2009) Codon usage is associated with the evolutionary age of genes in metazoan genomes. *BMC Evolutionary Biology* 9, 285. DOI: 10.1186/1471-2148-9-285.

Pratt, R.G. and Rowe, D.E. (1991) Differential responses of alfalfa genotypes to stem inoculations with *Sclerotinia sclerotiorum* and *S. trifoliorum*. *Plant Disease* 75, 188–191.

Preston, D.A. (1945) Host index of Oklahoma plant diseases. *Oklahoma Agricultural College Agricultural Experiment Station, Technical Bulletin, T* 21, 1–168.

Preston, D.A. and Dosdall, L. (1955) Minnesota plant diseases. *USDA, Plant Disease and Epidemiology and Identification Section, Special Publication* 8, 1–184.

Price, W.C. (1954) A phytopathological note. *Phytopathology* 44, 717–718.

Prova, A., Akanda, M.A.M., Islam, S., Sultana, F., Islam, M.T. *et al*. (2014) First report of stem and pod blight of hyacinth bean caused by *Sclerotinia sclerotiorum* in Bangladesh. *Journal of Plant Pathology* 96(3), 607.

Prova, A., Akanda, A.M., Islam, S. and Motaher Hossain, Md. (2017) First report of *Sclerotinia sclerotiorum* causing pod rot disease on okra in Bangladesh. *Canadian Journal of Plant Pathology* 39(1), 72–76. DOI: 10.1080/07060661.2017.1278723.

Prova, A., Akanda, A.M., Islam, S. and Hossain, M.M. (2018) Characterization of *Sclerotinia sclerotiorum*, an emerging fungal pathogen causing blight in hyacinth bean (*Lablab purpureus*). *The Plant Pathology Journal* 34(5), 367–380. DOI: 10.5423/PPJ.OA.02.2018.0028.

Punja, Z.K. and Ni, L. (2021) The bud rot pathogens infecting cannabis (*Cannabis sativa* L., *marijuana*) inflorescences: Symptomology, species identification, pathogenicity and biological control. *Canadian Journal of Plant Pathology* 43(6), 827–854. DOI: 10.1080/07060661.2021.1936650.

Purdy, L.H. (1979) *Sclerotinia sclerotiorum*: History, diseases and symptomatology, host range, geographic distribution, and impact. *Phytopathology* 69(8), 875. DOI: 10.1094/Phyto-69-875.

Putnam, M.L. (2004) First report of stem rot of rosemary caused by *Sclerotinia sclerotiorum* in the United States. *Plant Pathology* 53(2), 252–252. DOI: 10.1111/j.0032-0862.2004.00964.x.

Raabe, A. (1939) Investigations on parasitic fungal diseases of colza and rape. *Zentralblatt Fur Bakteriologie: International Journal of Medical Microbiology* 100, 35–52.

Raabe, R.D. (1966) Check list of plant diseases previously unreported in Hawaii. *Plant Disease Reporter* 50, 411–414.

Raabe, R.D. (1971) Sclerotinia blight of stephanotis flowers. *Phytopathology Journal* 61, 757–882. DOI: 10.1094/Phyto-61-1524.

Raabe, R.D. and Hendrix, F.F., Jr. (1966) Sclerotinia blight of *Gloxinia*. *Mycologia* 58(161), 166. DOI: 10.2307/3756999.

Raabe, R.D., Conners, I.L. and Martinez, A.P. (1981) *Checklist of plant diseases in Hawaii. College of Tropical Agriculture and Human Resources*, Text Series no. 22. University of Hawaii, Honolulu, Information, p. 313.

Raaymakers, T.M. and Van den Ackerveken, G. (2016) Extracellular recognition of oomycetes during biotrophic infection of plants. *Frontiers in Plant Science* 7, 906. DOI: 10.3389/fpls.2016.00906.

Rahman, M.M.E., Dey, T.K., Hossain, D.M., Nonaka, M. and Harada, N. (2015) First report of white mould caused by *Sclerotinia sclerotiorum* on jackfruit. *Australasian Plant Disease Notes* 10, 10. DOI: 10.1007/s13314-014-0155-9.

Rai, R.A. and Agnihotri, J.P. (1970) Foot rot of *Gaillardia pulchella*. *Mycologia* 58, 161–166.

Rajappan, K., Ramaraj, B. and Natarajan, S. (1999) Stem rot of *Salvia coccinea* caused by *Sclerotinia sclerotiorum*—A new host record. *Indian Phytopathology* 52, 425.

Ramsey, G.B. (1924) *Sclerotinia intermedia* n sp. a cause of decay of salsify and carrots. *Phytopathology* 14, 323–327.

Rascle, C., Dieryckx, C., Dupuy, J.W., Muszkieta, L., Souibgui, E. *et al.* (2018) The pH regulator pacc: A host-dependent virulence factor in *Botrytis cinerea*. *Environmental Microbiology Reports* 10(5), 555–568. DOI: 10.1111/1758-2229.12663.

Rashid, K.Y. (2001) Sclerotinia stem infection in flax in western Canada. In: Young, C.S. and Hughes, K.J.D. (eds) *Proceedings of Sclerotinia 2001, The XI International Sclerotinia Workshop*. Central Science Laboratory, York, UK, pp. 103–104.

Rathore, R.S., Rajpurohit, T.S., Solanki, J.S. and Bishnoi, H.R. (1993) Wild onion—A new record of *Sclerotinia sclerotiorum*. *Indian Phytopathology* 46(3), 261–262.

Redman, R.S., Dunigan, D.D. and Rodriguez, R.J. (2001) Fungal symbiosis from mutualism to parasitism: Who controls the outcome, host or invader. *New Phytologist* 151, 705–716.

Reinmuth, E. (1943) Further observations on the parasitic leaf desiccation of opium poppy. *Angewandte Botanik* 25, 300–302.

Richardson, M.J. (1990) *An Annotated List of Seed-Borne Diseases*, 4th edn. International Seed Testing Association, Zurich, p. 387.

Riddle, G.E., Burpee, L.L. and Boland, G.J. (1991) Virulence of *Sclerotinia sclerotiorum* and *S. minor* on dandelion (*Taraxacum officinale*). *Weed Science* 39, 109–118. DOI: 10.1017/S0043174500057969.

Riley, E.A. (1960) A revised list of plant diseases in Tanganyika Territory. *Mycology Paper* 75, 1–42.

Rossman, A.Y., Palm, M.E. and Spielman, L.J. (1987) *A literature guide for the identification of plant pathogenic fungi*. American Phytopathological Society, St. Paul, Minnesota, pp. 202–203.

Roy, A.K. (1973) Host range of *Sclerotinia sclerotiorum* and *Sclerotinia rolfsii* in Jorhat, Assam. *Science and Culture* 39, 319–320.

Roy, A.K. (1985) Fungal diseases of ornamentals and flowers in Assam. *Indian Horticulture* 30(1), 19–21.

Ruano-Rosa, D., Minutillo, S.A., Li Destri Nicosia, M.G., Agosteo, G.E. and Schena, L. (2017) First report of *Sclerotinia sclerotiorum* associated with olive fruit rot in Italy. *Plant Disease* 101(6), 1040–1040. DOI: 10.1094/PDIS-09-16-1306-PDN.

Russell, T.A. (1935) Diseases and pests observed during the year. *Report (Annual)* 18–23.

Russell, T.A. (1942) Report of director of agriculture. 1941. Report of the Bermuda, *Department of Agriculture and Fishries* 1–11.

Saharan, G.S. and Mehta, N. (2008) *Sclerotinia Diseases of Crop Plants: Biology, Ecology and Disease Management*. Springer Science, Dordrecht, the Netherlands, p. 485. DOI: 10.1007/978-1-4020-8408-9.

Saharan, G.S., Mehta, N. and Meena, P.D. (2021a) *Genomics of Crucifer's Host-Resistance*. Springer's Nature, Dordrecht, the Netherlands, p. 790. DOI: 10.1007/978-981-16-0862-9.

Saharan, G.S., Mehta, N. and Meena, P.D. (2021b) *Molecular Mechanism of Crucifer's Host Resistance*. Springer's Nature, Dordrecht, the Netherlands, p. 785.

Saharan, G.S., Mehta, N. and Meena, P.D. (2023) *Genomics of Crucifer's Host-Pathosystem*. Springer's Nature, Dordrecht, the Netherlands, p. 936. DOI: 10.1007/978-981-19-3812-2.

Saira, M., Rehman, A., Gleason, M.L., Alam, M.W., Abbas, M.F. *et al*. (2017) First report of *Sclerotinia sclerotiorum* causing stem and crown rot of berseem (*Trifolium alexandrinum*) in Pakistan. *Plant Disease* 101(5), 835–836. DOI: 10.1094/PDIS-10-16-1400-PDN.

Saito, I. (1997) *Sclerotinia nivalis*, sp. *nv*., the pathogen of snow mold of herbaceous dicots in northern Japan. *Mycoscience* 38(2), 227–236. DOI: 10.1007/BF02460857.

Salgado-Salazar, C., Beirn, L.A., Ismaiel, A., Boehm, M.J., Carbone, I. *et al*. (2018) *Clarireedia*: A new fungal genus comprising four pathogenic species responsible for dollar spot disease of turfgrass. *Fungal Biology* 122(8), 761–773. DOI: 10.1016/j.funbio.2018.04.004.

Sampson, P.J. and Walker, J. (1982) *An Annotated List of Plant Diseases in Tasmania*. Department of Agriculture, Hobart, Tasmania, p. 121.

Sanogo, S. and Puppala, N. (2007) Characterization of a darkly pigmented mycelial isolate of *Sclerotinia sclerotiorum* on Valencia peanut in New Mexico. *Plant Disease* 91(9), 1077–1082. DOI: 10.1094/PDIS-91-9-1077.

Sanogo, S., Lujan, P.A. and Baucom, D. (2015) First report of *Sclerotinia sclerotiorum* on cabbage in New Mexico. *Plant Disease* 99(6), 891. DOI: 10.1094/PDIS-12-14-1328-PDN.

Sarbhoy, A.K., Lal, G. and Varshney, J.L. (1971) *Fungi of India (1967–71)*. Navyug Traders, New Delhi, p. 148.

Sawada, K. (1959) Descriptive catalogue of Taiwan (Formosan) fungi. XI. *Special Publication College of Agriculture, National Taiwan University* 8, 1–268.

Saxena, V.C. and Rai, J.N. (1987) Survey of occurrence of white rot of crucifers caused by *Sclerotinia sclerotiorum* in Uttar Pradesh and Bihar. *Indian Journal of Mycology and Plant Pathology* 17, 89–91.

Schwartz, H.F. (1977) Epidemiology of white mold disease (*Sclerotinia sclerotiorum*) = (*Whetzelinia sclerotiorum*) of dry edible beans (*Phaseolus vulgaris*) with emphasis on resistance and host architectural disease avoidance mechanism. PhD. thesis, University of Nebraska, Lincoln.

Sehgal, S.P. and Agrawal, J.M. (1971) Drooping of fennel (*Foeniculum vulgare*) due to *Sclerotinia sclerotiorum*. *Indian Journal of Mycology and Plant Pathology* 14, 85–86.

Seifbarghi, S., Borhan, M.H., Wei, Y., Ma, L., Coutu, C. *et al*. (2020) Receptor-like kinases BAK1 and SOBIR1 are required for necrotizing activity of a novel group of *Sclerotinia sclerotiorum* necrosis-inducing effectors. *Frontier in Plant Science* 11, 1021. DOI: 10.3389/fpls.2020.01021.

Serdani, M. and Spotts, R.A. (2007) First report of blossom blight and green fruit rot of sweet cherry caused by *Sclerotinia sclerotiorum* in Oregon. *Plant Disease* 91(8), 1058. DOI: 10.1094/PDIS-91-8-1058B.

Sharma, P., Meena, P.D., Verma, P.R., Saharan, G.S., Mehta, N. *et al*. (2015) *Sclerotinia sclerotiorum* (lib) de Bary causing *Sclerotinia* rot in oilseed *Brassicas*: A review. *Journal of Oilseed Brassica* 6, 1–44.

Shavrova, L.A. (1967) Diseases of ornamental plants of Ranunculaceae family. *Bulletin Glavn. Botany Sada. Moshva* 66, 81–86.

Shaw, C.G. (1973) Host fungus index for the Pacific Northwest-I. *Hosts. Washington State Agricultural Experiment Station Bulletin* 765, 1–121.

Shen, Y.M., Chao, C.H., Wang, F.C., Liu, H.L. and Huang, T.C. (2012) First report of stem and leaf blight caused by *Sclerotinia sclerotiorum* on Eustoma in Taiwan. *Plant Disease* 96(6), 910. DOI: 10.1094/PDIS-02-12-0168-PDN.

Shi, Y.-X., Xie, X.-W., Song, J.-W., Jin, Z.-W., Chai, A.-L. *et al*. (2016) First report of Sclerotinia rot on *Andrographis paniculata* in China. *Canadian Journal of Plant Pathology* 38(4), 522–526. DOI: 10.1080/07060661.2016.1262450.

Shin, D.G. and Lee, J.T. (1987) Ecological studies on lettuce drop disease occurring under controlled cultivation conditions in drained paddy fields. *Korean Journal of Plant Pathology* 3, 252–260.

Shiraishi, T. and Sunaga, F. (1991) Chemical control of Sclerotinia rot of edible burdock. *Proceeding Kanto-Tosan Plant Protection Society* 38, 61–62.

Shivas, R.G. (1989) Fungal and bacterial diseases of plants in Western Australia. *Journal of Royal Society of Western Australia* 72, 1–62.

Shrestha, U., Swilling, K.J., Butler, D.M. and Ownley, B.H. (2018) First report of basal drop and white mold on lettuce, broccoli, and mustard caused by *Sclerotinia sclerotiorum* in Tennessee, USA. *Plant Disease* 102(1), 249. DOI: 10.1094/PDIS-07-17-1025-PDN.

Shukla, D.D. and Pavgi, M.S. (1964) Stem rot disease of some ornamental flowering plants. *Science and Culture* 30, 284–286.

Shumilenko, W.P. (1963) On the transfer of white rot together with the seeds of flowering plants. *Byull Gl Botany, Sada* 48, 84–85.

Siedel, D. and Daebeler, F. (1964) Sclerotinia infection of black salsify. *Nachrichtenbl. Dtsch. Pflanzenschutzdienst (Berlin)* 18, 75.

Simay, E.I. (1989) Fungi observed on chickpea (*Cicer arietinum* L.) in 1985–1988. *Novenytermeles* 38, 435–442.

Simmonds, J.H. (1966) *Host Index of Plant Diseases in Queensland*. Department of Primary Industries, Brisbane, Queensland, p. 111.

Singh, H.B. and Saha, I.R. (1989) Evaluation of soe fungicides against *Sclerotinia sclerotiorum*, the incitant of wilt and rot of knolkohl. *Pesticides* 23(2), 44–45.

Singh, S.B. and Singh, R. (2001) A new Sclerotinia stem rot of bishop weed. *Indian Phytopathology* 54, 140.

Singh, U.P. and Singh, R.B. (1983) The effect of soil texture, soil moisture, and depth of soil on carpogenic germination of *Sclerotinia sclerotiorum*. *Z. Pflanzenkr. Pflanzenschutz* 90, 662–669.

Singh, U.P. and Singh, R.B. (1986) Some new hosts of *Sclerotinia sclerotiorum* from India. *Indian Phytopathology* 39, 626.

Singh, U.P., Pathak, K.K., Khare, M.N. and Singh, R.B. (1979) Effect of leaf extract on *Fusarium oxysporum* f. sp. *ciceri*, *Sclerotinia sclerotiorum* and on gram seeds. *Mycologia* 71, 556–564.

Singh, S.B., Singh, A.K. and Singh, A.K. (2003) In *vitro* evaluation of fungi toxicants on developmental stages of *Sclerotinia sclerotiorum*. *Annals of Plant Protection Sciences* 11, 388–390.

Sinha, J.M. and Singh, A.P. (1991) New host for *Oidiopsis taurica* and *Whetzelinia sclerotiorum*. *Indian Phytopathology* 44, 558.

Skotland, C.B. and Menzies, J.D. (1957) Two pepper mint diseases found in the Yakima Valley of Washington. *Plant Disease Reporter* 41, 493.

Smith, R.E. (1900) *Botrytis* and *Sclerotinia*: Their relation to certain plant diseases and to each other. *Botanical Gazette* 29(6), 369–407. DOI: 10.1086/328001.

Smith, R.E. (1929) Life history of *Sclerotinia sclerotiorum* (Lib) Mass in connection with green rot of apricots. *Phytopathology* 19, 1136–1137.

Smith, R.E. (1931) The life history of *Sclerotinia sclerotiorum* with reference to the green rot of apricot. *Phytopathology* 21, 407–423.

Song, J.H. and Koh, Y.J. (1999) Sclerotinia twig blight on trees and cottony rot on fruits of *Satsuma mandarin* caused by *Sclerotinia sclerotiorum*. *Plant Pathology Journal* 15, 236–241.

Spaulding, P. (1961) Foreign diseases of forest trees of the world. *USDA Agriculture Handbook* 197, 1–361.

Spooner, B.M. (1987) Helotiales of Australasia: Geoglossaceae, Orbiliaceae, *Sclerotinia*, Hyaloscyphaceae. *Bibliotheca Mycologica* 116, 1–711.

Srivastava, U.K. and Diwakar, M.C. (1987) Note on stem rot and wilt of chillies in Haryana (India). *Plant Protection Bulletin, Faridabad* 39, 35.

Stevens, F.L. and Hall, J.G. (1911) A serious lettuce disease (*Sclerotinia)* and A method of control. *North Carolina Agricultural Experiment Station Technical Bulletin* 8, 85–145.

Stoner, W.N. (1951) A list of some of the plant diseases occurring in Everglades region of Florida during the 1949–1950 season. *Plant Disease Reporter* 35, 170–172.

Strauss, J. and Dillard, H.R. (2009) First report of Sclerotinia stem rot caused by *Sclerotinia sclerotiorum* on *Hibiscus trionum* in New York. *Plant Disease* 93(6), 673. DOI: 10.1094/PDIS-93-6-0673A.

Subbarao, K. (1998) Progress toward integrated management of lettuce drop. *Plant Disease* 82, 1068–1078.

Sugimoto, T. (1959) Studies on the Sclerotinia disease of bean plants in relation to mycelial development of the causal fungus. *Member Faculty, Agricultural Science Hokkaido University* 3, 114–120.

Sun, Z.X. and Hsiang, T. (2016) First report of *Sclerotinia sclerotiorum* on periwinkle (*Catharanthus roseus*) in Ontario, Canada. *Plant Disease* 100(8), 1789–1790.

Tai, F.L. (1979) *Sylloge Fungorum Sinicorum*. Scietific Press, Academic Sinica, Peking, p. 1527.
Tai, F.L. and Wei, C.T. (1933) Notes on chinese fungi III. *Sinensia* 4(5), 83–128.
Takeuchi, J. and Horie, H. (1996) Occurrence of Sclerotinia rot of *Brassica campestris* (Chinensis group), Angelica keiskei, beard-tongue [*Penstemon*], bladder campion [*Silene vulgaris*] and verbena [*V. officinalis*] caused by *Sclerotinia sclerotiorum*. *Proceedings of Kanto Tosan Plant Protection Society* 43, 67–70.
Takeuchi, J. and Horie, H. (1999) First occurrence of Sclerotinia rot in aster and strawflower in Japan. *Annual Report of the Kanto-Tosan Plant Protection Society* 46, 57–59.
Taubenhaus, J.J. (1920) *Diseases of greenhouse crops and their control. E.P.* Dutton, New York, p. 429.
Taubenhaus, J.J. and Ezekiel, W.N. (1931) A Sclerotinia limb blight of figs. *Phytopathology* 21, 1195–1197.
Terrones-Salgado, J., Ortega-Acosta, C., Sánchez-Ruiz, F.J., Ortega-Acosta, S.A., Palemon, F. *et al.* (2023) First report of white mold caused by *Sclerotinia sclerotiorum* on *Echeveria gigantea* in Mexico. *Plant Disease*. DOI: 10.1094/PDIS-01-24-0196-PDN.
Thomson, J.R. and Stelfox, D. (1983) *Sclerotinia* contamination of Alberta-produced rapeseed, from 1976–1981. *Canadian Plant Disease Survey* 63, 19–21.
Tian, B., Xie, J., Fu, Y., Cheng, J., Li, B. *et al.* (2020) A cosmopolitan fungal pathogen of dicots adopts an endophytic lifestyle on cereal crops and protects them from major fungal diseases. *The ISME Journal* 14(12), 3120–3135. DOI: 10.1038/s41396-020-00744-6.
Togawa, M. (2000) First occurrence of Sclerotinia rot (*Sclerotinia sclerotiorum*) in perilla (*Perilla frutescens*) and bupleurum *(Bupleurum rotundifolium*) in Japan. *Proceedings of Kanto Tosan Plant Protection Society* 42, 77–78. DOI: 10.4165/kapps1958.42.0_77.
Tok, F.M. (2008) First report of white mold caused by *Sclerotinia sclerotiorum* on sweet Basil in Turkey. *Plant Disease* 92, 1471. DOI: 10.1094/PDIS-92-10-1471C.
Toms, H.N.W. (1964) Plant diseases of southern British Columbia. A new host index. *Canadian Plant Disease Survey* 44, 143–225.
Tores, J.A. and Moreno, R. (1991) *Sclerotinia sclerotiorum*: Epidemiological factors affecting infection of greenhouse aubergine crops. *Journal of Phytopathology* 132, 65–74.
Trigiano, R.N., Boggess, S.L. and Ownley, B.H. (2018) First report of an aerial blight of *Chrysogonum virginianum* (green and gold) caused by *Sclerotinia sclerotiorum* in the United States. *Plant Disease* 102(2), 450–451.
Trinh, H.X., Quan, M.V., Groenewald, J.Z. and Burgess, L.W. (2012) First report of stub dieback of poinsettia (*Euphorbia pulcherrima*) caused by *Sclerotinia sclerotiorum* in Vietnam. *Australasian Plant Disease Notes* 7, 55–57.
Tripathi, A.N., De, R.K., Sharma, H.K. and Karmakar, P.G. (2015) Emerging threat of *Sclerotinia sclerotiorum* causing white/cottony stem rot of mesta in India. *New Disease Reports* 32, 19.
Tuset, J.J. and Marti, M.C. (1988) *Sclerotinia sclerotinia*—A new decay of citrus fruit in Spain. In: *Citriculture, Proceeding 6th International Citrus Congress*.Vol. 3. International Society of Citriculture, Tel Aviv, Israel, pp. 1451–1459.
Tziros, G.T., Bardas, G.A., Tsialtas, J.T. and Karaoglanidis, G.S. (2008) First report of oilseed rape stem rot caused by *Sclerotinia sclerotiorum* in Greece. *Plant Disease* 92(10), 1473. DOI: 10.1094/PDIS-92-10-1473C.
Uloth, M.B., Clode, P.L., You, M.P. and Barbetti, M.J. (2016) Attack modes and defense reactions in pathosystems involving *Sclerotinia sclerotiorum*, *Brassica carinata, B. juncea* and *B. napus*. *Annals of Botany* 117(1), 79–95. DOI: 10.1093/aob/mcv150.
Umemoto, S., Nagashima, K., Yoshida, S. and Tsushima, S. (2007) Sclerotinia rot of blueberry caused by *Sclerotinia sclerotiorum*. *Journal of General Plant Pathology* 73, 290–292.
Upadhyaya, J. (1994) Sclerotinia collar-rot of Ammi majus Linn. *Progressive Horticulture* 26, 77–78.
Urosevic, B. (1957) Mycoflora of stored acorns. *Pr Vysk Ustavu Lesn Csr* 13, 14–200.
Urtiaga, R. (1986) *Indice de enfermedades en plantas de Venezuela y Cuba. Impresos en Impresos Nuevo Siglo*. S.R.L, Barquisimeto, Venezuela, p. 202.
Valero-Jiménez, C.A., Veloso, J., Staats, M. and van Kan, J.A.L. (2019) Comparative genomics of plant pathogenic *Botrytis* species with distinct host specificity. *BMC Genomics* 20(1), 203. DOI: 10.1186/s12864-019-5580-x.
Valleau, W.D., Fergus, E.N. and Henson, L. (1933) Resistance of red clover to *Sclerotinia trifoliorum* Eriss and infection studies. *Kentucky Agriculture Experiment Station Bulletin* 341, 113–131.

Van de Wouw, A.P., Idnurm, A., Davidson, J.A., Sprague, S.J., Khangura, R.K. *et al*. (2016) Fungal diseases of canola in Australia: Identification of trends, threats and potential therapies. *Australasian Plant Pathology* 45(4), 415–423. DOI: 10.1007/s13313-016-0428-1.

Velastegui, J.R. and Ball, S.F.L. (1991) First record of *Sclerotinia sclerotiorum* on tree tomato in Ecuador. *Plant Pathology* 40, 476–477. DOI: 10.1111/j.1365-3059.1991.tb02406.x.

Vinod Kumar, S., Rajesh, P., Senthilraja, C., Nakkeeran, S. and Fernando, W.G.D. (2015) First report of *Sclerotinia sclerotiorum* causing stem rot of carnation (*Dianthus caryophyllus*) in India. *Plant Disease* 99, 1280–1281. DOI: 10.1094/PDIS-02-15-0240-PDN.

Voglino, P. (1922) The fungi most injurious to cultivated plants observed in the province of Turin and the vicinity in 1921. *Annals of Academy of Agriculture, Torino* 65, 53–64.

Wadsworth, D.F. (1979) Sclerotinia blight of peanuts in Oklahoma and occurrence of the sexual stage of the pathogen. *Peanut Science* 6, 77–79. DOI: 10.3146/i0095-3679-6-2-3.

Waipara, N.W. (2006) Isolation of white rot, *Sclerotinia sclerotiorum*, causing leaf necrosis on *Tradescantia fluminensis* in New Zealand. *Australasian Plant Disease Notes* 1(27), 28. DOI: 10.1071/DN06012.

Wakefield, E.N. (1924) On the names *Sclerotinia sclerotiorum* (Lib) Massee and *S. libertiana* Fuckel. *Phytopathology* 14, 126–127.

Walker, A.-S., Gautier, A., Confais, J., Martinho, D., Viaud, M. *et al*. (2011) *Botrytis pseudocinerea*, a new cryptic species causing gray mold in French vineyards in sympatry with *Botrytis cinerea*. *Phytopathology* 101, 1433–1445. DOI: 10.1094/PHYTO-04-11-0104.

Wallace, M.M. (1944) Sclerotinia disease of beans and other crops. *The East African Agricultural Journal* 9(3), 171–172. DOI: 10.1080/03670074.1944.11664370.

Wang, A.R., Lin, W., Chen, X., Lu, G., Zhou, J. *et al*. (2008) Isolation and identification of Sclerotinia stem rot causal pathogen in *Arabidopsis thaliana*. *Journal of Zhejiang University Science, B* 9(10), 818–822. DOI: 10.1631/jzus.B0860010.

Wang, C.H., Shang, W.J., Wang, Q., Fan, S.H., Subbarao, K.V. *et al*. (2021) White rot of *Panax quinquefolius* caused by *Sclerotinia nivalis*. *Plant Pathology* 70, 2034–2045. DOI: 10.1111/ppa.13439.

Waterson, J.M. (1939) *Annotated List of Diseases of Cultivated Plants in Bermuda*. Department of Agriculture, Bermuda, p. 38.

Waterston, J.M. (1939) *Report of the Department of Agriculture 1938*. Department of Agriculture and Fisheries, Bermuda, pp. 24–37.

Watson, A.J. (1950) Fungi associated with camellia flowers. *Plant Disease Reporter* 34, 186–187.

Watson, A.K., Copeman, R.J. and Renney, A.J. (1974) A first record of *Sclerotinia sclerotiorum* and *Macrosphaeropsis centaureae* on *Centaurea diffusa*. *Canadian Journal of Botany* 52, 2639–2640. DOI: 10.1139/b74-345.

Weimer, J.L. (1923) Two diseases of udo (*Aralia cordata* Thumb). *Journal of Agricultural Research* 26(6), 271–278.

Wellman, F.L. (1977) *Dictionary of Tropical American Crops and Their Diseases*. Scarecrow Press, Metuchen, New Jersey, p. 495.

Welty, R.E. (1977) *Lamium amplexicaule* (henbit): A new host for *Sclerotinia sclerotiorum* (*trifoliorum*). *Plant Disease Reporter* 61, 508–510.

Whetzel, H.H. (1945) A synopsis of the genera and species of the *Sclerotinia*, A family of stromatic inoperculate discomycetes. *Mycologia* 37, 648–714.

Wiehe, P.O. (1948) The plant diseases and fungi recorded from Mauritius. *Mycology Paper* 24, 1–39.

Willetts, H.J. (1997) Morphology, development and evolution of stromata *Sclerotinia* and macroconidia of the *Sclerotinia*. *Mycological Research* 101(8), 939–952. DOI: 10.1017/S0953756297003559.

Willetts, H.J. and Wong, J.A.L. (1980) The biology of *Sclerotinia sclerotiorum, S. trifoliorum*, and *S. minor* with emphasis on specific nomenclature. *Botanical Review* 46(2), 101–165. DOI: 10.1007/BF02860868.

Williams, L. and Hayne, S.C. (1982) *Index of Plant Diseases in West Virginia*. West Virginia Department of Agriculture and Plant Pest Control Division, Ripley, WI, p. 115.

Williams, J.R. and Stelfox, D. (1980) Occurrence of ascospores of *Sclerotinia sclerotiorum* in areas of central Alberta. *Canadian Plant Disease Survey* 60, 51–53.

Williams, B., Kabbage, M., Kim, H.J., Britt, R. and Dickman, M.B. (2011) Tipping the balance: *Sclerotinia sclerotiorum* secreted oxalic acid suppresses host defenses by manipulating the host redox environment. *PLoS Pathogens* 7, e1002107. DOI: 10.1371/journal.ppat.1002107.

Wilson, V.E. (1956) 1955 bean disease survey in southern Idaho. *Plant Disease Reporter* 40, 312.

Wilson, V.E. and Brandsberg, J. (1965) Fungi isolated from diseased lentil seedlings in 1963–1964. *Plant Disease Reporter* 49, 660–662.

Winstead, N.N. and Haasis, F.A. (1954) A flower blight of camellias in North Carolina incited by *Sclerotinia sclerotiorum*. *Phytopathology* 44, 717–718.

Wolcan, S.M. (2004) Occurrence of crown rot of gazania and gerbera caused by *Sclerotinia sclerotiorum* in Argentina. *Journal of Plant Pathology* 86, 263.

Wolcan, S.M. and Grego, P.J. (2005) Stem rot of *Trachellium caeruleum* and *Craspedia globosa* caused by *Sclerotinia sclerotiorum* in Argentina. *Journal of Plant Pathology* 87, 243.

Wolcan, S., Ronco, L., Dal Bo, E., Lori, G. and Alippi, H. (1996) First report of diseases on *lisianthus* in Argentina. *Plant Disease* 80, 223. DOI: 10.1094/PD-80-0223A.

Wolf, F.A. (1947) Twig blight of golden bell *Forsythia viridissima* Lindl. *Plant Disease Reporter* 31, 325.

Wolf, F.A. and Cromwell, R.O. (1919) Clover stem rot. *North Carolina Agriculture Experiment Station Technical Bulletin* 16, 22.

Woodard, K.E. and Newman, J.S. (1993) First report of *Sclerotinia minor* on Texas Blue bonnet. *Plant Disease* 77, 318.

Woodhall, J.W., Brown, L., Harrington, M., Olsen, N., Miller, J. *et al.* (2020) *Sclerotinia sclerotiorum* causes decay and forms Sclerotinia in potato tubers in Idaho. *Plant Health Progress* 21, 335–337. DOI: 10.1094/PHP-04-20-0032-RS.

Woodward, J.E., Brenneman, T.B., Kemerait, R.C., Jr., Culbreath, A.K. and Clark, J.R. (2006) First report of Sclerotinia blight caused by *Sclerotinia sclerotiorum* on peanut in Georgia. *Plant Disease* 90, 111.

Woodward, J.E., Nui, C., Wright, R.J., Batla, M.A. and Baughman, T.A. (2008) First report of *Sclerotinia sclerotiorum* infecting peanut in Texas. *Plant Disease* 92, 1468. DOI: 10.1094/PDIS-92-10-1468B.

Wright, E.R. and Palmucci, H.E. (2003) Occurrence of stem rot of chrysanthemum caused by *Sclerotinia sclerotiorum* in Argentina. *Plant Disease* 87, 98. DOI: 10.1094/PDIS.2003.87.1.98D.

Wright, E.R., Rivera, M.C., Chiesa, G. and Morisigue, D. (2005) Occurrence of Sclerotinia stem rot of *Osteospermum* sp., *Felicia amelloides*, and *Ranunculus asiaticus* in Argentina. *Plant Disease* 89(9), 1014. DOI: 10.1094/PD-89-1014B.

Wu, B.M. and Subbarao, K.V. (2006) Analyses of lettuce drop incidence and population structure of *Sclerotinia sclerotiorum* and *S. minor*. *Phytopathology* 96(12), 1322–1329. DOI: 10.1094/PHYTO-96-1322.

Wu, Y.S. and Wang, C.G. (1983) *Sclerotinia asari*: A new species of *Sclerotinia*. *Acta Phytopathological Sinica* 13, 9–14.

Wymore, L.A. and Lorbeer, J.W. (1986) Timing fungicidal control of *Sclerotinia minor* causing drop of lettuce grown on organic soils in New York. *Plant Disease* 70(12), 1134–1138. DOI: 10.1094/PD-70-1134.

Yan, L.Y., Kang, Y.P., Lei, Y., Huang, J.Q., Wan, L.Y. *et al.* (2014) First report of *Sclerotinia sclerotiorum* causing Sclerotinia blight on peanut (*Arachis hypogaea*) in Northeastern China. *Plant Disease* 98(1), 156. DOI: 10.1094/PDIS-05-13-0476-PDN.

Yang, S.M., Woi, S.W., Ge, C.F., Liang, K.Z. and Wang, L. (1988) Diseases of cultivated sunflower in Liaoning Province, People's Republic of China. *Plant Disease* 72(6), 546. DOI: 10.1094/PD-72-0546A.

Yang, X., Yang, J., Wang, Y., He, H., Niu, L. *et al.* (2019) Enhanced resistance to Sclerotinia stem rot in transgenic soybean that over expresses a wheat oxalate oxidase. *Transgenic Research* 28(1), 103–114. DOI: 10.1007/s11248-018-0106-x.

Yin, F., Song, Z., Liu, L., Xu, Q., Jiang, J. *et al.* (2024) Sclerotinia rot of *Zephyranthes candida* caused by *Sclerotinia sclerotiorum* and *sclerotinia minor*. *Fronters in Microbiology* 15, 1414141. DOI: 10.3389/fmicb.2024.1414141.

Young, P.A. (1936) *Sclerotinia* rot of squash and pumpkin. *Phytopathology* 26, 184–190.

Young, H.M., Srivastava, P., Paret, M.L., Dankers, H., Wright, D.L. *et al.* (2012) First report of Sclerotinia stem rot caused by *Sclerotinia sclerotiorum* on *Brassica carinata* in Florida. *Plant Disease* 96(10), 1581. DOI: 10.1094/PDIS-06-12-0525-PDN.

Yu, Y., Cai, J., Ma, L., Huang, Z., Wang, Y. *et al.* (2020) Population structure and aggressiveness of *Sclerotinia sclerotiorum* from rapeseed (*Brassica napus*) in Chongqing City. *Plant Disease* 104(4), 1201–1206. DOI: 10.1094/PDIS-07-19-1401-RE

Zhang, B. and Huang, Y. (1990) A list of important plant diseases in China. *Review of Plant Pathology* 69(3), 97–118.

Zhang, A.W., Hartman, G.L., Riccioni, L., Chen, W.D., Ma, R.Z. *et al.* (1997) Using PCR to distinguish *Diaporthe phaseolorum* and *Phomopsis longicolla* from other soybean fungal pathogens and to detect them in soybean tissues. *Plant Disease* 81(10), 1143–1149. DOI: 10.1094/PDIS.1997.81.10.1143.

Zhang, X., Xian, W., Qu, M., Xu, M., Guo, Z. *et al*. (2022) First report of Sclerotinia blight on peanut caused by *Sclerotinia sclerotiorum* in Qinghai Province, China. *Plant Disease* 106(4), 1301. DOI: 10.1094/PDIS-07-21-1503-PDN.

Zhou, F., Zhu, F.X., Zhang, X.L. and Zhang, A.S. (2014) First report of dimethachlon resistance in field isolates of *Sclerotinia sclerotiorum* on oilseed rape in Shaanxi Province of Northwestern China. *Plant Disease* 98(4), 568–568. DOI: 10.1094/PDIS-07-13-0730-PDN.

Zhou, L., Yu, L., Huang, P., Liu, W., Tang, Z. *et al*. (2019) First report of Sclerotinia stem rot caused by *Sclerotinia sclerotiorum* on *Macleaya cordata* in China. *Plant Disease* 103(3), 584–585. DOI: 10.1094/PDIS-08-18-1324-PDN.

Zimmer, R.C. and Campbell, C. (1990) First report of *Sclerotinia sclerotiorum* on *Lathyrus sativus*. *Canadian Plant Disease Survey* 70, 17–18.

Zimmer, D.E., Klisiewicz, J.M. and Thomas, C.A. (1963) Alternaria leaf spot and other diseases of safflower in 1962. *Plant Disease Reporter* 47, 643.

3 The Pathogen *Sclerotinia*

Abstract

The *Sclerotinia* genus has been designated as the type genus of the Sclerotiniaceae family. The classification and naming system relies on criteria such as type of stroma, color of ascospores, existence of and type of conidia, and a functional conidial state. The identification of *Sclerotinia* species is based on characters categorized on macroscopic, cultural, biological, microscopic, and microanatomical observations of sterile tissues. Different techniques—such as sclerotial ontogeny, mycelial interactions, cytology, and electrophoretic assays—have been employed to delineate species *Sclerotinia sclerotiorum*. The generic diagnosis is based on morphological characters of apothecia, stroma, substratal stroma, sclerotial stroma, ascus, and ascospores. *S. sclerotiorum* is the major pathogenic species affecting crucifers, apart from a new species, *Sclerotinia subarctica*, which attacks turnip rape and *Ranunculus acris*. The *Sclerotinia* genomics has been unveiled through genome sequencing, phylogenetic analysis, gene expression related to infection and pathogenesis, transcriptomic and secretomic investigation, genome architecture, genome assembly, gene annotations, and identification of effector genes responsible for the molecular mechanisms of infection and pathogenesis.

3.1 Introduction

Mycologists and plant pathologists have been interested in the taxonomic status of commercially relevant, plant pathogenic species of *Sclerotinia* for a long time, but the issue has remained unsolved. *Sclerotinia* is a genus that contains more than 250 species that exhibit a variety of ecological relationships. This genus contains both pathogenic and nonpathogenic species. The delineation of this genus has caused debate and misunderstanding. The outcomes of earlier attempts to identify the pathogenic species using conventional traits have been mixed, either identifying too few or too many species. Plant pathologists find it more difficult to correctly identify the diseases they come across because these methods do not align with the current knowledge of biology and microanatomy of Sclerotiniaceae group. The plant pathogenic species of *Sclerotinia* have been reclassified into a more narrowly defined genus by examining the preserved type specimens and analyzing living isolates taking into account the microanatomical and cultural traits utilized by modern discomycotina taxonomists.

3.2 Taxonomy and Nomenclature

The Sclerotiniaceae family was created by Whetzel (1945) to include discomycetes that do not have an operculum and that yield globose spermatia, stipital apothecia, ellipsoidal ascospores, and stromata. The use of keys and diagnoses based on important traits, such as stroma type, ascospore color, the existence of a functioning conidial state, and conidia type, has made it easier to identify new and revised genera within the Sclerotiniaceae. The Sclerotiniaceae was found to have a type genus in the genus *Sclerotinia*.

Scientist Whetzel distinguished between two basic forms of stroma: the substratal stroma and the distinct sclerotium. A portion of the stromatal surface of the substratal stroma, an unidentified type of stroma composed of a medulla—a portion of the host substrate invaded by hyphae—is covered in a thin black rind. Whetzel and later writers classified a number of unnamed genera in the *Sclerotinia*, together with *Lambertella*, *Ciboria*, *Ciboriopsis*, [= *Moleroduscus*], *Lanzia*, *Poculum*, and *Rutstroemia* because of this kind of stroma. The family was dubbed Sclerotiniaceae by

DOI: 10.1079/9781800629592.0003

Whetzel and other writers. The distinct sclerotium, however, can either develop within host tissues with remnants of these tissues remaining within the sclerotial medulla, as can be seen in *Ciborinia, Verpatinia, Monilinia, Myriosclerotinia, Phaeosclerotinia, Scleromitrula, Seaverinia, Botryotinia,* and *Streptotinia*, or it can develop independently from the suspect tissues, as demonstrated in *Sclerotinia* and *Martininia*, where susceptible tissues are not embedded in the sclerotial medulla. In stromatina, sclerotium occurs in two distinct forms: a thin, effuse mantling stroma with a medulla and rind, and minute, black 'sclerotules' that develop above the mantling stroma on aerial mycelium. Noviello and Korf (1961) established a straightforward laboratory technique to determine if susceptible tissues are integrated into the sclerotial medulla.

The existence of a functioning conidial state is another important characteristic that Whetzel and other writers have made use of. *Phaeosclerotinia* (*Monilia*), *Monilinia* (*Monilia*), *Pycnopeziza* (*Acarosporium*), *Scleromitrula* (unnamed), *Ovulina* (*Ovulitis*), *Botryotinia* (*Botrytis* and *Amphobotrys*), *Streptotinia* (*Streptobotrys*), *Seaverinia* (*Verrucobotrys*), *Gloeotinia* (unnamed), and *Septotinia* (Septotis) are among the genera that form sclerotia and have a recognized anamorphic (conidial) state. *Sclerotinia* is among the genera that do not, however, have a recognized functional conidial state. Only three genera—*Lambertella, Martininia,* and *Phaeosclerotinia*—from the Sclerotiniaceae family generate brown ascospores; the remaining genera are known to produce hyaline ascospores. Buchwald (1949), Dumont (1971), and Korf (1973) have taken into consideration the characteristics of the sterile tissues of the apothecium and sclerotium in the delimitation of genera, despite the fact that Whetzel generally ignored it within the Sclerotiniaceae. Moreover, Kohn (1979) further narrowed the definition of the genus *Sclerotinia* by limiting its inclusion to species whose globose cells are placed perpendicularly to the receptacle surface in the ectal excipulum, or the outermost layer of the apothecium. In the stems of sedges, rushes, and grasses, the *Myriosclerotinia* genus—which is different from *Sclerotinia*—forms sclerotia. Furthermore, it creates a *Myrioconium* condition (likely spermatia) in specific host tissue chambers. Conversely, *Sclerotinia* sensu Kohn (1979) exhibits the superficial development of both the sclerotia and the *Myrioconium* state. Furthermore, the anatomical composition of the apothecium varies across the two taxa.

3.3 The Correct Name for *Sclerotinia*

Among several studies, research conducted by Dennis (1978) classified *Sclerotinia* in a broader sense to include *Monilinia, Ovulinia, Ciborinia, Botryotinia, Myriosclerotinia,* and *Sclerotinia*; Kohn (1979) acknowledged *Sclerotinia* in a narrower sense. This taxonomic determination is interpreted subjectively. Fuckel (1870) defined the genus *Sclerotinia* to include *Sclerotinia candolleana, S. fuckeliana, S. tuberosa, S. baccata,* and *S. libertiana* (a required synonym of *S. sclerotiorum*, which Fuckel constructed to avoid a putative autonym). *S. baccata*, one of these early species, is no longer relevant to *Sclerotinia* and has been reclassified under Pezizales.

Based on taxonomic considerations, Whetzel (1945) classed *S. fuckeliana* as *Botryotinia* and *S. candolleana* as *Ciborinia*. If Whetzel's taxonomic determination is rejected (as it is by numerous plant pathologists and a small number of taxonomists), *Sclerotinia* could be seen more broadly as suggested by Fuckel, incorporating a large range of plant pathogenic taxa.While several studies, such as Dennis (1978), classified *Sclerotinia* in a broader sense to include *Monilinia, Ovulinia, Ciborinia, Botryotinia, Myriosclerotinia,* and *Sclerotinia*, Kohn (1979) acknowledged *Sclerotinia* in a narrower sense. This taxonomic determination is interpreted subjectively. Fuckel (1870) defined the genus *Sclerotinia* to include *Sclerotinia, Candolleana, S. fuckeliana, S. tuberosa, S. baccata,* and *S. libertiana* (a required synonym of *S. sclerotiorum*, which Fuckel constructed to avoid a putative autonym). *S. baccata*, one of these early species, is no longer relevant to *Sclerotinia* and has been reclassified under Pezizales. Based on taxonomic considerations, Whetzel (1945) classed *S. fuckeliana* as *Botryotinia* and *S. candolleana* as *Ciborinia*. Given that Whetzel's taxonomic determination is not recognized by many plant pathologists and taxonomists, *Sclerotinia* can be seen more broadly in the context of Fuckel's proposal, which includes a large range of plant pathogenic taxa.

When it comes to choosing the proper nomenclature for *Sclerotinia sclerotiorum*, there is no room for discretion with the narrower notion put forward by Whetzel (1945) and Kohn (1979), even though taxonomic reasons allow for some flexibility in the broad definition of Fuckel. With *S. candolleana*, the genus *Sclerotinia* was initially recognized and assigned as the lectotype by Honey (1928). In 1945, Whetzel renamed his own genus *Sclerotinia* with *S. sclerotiorum* and transferred *S. candolleana* to his newly formed genus *Ciborinia*, ignoring Honey's lectotypification. Accepting Whetzel's more narrow definition of genera (a taxonomic decision), *Sclerotinia*—lectotypified by *S. candolleana*—becomes the first name for *Ciborinia* that is now in use. Korf and Dumont (1972) presented evidence supporting this method when they introduced *Whetzelinia*, with *S. sclerotiorum* serving as an example. Due to the widespread acceptance of Whetzel's concept and naming of *Sclerotinia* with *S. sclerotiorum* among experts, particularly plant pathologists, it has been suggested that *S. sclerotiorum* remain the lectotype of *Sclerotinia*. The International Association of Plant Taxonomists General Committee and the special committee for lichens and fungi have approved this proposal. Since 1981, the International Botanical Congress has not yet made a definitive conclusion. Thus, from this moment on, it is proper to refer to *Sclerotinia sclerotiorum* and to drop the generic name *Whetzelinia*.

3.4 Species Characters in *Sclerotinia*

Observing and assessing various traits that fall into four primary categories—macroscopic, cultural, biological, and microscopic—help determine a taxonomic classification. Discomycetes were categorized and characterized in a very different way after Nannfeldt's Studien über die Morphologie und Systematik einer non lichenisierten inoperculaten Discomyceten was published in 1932. The use of sterile-tissue microanatomical research as an additional source of taxonomic features was first presented in this work. Before Nannfeldt's research, *Sclerotinia* species descriptions were restricted to a particular set of traits.

1. The macroscopic features, including the apothecium's color, size, and shape, as well as the stipe and sclerotium. Cultural characteristics often refer to the size and distribution of sclerotia on agar plates.
2. Biological traits encompass the host, season, and specific part of the substrate that has been invaded.
3. Microscopic characteristics typically focus on the size, shape, and color of the ascospores, asci, and paraphyses.

It is essential to acknowledge the significance of these characteristics and their impact on the decisions made regarding taxonomy. The zones of the apothecium, stipe, and sclerotium are further characterized according to Nannfeldt's classification of microanatomical features, in addition to the hymenium, which has traditionally drawn the greatest attention. The sterile zones of the sclerotium and apothecium have distinct tissue types (Korf, 1973). These zones include the ectal excipulum, medullary excipulum, and subhymenium. The edge, flank, stipe, and any hairs—or, in the case of the Sclerotiniaceae, the tomentum hyphae—are the three component zones that make up the ectal excipulum.

Different tissue types in the apothecial and sclerotial zones define the genus *Sclerotinia*. The dense layer of entangled prosenchyma, known as the subhymenium, is encased in gel and usually has brown walls. The medullary excipulum is composed of a loosely woven textura intricata arranged parallel to the surface of the apothecium. Textura prismatica makes up the zone of ectal excipulum, which is the most remarkable feature. It gradually changes into textura globulosa when it moves past the apothecial margin and becomes perpendicular to the apothecial surface. The cells undergo this change as they swell, round off, and partially disarticulate. Globose cells comprise the ectal excipulum of the stipe, and these cells frequently include tomentum hyphae. Usually, the walls of these cells and hyphae are brown. The sclerotial medulla in *Sclerotinia* is devoid of sensitive tissues. Rather, Saito (1977) observed that it is constituted of hyaline textura oblita with extensively gelatinized hyphal walls made of proteins, >β−1, and 3-glucans. The apices of these medullary cells grow perpendicular to the sclerotial surface and mature into textura prismatica, which forms the sclerotial rind. Additionally, these cells grow globose and somewhat disarticulated. There may be pigmentation up to six layers thick in the walls of the rind's outermost cells. The ascus pore channel wall reacts positively to Melzer's reagent (0.5g iodine, 1.5g potassium iodide, 20g chloral hydrate, and 20 ml distilled water) in all species of *Sclerotinia*. One species, Kohn (1979), showed dimorphism in spore size, like several species of *Sclerotinia allii* (Sawada, 1919) and *Monilinia* (Woronin, 1888), both of which are recognized as *Ciborinia* species.

3.5 Variability in Species Characters of *Sclerotinia*

Concern has been raised for a long time over the validity of taxonomic features. According to Purdy (1955), isolates that were provisionally classified as *Sclerotinia minor*, *Sclerotinia trifoliorum*, and *S. sclerotiorum* had differences in the diameters of their ascus, ascospore, and sclerotium. Purdy found that it was difficult to distinguish between different species by comparing the average and range of ascus and ascospore measures from two to three generations with those found in species diagnoses and other literature sources. He consequently came to the conclusion that distinguishing the asci or ascospores of one species from another was problematic, if not impossible. Purdy also noted a variety of cultural variations among species, from those that developed enormous masses resembling sclerotium to those that did not produce sclerotia. In addition, he observed a constant change in sclerotium size, progressing from *S. minor's* little sclerotia to *S. sclerotiorum's* larger ones. He was, therefore, unable to distinguish between species using the sclerotium size. Purdy combined *S. minor* and *S. trifoliorum* under *S. sclerotiorum*, along with *S. trifoliorum* var. *fabae*, *S. intermedia*, and *S. sativa*, based purely on this evidence. He contended that the variety of the features that have been employed makes it challenging to identify these species in real-world situations. Purdy's data may show the variety of the traits under study, but species identification can take into account a wide range of other features. It seemed at first that fewer species would be identified, which would make it simpler to identify the *Sclerotinia* species that are harmful to plants. However, it became clear that this broad definition of *S. sclerotiorum* looked to incorporate a number of taxa under a single species when more was learned about the biology of this group. Taxonomists and plant pathologists have researched particular cultural features in great detail, but perhaps less attention has been paid to the variations in apothecial structures between apothecia developed in nature and those created *in vitro*. More than 65 isolates of the species *Sclerotinia*, *Myriosclerotinia*, *Ciborinia*, and *Botryotinia*—which were obtained from various regions of the globe, including Europe, Australia, New Zealand, Asia, and North America—were studied by Kohn (1979). According to the results, 26% of these isolates showed signs of apothecial production and cultural traits. Interestingly, only those isolates that were finally determined to be *S. sclerotiorum* (10 out of 18 isolates), *S. trifoliorum* (two out of five isolates), and *S. minor* (five out of seven isolates) generated apothecia under *in vitro* conditions. While some isolates were acquired from individual ascospores or masses of ascospores, the majority were derived from sick tissue. Before being moved to 9 cm diameter Petri plates containing potato dextrose agar (PDA), the cultures were developed and maintained on PDA. After that, they were incubated for 3–4 days at room temperature. The cultures were then moved from the colony's growing edge to 500 ml Erlenmeyer flasks that contained autoclaved carrot discs and 25 ml of distilled water using a 5-mm-diameter cork borer. Next, these flasks were incubated at 15°C for a full 4 weeks without being exposed to light. After being gathered, the sclerotia were rinsed in sterile distilled water and placed in sterile Petri dishes with glass wool that had been wet with distilled water. The sclerotia were then exposed to a 'cold conditioning' procedure at 0°C for 4 weeks. After that, the dishes were placed in a growth chamber at a temperature of 15°C. The growth chamber had a 14 hour photoperiod and was lit with both fluorescent and incandescent lights, giving it an approximate illumination of 21,520 Lux (2000 ft-c). Four to 12 weeks after the placement in the growth chamber, apothecial initials began to form. As stated by Saito (1977), apothecial initiation in *Sclerotinia* has been the subject of extensive research by a number of different scientists. Fertile apothecia were generated when the illumination level reached 4304 Lux (400 ft-c) or greater, according to Kohn (1979). Only stipes were created when the illumination levels considerably dropped below 4304 Lux. A fertile hymenium was not developed by apothecia below this threshold level; instead, they merely developed a palisade layer of tomentum hyphae.

The apothecia and sclerotia produced in a controlled setting were contrasted with those produced spontaneously by *S. sclerotiorum*, *S. minor*, and *S. trifoliorum*. Multiple isolates of *S. tuberosa* and one strain of *S. trifoliorum* were obtained from a European collection of fresh apothecia, from which a mass ascospore shoot was produced. Sections of the apothecia and sclerotia, measuring 20 and 5 μm, were cut, respectively. Melzer's reagent, cotton blue in lacto phenol, and potassium hydroxide (KOH)/phloxine alone were used to analyze the sectioned apothecia. Groves and Bowerman (1955) found that "the apothecia in the Swedish specimen (topo type) were smaller and more delicate than those developed in cultures of Kohn (1979), who observed in other species of Sclerotiniaceae that apothecia produced in culture are usually more robust than those formed in nature.' This information was discovered during the investigation of the social behavior of Sclerotiniaceae. Before Christiansen (1966) examined sectioned apothecia to determine the histological relevance to this macroscopic variation, the variable apothecial structure of various species of *Ciboria*, *Ciborinia*, *Lambertella*, and *Rutstroemia* had not been explored further.

Melzer's reagent was used to investigate the sectioned apothecia. It has long been known that pathogenic *Sclerotinia* species found in field collections and laboratory cultures exhibit variation in the size, shape, and color of their apothecia as well as in their exterior texture and color. Kohn (1979) examined sectioned apothecia and found that changes in tissue types or cell sizes could not be the cause of this size variance. Christiansen's research has observed that the core problem is related to the various layers that make up the apothecia; these layers are all stable, but they differ in terms of compactness and the presence of admixtures. The proliferation and compactness of cells appear to have an impact on the variation in gross size of *Sclerotinia* apothecia. The subhymenium is the only zone of the apothecium that constantly has gel-bound cells; other zones can vary across isolates and may or may not contain gelatinous material. The presence or absence of gelatinous material in this group of species should be carefully considered as a taxonomic characteristic, since it may be involved in maintaining moisture within the apothecial tissue and is likely affected by different levels of moisture in the microenvironment during apothecium development.

Discomycetes taxonomists have long regarded the bluing of the ascus pore channel in the presence of Melzer's reagent as a significant taxonomic feature. According to Kohn and Korf (1975) and Nannfeldt (1976), pretreating with KOH can either start or improve the reaction. Gel-bound cells are only reliably found in the subhymenium. All *Sclerotinia* species, however, feature ascus pore channels that, in the absence of KOH pretreatment, turn blue in the presence of Melzer's reagent. When *S. minor* was pretreated with KOH, the majority of the collections exhibited a stronger reaction—if any at all. Remarkably, this reaction is not mentioned in Jagger's original species diagnosis (Jagger, 1920). Furthermore, all species exhibited bluing of the sterile zones of the apothecium in the presence of Melzer's reagent; however, the frequency of bluing varied among isolates. There was no difference seen in this variability between apothecia obtained from isolates grown *in vitro* and apothecia collected in the field. The apothecial zone that most frequently became blue in the presence of Melzer's reagent was the subhymenium, presumably because the layer contained gelatinous material. Not every apothecium with gel-bound tissues showed this reaction, though. Moreover, the pigmentation diversity in cell walls throughout all apothecium zones matched the macroscopic color variance of apothecia, including mottling. Though the color of the cell walls in this area can range from hyaline to light brown or dark brown, the pigmentation of the ectal excipulum in the apothecium and stipe is frequently observed. Because soil particles can stick to tomentum hyphae, their presence on the stipe and receptacle may be a factor in the outward mealiness. Tomentum hyphae are most commonly seen on the stipe, but their frequency and presence on the ectal excipulum of the apothecium and stipe can differ substantially. Tomentum appears to be a more dependable trait of the sclerotial rind, despite the fact that it is not a valid taxonomic diagnostic for species within *Sclerotinia*.

Comparing ascospore measurements with those provided in species diagnoses revealed variations in range and average values of up to 3 µm in various mounting media. Ascospores and dry specimens with collapsed cells are frequently rehydrated and inflated using diluted solutions of potassium hydroxide and chloral hydrate. It is noteworthy that the majority of writers do not identify the mounting material that is utilized for measurements. It is therefore important for researchers to investigate how mounting media affects ascospore size, which naturally varies within a range.

At temperatures between 15 and 20°C, cultures grown in PDA or autoclaved carrot discs consistently produce either massive sclerotia grouped in a radial pattern at the outer edge of the developing margin of the colony or little scattered sclerotia throughout the colony. This trend is consistent with the initial isolates being either big sclerotial forms (*S. sclerotiorum* and *S. trifoliorum*) or tiny sclerotial forms (*S. minor*) (Kohn, 1979). This observation was confirmed by Willetts and Wong (1971) through their study of sclerotial ontogeny, which revealed differences in the sclerotial initiation site and formation mechanism between these species. Several cultures of these three species have shown variations in the size and quantity of sclerotia formed as well as a deterioration in the ability to produce sclerotia. It has frequently been observed that an isolate may not always be able to produce sclerotia. However, it has been found that sclerotia form when the isolate is transferred to autoclaved carrot discs. Subcultures with distinct characteristics are obtained from the sclerotia formed on the carrot discs.

In order to improve morphological data and provide insights into the classification of biological species within this group, a number of approaches that had not been previously employed in comparative studies of economically significant *Sclerotinia* species have been investigated. These methods include mycelial interactions, sclerotial ontogeny, electrophoretic tests, and cytology. According to ; Bjorling's (1942, 1951) documentation, the ascospores of *S. sclerotiorum* and *S. trifoliorum* have two and four nuclear numbers, respectively, and six haploid chromosomes. According to Frandsen (1946), both species have eight haploid chromosomes. Subsequently, it was discovered by Wong and Willetts (1979) that *S. sclerotiorum* contained two nuclei per ascospore, whereas *S. minor* and *S. trifoliorum* had four nuclei each. Additionally, they reported that the haploid chromosome numbers in hyphal tips and germinating ascospores of *S. minor* were four, and those of *S. sclerotiorum* and *S. trifoliorum* were eight. Kohn (1979) used studies of recently created apothecia to validate these conclusions on nuclear numbers. However, Kohn (1979) found two to four nuclei per ascospore, while Berthet (1964) claimed that the nuclear number in *S. tuberosa* ascospores ranged from two to six.

Wong and Willetts (1975) divided their isolates of *Sclerotinia* species into four subgroups based on electrophoretic patterns for soluble proteins, aryl esterase, acid phosphatase, tetrazolium oxidase, glucose-6-phosphate dehydrogenate (Nicotinamide adenine dinucleotide phosphate (NADP) linked), and reduced nicotinamide adenine dinucleotide phosphate dehydrogenase. These isolates were obtained from several crops grown in New Zealand, Australia, Europe, and North America.

Group 1: Comprises tiny/small sclerotial isolates that were collected from a variety of host plants.

Group 2: Consists of big sclerotial isolates that have been isolated from legumes used as feed.

Group 3: Includes large sclerotial isolates from a range of host plants, including legumes and pasture.

Group 4: Contains only one isolate from an anemone.

According to Wong and Willetts (1975), *S. minor*, *S. trifoliorum*, *S. sclerotiorum*, and *S. tuberosa*, in that order, represent these groups. With *S. tuberosa* being 'characteristically different' from the other three species, they came to the conclusion that these four species are unique. Kohn (1979) recognized these species as well, placing *S. tuberosa* in the new genus *Dumontinia* and retaining the first three in *Sclerotinia* based on morphological characteristics.

3.6 Generic Diagnosis

The *Sclerotinia* Fuckel, Jahrb. Nassauischen Vereins *Naturk.* 23–24: 330. 1870 (Lectotype: *S. sclerotiorum* (Lib.) de Bary, typus conserv. proposition)

= *Ciboria* subg. *Sclerotinia* (Fuckel) Boud., Bull. Soc. Myc. France 1: 115. 1885
= *Whetzelinia* Korf and Dumont, *Mycologia* 64: 250. 1972 (nom Rejiclendum prop.)

Apothecia are stipulate, originating from a sclerotium, and range in breadth from 2 to 10 mm. A convex meniscus forms when a liquid's cohesive forces (attract itself) are stronger than its adhesive forces (attract to the container). This results in the liquid's surface curving upwards, like a dome or globule type at the edges. The sclerotium's medulla is devoid of any host tissues and is formed separately from the host tissues. The asci have thin walls that thicken at the tip and are eight-spored, J^{+}. The ascospores are primarily elliptical in shape, contain two oil droplets, and are organized in a single row with smooth walls. The well-developed subhymenium is made up of textura intricata, which has light brown walls and is typically encased in gelatinous material. The medullary excipulum is composed of a hyaline, loosely woven textura intricata. The textura prismatica, which has walls ranging from hyaline to light brown, makes up the ectal excipulum and is positioned perpendicular to the apothecial surface. The ectal excipulum's cells are globose and frequently contain tomentum hyphae. On the surface of the hymenium and in cultures, microconidia are formed superficially. They originate from solitary phialides that are carried laterally on hyphae or aggregated in sporodochia, and they have a globose, hyaline form. One of the genus' unique characteristics within the Sclerotiniaceae family is its tuberous sclerotium, which does not absorb host tissues. It does not produce disseminative conidia and is carried superficially by the aerial mycelium. It also yields superficial microconidia and hyaline ascospores.

3.7 Morphology of Stroma

The stroma, which consists of a hyaline hyphal medulla surrounded by a layer of cells with melanized walls, acts as a storage and survival organ. Whetzel (1945) distinguished between two different stroma types.

3.7.1 Substratal stroma

This kind of stroma is distinguished by a narrow black ring that covers at least a portion of the stromatal surface and an indiscriminate medulla composed of host tissue infiltrated by hyphae.

3.7.2 Sclerotial stroma

This type of stroma can grow independently of the host tissues or inside them, incorporating remnants of those tissues in the medulla. The sclerotial stroma of the *Sclerotinia* species develops independently of the host tissues, excluding any vestiges of the host tissue in the medulla, and is frequently collected without any identifiable host organ remnants.

The sclerotial medulla of *Sclerotinia* is formed by the hyaline textura oblita, which has walls that are 2–3 μm thick and highly gelatinized. The apices of these medullary cells, however, make up the sclerotial rind. These apices become textura prismatica and extend perpendicular to the sclerotia's surface. The upshot is that the cells enlarge, get globose, and frequently become disarticulated. The colour of the layers of the wall changed to brown,

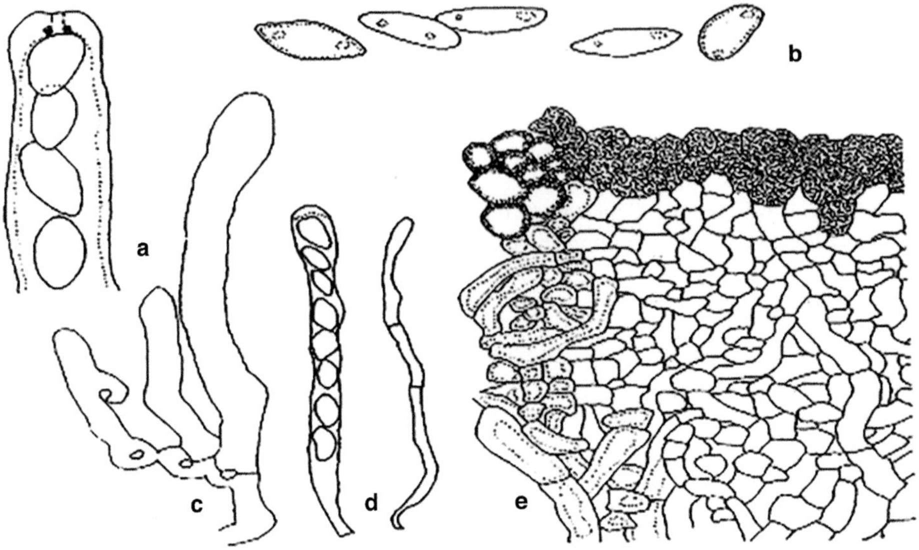

Fig. 3.1. (a) The ascus of *Sclerotinia sclerotiorum* with a J+ pore channel wall (magnification 1500×); (b) ascospores of *Sclerotinia sclerotiorum* (magnification 1500×); (c) young asci emerging from staves of *Sclerotinia sclerotiorum* (magnification 1500×); (d) ascus and paraphyses of *Sclerotinia sclerotiorum* (magnification 500×); and (e) cross section of the sclerotial rind and medulla of *S. sclerotiorum*. (Adapted from publication of Kohn, 1979, with permission.)

which is two to six depths within the outermost cells (Figs 3.1 and 3.2). It is known that the *Sclerotinia* species may produce sclerotia in large quantities when cultured; however, older isolates may no longer be able to do so. Sclerotial growth takes place on aerial mycelium just above the agar surface, according to Kohn (1979).

3.8 Molecular Mechanism of Sclerotial Development

Even though the synthesis of sclerotium is important for many fungi and has economic value, our knowledge of the molecular mechanisms underlying sclerotial development is still limited. Based on recent data, it appears that elements involved in reversible phosphorylative regulation, which regulates various aspects of fungal growth, proliferation, and morphogenesis, also regulate sclerotial development. Furthermore, it is now clear that Reactive Oxygen Species (ROS), primary and secondary metabolism, and sclerotial development are all critically related. Despite possible differences in their manner of action, it is plausible to anticipate that some of the molecular modules driving sclerotial development will be comparable, given the similarities in sclerotial structure, anatomy, physiology, and biochemistry among various fungi (such as cAMP – cyclic adenosine monophosphate). However, notable variations are also to be expected given the different kinds of fungi that produce sclerotium. This argument can be demonstrated by drawing a comparison between the conidiation processes of *A. nidulans* and *Neurospora crassa*. Significant conservation may be seen in the upstream signaling components that control conidiation, suggesting that these fungi respond to environmental cues in a similar manner to initiate macroconidial development. However, there is no evidence of conservation among the downstream signaling components, indicating that conidiation has evolved independently in these two taxa (Borkovich *et al.*, 2004; Fig. 3.3). There has been a rapid accumulation of study reports in recent years, despite the fact that the complicated processes involved in sclerotial development,

Fig. 3.2. (a) Ascospores of *S. trifoliorum* (magnification 1500×during germination); (b) magnified view of ascospores of *S. trifoliorum* at 1500×; (c) ascus of *S. trifoliorum* with J^+ pore channel wall magnification 1500×; (d) magnified view of ascus and paraphyses of *S. trifoliorum* at 500×; (e) a small conidiophore in the developing sporodochium formed on aerial hyphae in culture is visible in the *Myrioconium* microconidial state at a magnification of 1500×; and (f) *S. trifoliorum* cross section of the sclerotial rind and medulla. (Adapted from publication of Kohn, 1979, with permission.)

the activities within the resting structure, and germination are still not fully understood. This indicates that molecular analysis of sclerotial biology is far from dormant (Erental *et al.*, 2008).

Based on the macroscopic features that may be observed, sclerotial development has historically been divided into three separate stages (Fig. 3.4, upper portion): The process

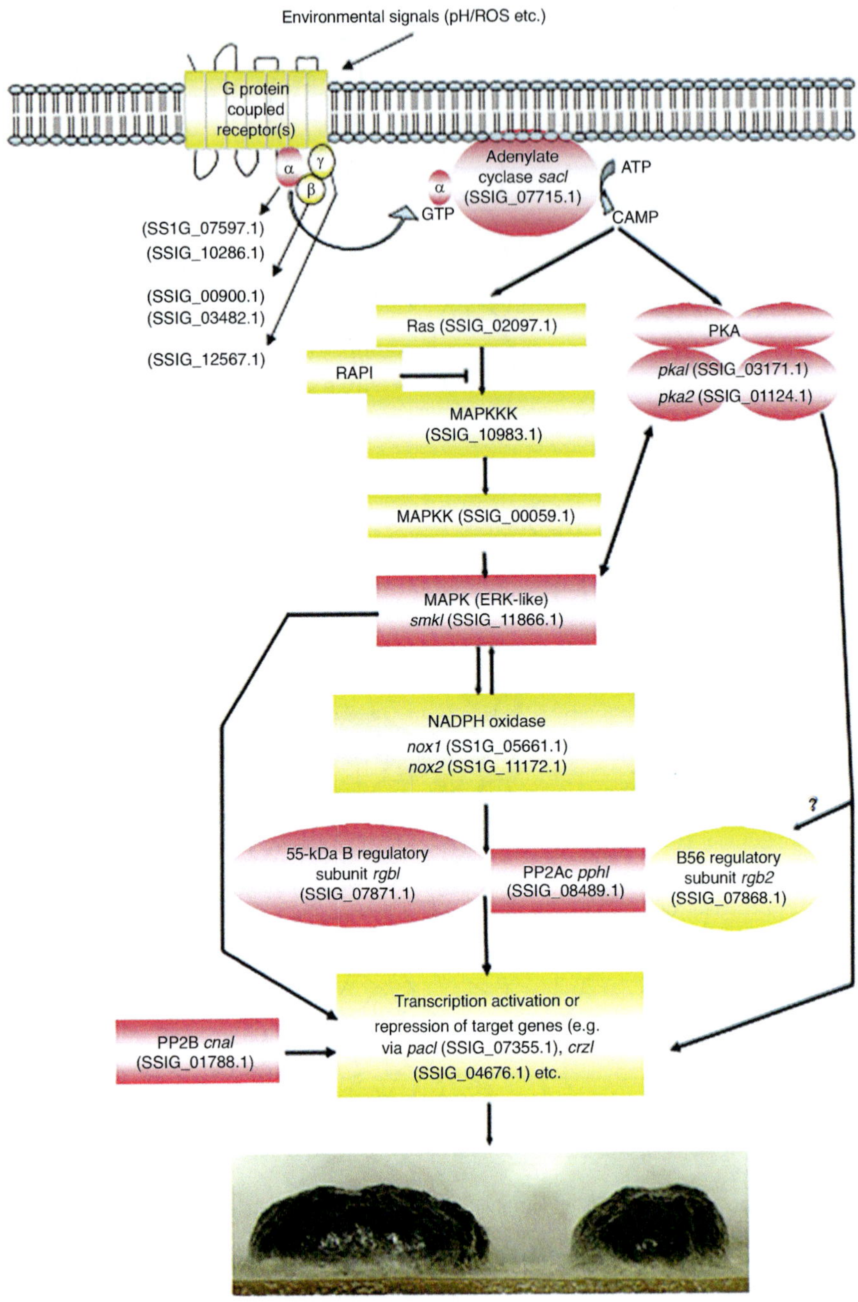

Fig. 3.3. This model elucidates the role of signal-transduction components involved in sclerotial development in *Sclerotinia sclerotiorum*, with a focus on the components that have been analyzed so far (highlighted in red color). However, there are still unresolved aspects regarding the hierarchical positioning of these components and the nature of their interactions, including whether they have positive, negative, or cooperative regulatory effects. The numbers in parentheses indicate the current genome annotation references, which can be found at http://www.broad.mit.edu/annotation/genome/sclerotinia_sclerotiorum/Home.html. (Adapted from publication of Erental *et al*., 2008, with permission.)

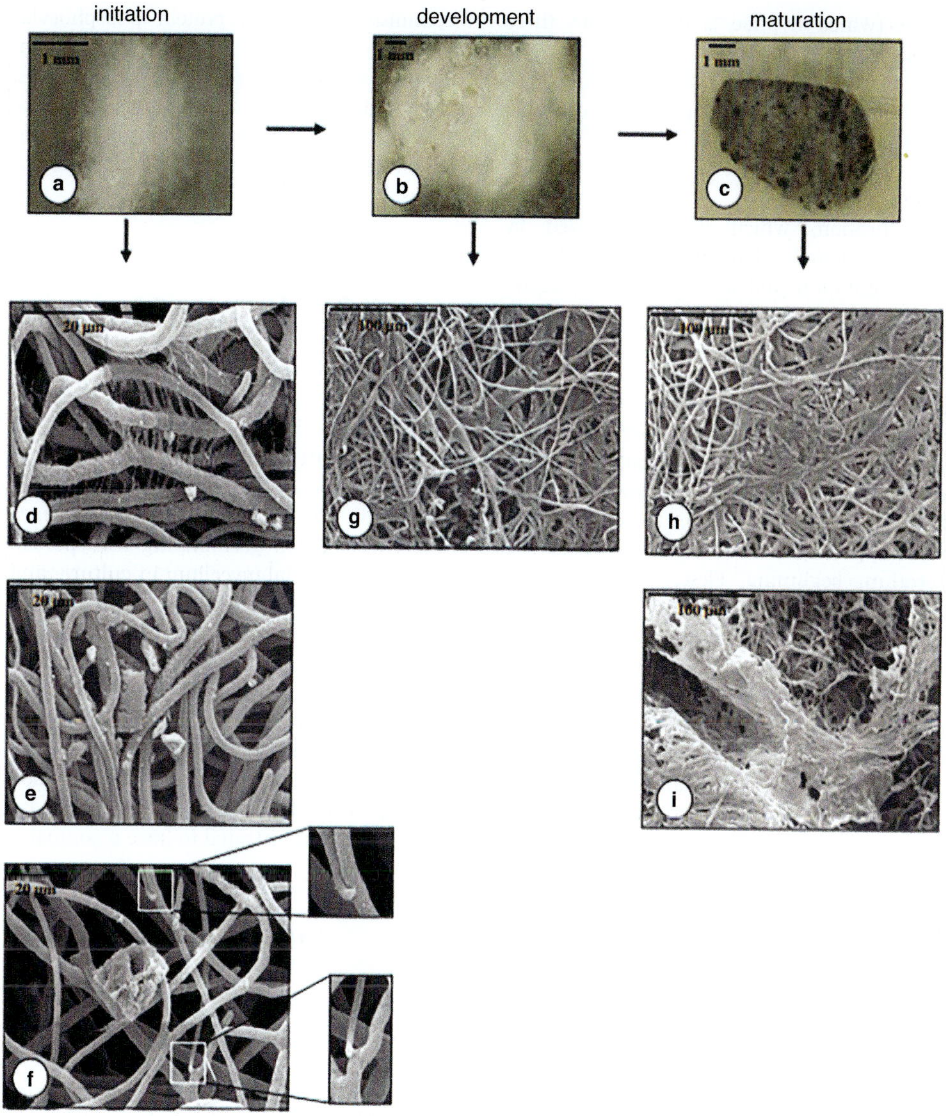

Fig. 3.4. The upper portion shows the macroscopic development of *Sclerotinia sclerotiorum* sclerotia, while the lower portion depicts the microscopic development. The process can be divided into following stages: (a) initiation, (b) development, (c) maturation, (d) secretion of a mucilage-like substance, (e) curling of hyphae and formation of localized crystalline structures, (f) crystalline structures and hyphal fusion, (g) formation of a continuous outer sclerotium layer, and (h–i) a wider and more highly condensed secreted layer. To prepare the samples for scanning electron microscopy, they were fixed for 4 hour using a 5% (v/v) glutaraldehyde solution in 0.1 M phosphate buffer with a pH of 7.2. The samples were then washed five times with the same buffer and dehydrated using a series of ethanol washes ranging from 25 to 100%. Following this, the fixed samples were dried for 1 min using a CPD750 drier and coated with gold using an E5150 Polaron SEM coating system apparatus. Finally, the samples were observed under a Jeol JSM 35 microscope. (Adapted from publication of Erental *et al.*, 2008, with permission.)

of intertwined hyphae is divided into three stages:

1. initiation, which is the appearance of small, distinct initial forms (Fig. 3.4a);
2. development, which is a growth in size (Fig. 3.4b); and
3. maturation, which is characterized by surface demarcation, internal consolidation, pigmentation, and, frequently, droplet secretion (Fig. 3.4; Townsend and Willetts, 1954).

Moreover, the microscopic analysis provides more information on the substages of the ongoing sclerotium formation process, including hyphal adhesion, early densification, and secretion along filaments (Fig. 3.4, lower portion). Multiple steps are involved in the change from linear hyphal growth to a more condensed sclerotium beginning. First, a material that resembles mucilage is secreted and serves as an adhesive for the hyphae (Fig. 3.4d). The next steps are the creation of localized crystalline structures (Fig. 3.4e,f), hyphal fusing (Fig. 3.4f), and hyphal curling (Fig. 3.4e). The continuous outer sclerotium layer that forms in the mature sclerotium has these crystalline structures as its nucleus (Fig. 3.4g). The secreted layers widen and condense as the sclerotium ages, resulting in the change from immature white to mature black sclerotia (Fig. 3.4h,i). Because of its capacity to generate multiple, large sclerotia under laboratory conditions, its ease of transformation and gene cloning via protoplast and *Agrobacterium*, the availability of its full genome sequence, and the extensive physiological and morphological analyses carried out over the past 50 years, *S. sclerotiorum* is an ideal model system for studying sclerotial development (Bolton *et al.*, 2006; Weld *et al.*, 2006; Erental *et al.*, 2008). The control of phosphorylation has been the main area of study in the present understanding of sclerotial development in *S. sclerotiorum*. Numerous molecular elements have been discovered as regulators of sclerotium growth, including Ser/Thr phosphatases type 2A and 2B, ERK-like mitogen-activated protein kinase, and cAMP-dependent protein kinase. It has previously been demonstrated that these elements are involved in controlling other fungal developmental stages. Sclerotial development in different fungi is likely regulated by a combination of regulatory mechanisms, including protein phosphorylation, DNA methylation, ribosome inactivation, secondary metabolites, reactive oxygen species, pH, and light. The diversity of fungi that produce sclerotium may reveal various molecular pathways involved in this phase of development in various species (Erental *et al.*, 2008).

3.9 Microconidia

The *Sclerotiniaceae* family, like other genera, uses flask-shaped phialidic conidiophores to create globose, hyaline microconidia. These microconidia are found alone or in clusters on the hyphae margins, where they form sporodochia (Fig. 3.2). The hymenial surface of apothecia (from germinating ascospores), the surface of aerial mycelium in culture, and the surface of sclerotia are where *Sclerotinia* microconidia are formed. However, it is unknown how homothallic *Sclerotinia* species, or microconidia, act as functional spermatia (Keay, 1939). While it has been demonstrated that microconidia in *Botrytis fabae* during overwintering serve as germinable spores (Harrison and Hargreaves, 1977), *Sclerotinia* species have not yet been demonstrated to have a comparable function.

3.10 Ascocarp

Apothecia are developed from a sclerotium and show sexual reproduction and stipulation (Figs 3.5 and 3.6). According to Saito (1977), sclerotium medulla is the birthplace of stripe primordia, which subsequently burst the outer layer. The diameter of the apothecia ranges from 2 to 10 mm, and they are usually uniformly colored from cinnamon to amber, with some mottling and darkening at the borders. The tomentum hyphae present in the receptacle and stipe surface may give them a slightly powdery quality. Repeating croziers positioned vertically form the eight-spored asci (Figs 3.1 and 3.2). The asci have a cylindrical form that tapers at the point where they connect to the crozier. The thickness of the ascus walls is 1–2 μm (Figs 3.1, 3.2, 3.5 and 3.6). When *Sclerotinia* species are exposed to Melzer's reagent (J^+), the ascus pore

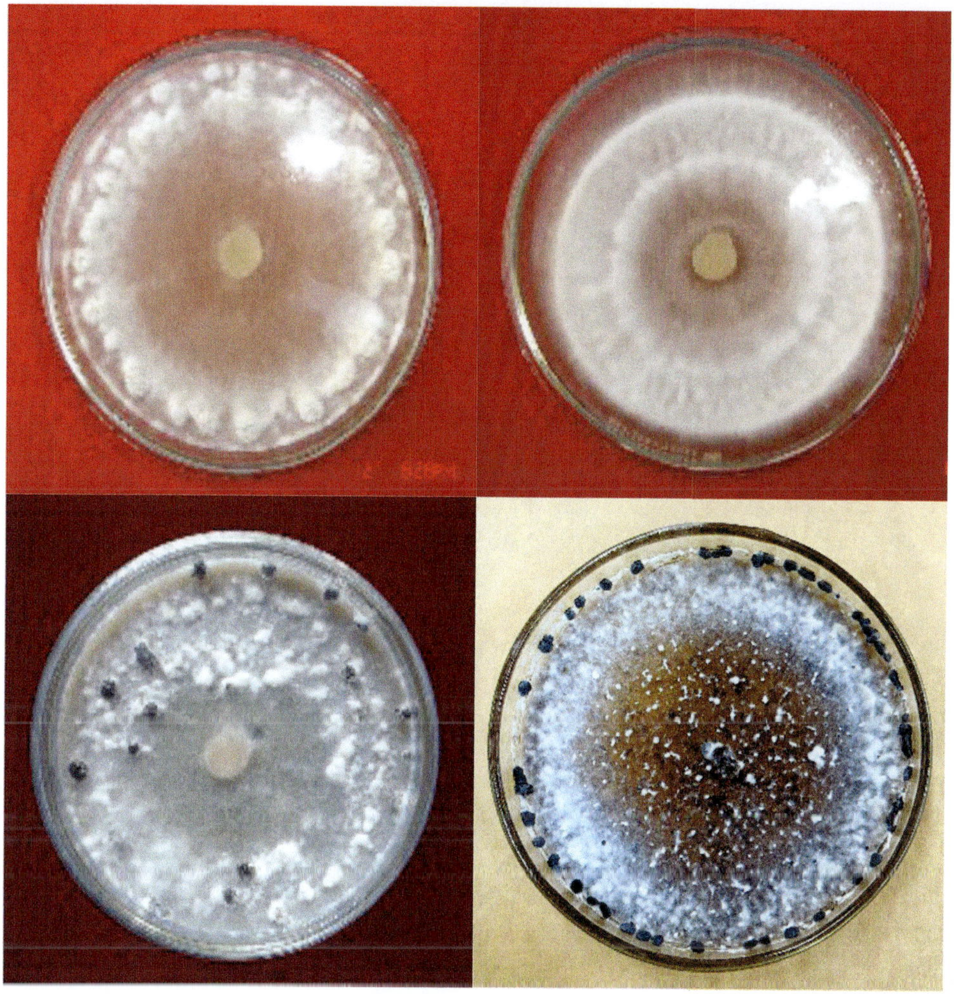

Fig. 3.5. The growth pattern of *Sclerotinia sclerotiorum* described as (a) mycelial growth, (b) initiation of sclerotia formation, (c) sclerotia formation at the outer periphery, and (d) mature sclerotia at the outer periphery.

channel wall exhibits a blue coloration. One species has a mild reaction initially, but it can be made stronger by pretreating with a 2% KOH solution, according to the protocol (Kohn and Korf, 1975). This species' translucent ascospores are elliptical in shape, but they can exhibit a slight flattening on one side (Fig. 3.8). This species has occasionally been found to exhibit dimorphism or variations in spore size. Within the ascus, small and big ascospores are usually distributed in a ratio of 4:4 (Fig. 3.7).

Woronin (1888) and Buchwald (1949) were the first to record and illustrate this occurrence in the genera *Ciborinia allii* and *Monilinia*, respectively. Within the ascus, the ascospores are organized in a single row (Fig. 3.8). Ascospores can grow in either a bipolar or unipolar manner during germination, with germination frequently starting in the center of the spores (Fig. 3.7). Mature ascospores in a lab setting have two oil droplets. Depending on the species, ascospores might have two to four

Fig. 3.6. Apothecia formation in *Sclerotinia sclerotiorum*.

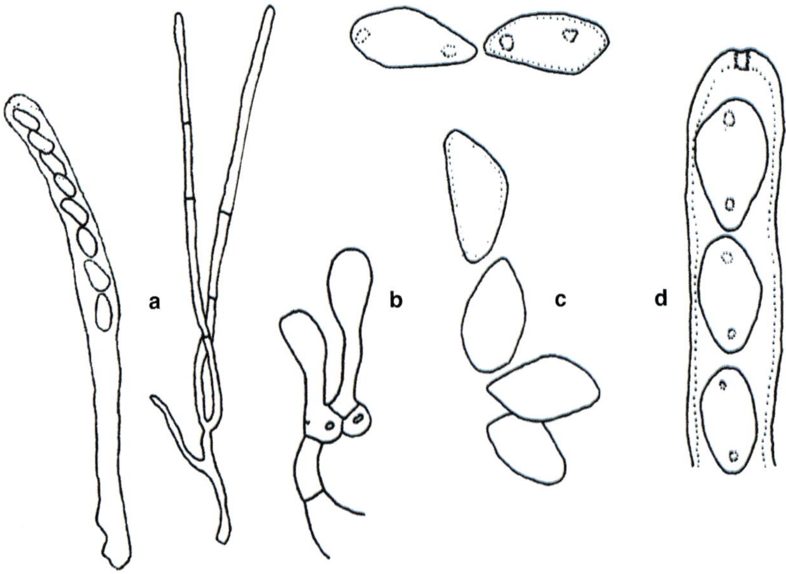

Fig. 3.7. The various structure of the pathogens are: (a) Ascus and paraphyses × 500, (b) young asci originating from crosiers × 1500, (c) ascospores ×1500, and (d) ascus with J + pore channel wall ×1500 describe different microscopic structures (ascus, paraphyses, crosiers, and ascospores) of *Sclerotinia*. (Adapted from publication of Kohn, 1979, with permission.)

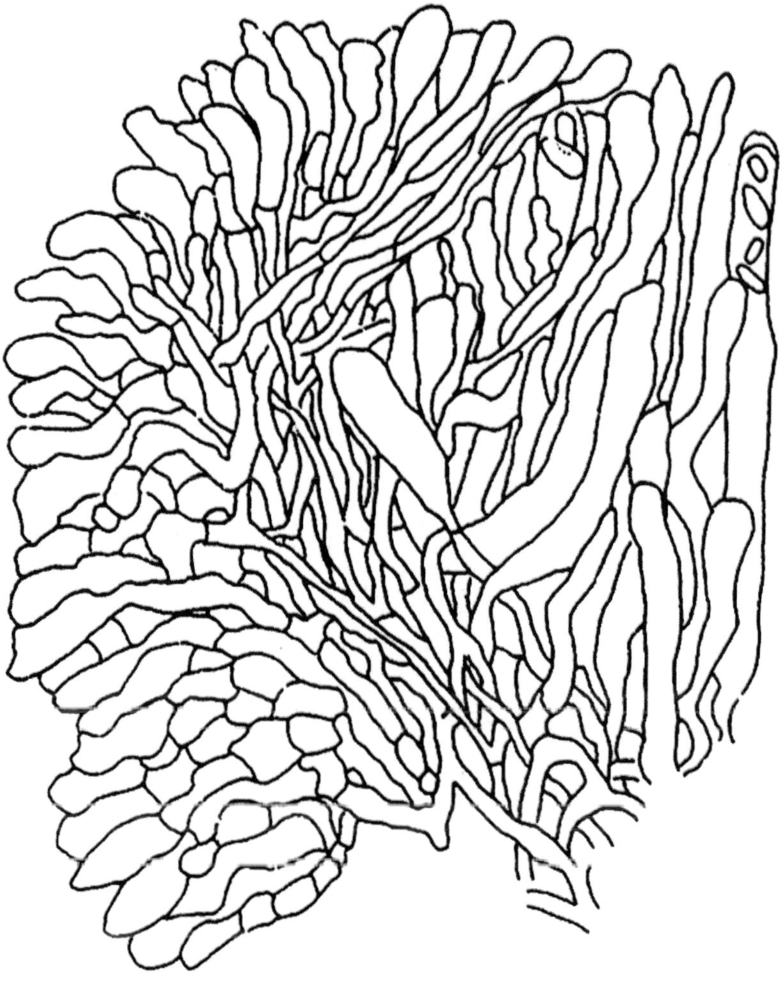

Fig. 3.8. Cross section taken at the apothecial margin at a magnification of 500×. *Sclerotinia* globose cells make up the ectal excipulum at the margin. (Adapted from publication of Kohn, 1979, with permission.)

nuclei (Kohn, 1979; Wong and Willetts, 1979). Figure 3.7 shows that the paraphyses are thin, sparsely segmented, and sometimes branching. The subhymenium, the layer beneath, is a tight, compact structure with intricately textured walls that is pale brown in color and usually submerged in a gel-like fluid. This specific layer frequently takes on a blue tint when exposed to Melzer's reagent.

The medullary excipulum is composed of a loosely woven textura intricata arranged parallel to the surface of the apothecium. The ectal excipulum, which is the outer layer of the apothecium and stipe, is made up of cells that are organized in a prism-like pattern perpendicular to the apothecial surface. These cells have walls that range from hyaline to pale brown. They have a tendency to swell and globulate, eventually disarticulating to some extent.

In one species, all of the spherical cells that make up the apothecium's border are present (Fig. 3.8). In contrast, the edge of the other species is made up of parallel textura porrecta cells that are oriented in line with the

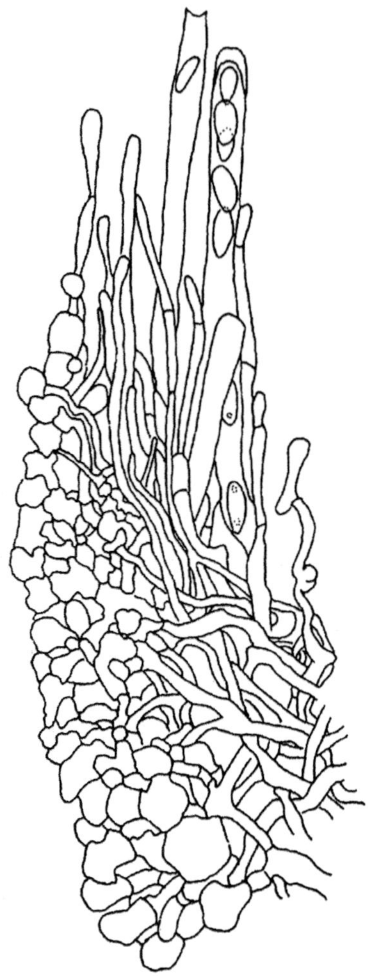

Fig. 3.9. The margin of the apothecia, when observed (magnification 500×), reveals a cross section. In this cross section, the ectal excipulum of *Sclerotinia sclerotiorum* is composed of prosenchyma. Prosenchyma is oriented perpendicularly to the apothecial surface. (Adapted from publication of Kohn, 1979, with permission.)

asci. The textura porrecta cells are positioned perpendicular to the apothecial surface and grow shorter as one approaches the sides of the ectal excipulum. Gelatinous material may or may not be present in this layer, and when the gelatinous material is treated with Melzer's reagent, bluing may occasionally be seen. The ectal excipulum of the apothecium, stipe, and sclerotium frequently contains one to two-cell-long tomentum hyphae. On the apothecium, these hyphae usually have hyaline-to-pale brown walls, and on the stipe, they frequently form darker brown clusters. Tomentum appears to vary in presence; however, it can be a helpful taxonomic feature of the sclerotial rind (Fig. 3.9).

3.11 Key to the Sclerotium-Forming Genera of the Sclerotiniaceae

Important sclerotium-forming genera of the family Sclerotiniaceae have been recommended as per details of Kohn (1979), which is as follows:

1. Apothecia engage in sexual reproduction or resemble verpoids, being stipitate and growing on a clearly defined sclerotium that possesses a well-differentiated rind and medulla. The production of conidia is not characteristic of *Sclerotinia* (cfr. *Phaeosclerotinia, Monilinia, Pycnopeziza, Scleromitrula, Botryotinia, Gloeotinia, Septotinia*.	
1. Apothecia engage in sexual reproduction or resemble verpoids, with a stipitate structure, growing on a clearly defined sclerotium that possesses a well-differentiated rind and medulla. Conidia are not present, except for the presence of phialidic microconidia (*Myrioconium*)	2
2. Ascospores brown	*Martininia*
2. Ascospores hyaline	3
3. Sclerotial medulla enveloping susceptible tissues or, in culture, sclerotia at least partially immersed in the agar	4
3. Sclerotial medulla free of susceptible tissues or, in culture, sclerotia formed above the agar surface	7
4. Ectal excipulum of apothecium composed of prosenchymatous cells	*Elliottinia*
4. Ectal excipulum of apothecium composed of globose cells	5
5. Sclerotia and microconidia produced within the stems of gramineous, juncaceous, and cyper aceoushosts	*Myriosclerotinia*
5. Sclerotia and microconidia produced within the stems of other monocot and dicot families	6
6. Apothecia verpoid	*Verpatinia*
6. Apothecia copulate	*Ciborinia*
7. Stroma consisting of mantling stroma of indefinite dimensions and smaller sclerotia (sclerotules) formed on aerial mycelium above the mantle; apothecia occurring on the mantling stroma only	*Stromatinia*
7. Not as above	8
8. Apothecia verpoid	*Mitrula sclerotiorum*. (a presumed parasite on sclerotia of *Sclerotinia* spp.)
8. Apothecia cupulate	9
9. Outer layer of apothecial ectal excipulum composed of prosenchymatous cells usually embedded in a gelatinous material	*Dumontinia*
9. Outer layer of apothecial ectal excipulum composed of globose cells	*Sclerotinia*

3.12 Key Leading to Plant Pathogenic *Sclerotinia* Species on the Basis of Sclerotia Production in Cultures

The identification of the plant pathogenic species of *Sclerotinia* can be determined by examining the production of sclerotia in cultures grown on PDA at temperatures ranging from 15 to 20°C, as well as by studying field-collected sclerotia (Kohn, 1979).

1.The sclerotia and mycelium possess one or both of the following characters:	
(a) Clamp connections; (b) dolipore septa (use phase microscopy or mounts in aniline blue/glycerine (Tu and Kimbrough, 1973)	Basidiomycetes
1. Sclerotia and mycelium with neither of the above characters	2
2. Conidia present (Cfr. *Verticillium, Phaeosclerotinia, Monilinia. Pycnopeziza, Scleromitrula, Botryotinia, Gloeotinia, Septotinia, Cristulariella,* etc.)	Not *Sclerotinia*
2. Conidia absent, except for phialidic 'spermatia' (*Myrioconium*)	3
3. A mantling sclerotial stroma of indefinite dimensions present and smaller sclerotia ('sclerotlutis' formed on aerial mycelium above the mantle	*Stromatinia*.
3. Not as above	4
4. Sclerotial medulla containing host cells (n.b. vessel elements with spiral cell-wall thickenings), or in culture, sclerotia at least partially immersed in the agar	*Ciborinia, Myriosclerotinia, 'Sclerotinia' kerneri*
4. Sclerotial medulla free of susceptible tissues, or in culture, sclerotia formed above the agar surface	5
5. Sclerotial rind composed of a single layer of dark-walled, clavate cells	*S. tuberosa*
5. Sclerotial rind is made up of 2–6 layers of dark walled globose cells	6
6. Sclerotia formed abundantly, scattered throughout colony, sometimes adhering to form an aggregate crust in culture; individual sclerotia 0.5–2 mm long	*S. minor*
6. Sclerotia produced at growing margins of colony only, forming concentric rings, radial lines, and other patterns; individual sclerotia 2–20 mm long	7
7. Sclerotial rind composed of textura prismatica with cells becoming globose, continuing beyond the rind as erect, tomentum hyphae	*S. trifoliorum*
7. Sclerotial rind composed of textura prismatica with cells becoming globose, no tomentum hyphae present	*S. sclerotiorum*

3.13 Key Leading to Pathogenic Species of *Sclerotinia* on the Basis of Apothecia with Sclerotia Production *In Vitro* and in Nature

Key leading to the sclerotium-forming plant pathogenic species of *Sclerotinia* based on apothecia with sclerotia produced *in vitro* or in nature (Kohn, 1979) as follows:

1. Apothecia cupulate, stipitate, on a distinct sclerotium with a well-differentiated rind and medulla, conidia present	Not *Sclerotinia*
1. Apothecia cupulate, stipitate on a distinct sclerotium with a well-differentiated rind and medulla, conidia absent, except for phialidic spermatia (*Myrioconium*)	2
2. Sclerotial medulla containing host cells, or, in culture, sclerotia at least partially immersed in the agar	*Ciborinia, Myriosclerotinia, 'Sclerotinia' kerneri*
2. Sclerotial medulla containing host cells, or in culture, sclerotia formed above the agar surface	3
3. Stroma consisting of a mantling stroma of indefinite dimensions and smaller sclerotia formed on aerial mycelium above the mantle; apothecia occurring on the mantling sclerotial stroma only	*Stromatinia*

3. Not as above	4
4. Outer layer of ectal excipulum of apothecia composed of prosenchymatous cells usually embedded in gelatinous material	*Sclerotinia tuberosa*
4. Ectal excipulum composed of globose cells, gelatinous material present or absent	*Sclerotinia* 5
5. Ascospores dimorphic in size, showing segregation in ascus, tetra-nucleate, length/width ratio of ascospores < 2.0	*S. trifoliorum*
5. Ascospores uniform in size, no segregation in ascus	6
6. Ectal excipulum at margin of apothecium composed of globose cells, ascospores tetra-nucleate	*S. minor*
6. Ectal excipulum at margin of apothecium composed of prosenchyma 'turning out' perpendicularly to the apothecial surface; ascospores binucleate, length/width ratio of ascospores > 2.0	*S. sclerotiorum*

3.14 Key for Leading Plant Pathogenic Species Included in *Sclerotinia*

Key to the plant pathogenic species included in Sclerotinia as per Kohn (1979) are as follows:

1. Ascospores dimorphic in size, showing segregation in ascus, tetra-nucleate, length/width ratio of ascospores < 2.0; ectal excipulum at margin of apothecium composed of prosenchyma 'turning out' perpendicularly to the apothecial surface; sclerotia produced at growing margins of colony only, forming concentric rings, radial lines and other patterns, individual sclerotia 2–20 mm long	*S. trifoliorum*
1. Ascospores uniform in size, no segregation in ascus	2
2. Ectal excipulum at margin of apothecium composed of globose cells; ascospores tetra-nucleate length/width ratio of ascospores < or > 2.0; sclerotia formed abundantly throughout the colony, sometimes adhering to form an aggregate crust in culture, individual sclerotia 0.5–2 mm long	*S. minor*
2. Ectal excipulum at the margin of apothecium, which is composed of prosenchyma 'turning out' perpendicularly to the apothecial surface, where length/width ratio of ascospore is 2.0; ascospores are binucleate. Sclerotia are produced at the growing margins of colony diameter only, showing concentric rings, radial lines, and other patterns, individual sclerotia are 2–20 mm long	*S. sclerotiorum*

3.15 Accepted Species of *Sclerotinia*

3.15.1 *Sclerotinia sclerotiorum*

Sclerotinia sclerotiorum (Lib.) de Bary, Vergl. der Pilze, Mycet. Bact., p. 22. c. i.e., 1884.

=*Peziza sclerotiorum* Lib., Plant. Crypt. Ard. 326. 1837. (!!)

=*Helotium sclerotiorum* (Lib.) Fuckel, Fung. Rhen. Exsicc. Suppl. IV. 1861.1866.

=*Sclerotinia libertiana* Fuckel, Jahrb. 23–24: 331. 1870 (name change).

=*Phialea sclerotiorum* (Lib.) Gillet, Chamb. .98, c. i.e., 3(1). 1881.

=*Hymenoscypha sclerotiorum* (Lib.) Phillips, Man. Brit. Discomyc., p.115.1887.

=*Whetzelinia sclerotiorum* (Lib.) Korf and Dumont, Mycologia 64:250. 1972;

[*Whetzelinia sclerotiorum* Wellman, Dictionary of Tropical American Crops and Their Diseases, p. 10, 1977 (Lapsuscalami)];

[*Sclerotinia whetzelinia* Wellman, Dictionary of Tropical Plant Crops and their Diseases, p. 395. 1977 (Lapsus calami).]

=[*Peziza sclerotii* Fuckel, Bot. Zeit. (Berlin) 35:249, c. i.e.,-x (a-d).1861 (nomennudum).]

=*Peziza kauffmanniana* Techomirow, 4: 295, c. i.e., IV–VII. 1868.

=*Rutstroemia homoeocarpa* Karst., Bildrag. Kannedom Finlands, Natur Folk 19: 107. 1871.(!!).

=*Sclerotinia postuma* Berk. and Wilson, Gard. Chron. 20: 333, c. i.e., 1883.

=*Sclerotinia ficariae* Rehn in Rabenh., Krypt. Fl. Deutschl. 11. 1 (3): 815. 1893. (!!)

=*Sclerotinia opuntiarum* Speg., Anales Soc. C i. Argent. 50: 37. July, 1900 (!!),

=*Sclerotinia sclerotiorum* (Lib.) opuntiarum (Speg.) Allipi, Revista Fac. 36: 149.1960.

=*Sclerotinia moelleriana* P. Henn., Hedwigia 41: 27.1902 (!!)

=*Sclerotinia wisconsinensis* Rehm., Ann. Mycol. 6: 317.1908. (!!)

=*Sclerotinia matthiolae* Lendn., Bull. Soc. Bot. Geneve 9: 21, c. i.e. 1–3, 1917; Bull. Soc. Bot. Geneva 9: 221. 1918.

=*Sclerotinia henningsiana* Kirschst., Verb. Bot. Provo Brandenburg 40: XXVII. 1918. (!)

=*Sclerotinia riograndensis* Rick, Broteria, Ser. Bot. 25: 99.1931 (!)

=*Sclerotinia galeopsidis* Velen., Monogr. Discomyc. Bohem. 1: 227.1934 (!!)

=*Sclerotinia caudata* Velen., Novitates Mycol. Novissimae P. 129. 1947. (!!)

=*Sclerotinia sclerotiorum* (Lib.) Narasimhan and Thirumalachar, 22: 426, c. i.e., 1954.

=*Sclerotinia xanthorrhoeae* Beaton and Weste, *Trans. Br. Mycol. Soc.* 68: 73, c. i.e. 1977. (!!)

Sclerotia can have a variety of shapes, although they usually appear as rounded to cylindrical forms on dense, white mycelium. They measure (2–)5–15 (–30) × 2–8 (–15) mm, with a white inner context and a black outer rind. They emerge in culture at the periphery of the colony, frequently creating radiating lines, concentric rings, and other patterns. The brainstem sclerotium is made up of hyaline textura oblita cells that are closely braided and range in width from 5 to 10 μm. with walls that are 2–3 μm thick and highly gelatinized. A thick layer of texture makes up the sclerotized rind. Sclerotized rind, a 2–6 cell deep layer of textura prismatica originating from medullary cells, turning out perpendicularly to the surface, becoming brown walled, inflated to globose, 5–15 μm in diameter. One to many apothecia, which range in color from ochraceous (cinnamon) to amber and are frequently darker around the base of the stipe, can be produced by each sclerotium. When the apothecia are young, its receptacle is flat to slightly concave; as it ages, it becomes flat to convex. Its width ranges from 2 to 8 (–10) mm. It frequently has an undulating edge with a central depression, tapering to a stipe that is 1–2 mm wide and 3–20 (–30) mm long. The thin-walled hyaline-to-pale brown textura prismatica makes up the 35–140 μm-wide ectal excipulum. The excipulum's inflated cells are globose and range in width from 5 to 30 μm. These cells may occasionally be joined by a gel-like material. The apothecial surface is perpendicular to the direction in which the cells are oriented. A textura porrecta with inflated apices and brown walls can be found at the edge of the excipulum. One- or two-celled tomentum hyphae may arise from the outermost cells of the excipulum. Light-brown-walled textura porrecta makes up the ectal excipulum of the stipe, and one-celled inflated tomentum hyphae that extend perpendicularly from the stipe axis originate from the outermost cells. The medullary excipulum has thin walls, a hyaline textura intricata, and is loosely woven. The 5–15 μm-diameter hyphae in the medullary excipulum occasionally display a blue coloring in the presence of Melzer's reagent. Sometimes it is seen in two layers: textura intricata on the inside and textura porrecta on the outside, both of which are parallel to the apothecial surface. The stipe's medullary excipulum can have walls that are 1–2 μm thick and be made of either textura porrecta or textura oblita.

Rhomboidal crystals can be found in large quantities in the stipe or receptacle. Compact or loosely woven light-brown-walled textura intricata makes up the subhymenium. The addition of Melzer's reagent may also reveal a blue coloring in it, and it is most likely bonded in gelatinous material. The subhymenium contains hyphae that are 2–3 µm in diameter. The eight-spored, cylindrical asci emerge from croziers. They have a thickened apex and measure (110–)130–150 (–160) × 6–10 µm. It can occasionally be seen in two layers, with the outer layer staining positively with J$^+$ by Melzer's reagent due to the asci's pore channel wall. Ascospores are translucent, oval-shaped, have two oil droplets, are placed in a single row, and have two nuclei. They measure between 9 and 14 µm in length and 4–5 µm (occasionally 6 µm) in width, with a length-to-width ratio typically exceeding 2.0. The paraphyses, which are transparent and thread-like, are 2 µm wide, segmented, and either simple or sparsely branched. The microconidial state, known as *Myrioconium*, consists of spherical, transparent microconidia measuring 2–4 µm in diameter. These microconidia are produced from phialides located in sporodochia or laterally on hyphae. They are found on the surface of the hymenium or in culture. For more detailed information regarding the habitat, type locality, type specimen, typical host of the basionym, typical hosts of synonyms, as well as the examination of exsiccate specimens with and without apothecia of this species, refer to Kohn's publication from 1979. Please refer to Chapter 7 of the book '*Sclerotinia Diseases of Crop Plants: Biology, Ecology and Disease Management*' by Saharan and Mehta (2008) for more information on the other two species, namely *S. minor* and *S. trifoliorum*.

3.16 Taxa Imperfectly Known

Kohn (1979) has documented 259 taxa within this category. As stated by Kohn (1979), the provided diagnoses are inadequate for species identification, and in certain instances, the correct generic classification remains uncertain. The comprehensive list of taxa can be found in *Mycotaxon* 9: 365–444.

3.17 Economically Important and Frequently Mentioned Species Not Included in *Sclerotinia*

According to the 1919 Sawada Government publication, Formosa Agricultural Experimentation Station, Special Bulletin 19: 206, *Sclerotinia allii* was renamed *Ciborinia allii* (Saw.) Kohn. However, there was an error by Yamamoto *et al.* (1956) by error thought that Sawada's species produces a *Botrytis* state, which is identified as *B. byssoidea* Walker, calling the species *Botryotinia allii* (Saw.) Yamam. Nevertheless, Sawada's species lacks a conidial state, according to unpublished cultural research by R. P. Korf and G. L. Hennebert. Regretfully, *Botryotinia* combo has gained traction in the literature. Unfortunately, the combination in *Botryotinia* has become well-established in the literature.

Sclerotinia camelliae Hara non-Hansen and Thomas belongs either in an unnamed new genus or may be a species of *Moellerodiscus*.

Sclerotinia camelliae Hansen and Thomas non-Hara is a *Ciborinia*. This epithet was published after 1935 without a Latin diagnosis and is, therefore, not valid. However, Kohn (1979) provided a Latin diagnosis for it in 1979.

Sclerotinia fructicola (Wint.) Rehm is *Monilinia fructicola* (Wint.) Honey.

Sclerotinia fructigena Aderh. and Ruhl. is *Monilinia fructigena* Honey.

Sclerotinia fuckeliana (de Bary.) Whetz. is *Botyotinia fuckeliana* (de Bary.) Whetz.

Sclerotinia gladioli Drayton is *Stromatinia gladioli* (Drayt.) Whetz.

Sclerotinia homoeocarpa Bennett has no existing type specimen and is not a *Sclerotinia*. The epithet has been applied to species with apothecia identified as belonging to *Lanzia* sp. and to *Moellerodiscus* sp. The symptoms attributed to "dollar spot" probably are caused by more than one species.

Sclerotinia kerneri Wettstein belongs in a new genus (Kohn, 1979).

Sclerotinia laxa Aderh. and Ruhl. is *Monilinia laxa* (Aderh. and Ruhl.) Honey.

Sclerotinia narcissicola Gregory is *Botryotinia narcissicola* (Qreg.) Buchw.

Sclerotinia panacis Rankin is transferred to *Stromatinia* as *Stromatinia panacis* (Rank.) Kohn, comb. nov. (Basionym: *Sclerotinia panacis* Rankin, Phytopathology 2: 30. 1912)

Sclerotinia perplexa Lawrence is *Ovulinia perplexa* (Lawr.) Seav.

Sclerotinia ricini Godfrey is *Botryotinia ricini* (Godfr.) Whetz.

Sclerotinia serica is transferred to *Stromatinia* as *Stromatinia serica* (Keay) Kohn. comb. Nov. (Basionym: *Sclerotinia serica* Keay. J. Bot. 75: 132. 1937).

Sclerotinia tuberosa (Hedw. ex-Merat) Fekl. belongs to a new genus (Kohn, 1979).

3.18 Description of Species

The present text provides a detailed account of the characteristics of *S. sclerotiorum*. However, for information on different species of *Sclerotinia*, please refer to Chapter 7 of the book titled " *Sclerotinia Diseases of Crop Plants: Biology, Ecology and Disease Management*' authored by Saharan and Mehta (2008), specifically dedicated to the pathogen–*Sclerotinia*.

3.18.1 The *Sclerotinia sclerotiorum*

3.18.1.1 Hosts

The fungus has a very wide host range and is capable of infecting over 900 different plant species worldwide (Chapter 2, Table 2.1, this volume). The specific details regarding the plants affected by this pathogen, which includes information on geographical distribution and host range. Notably, the primary hosts targeted by this pathogen are *Arachis hypogaea, Brassica oleracea* var. *capitata, B. chinensis, Carthamus tinctorius, Citrus, Coriandrum sativum, Cucumis melo, Curcurbita pepo, Cucumis sativus, Glycine max, Helianthus annuus, Hibiscus sabdariffa, Lycopersicon esculentum, Lactuca sativa, Nicotiana tabacum*, and *Phaseolus vulgaris*. Among the other crops that are affected are cabbage (*B. oleracea* var. *capitata* and *B. chinensis*), common bean (*P. vulgaris*), citrus (*Citrus* spp.), coriander (*Coriandrum sativum*), cucumber (*C. sativus*), groundnut (*A. hypogaea*), lettuce (*L. sativa*), melon (*C. melo*), roselle (*H. sabdariffa*), safflower (*C. tinctorius*), squash (*C. pepo*), soybean (*G. max*), sunflower (*H. annuus*), tomato (*L. esculentum*), and tobacco (*N. tabacum*).

3.18.1.2 Disease

The fungus is responsible for causing various diseases, including cottony soft rot, white mold, and watery soft rot. It primarily affects crucifers and vegetables in the field, during transit, and in storage, making it a significant pathogen. Cereals (monocots), grasses, and woody plants are rarely the targets of this pathogen. At any stage of growth, almost every portion of herbaceous crops is vulnerable to assault, both above and below the soil. The initial symptoms often involve plant collapsing due to stem infection near the soil level. This is followed by the formation of a soft rot and the obvious appearance of sclerotia, which are often generated in the pith, as well as exterior white mycelium. Ascospore-infected withering or fallen petals are typically the source of infections that appear at a specific height above the ground.

3.18.1.3 Geographical distribution

The disease is highly prevalent but it is primarily found in areas with relatively cool and moist conditions. In the Mediterranean regions, the disease is rarely or not at all present during the summer season. The main mode of transmission is through air-borne ascospores, which play a crucial role in spreading the disease. The primary structures for survival are the sclerotia, which are found in the soil and host debris and give rise to apothecia. The survival time of the disease varies greatly, but it can remain viable in the soil for up to 3 years (Hoes and Huang, 1975). In addition to ascospores, mycelium from sclerotia can also cause infection. The disease can also be transmitted through infected seeds, either through contaminated sclerotia or internal mycelium. This information is sourced from IMI Descriptions of Fungi and Bacteria, 1998, Sheet 513 (Mordue and Holliday, 1998).

3.18.2 New species of *Sclerotinia* on crucifers

A recent discovery has identified a new species of *Sclerotinia* known as *S. subarctica* nom. Prov.

(*Sclerotinia* sp.1). This species has been found to cause stem rot on turnip rape and *Ranunculus acris* in Norway and the UK, respectively (Clarkson *et al.*, 2010; Brodal *et al.*, 2017). In August 2013, stems of the turnip rape "*Pepita*" were found to have white-grayish lesions around the leaf axils, a characteristic of *Sclerotinia* stem rot. The observation was made in a field at the Norwegian Institute of Bioeconomy Research (NIBIO) research station Apelsvoll in Oppland County, Norway. After gathering the sclerotia from the infected turnip rape stump and harvesting the seeds, they were surface-sterilized. After being cut in half, they were spread out on PDA. *Sclerotinia's* fast-growing white mycelium was seen after 1–2 days of incubation at 20°C. Five to 7 days later, fresh sclerotia had begun to form. Even though they varied, the sclerotial size and growth patterns were distinctive to *S. sclerotiorum*. After that, 20 isolates underwent DNA extraction, PCR amplification, and sequencing of the Internal Transcribed Spacer (ITS) sections of the rDNA. Five isolates were identified as *S. subarctica* (formerly known as *Sclerotinia* sp. 1), and 15 isolates were recognized as *S. sclerotiorum* based on BLASTn (Nucleotide Basic Local Alignment Search Tool) analysis of the 475 bp amplicons (Hoist-Jensen *et al.*, 1998; Winton *et al.*, 2006; Winton *et al.*, 2007). According to Clarkson *et al.* (2010), five isolates matched 100% with the isolates of *S. subarctica* collected from the UK. GenBank received a typical ITS region sequence, entry number KX929095. The absence of a 304 bp intron in the Large Sub Unit (LSU) rDNA, which was compared to *S. sclerotiorum* by PCR amplification and gel electrophoresis, allowed for the validation of the *S. subarctica* isolates. Two *S. subarctica* isolates were used, and their sclerotia were cultured for 7 days on PDA in order to perform the pathogenicity test. Later, at growth stages BBCH (Biologische Bundesanstalt, Bundessortenamt, and CHemische Industrie scale) 57/59 and BBCH 64, spring oilseed rape plants ('Mosaik') were inoculated in the greenhouse using agar plugs containing actively growing mycelium. Two PDA plugs were affixed with tiny needles on the main stems of four plants for every treatment. PDA agar plugs that had not been inoculated were used to attach control plants. Three iterations of this experiment were conducted. After 1–2 weeks of incubation at 16–20°C with 100% relative humidity, the telltale signs of stem rot appeared. Around the sites of the inoculation, the stems started to show fluffy mycelium and white blemishes. The typical signs of a *Sclerotinia* infection, however, were absent from the control plants. Sclerotia were removed from the stems of the plants when they reached senescence and cultivated on PDA. White mycelium began to grow in 1–2 days, and new sclerotia resembling the ones used to create the original isolate formed in 7 days. In Norway's cereal-based agricultural system, *Brassica* oilseed crops are grown as essential break crops; however, they are very vulnerable to *Sclerotinia* stem rot. Every area where *Brassica* oilseed crops are farmed has seen the disease, and in extreme circumstances, untreated control treatments of fungicide experiments have shown a 25% decrease in oilseed output. Although *S. subarctica* has been documented on wild hosts earlier (Hoist-Jensen *et al.*, 1998), this is the first time the pathogen has been identified on a crop plant in Norway. Clarkson *et al.* (2010) in the UK showed that *S. subarctica*, which was isolated from *Ranunculus acris*, was pathogenic on oilseed rape. Given the similarity in symptoms between *S. subarctica* and *S. sclerotiorum*, it is probable that *S. subarctica* is hiding in many farmers' fields (Brodal *et al.*, 2017).

In the UK, Clarkson *et al.* (2010) reported the discovery of *Sclerotinia subarctica* nom. prov., commonly referred to as *Sclerotinia* species 1, on *Ranunculus acris*. This was a groundbreaking discovery. Their investigation concentrated on populations of *S. sclerotiorum* on *R. acris*, namely in a meadow in Herefordshire. They removed withered and dead blossoms from the plants and kept them moist at 20°C in order to conduct additional research. They saw the appearance of fungal growth from these blooms that showed *S. sclerotiorum*-specific traits such as sclerotia and white, fluffy mycelium. They moved the fungal growth to PDA and collected DNA from the cultures in order to verify the identity of the isolates. After that, the rDNA's ITS sections were amplified and sequenced. They found that of the 32 isolates that were randomly chosen, only 17 were identified as *S. sclerotiorum* and the remaining 15 as *S. subarctica* nom. prov. through BLAST analysis. According to earlier research by Hoist-Jensen *et al.* (1998) and Winton *et al.*, 2006, 2007) this result is consistent with the occurrence of *Sclerotinia* species 1. It was discovered that all 15 of the *S. subarctica* ITS sequences were identical (Gen Bank Accession No. GU018183). By comparing the lack of an intron in the LSU rDNA (large subunit ribosomal DNA) with *S. sclerotiorum*, the *S. subarctica* isolates were verified (Hoist-Jensen *et al.*, 1998). *S. subarctica* is believed to be morphologically identical to *S. sclerotiorum*, despite the lack of a formal description and host range (Hoist-Jensen *et al.*, 1998). In contrast to *S. sclerotiorum* (about 4.1 ± 0.9 mm), the cultures of *S. subarctica* in this investigation developed fewer but larger sclerotia *in vitro* (roughly 5.8 ± 1.0 mm diameter). Reports of *S.* been limited to wild hosts in Norway (Hoist-Jensen *et al.*, 1998) and white mold disease on vegetable crops in Alaska (Winton *et al.*, 2006) have been made. The susceptibility of oilseed rape to *S. subarctica* was examined using plant and detached leaf inoculation, as this crop serves as the primary host for *S. sclerotiorum* in the UK. Though the stem lesions of *S. subarctica* took longer to develop than those of *S. sclerotiorum*, the stem and leaf lesions produced by the former were the same. For the first time, this work shows that *S. subarctica* can infect oilseed rape. The discovery of *S. subarctica* in the UK suggests that, contrary to earlier claims, this pathogen is not exclusive to the 'High North' (Winton *et al.*, 2007). Importantly, it is possible that *S. subarctica* is now going unnoticed in crops infected by white mold in the UK because it exhibits symptoms that are similar to those of *S. sclerotiorum*. Therefore, more investigation is required to ascertain its distribution among wild and crop hosts. According to Clarkson *et al.* (2010), the fact that *S. subarctica* and *S. sclerotiorum* are found on *R. acris* implies that this wild host acts as a reservoir for inoculum intended for use in crop plants.

3.19 Biochemical and Cultural Traits to Differentiate Between *Sclerotinia* Species

The revised classification reduces the number of species in the genus *Sclerotinia* to just three: *S. trifoliorum* Erikss, *S. minor* Jagger, and *S. sclerotiorum* (Lib.) de Bary. The primary factors that determine the specific naming of these species are their morphological and physiological characteristics, such as the mycelium growth rates and the dimensions, morphology (both macro- and microanatomical), and growth characteristics of *Sclerotinia*, asci, and ascospores under laboratory conditions. The way a particular isolate has been linked to its host in the field has also proven to be a crucial factor for species identification in some circumstances, especially when distinguishing between *S. trifoliorum* and *S. sclerotiorum*. A small number of researchers, including Jagger (1920), Keay (1939), Whetzel (1945), Bjorling (1951), Williams and Western (1965), Buchwald and Neergaard (1973), Kohn (1979), and Willetts and Wong (1980), have supported the separation of *S. sclerotiorum*, *S. minor*, and *S. trifoliorum* using traditional methods. Some researchers, however, such as Purdy (1955), Walker (1969), Price and Colhoun (1975), and Morrall *et al.* (1972), believed that these taxa should be considered as a single species, *S. sclerotiorum*. Their conclusion was based on the discovery that the information provided by traditional taxonomic methods was insufficient and unreliable. Many researchers have researched *S. sclerotiorum* and closely related species taxonomy in great detail (Willetts and Wong, 1971; Wong and Willetts, 1973, 1974, 1979; Scott, 1981; Petersen *et al.*, 1982; Russo *et al.*, 1982; Cruickshank, 1983). These studies have used serological and electrophoretic data along with novel criteria, including ontogenetic and cytological variation. The results of these studies have supported the hypothesis that *S. minor*, *S. sclerotiorum*, and *S. trifoliorum* are distinct species.

In order to distinguish different cultural and biological characteristics that could differentiate various species of Sclerotinia, Tariq *et al.* (1985) compared these factors and evaluated a total of 27 *Sclerotinia* isolates, representing the three species groups recognized in taxonomic treatments, were included in the investigation. Mycelial properties were compared, and the results showed that *S. trifoliorum* may be distinguished from *S. sclerotiorum* and *S. minor* based on their growth rate at 25°C. Nevertheless, analyses of mycelial interactions and relative hyphal densities produced data with little taxonomic significance (Tables 3.1 and 3.2). However, by comparing the sclerotial features, it was possible to distinguish *S. minor*

Table 3.1. List of the general properties of mycelium characteristics of *Sclerotinia*. (Adapted from publication of Tariq *et al.* (1985) with permission.)

	On solid media (PDA)		In liquid media (CAT)		
Isolate	Colony radius after 48 hour (mm)	Relative density of aerial mycelium	Pellet diam. after 7 days (mm)	Dry wt. of mycelium after 7 days (mg)	General appearance of pellet after 7 days
Ss1 to Ss18	26.1 ± 1.7 (55)	Sparse–very dense	38.3 ± 1.7 (29)	527.3 ± 70.2 (17)	White to cream; no *Sclerotinia*
Ss19 and Ss20	15.8 ± 2.6 (8)	Dense–very dense	13.8 ± 3.2 (4)	120.5 ± 65.5 (2)	Creamy-white to grayish white; single *Sclerotinia* observed on one pellet
St21 to St24	6.3 ± 1.0 (15)	Sparse	10.0 ± 1.3 (6)	87.8 ± 5.5 (4)	White to cream; immature *Sclerotinia*
Sm25 to Sm27	17.6 ± 1.0 (11)	Moderately dense	21.0 ± 1.9 (5)	216.3 ± 32.9 (3)	White to cream; mature and immature Sclerotinia

Mean ± SEM; the mean is calculated using the values in parentheses.

Table 3.2. The characteristics of *Sclerotinia* are summarized (3 weeks on PDA, at 25°C in the dark). (Adapted from publication of Tariq *et al.* (1985) with permission.)

Isolate	Diam. of *Sclerotinia* (mm)	No. of *Sclerotinia* per 9-cm Petri dish	Wt. of dried *Sclerotinia* (mg)	*Sclerotinia* patterns on solid media	Conc. of protein in *Sclerotinia* (mg/g fresh wt.)
Ss1 to Ss18	3.04 ± 0.04(1.0–9.5)	27.0 ± 1.2	5.4 ± 0.8	Concentric circles throughout colony and/or at the end of the plate	30.8 ± 1.3
Ss19 and Ss20	4.2 ± 0.1(2.0–7.8)	9.8 ± 1.3	7.9 ± 0.6	Concentric circles	38.9 ± 7.3
St21 to St24	3.67 ± 0.12(10–16.0)	26.6 ± 2.4	6.7 ± 2.3	Concentric circles within colony or radiating lines from inoculum plug	19.7 ± 1.1
Sm25 to Sm27	1.48 ± 0.04(1.0–3.0)	411.2 ± 1.3	0.4 ± 1.0	Scattered throughout colony. No distinct pattern	30.3 ± 3.1

Mean ± SEM; numbers in parentheses represent range of values.
Ss- *Sclerotinia sclerotiorum*; St- *Sclerotinia trifoliorum*; Sm- *Sclerotinia minor*

from the other two species according to the size and pattern of the sclerotia inside the colony. It did not, however, differentiate between *S. trifoliorum* and *S. sclerotiorum*. The study also discovered that because ascocarp formation is unpredictable and time-consuming, spore features are not helpful for quick identification. By comparing the main soluble proteins from *Sclerotinia* using polyacrylamide gel electrophoresis and discriminant analysis of pyrolysis mass spectra of *Sclerotinia* preparations, more support for species groups was discovered. It would be difficult to differentiate between the three species using only biochemical criteria, though. According to Tariq *et al.* (1985), *S. sclerotiorum* and *S. trifoliorum* shared more similarities than did either fungus and *S. minor*. This suggests that the two species should be categorized as subspecies.

3.20 Cytology

Wong and Willetts (1979) studied the cytological parameters and reported that *S. minor* has four nuclei in each ascospore and a haploid chromosome count of four. These findings are based on cytological research. *S. trifoliorum*, on the other hand, contains eight haploid chromosomes but four nuclei in each ascospore. By comparison, *S. sclerotiorum* has eight haploid chromosomes and two nuclei per ascospore. Strong evidence supporting the unique species status of *S. minor*, *S. trifoliorum*, and *S. sclerotiorum* is provided by these consistent cytological distinctions.

3.21 Genomics of *Sclerotinia Sclerotiorum*

Genomic studies have been carried out on various aspects, *viz.*, sequencing of pathogen genome, phylogenetic analysis, expression of genes for infection and pathogenesis, transcriptomic and secretomic analysis, genome architecture, complete genome assembly, gene annotations, and effector genes for molecular mechanisms of infection and pathogenesis.

3.21.1 The *S. sclerotiorum* genome in its early stages

A group 1-intron, or intervening region of 1380 bp, was found when *S. sclerotiorum's* mitochondrial small subunit rRNA (nst ssu rRNA) gene was analyzed. The subgroup-IC2 introns of *Podospora* (ND 3i1, ND 5i2, and CO Ii5) and *Neurospora* (ND 5i1) share a secondary structure with this intron. This intron was discovered in the mt ssu rRNA gene of *S. minor*, *S. trifoliorum*, and *Sclerotinia cepivorum*, but not in other *Sclerotinia* species, using PCR amplification and Southern hybridization (Carbone *et al.*, 1995). Despite the fact that *S. sclerotiorum* may be changed using polyethylene glycol-mediated protoplast transformation, the multinucleate nature of hyphal cells and the instability and lack of reproducibility of transformants have hindered molecular studies on this pathogen (Rolland *et al.*, 2003). Nevertheless, co-cultivation with ascospores led to the first successful *Agrobacterium tumefaciens*-mediated transformation of *S. sclerotiorum* (Weld *et al.*, 2006). By utilizing the benefits of *Agrobacterium*-mediated transformation, future research on this pathogen incorporating gene transfer, gene knockdown, and insertional mutagenesis should be straightforward. Furthermore, *S. sclerotiorum* reporter genes, such the green fluorescent protein gene, have already been employed to learn more about the fungus' biology *in vivo* and *in vitro* (Lorang *et al.*, 2001; Guimaraes and Stotz, 2004; De Silva *et al.*, 2005).

The Massachusetts Institute of Technology and the Harvard Broad Institute sequenced the 38-megabyte genome of *Sclerotinia*. The assembly is based on shotgun sequencing and provides 8 × sequencing coverage of the genome. It consists of 679 sequence contigs in total. The trace sequence is currently in the possession of the National Centre for Biotechnology Information (NCBI). The publishing of the complete genome sequence has marked a significant turning point in the research on this hazardous virus (Bolton *et al.*, 2006).

3.21.2 Phylogeny of *Sclerotinia* and related genera

Based on nuclear ribosomal internal transcribed spacer (ITS) DNA sequences from an out-group made up of five related taxa within the same family and an in-group made up of 50 isolates representing 24 species of the discomycetes family *Sclerotiniaceae*, phylogenies have been built. The following taxa are included in the in-group: six *Myriosclerotinia* spp., nine *Sclerotinia* spp., one *Sclerotium* sp., one *Ciborinia* sp., one *Dumontinia* sp., three *Botrytis* spp., three *Monilinia* spp., one *Encoelia* sp., and one *Ciboria* sp. are included in the out-group taxa. With the exception of *Ciborinia* and *Encoelia*, all taxa include the type species. Numerous taxa mentioned in the list are significant plant pathogens. The morphology, life history, and taxonomy of the derived phylogenies are examined. While *Ciborinia ciborium* and *Myriosclerotinia* are thought to be related, a possible relationship between *Sclerotinia borealis* and *S. tetraspora* and *Myriosclerotinia* is firmly supported. Reinstated as a separate species of *Dumontinia*, *Sclerotinia ulmariae* was previously synonymized with *Dumontinia tuberosa*. There are two new combinations that are suggested: *Myriosclerotinia ciborium* and *Dumontinia ulmariae*. *Dumontinia* and the poorly understood taxon *S. cepivorum* appear to be most closely related. It is concluded that the existing taxa of *Dumontinia* and *Myriosclerotinia* are monophyletic, and that *Botryotinia* and the anamorphs of *Botrytis* most likely form a monophyletic lineage as well. It is likely that the genus *Sclerotinia* is polyphyletic, with simple isomorphies rather than synapomorphies being its defining characteristic. According to Hoist-Jensen *et al.* (1998), *Sclerotinia* sp.1 and *Sclerotinia* sp.2, two putatively new taxa, are most closely related to *S. minor*, *S. sclerotiorum*, and *S. trifoliorum*, and to *S. borealis*, respectively.

3.21.3 Gene expression related to the activity of enzymes

The mature *Sclerotinia* exhibits nil trehalase or amylase activity whereas it shows a modest but detectable of >β−1,3 glucanase and >β-glucosidase activity. The >β−1,3 glucanase activity seems to rise as germination progresses, whereas >β-glucosidase activity stays unchanged. >β−1,3 glucanase activity is found to increase most significantly during the active formation of apothecial stipes. A comparison was made between the activity of the enzymes responsible for the metabolism of carbohydrates in mature apothecia, immature apothecia referred to as stipes, and sclerotia, which are the different parts of germinating sclerotia. The highest specific activity of >β−1, >β−3, and >β-glucosidase is found in *Sclerotinia*. The relevant regions exhibit trehalase and amylase activity as germination progresses and the majority of stipes mature into apothecia. At this stage, amylase-specific activity is highest in mature apothecia, whereas trehalase activity is nearly equal in all three divisions. In apothecia, glucose-6-phosphate dehydrogenase activity is at its peak. However, compared to nongerminating sclerotia, activity in germinating sclerotia is comparatively modest, but slightly increased. These results unequivocally demonstrate the physiological distinctions between sclerotia and apothecia. The synthesis of elements needed for cellular development is enhanced in stipes and apothecia. Nonetheless, the medullary cells of the sclerotia are located primarily where internal reserves break down (Saito, 1977). The *S. sclerotiorum* is capable of producing a range of pectin-degrading enzymes both inside and outside of the cell. The gene pg1, which codes for an endopolygalacturonase, has been cloned and sequenced (pg1). The coding area consists of an open reading frame (ORF) with 1143 bp. In contrast to other fungal PG-encoding genes, essential transcriptional regulatory sequences were identified in the noncoding regions of *S. sclerotiorum* pg1. The projected amino acid sequence (380 aa) of the enzyme showed a significant degree of similarity (41.5–59.8%) when compared to seven fungal PG sequences. A similar protein structure, most likely composed of beta sheets organized antiparallel to one another, was discovered by predicted secondary structure analysis. A pg1 0.65 kb Bam HI fragment was used as a probe in hybridization analysis to successfully identify seven distinct recombinant phages from a genomic library. Analysis of the hybridizing restriction segments suggests that the PG-encoding genes are organized as a family (Reymond *et al.*, 1994).

3.21.4 Pathogenesis-related expression of calcineurin genes

The necrotrophic, omnivorous plant pathogen *S. sclerotiorum* is noted for its worldwide distribution. Sclerotia, pigmented structures composed of many hyphae, are produced by *S. sclerotiorum* and are essential to the pathogen's life cycle and cycles of infection. The fungal Ser/Thr phosphatase calcineurin is involved in the control of morphogenesis, cell wall integrity, pathogenicity, and cation homeostasis. Numerous signal-transduction pathways are connected to it. *S. sclerotiorum* shows phase-specific alterations in calcineurin expression during sclerotia formation. When calcineurin is inhibited by FK506, cysclosporin A, or inducible antisense calcineurin expression, the prematuration phase of sclerotium formation is impeded, and the germination of preformed sclerotia is accelerated. Reducing the pathogenesis on *Arabidopsis* and tomatoes involves inducing antisense calcineurin expression in *S. sclerotiorum*. However, one of *S. sclerotiorum's* main virulence factors, the release of oxalic acid, remains unchanged. Cell wall >β−1, 3-glucan content, susceptibility to enzymes that tear down cell walls, and the glucan synthase inhibitor caspofungin rise when calcineurin is suppressed. Consequently, calcineurin is involved in the pathogenesis of *S. sclerotiorum* as well as the development of sclerotia, and it is probably involved in other phytopathogens as well (Harel *et al.*, 2006).

3.21.5 The primary characteristics of the genome of *S. sclerotiorum*

The sequencing of the *S. sclerotiorum* genome has been accomplished by analyzing the transcriptome and secretome of pathogen isolates.

3.21.5.1 Genome sequences

Sanger technique was used in 2011 to sequence the genome of *S. sclerotiorum* strain 1980, which was originally isolated from beans in western Nebraska, USA. The sequencing had a coverage of 9.1×. These regions were removed from the assembly because they comprised around 1.6 Mb of sequences that were not covered by the 38.0-Mb scaffolds. The total size of the *S. sclerotiorum* genome is estimated to be approximately 38.3 Mb, with 16 linkage groups that most likely correspond to 16 chromosomes (Gen Bank accession numbers: AAGT01000000). The majority of the exposed regions are assumed to be centromeres, which are found at the chromosomes' center. A total of 14, 522 genes were predicted to be present in the genome, with an average GC content of 41.8%. After excluding genes that encode small proteins with fewer than 100 amino acids in length or those that do not exhibit detectable expression (from Expressed sequence tags (ESTs) and/ or microarray data), a total of 11,860 predicted proteins with high confidence were discovered. Many genes associated with virulence in the *S. sclerotiorum* 1980 strain were found when the genome sequences were first examined. These genes included those that make phytotoxins and other secondary metabolites, as well as those that code for the enzymes known as Cell wall-degrading enzymes (CWDEs), which break down cell walls. Furthermore, a substantial presence of repeated transposable elements (TEs) was found by Amselem *et al.* (2011); these TEs made up roughly 7.7%of the overall genome. This discovery gave rise to the theory that the *S. sclerotiorum* genome recently experienced significant remodeling, leading to a significant expansion of TEs. The arrangement of *S. sclerotiorum's* genome and the efficiency of gene inactivation, alteration, or expression regulation may also be affected by this increase of TEs (Amselem *et al.*, 2011). Using Pac Bio technology, Derbyshire *et al.* (2017) nearly finished sequencing the genome of the identical *S. sclerotiorum* 1980 strain. This genome is the preferred option for *Sclerotinia* since it was thoroughly annotated using RNA-Seq data and manual curation. They discovered a strong association between the locations of these released proteins and areas with a high relative repeat-induced point mutation (RIP) index while also identifying 70 putative effector genes. According to this correlation, transposition and RIP activity may be the cause of *S. sclerotiorum's* higher rate of secreted protein mutation in particular genomic compartments. By sequencing the genomes of 25 field isolates of *S. sclerotiorum* that were collected from different continents, including Australia,

Africa (north and south), Europe, and North America (Canada and Northern United States), Derbyshire *et al.* (2017) furthered their investigation. To investigate the population structure and selective sweeps, single nucleotide polymorphism (SNP)-based analysis was done. Two primary populations were formed from the 25 isolates. Eleven isolates from France, the USA, and Canada made up population 1, and nine isolates from Australia and one from Morocco made up population 2. Less than 0.001% of the genome was covered by a single potential selective sweep seen in the Australian and Moroccan groups. This study raises the possibility that the lack of powerful selective sweeps was caused in part by *S. sclerotiorum's* widespread self-fertilization and sluggish rate of evolution.

In the later stages of *Brassica napus* infection, it is important to note that one gene within a 10-kb region—encoding a key facilitator superfamily transporter—was detected in this selective sweep. Tropical and temperate *S. sclerotiorum* strains showed similar genetic patterns while being segregated in distinct populations, suggesting genome conservation (Del Ponte *et al.*, 2017; Xia *et al.*, 2020a, b).

3.21.5.2 Analysis of transcriptome and secretome

The high-quality genome sequences made this infamous disease easier to study through comparative transcriptomics and secretomic analysis, which in turn made biological research easier. RNA-Seq analysis focused on events occurring through the early (1 hour) to the mid (48 hour) phases of infection, allowing Seifbarghi *et al.* (2017) to identify gene expression patterns that signified transitions between pathogenic phases of *S. sclerotiorum*, namely, host penetration, ramification, and necrotrophic stages. These expression findings support the hypothesis that a brief biotrophic phase begins shortly after the host is punctured. An interspecies comparison analysis of the predicted secretomes of *Botrytis cinerea* and *S. sclerotiorum* revealed the presence of 432 proteins. Of these, small gene clusters containing 16% of the encoding genes are dispersed over 13 of the 16 predicted chromosomes. Further reverse genetics research, aimed at understanding the pathogenic mechanisms involved in the interactions between *S. sclerotiorum* and its host plants, could benefit greatly by examining the aforementioned potential genes. Because omics investigations only produce imprecise connections, mutant analysis is necessary to establish a causal association between a gene and a specific biological function. A summary of the examined genes of *S. sclerotiorum* has been provided through the use of credible mutant analysis techniques. For the purpose of providing readers with an overview, a Venn diagram depicting the roles of these encoded proteins is included (Fig. 7.3, Tables 7.1, 7.2 and 7.3, Chapter 7, this volume). For these genes, Xia *et al.* (2020a, b) have also produced a chromosomal map in order to find potential clustering events that are frequently linked to virulence factors (Fig. 7.4, Chapter 7, this volume). But just a small amount of clustering is seen, with the exception of genes related to mating and melanization, which is indicative of the scant study done on this fungus.

3.21.6 Genome architecture of *S. sclerotiorum*

The availability of fungal genome sequences has greatly aided the study of predicted secretome and effectors. However, despite the large number of sequenced fungal genomes (approximately 1509), many of them suffer from fragmentation and inadequate annotation. These gaps in the genomes are primarily due to the presence of repetitive sequences, which are not effectively captured by the sequencing technologies used. For instance, Illumina sequencing frequently provides reads of only 300 bp, but Sanger sequencing can produce reads of up to 1000 bp (Goodwin *et al.*, 2016). Because repeats frequently combine into a single, condensed sequence during graph-based assembly, sequence assemblies that depend on shorter reads relative to repeated components are unable to produce a consistent and continuous sequence (Ekblom and Wolf, 2014). As previously noted, fungal effectors have a tendency to cluster in repetitive genomic areas, which poses a unique obstacle for researchers investigating them (Thomma *et al.*, 2016). Reads up to 60 kilobases in length can be produced using the PacBio platform, which was created by Pacific

Biosciences (Goodwin *et al.*, 2016). Two gapless fungal plant pathogen genomes were assembled using optical mapping with PacBio data. Based on patterns of restriction sites across chromosomes, optical mapping uses restriction enzymes and fluorescence labeling to infer genomic structure (Faino *et al.*, 2015; van Kan *et al.*, 2017). When compared to other fungal genomes, the existing genome assembly of *S. sclerotiorum* is of excellent quality, although it is incomplete; an optical map indicates that 1.6 million bp of sequence are missing (Amselem *et al.*, 2011). Using the available optical map in conjunction with PacBio and Illumina sequencing, Derbyshire *et al.* (2017) improved the previous genome, its annotations, and the effector analysis that followed. They discovered numerous additional effector-like protein-producing genes using this enhanced genome sequence, some of which showed notable upregulation during infection. After TEs in the *S. sclerotiorum* genome were recharacterized, it was discovered that the five most prevalent TEs had undergone RIP and had a higher RIP index than a set of control sequences that were chosen at random. A comparative study comparing the genomes of different filamentous plant diseases that are exclusive to specific hosts revealed disparities in their in-planta expression profiles and revealed a weak correlation between them. Additionally, a comparison of *S. sclerotiorum*'s genome architecture to that of plant pathogens possessing "two-speed" genomes was conducted. Although the fraction of released proteins and the GC content did not correlate, it was shown that the secreted proteins of *S. sclerotiorum* exhibit considerable correlations with regions exhibiting a high RIP index. On the other hand, there was no negative correlation observed between the amount of released protein and the Cluster of Differentiation (CDs) content in the *S. sclerotiorum* population. It was found that although *S. sclerotiorum* does not exhibit these signatures throughout the whole genome, it does exhibit faint indications of increased secreted protein mutation in specific genomic compartments as a consequence of transposition and RIP activity (Fig. 8.1, Chapter 8, this volume; Derbyshire *et al.*, 2017).

3.21.7 *S. sclerotiorum* genome assembly

Derbyshire *et al.* (2017) have provided a thorough and precise annotation of the genome of the harmful plant disease pathogen *Sclerotinia*. The prior genome assembly used optical mapping in conjunction with 9.1 × Sanger sequencing. Despite the relatively high quality of this assembly, an estimated 1.6 million bp were missing. An additional 805,146 bp can be found in the current assembly of *S. sclerotiorum*—roughly half of the projected missing sequence noted during the study. Nonetheless, given that 14 of the 16 chromosomes have been sequenced from telomere to telomere with little to no gaps, it was determined that the new *S. sclerotiorum* genome is essentially complete (Fig. 3.10).

The sole gap in the structure is the nucleolus organizing complex, which contains repeats of rDNA. Similar results were noted in the PacBio-sequenced genome of the plant pathogenic fungus *Verticillium dahliae*. The two full fungal plant pathogen genomes, aside from *V. dahliae*, are those of the broad host range necrotroph *B. cinerea* (van Kan *et al.*, 2017) and the wheat head blight fungus *Fusarium graminearum* (King *et al.*, 2015). In contrast to the latter, which was put together using PacBio sequencing and optical mapping, the former was put together using Illumina mate pair technology. As a result, the new assembly of *S. sclerotiorum* will be useful for upcoming comparative genomics research and contribute to the very small pool of whole genomes of fungal plant diseases. Most of the extra constructed bases (78.7%) are found in areas that are expected to repeat. TEs make up 12.96% of the genome, which is more than the 7.7–9.5% range that was previously calculated (Amselem *et al.*, 2011, 2015a). Nevertheless, some of these sequences can be viewed as questionable in the future because TEs were not carefully curated following automated discovery. The new repeating sequence, while not thoroughly investigated, now enables a more thorough examination of TEs in *S. Sclerotinia*, expanding on the conclusions of Santana *et al.* (2014) and Amselem *et al.* (2015a).

3.21.8 Annotations of genes in the novel genome of *S. sclerotiorum*

The new gene annotations showed the most progress, even with minor changes to the *S. sclerotiorum* genome assembly. A total of 14,522 genes were predicted in the last genome assembly by ab initio gene finders that were trained on a carefully selected sample of 542 *s. sclerotiorum* genes. A total of 11,860 trustworthy gene

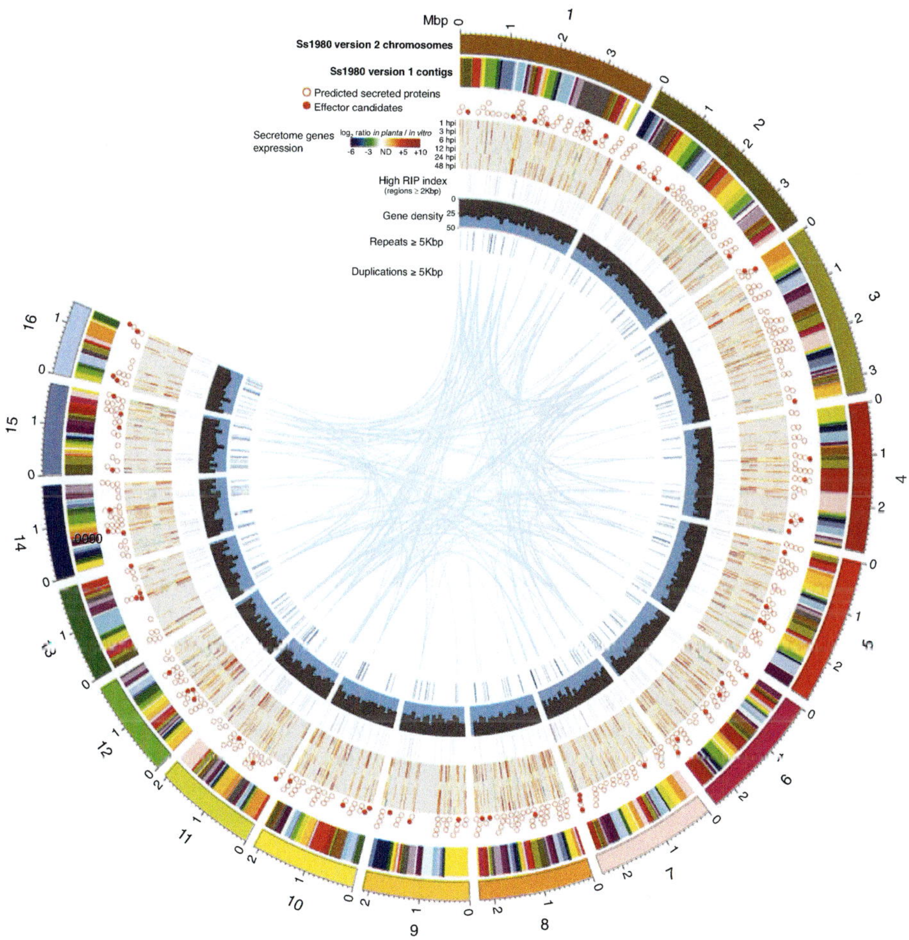

Fig. 3.10. The features of *S. sclerotiorum* and the locations of the relevant version 1 contigs are depicted in the Circos plot. The circular tracks are positioned as follows, going from outer to inner: To distinguish one contig from another, (1) the 16 contigs from the version 2 assembly (Ss1980 version 2. are color-coded. Contigs from the Ss1980 version 1 assembly are similarly color-coded to differentiate them from one another. The plot's circles—clear circles for secreted proteins and colorful circles for secreted small protein (SsPEs)—represent the locations of the two types of proteins. The color gradient represents the change in secreted protein (log fold) expressed *in vitro* for each of the tested *B. napus* with different time points. Blue represents down-regulated expression and red represents up-regulated expression. Furthermore, areas with high RIP index predicted by RipCal are highlighted; these include the repeat sequences longer than 5 kb from the REPET de novo annotation. (Adapted from Derbyshire *et al.*, (2017) publication with permission).

calls were obtained after these predictions were assessed using length, homology to TEs, EST, and BLAST evidence, along with additional tools (Amselem *et al.*, 2011). Only the 11,860 'non-dubious' genes from the previous version of the *S. sclerotiorum* gene annotation set were taken into consideration. With 11,130 predictions, the new version still has 730 fewer genes than the previous one. Additionally, only 10,528 of these sequences matched the sequences from version 1 in terms of reciprocal top BLAST hits. This indicates that 602 sequences were either absent from the prior genome or had undergone major alterations in the version 2 prediction set, preventing BLAST analysis from locating their associated loci. After more investigation, it was shown that only 279 of the gene predictions in version 2 were the outcome of splitting predictions from version 1, whereas 1301 loci in version 2 came from the fusion of prior loci. Therefore, the primary reason for the decrease in filtered gene predictions from the version 1 assembly to the version 2 assembly can be attributed to the fusion of previously different loci. Investigations later revealed that the gene predictions made by version 2 were more accurate than those made by version 1. When the amino acid sequences were blasted against the NCBI nr database, it was discovered that the version 2 sequences had, on average, better coverage and identity to their highest BLAST hits. This suggests that the new gene predictions produced more BLAST hits from precise protein predictions, displaying regions that are genuinely conserved. The *Parastagonospora nodorum* annotation collection was compared with an earlier version using a similar methodology (Syme *et al.*, 2016). Furthermore, a higher quantity of predicted functional domains and gene ontology (GO) keywords may also be a sign of improved protein-level accuracy. In addition, the second set of predictions showed a greater similarity to their corresponding Cufflinks transcripts when analyzed using BLAST. Although the median coverage only improved by 1% in the second set of predictions compared to the first set, there was a significant decrease in the degree of variability between the two sets. The interquartile range for the first set was 32%, while for the second set, it was only 1%, indicating a higher level of consistency with the Cufflinks transcripts in the second set of predictions. Furthermore, the RNA-Seq data further supports the accuracy of the second set of predictions, as the median coverage of the CDs in the second set is higher than that of the first set (Derbyshire *et al.*, 2017).

3.21.9 *S. sclerotiorum* effector gene family identification

Numerous researchers have reported a clear relationship between effector-like genes and TEs in a variety of plant pathogenic fungi (Rouxel *et al.*, 2011; Grandaubert *et al.*, 2014; Amselem *et al.*, 2015a; Asman *et al.*, 2016; Faino *et al.*, 2016; Selin *et al.*, 2016). According to some theories, this arrangement makes it easier for effector gene families to grow and then diversify through mutation (Dong *et al.*, 2015). In theory, this may improve a fungal capacity to swiftly adjust to recently developed *R*-genes or lost susceptibility loci in host plants. Furthermore, it is thought that effectors found in TE-rich locations experience higher rates of mutation due to RIP, a fungal defense mechanism against TEs. In duplicate sequences, this step entails the transformation of C bases to T bases. Compared to the rest of the genome, repeat-rich areas containing effector-like proteins are usually gene-poor. This guarantees that evolutionary processes linked with TE do not impact housekeeping genes.

It has been discovered that the majority of the TEs in the pathogen *Leptosphaeria maculans* are inactive in *B. napus*. The cause of this is thought to have been a prior invasion by ancestral TEs, which was followed by widespread RIP. In *L. maculans*, almost 20% of effector-like sequences are linked to areas that have TEs impacted by RIP, but only 4.2% are linked to regions that are rich in genes. It has been shown that repetitive region sequences most likely feature RIP-like mutations even if they do not typically belong to families that have grown through TE activity (Rouxel *et al.*, 2011). In order to investigate the possibility that Secreted Small Proteins (SsPEs) run in families, Derbyshire *et al.* (2017) carried out a number of analyses. To find recent paralogs, they employed Ortho MCL (Markov clustering) analysis using the gene predictions of *S. sclerotiorum* and *B. cinerea*. Furthermore, they aligned numerous sequences of effector-containing areas and checked if they included TE sequences, as REPET had predicted, to see if SsPE families were probably produced by TE activity. They discovered a solitary SsPE family, comprising seven proteins with more than 90% amino acid content.

The sequence family in question had only one ortholog in *B. cinerea*, indicating that they most likely diverged from *S. sclerotiorum* and *B. cinerea* after that. Six sequences from this family had previously been found by Guyon *et al.* (2014), but their relationship to transposition activity has not yet been thoroughly investigated. The absence of any functional domains that were expected for these sequences suggests that they may have specialized roles as effectors. The areas around these sequences also showed that they were part of a 4–5 kb genomic tract that was duplicated on six distinct chromosomes, according to many alignments of the surrounding regions. Remarkably, every single one of these SsPEs was discovered inside a full or partial Gypsy LTR-retro element, indicating that transposable element (TE) activity is likely the cause of these duplicated sequences. Consequently, *S. sclerotiorum* provides evidence of effector evolution through transposition, marking the first occurrence of this process in this species. This finding led to the hypothesis that *S. sclerotiorum* might have effector protein-enriched, TE-rich, gene-sparse, and possibly RIP (repeat-induced point mutation) affects areas with repeated or duplicated DNA sequences. A comparative genome analysis between *S. sclerotiorum* and the "two speed" genomes of *Blumeria graminis*, *L. maculans*, and *Phytophthora infestans* was carried out in order to evaluate this hypothesis. Only *L. maculans*' genome was found to have a bimodal GC content among the four microbial genomes. This discovery is consistent with the findings of a related investigation carried out with the earlier *S. sclerotiorum* genome version by Testa *et al.* (2016). Therefore, it may be assumed that *B. graminis*, *P. infestans*, and *S. sclerotiorum* have constant GC content across their whole genomes.

The obtained results have two possible explanations. First, the genomes under study might not have any RIP at all, or they might have very little RIP. On the other hand, it is possible that the genomes exhibit a constant degree of RIP throughout, lacking any particular RIP-affected areas with low GC content. Prior research has shown that *B. graminis* does not actively undergo RIP, although *S. sclerotiorum* and *L. maculans* do. Since RIP is unique to fungi, it is thought that *P. infestans* does not possess it (Hane *et al.*, 2015). In particular, RIP is displayed at a moderate frequency by *S. sclerotiorum* and at a high frequency by *L. maculans*. Conversely, whereas *B. graminis* exhibits a RIP-like signature in certain TE copies, this is probably not the result of RIP-related genes being present, but rather of ancestral activity. It has been demonstrated that *S. sclerotiorum*, in contrast to *L. maculans*, exhibits two different forms of RIP: CpT=>TpT and CpA=>TpA transitions (Amselem *et al.*, 2015b). The conclusions of the analysis are supported by the fact that *S. sclerotiorum* mostly exhibits these two kinds of RIP in the majority of its TE copies. Furthermore, compared to random control sequences, the TpA/ApT RIP indices of the five most prevalent TEs were substantially greater. TpA/ApT and CpA+TpG/ApC+GpT are two RIP indices used to measure the frequency of repeat-induced point (RIP) mutation in fungal genomes. Although *S. sclerotiorum* exhibits a comparatively high degree of RIP, the lack of bimodality in the GC content of its genome would indicate the absence of particular RIP-affected compartments, as seen in *L. maculans*.

Remarkably, a significant relationship between RIP affected regions and secreted proteins was found in *S. sclerotiorum*. RIP-affected sequences and effector-like sequences did not, however, appear to be related. This implies that the diversification of secreted proteins may be closely associated with RIP in particular regions of the *S. sclerotiorum* genome. A number of areas show clusters of secreted and effector-like proteins in areas with a high RIP index and numerous repeat sequences, especially at the distal ends of chromosomes 7, 15, and 16. However, further research is still needed to determine how these areas affect evolution and host specificity. These results led to the hypothesis that *S. sclerotiorum*-secreted proteins are connected to locations that are rich in adenine and thymine (AT). Furthermore, the possibility that these proteins are associated with gene-sparse regions—a marker of substantial compartmentalization away from critical genes—was investigated. The data showed no correlation between the GC content and the proportion of released proteins in *S. sclerotiorum*, *B. graminis*, and *P. infestans*. In contrast, a substantial negative correlation was seen between the number of released proteins and the GC level in *L. maculans*. This implies that *S. sclerotiorum*-secreted proteins do not substantially split into AT-rich (RIP-affected) sequence regions.

Additionally, the genomes of *B. graminis*, *P. infestans*, and *L. maculans* demonstrated a significant inverse relationship between the fraction of secreted proteins and CDs content, which is used to assess gene richness. This conclusion supports previous observations that effector-like and secreted proteins are generally located in regions of these animals where gene density is low. On the other hand, in *S. sclerotiorum*, there was no relationship found between the number of released proteins and the gene content. The distal end of chromosome 7 is one example of a tiny region that may have low gene density but significant diversity of secreted proteins, as seen in Fig. 3.10. On a bigger scale, though, *S. sclerotiorum* did not exhibit this phenomenon, as observed in other species (Raffaele and Kamoun, 2012).

3.21.10 Multifaceted role of small RNAs in molecular biology of *Sclerotinia sclerotiorum*

The discovery that small RNAs, specifically 20–25-nucleotide-long noncoding RNAs, can be transferred between pathogens and hosts, also known as cross-kingdom RNAi, has opened up exciting possibilities in the field of molecular plant pathology (Weiberg *et al.*, 2013; Cai *et al.*, 2018). These specific molecules function as guides for a group of enzymes that lead to complementary mRNAs in the RNAi pathway. This leads to the degradation or inhibition in the translation of these mRNAs (Weiberg and Jin, 2015). Cross-kingdom RNAi has been reported in the specialist pathogens *Fusarium oxysporum* f. sp. *lycopersici* and *Puccinia striiformis*, as well as in the generalist *B. cinerea*, which is closely related to *S. sclerotiorum* (Wang *et al.*, 2017; Ji *et al.*, 2021). But when it comes to the specialized pathogen *Zymoseptoria tritici*, there is a conspicuous lack of evidence supporting these phenomena (Kettles *et al.*, 2019; Ma *et al.*, 2019).

According to preliminary research, cross-kingdom RNAi is essential to the pathophysiology of *S. sclerotiorum*. Derbyshire *et al.* (2022) report that *S. sclerotiorum* produces a large number of short RNAs while infecting *Arabidopsis* and *P. vulgaris*. These small RNAs seem to target particular genes in *Arabidopsis*, resulting in decreased expression during infection. This could be a sign that fungal small RNAs are silencing these genes. Interestingly, in a wild population, these targeted genes also contain polymorphisms linked to quantitative resistance to *S. sclerotiorum*. A study found that two of the three knockout lines for the predicted *S. sclerotiorum* small RNA target genes in *Arabidopsis* were more prone to infection than the wild type. This finding offers an evolutionary explanation for the relationship between these specific mRNAs and pathogen small RNAs. The data, however, did not offer concrete proof of the interaction between plant mRNAs and short RNAs from *sclerotiorum*. Future studies may clarify the precise roles of short RNAs and address pertinent issues as the significance of these molecules in *S. sclerotiorum* pathogenesis remains largely unexplored. Furthermore, short RNAs have the ability to control endogenous gene expression in addition to spanning species' boundaries. Endogenous short RNAs may regulate sclerotia development-related genes in *S. sclerotiorum* (Xia *et al.*, 2020a). The function of short RNAs in defense of *S. sclerotiorum* against viruses has also been examined in another research (Neupane *et al.*, 2019).

The RNAi pathway in *S. sclerotiorum* provides opportunities for novel disease control strategies, such as host-induced and spray-induced gene silencing. The production of double-stranded RNAs reduced the susceptibility of *G. max* to illness by targeting the oxalic acid biosynthesis gene *oah* in *S. sclerotiorum* (McCaghey *et al.*, 2021). The results of *in vitro* investigations suggested that these double-stranded RNAs may be transferred from host to pathogen, as they could be absorbed by fungal hyphae. These double-stranded RNAs probably enter the endogenous RNAi pathway and produce short RNAs that control the expression of target genes, given the known function of RNAi in fungi (Dong *et al.*, 2011). Double-stranded RNAs can be applied topically to target virulence genes in *S. sclerotiorum*, since it is especially responsive to their absorption (Qiao *et al.*, 2021; Derbyshire *et al.*, 2022).

3.21.11 The effector repertoire diversity unveiled through the genome sequencing and secretome profiling of *S. sclerotiorum*

A highly aggressive strain of *S. sclerotiorum*, known as "ESR-01," was sequenced, resulting in an assembly size of approximately 41 Mb. This assembly consisted of 328 scaffolds, with an N50 value of 447,128. Through this sequencing, 27,450 SNPs were identified across 155 scaffolds when compared to the S. *Sclerotinia* 1980 isolate. On an average, there were approximately 1.5 SNPs per kilobase of the genome. Additionally, 667 repetitive elements were discovered, accounting for roughly 7% of the annotated genes. Among these repetitive elements, the DDE_1 family was the most prevalent, making up 68% of the total predicted repetitive elements, with a count of 454. Furthermore, a total of 3844 simple sequence repeats (SSR) were found within the 328 scaffolds. The whole-genome assembly predicted a total of 9469 protein-coding genes, with an average length of 1587 bp. In the genome, these genes were dispersed at a rate of 230.95 genes per mega base. It was discovered that 329 of the projected protein-coding genes encoded CAZymes, or carbohydrates-active enzymes, which are able to degrade complex polysaccharides. Glycosyltransferase (GT) families were the most prevalent CAZyme families, making up 497.1% of the predicted CAZymes. Specifically, the major families engaged in breaking down the cell wall were GT2 (23%), GT4 (20%), and glycoside hydrolase (GH) (23%), with GH18 (11%) making up the remaining 5%. Furthermore, 156 genes critical for the interactions between the virus and the host were found. A comprehensive examination of the whole-genome proteomics dataset yielded a total of 57 effector candidates (ECs). Thirty of these ECs were new, and the remaining 27 had analogs. In order to experimentally validate 11 chosen ECs, the expression profile of the *S. sclerotiorum* isolate ESR-01 was examined. Our knowledge of *S. sclerotiorum's* interactions with *Brassica* and its hemibiotrophic dual lifestyle as a pathogen is improved by these studies, which offer insightful information about the genome, secretome, and effector repertoire of the pathogen (Table 3.3; Gupta *et al.*, 2022).

S. sclerotiorum lives in hemi-biotrophic environments, which are defined as a brief biotrophic phase followed by an extended necrotrophic phase. *S. sclerotiorum* has a more complex pathogenic mechanism than a straightforward necrotrophic or biotrophic fungus. Recent discoveries about the life cycle and pathogenic mechanism of some necrotrophic fungi, however, suggest that the pathogen's effector repertoire is essential for the establishment of infection. This is accomplished by avoiding the hypersensitive response and suppressing the host defense, which is comparable to the early stages of the biotrophic mode of infection (Weiberg *et al.*, 2013; Zhu *et al.*, 2013; Rodriguez-Moreno *et al.*, 2018).

The effector repertoire of the pathogen both causes host cell death and acts as a defense mechanism against the oxidative bursts that the host experiences after infection. Repeat elements have the potential to create required genetic variation, which may also contribute to the organism's capacity for rapid adaptation to a wide range of hosts. A viable and effective technique for deciphering molecular recognition and determining genetic variations among pathogen isolates with differing degrees of virulence is the use of microsatellite (SSR) markers. Furthermore, using SNPs to conduct a genome-wide examination of genetic variation, it is possible to identify a variety of conserved molecular pathways that control the pathogen's ability to spread (Sexton and Howlett, 2004; Mert-Turk *et al.*, 2007; Singh *et al.*, 2012).

It was necessary to thoroughly examine the genetic elements and genomic properties of the pathogen in order to understand the host's reaction to geographically diverse isolates. Researchers were able to fully comprehend the genomic features of the *S. sclerotiorum* "ESR-01" isolate, including its universality and broad host range infection capacities, by performing whole-genome sequencing and de novo assembly of the isolate. The "ESR-01" isolate has a total assembly size of 40.98 Mb, 2.68 Mb more than the presequenced "1980" isolate (38.3 Mb). However, in comparison to the "1980" isolate (14,522) and the

Table 3.3. Features of *Sclerotinia sclerotiorum* assembly and gene statistics in the "ESR-01" isolate. (Adapted from publication of Gupta *et al*. (2022) with permission.)

Features	*S. sclerotiorum* isolate "ESR-01"
Coverage	129×
Assembly size (Mb)	40.98
Scaffold N50	447,128
GC content (%)	37.71
Predicted protein-coding genes	9469
Average gene length (bp)	1587
Gene density (number of genes per Mb)	231
Unique genes	57
GC content in predicted genes (%)	45.88
tRNA genes	74
rRNA gene	11
Transposable elements	647
Secreted proteins	944
SNPs	27,450
SSRs	3844

resequenced genome of "1980" (11,130), the number of protein-coding genes in the "ESR-01" isolate (9469) was comparatively smaller (Amselem *et al*., 2011; Derbyshire *et al*., 2017). The disparity in the number of genes might be ascribed to variations in the origin of the isolates or the application of a strict gene prediction approach aimed at reducing redundancy (Table 3.3; Fig. 3.11).

An important fraction of the predicted genes is critical for many cellular and metabolic activities, as well as for the formation of cells and cellular components, according to the examination of annotated data for the predicted genes. These processes are the essential building elements that allow the virus to develop quickly. Furthermore, KEGG (Kyoto Encyclopedia of Genes and Genomes) pathway analysis revealed a comparable distribution pattern, with a significant number of genes participating in the machinery in charge of protein synthesis. Moreover, it was shown that a large percentage of the genes were involved in pathways connected to the metabolism and transport of carbohydrates, which are crucial for fundamental biological processes (Table 3.4). A key component of the genome that enables infections to quickly adapt to new hosts and shifting environments is repeat elements. Repetitive DNA typically makes up 1–25% of fungal genomes (Castanera *et al*., 2016). The percentage of repetitive elements and secretory effector molecules in the genome affects how pathogens and effectors change over time. The most prevalent TE in the *S. sclerotiorum* ESR-01 isolate is DDE-1. Three amino acids—D, D, and E—which make up the conserved motif of this family of repeat components create an active core of the protein. Aspartic acid is represented by D in this motif, whereas glutamic acid is represented by E (Ohta *et al*., 2002).

Microsatellites used in the research of the *S. sclerotiorum* "ESR-01" isolate showed that trinucleotide repeats were the most prevalent, making up 66% of all SSRs found. With 107, 106, and 105 occurrences, respectively, the AAT/GAT/TCA element was the most frequently observed trinucleotide repeat, while GTT only appeared 27 times. According to the results of the genome-wide analyses, there are 27,450 SNPs in total, and T is replaced roughly 14,081 times. With 4927 occurrences, the most frequent SNP was the transversion from T to C. This was followed

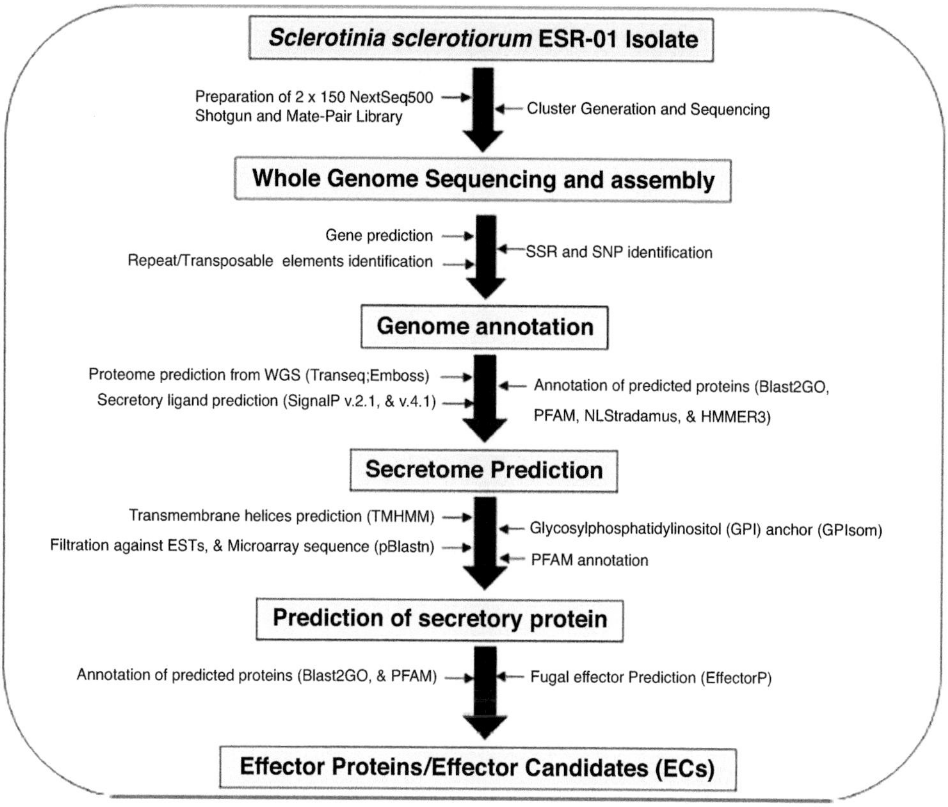

Fig. 3.11. Diagrammatic illustration of the *S. sclerotiorum* whole-genome sequencing and bioinformatic analysis protocol of isolate ESR-01. (Adapted from a publication by Gupta *et al*., 2022, with permission.)

by T/G (961 occurrences) and T/A (732 occurrences). The connection between the host and the production of different proteins, enzymes, and metabolites is the first step in a fungal pathogen's effective infection of its host (Heard *et al.*, 2015). Secretory proteins help the pathogen colonize during the host–pathogen contact, which is a critical function they play in the early stages of infection. Based on their distinguishing characteristics, 57 of the 554 projected secretory proteins were determined to be viable ECs. Guyon *et al.* (2014) conducted a secretome analysis of the *S. sclerotiorum* "1980" strain, which yielded 78 possible ECs out of 745 predicted secretory proteins. In a similar vein, 70 of the 523 secretory proteins predicted by Derbyshire *et al.* (2017) were thought to be potential effectors. The existence of CAZymes in 164 proteins further substantiated the functional linkage of the expected 369 secretory proteins in pathogenesis in the S. sclerotiorum "ESR-01" isolate. The secretome of the "ESR-01" isolate is composed of a wide variety of proteins other than CAZymes, according to the GO annotation of the secretory proteins. A noteworthy proportion of these projected ECs are involved in membrane-bound functions and oxidoreductase activities. The effector repertoire of the *S. sclerotiorum* "ESR-01" isolates comprises pathogenesis-related hydrolases, oxidoreductases, peptidases, transferases, and isomerases.

These proteins are responsible for the majority of the pathogen's pathogenicity. Three ECs (genes 70, 3672, and 5446)—isochorismatase, alpha-beta-hydrolase, and a different class of hydrolase—exhibited hydrolytic activity among the 27 identified effectors. Differential

Table 3.4. Predicted genes from the *S. sclerotiorum* "ESR-01" isolate analyzed using KEGG pathway. (Adapted from publication of Gupta *et al.* (2022) with permission.)

Pathway categories	Number of genes
Metabolism	
Overview	207
Carbohydrate metabolism	282
Energy	143
Lipid	140
Nucleotide	105
Amino acid	225
Metabolism of other amino acids	73
Glycan biosynthesis and metabolism	67
Metabolism of cofactors and vitamins	124
Metabolism of terpenoids and polyketides	28
Biosynthesis of other secondary metabolites	39
Xenobiotics, biodegradation, and metabolism	66
Genetic information processing	
Transcription	138
Translation	292
Folding, sorting, and degradation	233
Replication and repair	96
Environmental information processing	
Membrane transport	7
Signal transduction	231
Signaling molecules and interactions	2
Cellular process	
Transport and catabolism	261
Cell motility	23
Cell growth and death	154
Cellular community—eukaryotes	37
Organismal system	
Environmental adaptation	21

expression of this hydrolytic activity in the *S. sclerotiorum* transcriptome during *B. napus* infection has demonstrated its significance for virulence (Mindrebo *et al.*, 2016; Liang and Rollins, 2018). An oxalate decarboxylase is a protein belonging to the cupin family that is one of the secretory effectors (gene 2387). Liang *et al.* (2015) have provided evidence about the function of oxalate decarboxylase in aiding the initial stages of *S. sclerotiorum* infection through the creation of compound appressoria. Peptidyl-Prolyl *cis/trans* Isomerases (PPIases), a group of secreted effector proteins in plant pathogens like *Sclerotinia sclerotiorum*, exhibit characteristics similar to cysteine-rich secretory proteins (CRISPs) and possess peptidyl-prolyl cis/trans isomerase (PPIase) activity. These proteins, often termed SCP-like, play roles in virulence and interacting with the host plant's defense mechanisms. The defense response to fungus is associated with gene 6207, a CAZyme with a CBM-5 domain, whereas gene 7275 is involved in protein folding and has demonstrated a mixed response to pathogenicity against *B. cinerea* in the PHI database. It has been discovered that *Magnaporthe oryzae* infection in rice induces the tiny secreted protein SCP-like extracellular peptide (Wang *et al.*, 2020). Unal and Steinert (2014) have established the importance of PPIases in virulence and as a possible target for drugs, in addition to their recognized participation in biological processes such as gene expression, protein secretion, and signal transduction. PPIases play an important role in *S. sclerotiorum* pathogenicity, as evidenced by the markedly altered protein expression in the GATA-type IVb transcription factor ΔSsNsd1 mutant, which eliminated appressorium formation (Li *et al.*, 2018a).

Examples of secreted proteins with oxidoreductase activity are those encoded by genes 7790, 3331, 3900, and 2441. These proteins are categorized as 2-deoxy-D-gluconate-3-dehydrogenase, Rieske domain-containing proteins, members of the short-chain dehydrogenase reductase family, and di-copper center-containing proteins with esterase activity (particularly targeting RNA phosphodiester bonds), in that order. Laccases are a group of enzymes that belong to the oxidoreductase class with di- or multicopper form in *B. cinerea*. These laccases have been observed during the colonization of host tissues and are involved in the manufacture of melanin. Moreover, the hyphae grown in PDB contained a number of additional proteins like laccases, oxalate decarboxylase, peroxidase, polysaccharide deacetylase, and several proteins unique to *Sclerotinia* and *Botrytis* infection that might facilitate growth on a plant host (Liu and Free, 2016; Buddhika *et al.*, 2021).

The GT 51 domain-containing CAZyme 2-deoxy-D-gluconate-3-dehydrogenase (gene 2441) has been identified as deadly against *Ralstonia solanacearum* in the PHI basis. However, research on *Penicillium expansum* has led to the classification of the short-chain dehydrogenase reductase family protein (gene 3900) as having decreased virulence. The pentose phosphate pathway enzymes 2-deoxy-D-gluconate-3-dehydrogenases are upregulated in the presence of xylan in the growth media of *N. crassa*, suggesting their relationship with hemicellulose degradation. Based on research on *Burkholderia cenocepacia*, the predicted secretory effector phosphotyrosine phosphatase (gene 2590) has been classified as having reduced virulence in the PHI (pathogen–host interactions) database and has been reported to play a role in the regulation of virulence and multistress tolerance in *B. cinerea*. An effector 'ESR-01' of *S. sclerotiorum* is known to contain small coat protein complex II (COPII) coat GTPase SAR1 (gene 8968) for virulence. By enlisting the COPII assembly to create transport vesicles, secretion-associated Ras-related proteins, or SAR GTPases, control vesicular transport (Sun *et al.*, 2012; Yorimitsu *et al.*, 2014; Yang *et al.*, 2016). Proteomic examination of these extracellular vesicles has revealed the presence of CAZymes induced by cellulose in the growth conditions. GTPase SAR1 recruits COPII vesicles in *Trichoderma reesei* for the transfer of secretory proteins (de Paula *et al.*, 2019).

Protease propeptide inhibitor (gene 7551), d-tyrosyl-tRNA deacetylase (gene 2609), phospholipase A2 (gene 4192), phosphatidylglycerol phosphatidylinositol transfer (gene 5993), Sm-like ribonucleoprotein (gene 7451), proteasome subunit beta type 2 (gene 4766), RHA1a-like RING-H2 zinc finger protein (gene 8106), Mpv17 PMP22 family protein (gene 2343), and kinases (gene 6535) are among the other secretory effectors that have been annotated. Thirty of the 57 anticipated effectors were recently identified and had not been previously annotated. These comprise proteins containing the BYS1 and PAN domain (genes 7541 and 9070), necrosis and ethylene-inducing peptides (genes 7020, 7092), phospholipases (gene 6283), epimerases (gene 4208), dehydrogenases (gene 4195), and kinases (gene 1761) with a wide range of enzymatic activity. It was also discovered that BYS1 domain-containing proteins (gene 7541) and ethylene-inducing peptides (genes 7020, 7092) had CBM-13 and AA11 domains of CAZymes. PHI prediction revealed that among the unannotated proteins, necrosis-/ethylene-inducing peptides (gene 7020, gene 7092) and putative short-chain dehydrogenases (gene 4195) were found to be unaffected in pathogenicity against

F. graminearum and *Botrytis elliptica*, respectively. Furthermore, three additional annotated protein genes (47,151, and 8329) were identified as having decreased virulence against *M. oryzae* based on predictions made in the PHI database. The transcript profiles of specific ECs were analyzed in order to verify their presence within the *S. sclerotiorum* genome. Among the expressed SSEPs, genes 6207 and 2441 (SCP-like extracellular, 2-deoxy-D-gluconate-3 -dehydrogenase) contain the CBM5 and GT51 classes of CAZymes, respectively. *P. expansum* and *B. cenocepacia* were similar to phosphotyrosine phosphatase gene 2590 and short-chain dehydrogenase reductase gene 3900, respectively, suggesting decreased pathogenicity in PHI base. However, it was determined that 2-deoxy-D-gluconate 3-dehydrogenase (gene 2441) was fatal to *R. solanacearum*. Secondary metabolites are produced by pathogenic fungus for a variety of purposes, including virulence, defense, nutrition absorption, and signaling. The host cell's structure and functioning are disrupted by these metabolites. In fungal infections, the biotrophic life cycle is linked to the loss of the SM biosynthesis pathway. Secondary metabolites, such as phytotoxins, siderophores, and pigments, alter the structure and function of host cells, hence augmenting the virulence of pathogenic fungus. In biosynthetic gene clusters (BGCs), genes encoding secondary metabolites are frequently co-regulated (Keller and Hohn, 1997; Collemare and Lebrun, 2011; Chooi and Solomon, 2014). By applying bioinformatics techniques to examine the similarities between important 'backbone' enzymes and the "decorating" enzymes in charge of methylation, oxidation, reduction, or glycosylation, BGCs (biosynthetic gene clusters) can be predicted. Polyketide synthases (PKSs) and nonribosomal peptide synthases (NRPSs) are the primary enzymes implicated in BGCs. Precursor enzymes, transporters, and transcriptional regulators are among the other genes found in these clusters (Fedorova *et al.*, 2012). Two PKS (polyketide synthases genes) genes producing botcinic acid were shown to be increased in *B. cinerea* during host infection in *B. napus*; one of these genes was also found to be upregulated in *S. sclerotiorum* (Liang and Rollins, 2018). Graham-Taylor *et al.* (2020) examine the functional characterization of 11 secondary metabolite genes in *S. sclerotiorum*, which demonstrated homology to *B. cinerea* by analyzing their expression at six different time points during infection. The compound 1,3,6,8-tetrahydroxynaphthalene (T4HN), which is subsequently transformed into dihydroxy naphthalene (DHN) melanin during the production of sclerotia, is a product of naphthalene biosynthesis. Melanin is essential for virulence because it shields cells from harmful environments and aids in maintaining the proper turgor pressure in appressoria to allow penetration of the host tissue (Kimura and Tsuge, 1993; Eisenman and Casadevall, 2012). Upon infection of *B. napus*, it was also discovered that the homologs of phytoene synthase were increased in *S. sclerotiorum* and *B. cinerea* (Figs 3.12 and 3.13). Together with other genomic features, including SSRs, SNPs, and TEs, the ordered functioning of functionally varied secretory effector molecules, CAZymes, and PHI-related genes establishes the *S. sclerotiorum* pathosystem complex (Gupta *et al.*, 2022).

3.21.12 Proteomics analysis of *S. sclerotiorum*

Proteomics research by Tian *et al.* (2021) revealed important virulence variables, including acidic pH, oxalic acid, effectors, and extracellular exudates that break down the cell wall. The study particularly looked into the role exudates play in the sclerotial growth of the organism. A total of 258 proteins were found; of these, 193 were GO-annotated, and 54 were categorized using an analysis of the Kyoto Encyclopedia of Genes and Genomes. By assessing gene expression during sclerotial growth and/or quantifying the enzymatic activity of sclerotial exudates, additional proof was provided for four proteins linked to the breakdown of plant cell walls. The results imply that proteins secreted by sclerotia are involved in the formation of sclerotia and in the necrosis of host cells brought about by *S. sclerotiorum*. It was hypothesized that sclerotial exudates may break down plant cell walls and produce

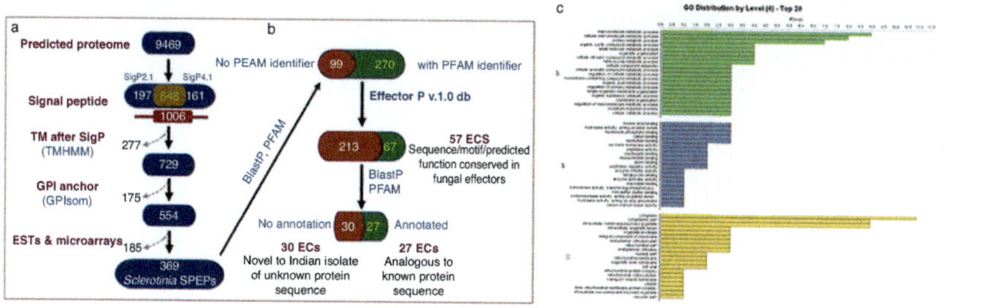

Fig. 3.12. The WEGO (Website-based Gene Ontology Annotation) map visualizes the identification and distribution of effector protein candidates (ECs) from *Sclerotinia sclerotiorum*, with a focus on the top 27 hits out of 554 predicted secreted proteins, of which 369 are confirmed to be expressed in planta as *S. sclerotiorum* secreted proteins expressed in planta (SPEPs). The map uses dot arrows for filtered-out proteins and boxes for selected proteins and computational tools, filtering for ECs based on conserved fungal effector features. Category (c) of the map categorizes the selected proteins using gene ontology (GO) into molecular function (MF), cellular components (CC), and biological processes (BP), showing the percentage of genes in each sub-group on the left y-axis. (Adapted from publication of Gupta *et al.*, 2022, with permission.)

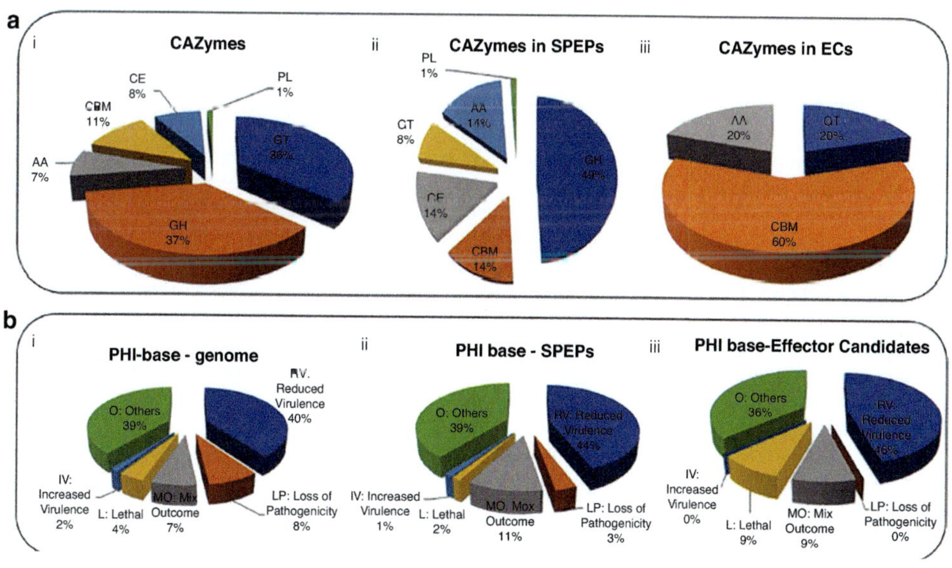

Fig. 3.13. (a) The distribution pattern of the predicted CAZymes and pathogen–host interaction factors in *S. sclerotiorum* 'ESR-01' isolates is briefly described as follows. (a–i) The six types of CAZymes are carbohydrate esterase (CE), polysaccharide lyases (PLs), glycoside hydrolases (GHs), carbohydrate-binding modules (CBMs), and auxiliary activities (AA). (a–ii) The sequestered proteins made in plants (SPEP) contain various summaries of the CAZymes. (a–iii) Additionally, each CAZyme has an own summary in the ECs. (b) Based on the pathogen–host interactions observed in the genome, SPEPs, and ECs, the ortholog genes of the *S. sclerotiorum* "ESR-01" isolate can be categorized into a number of phenotypic groupings. This info comes from the PHI base. (Adapted from publication of Gupta *et al.*, 2022, with permission.)

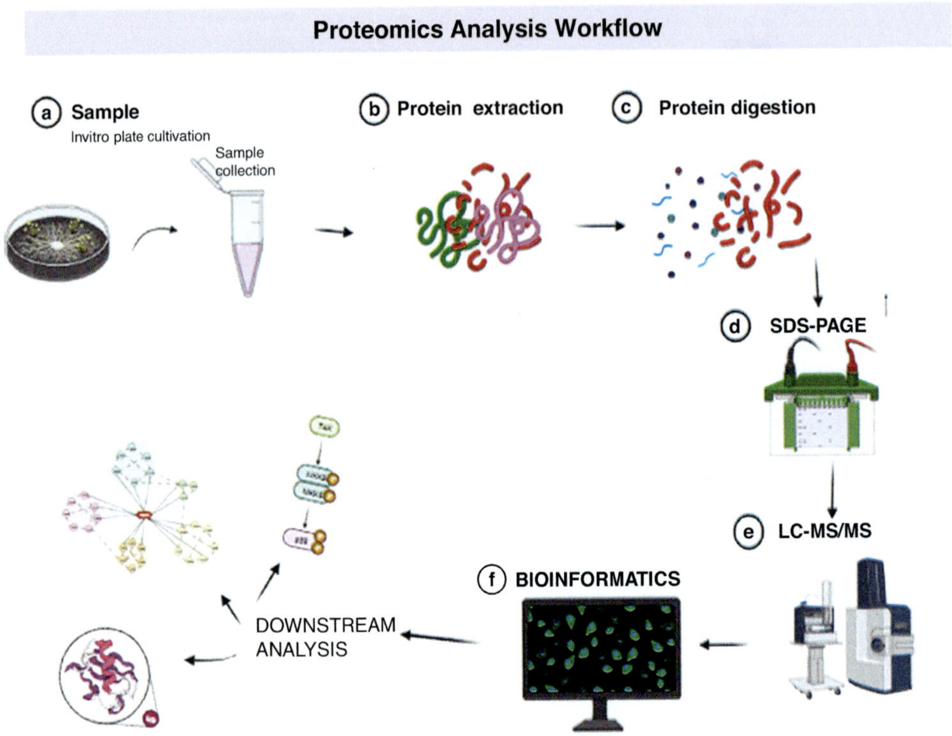

Fig. 3.14. The workflow for proteome analysis of *S. sclerotiorum* contains multiple phases. (a) First, mycelial samples are taken from a 5-day-old culture. (b) Following that, the proteins are extracted by means of chemical techniques like detergent cell lysis as well as mechanical techniques like grinding. (c) Next, sequence-specific enzymes, like trypsin and lysine, are used to break down the proteins into peptides. (d) SDS-PAGE is used to separate these peptides. (e) The separated peptides are ionized, for example, through electrospray ionization, and then measured using a mass spectrometer, specifically liquid chromatography-tandem mass spectrometry (LC-MS/MS). (f) The obtained data is then processed utilizing various bioinformatics tools, such as MaxQuant and Perseus. Subsequent uses of this data include pathway analysis, protein–protein interaction networks, functional characterization, and the creation of shared data repositories. Liquid chromatography with tandem mass spectrometry and sodium dodecyl-sulfate polyacrylamide gel electrophoresis are represented by the acronyms LC-MS/MS (liquid chromatography with tandem mass spectrometry) and SDS-PAGE (sodium dodecyl-sulfate polyacrylamide gel electrophoresis), respectively. (Adapted from publication of Otun *et al.*, 2022, with permission.)

carbohydrates that help the fungus thrive, possibly even aiding in the development of sclerotial cell walls (Tian *et al.*, 2021). The proteome of *S. sclerotiorum* that germinated *in vitro* was analyzed by Otun and Ntushelo (2020; Fig. 3.14). Proteins isolated from 5-day-old *S. sclerotiorum* were separated using sodium dodecyl-sulfate polyacrylamide gel electrophoresis (SDS-PAGE), and peaks were annotated using an AB Sciex TripleToF 6600 mass spectrometer. Next, we used Proteome Discoverer, MASCOT, and SEQUEST to explore and examine the resultant data. There were 1471 proteins found in all, 45 of which were kinases. Furthermore, 62% of the 78 proteins (with >50% similarity to the *S. sclerotiorum* protein database) were classified as having molecular functions, 25% as having BPs, and 13% as having cellular functions, according to the functional classification. The finding that *S. sclerotiorum* produces the potentially cancer-causing proteins tripeptidyl-peptidase and ochratoxin is significant. Mass spectrometry (MS) was employed in an independent proteome analysis of *S. sclerotiorum* to investigate the effect of SsNsd1 on appressorium formation. The SsNsd1 mutant and the wild type were subjected to a proteomic study using two-dimensional electrophoresis (2-DE). Forty-three proteins showed a threefold change in differential expression, according to the data. Of these proteins, four protein sites were entirely absent from the mutants, 77% of the proteins were downregulated, and 14% were elevated. Forty functionally annotated proteins were identified as a consequence of using MS to examine the peptide mass of these protein sequences. Just 17 (38%) of these proteins were known to play roles in cell organization, stress response, energy production, metabolism, protein synthesis, and cell growth and division. Four proteins (17%) had signal peptides, and the remaining 23 proteins (56%) were classified as hypothetical proteins. In the end, the differentially expressed proteins in our study illuminated the appressorium insufficiency caused by the SsNsd1 mutant, opening up new possibilities for investigating SsNsd1 signaling in *S. sclerotiorum* (Li *et al.*, 2018b; Otun *et al.*, 2022).

3.22 Electron Microscopy

Microanatomy allows the *S. sclerotiorum* to be identified from two closely related species. The size of the spores in the ascus and the quantity of rind layers in the *Sclerotinia* are used to make this differentiation. In *S. sclerotiorum*, the sclerotial rind comprises of two layers of globular cells. In addition, the asci generate eight equal-sized ascospores. Ziman (1997) reports that there are no other notable distinctions in the anatomy of the reproductive organs between *S. sclerotiorum* and comparable species. Arseniuk and Macewicz (1994) studied the microanatomy of mature, 1-month-old 'black' sclerotia of *S. trifoliorum*, *S. sclerotiorum*, and *S. minor*, as well as immature "white," mildly pigmented sclerotia. They noticed dried exudate on the surface of the sclerotia cells and a surface mycelial network covering the mature sclerotia. The outer layer of the medullary hyphal wall is formed at this stage of morphogenesis. The two distinct zones that could be seen in the sections of mature sclerotia were the rind and the medulla of hyphal tissue. The wall of rind cells is thick and single-layered, in contrast to the thick and bilayered wall of medullary hyphae. All three species appeared to have sclerotial medullas with lacunae, despite the absence of intercellular gaps in the sclerotial rind. At the SEM level, it was found that *S. trifoliorum* and sclerotia shared the same structural organization as *S. sclerotiorum* and *S. minor*. As a result, no unique structural characteristic of taxonomic relevance was discovered to differentiate *S. trifoliorum* from other *Sclerotinia* species. The only interspecific difference identified by SEM analysis in the sterile tissue zones of apothecia across all three species, however, was the form of the apothecial edge of small and big sclerotia. The ectal excipulum in the vicinity of *S. minor's* apothecial margin appeared to be made up of textura porrecta and prosenchymatous cells. Another difference between the several species was the dimorphism in spore size (ratio 4:4) on *S. trifoliorum* asci. Apart from these differences, it was found that *S. minor's* ascospore microanatomy was the same as that of *S. trifoliorum* and *S. sclerotiorum*. The hyphal density in the medullary excipulum in the apothecial discs of the three species did not differ noticeably. The occurrence of a membrane film covering the hymenium of the ascocarp of *S. trifoliorum* or any other species, as well as the bundle-like shape of apothecial stipes, as previously described, were not reported by Arseniuk and Macewicz (1994).

3.23 Characterizing and Creating New *Sclerotinia* Taxonomy Characters

This marks a significant era in the field of fungal systematics. Only a limited number of mycologists are currently engaged in taxonomy, which involves the creation of classification systems, expert identification, description of taxonomically significant organisms, and the preparation of comprehensive studies on fungi. To address these questions, Kohn (1992) suggested multiple possible sources of novel features in fungal systematics that may be applied to the taxonomy of *Sclerotinia*.

New characters	Expected resolution level
Morphological	
Histochemistry	Species, Genus, family
Ultrastructure	Any level
Anamorph connections	Genus, family
Anamorph morphology	Species, Genus
Genetic	
Ability to mate and form viable F1	Species
Vegetative incompatibility	Intraspecific
Mycelial intersterility	Species
Biological	
Host or substrate	Interspecific, species, Genus
Biogeography	Any level
Molecular—Proteins—Immunology	Any level
- Sequencing	Any level
- Isozyme electrophoresis	Population, intraspecific, species
DNA—Restriction analysis—RFLPs	Intraspecific, species (any level)
- Restriction mapping	Any level
PCR—Length polymorphism	Any level
- Restriction analysis	Any level
- Direct sequencing	Any level
- RAPD	Intraspecific (genetic)

Nonetheless, in order to differentiate *S. minor* from *S. sclerotiorum* and *S. trifoliorum*, Ekins *et al.* (2005) suggested using Restriction Fragment Length Polymorphism (RFLP) probes, contrasting host species, sclerotial diameter, ascosporic morphism, and breeding type.

3.24 Hypovirulent Viruses Infecting *Sclerotinia sclerotiorum*

Hypovirulent viruses infecting *S. sclerotiorum* have the potential to be useful as biocontrol agents. Two separate viruses, namely SsHV1 and SsDAHV-1, appear to individually greatly diminish *S. sclerotiorum*'s pathogenicity. It is critical to investigate and find other important biocontrol elements. Additionally, reports of +ssRNA, −ssRNA, dsRNA, and ssDNA viruses in *S. sclerotiorum* have identified about 16 families or genera, showing a wide variety of viruses. Studying the diversity and quantity of viruses in *S. sclerotiorum* is a great way to learn about the evolution of viruses. *Sclerotinia sclerotiorum* endornavirus 1 (SsEV1) is a globally widespread virus that can be found in Asia, North America, and Oceania. Endornaviruses have also been discovered from *S. sclerotiorum* (Mu *et al.*, 2021).

To enhance our understanding of the variety of fungal viruses and investigate potential resources for biological control, a study was conducted using RNA-Seq analysis on strains obtained from *Sclerotinia* found in infected rapeseed plants in a small field in Xinyang City, Henan Province, China. The purpose of this analysis was to identify new mycoviruses. Following sequencing and assembly, discovered a contig 2 encoding a putative protein with a length of 11,864 nt. This protein exhibited a significant resemblance to endornaviral RNA polymerase, suggesting the presence of an unidentified endornavirus in the debilitating strain XY79 (Luo *et al.*, 2022).

3.25 Characterization of Transcriptional Responses to Genomovirus Infection and Pathogenesis of *Sclerotinia*

Soybean leaves are linked to SlaGemV-1, a circular rep-encoding single-stranded DNA (CRESS-DNA) virus of the *Genomoviridae* family. It is well known that this virus causes hypovirulence in fungal infections belonging to the *Sclerotinia* family and inhibits the production of sclerotia. To comprehend the mechanisms behind this hypovirulence, RNA-Seq data from *S. sclerotiorum* strain DK3—both infected with SlaGemV-1 and virus-free—were compared. A total of 4639 genes exhibited differential expression, with 49.5% of them downregulated and 50.5% of them elevated, according to the research. Virus-like particles were visible close to the inner cell membrane in transmission electron microscopy pictures, and GO enrichments revealed alterations in integral membrane components. Genes associated with sclerotial development, ubiquitin proteolysis, cell cycle regulation, DNA replication and repair, methylation, gene silencing, pathogenesis-related activities, carbohydrate metabolism, and oxalic acid production were the subject of additional investigation. The metabolism of carbohydrates was found to be most affected, with two GH genes showed the higher downregulation by a combined −2396.1- and −648.6-fold, respectively. Pathogenesis-related genes were consistently downregulated; SsNcp1, SsSSVP1, and Endo2 showed the largest decreases, with fold changes of −4555,−14.7, and −12.3, respectively. Conversely, the pathways involved in DNA replication and repair and the cell cycle were largely upregulated, as evidenced by the 8.3- and 5.2-fold increases in the expression of a putative cyclin and separase, respectively. Moreover, there were downregulations of −12.1- and −17.2-fold alterations in the genes related to oxalic acid metabolism, oxalic acid precursor biosynthesis, and oxalic acid catabolism. Additionally, differential regulation was noted in the genes involved in the creation of sclerotia, such as *Pks1*, a gene linked to the manufacture of melanin, and *Sl2*, a gene implicated in the formation of sclerotia, with fold changes of 3.8 and 2.9, respectively (Pedersen and Marzano, 2022).

Numerous studies have demonstrated the potential of using mycoviruses as a biological control method—also known as "*in vitro* control"—to neutralize pathogenic fungus. Chestnut blight disease is well-managed by the natural spread of Cryptomycetes hypovirus 1 throughout Europe, a historical demonstration of the efficacy of *in vitro* control. This approach could be expanded to address other diseases that affect annual crops. A growing number of mycoviruses are being found to cause hypovirulence, or decreased virulence, in the fungal hosts they infect, despite the fact that some mycoviruses cause asymptomatic and dormant infections (Nuss, 2011). Kondo *et al.* (2022) reported that it has been possible to successfully rescue or purify a large number of RNA and DNA viruses in order to definitively determine the cause and consequence of virus infection. Furthermore, a thorough grasp of the genetic effects of virus infection can be obtained by comparing the gene expression levels of virus-transfected and virus-free cultures. Two novel *S. sclerotiorum*-infecting mycoviruses have been identified as a result of this analysis: one that modifies sRNA accumulation (Lee-Marzano *et al.*, 2018) and the other that inhibits growth and interferes with DNA replication/repair, carbohydrate metabolisms, translation, and virulence factors (Qu *et al.*, 2021). According to Xie and Jiang (2014), the mycoviruses that have been discovered thus far display a variety of genome types, such as single-stranded RNA, double-stranded RNA, negative-sense RNA, and CRESS-DNA forms. According to Zhang *et al.* (2020), CRESS-DNA mycoviruses, including *S. sclerotiorum* hypovirulence-associated virus-1 (SsHADV-1), have demonstrated potential as biocontrol pesticides because of their capacity to spread extracellularly and preserve environmental stability. It has been discovered that SsHADV-1, which was first identified from *B. napus* tissue, causes *S. sclerotiorum*, a fungal pathogen, to become hypovirulent. Furthermore, it has been shown that *S. sclerotiorum* infected with SsHADV-1 can colonize plants as an endophyte, which may improve plant resistance to disease (Zhang *et al.*, 2020; Zhou *et al.*, 2021). Changes in RNAi systems, virulence factor expression, and DNA replication and repair have been identified by a transcriptome investigation of *S. sclerotiorum* infected with SsHADV-1 (Qu *et al.*, 2021). It is interesting to note that these transcriptome alterations are similar to those seen in plants afflicted with the gemini-virus family of CRESS-DNA viruses (Hanley-Bowdoin *et al.*, 2013; Pedersen and Marzano, 2022).

3.26 Phylogenetic Investigation and Genome Characterization of the Endornavirus that Infects *S. sclerotiorum*

A class of viruses without virion and with a linear, (+) ssRNA genome are known as endornaviruses. The genome of endornaviruses ranges in size from 9.7 to 17.6 kb. The majority of endornaviruses have a large ORF that codes for a polyprotein made up of RNA-dependent RNA polymerase (RdRp), glycosyl-transferase (GTR), helicase (Hel), and methyltransferase (MTR). Endornaviruses, belonging to the family *Endornaviridae*, are phylogenetically related to alpha viruses. Currently, the family comprises two genera: Alpha-endornavirus and Beta-endornavirus. The establishment of these genera is based on factors such as genome size, host range, and the presence of unique domains. Alpha-endornaviruses have been identified in plants, fungi, and oomycetes, while Beta-endornaviruses exclusively infect fungi. More recently, endorna-like viruses have been discovered in animals, including nematodes, mites, and insects. However, their taxonomic classification remains undetermined (Valverde *et al.*, 2019; Luo *et al.*, 2022).

Endornaviruses are capsid-less linear (+) ssRNA viruses that are members of the *Endornaviridae* family. *S. sclerotiorum* endornavirus 11 (SsEV11), a recently identified endornavirus that infects the hypovirulent *S. sclerotiorum* strain XY79, was the subject of a recent investigation. Virome sequencing analysis and rapid amplification of cDNA ends (RACE) methods were used to identify and clone SsEV11. SsEV11's whole genome is 11,906 nucleotides long and has a large ORF that codes for a polyprotein with 3928 amino acid residues. A putative DEADc, an RNA-dependent RNA polymerase (RdRp) 2 domain, a cysteine-rich region, a viral helicase domain, a viral methyltransferase domain, and other domains are all present in this polyprotein. The lengths of SsEV11's 5′ and 3′ untranslated regions (UTRs) are 31 and 90 nucleotides, respectively. Using BLAST to compare the nucleotide sequence, it was discovered that SsEV11 and *S. minor* endornavirus 1 (SmEV1) have the highest degree of identity (45%). Further, SsEV11 clusters with other endornaviruses and shows a close relationship with Beta-endornavirus, according to phylogenetic study based on the amino acid sequence of RdRp. Three significant clusters were found in the phylogenetic tree based on the endornaviral RdRp domain sequence, according to the phylogenetic study. The alignment analysis supports Cluster I, which is made up of five subclusters that include different unclassified endornaviruses isolated from fungi, oomycetes, algae, and insects, as well as typical Alpha-endornavirus members. In addition, Cluster II consists of five subclusters, which include SsEV11, typical Beta-endornavirus members, and other unidentified viruses that infect fungus. However, several endorna-like viruses that infect worms, mites, and insects are included in Cluster III. While the viruses in Clusters I and II are reasonably near to one another, the viruses in Cluster III are more spread out. The fungal pathogen *S. sclerotiorum* is infected by a novel Beta-endornavirus known as SsEV11, which was recently described in a study by Luo *et al.* (2022). According to the study, endornaviruses exhibit considerable evolutionary diversity. Moreover, it is suggested that the endorna-like viruses in Cluster III be placed in a new genus called Gammaendorna virus (Figs 3.15 and 3.16).

3.27 *Sclerotinia*-induced Production of Phytotoxins and Phytoalexins

The *Sclerotinia* fungus produces a phytotoxin known as "sclerotin," which is rather selective and harmful to three species of cruciferous plants (*B. napus*, *B. juncea*, and *Sinapsis alba*) that are prone to the disease *Sclerotinia* stem rot. In these plants, this toxin results in severe chlorosis and necrosis. *Erucastrum gallicum*, a resistant species, is unaffected by it. The main fatty acid present in *S. sclerotiorum* sclerotia, oleic acid, is what gives extracts of these sclerotia their poisonous qualities when tested on brine shrimp larvae (Artemiasalina). After *e.g. allicum* leaves are stimulated, three known phytoalexins are

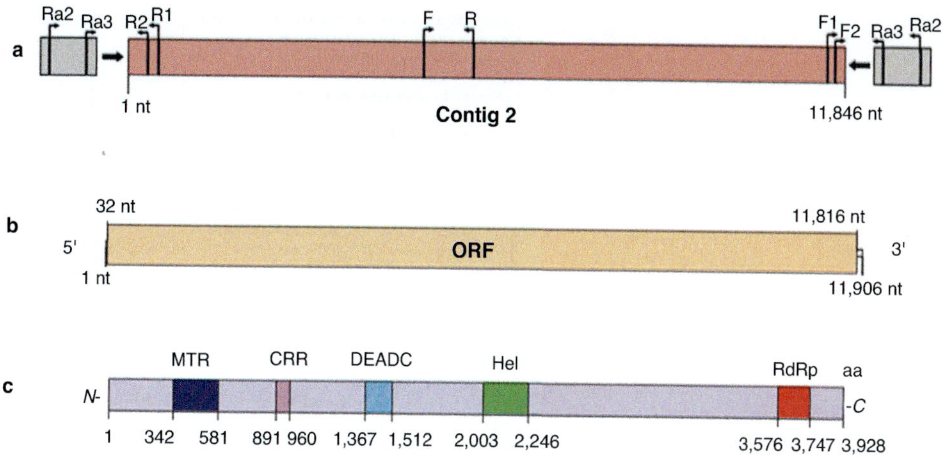

Fig. 3.15. This figure describes the genome architecture of SsEV11, a novel endornavirus. It details (a) the contig2 data from virome sequencing, highlighting primers (F, R, Ra) and RACE-OLOGO ligations used for identification and cDNA end amplification, and (b) the basic genomic structure including the open reading frame (ORF) and 5' and 3' untranslated regions (UTRs). (c) It also explains that the ORF encodes a polyprotein containing conserved functional domains: MTR, CRR, DEADc, Hel, and RdRp, with amino acid locations for each domain indicated. (Adapted from publication of Luo *et al*., 2022, with permission.)

isolated: 1 methoxyspirobrassinin, arvelexin, and indole-3-acetonitrile. Notably, *B. rapa*, a susceptible rape species, has the potential to develop *e.g. allicum's* resistance to *S. sclerotiorum*. Furthermore, *B. rapa* does not produce arvelexin or 1-methoxyspirobrassinin, indicating that these phytoalexins might be markers of resistance to *S. sclerotiorum* (Pedras and Ahiahonu, 2004). Nelson *et al.* (2002) found that the synthesis of phytoalexin in soybean plants is greater in glyphosate-resistant lines S20-B9 and P 93 B01 compared to glyphosate-susceptible lines S19-20 and P 9281.

3.28 Volatile Compounds Released From *S. Sclerotia*

Gas chromatography, MS, and solid phase micro-extraction have all been used to identify the volatile compounds released by *S. minor*, *S. sclerotiorum*, and *Sclerotium rolfsii* sclerotia. *S. minor* and *S. sclerotiorum* release 2-methyl-isoborneol and 2-methyl-enebornane, respectively. *S. minor* also releases gamma butyrolactone, transnerolidol, linalool, cis- and translinalool oxide, and mesityl oxide. On the other hand, *S. sclerotiorum* secretes two diterpenes, 1-methylcamphene and 2-methyl-2-bornene, with a molecular weight of 272. Fascinatingly, *S. rolfsii* releases just delta cadinene and ciscalamenene, not any other of these compounds (Fravel *et al.*, 2002).

3.29 The *Sporigermin* from Sclerotia of *Sclerotinia*

The outer layer of the sclerotia of *S. minor* contains a chemical that specifically causes the formation of macroconidia of *Sporidesmium sclerotivorum* (*Teratosperma sclerotivora*). This material is a tiny, hydrophobic molecule that is heat-sensitive when it comes into contact with water and can flow past a 3000-Da filter. Reversed-phase HPLC has been used to identify three distinct fractions. The portion of soil called "sporigermin," which is thought to have biological activity is slightly soluble in water. This is the first time that a substance that functions as a particular signal to promote the germination of a mycoparasitic fungus has been identified from a host fungus (Mischke *et al.*, 1995).

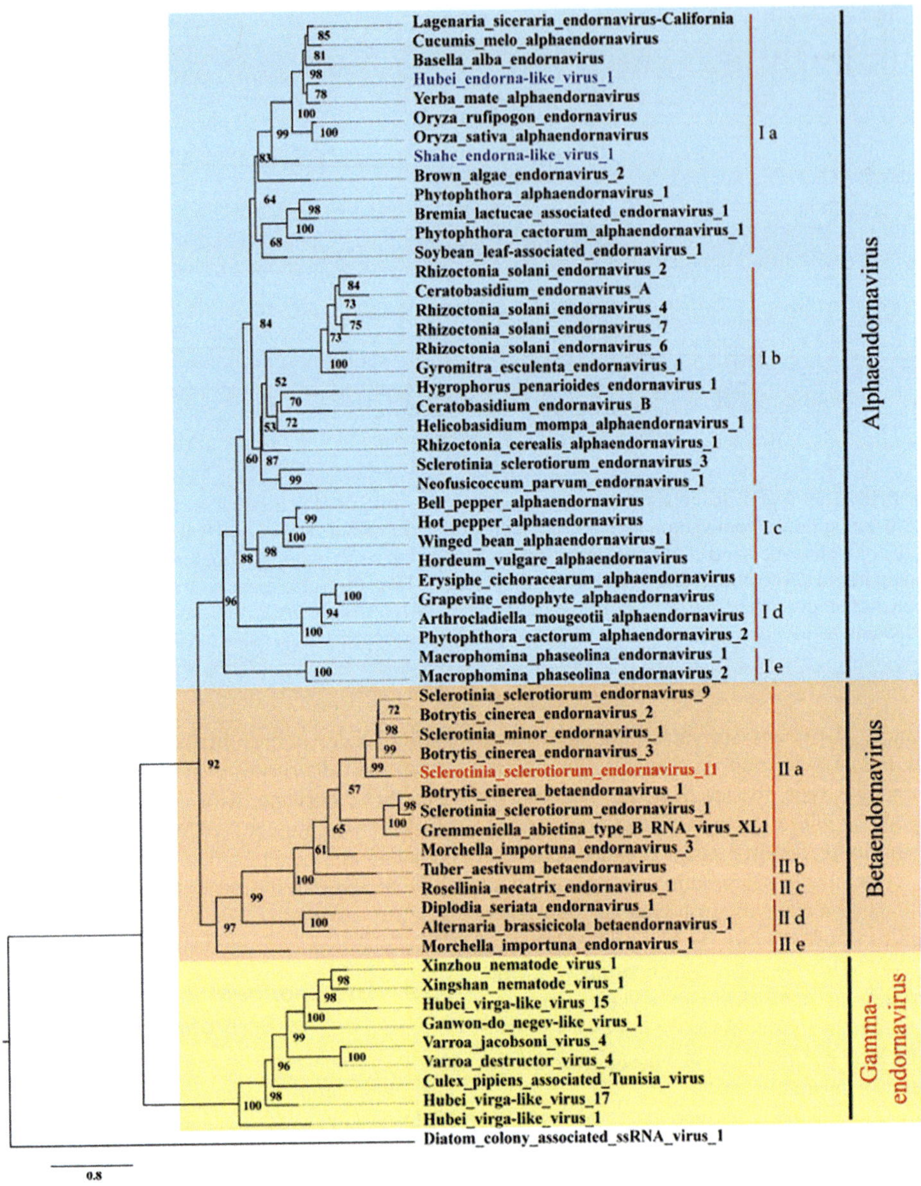

Fig. 3.16. SsEV11 and other selected endornaviruses and endorna-like viruses were subjected to a phylogenetic study with an emphasis on the RdRP domain using a Maximum Likelihood software with 1000 bootstrap replicates. A newly discovered virus is indicated in red in this analysis, while viruses that infect or are linked to insects are indicated in blue. (Adapted from publication of Luo *et al*., 2022, with permission.)

3.30 The Health Risks Associated with Sclerotinia Diseases

The poisonous qualities of the sclerotia may have economic implications. Researchers Ruddick and Harwig (1975) found that feeding rats meals containing 2–8% sclerotia from sunflower head rots caused delayed ossification, decreased maternal weight, and decreased food consumption. It is improbable that adequately cleaned sunflower seeds have sclerotial concentrations that are harmful to human health. Feeding diets containing 0, 1, and 5% ground sclerotia of *S. sclerotiorum* derived from infected rapeseed (*B. napus*) resulted in weight loss and increased feed waste in the group receiving the 5% ground sclerotia. Additionally, over half of the rats showed poor body condition, with alopecia and hyperkeratosis of the tail. The rats were weanling Wistar rats of both genders. The unappealing character of the cuisine and nutritional variables are probably to blame for these impacts. Additionally, there was a notable drop in serum glutamic pyruvic transaminase activity following consumption of sclerotia (Morrall *et al.*, 1978). Moreover, workers who handle celery contaminated with *S. sclerotiorum*, the source of pink rot illness, frequently develop dermatitis (Scheel *et al.*, 1963). Two distinct photosensitive furocoumarins, 8-methoxypsoralen and 4,5,8-trimethylpsoralen, are responsible for this skin irritation. They are generated by *S. sclerotiorum*-infested celery (Perone *et al.*, 1964).

References

Amselem, J., Cuomo, C.A., van Kan, J.A.L., Viaud, M., Benito, E.P. *et al.* (2011) Genomic analysis of the necrotrophic fungal pathogens *Sclerotinia sclerotiorum* and *Botrytis cinerea*. *PLoS Genetics* 7(8), e1002230. DOI: 10.1371/journal.pgen.1002230.

Amselem, J., Lebrun, M.H. and Quesneville, H. (2015a) Whole genome comparative analysis of transposable elements provides new insight into mechanisms of their inactivation in fungal genomes. *BMC Genomics* 16(1), 141. DOI: 10.1186/s12864-015-1347-1.

Amselem, J., Vigouroux, M., Oberhaensli, S., Brown, J.K.M., Bindschedler, L.V. *et al.* (2015b) Evolution of the EKA family of powdery mildew avirulence-effector genes from the ORF 1 of a LINE retrotransposon. *BMC Genomics* 16(1), 917. DOI: 10.1186/s12864-015-2185-x.

Arseniuk, E. and Macewicz, J. (1994) Scanning electron microscopy of sclerotia of *Sclerotinia trifoliorum* and related species. *Journal of Phytopathology* 141(3), 275–284. DOI: 10.1111/j.1439-0434.1994.tb01471.x.

Asman, A.K.M., Fogelqvist, J., Vetukuri, R.R. and Dixelius, C. (2016) *Phytophthora Infestans* Argonaute 1 binds microRNA and small RNAs from effector genes and transposable elements. *New Phytology* 211, 993–1007.

Berthet, P. (1964) *Essai biotaxonomique sur les discomycetes*. These Faculte des Sciences de l' Universite de Lyon, Berthet, Lyon, France, p. 157.

Bjorling, K. (1942) Undersokninger rorande kioverrotan. II studier av utvecklingshistoria och variation hos *Sclerotinia trifoliorum*. *Meddland Statens Vaxtskyddsanst* 37, 1–154.

Bjorling, K. (1951) Uber die entwicklungsgeschiehte, variabilitat and pathogenitat von *Sclerotinia trifoliorum* Erikss. *Phytopathologische Zeitschrift* 18, 129–156.

Bolton, M.D., Thomma, B.P.H.J. and Nelson, B.D. (2006) Pathogen profile: *Sclerotinia sclerotiorum* A (Lib.) de Bary, biology and molecular traits of a cosmopolitan pathogen. *Molecular Plant Pathology* 7, 1–16.

Borkovich, K.A., Alex, L.A., Yarden, O., Freitag, M., Turner, G.E. *et al.* (2004) Lessons from the genome sequence of *Neurospora crassa*: Tracing the path from genomic blueprint to multicellular organism. *Microbiology and Molecular Biology Reviews* 68, 1–108.

Brodal, G., Warmington, R., Grieu, C., Ficke, A. and Clarkson, J.P. (2017) First report of *Sclerotinia subarctica* nom. prov. (*Sclerotinia* sp. 1) causing stem rot on turnip rape (*Brassica rapa* subsp. *oleifera*) in Norway. *Plant Disease* 101, 386. DOI: 10.1094/PDIS-06-16-0785-PDN.

Buchwald, N.F. (1949) Studies in the *Sclerotinia* 1. Taxonomy of the *Sclerotinia*. *K Vet Landbohjsk Arsskr* 75–191.

Buchwald, N.F. and Neergaard, P. (1973) A plea for the retention of *Sclerotinia sclerotiorum* as type species for the *Sclerotinia* Fekl. *Emend Friesia* 10, 96–99.

Buddhika, U.V., Savocchia, S. and Steel, C.C. (2021) Copper induces transcription of the BcLCC2 laccase gene in phytopathogenic fungus, *Botrytis cinerea*. *Mycology* 12(1), 48–57.

Cai, Q., Qiao, L., Wang, M., He, B., Lin, F.-M. *et al.* (2018) Plants send small RNAs in extracellular vesicles to fungal pathogen to silence virulence genes. *Science* 360(6393), 1126–1129. DOI: 10.1126/science.aar4142.

Carbone, I., Anderson, J.B. and Kohn, L.M. (1995) A group-I intron in the mitochondrial small subunit ribosomal RNA gene of *Sclerotinia sclerotiorum*. *Current Genetics* 27(2), 166–176. DOI: 10.1007/BF00313431.

Castanera, R., López-Varas, L., Borgognone, A., LaButti, K., Lapidus, A. *et al.* (2016) Transposable elements versus the fungal genome: Impact on whole-genome architecture and transcriptional profiles. *PLoS Genetics* 12(6): e1006108. DOI: 10.1371/journal.pgen.1006108.

Chooi, Y.H. and Solomon, P.S. (2014) A chemical ecogenomics approach to understand the roles of secondary metabolites in fungal cereal pathogens. *Frontiers in Microbiology* 5, 640. DOI: 10.3389/fmicb.2014.00640.

Christiansen, M.A. (1966) Variation in apothecial structures of some sclerotia matured under natural and artificial conditions. Thesis, Cornell University, Ithaca, New York, 77p.

Clarkson, J.P., Carter, H.E. and Coventry, E. (2010) First report of *Sclerotinia subarctica* nom. prov. (*Sclerotinia* species 1) in the UK on *Ranunculus acris*. *Plant Pathology* 59, 1173–1173. DOI: 10.1111/j.1365-3059.2010.02271.

Collemare, J. and Lebrun, M.H. (2011) Fungal secondary metabolites: Ancient toxins and novel effectors in plant-microbe interactions. In: *Effectors in Plant-Microbe Interactions*. Wiley, Hoboken, New Jersey, pp. 377–400. DOI: 10.1002/9781119949138.

Cruickshank, R.H. (1983) Distinction between *Sclerotinia* species by their pectic zymograms. *Transactions of the British Mycological Society* 80(1), 117–119. DOI: 10.1016/S0007-1536(83)80171-5.

de Paula, R.G., Antoniêto, A.C.C., Nogueira, K.M.V., Ribeiro, L.F.C., Rocha, M.C. *et al.* (2019) Extracellular vesicles carry cellulases in the industrial fungus *Trichoderma reesei*. *Biotechnology for Biofuels* 12(1), 146. DOI: 10.1186/s13068-019-1487-7.

de Silva, A., Bolton, M.D. and Nelson, B.D. (2005) Transformation of *Sclerotinia sclerotiorum* with the green fluorescent protein gene and expression of fluorescence of host tissues. *Phytopathology* 95, 523.

Del Ponte, E.M., Eduardo, S., Lehner, M.S., Ju, T.J.D.P., Mizubuti, G. *et al.* (2017) Independently founded populations of *Sclerotinia sclerotiorum* from a tropical and a temperate region have similar genetic structure. *PLoS ONE* 12(3), e0173915. DOI: 10.1371/journal.pone.0173915.

Dennis, R.W.G. (1978) *British Ascomycetes*, Vol. xxvi. Liechtenstein, Cramer, Vaduz, p. 585.

Derbyshire, M., Denton-Giles, M., Hegedus, D., Seifbarghy, S., Rollins, J. *et al.* (2017) The complete genome sequence of the phytopathogenic fungus *Sclerotinia sclerotiorum* reveals insights into the genome architecture of broad host range pathogens. *Genome Biology and Evolution* 9(3), 593–618. DOI: 10.1093/gbe/evx030.

Derbyshire, M.C., Newman, T.E. and Khentry, T.A.O. (2022) Pathogen profile: The evolutionary and molecular features of the broad host range plant pathogen *Sclerotinia sclerotiorum*. *Molecular Plant Pathology* 23, 1–16. DOI: 10.1111/mpp.13221.

Dong, Y., Yang, Q., Xue, Z. and Liu, Y. (2011) RNA interference in fungi: Pathways, functions, and applications. *Eukaryotic Cell* 10(9), 1148–1155. DOI: 10.1128/EC.05109-11.

Dong, Y., Li, Y., Zhao, M., Jing, M., Liu, X. *et al.* (2015) Global genome and transcriptome analyses of *Magnaporthe oryzae* epidemic isolate 98-06 uncover novel effectors and pathogenicity-related genes, revealing gene gain and lose dynamics in genome evolution. *PLoS Pathogens* 11(4), e1004801. DOI: 10.1371/journal.ppat.1004801.

Dumont, K.P. (1971) *Sclerotinia.11. Lambertelia. Memories of the New York Botanical Garden*, Vol. 22. New York Botanical Garden, New York, pp. 1–178.

Eisenman, H.C. and Casadevall, A. (2012) Synthesis and assembly of fungal melanin. *Applied Microbiology and Biotechnology* 93(3), 931–940. DOI: 10.1007/s00253-011-3777-2.

Ekblom, R. and Wolf, J.B.W. (2014) A field guide to whole-genome sequencing, assembly and annotation. *Evolutionary Applications* 7(9), 1026–1042. DOI: 10.1111/eva.12178.

Ekins, M.G., Aitken, E.A.B. and Goulter, K.C. (2005) Identification of *Sclerotinia* species. *Australasian Plant Pathology* 34, 549–555.

Erental, A., Dickman, M.B. and Yarden, O. (2008) Sclerotial development in *Sclerotinia sclerotiorum*: Awakening molecular analysis of a "dormant" structure. *Fungal Biology Reviews* 22(1), 6–16. DOI: 10.1016/j.fbr.2007.10.001.

Faino, L., Seidl, M.F., Datema, E., van der Berg, G.C.M., Janssen, A. *et al.* (2015) Single-molecule real-time sequencing combined with optical mapping yields completely finished fungal genome. *mBio* 6(4), e00936–15. DOI: 10.1128/mBio.00936-15.

Faino, L., Seidl, M.F., Shi-Kunne, X., Pauper, M., van den Berg, G.C.M. *et al.* (2016) Transposons passively and actively contribute to evolution of the two-speed genome of a fungal pathogen. *Genome Research* 26(8), 1091–1100. DOI: 10.1101/gr.204974.116.

Fedorova, N.D., Moktali, V. and Medema, M.H. (2012) Bioinformatics approaches and software for detection of secondary metabolic gene clusters. In: Keller, N.P. and Turner, G. (eds) *Fungal Secondary Metabolism*. Humana Press Publishing Company, Totowa, New Jersey, pp. 23–45.

Frandsen, K.J. (1946) *Studies Over Sclerotinia Trifoliorum*. Erikss. Danske Forlag, Copenhagen, p. 222.

Fravel, D.R., Connick, W.J., Jr., Grimm, C.C. and Lloyd, S.W. (2002) Volatile compounds emitted by sclerotia of *Sclerotinia minor*, *Sclerotinia sclerotiorum* and *Sclerotium rolfsii*. *Journal of Agriculture Food Chemistry* 50, 3761–3764. DOI: 10.1021/jf0117098.

Fuckel, L. (1870) Symbolae mycologicae. Beiträge zur kenntniss der rheinischen pilze. *I JhrbNassauischen Vercins Naturk* 23, 1–459. DOI: 10.5962/bhl.title.47117.

Goodwin, S., McPherson, J.D. and McCombie, W.R. (2016) Coming of age: Ten years of next-generation sequencing technologies. *Nature Reviews Genetics* 17, 333–351.

Graham-Taylor, C., Kamphuis, L.G. and Derbyshire, M.C. (2020) A detailed *in silico* analysis of secondary metabolite biosynthesis clusters in the genome of the broad host range plant pathogenic fungus *Sclerotinia sclerotiorum*. *BMC Genomics* 21(1), 1–20.

Grandaubert, J., Lowe, R.G.T., Soyer Jessica, L., Schoch, C.L., van de Wouw, A.P. *et al.* (2014) Transposable element-assisted evolution and adaptation to host plant within the *Leptosphaeria maculans–Leptosphaeria biglobosa* species complex of fungal pathogens. *BMC Genomics* 15, 891.

Groves, J.W. and Bowerman, G.A. (1955) *Sclerotinia borealis* in Canada. *Canadian Journal of Botany* 33, 591–594.

Guimaraes, R.L. and Stotz, H.U. (2004) Oxalate production by *Sclerotinia sclerotiorum* deregulates guard cells during infection. *Plant Physiology* 136, 3703–3711.

Gupta, N.C., Yadav, S., Arora, S., Mishra, D., Budhlakoti, N. *et al.* (2022) Draft genome sequencing and secretome profiling of *Sclerotinia sclerotiorum* revealed effector repertoire diversity and allied broad-host range necrotrophy. *Scientific Reports* 12, 21855. DOI: 10.1038/s41598-022-22028-z.

Guyon, K., Balagué, C., Roby, D. and Raffaele, S. (2014) Secretome analysis reveals effector candidates associated with broad host range necrotrophy in the fungal plant pathogen *Sclerotinia sclerotiorum*. *BMC Genomics* 15(1), 336. DOI: 10.1186/1471-2164-15-336.

Hane, J.K., Williams, A.H., Taranto, A.P., Solomon, P.S. and Oliver, R.P. (2015) Repeat-induced point mutation: A fungal-specific, endogenous mutagenesis process. In: Berg, M.A. and Maruthachalam, K. (eds) *Genetic Transformation Systems in Fungi*, Vol. 2. Springer International, Cham, Switzerland, pp. 55–68.

Hanley-Bowdoin, L., Bejarano, E.R., Robertson, D. and Mansoor, S. (2013) Geminiviruses: Masters at redirecting and reprogramming plant processes. *Nature Reviews Microbiology* 11(11), 777–788. DOI: 10.1038/nrmicro3117.

Harel, A., Bercovich, S. and Yarden, O. (2006) Calcineurin is required for *Sclerotial* development and pathogenicity of *Sclerotinia sclerotiorum* in an oxalic acid-independent manner. *Molecular Plant-Microbe Interactions* 19(6), 682–693. DOI: 10.1094/MPMI-19-0682.

Harrison, J.G. and Hargreaves, A.J. (1977) Production and germination of *Botrytis fabae* microconidia. *Transactions of the British Mycological Society* 69(2), 332–335. DOI: 10.1016/S0007-1536(77)80061-2.

Heard, S., Brown, N.A. and Hammond-Kosack, K. (2015) An interspecies comparative analysis of the predicted secretomes of the necrotrophic plant pathogens *Sclerotinia sclerotiorum* and *Botrytis cinerea*. *PLoS ONE* 10(6), e0130534. DOI: 10.1371/journal.pone.0130534.

Hoes, J.A. and Huang, H.C. (1975) *Sclerotinia sclerotiorum*: Viability and separation of *sclerotia* from soil. *Phytopathology* 65(12), 1431. DOI: 10.1094/Phyto-65-1431.

Hoist-Jensen, A., Vaage, M. and Schumacher, T. (1998) An approximation to the phylogeny of *Sclerotinia* and related genera. *Nordic Journal of Botany* 18(6), 705–719. DOI: 10.1111/j.1756-1051.1998.tb01553.x.

Honey, E.E. (1928) The monilioid species of *Sclerotinia*. *Mycologia* 20(3), 127–157. DOI: 10.1080/00275514.1928.12016909.

Jagger, I.C. (1920) *Sclerotinia minor* n. sp., the cause of a decay of lettuce, celery and other crops. *Journal of Agriculture Research* 20, 331–334.

Ji, H.M., Mao, H.Y., Li, S.J., Feng, T., Zhang, Z.Y. *et al.* (2021) Fol-milR1, a pathogenicity factor of *Fusarium oxysporum*, confers tomato wilt disease resistance by impairing host immune responses. *The New Phytologist* 232(2), 705–718. DOI: 10.1111/nph.17436.

Keay, M.A. (1939) A study of certain species of the genus *Sclerotinia*. *Annals of Applied Biology* 26(2), 227–246. DOI: 10.1111/j.1744-7348.1939.tb06968.x.

Keller, N.P. and Hohn, T.M. (1997) Metabolic pathway gene clusters in filamentous fungi. *Fungal Genetics and Biology* 21(1), 17–29. DOI: 10.1006/fgbi.1997.0970.

Kettles, G.J., Hofinger, B.J., Hu, P., Bayon, C., Rudd, J.J. *et al.* (2019) SRNA profiling combined with gene function analysis reveals a lack of evidence for cross-kingdom RNAi in the wheat—*Zymoseptoria tritici* pathosystem. *Frontiers in Plant Science* 10, 892. DOI: 10.3389/fpls.2019.00892.

Kimura, N.O. and Tsuge, T.A. (1993) Gene cluster involved in melanin biosynthesis of the filamentous fungus *Alternaria alternata*. *Journal of Bacteriology* 175(14), 4427–4435. DOI: 10.1128/jb.175.14.4427-4435.1993.

King, R., Urban, M., Hammond-Kosack, M.C.U., Hassani-Pak, K. and Hammond-Kosack, K.E. (2015) The completed genome sequence of the pathogenic ascomycete fungus *Fusarium graminearum*. *BMC Genomics* 16(1), 544. DOI: 10.1186/s12864-015-1756-1.

Kohn, L.M. (1979) A monographic revision of the genus *Sclerotinia*. *Mycotaxon* 9, 365–444.

Kohn, L.M. (1992) Developing new characters for fungal systematics: An experimental approach for determining the rank of resolution. *Mycologia* 84(2), 139–153. DOI: 10.1080/00275514.1992.12026120.

Kohn, L.M. and Korf, R.P. (1975) Variation in ascomycetes iodine reactions: KOH pretreatment explored. *Mycotaxon* 3, 165–172.

Kondo, H., Botella, L. and Suzuki, N. (2022) Mycovirus diversity and evolution revealed/inferred from recent studies. *Annual Review of Phytopathology* 60, 307–336. DOI: 10.1146/annurev-phyto-021621-122122.

Korf, R.P. and Dumont, K.P. (1972) *Whetzelinia*, A new generic name for *Sclerotinia sclerotiorum* and *Sclerotinia tuberosa*. *Mycologia* 64(2), 248–251. DOI: 10.1080/00275514.1972.12019259.

Korf, R.P. (1973) Discomycetes and tuberales. In: Ainsworth, G.G., Sparrow, F.K. and Sussman, A.S. (eds) *The Fungi—An Advanced Treatise*, Vol. 4A. Academic Press, New York, pp. 249–319.

Lee-Marzano, S.-Y., Neupane, A. and Domier, L. (2018) Transcriptional and small RNA responses of the white mold fungus *Sclerotinia sclerotiorum* to infection by a virulence-attenuating hypovirus. *Viruses* 10(12), 713. DOI: 10.3390/v10120713.

Li, J., Mu, W., Veluchamy, S., Liu, Y., Zhang, Y. *et al.* (2018a) The GATA-type IVb zinc-finger transcription factor SsNsd1 regulates asexual-sexual development and appressoria formation in *Sclerotinia sclerotiorum*. *Molecular Plant Pathology* 19(7), 1679–1689. DOI: 10.1111/mpp.12651.

Li, J., Zhang, X., Li, L., Liu, J., Zhang, Y. *et al.* (2018b) Proteomics analysis of SsNsd1-mediated compound appressoria formation in *Sclerotinia sclerotiorum*. *International Journal of Molecular Sciences* 19(10), 2946. DOI: 10.3390/ijms19102946.

Liang, X. and Rollins, J.A. (2018) Mechanisms of broad host range necrotrophic pathogenesis in *Sclerotinia sclerotiorum*. *Phytopathology* 108(10), 1128–1140. DOI: 10.1094/PHYTO-06-18-0197-RVW.

Liang, X., Moomaw, E.W. and Rollins, J.A. (2015) Fungal oxalate decarboxylase activity contributes to *Sclerotinia sclerotiorum* early infection by affecting both compound appressoria development and function. *Molecular Plant Pathology* 16(8), 825–836. DOI: 10.1111/mpp.12239.

Liu, L. and Free, S.J. (2016) Characterization of the cell wall proteome. *Molecular Plant Pathology* 17(6), 985–995.

Lorang, J.M., Tuori, R.P., Martinez, J.P., Sawyer, T.L., Redman, R.S. *et al.* (2001) Green fluorescent protein is lighting up fungal biology. *Applied and Environmental Microbiology* 67(5), 1987–1994. DOI: 10.1128/AEM.67.5.1987-1994.2001.

Luo, X., Jiang, D., Xie, J., Jia, J., Duan, J. *et al.* (2022) Genome characterization and phylogenetic analysis of a novel endornavirus that infects fungal pathogen *Sclerotinia sclerotiorum*. *Viruses* 14(3), 456. DOI: 10.3390/v14030456.

Ma, X., Wiedmer, J. and Palma-Guerrero, J. (2019) Small RNA bidirectional crosstalk during the interaction between wheat and *Zymoseptoria tritici*. *Frontiers in Plant Science* 10, 1669. DOI: 10.3389/fpls.2019.01669.

McCaghey, M., Shao, D., Kurcezewski, J., Lindstrom, A., Ranjan, A. *et al.* (2021) Host-induced gene silencing of a *Sclerotinia sclerotiorum* oxaloacetate acetylhydrolase using bean pod mottle virus as a vehicle reduces disease on soybean. *Frontiers in Plant Science* 12, 677631. DOI: 10.3389/fpls.2021.677631.

Mert-Turk, F., Ipek, M., Mermer, D. and Nicholson, P. (2007) Microsatellite and morphological markers reveal genetic variation within a population of *Sclerotinia sclerotiorum* from oilseed rape in the Canakkale Province of Turkey. *Journal of Phytopathology* 155(3), 182–187. DOI: 10.1111/j.1439-0434.2007.01223.x.

Mindrebo, J.T., Nartey, C.M., Seto, Y., Burkart, M.D. and Noel, J.P. (2016) Unveiling the functional diversity of the alpha/beta hydrolase superfamily in the Plant Kingdom. *Current Opinion in Structural Biology* 41, 233–246. DOI: 10.1016/j.sbi.2016.08.005.

Mischke, S., Mischke, C.F. and Adams, P.B. (1995) A rind-associated factor from sclerotia of *Sclerotinia minor* stimulates germination of a mycoparasite. *Mycological Research* 99(9), 1063–1070. DOI: 10.1016/S0953-7562(09)80774-0.

Mordue, J.E.M. and Holliday, P. (1998) *Sclerotinia sclerotiorum, IMI Descriptions of Fungi and Bacteria*, Vol. 52. CABI Bioscience, Egham, Surrey, UK, p. 513.

Morrall, R.A.A., Duczek, L.J. and Sheard, J.W. (1972) Variations and correlations within and between morphology, pathogenicity, and pectolytic enzyme activity in *Sclerotinia* from Saskatchewan. *Canadian Journal of Botany* 50(4), 767–786. DOI: 10.1139/b72-095.

Morrall, R.A.A., Loew, F.M. and Hayes, M.A. (1978) Subacute toxicological evaluation of *sclerotia* of *Sclerotinia sclerotiorum* in rats. *Canadian Journal of Comparative Medicines* 42, 4.

Mu, F., Li, B., Cheng, S., Jia, J., Jiang, D. *et al.* (2021) Nine viruses from eight lineages exhibiting new evolutionary modes that co-infect a hypovirulent phytopathogenic fungus. *PLoS Pathogens* 17(8), e1009823. DOI: 10.1371/journal.ppat.1009823.

Nannfeldt, J.A. (1976) Iodine reactions in asens plugs and their taxonomic significance. *Transactions of the British Mycological Society* 67(2), 283–287. DOI: 10.1016/S0007-1536(76)80135-0.

Nelson, K.A., Renner, K.A. and Hammerschmidt, R. (2002) Effects of protoporphyrinogen oxidase inhibitors on soybean (*Glycine max* L.) response, *Sclerotinia sclerotiorum* disease development, and phytoalexin production by soybean. *Weed Technology* 16(2), 353–359.

Neupane, A., Feng, C., Mochama, P.K., Saleem, H. and Lee Marzano, S.Y. (2019) Roles of argonautes and dicers on *Sclerotinia sclerotiorum* antiviral RNA silencing. *Frontiers in Plant Science* 10, 976. DOI: 10.3389/fpls.2019.00976.

Noviello, C. and Korf, R.P. (1961) A simple technique for investing stomatal formation in the *Sclerotinia*. *Mycologia* 37, 648–714.

Nuss, D.L. (2011) Mycoviruses, RNA silencing, and viral RNA recombination. *Advances in Virus Research* 80, 25–48. DOI: 10.1016/B978-0-12-385987-7.00002-6.

Ohta, S., Tsuchida, K., Choi, S., Sekine, Y., Shiga, Y. *et al.* (2002) Presence of a characteristic DDE motif in IS1 transposase. *Journal of Bacteriology* 184(22), 6146–6154.

Otun, S. and Ntushelo, K. (2020) Proteomic analysis of the phytogenic fungus *Sclerotinia sclerotiorum*. *Journal of Chromatography B, Analytical Technologies in the Biomedical and Life Sciences* 1144, 122053.

Otun, O.S., Achilonu, I. and Ntushelo, K. (2022) Proteomics analysis of the phytopathogenic fungus *Sclerotinia sclerotiorum*: A narrative review. *Journal of Bio-X Research* 5(3), 112–117. DOI: 10.1097/JBR.0000000000000130.

Pedersen, C.J. and Marzano, S.-Y.L. (2022) Characterization of transcriptional responses to genomovirus infection of the white mold fungus *Sclerotinia Sclerotiorum*. *Viruses* 14(9), 1892. DOI: 10.3390/v14091892.

Pedras, M.S.C. and Ahiahonu, P.W.K. (2004) Phytotoxin production and phytoalexin elicitation by the phytopathogenic fungus *Sclerotinia sclerotiorum*. *Journal of Chemical Ecology* 30(11), 2163–2179.

Perone, V.B., School, L.D. and Meitus, R.J. (1964) A bioassay for the quantitation of cutaneous reactions associated with pink-rot celery. *The Journal of Investigative Dermatology* 42, 267–271. DOI: 10.1038/jid.1964.60.

Petersen, G.R., Russo, G.M. and Etten, J.L. (1982) Identification of major proteins in sclerotia of *Sclerotinia minor* and *Sclerotinia trifoliorum*. *Experimental Mycology* 6(3), 268–273. DOI: 10.1016/0147-5975(82)90119-0.

Price, K. and Colhoun, J. (1975) Pathogenicity of isolates of *Sclerotinia sclerotiorum* (Lib.) de Bary to several host. *Journal of Phytopathology* 83(3), 232–238. DOI: 10.1111/j.1439-0434.1975.tb03534.x.

Purdy, L.H. (1955) A broader concept of the species *Sclerotinia sclerotiorum* based on variability. *Phytopathology* 45, 421–427.

Qiao, L., Lan, C., Capriotti, L., Ah-Fong, A., Nino Sanchez, J. *et al.* (2021) Spray-induced gene silencing for disease control is dependent on the efficiency of pathogen RNA uptake. *Plant Biotechnology Journal* 19(9), 1756–1768. DOI: 10.1111/pbi.13589.

Qu, Z., Fu, Y., Lin, Y., Zhao, Z., Zhang, X. *et al.* (2021) Transcriptional responses of *Sclerotinia sclerotiorum* to the infection by SsHADV-1. *Journal of Fungi* 7(7), 493. DOI: 10.3390/jof7070493.

Raffaele, S. and Kamoun, S. (2012) Genome evolution in filamentous plant pathogens: Why bigger can be better. *Nature Reviews Microbiology* 10(6), 417–430. DOI: 10.1038/nrmicro2790.

Reymond, P., Deléage, G., Rascle, C. and Fevre, M. (1994) Cloning and sequence analysis of a polygalacturonase-encoding gene from the phytopathogenic fungus *Sclerotinia sclerotiorum*. *Gene* 146(2), 233–237. DOI: 10.1016/0378-1119(94)90298-4.

Rodriguez-Moreno, L., Ebert, M.K., Bolton, M.D. and Thomma, B.P.H.J. (2018) Tools of the crook—Infection strategies of fungal plant pathogens. *The Plant Journal* 93(4), 664–674. DOI: 10.1111/tpj.13810.

Rolland, S., Jobic, C., Fèvre, M. and Bruel, C. (2003) Agrobacterium-mediated transformation of *Botrytis cinerea*, simple purification of monokaryotic transformants and rapid conidia-based identification of the transfer-DNA host genomic DNA flanking sequences. *Current Genetics* 44(3), 164–171. DOI: 10.1007/s00294-003-0438-8.

Rouxel, T., Grandaubert, J., Hane, J.K., Hoede, C., Van de Wouw, A.P. *et al.* (2011) Effector diversification within compartments of the *Leptosphaeria maculans* genome affected by repeat-induced point mutations. *Nature Communications* 2, 202. DOI: 10.1038/ncomms1189.

Ruddick, J.A. and Harwig, J. (1975) Prenatal effects caused by feeding *sclerotia* of *Sclerotinia sclerotiorum* to pregnant rats. *Bulletin of Environmental Contamination and Toxicology* 13(5), 524–526. DOI: 10.1007/BF01685173.

Russo, G.M., Dahlberg, K.R. and Van Etten, J.L. (1982) Identification of a development-specific protein in *sclerotia* of *Sclerotinia sclerotiorum*. *Experimental Mycology* 6(3), 259–267. DOI: 10.1016/0147-5975(82)90118-9.

Saharan, G.S. and Mehta, N. (2008) Sclerotinia *Diseases of Crop Plants: Biology, Ecology and Disease Management*. Springer Science, Dordrecht, p. 485.

Saito, I. (1977) *Sclerotinia sclerotiorum* (Lib.) de Bary. studies on the maturation and germination of sclerotia of *Sclerotinia sclerotiorum* (Lib.) de Bary, a causal fungus of bean stem rot. Report Hokkaido Prefectural Agricultural Experiment Station, No. 26, March, 1977. Hokkaido, Japan: Hokkaido Central Agricultural Experiment Station, 069–13.

Santana, M.F., Silva, J.C.F., Mizubuti, E.S.G., Araújo, E.F. and Queiroz, M.V. (2014) Analysis of tc1-mariner elements in *Sclerotinia sclerotiorum* suggests recent activity and flexible transposases. *BMC Microbiology* 14, 256. DOI: 10.1186/s12866-014-0256-9.

Sawada, K. (1919) Descriptive catalogue of the fonnosan fungi. Part I. *Agricultural Experiment Station, Taihoku, Government Formosa, Bulletin* 19, 1–695. DOI: 10.5962/bhl.title.3911.

Scheel, L.D., Perone, V.B., Larkin, R.L. and Kupel, R.E. (1963) The isolation and characterization of two phototoxic furanocoumarins (Psoralens) from diseased celery. *Biochemistry* 2(5), 1127–1131. DOI: 10.1021/bi00905a038.

Scott, S.W. (1981) Serological relationships of three *Sclerotinia* species. *Transactions of the British Mycological Society* 77(3), 674–676. DOI: 10.1016/S0007-1536(81)80129-5.

Seifbarghi, S., Borhan, M.H., Wei, Y., Coutu, C., Robinson, S.J. *et al.* (2017) Changes in the *Sclerotinia sclerotiorum* transcriptome during infection of *Brassica napus*. *BMC Genomics* 18(1), 266. DOI: 10.1186/s12864-017-3642-5.

Selin, C., de Kievit, T.R., Belmonte, M.F. and Fernando, W.G.D. (2016) Elucidating the role of effectors in plant-fungal interactions: Progress and challenges. *Frontiers in Microbiology* 7, 600. DOI: 10.3389/fmicb.2016.00600.

Sexton, A.C. and Howlett, B.J. (2004) Microsatellite markers reveal genetic differentiation among populations of *Sclerotinia sclerotiorum* from Australian canola fields. *Current Genetics* 46(6), 357–365. DOI: 10.1007/s00294-004-0543-3.

Singh, K., Nizam, S., Sinha, M. and Verma, P.K. (2012) Comparative transcriptome analysis of the necrotrophic fungus *Ascochyta rabiei* during oxidative stress: Insight for fungal survival in the host plant. *PLoS ONE* 7(3), e33128. DOI: 10.1371/journal.pone.0033128.

Sun, J., Tian, C., Diamond, S. and Glass, N.L. (2012) Deciphering transcriptional regulatory mechanisms associated with hemicellulose degradation in *Neurospora crassa*. *Eukaryotic Cell* 11(4), 482–493.

Syme, R.A., Tan, K.C., Hane, J.K., Dodhia, K., Stoll, T. *et al.* (2016) Comprehensive annotation of the *Parastagonospora nodorum* reference genome using next-generation genomics, transcriptomics and proteogenomics. *PLoS ONE* 11(2), e0147221. DOI: 10.1371/journal.pone.0147221.

Tariq, V.N., Gutteridge, C.S. and Jeffries, P. (1985) Comparative studies of cultural and biochemical characteristics used for distinguishing species within *Sclerotinia*. *Transactions of the British Mycological Society* 84(3), 381–397. DOI: 10.1016/S0007-1536(85)80001-2.

Testa, A.C., Oliver, R.P. and Hane, J.K. (2016) Occulter cut: A comprehensive survey of AT-rich regions in fungal genomes. *Genome Biology and Evolution* 8, 2044–2064.

Thomma, B.P.H.J., Seidl, M.F., Shi-Kunne, X., Cook, D.E., Bolton, M.D. *et al.* (2016) Mind the gap; seven reasons to close fragmented genome assemblies. *Fungal Genetics and Biology* 90, 24–30. DOI: 10.1016/j.fgb.2015.08.010.

Tian, J., Chen, C., Sun, H., Wang, Z., Steinkellner, S. *et al.* (2021) Proteomic analysis reveals the importance of exudates on *Sclerotial* development in *Sclerotinia sclerotiorum*. *Journal of Agricultural and Food Chemistry* 69(4), 1430–1440. DOI: 10.1021/acs.jafc.0c06685.

Townsend, B.B. and Willetts, H.J. (1954) The development of *sclerotia* of certain fungi. *Transactions of the British Mycological Society* 37(3), 213–221. DOI: 10.1016/S0007-1536(54)80003-9.

Tu, C.C. and Kimbrough, J.W. (1973) A rapid staining technique for *Rhizoctonia solani* and related fungi. *Mycologia* 65(4), 941–944.

Unal, C.M. and Steinert, M. (2014) Microbial peptidyl-prolyl cis/trans isomerases (PPIases): Virulence factors and potential alternative drug targets. *Microbiology and Molecular Biology Reviews* 78(3), 544–571. DOI: 10.1128/MMBR.00015-14.

Valverde, R.A., Khalifa, M.E., Okada, R., Fukuhara, T., Sabanadzovic, S. *et al.* (2019) ICTV virus taxonomy profile: *Endornaviridae*. *The Journal of General Virology* 100(8), 1204–1205. DOI: 10.1099/jgv.0.001277.

van Kan, J.A.L., Stassen, J.H.M., Mosbach, A., Van Der Lee, T.A.J., Faino, L. *et al.* (2017) A gapless genome sequence of the fungus *Botrytis cinerea*. *Molecular Plant Pathology* 18(1), 75–89. DOI: 10.1111/mpp.12384.

Walker, J.C. (1969) *Plant Pathology*, 3rd edn. McGraw-Hill, New York, p. 819.

Wang, B., Sun, Y., Song, N., Zhao, M., Liu, R. *et al.* (2017) *Puccinia striiformis* f. sp. *tritici* micro-RNA-like RNA 1 (*Pst-milR1*), an important pathogenicity factor of Pst, impairs wheat resistance to Pst by suppressing the wheat pathogenesis-related 2 gene. *New Phytologists* 215(1), 338–350. DOI: 10.1111/nph.14577.

Wang, P., Yao, S., Kosami, K.I., Guo, T., Li, J. *et al.* (2020) Identification of endogenous small peptides involved in rice immunity through transcriptomics-and proteomics-based screening. *Plant Biotechnology Journal* 18(2), 415–428. DOI: 10.1111/pbi.13208.

Weiberg, A. and Jin, H. (2015) Small RNAs—The secret agents in the plant-pathogen interactions. *Current Opinion in Plant Biology* 26, 87–94. DOI: 10.1016/j.pbi.2015.05.033.

Weiberg, A., Wang, M., Lin, F.-M., Zhao, H., Zhang, Z. *et al.* (2013) Fungal small RNAs suppress plant immunity by hijacking host RNA interference pathways. *Science* 342(6154), 118–123. DOI: 10.1126/science.1239705.

Weld, R.J., Eady, C.C. and Ridgway, H.J. (2006) *Agrobacterium*-mediated transformation of *Sclerotinia sclerotiorum*. *Journal of Microbiological Methods* 65(1), 202–207. DOI: 10.1016/j.mimet.2005.07.010.

Whetzel, H.H. (1945) A synopsis of the genera and species of the *Sclerotinia*, a family of stromatic inoperculate discomycetes. *Mycologia* 37(6), 648–714. DOI: 10.1080/00275514.1945.12024025.

Willetts, H.J. and Wong, J.A.L. (1971) Ontogenetic diversity of *sclerotia* of *Sclerotinia sclerotiorum* and related species. *Transactions of the British Mycological Society* 57(3), 515–524. DOI: 10.1016/S0007-1536(71)80066-9.

Willetts, H.J. and Wong, J.A.L. (1980) The biology of *Sclerotinia sclerotiorum S. trifoliorum* and *S. minor* with emphasis on specific nomenclature. *The Botanical Review* 46(2), 101–165. DOI: 10.1007/BF02860868.

Williams, G.H. and Western, J.H. (1965) He biology of *Sclerotinia trifoliorum* Erikss. and other species of *Sclerotinia* forming fungi. I. Apothecium formation from *Sclerotinia*. *Annals of Applied Biology* 56, 253–260. DOI: 10.1111/j.1744-7348.1965.tb01233.x.

Winton, L.M., Krohn, A.L. and Leiner, R.H. (2006) Genetic diversity of *Sclerotinia* species from Alaskan vegetable crops. *Canadian Journal of Plant Pathology* 28(3), 426–434. DOI: 10.1080/07060660609507316.

Winton, L.M., Krohn, A.L. and Leiner, R.H. (2007) Microsatellite markers for *Sclerotinia subarctica* nom. prov., a new vegetable pathogen of the high North. *Molecular Ecology Notes* 7, 1077–1079. DOI: 10.1111/j.1471-8286.2007.01782.x.

Wong, J.A.L. and Willetts, H.J. (1973) Electrophoretic studies of soluble proteins and enzymes of *Sclerotinia* species. *Transactions of the British Mycological Society* 61(1), 167–178. DOI: 10.1016/S0007-1536(73)80098-1.

Wong, J.A.L. and Willetts, H.J. (1974) Polyacrylamide-gel electrophoresis of enzymes during morphogenesis of *sclerotia* of *Sclerotinia sclerotiorum*. *Microbiology* 81(1), 101–109. DOI: 10.1099/00221287-81-1-101.

Wong, J.A.L. and Willetts, H.J. (1975) A taxonomic study of *Sclerotinia sclerotiorum* and related species: Mycelial interactions. *Journal of General Microbiology* 88(2), 339–344. DOI: 10.1099/00221287-88-2-339.

Wong, J.A.L. and Willetts, H.J. (1979) Cytology of *Sclerotinia sclerotiorum* and related species. *Journal of General Microbiology* 112(1), 29–34. DOI: 10.1099/00221287-112-1-29.

Woronin, M. (1888) *Uber die sclerotinia krankheit der vacciniunburer*. Mémoires de l'Académie Impériale Des Sciences de St 7 36 Nr.

Xia, S., Xu, Y., Hoy, R., Zhang, J., Qin, L. *et al.* (2020a) The notorious soilborne pathogenic fungus *Sclerotinia sclerotiorum*: An update on genes studied with mutant analysis. *Pathogens* 9(1), 27. DOI: 10.3390/pathogens9010027.

Xia, Z., Wang, Z., Kav, N.N.V., Ding, C. and Liang, Y. (2020b) Characterization of microRNA-like RNAs associated with sclerotial development in *Sclerotinia sclerotiorum*. *Fungal Genetics and Biology* 144, 103471. DOI: 10.1016/j.fgb.2020.103471.

Xie, J. and Jiang, D. (2014) New insights into mycoviruses and exploration for the biological control of crop fungal diseases. *Annual Review Phytopathology* 52, 45–68.

Yamamoto, W., Oyasu, N. and Iwasaki, A. (1956) Studies on the leaf blight disease of *Allium* spp. caused by *Botrytis* and *Botryotinia fungi*. *Science Reporter, Hyogo University of Agriculture, Agricultural Biological Services* 2, 17–22.

Yang, Q., Yu, F., Yin, Y. and Ma, Z. (2016) Correction: Involvement of protein tyrosine phosphatases BcPtpA and BcPtpB in regulation of vegetative development, virulence and multi-stress tolerance in *Botrytis cinerea*. *PLoS ONE* 11(3), e0151720.

Yorimitsu, T., Sato, K. and Takeuchi, M. (2014) Molecular mechanisms of Sar/Arf GTPases in vesicular trafficking yeast and plants. *Frontiers in Plant Science* 5, 411.

Zhang, H., Xie, J., Fu, Y., Cheng, J., Qu, Z. *et al.* (2020) A 2-kb mycovirus converts a pathogenic fungus into a beneficial endophyte for *Brassica* protection and yield enhancement. *Molecular Plant* 13(10), 1420–1433. DOI: 10.1016/j.molp.2020.08.016.

Zhou, L., Li, X., Kotta-Loizou, I., Dong, K., Li, S. *et al.* (2021) A mycovirus modulates the endophytic and pathogenic traits of a plant associated fungus. *The ISME Journal* 15(7), 1893–1906. DOI: 10.1038/s41396-021-00892-3.

Zhu, W., Wei, W., Fu, Y., Cheng, J., Xie, J. *et al.* (2013) A secretory protein of necrotrophic fungus *Sclerotinia sclerotiorum* that suppresses host resistance. *PLoS ONE* 8(1), e53901. DOI: 10.1371/journal.pone.0053901.

Ziman, L. (1997) Microanatomy of the fungus *Sclerotinia sclerotiorum*. *Biologia Bratislava* 52, 389–394.

4 Reproductive Structures and Reproduction

Abstract

Sclerotinia's development and morphogenesis have been studied *in vitro* and *in vivo* to learn more about its reproductive systems and mechanisms. The spherical formations known as sclerotia are essential to the disease cycle because they act as both survival structures and inoculum. The essential element of the system is the Ras/MAPK pathway, which is adversely controlled by Rap1 via a cAMP signaling route independent of *PKA*. Many environmental and dietary factors affect the development of sclerotia. A thorough analysis has been conducted of the stages involved in the creation of sclerotium, as well as its cytology, morphology, composition, metabolism, impact of host residues and soils, and inhibitors. Numerous variables, including soil moisture, texture, pH, temperature, nutrient availability, and the depth of sclerotia burial in the soil, affect sclerotial survival. The carpogenic and myceliogenic germination of sclerotia is influenced by a number of biotic and abiotic stimuli. Ascospores are dispersed wherever it is warm and breezy; they do not survive well in humid or wet weather.

4.1 Introduction

The pathogen's survival, growth, and dissemination are greatly influenced by its strong reproduction and reproductive mechanism. *Sclerotinia sclerotiorum* has the capability of reproducing during hemibiotrophic stage with a dual mode of survival through both necrotrophic and biotrophic phases. The major survival structures are the formation of sclerotia. Numerous publications have discussed various facets of the growth and morphogenesis of *Sclerotinia* spp., highlighting the ability of their life cycle to progress on a range of artificial and synthetic media. The development and sprouting of sclerotia, which are dense, dark-colored masses of fungal tissue, have been thoroughly examined across different fungal species. These investigations have revealed both commonalities and distinctions in sclerotium formation, categorizing the primary types as terminal, strand, or loose, based on the aggregation of hyphae.

4.2 Sclerotia as Survival Structures

Sclerotia are essential in the disease cycle due to their ability to produce inoculum and act as the main structures for prolonged survival (Willetts and Wong, 1980). These structures can remain viable in the soil for up to 8 years (Adams and Ayers, 1979). A sclerotium consists of a cluster of hyphae enveloped by an outer layer of dark melanin-containing cells. This structure is thought to play a crucial role in shielding fungi from harsh environmental conditions and microbial attacks (Bell and Wheeler, 1986; Henson *et al.*, 1999). In specific instances, melanin also serves a function in the pathogenicity of certain fungi like *Magnaporthe* and *Colletotrichum* species. Strains that are unable to produce melanin struggle to form effective appressoria and invade the host. It's interesting to note that mutants in *Venturia inaequalis* and *Alternaria alternata* that lack the ability to produce melanin nonetheless display virulence (Howard and Valent, 1996; Thomma, 2003; Fitzgerald *et al.*, 2004). No correlation has been discovered between melanin and virulence in *S. sclerotiorum*. The medulla, the innermost portion of the sclerotium, is mostly made up of proteins and carbohydrates, particularly β-glucans. It is embedded in a fibrous matrix (Le Tourneau, 1979). Sclerotia of *S. sclerotiorum* exhibit considerable size differences based on the host plant they infect. For example, when infecting sunflowers, the sclerotia can grow up to 1 cm thick and have a diameter larger than 35 cm, while on dry beans,

DOI: 10.1079/9781800629592.0004

the sclerotia tend to be spherical and measure between 2 and 10 mm in diameter.

4.2.1 Stages of sclerotial development

Townsend and Willetts (1954) identified three distinct stages in the development of sclerotia:

Initiation, where a white mass is formed by the clustering of hyphae to form sclerotial initials;

1. Development, where hyphae grow and continue to aggregate to increase the size of the sclerotium; and
2. Maturation, where a defined surface is formed, melanin is deposited in the outer cells, and internal consolidation takes place.

A thorough list of dietary and environmental elements that affect sclerotia development has been put together (Chet and Henis, 1975; Le Tourneau, 1979; Willetts and Wong, 1980; Willetts and Bullock, 1992). Sclerotia usually grow when mycelial development comes into contact with a low-nutrient environment (Christias and Lockwood, 1973). The development of sclerotia has been discovered to be greatly impacted by the pH of the culture media. Inhibition of sclerotial formation occurs under neutral or alkaline pH conditions (Rollins and Dickman, 2001). On the other hand, the fungus's accumulation of oxalic acid (OA) results in a decrease in the pH of the surrounding environment (Maxwell and Lumsden, 1970). This change in pH creates favorable conditions for the development of sclerotia (Rollins and Dickman, 2001). This theory is substantiated by the observation that a mutated form of *S. sclerotiorum*, which cannot produce oxalic acid (OA), is also incapable of forming sclerotia in a laboratory setting and, consequently, does not induce any disease symptoms in plants (Godoy *et al.*, 1990; Dickman and Mitra, 1992). Even though decreasing the ambient pH in these mutants does not allow them to generate sclerotia again (Rollins and Dickman, 2001), this suggests that pH plays a more nuanced role than just that. The pac1 gene in *S. sclerotiorum* is essential for growth at neutral pH, complete virulence, regular synthesis of OA, and sclerotial development. It also plays a critical role in controlling pH-regulated gene expression. This demonstrates unequivocally how important pH-responsive gene expression is to the growth and pathophysiology of *S. sclerotiorum* (Rollins, 2003).

4.2.2 Transition period in skeletal development

There is an early transitional phase in the development of sclerotia from mycelium to sclerotia that is regulated by biochemical substances including OA and cyclic adenosine monophosphate (cAMP) buildup. cAMP is essential for the change from mycelial growth to sclerotial development (Rollins and Dickman, 1998). Greater amounts of either endogenous or exogenous cAMP prevent sclerotia from forming but encourage OA buildup in *S. sclerotiorum*. Moreover, Smk1, a mitogen-activated protein kinase (MAPK) in *S. sclerotiorum* that is essential for sclerotiogenesis, is inhibited by elevated cAMP levels (Fig. 4.1). Notably, Smk1 transcripts exhibit maximal expression in acidic environments, suggesting that MAPK cascades function as regulatory components in sclerotial development, regulated by pH and cAMP-dependent signaling (Chen *et al.*, 2004). Harel *et al.* (2005) conducted a study to investigate the impact of cAMP pathways on sclerotiogenesis. The results showed that in strains of the wild type, there was an increase in cAMP-dependent protein kinase A (*PKA*) levels during the formation of sclerotia 1. Conversely, mutants devoid of the capacity to consistently produce sclerotia had low levels of *PKA*. Furthermore, the study demonstrated that environmental signals that increased *PKA* levels were linked to the production of sclerotia-like structures in mutants that did not produce sclerotium, as illustrated in Fig. 4.1. Notable is the fact that a pka1 knockout mutant exhibited equivalent pathogenic behavior, sclerotial development, and cAMP sensitivity to the wild-type strain

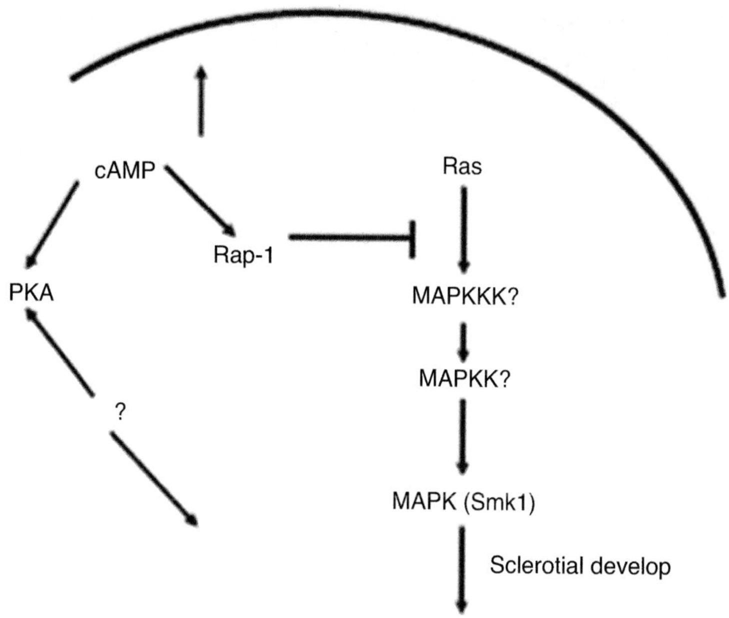

Fig. 4.1. Rasp –1 model. (Adapted from publication of Chen and Dickman, 2005, with permission.)

despite lacking the *PKA* catalytic subunit gene. This implies that additional *PKA*-related genes are involved in the sclerotiogenesis process, or that a distinct, *PKA*-independent route exists (Jurick *et al.*, 2004; Jurick and Rollins, 2007).

Chen and Dickman (2005) further suggested that *PKA* is not required for the suppression of cAMP-mediated sclerotial formation, since *PKA* inhibitors selectively inhibited *PKA* activity without affecting sclerotia development or cAMP-mediated MAPK inhibition.

4.2.3 Signaling pathways during sclerotial development

One idea states that Rap-1 uses cAMP to suppress the MAPK of *S. sclerotiorum*. Fig. 4.1 illustrates how *PKA* influences sclerotial development and that cAMP/Rap-1 suppresses the Ras/MAPK signaling pathway. *PKA* levels increase with the formation of sclerotia. Conversely, in mutants that do not produce sclerotium, environmental cues that raise *PKA* levels are associated with the formation of precursors that resemble sclerotia (Harel *et al.*, 2005). According to Chen and Dickman (2005), Rap-1 is activated by cAMP and most likely binds to components of the Ras/MAPK pathway that are unknown upstream of mitogen activated protein kinase kinase kinase (MAPKKK) and mitogen activated protein kinase kinase (MAPKK). This disease is brought on by the inhibition of sclerotial development by the downstream MAPK Smk1. Moreover, it has been observed that the small GTPase Ras, an activator upstream of the MAPK pathway, is necessary for appropriate sclerotial growth. The significance of the conserved Ras/MAPK system in sclerotiogenesis is highlighted by the fact that loss of Ras function prevents MAPK activation (Fig. 4.1). Interestingly, in the presence of cAMP, blocking Rap-1, a member of the Ras protein family, restores MAPK activation and promotes normal sclerotial growth. These results suggest that a mechanism involving the Ras/MAPK pathway, which is negatively regulated by Rap-1 in a cAMP signaling pathway that is *PKA*-independent, is responsible for skeletal development (Fig. 4.1).

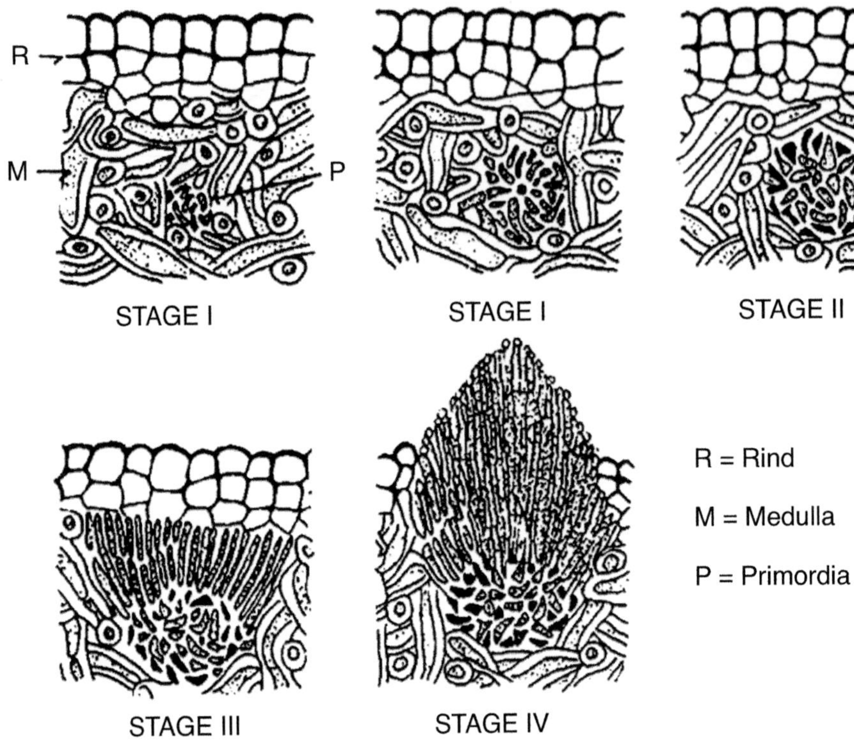

Fig. 4.2. The developmental stages of apothecial stipe primordium. (Adapted from Saito, 1977, published with permission.)

Research demonstrates the intricate relationship between ambient pH sensing and cAMP signaling in sclerotiogenesis; however, the precise function of *PKA* is still unknown.

4.3 Phases of Sclerotium Formation

Sclerotial formation passes through four main stages from mycelium to mature sclerotia to the young apothecial stage. Sclerotia formation is a prominent and easily observable process in *S. sclerotiorum* and similar species (Willetts and Wong, 1980). Numerous scholars have provided thorough explanations of the genesis of sclerotia (Chet and Henis, 1975; Saito, 1977; Le Tourneau, 1979; Willetts and Wong, 1980). The formation of sclerotia occurs in the following four primary stages (Figs 4.2 and 4.3; Trevethick and Cooke, 1973; Saito, 1977; Le Tourneau, 1979; Willetts and Wong, 1980):

1. In the initial phase of the process, the mycelium forms small sclerotial primordia.
2. The sclerotial primordia quickly develop into dense hyphal clusters throughout the development phase and reach their full size.
3. The maturity phase, which is marked by internal changes, desiccation, and pigmentation, starts when the sclerotia stop growing.
4. The primordia emerge from the outer layer and begin to grow as immature apothecial stipes during the last stage.

When completely developed, sclerotia serve as 'resting structures' in the pathogen's life cycle

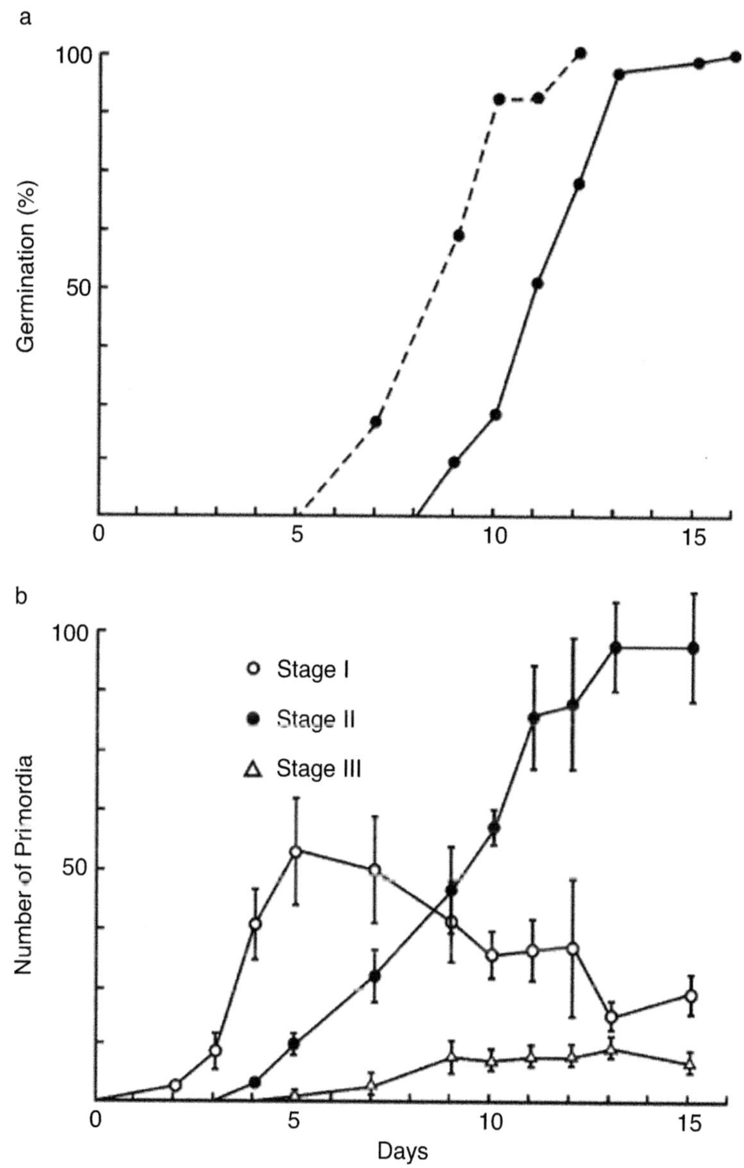

Fig. 4.3. The duration required for sclerotial germination and stipe primordium formation in sclerotial tissue. (a) Solid line represents the germination rates of sclerotia, while a dotted line shows the formation of Stage IV primordia in sclerotia. (b) The amount of stipe primordia at different developmental stages is shown. (Adapted from publication of Saito, 1977, with permission.)

and may withstand adverse environmental circumstances (Schwartz and Steadman, 1978; Adams and Ayers, 1979; Willetts and Wong, 1980). Studies conducted in laboratories have largely contributed to our understanding of how environmental conditions affect the genesis of sclerotium. Some of these research studies have demonstrated a positive association with field observations; however, others have not received this kind of validation. When the mycelium reaches a certain physiological condition, it produces certain chemicals that cause the production of sclerotium initials (Trevethick and Cooke, 1973). Initial sclerotium formation might result in either full development or no development at all. Seldom, if at all, do sclerotia remain partially developed. While non-nutritional factors have a significant impact on the onset of sclerotium, the nutritional environment of the medium mostly determines the course of growth or maturation (Trevethick and Cooke, 1973).

4.3.1 The impact of calcineurin on pathogenicity and sclerotial development

S. sclerotiorum is a globally prevalent necrotrophic/hemibiotrophic and omnivorous plant pathogen. Sclerotia, pigmented structures composed of many hyphae, are produced by *S. sclerotiorum* and are essential to the organism's life cycle. In fungus, calcineurin is a Ser/Thr phosphatase that is involved in several signal transduction pathways. It is essential for regulating cation balance, morphogenesis, cell wall strength, and pathogenicity. During the development of sclerotia, *S. sclerotiorum* experiences phase-specific alterations in the expression of calcineurin. Preformed sclerotia germination is increased when calcineurin is inhibited by FK506, cyclosporin A, or inducible antisense calcineurin expression. This hampers sclerotia formation during the prematuration phase. *S. sclerotiorum's* pathogenesis on tomato and *Arabidopsis* is reduced when antisense calcineurin expression is induced, but this has little effect on the pathogen's primary virulence factor, OA secretion. Reduced cell wall β−1,3-glucan concentration and enhanced susceptibility to enzymes that break down the cell wall and the glucan synthase inhibitor caspofungin are the outcomes of calcineurin inhibition. Thus, calcineurin is important for the development of sclerotia and for the pathogenesis of *S. sclerotiorum*. It is also probably important for other phytopathogens (Harel *et al.*, 2006).

4.3.2 The influence of missing exudates on the maturity of sclerotia

The removal of exudates from *S. sclerotiorum* sclerotia causes the sclerotia to not develop to their full potential, which reduces the size and compactness of the outer cell layers. Additionally, these exudates show antifungal qualities against particular saprophytic and parasitic fungi. The pure exudates showed the maximum sensitivity when tested against a variety of fungi, including *Fusarium udum*, *Colletotrichum capsici*, *Aspergillus flavus*, and *Cercospora blumea*. As a result, conidial germination was less than 40%. Through high-performance liquid chromatography a number of different acids, including tannic, gallic, ferulic, and cinnamic acids, were discovered in the ethyl acetate fraction of the exudate, along with a few other unidentified compounds. In a glasshouse experiment, the sclerotia that had run out of exudates could still infect eggplants (*Solanum melongena*). When eggplant leaves were treated with OA and sclerotial exudate of *S. sclerotiorum*, the result was the formation of phenolic acids, notably gallic and cinnamic acid. After 96 hour, leaves treated with 0.1% OA showed the highest induction of gallic acid, while after 48 hour, leaves treated with undiluted exudates showed the maximum induction of citric acid. The decreased synthesis of cinnamic acid when *S. sclerotiorum* mycelia are present indicates that resistance processes other than the phenylpropanoid pathway may be involved (Singh *et al.*, 2004).

4.3.3 Impact of damage to the outer layer and the healing process on sclerotial permeability

The *Sclerotinia minor* sclerotia possess the remarkable ability to regenerate a fresh outer layer in the event of damage occurring either before or after reaching maturity. This regenerative process entails the growth of an entirely new rind, effectively sealing off the exposed and compromised surface. The formation of the new rind and cortical cells is facilitated by the outgrowths of existing medullary hyphae, resulting in a more or less spherical structure. Over the course of several days, the newly formed rind cells gradually acquire pigmentation, while the cortical cells accumulate reserves once the pigmentation is complete. It is important to note that immature sclerotia experience the fastest rate of regeneration. Within 8 days, a fully regenerating rind is formed, even in old sclerotia.

When the rind is still intact, mature sclerotia effectively block the apoplastic tracer sulforhodamine G from entering. However, if the rind is injured, they do become permeable to this fluorochrome. Even when the regeneration process is finished, the permeability to sulforhodamine is not entirely restored, even though the regeneration of a new rind does reduce it. Therefore, even if they are regenerating, sclerotia that have sustained injury in the field might be more vulnerable to outside factors than those whose crust is intact (Young and Ashford, 1996).

4.4 Signaling Pathways Associated with the Start of Sclerotial Development

Sclerotia initiate their development from small primordia and rapidly transform into dense white masses of hyphae of *S. sclerotiorum*. These masses undergo dehydration and pigmentation as they mature and stop growing. The melanized external surface of the sclerotium plays a critical role in enduring unfavorable conditions. Mutants with broad regulatory functions not only show impaired growth and abnormal sclerotial formation but also consistently exhibit a loss of virulence and other pleiotropic defects. The melanization process of sclerotia is essential for the long-term survival of *Sclerotinia* species in their natural habitat. When the genes responsible for melanin biosynthesis, such as scytalone dehydratase (Scd1) and trihydroxy-naphthalene reductase (Thr1), are deleted, the mycelial growth becomes slower and the number, size, and pigmentation of sclerotia are reduced. This indicates that melanization plays a critical role in the maturation of sclerotia and potentially promotes growth through feedback mechanisms. Nevertheless, in contrast to pathogenic fungi like *Magnaporthe* species, the knockout strains of *Sclerotinia* do not exhibit a decrease in virulence. This indicates that melanin may not play a role in pathogenesis in *Sclerotinia* as it does in other fungi (Liang *et al.*, 2018). In *S. sclerotiorum*, knocking out distinct conserved genes can result in a variety of abnormalities. In *S. sclerotiorum*, the *Pth2* gene codes for the peroxisomal carnitine acetyltransferase (CAT). Acetyl-CoA transport, which is required for fatty acid metabolism, depends on the activity of CAT. According to Liberti *et al.* (2013), the pth2 knockout mutants exhibit aberrant growth patterns, a decreased capacity to form sclerotia, and decreased levels of pathogenicity and OA. Suppression of the integrin-like gene '*Itl*' is another example, wherein alterations in colony structure, smaller and irregularly shaped sclerotia, and decreased pathogenicity result. The cyanovirin-N homology (CVNH) domain is still preserved even though Cvnh homologs are unique to *Sclerotinia* and *Botrytis* (Lyu *et al.*, 2015). Most signal transduction pathways depend on phosphorylation and dephosphorylation processes, with several kinases and phosphatases being involved in sclerotial formation and growth. The two-component histidine kinase Shk1 mutation results in ineffective sclerotial development, as well as slower and altered hyphal growth. The mutant's pathogenicity is unaffected, but it exhibits sensitivity to osmotic stress and improved resistance to fungicides. It is believed that Shk1 controls these pathways by acting prior to the MAPK cascade. Smk1 and Smk3 are two MAPK kinases in *S. sclerotiorum* that have been studied in great detail. These kinases may work downstream of Shk1, as suggested by the comparable growth deficits and absence of sclerotial development seen in their mutants (Chen *et al.*, 2004; Bashi *et al.*, 2016). Furthermore, since the addition of exogenous cAMP reduces Smk1 expression and MAPK phosphorylation, cAMP might act prior to the MAPKs. It is believed that the Type 2A phosphoprotein phosphatase Pph1 functions subsequent to the MAPK. Pph1-deficient mutants experience stopped growth. According to Erental *et al.* (2007, 2008), mutations of the PP2A R2B regulatory subunit rgb1 exhibit aberrant sclerotial development and poor penetration during infection.

The study of the mutant Scat1, a type I catalase, shows the variety of roles that reactive oxygen species (ROS) play in the growth and pathogenicity of fungi. Reduced radial growth, an increase in tiny, improperly melanized sclerotia, and hypovirulence are the results of Scat1 deletion. Significant involvement of calcium is also seen in signaling pathways. Studies conducted on Cna1 and Caf1 demonstrate the role of calcium in *S. sclerotiorum* life. The conserved Ser/Thr protein phosphatase Cna1 is dependent on calmodulin and calcium to operate. Mutants carrying antisense cna1 exhibit altered growth, sclerotial development, and hypovirulence. On the other hand, fungi have conserved the putative Ca^{2+}-binding EF-hand motif found in the secreted protein Caf1. According to Xiao *et al.* (2013), mutants with transfer DNA (T DNA) insertion in caf1 show aberrant sclerotia, hypovirulence, and somewhat slower development.

Significant involvement of calcium is also seen in signaling pathways. A large class of genes involved in signaling are transcription factors that are conserved. GATA transcription factors are a family of zinc finger DNA-binding proteins that regulate gene expression during development and cell differentiation, playing critical roles in the formation and function of various tissues and organs. The SNF5 domain is present in the GATA transcription factor Sfh1. It is a component of the ATP-dependent chromatin remodeler known as the chromatin remodeling complex (RSC) (Remodels the Structure of Chromatin) complex, which is essential to the cell cycle. Its participation in growth, sclerotial development, and virulence is demonstrated by its varied functions. It is probable that the transcription factor Ste12 functions after the MAP kinase cascades. Hypovirulence was seen in ste12 RNAi lines, with smaller sclerotia, fewer appressoria, and sluggish development. In mammals and fungi, the transcription factors belonging to the Forkhead (FKH) box (FOX) family are essential for controlling primary metabolism, cell cycle, and morphogenesis. According to Fan *et al.* (2017), the fkh1 RNAi mutants show slower growth, non-sclerotial formation, hypovirulence, and reduced stress responses, suggesting that it functions as a common transcriptional regulator across several pathways.

It has been speculated that cell wall degrading enzymes (CWDEs) may play a significant role in fungal pathogenesis. This is corroborated by the observation that pathogenicity is lost upon deletion of Xyl1, which encodes an endo-β−1,4-xylanase. Furthermore, the mutant displays unusual traits such as accelerated hyphal branching, decreased sclerotial formation, and growth retardation, suggesting xylanase may also be involved in developmental processes (Yu *et al.*, 2016).

Conserved are the light-sensitive proteins known as opsins, which play a role in circadian rhythms. The sop1 RNAi lines show hypovirulence, non-sclerotial formation, decreased growth, and enhanced sensitivity to fungicide and osmotic agent stresses. These findings imply that different light-induced processes in sclerotia may be affected differently by opsins. In bacteria and eukaryotes, rearrangement hotspot (Rhs) repeat-containing proteins are extensively dispersed. Rhs1 is a secreted protein, and its orthologs seem to be specific to *Sclerotinia* and *Botrytis* species. Rhs1 RNAi mutants exhibit hypovirulence, fewer and larger sclerotial formations, and significantly slower colony expansion. The results suggest that Rhs1 may be important for *Sclerotinia* virulence and may also have a small effect on sclerotial development and growth. Fungal elicitors called cerato-platanins have the power to induce cell death and plant immunity. Sm1 has the capacity to cause the host to become hypersensitive. Interestingly, RNAi mutants of Sm1 show decreased growth, impaired sclerotial development, and increased susceptibility to other stimuli in addition to reduced virulence, indicating a broader function for this elicitor. Nevertheless, in another investigation using the deletion mutant cp1, which impacts the same gene, these defects were not noted. To resolve these discrepancies, careful transgenic complementation studies must be carried out in the future (Yang *et al.*, 2018).

Mutants that fail to create sclerotia but exhibit normal mycelial growth indicate that sclerotial development is distinct from growth. Although many of these mutants have decreased virulence, a few like sl2 and ggt1 preserve normal pathogenicity. RNA interference is used to silence Sl2, GAPDH (glyceraldehyde-3-phosphate dehydrogenase) (Gpd), or Hex1. This results in modifications to sclerotial development or melanization, which is similar to the effect observed in interacting proteins like Hex and glyceraldehyde-3-phosphate dehydrogenase

(GAPDH). Nevertheless, sl2's pathogenicity does not change, indicating that different downstream components may regulate sclerotial development and pathogenicity independently. It is unknown how the protein Sl2, which is only released by fungi and is found inside their cell walls, affects the growth of sclerotia. On the other hand, the highly conserved enzyme Ggt helps to keep the redox equilibrium in place. Elevated levels of GSH (tripeptide glutathione) + GSSG (oxidized glutathione) and hydrogen peroxide are observed in ggt1 deletion mutants, and in mature sclerotia, the cortical layer and medulla are separated. Additionally, these mutants struggle with producing compound appressoria, impacting the penetration phase of infection. These mutant characteristics indicate that GSH, GSSG, or the GSH/GSSG ratio may be involved in regulating sclerotia maturation and appressoria formation (Li *et al.*, 2012).

The genes that have been analyzed further affect sclerotial formation without influencing growth, showing reduced virulence. These genes are believed to function upstream of Sl2 and Ggt. In *S. sclerotiorum*, an increase in ROS was observed in the initial stages of sclerotial development and in the infection cushions. The combined effects of NADPH (nicotinamide adenine dinucleotide phosphate) oxidases Nox1 and Nox2 are probably responsible for the rise in sclerotia formation. Mutants of either Nox1 or Nox2, as well as the application of an NADPH inhibitor or ROS scavenger, failed to develop sclerotia, indicating the necessity of ROS for sclerotial initiation. It's interesting to note that only the nox1 mutant showed a decrease in OA synthesis and virulence, not the nox2 mutant. This finding suggests that these two genes may not be entirely redundant because of their different patterns of expression. Larger and fewer sclerotia are consistently generated when thioredoxin reductase (Trr1), the gene in charge of ROS detoxification, is silenced. Nonetheless, Trr1 silenced mutants exhibit reduced virulence, suggesting that virulence requires an ideal concentration of ROS. Low pH is known to encourage the growth of sclerotia. The transcription factor Pac1, which is sensitive to pH, is responsible for the expression of genes induced by ambient pH. The absence of Pac1 leads to irregular sclerotial development and decreased OA levels, causing reduced virulence. Maintaining a balanced interplay among ROS, OA, and pH is essentiConserved are the light-sensitiveal for the biology of this fungal pathogen, as elevated OA levels can affect pH and potentially disrupt ROS equilibrium (Rollins, 2003; Xia *et al.*, 2019).

4.5 Genetic Factors Influencing the Growth of Apothecia and Ascospores

The genus *Sclerotinia* is distinguished by the apothecium, comprising a stem and a disc that is saucer shaped. The upper surface of the apothecia is where the asci develop, and when they reach maturity, the ascospores are dispersed. The apothecial development in the homothallic *S. sclerotiorum* is influenced by the mating-type (Mat) genes, which are known to be key regulators of sexual development in ascomycetes. The Mat locus of *S. sclerotiorum* is home to a group of four Mat genes responsible for encoding transcription factors. The Mat1-1s encode an alpha-domain, while the Mat1-2s encode high-mobility group (HMG) proteins. The functions of each of the four Mat genes in mycelial growth, virulence, sclerotial formation, and apothecial development were investigated using deletion knockout studies. Mat1-1-5, Mat1-1-1, and Mat1-2-1 are critical for early sexual stages, and Mat1-2-4 is critical for late sexual stages, including apothecial initiation and ascospore formation. These Mat genes have little effect on vegetative growth, development, and pathogenesis. It is not unexpected that these Mat gene mutants, which are transcription factors, show altered expression of the Mat locus and suspected pheromone and pheromone receptor genes, since this provides insight into their possible regulatory mechanism and particular symptoms (Doughan *et al.*, 2016).

There is another gene, Fox E2, that is similar to the Mat genes and is primarily implicated in the development of apothecia. A member of the FOX transcription factor family, Fox E2 is well known for its roles in morphogenesis, development, pathogenicity, and stress responses in both fungi and animals. The results of the Fox E2 knockdown investigation for *S. sclerotiorum* indicate that there is no discernible effect on virulence, sclerotial development, or mycelial growth. Interesting issues about the interactions between these transcription factors are raised by the remarkable similarities in mutant phenotypes between Mat and Fox E2. It is plausible that a coordinated transcriptional network involving these factors is necessary for the proper development of the sexual stage in *S. sclerotiorum* (Wang *et al*., 2016). Cry1, identified as a UV-A photoreceptor, was first thought to play a role in apothecial development because it is activated by UV light during this particular phase of development. Nevertheless, no significant abnormalities in apothecial development were detected in the cry1 deletion mutant, indicating that it might not be a primary regulator or that there are other elements concealing its impact on photomorphogenesis (Veluchamy and Rollins, 2008).

Pleiotropic deficiencies involving conserved proteins are seen in mutants with apothecial development disorders, such as nsd1 (GATA-type IVb zinc-finger transcription factor) and sac1 (adenylate cyclase). In fungal diseases that affect plants, the cAMP molecule has multiple functions in biological processes. Despite the fact that the protein kinase A catalytic subunit gene (Pka1) was previously believed to be involved in cAMP signaling, the pka1 single mutant does not show any discernible abnormalities, which may be because Pka2 is genetically redundant. Due to a marked decrease in cAMP levels, the sac1 mutant exhibits slower growth, loss of pathogenicity, and failure in apothecial development. Conversely, Nsd1 (histone methyltransferases) enzymes functions in a variety of fungi as a transcription factor for development and environmental reactions. Knockout mutants of nsd1 display altered hyphal growth, smaller sclerotia, failure in appressorium formation leading to loss of pathogenicity, and failure in apothecial initiation. These observations suggest that Nsd1 is likely a general transcription factor that regulates multiple processes in *S. sclerotiorum*, or a specific regulator that controls a downstream factor shared by various pathways. One of the targets of Nsd1 is a gene encoding a NO homeostasis-regulating glutathione-dependent formaldehyde dehydrogenase (Fdh1). The incapacity of fdh1 deletion mutants to infiltrate host tissue is due to decreased mycelial development, smaller sclerotia, and impaired compound appressoria. These observations underscore the critical roles that NO plays in these pathways. The striking similarities between the traits of the fdh1 and nsd1 mutants suggest that Fdh1 may be the primary target of Nsd1 during growth and development (Li and Rollins, 2009; Pan and Rollins, 2017; Zhu *et al*., 2019).

4.6 Cytology and Morphology of Sclerotia

In *S. sclerotiorum*, sclerotium formation usually follows a general pattern, while isolates, substrates, and environmental factors can cause deviations. Ascospores, mycelia, or sclerotia placed on a solid nutritional medium cause hyphae to develop and form a thin layer on the surface. As the colony spreads and approaches the container's periphery or faces growth limits, the mycelium mass thickens, resulting in the creation of white mycelium mounds ornamented with little liquid droplets. Larger exudate droplets are visible and the surface gradually darkens as the sclerotium grows.

The sclerotium darkens gradually until it turns black, at which point exudate droplets disappear. The process takes over a week to finish, at which point a fully formed sclerotium can be detached from the substrate. Sclerotium can spread throughout the substrate in concentric rings or other regular patterns, even though they typically form near the plate's edge. Certain fungal strains' internal rhythms seem to have an impact on these patterns (Humpherson-Jones and Cooke, 1977a, b). Light transmission and scanning electron microscopy have been used to study the development of sclerotia in great detail (Willetts and Wong, 1971; Colotelo, 1974; Saito, 1974Saito, 1977). Similar to other Ascomycotina, the vegetative hyphae have a simple perforated septum and a variety of organelles. In addition, an unexplained tubular vesicle complex in the cytoplasm and a structure resembling lomasomes between the single-layered cell wall and plasma membrane were seen by Saito (1974).

In 1971, Willetts and Wong carried out research on how different *Sclerotinia* species produce their sclerotium. The growth process of bigger sclerotia species, such as *S. sclerotiorum* and *Sclerotinia trifoliorum*, follows a terminal pattern. The fusing of lengthy primary hyphae produces initials, and several initials combine to form a significant sclerotium. On the other hand, smaller sclerotia species, such as *Sclerotinia libertiana* and *S. minor*, produce initials by means of a different process of strand development. These initials originate from small, aerial hyphae that entwine with one another and emerge laterally from the major hyphae (Figs 4.4, 4.5 and 4.6). As the growth continues, the mycelial network on the surface is replaced by rind cells as the medullary cells within the structure get more closely spaced. Liquid fills the intercellular gaps during this phase and is connected to the surface droplets (Colotelo, 1974). Numerous ribosomes as well as multivesicular and multitubular lomasomes are present in the growing sclerotia cells. It is also observed that the plasma membrane regularly infolds. A thick layer of fibrillar tissue covers the interior hyphae's initial chitinous wall upon sclerotium maturation. This layer is made up of protein and β−1,3-glucan, according to histochemical analyses combined with enzyme techniques (Saito, 1974Saito, 1977). Inside the cells, the same hyphae are made up of a polysaccharide that is probably glycogen. The mitochondrial cristae become less defined and electron-dense molecules, possibly polyphosphates, accumulate in the vacuoles as the sclerotia grow. Melanin is abundant in fully grown rind cells, which also have several pores or apertures (Jones, 1970; Saito, 1977). When a sclerotium is fully formed, it consists of a core of prosenchymatous tissues encircled by a fibrous matrix, and an outer layer that is black and around three cells wide (Colotelo, 1974; Saito, 1974). Between the outer layer and the core, a cortical layer that is two to four cells thick has been described by Kosasih and Willetts (1975).

4.7 Chemical Composition of Sclerotia

Sclerotia obtained from bean- and pea-cleaning operations in commercial settings were subjected to proximate chemical analysis. The results revealed that the sclerotia contained less than 2% crude fat, 3.5–5.0% ash, and 20–25% protein. It was demonstrated that a sizable portion (20–25%) of the dried ground sclerotia may dissolve in ethanol. However, if the calculation was based on alcohol-insoluble nitrogen, the protein content was found to be between 10 and 15%. Sclerotia cultivated on synthetic agar medium exhibited reduced levels of crude fat, alcohol-soluble material, and protein in comparison to sclerotium obtained from the wild.

Further investigation has been conducted on the hydrocarbon components, free fatty acids, and amino acids found in various fractions of these sclerotia (Weete *et al.*, 1970; Le Tourneau, 1979). To stimulate the growth of mushrooms, it is important to highlight that sclerotia contain considerable amounts of unsaturated fatty acids, such as oleic and linoleic acids (Khalil and Ragab, 1990).

In mature sclerotia, carbohydrates comprise the majority of the dry matter. According to chemical research, around 25% of the sclerotia are composed of β-glucan (Le Tourneau, 1979). Histochemical data from ; studies demonstrate the presence of glycogen, chitin, and polysaccharides. On the other hand, cellulose is not suggested to be present by the infrared spectra research conducted by Michell and Scurfield (1967) Trehalose, a non reducing disaccharide, and mannitol, a sugar alcohol, are the two primary low-molecular-weight carbohydrates present in sclerotia. Normal contents of mycelia and sclerotia include glucose or fructose monosaccharides. The presence of extra sugars and sugar alcohols is dependent on the carbon source in the medium (Wang and Le Tourneau, 1971).

4.8 Metabolites Associated with Production of Sclerotium

Significant changes in cellular structure are necessary for the development of sclerotia, as is the transport and accumulation of various chemicals. But the variables that affect the beginning phase have received little attention. According to Willetts (1978), vegetative hyphae usually develop apart from one another. When initials form, there is an attraction between

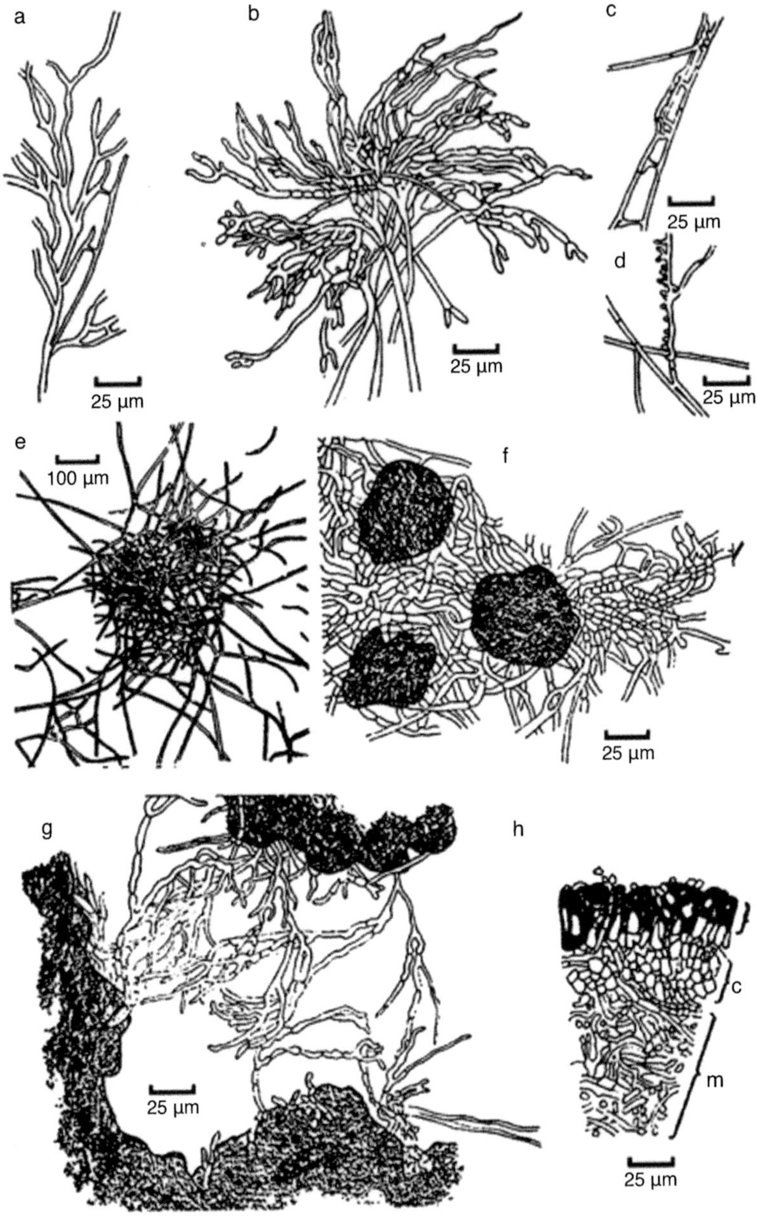

Fig. 4.4. (a) Sclerotinia trifoliorum and *Sclerotinia sclerotiorum* mycelium is growing regularly. (b) During the early stages of development, a large number of dichotomous branches have formed and the growth of the sclerotial initials has stopped. (c) Anastomosis is fusing nearby hyphae together. (d) Anastomosis is linked to a variety of protuberances. (e) A growing sclerotium appears tufted and has white hyphae. (f) Hyphae are weaving together in the space between three tiny colored initials. (g) Hyphae connect two substantial pigmented initials. The void left by mycelium could eventually cause a sizable sclerotium to grow when the initials combine. (h) When examined in cross section, a thick-walled cell composition forms the protective rind of the mature sclerotium of *S. sclerotiorum*. (i) The cortex contains cells that are pseudo-parenchymatous, while the medulla is characterized by hyphae arranged irregularly. (Adapted from publication of Willetts and Wong, 1971, with permission.)

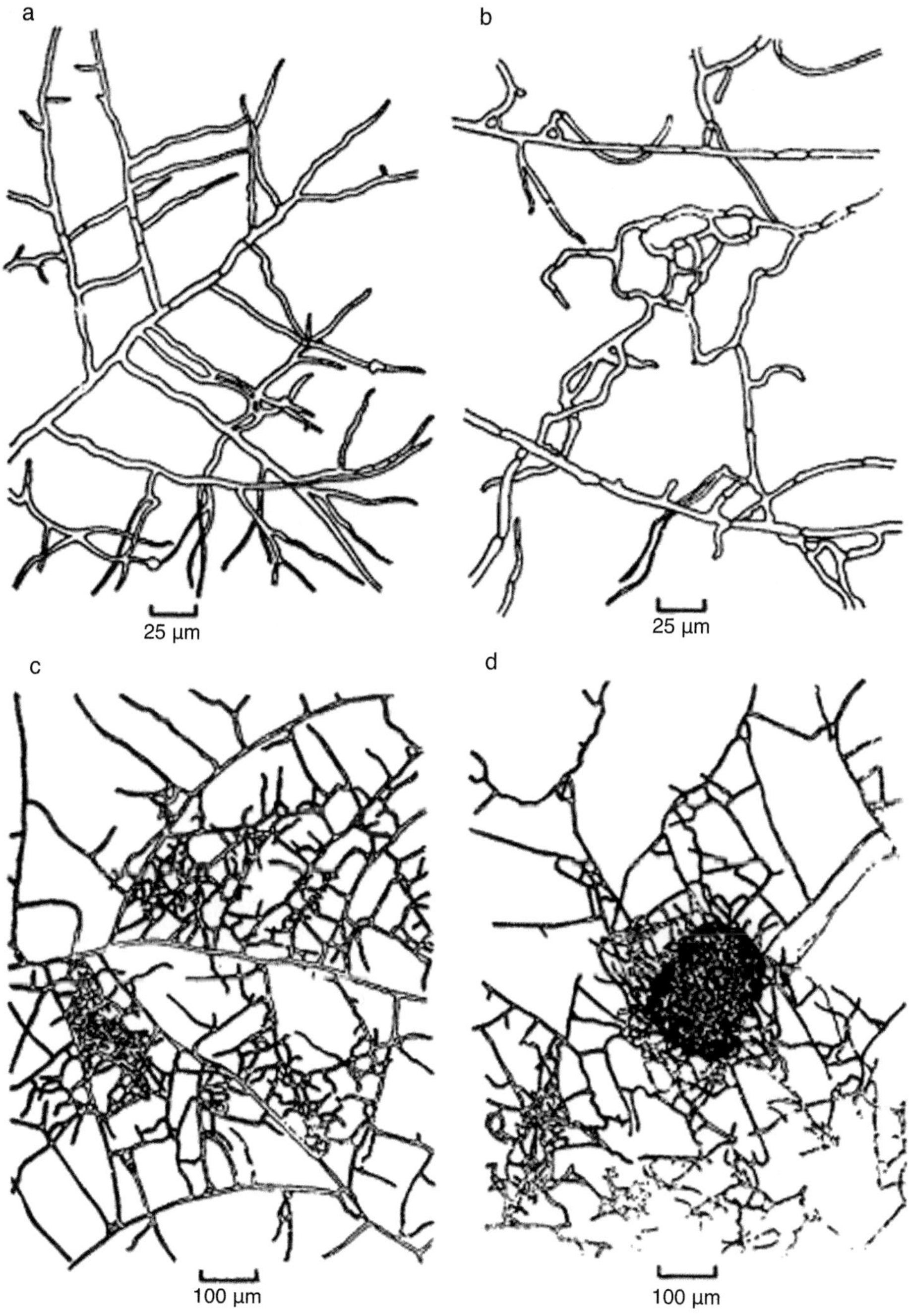

Fig. 4.5. Both *Sclerotinia minor* and *Sclerotinia libertiana* produce small-type sclerotium, which may be categorized as follows: (a) standard mycelium development; (b) several branches interlinked during the early stages of sclerotial creation; (c) subsequent phase in initial development; and (d) minute unique sclerotium. (Adapted from publication of Willetts and Wong, 1971, with permission.)

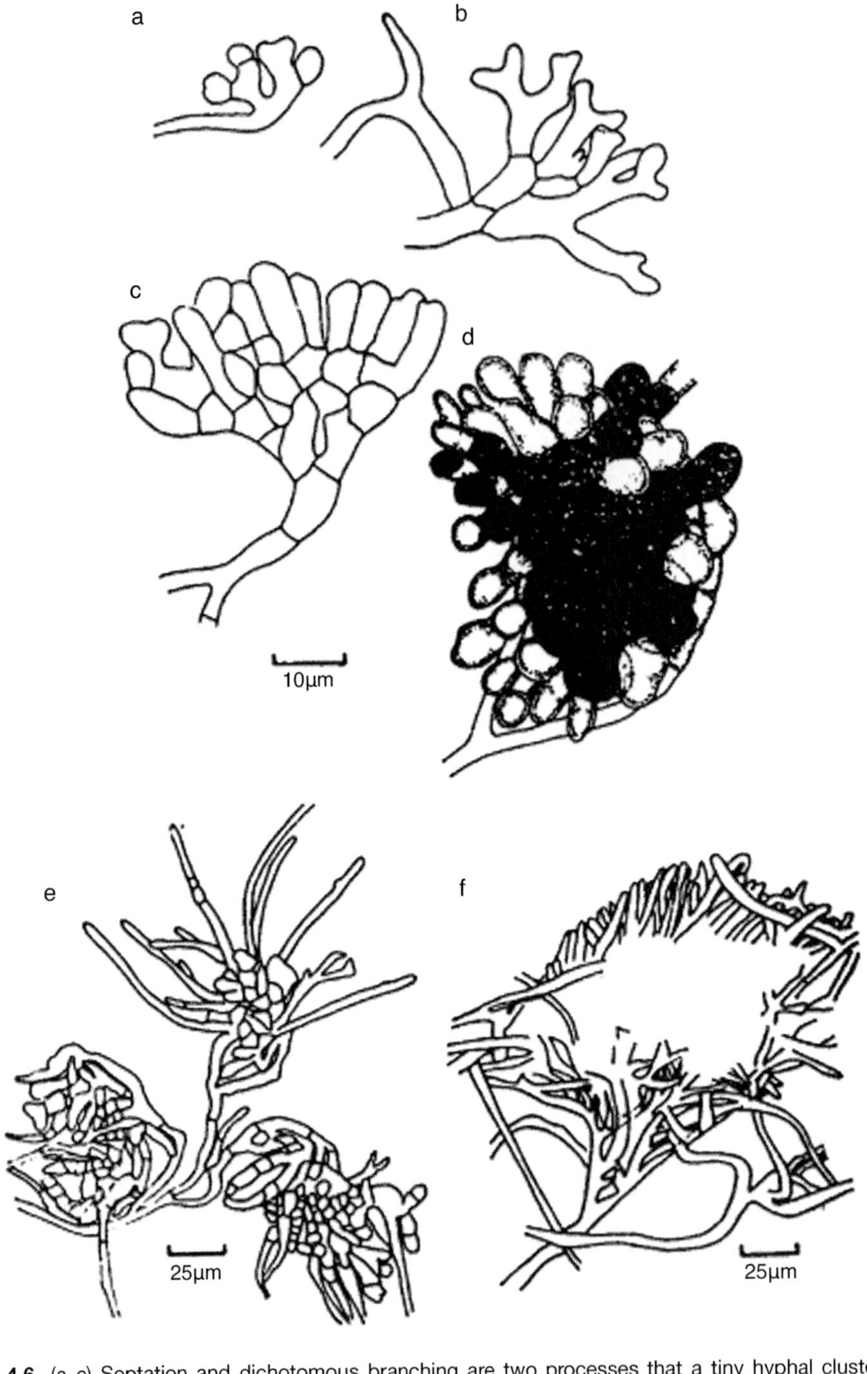

Fig. 4.6. (a–c) Septation and dichotomous branching are two processes that a tiny hyphal cluster of *Sclerotinia minor* goes through as it grows. (d) The final stage of the creation process. Almost all the cells have developed pigmentation. (e) Small clusters that resemble spores. There are certain cells that have sprung vegetative hyphae. (f) On the cellophane that covered the culture medium, a mass of hyphae formed. (Adapted from publication of Willetts and Wong, 1971, with permission.)

them that facilitates hyphal fusions. Variations in nutrition availability may cause initials to emerge; these events usually occur after the substrate has been completely colonized. This idea is supported by the finding that certain fungus, such as *S. sclerotiorum*, produce sclerotia when their mycelial mats are moved to a medium lacking in nutrients (Christias and Lockwood, 1973). The mycelium or media must create particular metabolic processes before initials may be produced. Known to produce organic acids, some *Sclerotinia* species may play a part in the establishment of sclerotium (Wang and Le Tourneau, 1971; Humpherson-Jones and Cooke, 1977b).

Although the amount and kinds of organic acids can vary depending on the isolate and the medium, *S. sclerotiorum* normally produces oxalic, fumaric, malic, and succinic acids (Corsini and Le Tourneau, 1973; Humpherson-Jones and Cooke, 1977b). There have also been discoveries of other acids, including citric, glycolic, glyoxylic, and α-ketoglutaric (Colotelo, 1973; Humpherson-Jones and Cooke, 1977b). Enzymes for the synthesis of OA and Krebs cycle acids are present in *S. sclerotiorum* (Corsini and Le Tourneau, 1973; Maxwell, 1973). The pH of the culture filtrate of *S. sclerotiorum* cultured in a liquid medium decreases to pH 3–4 and then gradually rises (Wang and Le Tourneau, 1971). The greatest acid generation and the highest specific activity of Krebs cycle enzymes are correlated with this pH reduction. When the fungus is grown in a medium that facilitates optimal growth, a lot of sclerotia emerge, which is consistent with the pH dropping. On the other hand, the pH of the culture filtrate usually increases when the fungus is grown in a medium that promotes limited growth or sclerotium development (Wang and Le Tourneau, 1971). The provision of resources essential for the production of reserves and other compounds present in mature sclerotia is crucial once the initial sclerotia have grown. Willetts (1978) and Chet and Henis (1975) have provided summaries of the solute movement research. The developing sclerotium's core is connected to droplets that cover the growing sclerotia (Colotelo, 1974). Enzymes, among other inorganic and organic materials, make up these membrane-enclosed droplets. The concentration of these components varies as the sclerotium develops (Colotelo, 1973). There is still room for debate on the exudate's precise purpose. It might contribute to the release of water from the sclerotium during maturation or function as an osmotic agent during translocation (Willetts, 1978).

Even though a wide variety of enzymes have been found in sclerotia, it is still unclear how these enzymes work together and whether any specific enzymes are necessary for morphogenesis. Wong and Willetts (1974) postulated that during the compaction and growth of sclerotia, the pentose phosphate pathway is increased and glycolysis and the Krebs cycle are repressed, based on experiments using polyacrylamide gel electrophoresis. Moreover, enzymes linked to phenol oxidation have been discovered (Jones, 1970; Colotelo, 1973; Wong and Willetts, 1974). Other phenoloxidases may contribute to the development of the darker rind, and tyrosinase may be involved in the onset of sclerotia (Wong and Willetts, 1974).

4.9 Variables Influencing the Development of Sclerotia

Sclerotia can form *in vitro*, *in vivo*, or a combination of both depending on a number of parameters, including physical, chemical, soil, nutrition, host contact, and particular characteristics.

4.9.1 Influence of temperature

There are three different temperatures at which sclerotia can form: minimal, optimal, and maximum. Different temperatures are ideal for the growth of sclerotia and mycelium. It has not been demonstrated that abrupt temperature changes induce the production of sclerotium initials in *S. sclerotiorum* or *S. minor* (Willetts and Wong, 1980). In culture, *S. sclerotiorum* can produce sclerotia at temperatures between 0 and 30°C (Bedi, 1962; Marukawa *et al.*, 1975; Le Tourneau, 1979). Nevertheless, the fungus grows slowly at temperatures below 5°C, without producing sclerotia (Bedi, 1962). Sclerotial production occurs at 15°C, where their formation rate is at its peak (Bedi, 1962). Research studies by Bedi (1962), Marukawa *et al.* (1975), and Le Tourneau (1979) have demonstrated that smaller sclerotia occur at lower temperatures. On the other hand, Bedi (1962) found that the tiniest sclerotia occur at higher temperatures. According to Kruger (1975a), field observations have also shown that the temperature at which more sclerotia are generated is roughly 17°C, as opposed to temperatures below 10°C. It is noteworthy that Bedi (1962) and Willetts and Wong (1980) found that the ideal temperature for mycelium growth is between 20 and 25°C, whereas the optimal temperature for sclerotium production is approximately 15°C.

This implies that when the environment is not conducive to mycelial growth, the pathogen responds by creating resistant propagules.

4.9.2 Consequence of light exposure

There is disagreement among researchers over how light affects the formation of sclerotium in *Sclerotinia* species (Le Tourneau, 1979). Generally speaking, sclerotia created in darkness are larger than those made in light (Trevethick and Cooke, 1973; Le Tourneau, 1979), although more sclerotia are produced in light than in darkness. When the mycelium is exposed to light prior to the formation of sclerotium initials, the number of sclerotia increases most effectively (Humpherson-Jones and Cooke, 1977a; Le Tourneau, 1979). An increase in light levels results in an increase in sclerotia (Trevethick and Cooke, 1973; Humpherson-Jones and Cooke, 1977a). However, this rise in quantity is accompanied by a decrease in the size of individual sclerotia, resulting in a consistent overall supply of sclerotium material (Humpherson-Jones and Cooke, 1977a). Beyond light levels of 30 lumens/ft^2, there is no further increase in the number of sclerotia (Trevethick and Cooke, 1973). Increased amounts of sclerotial primordia and mature sclerotia are observed up to around 500 lux; however, the dry weight of mature sclerotia is decreased. Short exposure periods (48 hour) to light levels up to 5000 lux promote the development of both primordia and mature sclerotia (Kim, 1976).

Light exposure appears to have a direct effect on the establishment of sclerotium initials on undifferentiated hyphae (Humpherson-Jones and Cooke, 1977a). Notably, sclerotium initials do not appear to be formed in response to sudden variations in light intensity (Willetts and Wong, 1980).

However, the length of the photoperiod can affect the number and size of sclerotia that form in a particular setting (Trevethick and Cooke, 1971). It's vital to remember that mycelia grow most robustly when exposed to wavelengths in the blue to ultraviolet portion of the light spectrum. On the other hand, no stimulating effects are shown by red, infrared, or green light (Humpherson-Jones and Cooke, 1977a; Le Tourneau, 1979). Surprisingly, red and blue light have been shown to produce a small number of large sclerotia in culture, despite UV irradiation having no effect on sclerotia development (Nagy and Fischl, 2002).

4.9.3 Impact of nutrition

The initiation and growth of sclerotia are influenced differently by various nutrients. Abrupt changes in nutrient levels do not seem to trigger the onset of sclerotium growth, in contrast to temperature and light (Willetts and Wong, 1980). Nonetheless, sclerotium growth and ripening are significantly influenced by the precise kind and quantity of nutrients supplied (Trevethick and Cooke, 1973; Grogan, 1979). Zinc is required for most fungi to grow and to produce sclerotium, according to Trevethick and Cooke (1971), Le Tourneau (1979), and Vega and Le Tourneau (1974). Additionally, it appears that potassium, phosphorus, magnesium, and sulfur are necessary nutrients (Purdy and Grogan, 1954; Le Tourneau, 1979). According to Le Tourneau (1979) and Wang and Le Tourneau (1971), the *Sclerotinia* spp. use a variety of chemicals to promote the growth of their mycelial cells and the creation of sclerotium. Both organic and inorganic molecules can provide nitrogen, and research by Le Tourneau (1979) and Marukawa *et al.* (1975) suggests that the type of nitrogen and the carbon/nitrogen ratio may have an effect on the production of sclerotium. Sclerotia in *S. sclerotiorum* can only grow and develop when inorganic macronutrients like P, K, Mg, and S are present in the media. Moreover, adding a synthetic micronutrient combination strengthens this mechanism (Purdy and Grogan, 1954). The absence of zinc causes a drop in dry mass yield and prevents the development of sclerotia when the mold is grown in a fluid solution containing refined materials in synthetic containers (Vega and Le Tourneau, 1974). The amount of sclerotia produced indicates how indirectly amino acids containing sulfur, metabolic suppressants, and chelating agents affect sclerotium development by altering unrelated metabolic pathways (Trevethick and Cooke, 1971). Generally speaking, more readily available nutrients lead to more sclerotia being produced (Trevethick and Cooke, 1973).

4.9.4 Effects of different osmotic pressures and pH values

On media with an initial pH range of 2.5–9.0, *S. sclerotiorum* has been shown to thrive and form sclerotia (Marukawa *et al.*, 1975). It is vital to understand that the outcomes of these research studies are modified by the type of medium utilized, specifically its buffering capability. Because organic acids are synthesized throughout growth, the medium's pH may fluctuate (Corsini and Le Tourneau, 1973). When salts or sucrose are added to an agar medium, the osmotic potential drops from −1 to −14 bars, which promotes *S. sclerotiorum* mycelial development. Growth slows down below this level, although it is still noticeable around −100 bars. According to Grogan and Abawi, sclerotia formation occurs at −65 bars but not at −73 bars (Grogan and Abawi, 1975).

4.9.5 The impact of specific components on the culture process

The absence of certain chemicals required for the development of sclerotium may cause the fungus *Sclerotinia* spp. to lose its capacity to create sclerotia. Numerous scientists have made an effort to identify these substances. Sclerotia production is measured in their tests by introducing aged culture filtrates into a new substrate. The identification of active ingredients is possible with this technique. Two unidentified acids have been found by Humpherson-Jones and Cooke (1977b) to increase sclerotium formation. Marukawa and Satomura (1977) postulated that melanogenesis and the formation of hyphal aggregates are mediated by sclerin, a known metabolite of *Sclerotinia* spp., in conjunction with other compounds such as phenols (Marukawa *et al.*, 1975).

4.9.6 Suppressor effect on sclerotial production

When inhibitors are present, some strains of *Sclerotinia* spp. may be unable to form sclerotia. Numerous substances have been found, and each has a specific concentration that must be met in order to prevent the formation of sclerotia. These substances include fluorophenylalanine at 0.1 mM (Dimopoulou, 1975), β-aminobenzoic acid at 6 mM (Marukawa *et al.*, 1975), phenylthiourea at 1 mM (Le Tourneau, 1976), and Al3 +at 0.6 mM (Orellana *et al.*, 1975). It is noteworthy that polyphenyl oxidase is known to be inhibited by β-aminobenzoic acid and phenylthiourea. To find more powerful inhibitors and comprehend the processes by which these inhibitors function, more investigation is required.

4.9.7 Effects of microorganisms, soil type, and residual matter from host

Many factors influence the production of sclerotia in *S. sclerotiorum*, both in the host plant and in the soil following the plowing of harvest residues. Soil sclerotia density increases when harvest remains are present because of their stimulating properties (Kruger, 1975b). Moreover, the nature of the soil influences this process as well (Kruger, 1975b). More sclerotia are developed in relatively damp to wet soils (Kruger, 1975a), perhaps as a result of the presence of water-soluble nutrients. Although *S. sclerotiorum*'s reaction to these bacteria is mostly dependent on soil temperature, the presence of other soil microorganisms also affects the production of sclerotium (Bedi, 1958).

4.9.8 The *SsAtg1* gene activates autophagy in *Sclerotinia* for sclerotia formation and pathogenicity

Autophagy, a cellular reaction mechanism for intracellular material turnover in eukaryotes, is widely present and evolutionarily conserved (Liu *et al.*, 2007; Parzych and Klionsky, 2014). When cells face external environmental stress like nutrient deficiency and hypoxia during growth and differentiation, they activate autophagy to break down organelles and generate essential energy for survival (Mizushima and Komatsu, 2011; Ravanan *et al.*, 2017; Zhu *et al.*, 2019). Autophagy in filamentous fungi is involved in maintaining nutrient balance, regulating secondary metabolism, influencing cell differentiation, and impacting pathogenicity (Liu *et al.*, 2016). Among yeast autophagy-related proteins, the serine/threonine protein kinase Atg1 stands out as the sole protein kinase. During the beginning phase of autophagy, the ATG1-ATG13-ATG17 complex, which includes Atg13 as the controlling subunit and Atg17 as the framework protein, is called upon to gather at the PAS (pre-autophagosomal structure) close to vacuoles in order to organize the necessary proteins (Mizushima *et al.*, 1998; Ichimura *et al.*, 2000). Although there has been thorough research conducted on the impact of MAPK and cAMP/PKA pathways on sclerotia formation and pathogenicity in *S. sclerotiorum* by various workers (Xu *et al.*, 2018; Cong *et al.*, 2022; Yu *et al.*, 2022), the role of the autophagy-related protein Atg1 in this fungus has yet to be investigated. RNA sequencing was utilized to investigate the transcriptomic dynamics of sclerotia development in order to address the existing gap. The analysis of differentially expressed genes (DEGs) at the sclerotial stage aimed to uncover the molecular mechanism underlying sclerotia development, including processes such as ribosomal biogenesis, melanin biosynthesis, cell wall/membrane/envelope biogenesis, autophagy and ubiquitination, reactivation of oxygen metabolism, and signal transduction. A homolog of serine/threonine kinase in *S. sclerotiorum*, known as SsAtg1, has been identified and investigated (Jiao *et al.*, 2022a). The impairment of autophagosome induction was observed upon the deletion of SsAtg1, highlighting its crucial role in sclerotia formation and the full virulence of *S. sclerotiorum*. These findings provide evidence for the essential involvement of the autophagy pathway in the asexual reproduction and pathogenicity of this necrotrophic fungus.

The *S. sclerotiorum* is a phytopathogenic fungus that follows a hemibiotrophic lifestyle and produces sclerotia, which are crucial for its survival and disease cycle. In a study by Jiao *et al.* (2022a), the comparative transcriptomics of hyphae and sclerotia were analyzed. The researchers discovered 1959 genes that were expressed differently, including 919 genes that were downregulated and 1040 genes that were upregulated. This transcriptomic information provides valuable insights into the development of sclerotia. Additional examination of the genes that are expressed differently in sclerotia uncovered crucial molecular processes that play a part in sclerotia development, such as ribosome formation and translation, melanin production, autophagy, and the restoration of oxygen metabolism. Notably, the gene *SsAtg1*, which is related to autophagy, was discovered to be overexpressed in sclerotia. Atg1 homologs have important functions in autophagy, a cellular process that is widely preserved and accountable for the breakdown of intracellular substances in eukaryotes.

In order to explore the role of the *SsAtg1* gene in the autophagy pathway of *S. sclerotiorum*, scientists investigated its impact on autophagosome accumulation in nitrogen-starved cells. They noted that the absence of SsAtg1 impeded autophagosome accumulation in the vacuoles. Moreover, the ΔSsAtg1 mutant strain exhibited stunted vegetative growth and failed to produce sclerotia in conditions with limited nutrients. The formation and penetration of compound appressoria were also abnormal in the ΔSsAtg1 mutant. Further analysis on pathogenicity demonstrated that SsAtg1 is crucial for the complete virulence of *S. sclerotiorum*. These results underscore the significant role of *SsAtg1* as a fundamental autophagy-related gene involved in nutrient utilization, sclerotia development, and pathogenicity in *S. sclerotiorum* (Figs 4.7, 4.8 and 4.9; Jiao *et al.*, 2022a).

Autophagy, a widely preserved mechanism, is essential for breaking down accumulated and improperly folded proteins, along with impaired organelles, in different types of eukaryotic organisms (Mizushima *et al.*, 1998; Abeliovich and Klionsky, 2001; Ravanan *et al.*, 2017; Pérez-Pérez *et al.*, 2021). By utilizing the degraded materials, autophagy contributes to energy production and the reconstruction of cellular structures, thereby maintaining cellular homeostasis (D'Arcy, 2019). Autophagy in filamentous fungi is strongly linked to pathogenicity, cellular differentiation, nutrient

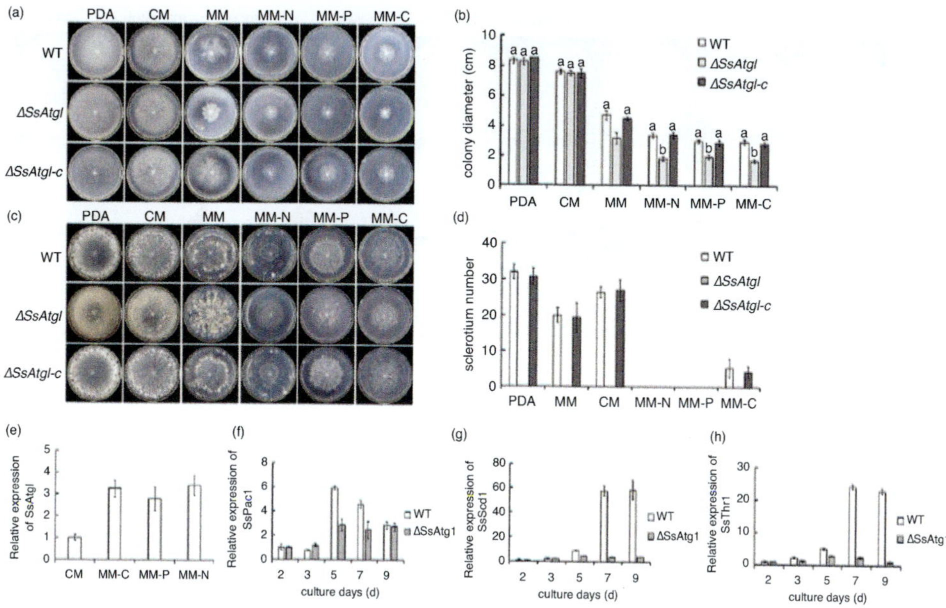

Fig. 4.7. The presence of *SsAtg1* is necessary for the formation of sclerotium. *sAtg1* is required for sclerotium formation. (a,b) Colony morphology and colony diameter of *ΔSsAtg1* and control strains. Strains were inoculated on PDA, CM, MM, MM-N, MM-C, and MM-P and colony diameters were measured by the cross method at 48 hour. Error bars represent the standard deviations, and the letters indicated significant differences ($p<0.05$). (c,d) *SsAtg1* mutation affected sclerotia development. Colony morphology and sclerotia number were recorded after 4 weeks. Error bars represent mean standard deviation, and statistical significance of data was analyzed using SPSS software ($p<0.05$). (c) The expression of *SsAtg1* was significantly increased in nutrient deficient medium. (f–h) Deficiency of *SsAtg1* affects the expression of key genes regulating sclerotia development. (Adapted from publication of Jiao *et al.*, 2022b, with permission.)

equilibrium, and secondary metabolic processes (Veneault-Fourrey *et al.*, 2006; Asakura *et al.*, 2009; Huang *et al.*, 2022; Yu *et al.*, 2022). The studies have demonstrated the impact of SsAtg8 on sclerotia formation, compound appressoria development, and pathogenicity in *S. sclerotiorum* (Zhang *et al.*, 2021; Jiao *et al.*, 2022a). The yeast ATG autophagy pathway relies on the autophagy-related protein Atg1 as the primary protein kinase.

Deletion of SsAtg1 significantly impacts the vegetative growth of *S. sclerotiorum*, resulting in a notable decrease in colony growth on minimal media (MM) and nutrient-poor media (MM-N). However, there is no observable difference in growth rate between ΔSsAtg1 and the wild type when provided with nutritionally adequate PDA and complete medium (CM) media. These results indicate that SsAtg1 may affect the nutrient utilization of *S. sclerotiorum*, as a lack of SsAtg1 impedes hyphal growth under nutrient-deficient conditions. Sclerotia formation in *S. sclerotiorum* is dependent on the autophagy pathway triggered by SsAtg1, as demonstrated by the inability of ΔSsAtg1 to form sclerotia even after prolonged cultivation on various media. This phenomenon is consistent with the effects observed in other filamentous fungi, such as *M. oryzae*, *F. graminearum*, *B. cinerea*, and *Aspergillus fumigatus*, where loss of Atg1 function also hinders the formation of asexual reproductive structures (Lv *et al.*, 2017; Ren *et al.*, 2017; Yin *et al.*, 2020). Furthermore, it has been observed that the wild type is unable to form sclerotia on N-starvation medium (MM-N), further emphasizing the importance of autophagy induction and termination in sclerotia development. Furthermore, by altering carbon and nitrogen sources in MM medium, researchers found that SsAtg1 is crucial for sclerotia formation under nutritional stress conditions. As a result, a deeper exploration of the connection

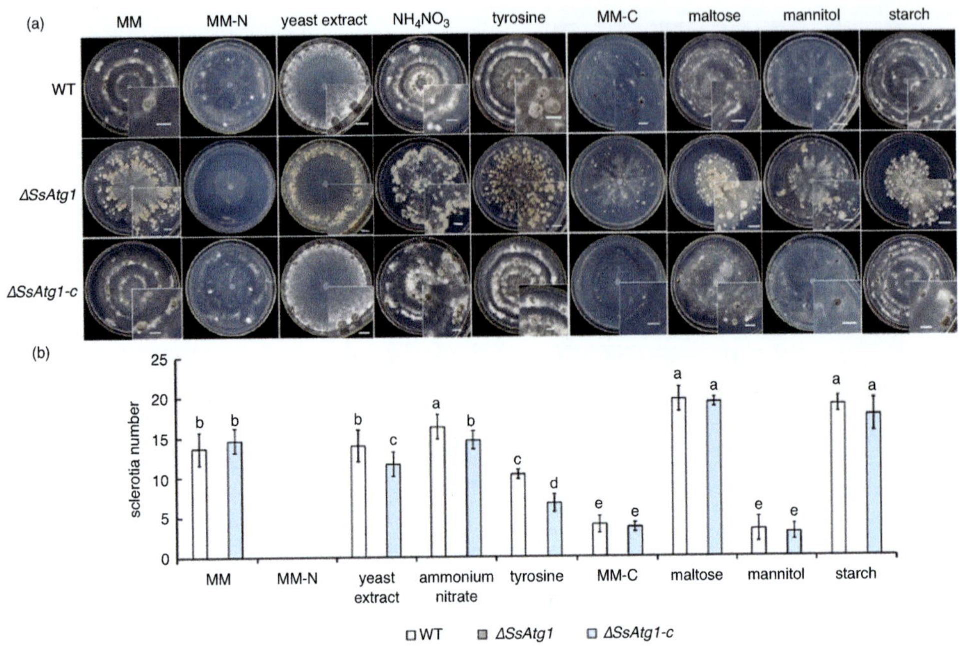

Fig. 4.8. Function of SsAtg1 in reaction to nutrient deprivation. (a) Comparison of colony appearance between wild type, ΔSsAtg1, and the strain with genetic complementation on various growth substrates. The strains were cultured on MM with specified carbon and nitrogen sources. Scale bar = 0.5 cm. (b) Quantification of sclerotia formation by ΔSsAtg1 on different growth media. Colony morphology and sclerotia count were assessed after 4 weeks. Statistical significance is denoted by letters ($P<0.05$), with error bars indicating standard deviation. (Adapted from publication of Jiao *et al.*, 2022b, with permission.)

between autophagy and carbon metabolism in *S. sclerotiorum* is necessary.

OA secretion by *S. sclerotiorum* triggers programmed cell death (PCD) in the host, leading to the production of and subsequent necrosis caused by *S. sclerotiorum* (Govrin and Levine, 2000; Kim *et al.*, 2008). However, disruption of SsAtg1 results in increased sensitivity to ROS, establishing a connection between autophagy and redox processes. Additionally, the redox environment also has a significant impact on fungal growth and development (Viefhues *et al.*, 2014; Rana, *et al.*, 2021). The ΔSsAtg1 strain shows increased sensitivity to 10 mM hydrogen peroxide compared to the wild type, and the expression levels of *Ssnox1*, *Ssnox2*, and *SsCat1* genes involved in reactive oxygen metabolism are reduced during sclerotia formation. These results indicate that *SsAtg1* affects the expression of genes related to ROS metabolism, and ROS is essential for sclerotia formation in *S. sclerotiorum*. The *SsAtg1* gene of *S. sclerotiorum* is also involved in responding to nutritional stresses that control mycelial growth and is crucial for sclerotia formation, contributing to virulence and the development of compound appressoria (Jiao *et al.*, 2022b).

Compound appressoria and OA play crucial roles as pathogenic factors in *S. sclerotiorum* (Liang and Rollins, 2018). The analysis of pathogenicity revealed a significant decrease in virulence in ΔSsAtg1 compared to the wild-type strain. Upon further analysis of the development and penetration of compound appressoria in ΔSsAtg1, it was noted that ΔSsAtg1 formed fewer compound appressoria compared to the wild type. Additionally, experiments conducted on onion epidermis revealed a notable decrease in the penetration efficiency of ΔSsAtg1. Moreover, the pH-indicator dye bromophenol blue was employed to identify the *in vitro* accumulation of OA, indicating that the mutant strain displayed similar accumulation levels as the wild type. These results indicate that the influence of SsAtg1 on the pathogenicity of *S. sclerotiorum* is complex, involving compound appressoria, nutrient utilization, and not solely OA accumulation.

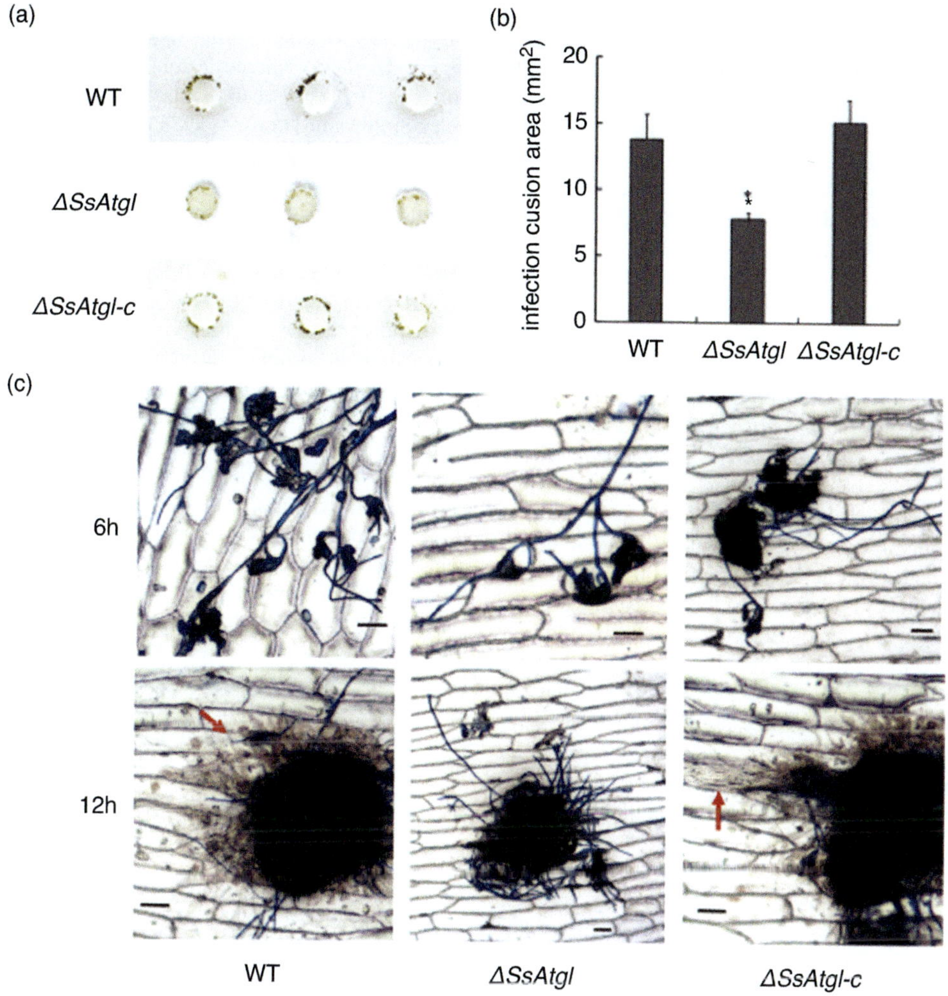

Fig. 4.9. The SsAtg1 gene mutation impacted the development of compound appressoria and their ability to penetrate. (a, b) Compound appressoria of various strains were observed on a hydrophobic surface. ImageJ software was utilized to analyze the compound appressoria by converting the image to 8-bit format, delineating the compound appressoria using adjust-threshold, and using the agar disc area as a reference to measure the compound appressoria size. The error bars in the graph represent the standard deviations, with significant differences denoted by an asterisk (*) ($P<0.05$). (c) Formation of compound appressoria on onion epidermal cells was examined. The red arrow highlights the invasion hypha, which was not visible after staining with lactophenol cotton blue solution. Samples were stained with lactophenol cotton blue solution for 30 s, rinsed with ddH_2O, and observed under a microscope. Scale bar = 50 µm. (Adapted from publication of Jiao *et al.*, 2022b, with permission.)

4.10 Survival Time of Sclerotium

Sclerotia are the primary structures required for survival in *Sclerotinia* spp. Several studies (Coley-Smith and Cooke, 1971; Cook *et al.*, 1975; Schwartz and Steadman, 1978; Adams and Ayers, 1979; Willetts and Wong, 1980) have demonstrated that these structures have the capacity to keep the fungus in a latent state and start germination when the proper circumstances are met. Since sclerotia populations in the soil continue to be rather stable, it is believed that only a small percentage of the sclerotia that have been observed to emerge during epidemics actually endure (Abawi and Grogan, 1979; Grogan, 1979). However, when non-host crops are cultivated, soil populations do not decrease for several years, suggesting that some sclerotia have a long lifespan (Dueck and Morrall, 1971; Schwartz and Steadman, 1978; Morrall and Dueck, 1982).

S. sclerotiorum sclerotia were either shallowly buried (buried in the soil at a depth of 3 cm) or deployed in mesh bags on the soil's surface in a field trial that was carried out in Geneva, New York. Bags were collected on a regular basis between 67 and 769 days. Sclerotia retrieval, or the quantity of sclerotia, was measured at each time by counting, and myceliogenic germination was used to determine viability. Soil depth had a major impact on sclerotial retrieval, which was greater in surface-dwelling specimens than in buried ones. The retrieval of sclerotia was also impacted by time; it was much less after 250 days. On the survivability of sclerotia, burial and time interacted significantly. After 769 days, about 15% of the sclerotia that were placed on the surface were still alive. The vitality of buried sclerotia was likewise much lower after 433 days than that of surface-dwelling sclerotia. None of the buried sclerotia remained viable after 670 days. To decrease primary inoculum, crops that are vulnerable to infection must be rotated for a minimum of 2 years. A longer cycle, however, would be advantageous because low inoculum densities are sufficient to start a white mold outbreak. It is recommended to promote degradation in farmed systems by timely tillage of agricultural residue to bury sclerotia after harvest (Pethybridge *et al.*, 2023).

4.10.1 Sclerotial longevity in the field

It can be difficult to ascertain the lifespan of sclerotia in the field, particularly in the absence of susceptible hosts. This is because, in the lack of appropriate substrates for colonization, 'primary' sclerotia give rise to 'secondary' or 'daughter' sclerotia. This lengthens the lifespan of sclerotia in addition to increasing their quantity. Sclerotia survival rates in the field vary greatly; estimates range from a few weeks to over 10 years. Different estimates have been published by a number of studies, including Coley-Smith and Cooke (1971), Cook *et al.* (1975), Adams and Ayers (1979), and Grogan (1979). Six to 8 years can pass before *S. trifoliorum* sclerotia is detected, according to Dillon Weston *et al.* (1946). *S. trifoliorum* sclerotia can live for around 8 years if buried deeper than 6.4 cm, according to Tribe (1957). After a year, *S. sclerotiorum* sclerotia near the soil's surface go dormant, according to Davis (1925). On the other hand, Young and Morris (1927) recommend waiting at least 4 years before planting sunflowers in a field where *Sclerotinia* wilt has previously occurred. A multitude of factors influence the survival of sclerotia in *Sclerotinia* spp. under outdoor circumstances. Starr *et al.* (1953) suggest rotating non-host crops every 3–5 years as a means of controlling bean white mold. However, Cook (1973) discovered that alternating beans every third year with maize and sugar beets is ineffective in Nebraska. According to study by Brown and Butler (1936), *S. sclerotiorum* sclerotia can survive for at least 10 years in optimum dry conditions. On the other hand, *S. sclerotiorum* can survive in the soil for up to 10 years, according to Quentin (2004), especially in areas with warmer temperatures and irrigation. This is in opposition to the estimated 4–5 years that sclerotia in field soil are expected to survive. Variations in environmental circumstances across places are more likely to be the cause of survival variability than observational mistakes. Sclerotium survival has been demonstrated to be influenced by environmental conditions, such as those examined by a number of researchers (Coley-Smith and Cooke, 1971; Orellana, 1975; Morrall *et al.*, 1976; Adams and Ayers, 1979; Willetts and Wong, 1980). It is important to note that environmental elements often interact, meaning that changes in one aspect can affect other factors as well as the frequency and intensity of drying (Willetts and Wong, 1980). Sclerotia leak nutrients when they are wet and dry, which reduces their nutritional reserves and encourages the growth of other soil organisms, both of which are detrimental to their survival (Coley-Smith and Cooke, 1971). The following variables affect *Sclerotinia* sclerotia's ability to survive in the field.

4.10.2 The impact of temperature, nutritional levels, composition, moisture content, acidity, and depth of sclerotial implantation in the soil

According to Coley-Smith and Cooke (1971) and Imolehin, and Grogan, (1980), wet soils are the least conducive to the survival of sclerotia, whereas dry soils offer the ideal circumstances for them. Research has demonstrated that bean white mold can be caused by sclerotia from *S. sclerotiorum*, which can contaminate surface irrigation water as it runs off from agricultural areas (Steadman *et al.*, 1975). Sclerotial bodies can survive in running water for 10–21 days, which can cause the white mold illness to proliferate among beans. Their incapacity to control water loss is the cause of this. It is not the physical characteristics of the sclerotium's exterior tissues that allow individual sclerotial cells to survive during dry times. Rather, it is their resistance to desiccation. Although this is usually not a big worry in most field conditions, it is important to note that rising soil moisture can accelerate the destruction of sclerotia and make them more prone to damage from sun radiation. Although prolonged exposure to soil temperatures above 35°C can reduce sclerotium survival, this is not a serious problem because natural field circumstances in temperate countries rarely involve such high soil temperatures (Adams, 1975; Adams and Ayers, 1979). Nonetheless, the combination of high soil moisture and temperature significantly amplifies the degree of sclerotium destruction (Cook *et al.*, 1975; Willetts and Wong, 1980). Moore (1949) found that nearly all sclerotia are killed in soil that is buried in water for 16–31 days. Flooding has been used in some places to remove sclerotia from fields since the higher moisture levels encourage the decomposition of sclerotium (Rotem and Palti, 1969; Coley-Smith and Cooke, 1971; Adams and Ayers, 1979; Steadman, 1979). However, since so much water is needed, this strategy is not as beneficial in areas that are not irrigated (Steadman, 1979). It has been demonstrated that burying sclerotia for 35 weeks at a depth of 4 cm entirely prevents sclerotia recovery in sandy clay loam soil and reduces recovery by 50% in sandy clay soil. In sandy clay loam soil, recovery is reduced by 50% and in sandy loam soil by 10% when sclerotia are present at the soil surface. It has been discovered that fewer than half of the recovered sclerotia are viable (Merriman, 1976). Nonetheless, Adams (1975) found *that S. sclerotiorum* sclerotia buried in soil can endure for almost 15 months at a depth of 1–12 inches, but that at 25 inches, their chances of survival drastically decrease. Cook *et al.* (1975) found that around 75% of the recovered sclerotia were able to germinate and develop apothecia in culture after 3 years of burial at depths of 5, 12, 15, and 20 cm below the soil surface. Under Indian conditions, *S. sclerotiorum* sclerotia can survive for more than 18 months at a soil depth of 20 cm; however, none of those at the surface were able to germinate after 6 months. Apothecia occur between November and March and produce ascospores, which are the main source of infection for cauliflower (Sharma and Sharma, 1986a, b). In comparison to deeper depths, *S. minor* sclerotia had higher survival rates at shallower soil depths. According to Kakoti and Saikia (1997), sclerotia can withstand 7 months at the soil's surface, 8 months at a depth of 5 cm, and 7 and 6 months when buried at depths of 10 and 15 cm, respectively. But within 5 months, sclerotia buried 20 cm below the surface lose vitality. Microorganism colonization is influenced by the depth at which sclerotia are buried in the soil (Adams, 1975; Willetts and Wong, 1980). Sclerotia are able to survive arid circumstances for over 10 years.

When discovered in field soil at depths of 5 mm and 10–20 mm, respectively, sclerotia in Japan have a 4 year lifespan (Akai, 1981). Furthermore, fields previously affected by lettuce decline caused by *S. minor* and *S. sclerotiorum* experience a decrease in the population of viable sclerotia when they are flooded for 2–3 weeks (Matheron and Porchas, 2005). The presence of sclerotium is influenced by soil nitrogen levels (Coley-Smith and Cooke, 1971). Because high-nutrient soils support a broad variety of microorganisms that break down organic matter, sclerotia in these soils are susceptible to microbial deterioration. However, it seems that the pH of the soil has no bearing on sclerotium survival (Coley-Smith and Cooke, 1971; Adams and Ayers, 1979).

4.10.3 Influence of alternative soil microorganisms

According to Adams and Ayers (1979), the main factor affecting sclerotium survival is assumed to be the biological component of soil. The numerous bacteria that function as antagonists or mycoparasites of *Sclerotinia* spp. are included in Table 8.1 of another article, *Sclerotinia Stem Rot of Crucifers: Epidemiology and Disease Management* Vol–II, which is located in Chapter 8 on *Biological Control. Sporidesmium sclerotivorum*, *Coniothyrium minitans*, and *Trichoderma* spp. are among the microorganisms that are thought to be the most significant in their natural environment (Tribe, 1957; Coley-Smith and Cooke, 1971; Dueck, 1977; Huang, 1977, 1979; Adams and Ayers, 1979; Willetts and Wong, 1980).

Variations in the population of soil microorganisms in different places can help to explain some of the known variability in sclerotia lifespan (Huang, 1977; Willetts and Wong, 1980). A number of variables, including temperature, nitrogen levels, pH, and soil moisture, can affect the makeup and quantity of soil microorganisms in a particular location, which can have an indirect effect on sclerotia survival (Steadman, 1979). Additionally, these elements affect how actively soil microorganisms break down organic matter in the soil. Furthermore, the germination of latent sclerotia can be induced by the presence of soil nutrients, making them more susceptible to microbial attacks (Coley-Smith and Cooke, 1971). Furthermore, even while the sclerotial outer layer can regenerate, damage to this layer increases the vulnerability of sclerotia to microbial attacks (Coley-Smith and Cooke, 1971).

4.10.4 The effect of feeding animals

For 10 days, the hybrid heifer was fed peanut hay tainted with *S. minor* and sclerotia. There were live sclerotia in the ruminal and fecal samples. Remarkably, ruminal samples had a greater sclerotia survival rate than fecal samples did. Melouk *et al.* (1989) found that *S. minor* cultures from ruminal and fecal sclerotia were pathogenic to peanuts or Tamnutzu in a greenhouse environment.

4.10.5 Effect on host tissues

The viability of sclerotia generated inside or outside of bean stems, with or without bean straw protection, was examined in a study. According to research, sclerotia that were protected by bean straw or that had grown inside of stems had the highest survival rate (11%) (Merriman *et al.*, 1979).

4.10.6 Impact of soil atmospheric conditions

It seems doubtful that soil aeration is the cause of the observed effect on sclerotium survival. It is unlikely that levels of carbon dioxide buildup or oxygen depletion, which could be fatal to dormant sclerotia, would occur for an extended period of time (Coley-Smith and Cooke, 1971).

4.10.7 Effect of mode of germination

It appears that there is no discernible relationship between the myceliogenically or carpogenically germination method of sclerotia and the survival rate (Coley-Smith and Cooke, 1971; Cook *et al.*, 1975). However, the longevity of sclerotia that germinate myceliogenically is shortened, most likely due to the depletion of nutritional stores (Saito, 1977).

4.11 Spread of Sclerotium

Sclerotia spreads by means of multiple agencies from the sources of survival to different areas, where it serves as the main source of inoculum to induce infections. Through a variety of techniques, *Sclerotinia* spp. establishes itself and expands throughout many fields and geographical locations. It spreads from one field to another by adhering to persons, animals, farm equipment, or plants. Within the diseased host tissue, sclerotia or mycelium may form (Dillon Weston *et al.*, 1946; Starr *et al.*, 1953). Farms that use diseased plant tissue for cow bedding or feed run the risk of transmitting diseases to otherwise pristine areas through the spread of manure (Dillon Weston *et al.*, 1946). According to research, fewer than 2% of *S. sclerotiorum* sclerotia survived after being eaten by sheep and passing through their digestive systems (Brown, 1937). Sheep and other animals may transfer the infection to areas free of *Sclerotinia* if they consume contaminated plant material and are then let back onto pasture. Additionally, studies have indicated that irrigation can aid in the dispersal of *Sclerotinia* species from one field to another. Steadman *et al.* (1975) discovered *S. sclerotiorum* sclerotia in streams and irrigation runoff from fields in Nebraska. These sclerotia can live for 10–21 days or longer in flowing water.

4.12 Sclerotia as Naturally Occurring Soil Inoculum Concentrations

The knowledge base on sclerotia's existence in native soil populations is currently small. In the field, sclerotia of *S. sclerotiorum* were discovered at depths of up to 5.1 cm, in amounts ranging from 0 to 20 per 929 cm^2 (1 ft^2) of soil (Henderson, 1962). This corresponds to an average of between 0 and 3 sclerotia kg^{-1} of soil. According to research by Hoes and Huang (1975), non-rhizosphere soil has between 2 and 3 sclerotia kg^{-1}. The amount of sclerotia in the rhizosphere of sick plants increased dramatically to about 211 kg^{-1} of soil. An investigation was carried out in 1975 by Abawi and Grogan to ascertain the *S. sclerotiorum* inoculum density in a New York bean field. Their results showed that, previous to plowing, the sclerotia count per kilogram of soil was around 7.2, 0.5, and nearly zero at depths of 0–2.5 cm, 2.5–10 cm, and 10–17.5 cm, respectively. At all depths, the quantity of sclerotia remained minimal following planting. The number of *S. sclerotiorum* inoculum varied from 0.1 to 6.2 sclerotia kg^{-1} of soil in bean fields in western Nebraska (Schwartz and Steadman, 1978). According to what little data there is, sclerotia levels in natural soils range from zero to less than 10 kg^{-1} in a field that is ready for planting. Adams and Ayers (1979) found that in fields with a history of severe lettuce decline in three New York States, the inoculum density of *S. minor* ranges from 160 to 820 sclerotia kg^{-1} of soil. In 11 mineral oil fields in New Jersey, the inoculum density of *S. minor* ranges from 0 to 230 sclerotia kg^{-1} of soil. The inoculum density in a Virginian field that had previously suffered from severe *Sclerotinia* peanut blight ranged from 35 to 100 sclerotia kg^{-1} of soil. Thus, in areas where *Sclerotinia* spp. have been blamed for losses, the inoculum density of *S. minor* has been demonstrated to be 10–100 times higher than that of *S. sclerotiorum*.

4.12.1 Elevation of soil sclerotial inoculum densities

The sclerotial inoculum density of *Sclerotinia* species can be increased in two ways:

1. by means of the production of secondary sclerotia in the soil and
2. by means of the development of sclerotia on their respective hosts.

The two techniques accessible to increase the sclerotial inoculum density of *Sclerotinia* species are the generation of secondary sclerotia in the soil and the creation of sclerotia on their respective hosts. The sclerotia of *S. trifoliorum* and *S. sclerotiorum* can produce secondary sclerotia in the soil in the absence of a host (Williams and Western 1965a, b). Moreover, *S. minor* has the ability to form secondary sclerotia, which multiply the organism. In their investigation into the host's function in sclerotia production, Stevens and Hall (1911) discovered that a diseased plant bed containing 2000 lettuce plants and measuring 2.7 × 61 m (9 × 200 ft) could yield up to 17,000 S. *minor* sclerotia, which would eventually sink into the soil. If these sclerotia were well mixed into the soil down to a depth of 15 cm, the inoculum density would rise by about 0.5 sclerotia kg^{-1} of soil. One ill romaine lettuce plant may produce up to 1000 *S. minor* sclerotia, according to Adams (1975). If every plant in the field filled an area of 929 cm^2 (1 ft^2) and produced 1000 sclerotia, the inoculum density of the field increased by 50 sclerotia kg^{-1} of soil when the diseased crop is absorbed into the soil to a depth of 15 cm. The importance of highlighting the function that various weed hosts play in raising the inoculum densities of *Sclerotinia* spp. sclerotia in field settings cannot be overstated.

4.13 Determinants of Sclerotial Viability

The viability of *S. sclerotiorum* sclerotia is greatly influenced by the soil moisture and duration. These sclerotia act as resting structures for the pathogen and aid in its continued existence. An investigation was conducted to examine the soil moisture impact on the viability of the sclerotia. Soil that is conducive to *S. sclerotiorum* was collected from a cabbage-growing field in the Kothagiri area of Nilgiri district, Tamil Nadu. A specific amount of this soil was placed in plastic containers measuring 14 cm in height and 11 cm in diameter. Nylon mesh pouches, each containing five sclerotia, were then placed in the upper 5 cm of soil, which was covered by an additional 5 cm of soil. Four different levels of soil moisture (60, 70, 80, and 90%) were maintained in the containers. The sclerotia were sampled every 15 days for a total of 60 days, and the percentage of germination was recorded. The results revealed that the sclerotia had higher survival rates at lower soil moisture levels, with the maximum germination occurring at a soil moisture level of 60%, which gradually decreased over time. This study demonstrated that higher soil moisture levels hindered the viability of the sclerotia, and the viability also decreased as the duration of the sclerotia in the soil increased (Krishnamoorthy *et al.*, 2017).

4.14 Sclerotium Germination Pathways

According to various workers, sclerotia of *Sclerotinia* spp. can germinate in two different ways. Since myceliogenic germination and infection are rare in natural settings, they are not regarded as important epidemiological factors for a variety of crops (Abawi and Grogan, 1975, 1979; Grogan, 1979; Akai, 1981). But it has a major impact on certain crops (Hoes and Huang, 1976; Huang and Hoes, 1980). Although its infectivity is lower than that of ascospores, which are the primary cause of epidemics (Kruger, 1975a; Dueck, 1977; Williams and Stelfox, 1980b; Scheibert-Bohm *et al.*, 1981), sclerotia mycelium is known to infect rapeseed crops (Dueck, 1977). An individual apothecium has the capacity to generate millions of ascospores (Schwartz and Steadman, 1978; Abawi and Grogan, 1979; Williams and Stelfox, 1980a). On the other hand, less than ten sclerotia are typically found per kilogram of soil in natural soils (Adams and Ayers, 1979). Ascospores have the ability to spread via air currents across great distances, up to several kilometers (Abawi and Grogan, 1979; Williams and Stelfox, 1979), which permits infections to happen distant from the initial inoculum source. However, mycelium derived from sclerotia is unlikely to infect plants more than 2 cm away from the parent sclerotium (Newton and Sequeira, 1972).

4.14.1 Preconditioning of sclerotia for germination

Apothecia production in sclerotia is only possible when they are preconditioned and functionally mature (Abawi and Grogan, 1979; Purdy, 1979). Although it is commonly accepted that freshly formed sclerotia must be exposed to cool, moist conditions for a specified amount of time before they may achieve optimal carpogenic germination, the precise criteria for preconditioning sclerotia are still unknown (Abawi and Grogan, 1979). According to Adams and Tate (1976), the duration of this phase differs among different isolates of *S. minor*, suggesting that this may also apply to distinct isolates of *S. sclerotiorum*. Sclerotia of *S. sclerotiorum* can commence apothecia production in as little as 13–208 days (Letham, 1975), indicating that there is no set amount of time that is necessary. According to field observations, apothecial production is significantly influenced by environmental conditions. In different rapeseed stands during the same year, sclerotia have been seen to develop apothecia at varied rates (Kruger, 1974, 1975a,b). Rather than being dependent on a single cause or physiological differences across isolates, the length of time needed for sclerotia preconditioning is most likely determined by the intricate interactions of multiple environmental variables. According to Willetts and Wong (1980), this mechanism makes sure that apothecia and ascospores are only produced in environments that allow ascospores to germinate and infect vulnerable hosts.

It is important to keep in mind that apothecium development and sclerotia germination are two different processes. While apothecium creation entails the extension of the carpophores' tips to form a disc-shaped ascocarp where asci and ascospores will eventually be generated, carpogenic germination refers to the growth of stipes or carpophores on the sclerotium. Sclerotia germinate by means of two distinct processes: (i) carpogenic germination and (ii) myceliogenic germination.

4.14.2 Carpogenic germination technique

Under the correct circumstances, sclerotia on a substrate devoid of nutrients, such as moistened sand, cotton, polyurethane, water agar, or water, may be able to form one or more stipes with an apothecium. Low osmotic potential inhibits the production of apothecia, as demonstrated by Morrall (1977) and Grogan and Abawi (1975). While Kosasih and Willetts (1975) reported that some researchers had observed carpogenic germination in sclerotia taken directly from host tissue or laboratory media, other researchers (Saito, 1973; Steadman and Nickerson, 1975) had success conditioning the sclerotia in a moistened state for several weeks or longer. Numerous histochemical assays, in addition to optical and electron microscopy, have been used to thoroughly examine the development of apothecia from sclerotia (Saito, 1973; Jones, 1974; Kosasih and Willetts, 1975). Apothecial initials are brown to hyaline-colored hyphae that have entwined into clusters or nests within the medulla or cortex. The energy produce leads to the hyphae proliferation, and a tightly packed knot full of cytoplasm is formed. Some of these primordia eventually pierce the sclerotium's surface. After that, clusters of microconidia become apparent; however, spermatization does not appear to be required for the development of apothecial structures (Kosasih and Willetts, 1975). Even while many primordia may form, only a small percentage of them are able to penetrate the surface, and if every primordium develops, the sclerotia's reserve supplies may run out (Saito, 1973). This implies that the sclerotia have intrinsic mechanisms that regulate the growth of these initial structures. The first formations are pointy and buff in color, and they pierce the surface before growing and maturing. The stipe exhibits positive phototropism; it is only in the presence of light that it changes into a disc. It has been demonstrated that apothecium production can be efficiently triggered by light with a wavelength shorter than 390 nm (Honda and Yunoki, 1977). At the peak of the apothecial initial structures, a depression appears, and paraphyses begin to take shape. The subhymenium produces ascogenous hyphae with crosiers. Eight ascospores are present in the ascus upon differentiation (Kosasih and Willetts, 1975). A medulla consisting of elongated, septate, sparingly branched hyphae and an outer layer of compacted hyphae makes up a fully grown stipe. A fully formed stipe is composed of an outer layer of compacted hyphae and a medulla of elongated, septate, sparingly branched hyphae. When the adult apothecium is fully grown, it has the following components, according to Kosasih and Willetts (1975):

1. an ectal excipulum,
2. a medullary excipulum,
3. the hymenium, and
4. the subhymenium.

The reserve materials go through metabolism during carpogenic germination to create necessary 'building blocks' that are then used to synthesize the stipe and apothecium components. High levels of hydrolytic and synthetic enzymatic activity are seen in the apothecial initials (Kosasih and Willetts, 1975). Furthermore, it has been reported that the β-glucanase enzyme is present during germination (Saito, 1974).

4.14.2.1 Diversity in carpogenic germinability of Sclerotinia populations

The management of the disease heavily relies on cultural and chemical methods of control.

However, accurately predicting the release and presence of ascospores on the plant remains a challenge, making the timing of fungicide application difficult. In order to determine the ideal temperature for the carpogenic germination of *S. sclerotiorum* populations, samples were collected from canola and lupin fields in southwestern Australia. These samples were then characterized for diversity using mycelial compatibility groupings (MCGs). Sclerotia were collected from four diseased canola fields and one diseased lupin field across southwestern Australia. Each population had 40 sclerotia, which were incubated at four different alternating temperatures: 30/15°C, 20/15°C, 20/4°C, and 15/4°C (with a 12 hour light/dark cycle). The sclerotia were assessed every 2–3 days over a period of 180 days. The MCG grouping for the populations was determined using 12 reference isolates. The results showed that the time for initial carpogenic germination decreased as the diurnal temperature fluctuations decreased, with the greatest decrease observed at a fluctuation of 5°C (20/15°C), followed by 16°C (20/4°C). The optimal germination temperature for all five populations was 20/15°C, but the responses to other diurnal temperature regimes varied significantly among the populations. No germination was observed at 30/15°C. The MCG results revealed extensive diversity within and between populations, with at least 40% of sclerotia in each population unable to be characterized. This suggests that the diversity has allowed *S. sclerotiorum* populations to adapt to the varying environmental conditions in southwestern Australia (Michael *et al.*, 2020).

4.14.2.2 Carpogenic germinability of Sclerotinia isolates

The *S. sclerotiorum* primarily exhibits homothallic characteristics, meaning that it possesses both male and female reproductive structures on the same thallus. After self-fertilization, it produces ascospores without generating any asexual spores (Kamal *et al.*, 2016). As a result, the population structure of this pathogen is predominantly clonal, with a few clones occurring at high frequencies and the rest found only once or a few times (Kohli *et al.*, 1995; Clarkson *et al.*, 2013). However, there have been reports of outcrossing and exchange of genotypes in *S. sclerotiorum* populations (Malvárez *et al.*, 2007; Attanayake *et al.*, 2014; Derbyshire *et al.*, 2019), which could potentially enhance the genetic diversity and adaptability of the pathogen. One method of assessing diversity in *S. sclerotiorum* populations is through mycelial compatibility groupings (MCGs), which is a phenotypic marker controlled by multiple loci. It is based on the ability of two isolates to form a single indistinct colony. This characterization method is often used in conjunction with DNA fingerprints or microsatellite alleles (Schafer and Kohn, 2006; Sexton *et al.*, 2006; Clarkson *et al.*, 2013; Aldrich-Wolfe *et al.*, 2015; Liu *et al.*, 2018). The correlation between phenotypic and genetic diversity has been established for various agronomic traits, including aggressiveness, fungicide sensitivity, and sclerotia production (Kohli *et al.*, 1995; Garg *et al.*, 2010b; Attanayake *et al.*, 2013; Denton-Giles *et al.*, 2018; Leyronas *et al.*, 2018). However, some studies have found no correlation between genetic diversity and phenotypic traits such as aggressiveness and crop host type (Durman *et al.*, 2003; Sexton and Howlett, 2004; Clarkson *et al.*, 2013; Aldrich-Wolfe *et al.*, 2015; Leyronas *et al.*, 2018). Therefore, understanding the phenotypic and genetic diversity of important agronomic traits in this pathogen is crucial for the development of effective disease management strategies.

Michael *et al.* (2020) conducted a study to determine the most suitable temperature for carpogenic germination of *S. sclerotiorum* populations collected from canola and lupin fields in southwestern Australia. They also aimed to assess the diversity using mycelial compatibility groupings (MCGs). The researchers collected sclerotia from four diseased canola fields and one diseased lupin field across southwestern Australia. Each population consisted of 40 sclerotia, which were then incubated at four different alternating temperatures: 30/15°C, 20/15°C, 20/4°C, and 15/4°C (12 hour light/dark cycle). The assessment was carried out every 2–3 days over a period of 180 days. The MCG groupings for the populations were determined using 12 reference isolates. The findings revealed that the time taken for initial carpogenic germination decreased as the diurnal temperature fluctuations decreased. The most rapid initial germination was observed at a fluctuation of 5°C (20/15°C), followed by 11°C (15/4°C), and then 16°C (20/4°C). The optimal germination temperature for all five populations was found to be 20/15°C. However, the responses of the populations to other diurnal temperature regimes varied significantly. No germination was observed at 30/15°C. The MCG results indicated a substantial diversity within and between populations, with at least 40% of sclerotia in each population remaining uncharacterized. The researchers suggested that this diversity has allowed *S. sclerotiorum* populations to adapt to the diverse environmental conditions in southwestern Australia (Michael *et al.*, 2020).

4.14.2.3 Scarification and environmental factors on carpogenic germination of Sclerotinia sclerotia

Seven different environmental treatments were used to identify the environmental factors that promote carpogenic germination. These treatments involved applying sterilized distilled water, aerated water, constant rinsing with tap water, cold-conditioning followed by transfer into moist unsterilized compost, incubation in sterilized river sand, air drying followed by transfer into sterilized moist river sand, and placement into 0.5% water agar and incubation. The results showed that scarified sclerotia had significantly higher carpogenic germination compared to un-scarified sclerotia. There was also a significant interaction between scarification and the environmental treatments in relation to carpogenic germination. Incubation in compost or sterilized river sand enhanced the carpogenic germination of scarified sclerotia. Additionally, constant rinsing with tap water resulted in the highest overall carpogenic germination for both scarified and un-scarified sclerotia. The study also monitored the progression of carpogenic germination over two consecutive years and found that the majority of germination occurred between June and September, which coincides with the cropping season for *Sclerotinia* stem rot in rainfed canola in Western Australia. These findings suggest the presence of a seasonal rhythm-like pattern in the carpogenic germination of this pathogen (Garg *et al.*, 2010a).

4.14.2.4 Effect of sclerotial water content on carpogenic germination of Sclerotinia

The study conducted in a controlled environment looked into the dynamics of sclerotial water imbibition and the correlation between moisture content and carpogenic germination of *S. sclerotiorum* sclerotia. Laboratory-produced sclerotia from seven *S. sclerotiorum* isolates were used in the study. The gravimetric determination of water imbibition was conducted on three different sizes of sclerotia in silty clay, sandy loam, and sandy soils, which were maintained at various levels of soil saturation (100, 75, 50, and 25%), as well as in distilled water. The results showed that smaller sclerotia absorbed water at a significantly faster rate compared to larger sclerotia, both in water and in soil at all saturation percentages. When buried in soil, small, medium, and large sclerotia reached full saturation within 5, 15, and 25 hour, respectively, in all three soil types and moisture percentages. The impact of sclerotia moisture content on carpogenic germination was assessed by maintaining sclerotia at different levels of water saturation capacity (95–100%, 70–80%, 40–50%, and 20–30%) using cool mist humidifiers. The study found that sclerotial moisture content had a significant influence on carpogenic germination, with maximum carpogenic germination observed in fully saturated sclerotia and no carpogenic germination observed below 70–80% saturation. These findings provide insights into the ability of *S. sclerotiorum* to produce apothecia in soils with relatively low moisture levels (Nepal and Del Río Mendoza, 2012).

4.14.2.5 Impact of preconditioning period on carpogenic germination of Sclerotinia isolates

Sclerotia conditioning typically occurs during the non-cropping season (Saharan and Mehta, 2008). In southern Australia, *S. sclerotiorum* isolates experience a warm temperate Mediterranean-type climate, characterized by hot fluctuating summer temperatures until the onset of opening rainfall in the winter growing period (Beck, *et al.*, 2018). Preconditioning of pasture and weed seeds adapted to this climate has been extensively documented by Taylor (1981), highlighting the requirement of hot dry temperatures over summer before germination can occur during milder temperatures associated with the onset of winter rains. Research has shown that there is ecotypic variation between genotypes in their temperature requirements for preconditioning, which is influenced by the environment in which the seeds form in spring (Smith *et al.*, 1996; Taylor, 1996). Additionally, there is variation within a population, as noted by Norman *et al.* (2002). While *S. sclerotiorum* primarily exhibits a clonal population structure, there are reports of outcrossing, potentially enhancing the genetic diversity and adaptability of the pathogen to different environments (Derbyshire *et al.*, 2019). Ecotypic diversity of *S. sclerotiorum* populations has been demonstrated by Michael *et al.* (2020), both within a single field and between fields in southwestern Australia, using mycelial compatibility groups (MCGs) as a phenotypic marker. This marker is controlled by multiple loci and is based on the ability of two isolates to form one indistinct colony. Furthermore, Michael *et al.* (2020) found that populations exhibited variations in their response to carpogenic germination under different fluctuating temperatures. Correlations between genetic diversity and phenotypic traits, such as infection aggressiveness, have also been observed within southern Australian isolates (Garg *et al.*, 2010b; Denton-Giles *et al.*, 2018).

Sclerotia, the melanized resting structure of the pathogen, must undergo preconditioning in order for carpogenic germination to take place. In a study conducted by Michael *et al.* (2021), the impact of preconditioning temperature (4, 20, 35, and 50°C and field conditions) and duration (0, 30, 60, 120, 179, 240, and 301 days) on the germination of *S. sclerotiorum* sclerotia collected from five canola fields in the southwestern Australia grain belt was investigated. The ecological diversity of each population was assessed using mycelial compatibility group (MCG) typing. The findings revealed that isolates conditioned at 4°C did not exhibit any response during any of the time periods, indicating that chilling is not a preconditioning requirement for these isolates. It was determined that sclerotia required a minimum of 60 days of preconditioning before any significant increase in germination was observed. No further increases in germination were recorded beyond 60 days of conditioning. The highest germination rates were observed in sclerotia that were conditioned at 50°C. The MCG results demonstrated significant diversity both within and between populations, suggesting local adaptation to different environments and the ability to respond to seasonal variations between years (Michael *et al.*, 2021).

4.14.3 Myceliogenic method of germination

Sclerotia have the ability to initiate growth by generating mycelia (Chet and Henis, 1975). It has been demonstrated that *Sclerotinia* spp. can produce mycelia through either hyphal or mycelial germination (Adams and Tate, 1976). Sclerotia located on a nutrient-deficient substrate usually exhibit limited mycelial growth when hyphal germination takes place, with emerging hyphae penetrating the outer layer. Smaller secondary sclerotia may occasionally form as a result of this mycelial growth (Steadman and Nickerson, 1975). Conversely, if hyphal germination takes place on a media rich in nutrients, the mycelia will proliferate and produce more sclerotia. Strains such as *S. minor* that produce tiny sclerotia have been reported to exhibit mycelial germination (Adams and Tate, 1976). Sclerotia placed on a moist surface will develop one or more projections after a period of dormancy. These protuberances progressively enlarge until they finally rupture, releasing a thick clump of mycelium or a visible plug. There are situations where several masses that come from the same sclerotia merge. The mycelial clusters extend 2–3 mm from the sclerotia.

4.14.3.1 Influences on myceliogenic germination of sclerotia

Myceliogenic development occurs when the temperature is between 20 and 25°C, the relative humidity (RH) is above 80%, and the primary medium pH is 5. Sclerotia exhibit a higher germination rate under 100% RH conditions when deprived of external nutrients, as opposed to 95% or lower, as demonstrated over 95% RH (Huang and Dueck, 1980; Huang *et al.*, 1998). With the exception of the flowering period, the application of root exudates from hybrid sunflower plants greatly increases myceliogenic germination.

However, amino acid supplementation alone has no discernible impact on myceliogenic germination. It certainly increases germination greatly when mixed with peptone. On the other hand, fungicide application reduces myceliogenic germination; benomyl treatment results in the greatest reduction in sclerotia germination (88%). Among the biological control agents, *Trichoderma viride* is the most effective, followed by *Bacillus* sp. After being cultured for 4–6 weeks in sand or vermiculite, sclerotia generated at 15°C and stored for 8 weeks at 4°C exhibit the maximum carpogenic germination (27.5%; Mosa *et al.*, 2000). Hyphal development and mycelium germination are impacted by sclerotia drying out. When exposed to damp sand or 100% RH, desiccant-dried sclerotia quickly germinate, show vigorous hyphal development, and frequently form colonies (Huang and Dueck, 1980; Huang *et al.*, 1998).

4.14.3.2 Abiotic variables influencing Sclerotinia sclerotia myceliogenic germination

The phytopathogenic fungus *S. sclerotiorum* forms dormant structures known as sclerotia, which undergo myceliogenic germination under specific environmental conditions. Basal infection occurs when these sclerotia produce hyphae that directly infect the leaves or stems of host plants from the ground. Various abiotic factors play a crucial role in promoting myceliogenic germination, including sclerotial maturity, rind melanization, temperature, moisture, and their combination. Optimal mycelial growth was observed when the sclerotium hydration level was high and the incubation temperature was low (15°C) in the presence of a nutrient source. Interestingly, sclerotia incubated without a nutrient source on moist sand exhibited the highest frequency of myceliogenic germination (63%) when they had been previously imbibed and conditioned at 20°C. However, the most consistent and vigorous myceliogenic germination (>75%) occurred when the sclerotia were heat-dried before being submerged in water. These germinated sclerotia were found to infect and proliferate on the leaves of intact *Brassica napus* plants. These findings provide valuable insights into the abiotic conditions that increase the risk of basal infection by *S. sclerotiorum* (Lane *et al.*, 2018, 2019).

4.15 Regulation of Stipe Formation From Sclerotia

Certain stipes that were established previously can cause dormant pigmented primordia in sclerotia to stay inactive. Researchers Kapoor (1994) and Saito (1977) examined how stipe removal affected sclerotia's ability to produce new stipes. The removal of stipes from sclerotia that are sprouting has been found to increase the yield of stipes relative to the control sclerotia. The total number of stipes created increases fourfold when stipes are removed. This implies that in sclerotia, stipes and pigmented primordia have a growth-inhibitory link. A comparable inhibitory link between a stipe and its lateral branches has been demonstrated before by Henderson (1962). We'll talk more about how various *in vitro* and *in vivo* conditions affect sclerotium germination and apothecium production.

4.15.1 Effect of food consumption

Even though sclerotia appear to be identical in color and shape, their germinability is greatly influenced by the type of growth medium on which they grow.

The two most effective media for achieving high germinability are potato dextrose broth and bean leaf decoction dextrose broth. On the other hand, sclerotia grown on Houston's solution, a synthetic medium, exhibit low germination rates even when they seem fully mature. This implies that traditional indicators of maturity, including dark coloring or the absence of liquid droplets on the sclerotia's surface, could not be accurate measures of their ability to proliferate. Consequently, in the context of *Sclerotinia*, a sclerotium is only deemed mature if it is able to germinate and create apothecia. Amino acids have been found to be the optimum nitrogen sources for functional maturation (Table 4.1).

Despite the externally formed sclerotia seeming mature, nitrate and ammonium nitrogen are not preferred (Table 4.2). The 'functional maturation' of sclerotia appears to be influenced by the carbon-to-nitrogen ratio as well. Using

Table 4.1. Effects of different nitrogen sources, with a focus on amino acids, on the development of sclerotia. (Adapted from publication of Saito, 1977, with permission.)

Nitrogen sources	Average number of sclerotia/Petri dish	Average dry weight of sclerotium (mg)
L-Alanine	22.4	7.0
γ-Amino butyric acid	17.8	9.5
V-Aminocaproic acid	0	0
L-Aspartic acid Na	19.8	6.8
L-Arginine HCl	14.1	5.7
L-Citrulline	17.7	5.5
L-Cysteine	5.6	7.9
L-Cystine	2.7	4.6
L-Glutamic acid Na	11.7	17.2
Glycine	18.4	5.2
L-Histidine HCl	10.7	9.6
L-Homoserine	9.2	6.1
L-Leucine	10.8	6.6
L-Lysine HCl	0	0
L-Methionine	15.8	5.2
L-Ornitine	12.5	9.5
L-Proline	13.8	8.3
L-Phenylalanine	8.9	9.3
L-Threonine	10.0	2.6
L-Tryptophane	21.3	3.3
L-Tyrosine	16.1	8.6
L-Valine	7.8	8.0
NH_4NO_3	19.8	6.2
Bean leaf decoction dextrose agar	6.4	10.0

Nitrogen content, 245 mg/1000 ml

Table 4.2. Variation in sclerotia formation significantly depending on ammonium salts or nitrates as nitrogen source. (Adapted from publication of Saito, 1977, with permission.)

Nitrogen sources	Average number of sclerotia/ Petri dish	Average dry weight of sclerotium (mg)
NH_4NO_3	11.6	9.0
$(NH_4)\ SO_4$	13.3	9.3
$(NH_4)\ HPO_3$	8.3	8.2
CH_3COONH_4	12.8	5.1
$(NH_4)\ C_2\ O_4H_2O$	14.6	4.3
$(NH_4)\ HC_6H_6O_7$	11.5	9.8
$(NH_4)\ _2C_4H_4O_6$	13.1	4.6
KNO_3	13.7	12.2
$Ca\ (NO_3)_2$	14.2	13.1
$NaNO_3$	14.8	9.4
Casamino acids	7.8	19.4

Nitrogen content, 245 mg/1000 ml

glutamic acid as the only source of nitrogen has allowed researchers to study the impact of carbon sources on sclerotial development (Tables 4.3, 4.4, 4.5, 4.6 and 4.7). A carbon supply is considered to be functionally mature if it promotes both mycelial development and sclerotial production. Consequently, sclerotia grow on a variety of polysaccharide-, disaccharide-, and monosaccharide-containing media, except for sorbase, which prevents the fungus from growing and germinating normally. Except for sorbitol, polyols do not support the development of fully developed, mature sclerotia. There is no relationship between the maximum germination percentage and the amount of time it takes for sclerotia to attain 'functional maturity.' There is a suggestion that the developmental stage of a particular plant or the diversity of host plants influences the shift in 'functional maturity' in natural environments. Although the exact meaning of 'functional maturity' is still unclear, it is most likely distinct from vegetative growth because vitamins usually promote mycelial growth (Tables 4.8, 4.9, 4.10, 4.11, 4.12 and 4.13) but not the 'functional maturation' of the sclerotia that are produced (Saito, 1977).

Table 4.3. Variations in sclerotia's capacity to germinate while using different nitrogen sources, especially amino acids. (Adapted from publication of Saito, 1977, with permission.)

Nitrogen sources	Germination (%)						
	Days after seeding of sclerotia						
	14	22	27	34	40	47	56
L-Alanine	0	7.0	11.0	13.0	18.0	37.0	42.0
γ-Aminobutyric acid	0	2.0	11.0	14.0	38.0	71.0	72.0
L-Aspartic acid Na	0	0	0	1.0	2.0	65.0	76.0
L-Arginine HCl	0	0	0	0	0	2.0	3.0
L-Citrulline	0	0	0	1.0	2.0	3.0	9.0
L-Cysteine	0	0	0	0	0	0	1.0
L-Cystine	0	0	0	0	0	0	0
L-Glutamic acid Na	0	4.0	6.0	15.0	47.0	67.0	70.0
Glycine	0	7.0	7.0	15.0	21.0	60.0	68.0
L-Histidine HCl	0	9.0	11.0	13.0	19.0	25.0	30.0
L-Homoserine	0	0	1.0	2.0	6.0	11.0	16.0
L-Methionine	0	2.0	3.0	10.0	16.0	44.0	54.0
L-Ornitine	0	0	0	0	1.0	1.0	1.0
L-Threonine	0	0	0	3.0	5.0	12.0	20.0
L-Tryptophane	0	1.0	5.0	8.0	11.0	20.0	24.0
L-Tyrosine	0	3.0	3.0	5.0	9.0	29.0	43.0
L-Valine	0	2.0	4.0	5.0	8.0	15.0	21.0
NH_4NO_3	2.0	8.0	14.0	16.0	21.0	44.0	53.0
Bean leaf decoction dextrose agar	5.0	10.0	16.0	25.0	45.0	75.0	100

4.15.2 Effect of low-temperature pretreatment on sclerotial germination

For a month, sclerotia that were isolated from bean stem cultures were stored at four different temperatures (−4,−10, and −20°C), with or without access to water. The germination of these sclerotia was then contrasted with that of control sclerotia that had been kept at room temperature. The findings showed that when the sclerotia are submerged in water and kept in a moist environment during the pretreatment phase, the chilling impact of low temperature at 4°C is evident. However, the chilling effect is minimal in the absence of moisture, and the germination rate at 15°C is almost the same as the control sclerotia. According to Saito (1977), the chilling period appears to shorten the time needed to attain maximal germination. It is important to note that carpogenic germination of sclerotia does not require exposure to low temperatures (Coley-Smith and Cooke, 1971; Willetts and Wong, 1980). According to Huang (1991), sclerotia of a Canadian isolate of *S. sclerotiorum* undergo a change in germination pattern from carpogenic to myceliogenic after being exposed to −10 or −20°C for 4 weeks.

4.15.3 Requirements for nutrition during mycelial development and germination

Mycelial production is frequently seen on the surface of the sclerotia, even though the germination of *Sclerotinia* spp. sclerotia usually

Table 4.4. Variations in sclerotia germination capacity that result from employing different nitrogen sources, particularly ammonium salts and nitrates. (Adapted from publication of Saito, 1977, with permission.)

Nitrogen sources	Germination (%)						
	Days after seeding of sclerotia						
	8	14	21	27	34	56	69
NH_4NO_3	0	0	2.0	8.0	18.0	35.0	35.0
$(NH_4)_2SO_4$	0	0	0	0	0	0	1.0
$(NH_4)_2HPO_3$	0	0	0	2.0	7.0	20.0	22.0
CH_3COONH_4	0	0	1.0	30	4.0	9.0	9.0
$(NH_4)\ C_2\ O_4H_2O$	0	0	0	0	0	0	2.0
$(NH_4)_2HC_6H_5O_7$	0	0	0	0	3.0	6.0	12.0
$(NH_4)_2C_4H_4O_6$	0	0	3.0	7.0	8.0	19.0	31.0
KNO_3	0	0	0	0	0	5.0	10.0
$Ca\ (NO_3)_2$	0	0	0	3.0	10.0	18.0	19.0
$NaNO_3$	0	0	0	0	0	0	0
Casamino acids	0	0	1.0	4.0	15.0	27.0	27.0
Bean leaf decoction dextrose agar	34.0	87.0	99.0	99.0	100	100	100

Table 4.5. Effects of nitrogen-containing amino acids on the initiation and outward development of sclerotia. (Adapted from publication of Saito, 1977, with permission.)

Nitrogen sources	Days of sclerotial initiation	Duration of exudation period (days)	Days of pigmentation
L-Alanine	6	8	15
L-Aspartica acid Na	5	3	8
L-Arginine HCl	4	5	11
L-Cysteine	8	6	13
L-Cystine	8	10	17
L-Glutamic acid Na	4	6	11
Glycine	8	6	15
L-Histidine HCl	5	8	13
L-Serine	7	7	15
L-Tryptophane	7	5	15
L-Tyrosine	6	6	15

Nitrogen content, 245 mg/1000 ml

culminates in the creation of apothecia. This phenomenon, which falls under the category of germination, is thought to be the consequence of sclerotial cells that were dormant before beginning to proliferate again. 'Myceliogenic germination' is the term originally used to describe the process by which fungal sclerotia germinate to develop into mycelia (Coley-Smith and Cooke, 1971). This specific type of germination happens when sclerotia are

Table 4.6. Variations in the rates of sclerotia germination using different sources of nitrogen on agar plates. (Adapted from publication of Saito, 1977, with permission.)

Nitrogen sources	Germination (%)					
	Days after seeding of sclerotia					
	6	8	11	13	15	23
L-Alanine	0	0	0	1.5	1.5	18.5
L-Aspartic acid	10.1	26.1	44.2	52.2	63.3	63.3
L-Arginine HCl	30.3	54.9	57.4	57.4	60.7	63.0
L-Cysteine	0	0	0	0	0	0
L-Cystine	0	0	0	0	0	0
L-Glutamic acid Na	5.7	20.6	21.6	34.3	40.2	47.1
Glycine	4.0	19.2	27.3	32.3	34.3	38.4
L-Histidine HCl	2.9	5.8	7.9	9.4	14.4	22.9
L-Serine	34.3	51.8	77.4	78.8	82.5	82.5
L-Tryptophane	6.6	15.6	18.0	20.5	24.6	25.4
L-Tyrosine	77.6	83.2	87.9	88.6	88.6	88.6

Table 4.7. Different carbon sources' effects on the onset, amount, dry weight, and outward development of sclerotia. (Adapted from publication of Saito, 1977, with permission.)

Carbon sources	Average number/Petri dish	Average dry weight of sclerotium (mg)	Days of initiation	Duration of exudation period (days)	Days of pigmentation
Arabinose	9.3	8.2	6	6	9
Fructose	8.0	11.9	4	8	10
Galactose	8.5	11.7	5	11	12
Glucose	7.0	13.5	4	10	10
Mannose	5.6	14.6	4	12	12
Rhamnose	20.4	3.2	4	8	7
Sorbose	–	–	–	–	–
Xylose	18.1	1.6	5	7	9

placed on damp sponges or sand; however, it is more commonly observed when nutrients, like glucose or powdered bean leaf, are given to the medium in order to inhibit the establishment of the apothecial layer (Table 4.14). These two germination mechanisms appear to be at odds with one another (Tables 4.15 and 4.16). Because of this, apothecia creation is inhibited in proportion to mycelial growth, and sclerotia that are actively creating apothecia show little to no mycelial growth. It is regarded as unusual when both germination modalities co-occur in a single sclerotium. Reserves are rapidly depleted during the 'myceliogenic germination' of sclerotia in *Sclerotinia* spp. (Saito, 1977).

4.15.4 Effect of soil moisture content for sclerotial germination

According to several studies (Partyka and Mai, 1962; Abawi and Grogan, 1975; Kruger, 1975a,

Table 4.8. Effect of different carbon sources on the sclerotial development, including initiation, quantity, dry weight, and external maturation. (Adapted from publication of Saito, 1977, with permission.)

Carbon sources	Average number/Petri dish	Average dry weight of sclerotium (mg)	Days of initiation	Duration of exudation period (days)	Days of pigmentation
Lactose	5.7	13.0	4	10	11
Maltose	12.9	7.7	4	8	10
Sucrose	5.4	17.0	6	5	12
Dextrin	11.4	8.7	4	8	10
Inulin	13.2	6.0	4	8	9
Soluble starch	5.1	11.4	5	7	11

Table 4.9. Impacts of different carbon sources on sclerotia-polyol development, mass multiplication, quantity, and visible expansion. (Adapted from publication of Saito, 1977, with permission.)

Carbon sources	Average number/Petri dish	Average dry weight of sclerotium (mg)	Days of initiation	Duration of exudation period (days)	Days of pigmentation
Galactitol	2.7	4.9	5	7	9
Erythritol	–	–	–	–	–
Glycerol	6.0	13.1	4	5	7
Inositol	1.8	1.7	6	6	9
Mannitol	1.7	2.1	4	6	7
Sorbitol	7.6	9.1	4	8	–

b, Kruger, 1976; Williams and Stelfox, 1979; Kruger, 1980), moisture is essential for the germination and growth of apothecia in sclerotia. Since stipes cannot form at even 100% RH, the availability of water is critical (Coley-Smith and Cooke, 1971; Grogan and Abawi, 1975). For germination and the production of apothecia, a soil moisture level of 30% is very favorable (Singh and Singh, 1983). Sclerotia can germinate in a range of moisture levels from 15 to 50% when buried in deep soils at 15°C (Morrall, 1977). Prior research has indicated that even low levels of moisture tension prevent germination; however, more recent investigations have demonstrated that carpogenic sclerotia can germinate at water potentials between 0 and −7.5 bars. Ascospore infections on rapeseed crops in Western Canada provide clear evidence that apothecia can form in the field at water potentials far lower than 0 bars. Methods used by Teo and Morrall (1985) showed that germination can only occur at matric potentials close to 0 bars when there is sufficient oxygen present and sclerotia on the soil surface. Many of the buried sclerotia decompose and do not germinate. The methods of using an inclined box, polyethylene glycol, and sealed jars cause fruiting bodies to begin growing at very low matric potentials, −4,−7, and −15 bars, respectively. According to Teo and Morrall (1985), this suggests that sclerotia can sprout in soil that is well below its field capacity.

4.15.5 Temperature's effect on apothecial development and sclerotial germination

Sclerotia germination and apothecium production are influenced by temperature (Coley-Smith and Cooke, 1971; Kruger, 1976; Saito, 1977; Willetts and Wong, 1980; Singh *et al.*, 1985).

Table 4.10. Variation in the ability of sclerotia to germinate based on the carbon sources used in agar plate production. (Adapted from publication of Saito, 1977, with permission.)

Carbon sources	Germination (%)					
	Days after seeding of sclerotia					
	6	8	10	12	14	16
Arabinose	27.5	59.5	79.0	94.0	98.5	99.0
Fructose	33.6	49.3	65.7	75.0	84.3	89.3
Galactose	30.2	43.6	49.7	57.7	73.8	85.9
Glucose	38.3	71.3	85.1	93.6	95.7	97.9
Mannose	39.1	60.0	79.1	82.6	88.7	97.4
Rhamnose	6.9	20.6	29.9	39.5	54.8	71.0
Xylose	18.8	34.8	46.5	61.5	79.4	96.0
Lactose	7.1	21.4	24.1	33.9	34.8	38.4
Maltose	1.8	10.1	15.1	26.0	33.8	41.6
Sucrose	18.6	34.9	66.3	90.7	96.5	97.7
Dextrin	54.4	59.7	75.2	86.9	92.7	97.1
Inulin	4.9	19.2	23.6	38.9	40.9	55.2
Soluble starch	21.5	45.6	64.6	72.2	84.3	88.6
Galactitol	1.2	–	4.9	4.9	–	25.6
Glycerol	0	0	0	0	–	7.5
Inositol	0	0	0	1.2	0	16.7
Mannitol	0	0	0	0	–	16.1
Sorbitol	4.9	11.8	32.4	43.1	54.9	58.8

Table 4.11. Impact of vitamins on sclerotia formation. (Adapted from publication of Saito, 1977, with permission.)

Vitamins	Concentration (ppm)	Average number of sclerotia/Petri dish	Average dry weight of sclerotia (mg)
Biotin	0.05	10.9	11.2
Pyridoxine	0.1	13.9	12.2
Thiamine	0.1	10.9	7.4
Control	–	12.0	9.2

According to Kruger (1976), the optimal temperature range for the production of apothecia is between 7 and 11°C. According to Singh *et al.* (1985), high temperatures cause a delay in seed germination, mycelial expansion, the start of sclerotial production, secretion, coloration/ripening, and a decrease in the amount of sclerotia on culture dishes. The amount and timing of crop foliage-induced shadowing on the soil surface have a significant effect on the development of apothecia. Apothecial development is significantly influenced by the microclimate on the soil's surface (Akai, 1981). When *S. minor* sclerotia are cultivated under a

Table 4.12. Sclerotia germination on the basal medium supplemented with vitamins. (Adapted from publication of Saito, 1977, with permission.)

Vitamins	Germination (%)						
	Days after seeding of sclerotia						
	8	14	21	27	34	56	69
Biotin	0	0	0	1.0	3.0	6.0	8.0
Pyridoxine	0	0	9.0	23.0	29.0	41.0	44.0
Thiamine	0	0	0	0	0	1.0	2.0
Control	0	0	2.0	7.0	20.0	35.0	38.0

Table 4.13. Effects of vitamins on mycelium growth. (Adapted from publication of Saito, 1977, with permission.)

Vitamins	Mycelial dry weight (mg)					
	Days of culture					
	3	6	9	12	16	22
Biotin	23.5	259.0	289.1	312.7	358.5	338.9
Pyridoxine	25.2	306.6	349.4	367.2	343.8	348.8
Thiamine	18.0	236.2	343.7	360.9	352.0	352.5
Control	36.0	49.9	34.6	104.4	227.9	246.8
Bean leaf decoction dextrose broth	36.0	234.9	219.0	223.3	215.6	215.8

Table 4.14. The development of apothecia (carpogenic germination) is inhibited by the growth of mycelium from sclerotia (myceliogenic germination). (Adapted from publication of Saito, 1977, with permission.)

Experiment	No. of sclerotia observed	Count of germinated sclerotia						
		Grade of mycelial growth						
		0	1		2		3	
		A[a]	M[b]	A+M	M	A+M	M	A+M
1	48	32	2	5	2	1	2	1
2	66	31	3	5	2	3	12	0
3	54	34	0	5	4	3	8	0
Total	168	97	5	15	8	7	22	1

[a]Number of apothecia
[b]Number of sclerotia

diurnal temperature cycle of 15°C for 8 hour and 10°C for 16 hour, abundant stipe development happens. On the other hand, stipes do not form during the 20/15°C and 25/20°C diurnal cycles (Hawthorne, 1973). Hawthorne (1976) discovered in follow-up controlled laboratory tests that apothecial stipes form when sclerotia are cultivated from 10 to 19°C, with 15°C being the most suitable temperature. While apothecial discs can form between 10 and 20°C, sclerotia can germinate and generate stipes anywhere between 5 and 25°C (Coley-Smith and Cooke, 1971; Saito, 1977; Willetts and Wong, 1980). It was shown that the ideal temperature range for

Table 4.15. Impact of sclerotial sand and soil supplemented with organic materials, both under sterilized and unsterilized conditions, on myceliogenic and carpogenic germination of sclerotia. (Adapted from publication of Saito, 1977, with permission.)

Media	Amendments	No. of sclerotia at different days after seedling									
		6		10		14		27		42	
		A[a]	M[a]	A	M	A	M	A	M	A	M
Sanitized sand	Nonglucose	0	11.0	2.0	27.0	9.0	30.0	15.0	29.0	41.0	34.0
	bean leaf	0	26.0	0	46.0	1.0	46.0	1.0	82.0	–[b]	–
	powder	0	62.0	0	64.0	0	67.0	0	93.0	–	–
Sanitized soil	Nonglucose	0	10.0	5.0	21.0	7.0	21.0	–	–	–	–
	bean leaf	0	13.0	0	19.0	0	35.0	–	–	–	–
	powder	0	28.0	2.0	36.0	2.0	–[b]	–	–	–	–
Nonsanitized soil	Nonglucose	0	6.0	12.0	13.0	15.0	23.0	50.0	19.0	66.0	12.0
	bean leaf	0	40.0	1.0	47.0	2.0	49.0	5.0	52.0	3.0	52.0
	powder	0	17.0	1.0	17.0	1.0	7.0	1.0	–[c]	1.0	–

[a]A. Carpogenic germination; M, Myceliogenic germination
[b]Additional examination was halted by the mycelial expansion across the substrate.
[c]The mycelium of *S. sclerotium* could not be differentiated from the other soil fungi present in the soil.

Table 4.16. Study on myceliogenic and carpogenic sclerotial germination (%) in unsterilized soil treated with organic amendments. (Adapted from publication of Saito, 1977, with permission.)

Amendments		Days after seeding of sclerotia							
		10		15		28		40	
	Dose (%)	A[a]	M[a]	A	M	A	M	A	M
Glucose	1	0	5.0	26.0	0	81.0	0	87.0	0
	2	0	29.0	12.0	17.0	71.0	10.0	84.0	7.0
	5	0	41.0	0	51.0	2.0	91.0	2.0	80.0
Bean leaf powder	1	0	–	5.0	–	70.0	–	89.0	–
	2	0	–	9.0	–	62.0	–	77.0	–
	5	0	–	0	–	31.0	–	46.0	–
Rice bean	1	0	–	23.0	–	75.0	–	79.0	–
	2	0	–	5.0	–	77.0	–	89.0	–
	5	0	–	4.0	–	54.0	–	70.0	–
Control		0	0	34.0	0	84.0	0	93.0	0

[a]A = Carpogenic germination; M = Myceliogenic germination

sclerotium germination was 12–18°C (Partyka and Mai, 1962) and 8–16°C, in that order (Dillard *et al.*, 1995). The number of germinating sclerotia and apothecial sclerotium increases with sclerotia size. Cold or high temperatures have the potential to prevent germination or reduce the pace of germination (Kruger, 1980). By decreasing sclerotium germination frequency and preventing apothecial disc expansion, soil temperatures below 10°C postpone the appearance of apothecia in the field. This assertion is supported by field research (Kruger, 1973, 1975a, 1980), which demonstrates a clear correlation between spring temperatures and the timing of apothecia's emergence in the soil (Kruger, 1975a).

Temperature and wetness have a strong link that affects a lot of different processes.

Sclerotial germination and apothecium production are frequently enhanced by cool, moist settings (Partyka and Mai, 1962; Walker, 1969; Kruger, 1975a). However, high temperatures and dry soil inhibit the germination of sclerotia and their subsequent development into apothecia (Partyka and Mai, 1962). This is due to the stipes' inability to break through the tough outer layer of the topsoil (Kruger, 1976). Sclerotia exposed to extreme dryness in the field show reduced carpogenic germination ability, yet they still show viability when exposed to optimal conditions later on (Abawi and Grogan, 1975; Grogan and Abawi, 1975). The reduced capacity could be the consequence of the stipe start cells' decreased ability to withstand desiccation in contrast to the myceliogenic germination cells (Abawi and Grogan, 1975). Since both mycelia and carpophores originate from cells in the sclerotial medulla, it is unclear if there are specific cell groups for each germination mechanism (Adams and Tate, 1976; Saito, 1977). When temperatures are extremely high, a large percentage (at least 50%) of the sclerotia in the soil disappear in a comparatively short amount of time. At 40°C, for instance, it takes around 39 hour to destroy more than half of the sclerotia. Similarly, it takes 6 hour at 45°C and 2 hour at 50°C, respectively (Adams, 1987). When soil is dried to a moisture content of −1516 bars or less for 7 days, then it is rehydrated to −0.2 bar. During the summer, the ability and persistence of sclerotia in soil decrease at the soil's surface and at depths of 0–2 cm. Conversely, at depths of 2–8 cm and 8–14 cm, there is a marginal rise in the number of sclerotia, which ultimately stays constant (Tables 4.17, 4.18 and 4.19; Adams, 1987). However, compared to those buried deeper, sclerotia positioned between 0 and 2 cm produce more apothecia, according to Mitchell and Wheeler (1990).

4.15.6 Impact of light on sclerotial germination and apothecial formation

Studies have demonstrated that sclerotia may develop in both dark and light settings, allowing them to germinate and form stipes regardless of the presence of light (Purdy, 1956; Bedi, 1962; Coley-Smith and Cooke, 1971; Natti, 1971; Letham, 1975; Willetts and Wong, 1980). However, the stipes show positive phototropism and bend in the direction of the light once they separate from the parent sclerotium (Le Tourneau, 1979; Willetts and Wong, 1980). On the other hand, light is necessary for the apothecial disc to fully develop and grow (Purdy, 1956; Bedi, 1962; Walker, 1969; Coley-Smith and Cooke, 1971; Natti, 1971; Letham, 1975; Le Tourneau, 1979; Willetts and Wong, 1980). This means that the disc needs a minimum of 8–12 hour of light in a 24 hour period. High light intensity (120–130 mol m^{-2} s^{-1}) is more likely to produce apothecial initials than low light intensity (80–90 mol m^{-2} s^{-1}). The apothecia produced under high light intensity are larger than those formed under low light intensity (Sun and Yang, 2000). For the fungus to form apothecial discs above the soil's surface and facilitate the efficient spread of ascospores, light is an essential component (Willetts and Wong, 1980). White light is seen to be the most beneficial of the various light

Table 4.17. The length of time and temperature at which three soil-borne fungi must be exposed in order to destroy 50% of their propagules (LD_{50}). (Adapted from publication of Adams, 1987, with permission.)

Temperature (°C)	LD_{50} (h)		
	Sclerotium cepivorum (sclerotia)	*Sporidesmium sclerotivorum* (macroconidia)	*Sclerotinia minor* (sclerotia)
35	129.6	266.4	624.0
40	9.5	37.2	38.4
45	1.7	5.2	3.4
50	0.8	0.8	1.5

Table 4.18. Sclerotia survival (in relation to inoculum density) of *Sclerotinia minor* and *Sclerotium cepivorum* in damp soil (–0.2 bar) 6 weeks after infestation. (Adapted from publication of Adams, 1987, with permission.)

Matric potential of dried soils	Survival based on percentage of control		
	Sporidesmium sclerotivorum	*Sclerotinia minor*	*Sclerotium cepivorum*
–0.2 bar (control)	100 a[a]	100 a	100 a
–1.160 bars	73 b	77 ab	90 ab
–1.516 bars	61 c	69 bc	80 ab
–2.915 bars	39 d	30 c	59 bc
–3.656 bars	24 c	19 c	28 c

[a]Values in a column that share the same letter are deemed not to be significantly different based on Duncan's multiple range tests.

Table 4.19. Sclerotial survival of *Sclerotinia minor* at different soil depths during summer 1985. (Adapted from publication of Adams, 1987, with permission.)

Sampling date	Survival of sclerotia at different depths (%)			
	Soil surface	0–2 cm	2–8 cm	8–14 cm
8 May	100	100	100	100
20 May	83 cdef[a]	114 abcde	120 abcd	124 abcd
30 May	99 abcdef	137 abc	140 ab	136 abc
20 June	85 bcdefg	109 abcde	126 abcd	126 abcd
5 July	53 fg	113 abcde	131 abc	144 a
18 July	83 cdefg	115 abcde	140 ab	138 abc
30 July	36 g	94 abcdef	140 ab	140 ab
16 August	32 g	62 efg	126 abcd	140 ab
27 August	44 fg	72 defg	113 abcde	142 ab
15 September	32 g	53 fg	130 abc	125 abcd

[a]Values that share the same letter do not exhibit significant differences ($P \geq 0.01$) based on Duncan's multiple range tests.

types, but red, green, and blue lights are thought to be less beneficial (Bedi, 1962).

4.15.7 Impact of sclerotium dimensions and burial depth in soil on germination and apothecial formation

Numerous studies have demonstrated a direct correlation between the number of apothecia produced and the size of sclerotia (Bedi, 1963; Coley-Smith and Cooke, 1971; Kruger, 1974; 1975a). The reason for this correlation is most likely that larger sclerotia store more nutrients in reserve, which is necessary for the growth of apothecium. Moreover, the depth of sclerotia in the soil affects apothecial growth. Sclerotia that germinate closer to the soil's surface tend to generate more apothecia (Kruger, 1975a; Willetts and Wong, 1980; Akai, 1981; Singh and Singh, 1983). Sclerotia can start to grow as low as 10 cm (Cook *et al.*, 1975; Willetts and Wong, 1980). However, they have a better probability of germinating when they are closer to the soil's surface (Kruger, 1980). If the sclerotia germinate too deeply, the stipes or stalks may not extend sufficiently to break through the soil's surface and expose the apothecial discs to the essential light for expansion (Willetts and Wong, 1980). According to field observations, apothecia are only formed by sclerotia in the top 2–3 cm of soil; stipes longer than 3 cm are rarely generated (Abawi and Grogan, 1979).

4.15.8 Effects of soil mixture, pH, fertility, and structure on apothecial development and sclerotial germination

The pH of the media on which sclerotia form has no effect on their ability to produce apothecia (Bedi, 1963). Nevertheless, the pH of the substrate that the sclerotia rest on during germination affects both germination and the growth of the apothecium. The pH range needed for apothecium formation is 6.0–9.7 (Bedi, 1963; Coley-Smith and Cooke, 1971). Only stipes are produced by sclerotia, or they do not germinate at all outside of this area (Bedi, 1963). However, Hau *et al.* (1982) discovered that the ideal soil pH range for sclerotial germination is between 6.0 and 6.5.

Unlike sandy loam, silt loam, and various combinations of sand and clay, pure sand has the ideal conditions for germination and the growth of adult apothecia. The greatest apothecia and sclerotia germination occur in sand–clay mixes with a ratio of three parts sand to one part clay. The formation and germination of sclerotia diminish in these mixes as the quantity of clay increases (Singh and Singh, 1983; Mitchell and Wheeler, 1990). In eastern Schleswig-Holstein, Kruger (1976) found that apothecial development is more prominent in marsh soils than in sandy loam and loam soils. The size and depth of the apothecia (*S. sclerotiorum* sclerotia) burial directly affects the apothecia diameter, number, stipe length, and germination period. For example, sclerotia placed 2 cm deep in clay loam soil take 38 days to germinate, but those buried 5 cm deep in river sand take 62 days, producing bigger 5.6 mm apothecia (Singh and Tripathi, 1996).

In comparison to control II (no irrigation but covered), the most successful methods for reducing the apothecial population were floods for two consecutive days once a week and optimum irrigation on the seventh day. This was demonstrated by the production of apothecia in connection to irrigation levels. On the tenth day, the treatment with optimal irrigation showed the highest amount of apothecia formation. A comparison of the various soil types revealed that sandy soil produced the fewest apothecia, whereas sandy loam soil produced the most (Table 4.20). Before seeding, weekly flooding of the inoculated pots produced the lowest disease incidence and the shortest lesion length, while daily irrigation produced the highest disease incidence when compared to control treatments. Plants grown in sandy soil produced the longest lesions, while plants grown in sandy loam soil produced the shortest lesions. Compared to control pots, disease intensity changed less in the treatments that provided optimum irrigation on the third and seventh day of each week and flooding once a week (Table 4.21). The highest disease intensity was found in sandy loam soil, whereas the least amount was found in sandy soil (Mehta *et al.*, 2009).

The nutrient quality of the soil affects the formation of sclerotia and their germination, which are both aspects of apothecia production. When sclerotia grow in nutrient-rich environments, they either produce relatively little apothecium or are sterile. In contrast, primitive or small apothecia arise from sclerotia that are created in nutrient-poor settings. In contrast, highly fertile sclerotia in terms of apothecium production are produced under ideal nutritional conditions (Bedi, 1963; Willetts and Wong, 1980). Sclerotia are more likely to germinate carpogenically under nutrient-poor environments (Bedi, 1963; Le Tourneau, 1979). On the other hand, nutrients tend to favor myceliogenic germination during sclerotia germination, which is the opposite of carpogenic germination (Bedi, 1963; Saito, 1977). According to Mitchell and Wheeler (1990), adding nitrogen to pots slows down the formation of sclerotia and decreases their production. Moreover, low concentrations of Mg^{2+} and HCO_3^-, along with high concentrations of Na^+, Ca^{2+}, and SO_4^{-2}, inhibit the carpogenic germination of sclerotia (Singh *et al.*, 1995a, b).

Table 4.20. Effect of different irrigation levels[a] on the number of apothecia and disease incidence of Indian mustard due to *Sclerotinia sclerotiorum* in various types of soil. (Adapted from publication of Hieu, 2007, with permission.)

Treatments (Irrigation levels)[a]	Number of apothecia appearance/pot				Disease incidence (%)			
	Sandy soil	Clay soil	Sandy loam soil	Mean	Sandy soil	Clay soil	Sandy loam soil	Mean
Flooding once in a week	22.7	29.2	31.7	27.9	67.5	62.5	54.3	61.4
Flooding for two consecutive days in a week	17.7	23.0	25.5	18.5	60.0	42.5	41.6	48.0
Optimum irrigation on every third day	15.7	33.2	37.0	28.7	65.0	46.9	44.3	52.1
Optimum irrigation on every seventh day	14.7	17.5	29.7	20.7	55.0	52.5	40.5	49.3
Optimum irrigation on every tenth day	30.2	32.0	33.0	31.7	62.5	45.0	50.0	52.5
Optimum irrigation daily	18.0	30.5	30.5	24.3	67.5	65.4	57.0	63.3
Control I (no irrigation and uncovered)	17.0	33.2	38.0	29.4	47.5	52.5	68.6	56.2
Control II (no irrigation but covered)	28.7	39.0	43.0	36.9	47.5	55.0	67.5	56.7
Mean	20.6	29.7	33.6	–	50.3	52.8	53.0	–
CD (*P*=0.05) Irrigation levels:		= 2.44				= 5.64		
Soil types:		= 1.49				= 3.45		
Irrigation × soil:		= 4.23				= 0.76		

[a]Irrigation levels applied before sowing in pots inoculated with sclerotia thereafter uniform irrigation applied.

4.15.9 Impact of inhibitors on sclerotial germination and apothecial formation

Differential inhibition is a phenomenon that Steadman and Nickerson (1975) have reported. It states that various stages of fungal growth require different amounts of an inhibitor to be prevented. For instance, to prevent apothecium development, myceliogenic germination, and stipe production, various inhibitor dosages can be needed. Certain commercial fungicides and transition metal ions (cadmium, in particular) have inhibitory effects between 1 μM and 1 mM. Germination is inhibited by low-molecular-weight carbohydrates such as glucose, mannitol, sucrose, and trehalose; cations such as sodium and lithium; anions such as nitrate and sulfate; and high concentrations (1–100 mM) of common buffers such as acetate, phosphate, and Tris. However, Singh *et al.* (1995b) have documented a decrease in the proportion of carpogenic germination of *S. sclerotiorum* sclerotia in soils with high concentrations of Na^+, Ca^{2+}, Cl^-, and SO_4^-. Similarly, limited carpogenic germination has been seen in soils with low amounts of Mg^{2+} and HCO_3^-.

4.15.10 The influence of growth regulators on sclerotial germination

To evaluate the effects of growth regulators, desiccated sclerotia were submerged in solutions with varying concentrations (1 to 100 ppm) of gibberellic acid, kinetin, auxins, antiauxins, and sclerin, allowing them to germinate. Among

Table 4.21. Effect of different irrigation levels[a] on lesion length and disease intensity of Indian mustard due to *Sclerotinia sclerotiorum* in various types of soil. (Adapted from publication of Hieu, 2007, with permission.)

	Lesion length (cm)				Disease intensity (%)			
Treatments (Irrigation levels)[a]	Sandy soil	Clay soil	Sandy loam soil	Mean	Sandy soil	Clay soil	Sandy loam soil	Mean
Flooding once in a week	1.6	2.1	3.1	2.3	34.1	38.6	50.5	41.4
Flooding for two consecutive days in a week	2.3	2.8	3.0	2.7	38.7	51.6	53.1	47.8
Optimum irrigation on every third day	2.1	2.1	3.1	2.4	38.7	37.5	53.6	43.0
Optimum irrigation on every seventh day	1.8	2.7	3.7	2.7	31.7	44.8	57.8	44.8
Optimum irrigation on every tenth day	1.8	2.8	3.9	2.9	31.2	47.5	64.6	47.7
Optimum irrigation daily	2.1	3.5	3.9	3.2	38.7	51.0	62.5	50.8
Control (no irrigation and uncovered)	2.8	3.3	4.0	3.4	35.9	56.6	66.2	52.9
Control (no irrigation but covered)	2.9	3.5	3.7	3.4	36.2	62.6	64.1	54.4
Mean	2.2	2.9	3.6	–	35.7	48.8	59.1	–
CD (P=0.05) Irrigation levels:		= 0.13					= 5.21	
Soil types:		= 0.08					= 3.19	
Irrigation × soil:		= 0.21					= 9.03	

[a]Irrigation levels applied before sowing in pots inoculated with sclerotia thereafter uniform irrigation applied.

these substances, only sclerin, a metabolite derived from *S. sclerotiorum*, exhibited the ability to induce germination, as indicated in Table 4.22. It is worth noting that sclerin is recognized as a biologically active compound (Saito, 1977).

4.15.11 Effect on apothecia and sclerotia dry weight

As apothecia form, the bulk of sclerotia gradually decreases (Saito, 1977). As apothecia are generated, there is a considerable drop in the sclerotan, a β−1,3-glucan. The primary cause of the decrease in sclerotia's dry weight is the apothecial manufacturing process's consumption of β−1,3-glucans.

4.15.12 Impact of enzymatic function on sclerotia during germination

As they mature, sclerotia show only weak β−1,3-glucanase and β-glucosidase activity, and

Table 4.22. Effects of plant growth regulators on sclerotial sprouting. (Adapted from publication of Saito, 1977, with permission.)

Substances	Conc. (ppm)	Germination (%)[a]					
		Days after seedling of sclerotia					
		17	19	21	23	27	31
β-(furyl) acrylic acid	1	0.0	1.3	8.0	14.7	20.0	26.7
	10	0.0	0.0	0.0	0.0	0.0	0.0
	100	0.0	0.0	0.0	0.0	1.3	2.7
Gibberellic acid	1	0.0	1.3	4.0	8.0	9.3	14.7
	10	1.3	1.3	2.7	8.0	10.7	13.3
	100	0.0	2.7	5.3	6.7	12.0	13.3
Indoll-3-acetic acid (IAA)	1	1.3	1.3	2.7	2.7	8.0	12.0
	10	0.0	0.0	0.0	0.0	0.0	0.0
	100	0.0	0.0	1.3	4.0	5.3	6.7
2,4-Dichloro-phen-oxyacetic acid (2,4D)	1	0.0	0.0	0.0	0.0	5.3	10.7
	10	0.0	0.0	2.7	4.0	6.7	6.7
	100	0.0	2.0	2.0	2.0	2.0	6.0
Maleic hydrazide (MH)	1	0.0	0.0	4.0	5.3	12.0	14.0
	10	0.0	0.0	0.0	0.0	0.0	0.0
	100	0.0	0.0	0.0	0.0	1.3	4.0
2,4,5-Trichlorophenoxyacetic acid (2,4,5-TPA)	1	0.0	0.0	0.0	1.3	1.3	2.7
	10	0.0	0.0	0.0	0.0	0.0	0.0
	100	0.0	0.0	0.0	0.0	0.0	0.0
Kinetin	1	0.0	0.0	0.0	1.3	8.0	8.0
	10	0.0	0.0	0.0	0.0	4.0	4.0
	100	0.0	0.0	0.0	0.0	0.0	0.0
Sclerin	1	1.3	10.7	12.0	21.3	29.3	37.3
	10	0.0	1.3	1.3	6.7	13.3	18.7
	100	22.7	34.7	44.0	54.7	60.0	66.7
Control	–	–	5.3	9.3	14.7	18.7	18.7

[a]Mean of three replications.

neither trehalase nor amylase activity. The activity of β−1,3-glucanase appears to increase when they begin to sprout, although the activity of β-glucosidase stays constant. The development of apothecial stipes happens when β−1,3-glucanase activity rises to the greatest extent. The activities of the enzymes involved in the metabolism of carbohydrates were compared in the mature apothecia, immature apothecia, and sclerotia, which are the three separate parts of germinating sclerotia (Figs 3.11 and 3.12; Chapter 3, this volume). The sclerotia include the highest specific activity of β−1, β-glucosidase, and 3-glucanase. There is observable amylase and trehalase activity as germination proceeds and the majority of stems mature into apothecia in the respective parts. At this stage, amylase-specific activity is highest in mature apothecia, whereas trehalase activity is nearly equal in all three divisions.

The most active form of glucose-6-phosphate dehydrogenase was found in apothecia. However, germinating sclerotia exhibit comparatively less activity in comparison to un-germinating sclerotia (Fig. 3.12; Chapter 3, this volume). While stipes and apothecia facilitate the production of resources required for cellular formation, medullary cells of sclerotia are principally responsible for the breakdown of endogenous reserves (Saito, 1977). These findings unequivocally highlight the physiological distinctions between sclerotia and apothecia.

4.15.13 Effects of conditioning medium and duration on apothecia and sclerotia

The amount of time that must pass after conditioning in the soil for *S. sclerotiorum* sclerotia to germinate and develop apothecia is shortened. Sunflower root sclerotia germinate more quickly than stem cavity sclerotia. The longer the conditioning period, the greater the ability to germinate.

It is important to remember that conditioning in soil produces superior outcomes than conditioning in damp vermiculite (Phillips, 1986).

4.15.14 Effects of host exudates and organs on sclerotial germination and apothecial formation rate

The number of apothecial formations is mostly determined by the root exudates of different soybean cultivars (Chaves *et al.*, 1996). The presence of remoistened peanut leaves consistently boosts *S. minor* sclerotia germination. Although more peanut tissue than 1g may inhibit germination, 0.25–0.50g of dried peanut leaves is the recommended quantity of stimulation (Hau *et al.*, 1982).

4.15.15 Cropping history's effect on sclerotial germination and apothecia formation

The germination of *S. sclerotiorum* sclerotia is greatly influenced by the previous cropping history. Rapeseed plots sown for 2 years show a greater amount of germinated sclerotia than those sown for just 1 year. Additionally, fertilization in the spring results in higher germination rates in early summer, in contrast to fall fertilization or no fertilizer application. The development of apothecia typically mirrors sclerotium germination, suggesting the possibility of *S. sclerotiorum* ascospores being generated as inoculum in non-host crops.

4.15.16 Impact of plant cover on sclerotial germination and apothecia formation

The development of apothecium and the germination of sclerotium are strongly impacted by temperature, moisture content, and light in the soil. Studies in rapeseed fields (Morrall and Dueck, 1982) and bean fields (Schwartz and Steadman, 1978) have revealed that the existence of a crop canopy affects all three parameters. The significance of the microclimate beneath the crop canopy in the epidemiology of Sclerotiorum is highlighted by these investigations. According to Natti (1971), the microclimate in this region is characterized by high water potentials in the upper soil layers (Abawi and Grogan, 1979) and an ambient temperature that is rather steady in relation to the surrounding environment (Morrall and Dueck, 1982). These circumstances are ideal for sclerotium germination and apothecium production.

4.15.17 Other microorganisms' effects on sclerotial germination and apothecia formation

The field is home to a number of contradictory microorganisms. Chapter 8, *Biological Control, Sclerotinia Stem Rot of Crucifers: Epidemiology and Disease Management, Vol. II*, extensively explores their effects on sclerotia survival and germination, as well as apothecia production.

4.15.18 Effect of fungicides and herbicides on sclerotial survival and germination

When applied on PDA, a number of fungicides, such as Bavistin (50 ppm), Topsin M (100 ppm), Sailaxyl MZ (100 ppm), mancozeb (1000 ppm), and neem extract (5000 ppm), can effectively stop the development of sclerotia (Zewain *et al.*, 2004). Nonetheless, carbendazim (100 μg/ml for 20 min) is the most effective at preventing carpogenic germination of sclerotia (Singh *et al.*, 2003). Additionally, at a concentration of 100 ppm, the fungicides Topsin M, Benlate, Tecto 60, Ridomil G, and Benlate can reduce mycelial development (Iqbal *et al.*, 2003). It is important to note that using fungicides and herbicides in the field can affect sclerotia's ability to survive, germinate, and develop apothecia. The specific effects have been discussed in the book *Sclerotinia Stem Rot of Crucifers: Epidemiology and Disease Management*, Vol. II, Chapter 7, *Disease Management*.

4.15.19 Effects of different irrigation regimes on sclerotial carpogenic germination

After a steady 25–30 day period of soil moisture, apothecia development starts, and irrigation technique has a big impact. If there is a halt in irrigation before the apothecia begin to form, their growth will be delayed. This delay is about the same length as the irrigation break. Compared to loamy soil, apothecia are produced more abundantly in sandy soil at greater irrigation levels. On the other hand, compared to sandy soil, loamy soil produces more apothecia at a lower irrigation level. Apothecia may grow in sand and require 5 mm of watering every day, with a maximum lifespan of 33 days. One can estimate the start of apothecia growth by knowing the required moist period for carpogenic germination and taking into account the impacts of dry spells. To precisely anticipate the number and duration of apothecia, it is also essential to take into account factors such as soil composition, irrigation or rainfall levels, and their dispersion (Twengström *et al.*, 1998).

4.15.20 Consequence of sclerotia development stage

When cultures are 6–8 day old, immature sclerotia that can range in color from light brown to grayish black and include liquid sacs on their surface germinate myceliogenically with ease, producing hyphae. However, mature sclerotia, which are dry and black, develop carpogenically, resulting in the direct generation of apothecia. They are harvested from 14-, 21-, and 42-day-old cultures. For two isolates, the germination responses of mature sclerotia from cultures ranging in age from 14 to 42 days are comparable (Huang and Kozub, 1994).

4.15.21 Dissemination and release of ascospores from apothecia

Ascospores form inside of them during the apothecium maturation phase. When apothecia are juvenile, they contain very few ascospores (Kruger, 1974, 1975a). However, in the right conditions, a mature apothecium can develop and release up to 2.32×10^6 ascospores over a period of 1 week or longer (Schwartz and Steadman, 1978). When one takes into account that a single sclerotium can produce up to 100 apothecia, it can produce up to 2.3×10^8 ascospores (Schwartz and Steadman, 1978). The rate of ascospore discharge, or 'puff,' increases rapidly to a maximum level during the full maturity of *S. sclerotiorum* apothecia, is sustained for a brief period of time, and then gradually decreases. Ascospores are released by each ascus for a few seconds following the primary discharge. *S. minor* apothecia, on the other hand, have a more transient discharge; the puff appears to start at the maximal discharge rate, and the discharge ceases suddenly (Hartill and Underhill, 1976).

Ascospores, which traverse at least 1 cm in the atmosphere, are ejected by fully formed asci with force. As a result, the ascospores are able to ascend to the more turbulent air above and escape the stagnant layer of air close to the soil surface (Abawi and Grogan, 1979). In turbulent air, ascospores are continuously released; in calm air, they are released in brief bursts that last for about 0.1 s. As a result, vast columns of ascospores begin to grow in the air above the apothecium. Since it is a more effective mechanism for ejecting ascospores through the still air layer, calm air causes a large quantity of ascospores to be released. This guarantees that any minute air currents that may exist will carry away some ascospores at the very least (Willetts and Wong, 1980).

4.15.21.1 Conditions for discharge of ascospore from apothecia

Apothecia require a minimum of 98% RH to maintain their firmness, and they will collapse if the RH falls below this threshold. However, when placed back in an environment with 100% RH, apothecia can recover. Dry apothecia are not a significant source of inoculum due to high temperatures and dry soils causing them to lose moisture and prevent the discharge of ascospores. Temperature alone determines the release of ascospores as long as there is sufficient moisture. The optimal temperature for ascospore discharge is 22°C, occurring within a range of 4–32°C. Ascospore release is significantly reduced at lower temperatures than the ideal range; at 4°C, hardly any ascospores are released. Overly warm temperatures also prevent the release of ascospores, which has a long-term detrimental effect on apothecia.

Field studies have demonstrated that, as long as the apothecia are still turgid, warm, windy weather causes a large number of ascospores to be released. However, ascospore discharge is limited in humid and rainy conditions (Kruger, 1975a). When it rains, ascospores are not released into the atmosphere but instead spread into the water films on the apothecia and are finally washed into the soil (Kruger, 1975a, 1980). Therefore, if strong rains coincide with ascospore release, they can effectively lower the disease level in a rapeseed crop. Studies on rapeseed (Williams and Stelfox, 1979, 1980a, b) and beans (Abawi and Grogan, 1975; Akai, 1981) have provided evidence for the phenomenon of ascospores being liberated from apothecia and being viable for several kilometers when carried by wind. Fields far from the apothecia have shown disease incidence, showing the important mechanism of wind-mediated ascospore dissemination. Ascospores can be discovered up to 150 cm above the soil's surface, according to spore trapping research, which further supports their long-distance dispersal. Ascospores are also present on pollen grains; however, the spread of illness is not significantly aided by bees' transfer of ascospores (Williams and Stelfox, 1979). Two primary components seem to be involved in ascospore dissemination. The ascospores first need wind to get to hosts that are vulnerable. Additionally, by blocking ascospores' access to the air currents above the canopy, the crop canopy may hinder their dissemination. It's likely that different crops can prevent ascospores from spreading in different ways (Williams and Stelfox, 1979).

4.16 Survival of Ascospores

Ascospores exhibit the highest likelihood of survival when they are dry, indicating that their ability to endure is affected by both temperature and moisture. In natural field conditions, ascospores can persist for up to 12 days (Grogan and Abawi, 1975). However, in greenhouse conditions, they can remain viable for a longer period of 2–5 months. Ascospores are more viable at 60% RH than they are at 80 and 98% RH, according to research (Partyka and Mai, 1962; Newton and Sequeira, 1972; Scheibert-Bohm *et al.*, 1981). *S. sclerotiorum* ascospore viability is especially harmed by temperatures of 25°C or higher and RH levels of more than 35%. Ascospore death rates typically rise in tandem with changes in temperature and RH. Ascospore survival rates beneath plastic sheets that transmit 10 and 50% UV radiation in the 300–400 nm band, respectively, were reported to be 49 and 13.4% in laboratory trials following exposure to two FS-40 sunlamps for 224 hour (Caesar and Pearson, 1983).

4.17 Ascospore Germination

Ascospore germination appears to be unaffected by temperature variations. In a controlled laboratory environment, the majority of ascospores will start to germinate in 6 hour at temperatures between 10 and 30°C (Abawi and Grogan, 1975). Between 20 and 25°C is the ideal temperature range for germ-tube growth and the start of infection; temperatures above and below this range are associated with a decrease in growth rate (Abawi and Grogan, 1975). The percentage or rate of germination is unaffected by water potentials as low as −56 bars. However, germination is diminished as water potentials drop even further, and it is entirely suppressed at a water potential of −91 bars (Grogan and Abawi, 1975). On the other hand, free water is necessary for host infection to happen. According to Abawi and Grogan (1975), ascospores require at least 48–72 hour of continuous leaf wetness in order to infect beans. Additionally, RH values near 100% are insufficient for infection to occur. The development of germ tubes, which are crucial for infection, is dependent less on ascospore germination and more on the presence of free water at the ascospore–host tissue interface. It's interesting to note that laboratory tests have shown that water potentials between −1 and −14 bars promote the formation of germ tubes, indicating that a water potential of 0 bars is not necessary (Grogan and Abawi, 1975). However, more investigation is required to get more details on this particular element.

4.18 Ascospore Penetration and Infection Process

It has been noted that ascospores at the germination stage directly infiltrate healthy tissues (Purdy, 1979). Energy is needed for both the ascospores' germination and the host's penetration. Although ascospores have sufficient stocks of stored nutrients for germination and the formation of germ tubes, they lack the energy necessary for the creation of appressoria (Lumsden, 1979). Therefore, an external source of nutrients must exist in the infection arena prior to host penetration (Purdy, 1958; Abawi and Grogan, 1975, 1979; Willetts and Wong, 1980). Dead floral components are crucial in providing these nutrients, according to multiple studies (Purdy and Bardin, 1953; McLean, 1958; Natti, 1971; Abawi *et al.*, 1975a, b ; Kruger, 1975b; Abawi and Grogan, 1979; Akai, 1981). Furthermore, pollen grains act as a rich source of nutrients and promote ascospore germination (Stelfox *et al.*, 1978). In addition, ascospore germination and subsequent host penetration are facilitated by wounds or senescent tissues that are still attached to the plant (Newton and Sequeira, 1972; Abawi and Grogan, 1979; Willetts and Wong, 1980). See the section on infection and pathogenesis in the book *Sclerotinia Stem Rot of Crucifers: Epidemiology and Disease Management Vol–II* for additional details on the host infection process.

References

Abawi, G.S. and Grogan, R.G. (1975) Source of primary inoculum and effects of temperature and moisture on infection of beans by *Whetzelinia sclerotiorum*. *Phytopathology* 65(3), 300.

Abawi, G.S. and Grogan, R.G. (1979) Epidemiology of diseases caused by *Sclerotinia* species. *Phytopathology* 69, 899–904.

Abawi, G.S., Polach, F.J. and Molin, W.T. (1975a) Infection of bean by ascospores of *Whetzelinia sclerotiorum*. *Phytopathology* 65(6), 673.

Abawi, G.S., Provvidenti, R., Grogan, R.G. and Hunter, J.E. (1975b) Predisposition of beans to infection by ascospores of *Whetzelinia sclerotiorum* prior to blossoming. *Proceeding of American Phytopathological Society* 2, 61.

Abeliovich, H. and Klionsky, D.J. (2001) Autophagy in yeast: Mechanistic insights and physiological function. *Microbiology and Molecular Biology Reviews* 65(3), 463–479.

Adams, P.B. (1975) Factors affecting survival of *Sclerotinia sclerotiorum* in soil. *Plant Disease Reporter* 59, 599–603.

Adams, P.B. and Tate, C.J. (1976) Mycelial germination of sclerotia of *Sclerotinia sclerotiorum* on soil. *Plant Disease Reporter* 60, 515–518.

Adams, P.B. and Ayers, W.A. (1979) Ecology of *Sclerotinia* species. *Phytopathology* 69, 896–899.

Adams, P.B. (1987) Effects of soil temperature, moisture, and depth on survival and activity of *Sclerotinia minor*, *Sclerotium cepivorum*, and *Sporidesmium sclerotivorum*. *Plant Disease* 71(2), 170.

Akai, J. (1981) Studies on the epidemiology and control of Sclerotinia disease of beans caused by *Sclerotinia sclerotiorum* (Lib.) de Bary. *Hokkaido Prefectural Agricultural Experiment Station Report* 36, 77–79.

Aldrich-Wolfe, L., Travers, S. and Nelson, B.D. (2015) Genetic variation of *Sclerotinia sclerotiorum* from multiple crops in the North central United States. *PLoS ONE* 10, e0139188.

Asakura, M., Ninomiya, S., Sugimoto, M., Oku, M., Yamashita, S.I. *et al.* (2009) Atg26-mediated pexophagy is required for host invasion by the plant pathogenic fungus *Colletotrichum orbiculare*. *The Plant Cell* 21(4), 1291–1304.

Attanayake, R.N., Carter, P.A., Jiang, D., Rio-Mendoza, L. and Chen, W. (2013) *Sclerotinia sclerotiorum* populations infecting canola from China and the United States are genetically and phenotypically distinct. *Phytopathology* 103, 750–761.

Attanayake, R.N., Tennekoon, V., Johnson, D.A., Porter, L.D., del Río-Mendoza, L. *et al.* (2014) Inferring out crossing in the homothallic fungus *Sclerotinia sclerotiorum* using linkage disequilibrium decay. *Heredity* 113(4), 353–363.

Bashi, Z.D., Gyawali, S., Bekkaoui, D., Coutu, C., Lee, L. *et al.* (2016) The *Sclerotinia sclerotiorum* Slt2 mitogen-activated protein kinase ortholog, SMK3, is required for infection initiation but not lesion expansion. *Canadian Journal of Microbiology* 62, 836–850. DOI: 10.1139/cjm-2016-0091.

Beck, H.E., Zimmermann, N.E., McVicar, T.R., Vergopolan, N., Berg, A. *et al.* (2018) Present and future Köppen-Geiger climate classification maps at 1-km resolution. [Published correction appears in Science Data; 7, 274]. *Scientific Data* 5, 180214.

Bedi, K.S. (1958) Effect of ultra-violet radiation of sclerotia of *Sclerotinia sclerotiorum* de Bary on speed of their germination and apothecial development. *Indian Phytopathology* 11, 37–39.

Bedi, K.S. (1962) Light, air and moisture in relation to the formation of apothecia of *Sclerotinia sclerotiorum* (Lib.) de Bary. *Proceedings/Indian Academy of Sciences* 55(5), 213–223.

Bedi, K.S. (1963) The age and the size of sclerotia of *Sclerotinia sclerotiorum* (Lib.) de Bary in relation to the formation of apothecia. *Indian Botanical Society* 42, 204–207.

Bell, A.A. and Wheeler, M.H. (1986) Biosynthesis and functions of fungal melanins. *Annual Review of Phytopathology* 24(1), 411–451.

Brown, J.G. (1937) Relation of livestock to the control of sclerotinosis of lettuce. *Phytopathology* 27, 1045–1050.

Brown, J.G. and Butler, K.D. (1936) Sclerotiniase of lettuce in Arizona. *Agricultural Experimental Station Technical Bulletin* 63, 475–506.

Caesar, A.J. and Pearson, R.C. (1983) Environmental factors affecting survival of ascospores of *Sclerotinia sclerotiorum*. *Phytopathology* 73(7), 1024.

Chaves, M.S., Martinelli, J.A. and Loch, L.C. (1996) Effect of root exudates of soybeans on carpogenic germination of *Whetzelinia sclerotiorum* sclerotia. *Summa Phytopathologica* 22, 256–258.

Chen, C. and Dickman, M.B. (2005) CAMP blocks MAPK activation and sclerotial development via rap-1 in a PKA-independent manner in *Sclerotinia sclerotiorum*. *Molecular Microbiology* 55(1), 299–311.

Chen, C., Harel, A., Gorovoits, R., Yarden, O. and Dickman, M.B. (2004) MAPK regulation of sclerotial development in *Sclerotinia sclerotiorum* is linked with pH and cAMP sensing. *Molecular Plant-Microbe Interactions* 17(4), 404–413.

Chet, I. and Henis, Y. (1975) Sclerotial morphogenesis in fungi. *Annual Review of Phytopathology* 13(1), 169–192.

Christias, C. and Lockwood, J.L. (1973) Conservation of mycelial constituents in four *Sclerotium* forming fungi in nutrient-deprived conditions. *Phytopathology* 63(5), 602.

Clarkson, J.P., Coventry, E., Kitchen, J., Carter, H.E. and Whipps, J.M. (2013) Population structure of *Sclerotinia sclerotiorum* in crop and wild hosts in the UK. *Plant Pathology* 62(2), 309–324.

Coley-Smith, J.R. and Cooke, R.C. (1971) Survival and germination of fungal sclerotia. *Annual Review of Phytopathology* 9(1), 65–92.

Colotelo, N. (1973) Physiological and biochemical properties of the exudate associated with developing sclerotia of *Sclerotinia sclerotiorum* (Lib.) de Bary. *Canadian Journal of Microbiology* 19(1), 73–79.

Colotelo, N. (1974) A scanning electron microscope study of developing sclerotia of *Sclerotinia sclerotiorum*. *Canadian Journal of Botany* 52(5), 1127–1130.

Cong, J., Xiao, K., Jiao, W., Zhang, C., Zhang, X. *et al.* (2022) The coupling between cell wall integrity mediated by MAPK kinases and SsFkh1 is involved in sclerotia formation and pathogenicity of *Sclerotinia sclerotiorum*. *Frontier in Microbiology* 13, 816091. DOI: 10.3389/fmicb.2022.816091.

Cook, G.E. (1973) Some aspects of the epidemiology of white mold on dry beans caused by *Whetzelinia sclerotiorum* in Western Nebraska. PhD. thesis, University of Nebraska, Lincoln.

Cook, G.E., Steadman, J.R. and Boosalis, M.G. (1975) Survival of *Whetzelinia sclerotiorum* and initial infection of dry edible beans in Western Nebraska. *Phytopathology* 65(3), 250.

Corsini, D.L. and Le Tourneau, D. (1973) Organic acid metabolism in *Sclerotinia sclerotiorum*. *Archiv Fur Mikrobiologie* 90(1), 59–64.

D'Arcy, M.S. (2019) Cell death: A review of the major forms of apoptosis, necrosis and autophagy. *Cell Biology International* 43(6), 582–592.

Davis, W.H. (1925) Drop of Chinese cabbage and our common cabbage caused by *Sclerotinia sclerotiorum* (Lib.) de Bary (*Sclerotinia libertiana Fckl*). *Phytopathology* 15, 249–259.

Denton-Giles, M., Derbyshire, M.C., Khentry, Y., Buchwaldt, L. and Kamphuis, L.G. (2018) Partial stem resistance in *Brassica napus* to highly aggressive and genetically diverse *Sclerotinia sclerotiorum* isolates from Australia. *Canadian Journal of Plant Pathology* 40(4), 551–561.

Derbyshire, M.C., Denton-Giles, M., Hane, J.K., Chang, S., Mousavi-Derazmahalleh, M. *et al.* (2019) A whole genome scan of SNP data suggests a lack of abundant hard selective sweeps in the genome of the broad host range plant pathogenic fungus *Sclerotinia sclerotiorum*. *PLoS ONE* 14(3), e0214201.

Dickman, M.B. and Mitra, A. (1992) *Arabidopsis thaliana* as a model for studying *Sclerotinia sclerotiorum* pathogenesis. *Physiological and Molecular Plant Pathology* 41(4), 255–263.

Dillard, H.R., Ludwig, J.W. and Hunter, J.E. (1995) Conditioning sclerotia of *Sclerotinia sclerotiorum* for carpogenic germination. *Plant Disease* 79(4), 411.

Dillon Weston, W.A.R., Loveless, A.R. and Taylor, R.E. (1946) Clover rot. *The Journal of Agricultural Science* 36(1), 18–28.

Dimopoulou, L. (1975) Influence de quelques substances *in vitro* sur la formation des sclerotes du *Sclerotinia sclerotiorum*. *Annals de l'Institut Phytopathologique Benaki* 11, 127–131.

Doughan, B., Rollins, J.A. and Avery, S.V. (2016) Characterization of *MAT* gene functions in the life cycle of *Sclerotinia sclerotiorum* reveals a lineage-specific *MAT* gene functioning in apothecium morphogenesis. *Fungal Biology* 120(9), 1105–1117. DOI: 10.1016/j.funbio.2016.06.007.

Dueck, J. (1977) Sclerotinia in rapeseed. *Canada Agriculture* 22, 7–9.

Dueck, J. and Morrall, R.A.A. (1971) *Sclerotinia* in Saskatchewan in 1970. *Canadian Plant Disease Survey* 51, 116–121.

Durman, S.B., Menéndez, A.B. and Godeas, A.M. (2003) Mycelial compatibility groups in Buenos Aires field populations of *Sclerotinia sclerotiorum* (Sclerotiniaceae). *Australian Journal of Botany* 51(4), 421.

Erental, A., Harel, A. and Yarden, O. (2007) Type 2A phosphoprotein phosphatase is required for asexual development and pathogenesis of *Sclerotinia sclerotiorum*. *Molecular Plant-Microbe Interactions* 20(8), 944–954. DOI: 10.1094/MPMI-20-8-0944.

Erental, A., Dickman, M.B. and Yarden, O. (2008) Sclerotial development in *Sclerotinia sclerotiorum*: Awakening molecular analysis of a "Dormant" structure. *Fungal Biology Reviews* 22(1), 6–16. DOI: 10.1016/j.fbr.2007.10.001.

Fan, H., Yu, G., Liu, Y., Zhang, X., Liu, J. *et al.* (2017) An atypical forkhead-containing transcription factor SsFKH1 is involved in sclerotial formation and is essential for pathogenicity in *Sclerotinia sclerotiorum*. *Molecular Plant Pathology* 18(7), 963–975.

Fitzgerald, A., Kan, J.A.L. and Plummer, K.M. (2004) Simultaneous silencing of multiple genes in the apple scab fungus, *Venturia inaequalis*, by expression of RNA with chimeric inverted repeats. *Fungal Genetics and Biology* 41(10), 963–971.

Garg, H., Sivasithamparam, K. and Barbetti, M.J. (2010a) Scarification and environmental factors that enhance carpogenic germination of sclerotia of *Sclerotinia sclerotiorum*. *Plant Disease* 94(8), 1041–1047.

Garg, H., Kohn, L.M., Andrew, M., Li, H., Sivasithamparam, K. *et al.* (2010b) Pathogenicity of morphologically different isolates of *Sclerotinia sclerotiorum* with *Brassica napus* and *B. juncea* genotypes. *European Journal of Plant Pathology* 126(3), 305–315.

Godoy, G., Steadman, J.R., Dickman, M.B. and Dam, R. (1990) Use of mutants to demonstrate the role of oxalic acid in pathogenicity of *Sclerotinia sclerotiorum* on *Phaseolus vulgaris*. *Physiological and Molecular Plant Pathology* 37(3), 179–191.

Govrin, E.M. and Levine, A. (2000) The hypersensitive response facilitates plant infection by the Necrotrophic pathogen Botrytis cinerea. *Current Biology* 10(13), 751–757.

Grogan, R.G. (1979) *Sclerotinia* species: Summary and comments on needed research. *Phytopathology* 69(8), 908.

Grogan, R.G. and Abawi, G.S. (1975) Influence of water potential on growth and survival of *Whetzelinia sclerotiorum*. *Phytopathology* 65(2), 122.

Harel, A., Gorovits, R. and Yarden, O. (2005) Changes in protein kinase A activity accompany sclerotial development in *Sclerotinia sclerotiorum*. *Phytopathology* 95(4), 397–404.

Harel, A., Bercovich, S. and Yarden, O. (2006) Calcineurin is required for sclerotial development and pathogenicity of *Sclerotinia sclerotiorum* in an oxalic acid-independent manner. *Molecular Plant-Microbe Interactions* 19(6), 682–693.

Hartill, W.F.T. and Underhill, A.P. (1976) Puffing in *Sclerotinia sclerotiorum* and *S. minor*. *New Zealand Journal of Botany* 14, 355–358.

Hau, F.C., Beute, M.K. and Smith, T. (1982) Effect of soil pH and volatile stimulants from remoistened peanut leaves on germination of sclerotia of *Sclerotinia* minor. *Plant Disease* 66(1), 223.

Hawthorne, B.T. (1973) Production of apothecia of *Sclerotinia minor*. *New Zealand Journal of Agricultural Research* 16, 559–560.

Hawthorne, B.T. (1976) Observations on the development of apothecia of *Sclerotinia minor* Jagg. in the field. *New Zealand Journal of Agricultural Research* 19(3), 383–386.

Henderson, R.M. (1962) An inhibitory growth correlation in the apothecial stipe of *Sclerotinia sclerotiorum*. *Nature* 195(4843), 826–826.

Henson, J.M., Butler, M.J. and Day, A.W. (1999) The dark side of the mycelium: Melanins of phytopathogenic fungi. *Annual Review of Phytopathology* 37, 447–471.

Hieu, N.T. (2007) Iintegrated management of sclerotinia rot (*Sclerotinia sclerotiorum* (Lib.) de Bary) of indian mustard (*Brassica juncea* L.) czern & coss, MSc.Thesis submitted to CCS Haryana Agricultural University, Hisar, India, p. 150.

Hoes, J.A. and Huang, H.C. (1975) *Sclerotinia sclerotiorum*: Viability and separation of sclerotia from soil. *Phytopathology* 65(12), 1431.

Hoes, J.A. and Huang, H.C. (1976) Control of *Sclerotinia* basal stem rot of sunflower: A progress report. In: *Proceedings of the Sunflower Forum*, National Sunflower Association, Fargo, North Dakota, p. 18.

Honda, Y.H. and Yunoki, T. (1977) Control of *Sclerotinia* disease of greenhouse eggplant and cucumber by inhibition of development of apothecia. *Plant Disease Reporter* 61, 1036–1040.

Howard, R.J. and Valent, B. (1996) Breaking and entering host penetration by the fungal rice blast pathogen *Magnaporthe grisea*. *Annual Review of Microbiology* 50, 491–512.

Huang, H.C. (1977) Importance of *Coniothyrium minitans* in survival of sclerotia of *Sclerotinia sclerotiorum* in wilted sunflowers. *Canadian Journal of Botany* 55, 289–295.

Huang, H.C. (1991) Induction of myceliogenic germination of sclerotia of *Sclerotinia sclerotiorum* by exposure to sub-freezing temperatures. *Plant Pathology* 40(4), 621–625.

Huang, H.C. (1979) Biological control of Sclerotinia wilt in sunflower. *Report Canada Agriclture* 24(3), 12–14.

Huang, H.C. and Hoes, J.A. (1980) Importance of plant spacing and sclerotial position to development of Sclerotinia wilt of sunflower. *Plant Disease* 64(1), 81.

Huang, H.C. and Dueck, J. (1980) Wilt of sunflower from infection by mycelial-germinating sclerotia of *Sclerotinia sclerotiorum*. *Canadian Journal of Plant Pathology* 2(2), 47–52.

Huang, H.C. and Kozub, G.C. (1994) Longevity of normal and abnormal sclerotia of *Sclerotinia sclerotiorum*. *Plant Disease* 78(12), 1164.

Huang, C., Li, L., Wang, L., Bao, J., Zhang, X. *et al*. (2022) The amino acid permease MoGAP1 regulates TOR activity and autophagy in *Magnaporthe oryzae*. *International Journal of Molecular Sciences* 23(21), 13663.

Huang, H.C., Chang, C. and Kozub, G.C. (1998) Effect of temperature during sclerotial formation, sclerotial dryness, and relative humidity on myceliogenic germination of sclerotia of *Sclerotinia sclerotiorum*. *Canadian Journal of Botany* 76(3), 494–499.

Humpherson-Jones, F.M. and Cooke, R.C. (1977a) Morphogenesis in sclerotium forming fungi II. Rhythmic production of sclerotia by *Sclerotinia sclerotiorum* (Lib.) de Bary. *New Phytologists* 78, 181–187.

Humpherson-Jones, F.M. and Cooke, R.C. (1977b) Induction of sclerotium formation by acid staling compounds in *Sclerotinia sclerotiorum* and *Sclerotium rolfsii*. *Transactions of the British Mycological Society* 68(3), 413–420.

Ichimura, Y., Kirisako, T., Takao, T., Satomi, Y., Shimonishi, Y. *et al.* (2000) A ubiquitin-like system mediates protein lipidation. *Nature* 408(6811), 488–492.

Imolehin, E.D. and Grogan, R.G. (1980) Factors affecting survival of sclerotia, and effects of inoculum density, relative position, and distance of sclerotia from the host on infection of lettuce by *Sclerotinia minor*. *Phytopathology* 70(12), 1162.

Iqbal, S.M., Ghafoor, A., Ahmad, Z. and Haqqani, A.M. (2003) Pathogenicity and fungicidal efficacy for *Sclerotinia* rot of brinjal. *International Journal of Agriculture and Biology* 5(4), 618–620.

Jiao, W., Yu, H., Chen, X., Xiao, K., Jia, D. *et al.* (2022a) The SsAtg1 activating autophagy is required for sclerotia formation and pathogenicity in *Sclerotinia sclerotiorum*. *Journal of Fungi* 8, 1314. DOI: 10.3390/jof8121314.

Jiao, W., Yu, H., Cong, J., Xiao, K., Zhang, X. *et al.* (2022b) Transcription factor SsFoxE3 activating SsAtg8 is critical for sclerotia, compound appressoria formation, and pathogenicity in *Sclerotinia sclerotiorum*. *Molecular Plant Pathology* 23(2), 204–217.

Jones, D. (1970) Ultrastructure and composition of the cell walls of *Sclerotinia sclerotiorum*. *Transactions of the British Mycological Society* 54(3), 351–360.

Jones, D. (1974) Ultrastructure of the stipe and apothecium of *Sclerotinia sclerotiorum*. *Transactions of the British Mycological Society* 63(2), 386–389.

Jurick, W.M., II and Rollins, J.A. (2007) Deletion of the adenylate cyclase (*sac1*) gene affects multiple developmental pathways and pathogenicity in *Sclerotinia sclerotiorum*. *Fungal Genetics and Biology* 44(6), 521–530. DOI: 10.1016/j.fgb.2006.11.005.

Jurick, W.M., II, Dickman, M.B. and Rollins, J.A. (2004) Characterization and functional analysis of a cAMP dependent protein kinase A catalytic subunit gene (*pka1*) in *Sclerotinia sclerotiorum*. *Physiological and Molecular Plant Pathology* 64(3), 155–163.

Kakoti, R.K. and Saikia, U.N. (1997) Sclerotial survival of *Sclerotinia sclerotiorum* (Lib.) de Bary in soil. *Journal of Agricultural Scientific Society North-East India* 10, 265–267.

Kamal, M., Savocchia, S., Lindbeck, K. and Ash, G.J. (2016) Biology and biocontrol of *Sclerotinia sclerotiorum* (Lib.) de Bary in oilseed *Brassicas*. *Australasian Plant Pathology* 45(1), 1–14.

Kapoor, K.S. (1994) Stipe regeneration potential of sclerotia of *Sclerotinia sclerotiorum* (Lib.) de Bary. *Plant Disease Research* 9, 98–100.

Khalil, F.A. and Ragab, M.M. (1990) The behavior and survival of *Sclerotinia sclerotiorum* in relation to sclerotial lipid contents. *Egypt Journal of Phytopathology* 20, 107–116.

Kim, K.G. (1976) Effect of light on the sclerotial formation of *Sclerotinia sclerotiorum* (Lib.) de Bary. *Korean Journal of Plant Protection* 15, 153–159.

Kim, K.S., Min, J.Y. and Dickman, M.B. (2008) Oxalic acid is an elicitor of plant programmed cell death during *Sclerotinia sclerotiorum* disease development. *Molecular Plant-Microbe Interactions* 21, 605–612.

Kohli, Y., Brunner, L.J., Yoell, H., Milgroom, M.G., Anderson, J.B. *et al.* (1995) Clonal dispersal and spatial mixing in populations of the plant pathogenic fungus *Sclerotinia sclerotiorum*. *Molecular Ecology* 4, 69–77.

Kosasih, B.D. and Willetts, H.J. (1975) Types of abnormal apothecia produced by *Sclerotinia sclerotiorum*. *Mycologia* 67, 89.

Krishnamoorthy, K.K., Sankaralingam, A. and Nakkeeran, S. (2017) Effect of soil moisture and duration on the viability of sclerotia of *Sclerotinia sclerotiorum* causing head rot of cabbage. *Journal of Soil Biology and Ecology* 37(1), 7–14.

Kruger, W. (1973) Control measures for *Sclerotinia sclerotiorum* in rape. *Phytopathology Z* 77, 125–137.

Kruger, W. (1974) Investigation concerning the epidemiology of rape cankers giving rise to *Sclerotinia sclerotiorum* (Lib.) de Bary. In: *Proceedings of the 4th International Rapeseed Conference*, Global Council for Innovation in Rapeseed and Canola (GCIRC), Giessen, Germany, pp. 595–603.

Kruger, W. (1975a) Influence of more environmental factors on the development of apothecia and ascospores of *Sclerotinia sclerotiorum* (Lib.) de Bary. *Z Pflimzenkaran Kh Planzensehtz* 82, 101–108.

Kruger, W. (1975b) Influence of weather on the attack of rape by *Sclerotinia sclerotiorum* (Lib.) de Bary. *Nachrichtenble Devet. Pflanzenschutzd (Braunschweig)* 27, 1–6.

Kruger, W. (1976) The influence of environmental factors on the apothecial development of *Sclerotinia sclerotiorum* (Lib.) de Bary. *Nachrichtenblatt Deut. Pflanzensehtzd. (Braunschweig* 28, 130–135.

Kruger, W. (1980) On the effect of calcium cyanamide on the development of apothecia of *Whetzelinia sclerotiorum* (Lib.) Korf and Dumont, the agent of stalk rot of rape. *Review of Plant Pathology* 59, 5438.

Lane, D., Kamphuis, L.G., Derbyshire, M.C. and Denton-Giles, M. (2018) Heat-dried sclerotia of *Sclerotinia sclerotiorum* myceliogenically germinate in water and are able to infect *Brassica napus*. *Crop and Pasture Science* 69(8), 765. DOI: 10.1071/CP18109.

Lane, D., Denton-Giles, M., Derbyshire, M. and Kamphuis, L.G. (2019) Abiotic conditions governing the myceliogenic germination of *Sclerotinia sclerotiorum* allowing the basal infection of *Brassica napus*. *Australasian Plant Pathology* 48(2), 85–91. DOI: 10.1007/s13313-019-0613-0.

Le Tourneau, D. (1976) Effect of phenylthiourea on growth and sclerotial formation of *Sclerotium rolfsii* and *Whetzelinia sclerotiorum*. *Canadian Journal of Microbiology* 22, 379–383.

Le Tourneau, D. (1979) Morphology, cytology, and physiology of *Sclerotinia* species in culture. *Phytopathology* 69(8), 887.

Letham, D.B. (1975) Stimulation by light of apothecial initial development of *Sclerotinia sclerotiorum*. *Transactions of the British Mycological Society* 65(2), 333–335.

Leyronas, C., Bardin, M., Berthier, K., Duffaud, M., Troulet, C. *et al.* (2018) Assessing the phenotypic and genotypic diversity of *Sclerotinia sclerotiorum* in France. *European Journal of Plant Pathology* 152(4), 933–944.

Li, M. and Rollins, J.A. (2009) The development-specific protein (Ssp1) from *Sclerotinia sclerotiorum* is encoded by a novel gene expressed exclusively in sclerotium tissues. *Mycologia* 101(1), 34–43.

Li, M., Liang, X. and Rollins, J.A. (2012) *Sclerotinia sclerotiorum* γ-glutamyl transpeptidase (Ss-Ggt1) is required for regulating glutathione accumulation and development of sclerotia and compound appressoria. *Molecular Plant-Microbe Interactions* 25(3), 412–420.

Liang, X. and Rollins, J.A. (2018) Mechanisms of broad host range necrotrophic pathogenesis in *Sclerotinia sclerotiorum*. *Phytopathology* 108(10), 1128–1140.

Liang, Y., Xiong, W., Steinkellner, S. and Feng, J. (2018) Deficiency of the melanin biosynthesis genes SCD1 and THR1 affects sclerotial development and vegetative growth, but not pathogenicity, in *Sclerotinia sclerotiorum*. *Molecular Plant Pathology* 19(6), 1444–1453. DOI: 10.1111/mpp.12627.

Liberti, D., Rollins, J.A. and Dobinson, K.F. (2013) Peroxysomal carnitine acetyl transferase influences host colonization capacity in *Sclerotinia sclerotiorum*. *Molecular Plant-Microbe Interactions* 26, 768–780. DOI: 10.1094/PHYTO-06-18-0197-RVW.

Liu, J., Meng, Q., Zhang, Y., Xiang, H., Li, Y. *et al.* (2018) Mycelial compatibility group and genetic variation of sunflower *Sclerotinia sclerotiorum* in northeast China. *Physiological and Molecular Plant Pathology* 102, 185–192.

Liu, X.H., Lu, J.P. and Lin, F.C. (2007) Autophagy during conidiation, conidial germination and turgor generation in *Magnaporthe grisea*. *Autophagy* 3(5), 472–473.

Liu, X.H., Xu, F., Snyder, J.H., Shi, H.B., Lu, J.P. *et al.* (2016) Autophagy in plant pathogenic fungi. *Seminars in Cell and Developmental Biology* 57, 128–137.

Lumsden, R.D. (1979) Histology and physiology of pathogenesis in plant diseases caused by *Sclerotinia* species. *Phytopathology* 69(8), 890.

Lv, W., Wang, C., Yang, N., Que, Y., Talbot, N.J. *et al.* (2017) Genome-wide functional analysis reveals that autophagy is necessary for growth, sporulation, deoxynivalenol production and virulence in *Fusarium graminearum*. *Scientific Reports* 7(1), 11062.

Lyu, X., Shen, C., Fu, Y., Xie, J., Jiang, D. *et al.* (2015) Comparative genomic and transcriptional analyses of the carbohydrate-active enzymes and secretomes of phytopathogenic fungi reveal their significant roles during infection and development. *Scientific Reports* 5, 15565. DOI: 10.1038/srep15565.

Malvárez, G., Carbone, I., Grünwald, N.J., Subbarao, K.V., Schafer, M. *et al.* (2007) New populations of *Sclerotinia sclerotiorum* from lettuce in California and peas and lentils in Washington. *Phytopathology* 97(4), 470–483.

Marukawa, S. and Satomura, Y. (1977) Conditions for reversion of sclerotial formation in a mutant of *Sclerotinia libertiana* fuckel. *Agricultural and Biological Chemistry* 41(4), 655–660.

Marukawa, S., Funakawa, S. and Satomura, Y. (1975) Some physical and chemical factors on formation of sclerotia in *Sclerotinia libertiana* fuckel. *Agricultural and Biological Chemistry* 39(2), 463–468.

Matheron, M.E. and Porchas, M. (2005) Influence of soil temperature and moisture on eruptive germination and viability of sclerotia of *Sclerotinia minor* and *S. sclerotiorum*. *Plant Disease* 89(1), 50–54.

Maxwell, D.P. (1973) Oxalate formation in *Whetzelinia sclerotiorum* by oxaloacetate acetylhydrolase. *Physiological Plant Pathology* 3(2), 279–288.

Maxwell, D.P. and Lumsden, R.D. (1970) Oxalic acid production by *Sclerotinia sclerotiorum* in infected bean and in culture. *Phytopathology* 60(9), 1395.

McLean, D.M. (1958) Role of dead flower parts in infection of certain crucifers by *Sclerotinia sclerotiorum* (Lib.) de Bary. *Plant Disease Reporter* 42, 663–666.

Mehta, N., Hieu, N.T. and Sangwan, M.S. (2009) Influence of soil types and frequency of irrigation on the development of Sclerotinia stem rot of mustard. *Journal of Mycology and Plant Pathology* 39(3), 506–510.

Melouk, H.A., Singleton, L.L., Oweres, F.N. and Akem, C.N. (1989) Viability of sclerotia of *Sclerotinia sclerotiorum* after passage through the digestive track of a crossbred heifer. *Plant Disease* 73(1), 68.

Merriman, P.R. (1976) Survival of sclerotia of *Sclerotinia sclerotiorum* in soil. *Soil Biology and Biochemistry* 8(5), 385–389.

Merriman, P.R., Pywell, M., Harrison, G. and Nancarrow, J. (1979) Survival of sclerotia of *Sclerotinia sclerotiorum* and effects of cultivation practices on disease. *Soil Biology and Biochemistry* 11(6), 567–570.

Michael, P.J., Lui, K.Y., Thomson, L.L., Stefanova, K. and Bennett, S.J. (2020) Carpogenic germinability of diverse *Sclerotinia sclerotiorum* populations within the south western Australian grain belt. *Plant Disease* 104(11), 2891–2897. DOI: 10.1094/PDIS-12-19-2575-RE.

Michael, P.J., Lui, K.Y., Thomson, L.I., Lamichhane, A.R. and Bennett, J. (2021) Impact of preconditioning temperature and duration period on carpogenic germination of diverse *Sclerotinia sclerotiorum* populations in southwestern Australia. *Plant Disease* 105(6), 1798–1805.

Michell, A.J. and Scurfield, G. (1967) Composition of extracted fungal cell walls as indicated by infrared spectroscopy. *Archives of Biochemistry and Biophysics* 120(3), 628–637.

Mitchell, S.J. and Wheeler, B.E.J. (1990) Factors affecting the production of apothecia and longevity of sclerotia of *Sclerotinia sclerotiorum*. *Plant Pathology* 39(1), 70–76.

Mizushima, N. and Komatsu, M. (2011) Autophagy: Renovation of cells and tissues. *Cell* 147(4), 728–741.

Mizushima, N., Noda, T., Yoshimori, T., Tanaka, Y., Ishii, T. *et al.* (1998) A protein conjugation system essential for autophagy. *Nature* 395(6700), 395–398.

Moore, W.D. (1949) Flooding as a means of destroying sclerotia of *Sclerotinia sclerotiorum*. *Phytopathology* 39, 920–927.

Morrall, R.A.A. (1977) A preliminary study of the influence of water potential on sclerotium germination in *Sclerotinia sclerotiorum*. *Canadian Journal of Botany* 55(1), 8–11.

Morrall, R.A.A. and Dueck, J. (1982) Epidemiology of Sclerotinia stem rot of rapeseed in Saskatchewan. *Canadian Journal of Plant Pathology* 4(2), 161–168.

Morrall, R.A.A., Duczek, I.J., Mc Kenzie, D.L. and McGee, D.C. (1976) Some aspects of *Sclerotinia sclerotiorum* in Saskatchewan, 1970–75. *Canadian Plant Disease Survey* 56, 56.

Mosa, A.A., Abd Allah, S.M. and Ibrahim, M.M. (2000) Factors affecting germination of sclerotia of *Sclerotinia sclerotiorum*. *Annals of Agricultural Science Cairo* 4(S), 1353–1371.

Nagy, P. and Fischl, G. (2002) Effect of UV and visible light irradiation on mycelial growth and sclerotium formation of *Sclerotinia sclerotiorum*. *Acta Phytopathologica et Entomologica Hungarica* 37(1–3), 83–89.

Natti, J.J. (1971) Epidemiology and control of bean white mold. *Phytopathology* 61(6), 669.

Nepal, A. and Del Río Mendoza, L.E. (2012) Effect of sclerotial water content on carpogenic germination of *Sclerotinia sclerotiorum*. *Plant Disease* 96(9), 1315–1322.

Newton, H.K. and Sequeira, L. (1972) Ascospores as the primary infective propagules of *Sclerotinia sclerotiorum* in Wisconsin. *Plant Disease Reporter* 56, 798–802.

Norman, H.C., Galwey, N.W. and Cocks, P.S. (2002) Hardseededness in annual clovers: Variation within populations and subsequent shifts due to environmental changes. *Australian Journal of Agricultural Research* 53(7), 831.

Orellana, R.G. (1975) Photoperiod influence on the susceptibility of sunflower to Sclerotinia stalk rot. *Phytopathology* 65(11), 1293.

Orellana, R.G., Foy, C.D. and Fleming, A.L. (1975) Effect of soluble aluminum on growth and pathogenicity of *Verticillium albo-atrum* and *Whetzelinia sclerotiorum* from sunflower. *Phytopathology* 65(2), 202.

Pan, H. and Rollins, J.A. (2017) The GATA-type IVb zinc-finger transcription factor SsNsd1 regulates asexual and sexual development and appressoria formation in *Sclerotinia sclerotiorum*. *Molecular Plant Pathology* 9, 1679–1689.

Partyka, R.E. and Mai, W.F. (1962) Effect of environment and some chemicals on *Sclerotinia sclerotiorum* in laboratory and potato field. *Phytopathology* 52, 766–770.

Parzych, K.R. and Klionsky, D.J. (2014) An overview of autophagy: Morphology, mechanism, and regulation. *Antioxidants and Redox Signaling* 20(3), 460–473.

Pérez-Pérez, M.E., Lemaire, S.D. and Crespo, J.L. (2021) The ATG4 protease integrates redox and stress signals to regulate autophagy. *Journal of Experimental Botany* 72(9), 3340–3351.

Pethybridge, S.J., Murphy, S., Margie, L. and Kikkert, J.R. (2023) Survival of *Sclerotinia sclerotiorum* in central New York. *Plant Disease* 108(5), 1165–1168. DOI: 10.1094/PDIS-10-23-2126-SC.

Phillips, A.J.L. (1986) Carpogenic germination of sclerotia of *Sclerotinia sclerotiorum* after periods of conditioning in soil. *Journal of Phytopathology* 116(3), 247–258.

Purdy, L.H. (1956) Factors affecting apothecial production by *Sclerotinia sclerotiorum*. *Phytopathology* 46, 409–410.

Purdy, L.H. (1958) Some factors affecting penetration and infection by *Sclerotinia sclerotiorum*. *Phytopathology* 48, 605–609.

Purdy, L.H. (1979) *Sclerotinia sclerotiorum*: History, diseases and symptomatology, host range, geographic distribution, and impact. *Phytopathology* 69(8), 875.

Purdy, L.H. and Bardin, R. (1953) Mode of infection of tomato plant by the ascospores of *Sclerotinia sclerotiorum*. *Plant Disease Reporter* 37, 361–362.

Purdy, L.H. and Grogan, R.L. (1954) Physiological studies of *Sclerotinia sclerotiorum* in liquid and agar culture. *Phytopathology* 44, 36–38.

Quentin, U. (2004) *Sclerotinia sclerotiorum*, occurrence and control. *Kartoffelbau* 8, 318–319.

Rana, K., Ding, Y., Banga, S.S., Liao, H., Zhao, S. *et al.* (2021) *Sclerotinia sclerotiorum* Thioredoxin1 (*SsTrx1*) is required for pathogenicity and oxidative stress tolerance. *Molecular Plant Pathology* 22(11), 1413–1426.

Ravanan, P., Srikumar, I.F. and Talwar, P. (2017) Autophagy: The spotlight for cellular stress responses. *Life Sciences* 188, 53–67.

Ren, W., Zhang, Z., Shao, W., Yang, Y., Zhou, M. *et al.* (2017) The autophagy-related gene BcATG1 is involved in fungal development and pathogenesis in *Botrytis cinerea*. *Molecular Plant Pathology* 18(2), 238–248.

Rollins, J.A. (2003) The *Sclerotinia sclerotiorum pac1* gene is required for sclerotial development and virulence. *Molecular Plant-Microbe Interactions* 16(9), 785–795.

Rollins, J.A. and Dickman, M.B. (1998) Increase in endogenous and exogenous cyclic AMP levels inhibits sclerotial development in *Sclerotinia sclerotiorum*. *Applied and Environmental Microbiology* 64(7), 2539–2544.

Rollins, J.A. and Dickman, M.B. (2001) PH signaling in *Sclerotinia sclerotiorum* identification of a PacC/RIM1 homolog. *Applied and Environmental Microbiology* 67(1), 75–81.

Rotem, J.P. and Palti, J. (1969) Irrigation and plant diseases. *Annual Review of Phytopathology* 7(1), 267–288.

Saharan, G.S. and Mehta, N. (2008) *Sclerotinia Diseases of Crop Plants: Biology, Ecology and Disease Management*. Springer, Dordrecht, p. 485.

Saito, I. (1973) Initiation and development of apothecial stipe primordia in sclerotia of *Sclerotinia sclerotiorum*. *Transactions of Mycological Society of Japan* 14, 343–351.

Saito, I. (1974) Ultra-structural aspects of the maturation of sclerotia of *Sclerotinia sclerotiorum* (Lib.) de Bary. *Transactions of Mycological Society* Japan 15, 384–400.

Saito, I. (1977) *Sclerotinia sclerotiorum (Lib.) de Bary. Studies on the maturation and germination of sclerotia of Sclerotinia sclerotiorum (Lib.) de Bary, a causal fungus of bean stem rot*. Report Hokkaido Prefectural Agricultural Experiment Station, National Agriculture and Food Research Organization (NARO), Hitsujigaoka, Sapporo, Japan, p. 106.

Schafer, M.R. and Kohn, L.M. (2006) An optimized method for mycelial compatibility testing in *Sclerotinia sclerotiorum*. *Mycologia* 98(4), 593–597.

Scheibert-Bohm, F., Schlosser, E. and Schuster, W. (1981) Studies on the infection of rape by *Sclerotinia sclerotiorum*. *Angew Botanik* 55, 37–47.

Schwartz, H.F. and Steadman, J.R. (1978) Factors affecting sclerotium populations of, and apothecium production by *Sclerotinia sclerotiorum*. *Phytopathology* 68(3), 383.

Sexton, A.C. and Howlett, B.J. (2004) Microsatellite markers reveal genetic differentiation among populations of *Sclerotinia sclerotiorum* from Australian canola fields. *Current Genetics* 46(6), 357–365.

Sexton, A.C., Whitten, A.R. and Howlett, B.J. (2006) Population structure of *Sclerotinia sclerotiorum* in an Australian canola field at flowering and stem-infection stages of the disease cycle. *Genome* 49(11), 1408–1415.

Sharma, R.C. and Sharma, S.L. (1986a) Survival of *Sclerotinia sclerotiorum* and initial infection of cauliflower in Himachal Pradesh. *Indian Journal of Ecology* 13, 367–369.

Sharma, R.C. and Sharma, S.L. (1986b) Effect of soil amendment and artificial defoliation of infected leaves on *Sclerotinia* stalk rot of cauliflower. *Indian Journal of Plant Pathology* 4, 126–128.

Singh, H. and Singh, H. (1983) Sclerotia of *Sclerotinia trifoliorum* on seed as source of primary infection of stem rot of berseem. *Indian Phytopathology* 46, 93–94.

Singh, H., Kalda, T.S.S. and Singh, H. (1995a) Screening of cauliflower germplasm against Sclerotinia rot. *Indian Journal Genetics and Plant Breeding* 55, 98–102.

Singh, K.P., Shahi, D.K. and Singh, U.P. (1995b) Carpogenic germination of sclerotia of *Sclerotinia sclerotiorum* in some soil samples differing in ion exchange and soil nutrient properties. *Tropical Science* 35, 354–358.

Singh, R. and Tripathi, N.N. (1996) Effect of soil type on apothecial production of *Sclerotinia sclerotiorum* (Lib.) de Bary. *Indian Journal of Agricultural Research* 30, 191–196.

Singh, S.B., Singh, A.K. and Singh, A.K. (2003) *In vitro* evaluation of fungitoxicants on developmental stages of *Sclerotinia sclerotiorum*. *Annals of Plant Protection Sciences* 11, 388–390.

Singh, U.P., Sarma, B.K., Singh, D.P., Maurya, S., Mishra, P.K. *et al.* (2004) Effects of exudates depletion on sclerotial development of *Sclerotinia sclerotiorum* and of oxalic acid on the synthesis of phenolic acids in eggplant (*Solanum melongena* L.). *Zeitschrift Fur Pflanzenkrankheiten Und Pflanzenschutz* 111, 371–379.

Singh, V.P. and Singh, R.B. (1983) The effect of soil texture, soil moisture, soil mixture and depth of soil on carpogenic germination of *Sclerotinia sclerotiorum*. *Journal of Plant Diseases and Protection* 90, 662–669.

Singh, V.P., Singh, R.B. and Gupta, S. (1985) Effect of temperature on sclerotial germination, growth and sclerotial formation in *Sclerotinia sclerotiorum*. *Journal Plant Disease Protection* 92, 281–287.

Smith, F.P., Cocks, P.S. and Ewing, M.A. (1996) Short-term patterns of seed softening in *Trifolium subterraneum*, *T. glomeratum* and *Medicago polymorpha*. *Australian Journal of Agricultural Research* 47(5), 775.

Starr, G.H., Walters, H.J. and Bridgmon, G.H. (1953) White mould (*Sclerotinia*) of beans. *Wyoming Agricultural Experiment Station Technical Bulletin* 322, 11.

Steadman, J.R. and Nickerson, K.W. (1975) Differential inhibition of sclerotial germination in *Whetzelinia sclerotiorum*. *Mycopathologia* 57(3), 165–170.

Steadman, J.R., Maier, L.R., Schwartz, H.F. and Kerr, B.D. (1975) Pollution of surface irrigation waters by plant pathogenic organisms. *JAWRA Journal of the American Water Resources Association* 11(4), 796–804.

Steadman, J.R. (1979) Control of plant diseases caused by *Sclerotinia* species. *Phytopathology* 69, 904–907.

Stelfox, D., Williams, J.R., Soehngen, U. and Topping, R.C. (1978) Transport of *Sclerotinia sclerotiorum* ascospores by rapeseed pollens in Alberta. *Plant Disease Reporter* 62, 576–579.

Stevens, F.L. and Hall, J.G. (1911) A serious lettuce disease (*Sclerotiniaceae*) and a method of control. *North Carolina Agricultural Experiment Station Technical Bulletin* 8, 85–145.

Sun, P. and Yang, X.B. (2000) Light, temperature, and moisture effects on apothecium production of *Sclerotinia sclerotiorum*. *Plant Disease* 84(12), 1287–1293.

Taylor, G.B. (1981) Effect of constant temperature treatments followed by fluctuating temperatures on the softening of hard seeds of *Trifolium subterraneum* L. *Functional Plant Biology* 8(6), 547–558.

Taylor, G.B. (1996) Effect of the environment in which seeds are grown and softened on the incidence of autumn seed softening in two species of annual medics. *Australian Journal of Agricultural Research* 47(1), 141.

Teo, B.K. and Morrall, R.A.A. (1985) Influence of matric potentials on carpogenic germination of sclerotia of *Sclerotinia sclerotiorum*. II. A comparison of results obtained with different techniques. *Canadian Journal of Plant Pathology* 7(4), 365–369.

Thomma, B.P.H.J. (2003) Alternaria spp. from general saprophyte to specific parasite. *Molecular Plant Pathology* 4(4), 225–236.

Townsend, B.D. and Willetts, H.J. (1954) The development of sclerotia of certain fungi. *Transactions of the British Mycological Society* 37(3), 213–221.

Trevethick, J. and Cooke, R.C. (1971) Effect of some metabolic inhibitors and sulfur containing amino acids on sclerotium formation in sclerotium rolfsii *S. Delphini* and *Sclerotinia sclerotiorum*. *Transactions of British Mycological Society* 57, 340–341.

Trevethick, J. and Cooke, R.C. (1973) Non-nutritional factors influencing sclerotium formation in some *Sclerotinia* and *Sclerotium* species. *Transactions of the British Mycological Society* 60(3), 559–566.

Tribe, H.T. (1957) On the parasitism of *Sclerotinia trifoliorum* by coniothyrium minitans. *Transactions of the British Mycological Society* 40(4), 489–499.

Twengström, E., Köpmans, E., Sigvald, R. and Svensson, C. (1998) Influence of different irrigation regimes on carpogenic germination of sclerotia of *Sclerotinia sclerotiorum*. *Journal of Phytopathology* 146(10), 487–493.

Vega, R.R. and Le Tourneau, D. (1974) The effect of zinc on growth and sclerotial formation in *Whetzelinia sclerotiorum*. *Mycologia* 66(2), 256–264.

Veluchamy, S. and Rollins, J.A. (2008) A CRY-DASH-type photolyase/cryptochrome from *Sclerotinia sclerotiorum* mediates minor UV-a-specific effects on development. *Fungal Genetics and Biology* 45(9), 1265–1276.

Veneault-Fourrey, C., Barooah, M., Egan, M., Wakley, G. and Talbot, N.J. (2006) Autophagic fungal cell death is necessary for infection by the rice blast fungus. *Science* 312(5773), 580–583.

Viefhues, A., Heller, J., Temme, N. and Tudzynski, P. (2014) Redox systems in *Botrytis cinerea*: Impact on development and virulence. *Molecular Plant-Microbe Interactions* 27(8), 858–874.

Walker, J.C. (1969) *Plant Pathology*, 3rd edn. McGraw-Hill, New York, p. 819.

Wang, L., Liu, Y., Liu, J., Zhang, Y., Zhang, X. *et al.* (2016) The *Sclerotinia sclerotiorum foxE2* gene is required for apothecial development. *Phytopathology* 106(5), 484–490.

Wang, S.C. and Le Tourneau, D. (1971) Carbon sources, growth, sclerotium formation and carbohydrate composition of *Sclerotinia sclerotiorum*. *Archives of Microbiology* 80(3), 219–233.

Weete, J.D., Weber, D.J. and Le Tourneau, D. (1970) Hydrocarbons, free fatty acids, and amino acids of sclerotia of *Sclerotinia sclerotiorum*. *Archives of Microbiology* 75(1), 59–66.

Willetts, H.J. (1978) Sclerotium formation. In: Smith, J.E. and Berry, D.R. (eds) *The Filamentous Fungi*, Vol. 3. Wiley, New York, pp. 197–213.

Willetts, H.J. and Bullock, S. (1992) Developmental biology of sclerotia. *Mycological Research* 96(10), 801–816.

Willetts, H.J. and Wong, J.A.L. (1971) Ontogenetic diversity of sclerotia of *Sclerotinia sclerotiorum* and related species. *Transactions of the British Mycological Society* 57(3), 515–524.

Willetts, H.J. and Wong, J.A.L. (1980) The biology of *Sclerotinia sclerotiorum*, *S. trifoliorum*, and *S. minor* with emphasis on specific nomenclature. *The Botanical Review* 46(2), 101–165.

Williams, G.H. and Western, J.H. (1965a) The biology of *Sclerotinia trifoliorum Eriks*s. and other species of sclerotium forming fungi I. Apothecium formation from sclerotia. *Annals of Applied Biology* 56, 253–260.

Williams, G.H. and Western, J.H. (1965b) The biology of *Sclerotinia trifoliorum Erikss*. and other species of sclerotium forming fungi II. The survival of sclerotia in soil. *Annals of Applied Biology* 56, 261–268.

Williams, J.R. and Stelfox, D. (1979) Dispersal of ascospores of *Sclerotinia sclerotiorum* in relation to sclerotinia stem rot of rapeseed. *Plant Disease Reporter* 63, 395–399.

Williams, J.R. and Stelfox, D. (1980a) Occurrence of ascospores of *Sclerotinia sclerotiorum* in areas of central Alberta. *Canadian Plant Disease Survey* 60, 51–53.

Williams, J.R. and Stelfox, D. (1980b) Influence of farming practices in alberta on germination and apothecium production of sclerotia of *Sclerotinia sclerotiorum*. *Canadian Journal of Plant Pathology* 2(3), 169–172.

Wong, J.A.L. and Willetts, H.J. (1974) Polyacrylamide-gel electrophoresis of enzymes during morphogenesis of sclerotia of *Sclerotinia sclerotiorum*. *Microbiology* 81(1), 101–109.

Xia, S., Xu, Y., Hoy, R., Zhang, J., Qin, L. *et al.* (2019) The notorious soilborne pathogenic fungus *Sclerotinia sclerotiorum*: An update on genes studied with mutant analysis. *Pathogens* 9(1), 27.

Xiao, X., Xie, J., Cheng, J., Li, G., Yi, X. *et al.* (2013) Novel secretory protein *Ss-caf1* of the plant-pathogenic fungus *Sclerotinia sclerotiorum* is required for host penetration and normal sclerotial development. *Molecular Plant-Microbe Interactions* 27(1), 40–55.

Xu, T., Li, J., Yu, B., Liu, L., Zhang, X. *et al.* (2018) Transcription factor SsSte12 was involved in mycelium growth and development in *Sclerotinia sclerotiorum*. *Frontiers in Microbiology* 9, 2476.

Yang, G., Tang, L., Gong, Y., Xie, J., Fu, Y. *et al.* (2018) A cerato-platanin protein SsCP1 targets plant PR1 and contributes to virulence of *Sclerotinia sclerotiorum*. *The New Phytologist* 217(2), 739–755.

Yin, Z., Feng, W., Chen, C., Xu, J., Li, Y. *et al.* (2020) Shedding light on autophagy coordinating with cell wall integrity signaling to govern pathogenicity of *Magnaporthe oryzae*. *Autophagy* 16(5), 900–916.

Young, N. and Ashford, A.E. (1996) The effects of rind damage and regeneration on permeability of the apoplast in sclerotia of *Sclerotinia minor*. *New Phytologist* 134(1), 13–24.

Young, P. and Morris, H.E. (1927) Sclerotinia wilt of sunflower. *University of Montana Agricultural Experiment Station Bulletin* 208, 32.

Yu, H., Jia, W., Xiao, K., Jiao, W., Zhang, X. *et al.* (2022) The autophagy genes *ChATG4* and *ChATG8* are required for reproductive development, virulence, and septin assembly in *Cochliobolus heterostrophus*. *Phytopathology* 112, 830–841.

Yu, Y., Xiao, J., Zhu, W., Yang, Y., Mei, J. *et al.* (2016) Ss-Rhs1, a secretory Rhs repeat-containing protein, is required for the virulence of *Sclerotinia sclerotiorum*. *Molecular Plant Pathology* 18(8), 1052–1061.

Zewain, Q.K., Bahadur, P. and Sharma, P. (2004) Effect of fungicides and neem extract on mycelial growth and myceliogenic germination of *Sclerotinia sclerotiorum*. *Indian Phytopathology* 57, 101–103.

Zhang, H., Li, Y., Lai, W., Huang, K., Li, Y. *et al.* (2021) SsATG8 and SsNBR1 mediated-autophagy is required for fungal development, proteasomal stress response and virulence in *Sclerotinia sclerotiorum*. *Fungal Genetics and Biology* 157, 103632.

Zhu, X.M., Li, L., Wu, M., Liang, S., Shi, H.B. *et al.* (2019) Current opinions on autophagy in pathogenicity of fungi. *Virulence* 10(1), 481–489.

5 *Sclerotinia* Ultrastructures

Abstract

The use of light and electron microscopy made it possible to observe changes in constituent cells, tissue differentiation, and sclerotial growth in host tissues. At various stages of sclerotial development, alterations in the cytoplasm have been observed. These include modifications to the numerous vegetative hyphal organelles, endoplasmic reticulum, vacuoles, mitochondria, ribosomes, lomasomes, septal apertures, and medullary cell cytoplasmic characteristics. Juvenile sclerotial hyphal ultrastructures are similar to those of vegetative hyphae that are growing quickly. The presence of β-1,3 glucans and protein components was detected in the fibrous layer and medullary cell wall. Upon closer inspection, the stipe and apothecium's ultrastructures are revealed to be made up of a group of hyphae that are arranged longitudinally and loosely packed. The analysis of ascospores ultrastructure has revealed the existence of nuclei, vacuoles, microconidia, and stroma in the fungal cell. Ascospore ultrastructures have revealed nuclei, vacuoles, microconidia, and stroma. Histochemical analyses at the penetration sites demonstrated biological changes in the cutin layer as well as esterase activity. Enzymes activity initially manifests at the interface between the appressorium and the host's surface. At the host penetration site, adjacent to the penetration pore, electron-opaque deposits are seen as the infection worsens.

5.1 Introduction

Throughout their life cycles, all pathogens, including fungi, produce ultrastructures. After obtaining ultrathin sections of each component and process, the ultrastructures of *Sclerotinia's* morphological parts—such as sclerotia, stipe, apothecia, microconidia, stroma, germ tube, infection cushions, appressoria—and infection process have been observed using electron microscopy. Electron microscopy has provided important insights into the complex host–parasite dynamics throughout different infection stages and pathogenesis, in conjunction with physiological, biochemical, and genetic studies. This comprehensive approach provides insight into the life cycle of the pathogen and enables the identification of vulnerable locations to guide the development of suitable management plans.

5.2 Ultrastructures of Sclerotial Development and Maturation

Sclerotium production by pathogens follows a sequence of three stages, starting from vegetative hyphae, enlargement, and maturation. The variations in development documented among different *Sclerotinia* species are worth mentioning (Townsend and Willetts, 1954; Butler, 1966). Fully formed sclerotia of *Sclerotinia sclerotiorum* comprise of a central region made up of prosenchymatous hyphae and a clearly defined outer layer. The initiation of apothecial stipe growth is attributed to the core hyphae, which are situated beneath the outer layer of the spreading sclerotia, as revealed by a previous study utilizing light microscopy (Saito, 1973). The development of these sclerotia, known as 'carpogenic germination,' is a result of cellular differentiation (Coley-Smith and Cooke, 1971). It is important to note that this process differs from the vegetative development of specific core hyphae.

5.2.1 Ultrastructural alterations in constituent cells and cellular variety of sclerotia

The juvenile sclerotia undergo a process where the marginal cells lose their filamentous structure due to repetitive septation, resulting in the formation of white sclerotia. Furthermore, the terminal cells differentiate into cells with slightly thicker walls, resulting in a globose shape. This transition produces moderately pigmented sclerotia.

DOI: 10.1079/9781800629592.0005

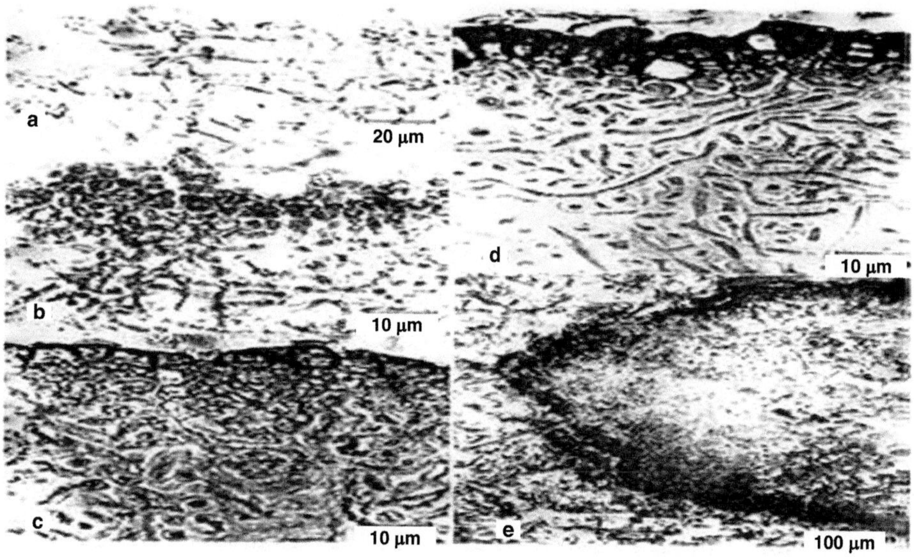

Fig. 5.1. Vertical sections of sclerotia were captured on light micrographs. (a) An immature sclerotium with a trace of color, (b) a white sclerotium, (c) a sclerotium with a hint of color, (d) a mature sclerotium, and (e) a sclerotium that sprouted and turned into a mycelium. There has been partial destruction of the outer layer (arrows). (Adapted from publication of Saito, 1977, with permission.)

Following that, a rind made of dark-colored, thick-walled cells forms, resulting in the maturation of the sclerotia. During this period, mature sclerotia are easily detached from the underlying colony. Simultaneously, the core cells of the sclerotia remain filamentous, while a nonstained layer eventually grows and surrounds them. This process is depicted in Fig. 5.1. Finally, the bulky, dominant zone of the sclerotia becomes filled with these cells, forming a medullary tissue ().

5.2.2 Cell wall modifications

The microscopic detail in Fig. 5.2 reveals that ascomycetous fungi are distinguished by the presence of a simple, perforated septum. The vegetative hyphae's cell wall (CW) displays a consistent, single-layered zone with a width that varies from 0.09 to 0.18 μm. The majority of the inner cells (medullary cells) in white sclerotia have CWs which are similar in thickness to the vegetative hyphae. Nevertheless, these cells remain surrounded by varying amounts of fibrous material, and in certain instances, this fibrous material creates a visible layer (Fig. 5.3). The increase in width of the cellular barrier is due to the development of the fibrous layer. The formation of such a fibrous layer also stops in poorly pigmented sclerotia, and in certain cases, the CW thickens without interruption, but an incredibly thin, electron-dense layer coats the outside (Figs 5.4 and 5.5). As a consequence of these alterations, fully developed sclerotial medullary cells possess two primary layers: a fibrous layer, which is significantly thicker than the CW and appears as an unstainable layer under a light microscope (Fig. 5.6). A noncellular matrix is absent; hence, intercellular spaces are discernible (Colotelo, 1974; Saito, 1974; Saito, 1977).

5.2.3 Cytoplasmic modifications

The cytoplasm of vegetative hyphae frequently contains a variety of fungus-specific organelles. These organelles consist of the endoplasmic reticulum, mitochondria, ribosomes, vacuoles,

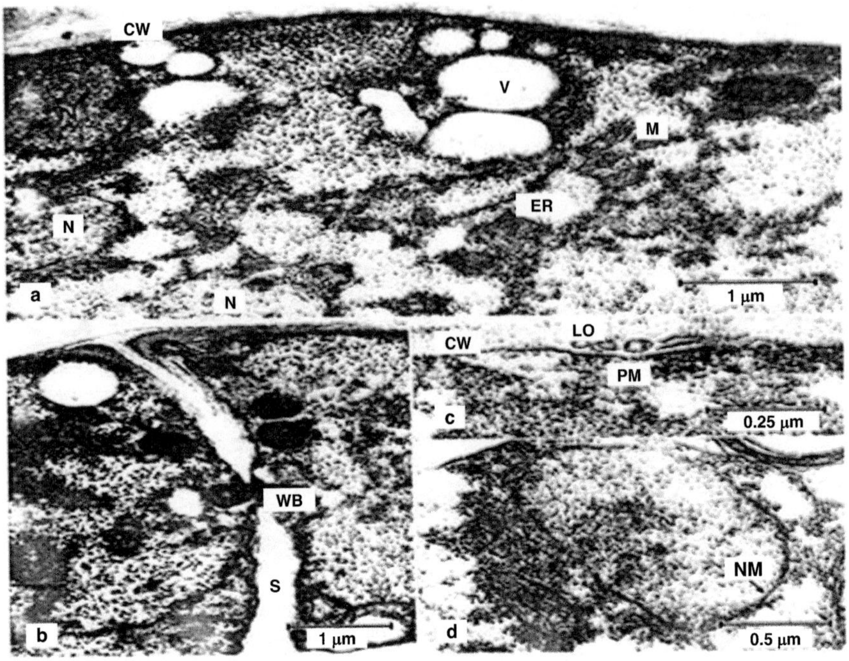

Fig. 5.2. Fine sections of mature hyphae. (a) A segment of hyphal cells, (b) Woronin bodies (WB) coupled to a septum (S), (c) lomasomes (LO) between plasma membrane (PM) and cell wall (CW) of the cell wall, (d) the nucleus (N), containing the mitochondrion (M) and binary nuclear membrane (NM). (Adapted from publication of Saito, 1977 with permission.)

and lomasomes (Fig. 5.2). Furthermore, it is common to observe a slight buildup of shapeless, electron-dense substance surrounding the edge of the septal pore. Moreover, Woronin bodies, which are electron-dense, are present in the vicinity of the septal pore (Figs 5.2b and 5.3a). Medullary cells in white sclerotia usually share cytoplasmic characteristics with vegetative hyphae. The lomasomes, on the other hand, have a more intricate structure and resemble multivesicular and multitubular entities, while the electron-dense materials disappear from the pore rim (Fig. 5.7). During sclerotial development, the mitochondria and vacuoles undergo significant modifications. Numerous mitochondria with well-defined cristae can be found in the medullary cells of lightly pigmented sclerotia, despite their tendency to be more elongated (Fig. 5.4c). However, in mature sclerotia, mitochondrial cristae become unclear, making identification challenging (Fig. 5.5b). The mitochondrial structural shift seen in mature sclerotia (Fig. 5.8) is a direct result of reduced oxygen intake. Within the vacuoles, electron-dense, amorphous materials are deposited; as the sclerotia develop, their volume progressively increases (Figs 5.4 and 5.5). It is clear that mature sclerotia have several vacuoles packed with these components (Figs 5.5b and 5.9; Colotelo, 1974; Saito, 1977).

The production and contents of sclerotia in *Sclerotinia minor*, which are smaller than those in *S. sclerotiorum* and *S. trifoliorum*, were investigated using light and electron microscopy (Bullock *et al.*, 1980a). The differentiation of sclerotial hyphae into distinct regions, including the rind, cortex, and medulla, was observed by many researchers to commence approximately 12–24 hour following the initiation of sclerotial development. The rind, consisting of densely packed cells, surrounds the sclerotium, while the cortex, composed of pseudo-parenchymatous tissue, is about three cells wide. The medulla,

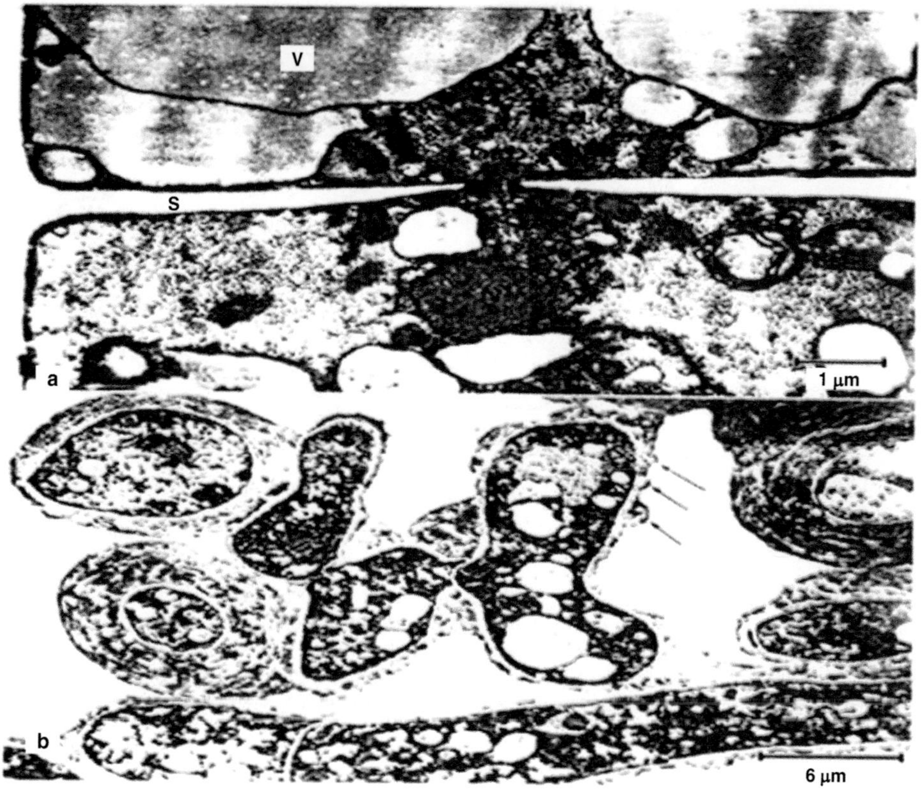

Fig. 5.3. (a) Fine segment of growing hyphae with an electron-dense material (DED) deposition surrounding the edge and a septum (S) with a basic hole, (b) there is a distinct, ultrathin portion of a white sclerotium visible, and arrows show how the fibrous layer is progressing. (Adapted from publication of Saito, 1977, with permission.)

which is prosenchymatous in nature, forms the bulk of the sclerotium. Protein bodies and polyphosphate granules, which are essential cytoplasmic reserves, are predominantly found in the cortical and medullary hyphae. Extracellular material quickly builds up, surrounding the cortex and medullary hyphae, finally filling the spaces between them entirely when fully grown. The ultrastructural characteristics of juvenile sclerotial hyphae closely resemble those of actively growing vegetative hyphae. With maturation of the sclerotia, there is a rise in the quantity of endoplasmic reticulum cisternae profiles, as the number of nuclei and mitochondrial profiles declines. The appearance of the cytoplasm stays granular during the differentiating process. In general, the adult sclerotial morphology of *S. minor* bears similarity to that of other species of *Sclerotinia*.

5.2.4 Sclerotial histochemistry

Histochemical observations of the sclerotia have provided information on the biochemical compounds present within sclerotia and their role in its development, maturation, and germination. The medullary cells of *S. sclerotiorum* do not exhibit any response to the periodic acid–Schiff (PAS) test and completely disintegrate when autoclaved for 2 hour in approximately 23 M KOH (potassium hydroxide) (Fig. 5.10). However, the PAS test demonstrates a favorable

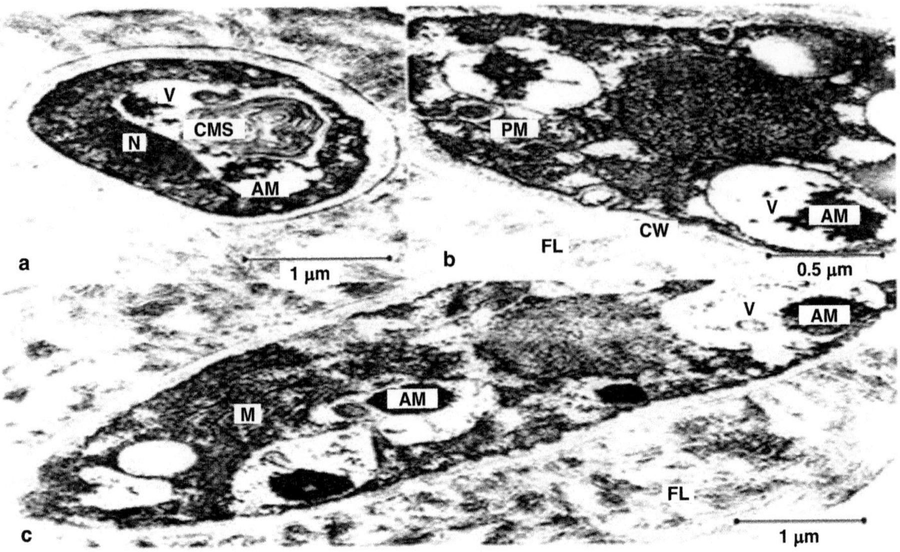

Fig. 5.4. Ultrathin slices reveal moderately colored sclerotium medullary cells. (a) The large vacuole contains electron-dense amorphous components and a concentric membrane structure; (b) the cell is presently thickening its wall; (c) cell with extended mitochondria (M) and unstructured elements, including vacuoles (V); the observation of abundant plasma membrane folds as well as the accumulation of electron-dense, shapeless substances contained by vacuoles (V) ought to be duly acknowledged. (Adapted from publication of Saito, 1977 with permission.)

response in the cellular layer and their septa, which remain intact even following sterilization using a KOH solution. Chitosan is detected in remaining cell walls (CWs) and septa by using I Ki (potassium iodide) in 1% solution of H_2SO_4. Both the fibrous layer and the CW exhibit yellow fluorescence when treated with aniline blue and examined under an ultraviolet light source. The breakdown of medullary cells using moderately refined β−1,3 glucanase and/or papain is illustrated in Fig. 5.11, demonstrating the existence of β−1,3 glucans in both layers with a slightly decreased fluorescence in the latter layer. Fig. 5.11 illustrates the breakdown of medullary cells through the use of moderately refined β−1,3 glucanase and/or papain, showing the presence of β−1,3 glucans in both layers, with a slightly reduced fluorescence in the latter layer. Inadequate breakdown of these layers is only observed when both enzymes are used together, as previously mentioned. The findings of the study suggest that the fibrous layer primarily comprises chitin, β−1,3 glucans, and protein constituents present in the cell membranes of medullary cells (Jones, 1970).

The cytoplasm of fully developed sclerotia in *S. sclerotiorum* medullary cells has been discovered to have substances that show positive results with PAS staining (Fig. 5.10). These substances are believed to be glycogen since their ability to be stained with PAS decreases or vanishes following an amylase digestion of cells (Fig. 5.10). As sclerotia mature, the medullary cells inside them have numerous granules that increase in number. It is possible that these granules resemble vacuoles which hold electron-dense, shapeless components, indicating the buildup of polyphosphates in vacuoles of developed sclerotial structures (Saito, 1977).

The substance, cellular makeup, and distribution of store reserves in *S. minor* sclerotia were analyzed using a histochemical approach (Bullock *et al.*, 1980a,b). This work added to a previous comprehensive ultrastructural analysis of the genesis of sclerotia. The walls and septa of all hyphae in the cell contain chitin and β−1, 3 glucans, whereas those of the rind contain in addition, a melanin-like

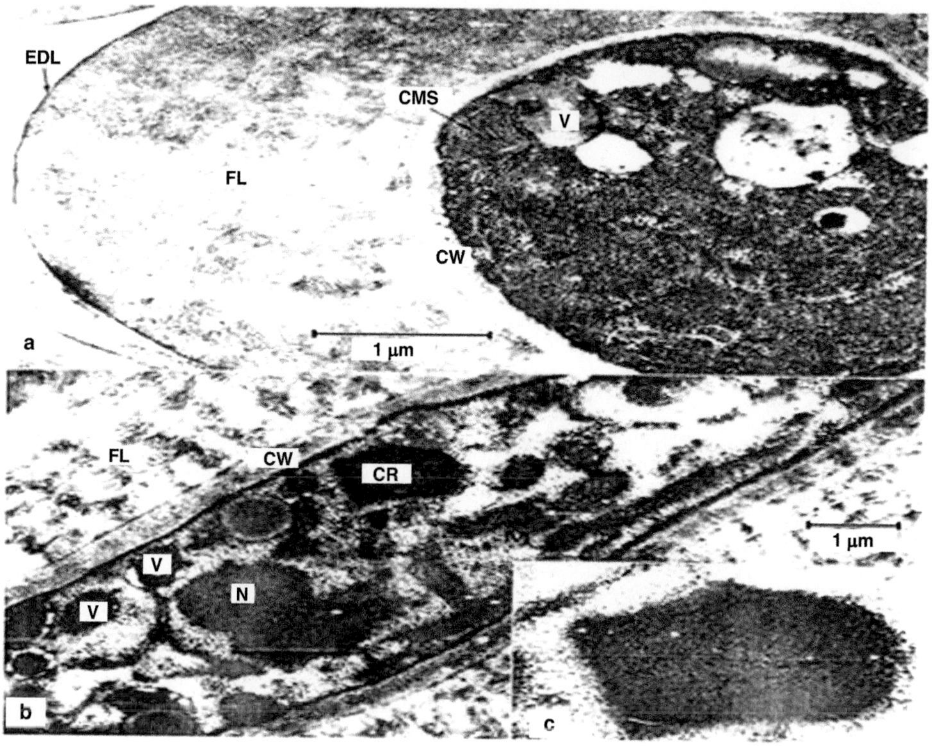

Fig. 5.5. (a) In a moderately pigmented sclerotial cell, an advanced fibrous layer (FL) is surrounded by an extremely thin layer of electron-dense material (EDL). (b) In an ultrathin section of the medullary cell of a developed sclerotium, fragmented mitochondria (M), vacuoles (V carrying electron-condensed amorphous substances, and the presence of a microbody with a crystal-like assembly (CR) are shown. (c) Enlarged duplicate of a little insertion in 2. (Adapted from publication of Saito, 1977, with permission.)

substance. An extracellular matrix, which accumulates around cortical and medullary hyphae, consists of β−1, 3 glucans, however, some polysaccharide that could not be identified by histochemical analysis is also present. Throughout the entire developmental process, phenolic material was accumulated in both the extracellular matrix and the inter-hyphal gaps as the hyphae reached maturity. Furthermore, glycogen was consistently detected in the cytoplasm of both cortex and medullary hyphae.

Polyphosphate granules are stored in tiny vacuoles, and as sclerotia reach maturity, they predominantly accumulate in the cortical region. Upon reaching full maturity, a crucial characteristic of inter-hyphal structures is the swift creation of protein bodies within the cortical and medullary hyphae. These bodies can be spherical or elongated, with the latter commonly aligning parallel to the hyphae's long axis and being closely attached to endoplasmic reticulum strands. No lipid reserves have been discovered.

5.2.5 Histology of ordinary and irregular sclerotia

Sclerotinia produces normal as well as abnormal sclerotia. Histological observations have provided insight into their tissue differences and how they behave during the life cycle. The *S. sclerotiorum*, a fungus that affects sunflower

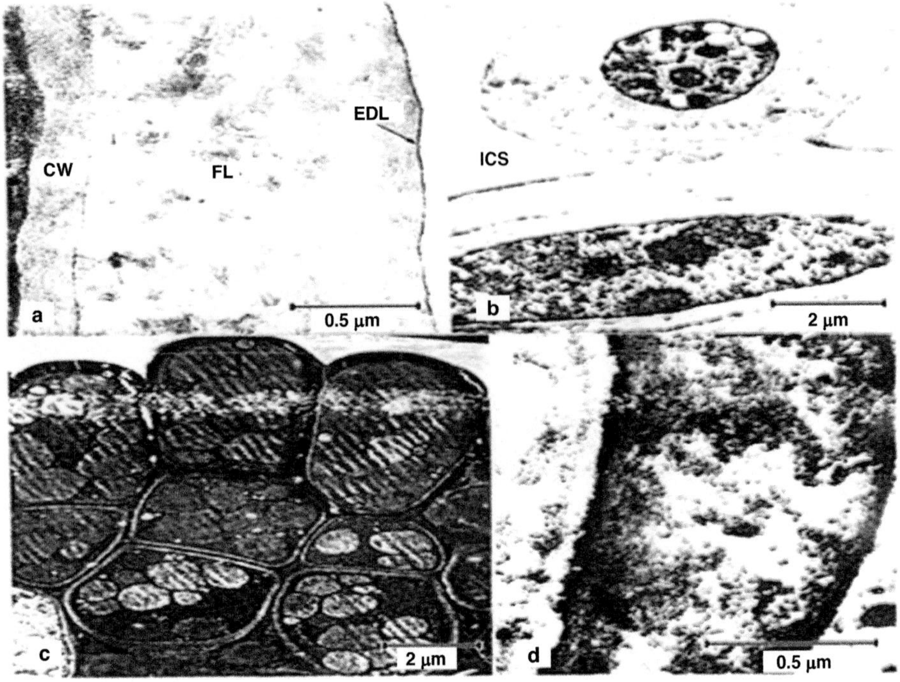

Fig. 5.6. Ultrathin slices of developing sclerotium. (a) A piece of the medullary cell wall displaying the various layer thicknesses; (b) the intercellular space (ICS) of the medulla; (c) the outermost layer; (d) a portion of the cell wall that has two electron-dense layers inside of it and a large, microfibril-filled central region. (Adapted from publication of Saito, 1977, with permission.)

plants, has been found to produce morphologically abnormal sclerotia. These abnormal sclerotia differ from the normal ones in terms of their rough appearance and brown medullary tissue, as opposed to the smooth surface and white medulla of the normal ones. It is important to note that this abnormality is not inherited. However, the abnormal sclerotia exhibit reduced variability and pathogenicity compared to the normal ones. A microscopic analysis of tissue showed that the typical sclerotia of *S. sclerotiorum* recovered after infection of sunflower soft tissue are composed of rind, cortex, and medulla layers, which is consistent with previous accounts of their composition (Kosasih and Wllletts, 1975; Saito, 1977; Huang, 1982).

A typical sclerotium presents a ring layer that is well-preserved and made up of two or three cells. The cortex varies in width from two to four cells, while the medullary region is broad and white, containing numerous loosely intertwined, darkly pigmented hyphae embedded in an undefined matrix. On the other hand, an atypical sclerotium shows a damaged outer layer and a brown medullary area. The medullary section of the anomalous sclerotium contains sparse, lightly pigmented filamentous hyphae within a frequently vacuolated matrix. Within abnormal sclerotia, viable cells are primarily concentrated in the white medullary area, while the brown medullary tissue does not exhibit any hyphal growth. Abnormal sclerotia secrete increased levels of amino acids in comparison to normal sclerotia. Through chemical analyses, it has been determined that the protein, alcohol-soluble compounds, and oil levels exhibit a constant composition in both

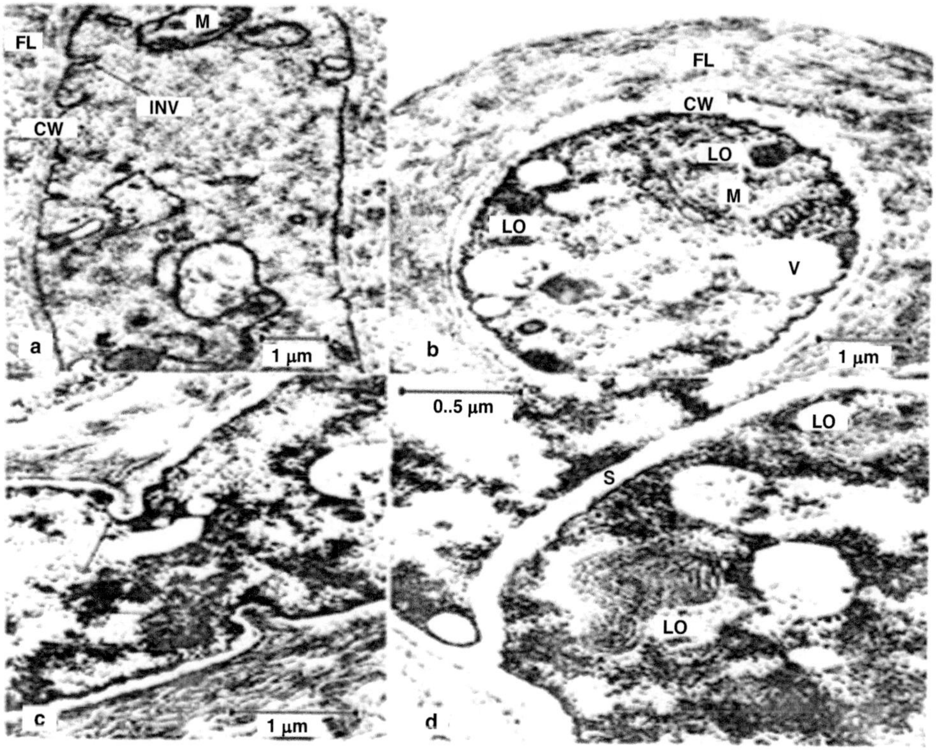

Fig. 5.7. Thin slices of white sclerotial medullary cells. (a) The plasma membrane invades; (b) a cell appears in the cross section; (c) the cross wall expands to produce a new septum (arrows); (d) lomasomes (LO) are found close to the septum (S). (Adapted from publication of Saito, 1977, with permission.)

regular and irregular sclerotia (Table 5.1). Nevertheless, abnormal sclerotia exhibit notably higher ash content (Huang, 1983). The sclerotia commonly seen in *S. sclerotiorum*, derived from sunflower plants that are infected, demonstrate similar levels of oil, protein, and alcohol-soluble compounds, as obtained from bean or pea foliage affected by the same disease (Le Tourneau, 1966; Weete *et al.*, 1970). As the primordia break through the rind, they start developing into juvenile apothecial stipes (Stage IV), marking the beginning of Stage III. The formation of regular apothecia is initiated when the surface darkens or the rind regenerates, triggering the germination of medullary tissue cubes extracted from mature sclerotia (Fig. 5.12). Primordia are also observed in these medullary tissue cubes close to the redeveloped outer layer. This suggests that any section of the medulla has the ability to start primordium formation, with the starting point being closely linked to the area where outer layer differentiation occurs (Saito, 1977).

5.3 Ultrastructures of the Germination Process of Sclerotia

The appearance of an apothecial stipe from the sclerotium suggests carpogenic germination of *S. sclerotiorum*. Figures 4.2 (Chapter 4, this volume) and 5.11 depict the four steps involved in the development of stipes. The microscopic detail in Fig. 5.12 as well as in Fig. 4.2 (Chapter 4, this volume) depict the four stages involved in the development of stipes. The intensely pigmented patches of the medulla close to the outer layer of the sclerotium are the first indications of primordium formation. The regions consist

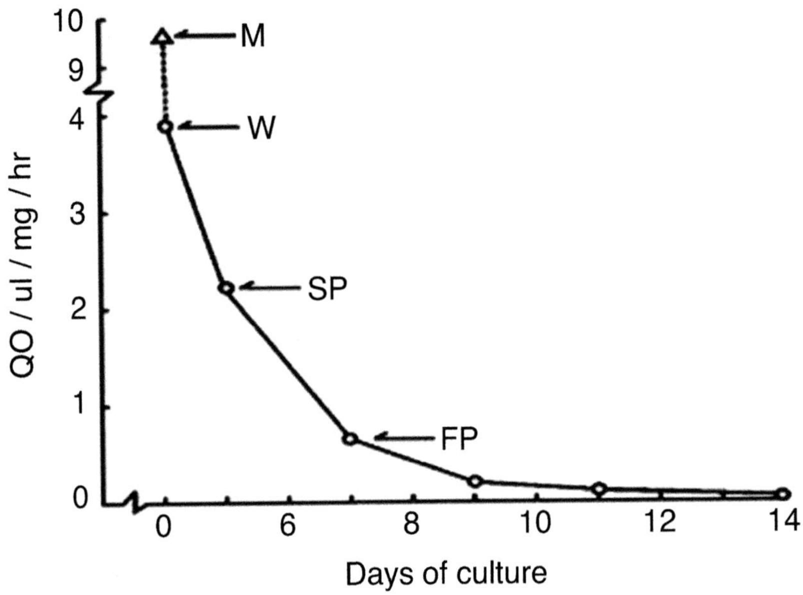

Fig. 5.8. This graph demonstrates how the respiration rate in sclerotia varies with the developmental stage. (FP: totally pigmented sclerotium; M: mycelium; SP: slightly pigmented sclerotium; W: white sclerotium). (Adapted from publication of Saito, 1977, with permission.)

of a mixture of small and slightly broader, condensed cytoplasmic cells with thin walls (Phase I). Then, during Phase II, black pigments start to show up outside or inside the primordia, possibly as a result of the development of dark-colored, thick-walled cells. Under the outer layer of abnormal sclerotia, which is different from that of normal sclerotia, the Phase II primordia give rise to a cluster of thin-walled hyphae with copious amounts of cytoplasm. This cluster develops into a distinctive tissue arrangement that is easily visible from the medulla. Analytical research indicates that normal and diseased sclerotia have similar protein, alcohol-soluble chemical, and oil contents (Table 5.1). Nonetheless, aberrant sclerotia have a noticeably greater ash content (Huang, 1983). The oil, protein, and alcohol-soluble chemicals detected in the sclerotia of diseased sunflower plants are similar to those reported in infected pea or bean plants (Le Tourneau, 1966; Weete *et al.*, 1970). Nonetheless, the concentration of linoleic acid in the sclerotia of infected sunflower plants seems to be higher when compared to infected pea plants (Weete *et al.*, 1970). As the primordia break through the rind, they start developing into juvenile apothecial stipes (Stage IV), marking the beginning of Stage III. The formation of regular apothecia is initiated when the surface darkens or the rind regenerates, triggering the germination of medullary tissue cubes extracted from mature sclerotia (Fig. 5.13). Primordia are also observed in these medullary tissue cubes close to the redeveloped outer layer. This suggests that any section of the medulla has the ability to start primordium formation, with the starting point being closely linked to the area where outer layer differentiation occurs (Saito, 1977).

5.3.1 Medullary cell ultrastructures

The cytoplasm of developing sclerotia contains a majority of medullary cells that retain vacuoles filled with polyphosphate-like compounds. These cells also show signs of mitochondrial degeneration (Fig. 5.14). Consequently, metabolic functions show little variance among medullary cells in germinating sclerotia compared to those in mature but nongerminating sclerotia. Nevertheless, unique characteristics are evident

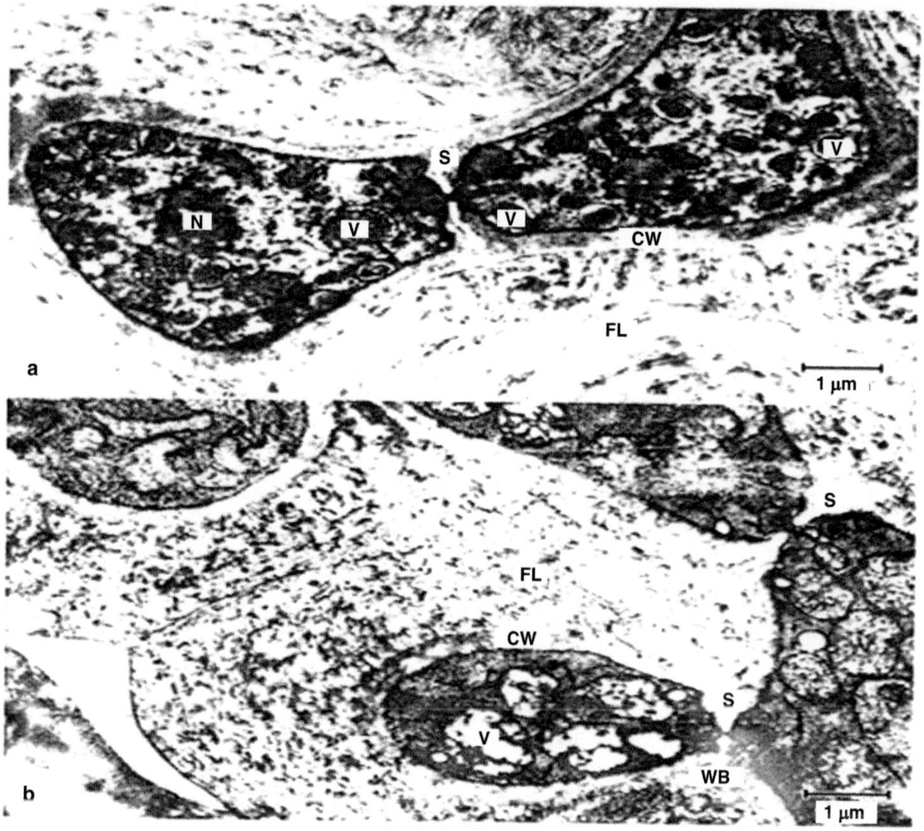

Fig. 5.9. The preservation of mature sclerotium results in the formation of ultrathin sections of medullary cells with varying thickness. (a) Septate cells with O_8SO_4, acrolein, and glutaraldehyde; (b) septate cells with $KMnO_4$-containing septum. (Adapted from publication of Saito, 1977, with permission.)

in the former, including a marked rise in endoplasmic reticulum and the CW zonation. These characteristics are unique to the medullary cells in the germinating sclerotia. Furthermore, the undifferentiated medullary cells can be easily distinguished from the cells that make up the stipe primordia (Figs 5.15c and 5.16). The primordial cells (PC) exhibit irregular shapes and sizes, with thin and uniform walls lacking fibrous layers. Notably, these cells have many ribosomes and mitochondria having distinct cristae, indicating significant metabolic activity (Fig. 5.17). Polyphosphate-like substances are not found in primitive cell vacuoles. The size reduction observed in the chemicals within the vacuoles of nearby medullary cells suggests that they are used as energy sources. Electron-dense deposits are occasionally seen in formations that resemble Stage II primordial environments (Fig. 5.18). However, these deposits are not located in the CWs of the primordia but rather in their intercellular matrix. They seem to be present in fibrous materials that cling to neighboring medullary cells or in the fibrous layer between them. Thus, these fibrous components may have contributed to the observed pigmentation, perhaps by melanization (Saito, 1977).

Primary cells differ from medullary cells in that they have thin walls and no fibrous layers. There is no indication of direct growth originating from the medullary cells. Nevertheless, there are specific cells that resemble endo-hyphae, appearing to be created by the separation of binary distinct layers within the hardened wall

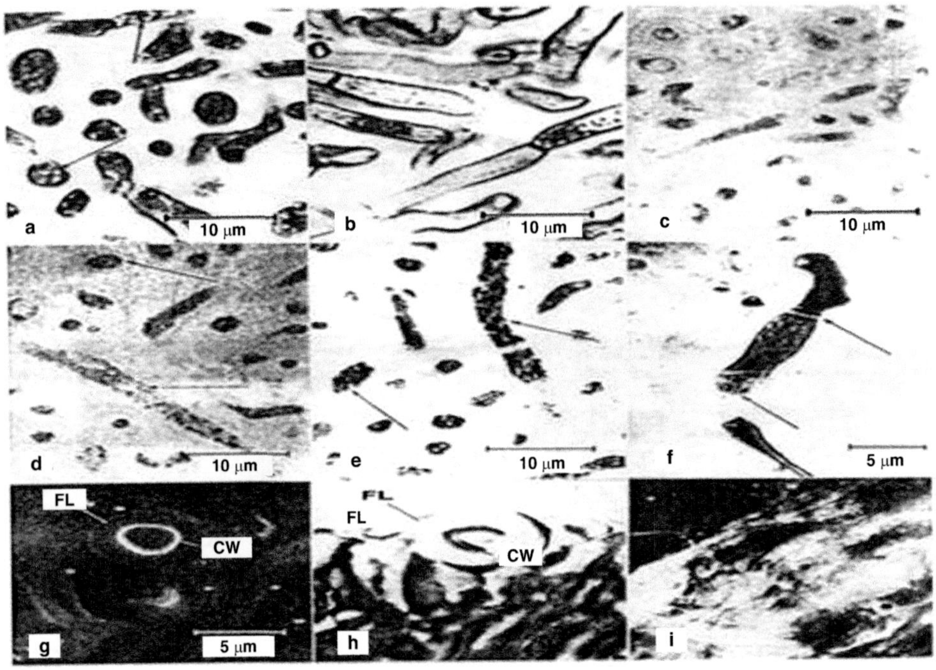

Fig. 5.10. (a) A naturally occurring slice of a matured sclerotium stained with PAS chemical with arrows pointing to PAS-negative granules; (b) the same tissue section subjected to absorption of alpha-amylase; (c) aniline blue has been used to stain the white sclerotium's HPMA segment at a pH of 4.4; (d) a similarly colored, light-hued sclerotium; (e) a similarly colored, developed sclerotium. Note the arrow-designated meta-chromatic granules; (f) the arrows show that the mature sclerotium exhibits a chitosan-specific response in the septa and cell walls; (g) a fluorescence microscopy picture displaying medullary cells treated with a diluted aniline blue solution (pH 9.0); (h) the micrograph is identical to (g) but with normal brightness removed; (i) a diluted aniline blue stain was used to visualize a fluorescence micrograph of isolated β−1,3 glucans. (Adapted from publication of Saito, 1977, with permission.)

of the medullary cells (Fig. 5.19). The enhanced metabolic activity in endo-hyphae-like cells can be attributed to the higher presence of ribosomes and mitochondria in their cytoplasm, as opposed to undifferentiated medullary cells. The emergence of these cells resembling endo-hyphae marks the initiation of cellular differentiation. Upon the disintegration of the fibrous layer and mother CW, these cells are produced from quiescent medullary cells. The enzyme β−1,3 glucanase is indispensable for the medullary CW zonation, a crucial process for cellular differentiation, as previously explained. Endo-hyphae are formed within a deceased cell by extending through the septal opening from an adjacent living cell, a phenomenon that has been observed in various fungi. Therefore, instead of forming endo-hyphae, this process can be referred to as the 'ecdysis' of cells. 1. Within the medulla of decaying sclerotia, CWs break down and fibrous layers disintegrate, suggesting the consumption of resources such as β−1,3 glucans or proteins in the process of apothecial formation (Fig. 5.20). However, complete fibrous layers are occasionally seen. The cells in the stipe tissue contain numerous mitochondria and ribosomes, indicating active ribonucleic acid synthesis and high metabolic activity (Fig. 5.21; Saito, 1977). Bullock and Willetts (1996) found that when mycelia sprout from *S. minor* sclerotia, germ hyphae emerge from the sclerotium's outer regions and pass through the degrading outer medullary layer. The mycelium develops when the cortical hyphae develop separately through the outer layer and cluster around the sclerotium. This

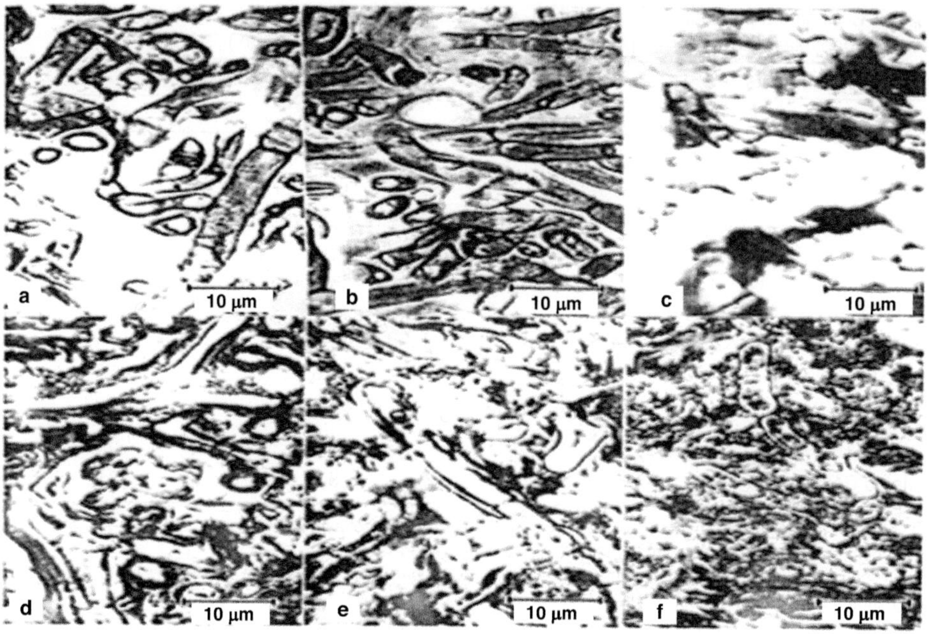

Fig. 5.11. Enzymes were applied to the sections of the mature sclerotia. Sections (1)–(3) were stained using the PAS staining procedure; (a) untreated portion; (b) portion treated with β–1,3 glucanase; (c) β–1,3 glucanase+papain-treated section; (d) untreated section; (e) β–1,3 glucanase-treated section; (f) β–1,3 glucanase+papain. Sections (d)–(f) were left unstained and analyzed using phase contrast microscopy. (Adapted from publication of Saito, 1977, with permission.)

particular process of mycelial growth is known as 'noneruptive.' Germ hyphae are formed almost exclusively from intracellular and extracellular sclerotial reserves.

5.3.2 Histochemistry of primordia

An intense presence of a protein-associated substance is observed in the initial phases of stipe growth (Fig. 5.22). At the outset, the protein concentration in these initial phases, as evidenced by mercuric bromophenol blue and the Mullion reaction, is not elevated. As the primordia progress in their development, the reaction becomes increasingly noticeable, especially in the apical regions of the growing stipes (Figs 5.22 and 5.23). The RNA content varies greatly between the primordial and nonprimordial sections of the medulla, suggesting that RNA synthesis is highly active in PC (primordial cell). RNA synthesis, or transcription, is a continuous process that occurs in the nucleus of eukaryotic cells during interphase, but it is most active in the G1 phase of the cell cycle, not a specific cellular structure called "PC". Mostly, the pigmented primordia (Stage II) and rind cells exhibit discernible black deposits (Fig. 5.24). These deposits are attributed to the presence of phenolic compounds, which undergo oxidative polymerization to form melanin, resulting in pigmentation. Fibrous materials are likely to contain phenolic chemicals. The presence of succinate dehydrogenase activity was confirmed through the use of nitro blue tetrazolium (Fig. 5.25). Blue-black formazan deposits were detected in the cellular composition of the dominant area of immature apothecial stipes, indicating the existence of succinate dehydrogenases. In contrast, there was minimal activity detected in the nonprimordial section of the medulla during sclerotial development (Saito, 1977).

Table 5.1. Chemical constituents of typical and unusual sclerotia of *S. sclerotiorum* derived from sunflower capitula. (Adapted from publication of Huang, 1983 with permission.)

Chemical constituents[a]	Sclerotia	
	Typical	Unusual
Oil	1.02	0.92
Protein	12.8	13.1
Alcohol-soluble materials	12.1	12.1
Ash[b]	1.7	2.1
Fatty acid		
Palmitic C16:0	13.1	13.8
Stearic C18:0	1.8	1.7
Poleic C18:1	22.6	19.1
Linoleic C18:2	53.2	50.9
Linolenic C18:3	8.2	12.0

[a]The values of the dry weight matter are represented by the percentages of ash, protein, oil, and alcohol-soluble substances. The fatty acids percentage reveals the number of oils in seed.
[b]There was a notable contrast in the ash quantity observed between typical and unusual sclerotia ($P < 0.05$). However, nonsignificant variations were found in the values of the other components.

5.4 Ultrastructures of Stipe and Apothecium Development and Their Components

Following germination, sclerotia produce morphological structures, such as stipe and apothecia, to continue their life cycle under *in vitro* as well as *in vivo* conditions. Developing sclerotia displayed an interesting characteristic when kept in pots at a temperature of 20°C for 8–10 weeks (Fig. 5.26a). It is worth mentioning that some of these sclerotia have multiple stalks, and the apothecia are at different stages of growth. The stalk is composed of tightly packed hyphae aligned in a longitudinal manner (Fig. 5.26b). Young apothecia gradually develop into a trumpet-like shape (Fig. 5.26e). Within this structure, asci with inoperculate apices can be observed (Fig. 5.26f). The immature apothecia (Fig. 5.26c) and undifferentiated hyphae (Fig. 5.26d) can be seen. A longitudinal section of an ascus is presented, illustrating both ascospores that have been discharged and those that remain within the ascus (Fig. 5.27g). It is important to note that the aperture through which the spores are ejected is not visible in this cross section (Fig. 5.27h), likely due to the oblique sectioning of the empty ascus. The young ascus contains nuclei that have unique double-layered membranes (Fig. 5.27i). The ascospores are enclosed by walls that develop in a subsequent phase (Fig. 5.27j). In some cases, mature ascospores may contain more than one nucleus (Fig. 5.27k). Additionally, both unidentified inclusions and vacuoles can be observed in the ascospore (Fig. 5.27l). Around the ascospores, there are other vacuoles that resemble those in the ascus cavity. Nuclei in freeze-etched ascospores can be identified thanks to holes in the NM (nuclear membrane) (Fig. 5.28a). These pores can be observed in Fig. 5.28a where the nucleus is broken in such a way that the outer surface of the membrane is exposed and where the inner surface is revealed in Fig. 5.28a spore 4. Moreover, NM pores in cross-fractured nuclei are visible as holes in Fig. 5.27h. Ascospores that have been freeze-etched (Fig. 5.27i) display vacuoles; some of these show the vacuolar membrane's outer surface, while others show its inner surface. The inset pic shows the inner surface of the plasmalemma with characteristic ridges, where the outer

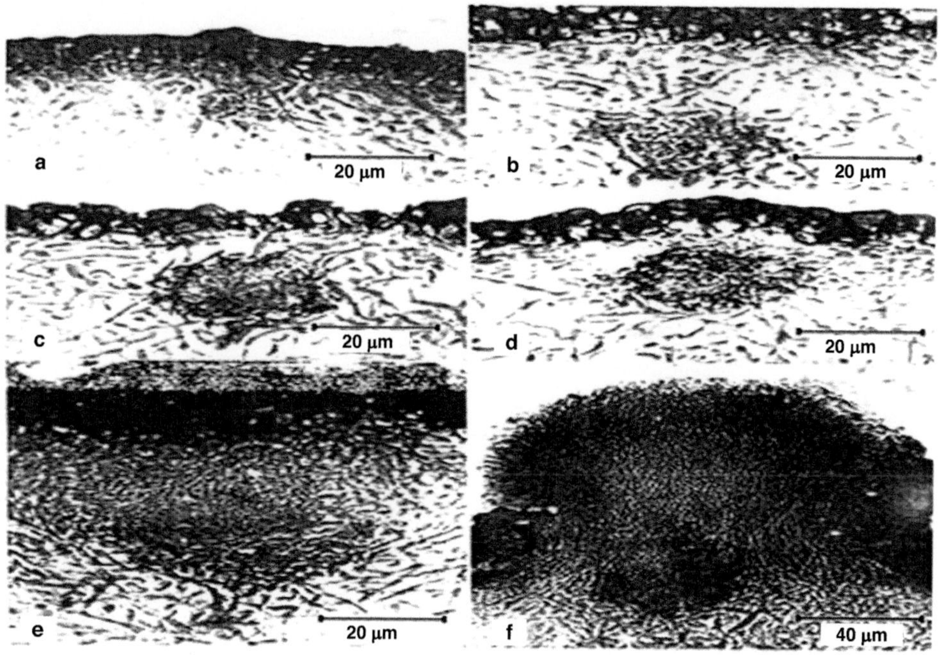

Fig. 5.12. The vertical sections of the sclerotium depict the various stages of apothecial stipe growth. (a) A meristematic structure in Stage I-primordium that is highly pigmented; (b) the primordium has gotten bigger in Stage I, but pigmentation has not yet appeared; the primordium is covered in dark pigments in Stage II, pigmentation penetrates Stage II, the primordium is in Stage III, and the primordium is in Stage IV. (Adapted from publication of Saito, 1977, with permission.)

surface is exposed during invaginations. The distinct surfaces of the spore organelles previously mentioned are highlighted in a detailed image of a broken and etched ascus containing ascospores (Fig. 5.27h) (seven out of eight spores displayed). Furthermore, the broken apothecium is made up of a variety of structures, possibly including paraphyses and juvenile asci. Additionally, a semi-diagrammatic representation of an ascus rupture is shown in Fig. 5.28. The apothecial walls and stipe's infrared spectra have a striking resemblance to *S. sclerotiorum's* hyphal walls. These walls are made up of chitin and β−1,3 glucans, with chitin making up the majority of the wall's components (Jones, 1970; Jones, 1974).

Formation of sporogenous area during ascus development heavily relies on the ascus vesicle, a membrane system that takes the shape of a muff with open ends. Interestingly, despite the near resemblance between the vesicle's component membranes and the plasmalemma, this vesicle displays a substantially higher reactivity to polysaccharides than the ascus plasmalemma. Additionally, significant alterations in the quantity and arrangement of glycogen storage, along with the cellular cytoplasm's composition, have been recorded throughout the progression of *Sclerotinia* asci. Moreover, the intricate phenomenon of carpogamy, the creation of the intricate apical device, and the growth of ascospore walls have been thoroughly examined in great detail (Codron, 1974).

5.5 Ultrastructures of Microconidia and Stroma

The small conidia produced by *S. sclerotiorum* contain identical cell organelles, as previously described. These consist of a well-defined nucleus, a noticeable lipid body, two mitochondria, and a limited endoplasmic reticulum. The

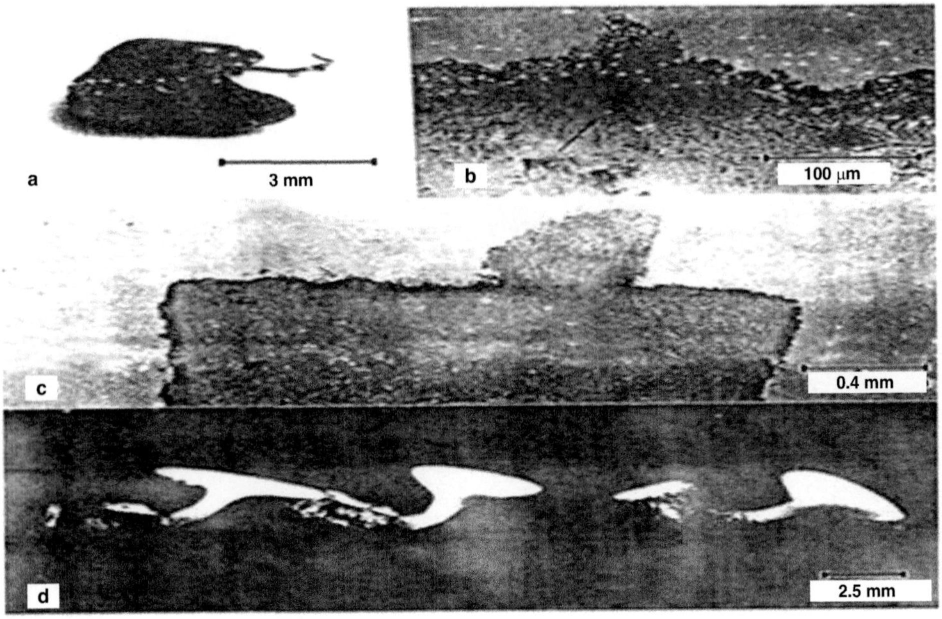

Fig. 5.13. (a) The apothecial stipe emerges from the sclerotium incision; (b) a portion of cubed medullary tissue illustrating redeveloped peel that forms the apothecial stipe primordia, as shown by the arrows; (c) a segment of a cubed medullary tissue illustrating the regrowth of the outer layer and its sprouting; (d) during growth, developed apothecia emerge from cubed medullary tissues. (Adapted from publication of Saito, 1977, with permission.)

stroma contains a plentiful number of hyphae that are rich in lipid bodies and storage vacuoles, serving as a valuable source of nutrition. Occasionally, there are basic septa with blocked pores. There are two sorts of hyphae: one full of nutrients and the other showing signs of deterioration. Both forms of hyphae have identical CW thickness of roughly 0.2 μm and widths varying from 1.5 to 3 μm (Calonge, 1970).

5.6 The Host–Pathogen Interface

The mechanism of host-surface penetrations and their infection by the pathogen observed through scanning electron microscopy and explained how the pathogen penetrates host cells and became inter and intracellular, ramify host tissues, emerge out from host surface and produce mycelium and sclerotia. Pathogenic fungi commonly invade intact host surfaces during host–parasite interactions. The cuticle on the plant surfaces acts as the first line of defense against infection. The method by which the fungi penetrate the cuticle, whether through physical force or enzymatic digestion, has been a subject of debate for a long time (Kolattukudy, 1985). Several studies have provided evidence that *Sclerotinia* usually infiltrate the tissues of host by directly penetrating the cuticle. However, there have also been instances where pathogen entry through open stomata has been observed (Purdy, 1958; Prior and Owen, 1964; Lumsden and Dow, 1973; Abawi *et al.*, 1975; Jones, 1976; Tariq, 1984; Tariq and Jeffries, 1984; Tariq and Jeffries, 1986). The scanning electron microscopy analysis (Figs 5.29 and 5.30) demonstrated that the *S. sclerotiorum* hyphae have the ability to enter through stomatal openings located on the upper surfaces of potato leaf (Jones, 1976). This technique was also employed to observe the infection progression of *S. sclerotiorum* and *S. minor*, which further supported the outcomes of previous investigations conducted using light microscopy on *S. sclerotiorum* (Boyle, 1921; Purdy, 1958; Lumsden and Dow, 1973;

Abawi *et al.*, 1975), *S. minor* (Lumsden and Dow, 1973), and *S. trifoliorum* (Prior and Owen, 1964). In order to promote the onset of an infection, a sequence of procedures is carried out. First, infection cushions envelop the surface of the host. Subsequently, the cuticle is mechanically breached. Following that, the infection hyphae expand and disseminate between host cells. The host tissue is subsequently invaded by the branching hyphae, both internally and externally. Hyphae appear through the stomata, leading to the sclerotium formation and compact mycelial covering on the surface of host (Lumsden and Wergin, 1980).

5.6.1 Appressorial development for infection

Appressoria play a crucial role in invading intact host tissue by penetrating the cuticle. It is theorized that they establish a strong connection with the host's surface and are thought to be connected by an adhesive substance. This material, along with the dome-shaped structure of the appressorium, helps apply pressure on the cuticle, facilitating the entry into the host tissue through infection pegs. The importance of mechanical force in the pathogen's penetration is strongly backed by existing evidence. This is demonstrated by the finding that the infection cushions, or appressoria, have a tendency to compress the host tissue around them, creating a mound that is somewhat elevated. Furthermore, at the penetration sites, the cuticle is forced inward, suggesting that mechanical force played a part in the pathogen's entrance (Boyle, 1921; Lumsden and Dow, 1973). Before the fungus infiltrates, the cuticle does not dissolve, soften, or change (Boyle, 1921). Despite the advancement of the disease, the cuticle retains its initial staining reactivity and effectively prevents the fungus from infiltrating (Boyle, 1921; Lumsden and Dow, 1973). It has been observed that the cuticle and epidermal cells are undergoing destruction (Prior and Owen, 1964). The researchers believed that this destruction was caused by the enzyme activity of the appressoria. They also determined that *S. trifoliorum* has the ability to penetrate clover leaves as a result of the mechanical pressure exerted by the infection cushions. There have been additional reports of plant pathogenic fungi breaking down cutin in laboratory settings (Kolattukudy, 1985).

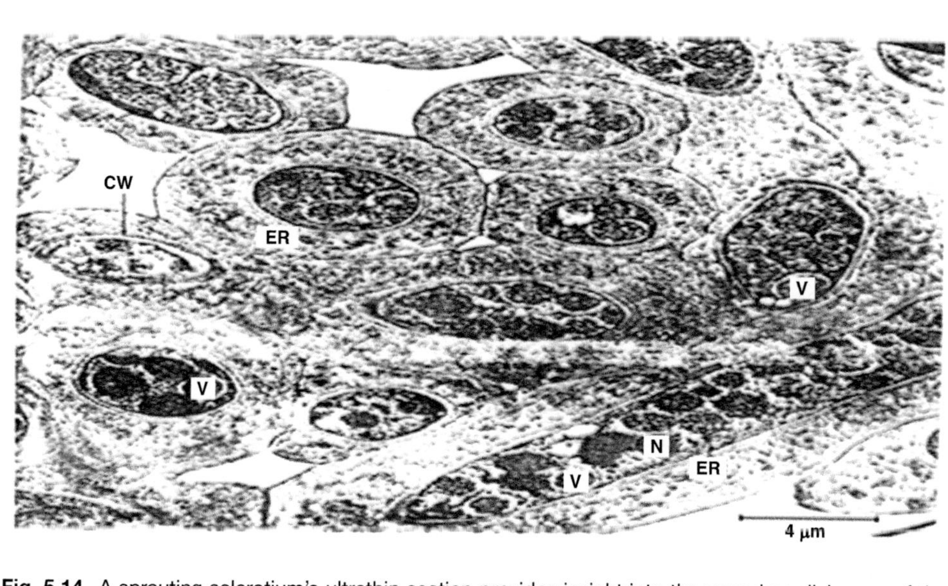

Fig. 5.14. A sprouting sclerotium's ultrathin section provides insight into the nonprimordial areas of the medulla's cellular state. Vacuoles (V) with electron-dense, amorphous material occupy the majority of the cytoplasm, but the cell wall (CW) has a well-developed endoplasmic reticulum (ER) zonation. (Adapted from publication of Saito, 1977, with permission.)

Fig. 5.15. (a) A damaged cell that is growing from the medulla of sclerotia; (b) A nearly fully degraded cell adjacent to a healthy one, with a Woronin body (WB) covering the entrance; (c). The initial cell cluster of the medulla. Observe the changes in size between medullary cells (MC) and starting cells (PC). (Adapted from publication of Saito, 1977, with permission.)

5.6.2 Histology and histochemical observations during infection

Histological investigations conducted at an ultrastructural level have revealed that fungal pathogens may employ enzymes to facilitate the penetration of plant tissues. Furthermore, the utilization of light microscopy has facilitated histochemical investigations to reveal esterase action and chemical alterations in cutin at the penetration site. A study conducted on *Phaseolus* plants demonstrated the invasion of *Sclerotinia* and observed various alterations in the chloroplast ultrastructure (Tariq and Jeffries, 1986). Additionally, Tariq (1984) suggested that the penetration process requires the presence of enzymes that break down cell walls. Additionally, Tariq and Jeffries (1986) successfully induced non-specific esterase activity in the liquid filtrate of *S. sclerotiorum* through the addition of cutin substrates. It was identified that *S. sclerotiorum* displayed lipolytic enzyme activity as it invaded *Phaseolus* leaf tissues. Initially, this activity was detected at the locations where an appressorium closely interacted with the host surface. In the fungal cytoplasm, an extra-plasmalemma zone was present, along with electron-opaque deposits directly below a vesicular region. The deposition was restricted to the region between the host cuticle and the plasma membrane of the fungus. Interestingly, during the later stages of infection, the electron-dense deposition was specifically perceived at the site of penetration on the host or near the entry aperture.

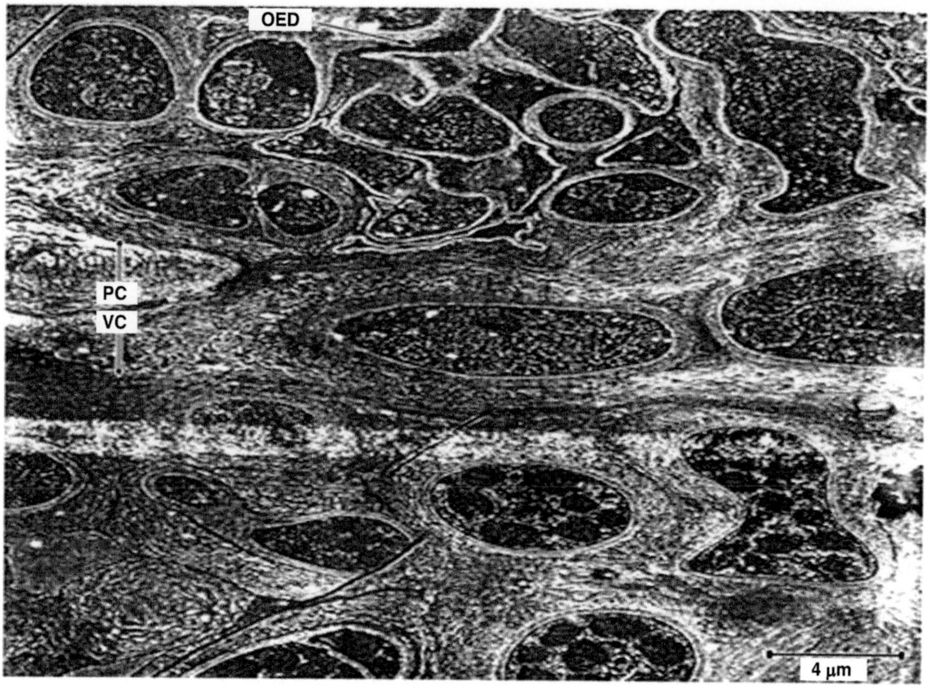

Fig. 5.16. An incredibly thin slice of medullary tissue allows for the obvious separation of primordial cells (PC) and medullary cells (MC). The buildup of electron-dense materials inside the PC and the reduction in vacuole contents (V) in medullary cells close to PC are noteworthy. (Adapted from publication of Saito, 1977, with permission.)

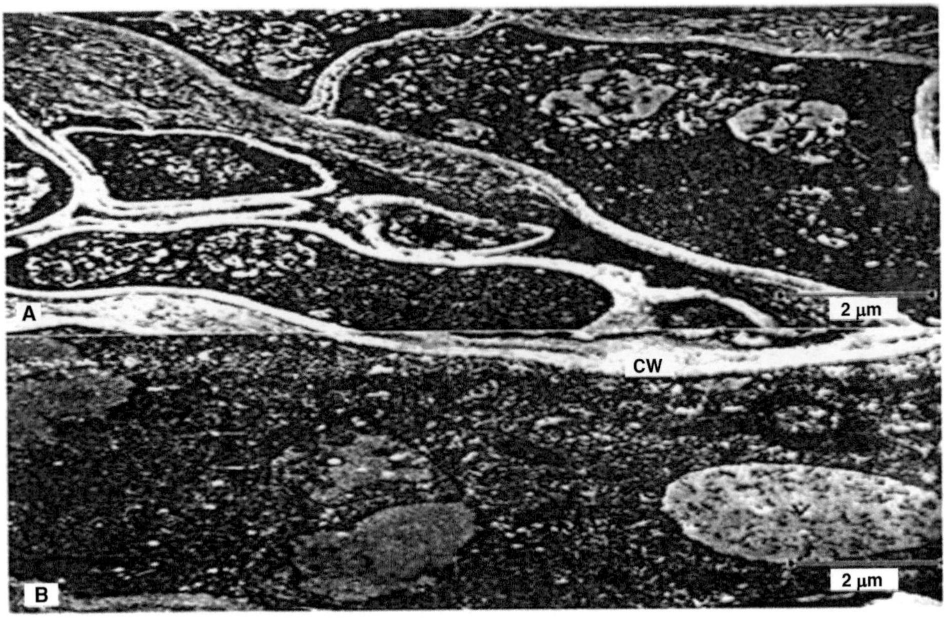

Fig. 5.17. (a) A primordial structural element; (b) a primitive cell's cytoplasmic region with a large number of mitochondria and ribosomes. (Adapted from publication of Saito, 1977 with permission.)

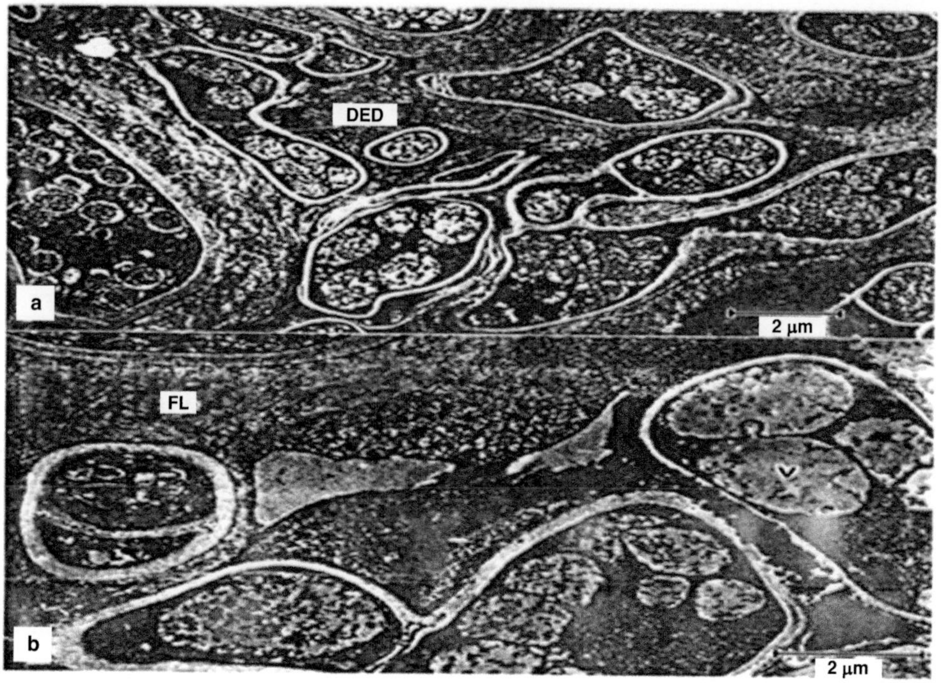

Fig. 5.18. (a) A thin section of a primordium demonstrating the presence of primordial cells with thin walls, irregular shapes, and a range of sizes, as well as electron-dense materials (DED). (b) The primordium's exterior displays the accumulation of electron-dense elements in a nonfibrous layer surrounding the medullary canal. (Adapted from publication of Saito, 1977, with permission.)

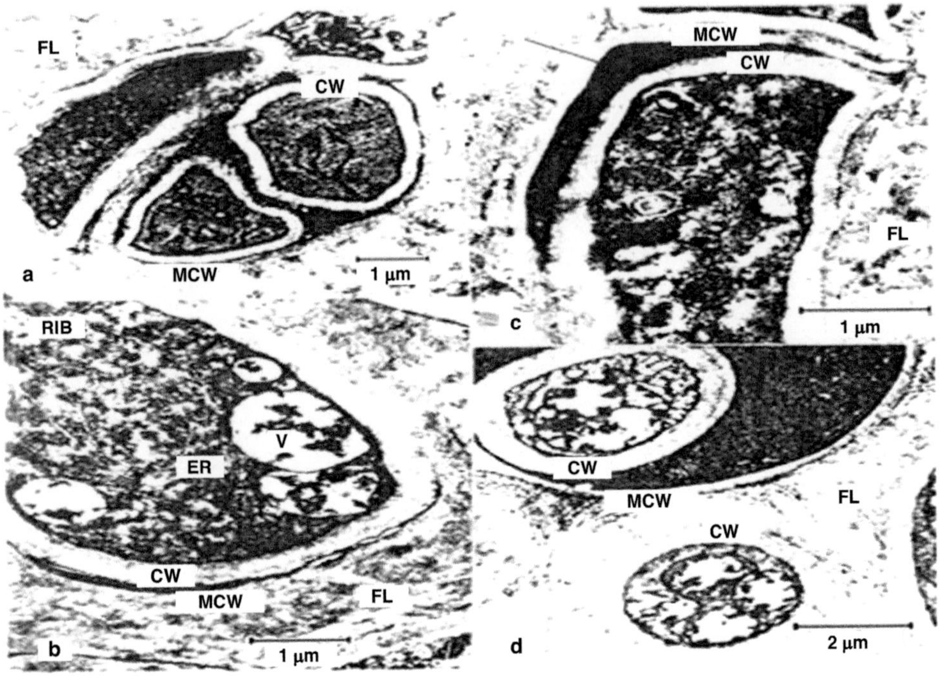

Fig. 5.19. (a) The medullary cells resemble endo-hyphae. We use $KMnO_4$ to color these cells. (b) Discrete, separate layers in the cell wall (CW) are seen in a medullary cell's cross section. Endoplasmic reticulum and many ribosomes may be observed. (c) A medullary cell's ultra-thin cross section displays different CW layers, just like the previous stage. Notably, there is a zone of high electron density located in between these dividing layers. (d) The medulla of a sprouting *Sclerotium borealis* sclerotium contains cells that resemble endo-hyphae. (Adapted from publication of Saito, 1977, with permission.)

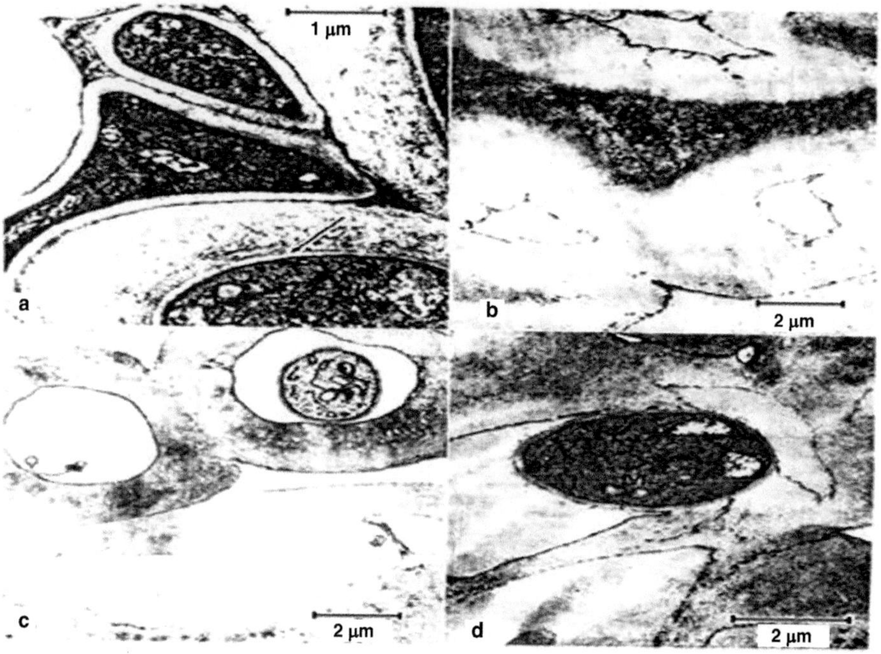

Fig. 5.20. (a) The medullary cells display distorted cells of the stipe fundament. Zonation can be observed in the medullary cell wall, indicated by the arrow, when treated with $KMnO_4$. (b) Within the decayed sclerotium, a portion of the medullary tissue can be observed. This tissue has given rise to numerous apothecia. The cell walls start to disappear, and the fibrous layers lose their structure to a large extent when treated with $KMnO_4$. (c) The fibrous layers in the decayed medullary tissue maintain their structure even after the decay. It was observed in $KMnO_4$ treated cells. (d) Cross section of a secondary developed hyphal cell in degraded medullary tissue exhibits its properties after $KMnO_4$ treatment. (Adapted from publication of Saito, 1977, with permission.)

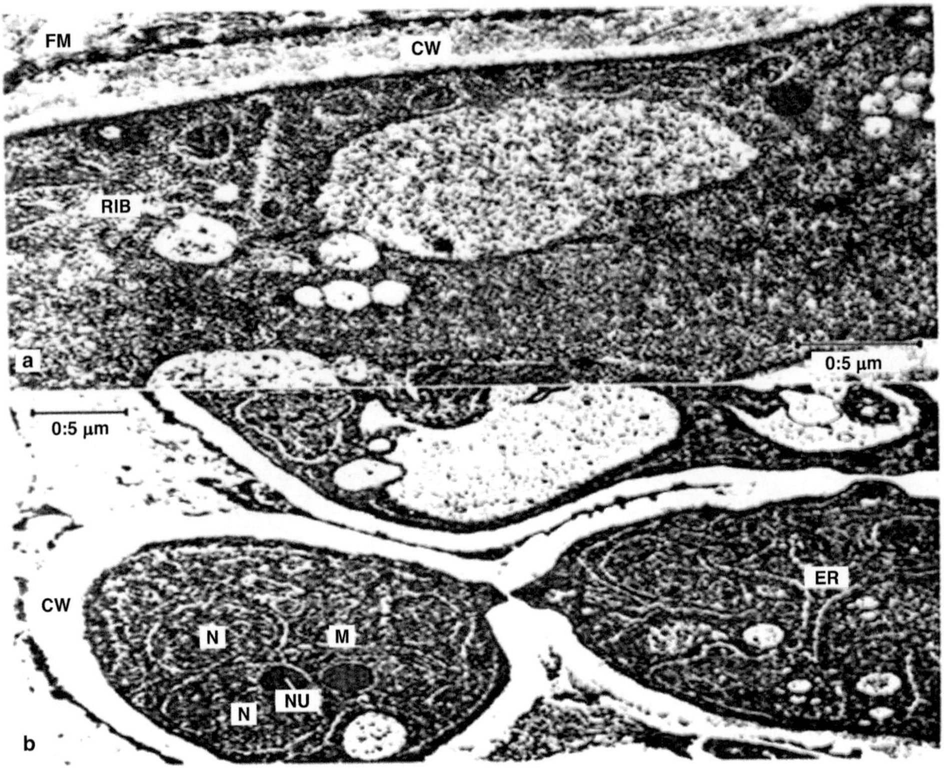

Fig. 5.21. (a) A cell that is located near the stipe's base; (b) the different cytoplasmic ribosomes, many mitochondria (M), and cell wall (CW) layers (shown by an arrow). (Adapted from publication of Saito, 1977, with permission.)

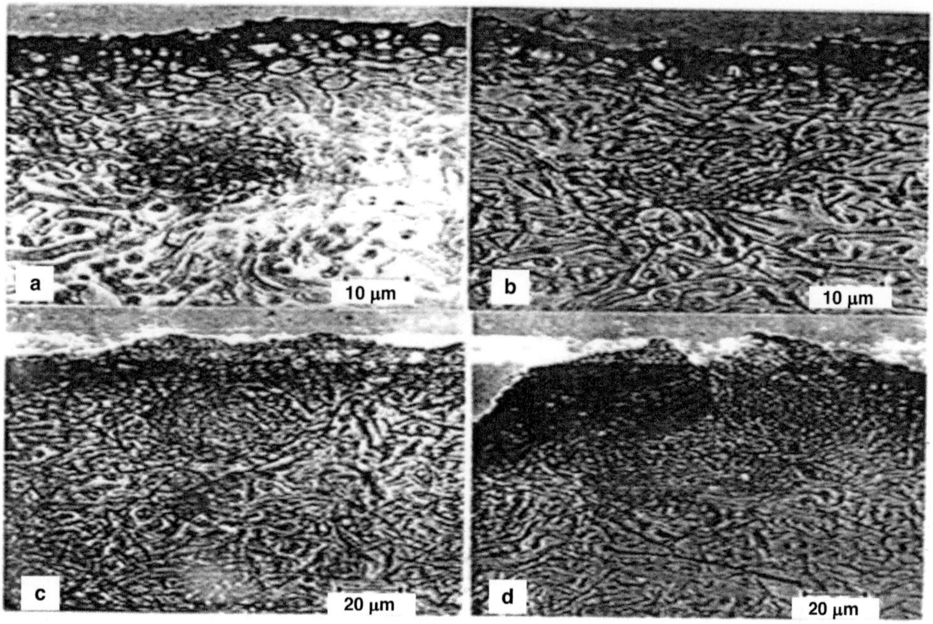

Fig. 5.22. Light micrographs were used to see the histochemical reactions in sclerotium slices. (a) The first micrograph showed the PAS response in a primordium and the surrounding medullary tissue. (b) The second micrograph showed the proteins in a primordium and the surrounding medullary tissue stained with mercuric bromophenol blue. The second micrograph's staining was the same as that of the third and fourth (c and d). (d) The fourth micrograph displayed a mature Stage III primordium. (Adapted from publication of Saito, 1977, with permission.)

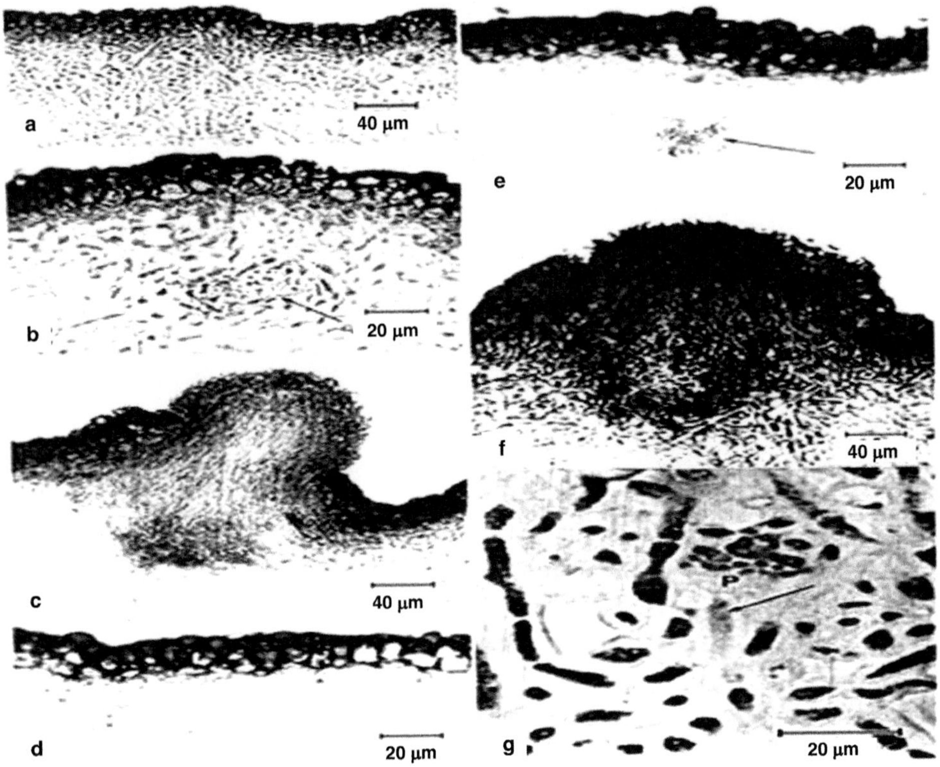

Fig. 5.23. (a) The histochemical reactions in various sclerotium sections were observed using light micrographs. The Millipore response in the nonprimordial area of the medullary tissue is depicted in the first micrograph. (b) Inside a primordium and the surrounding medullary tissue, the Millon reaction is observable. (c) The Millon reaction in a Stage IV primordium is depicted in the third micrograph. (d) The pyronin dye allows the nonprimordial portion of the medullary tissue to be seen in the fourth micrograph. An arrow indicates a substantial staining in the primordium of the (e) Micrograph, which shows pyronin staining in the primordium and the surrounding medullary tissue. (f) A Stage IV primordium stained with toluidine blue is seen in micrograph. (g) The seventh micrograph displays a colored HPMA segment. A blue-stained primordium (P) and a nearby medullary cell devoid of polyphosphate-like structures are visible indicated by an arrow. (Adapted from publication of Saito, 1977, with permission.)

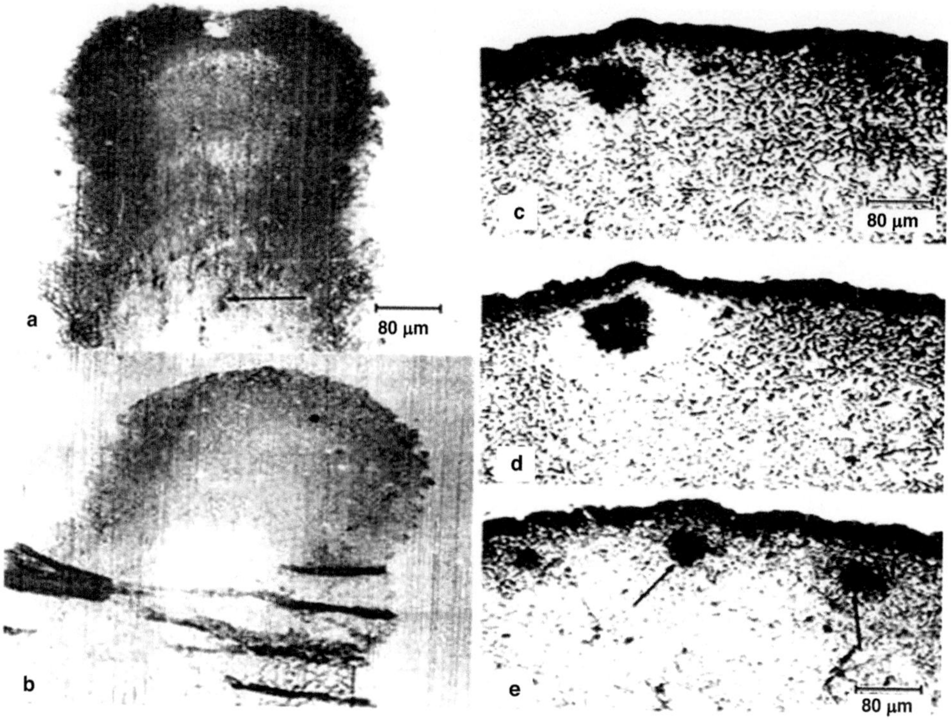

Fig. 5.24. (a) A characteristic pyronin stain is shown on a section of the immature apothecial stipe. The apical region is where the staining is most dense; however, some cells also show distribution in the middle zone (shown by an arrow). (b) The same slice was treated with ribonuclease and then stained in a pattern that mirrored that of the staining reported in section (a). (c) A portion of sclerotia that includes a Stage II primordium has been toluidine blue-stained. Important features within the primordium are shown by the staining, which also covers the rind. (d) Like section (c), this one has been subjected to ribonuclease treatment and then stained with toluidine blue, yielding a staining pattern that matches the explanation given in section (c). (e) A section subjected to Masson's ammoniacal silver nitrate staining exhibits intense staining in both the primordial region (indicated by arrows) and the surrounding rind. (Adapted from publication of Saito, 1977, with permission.)

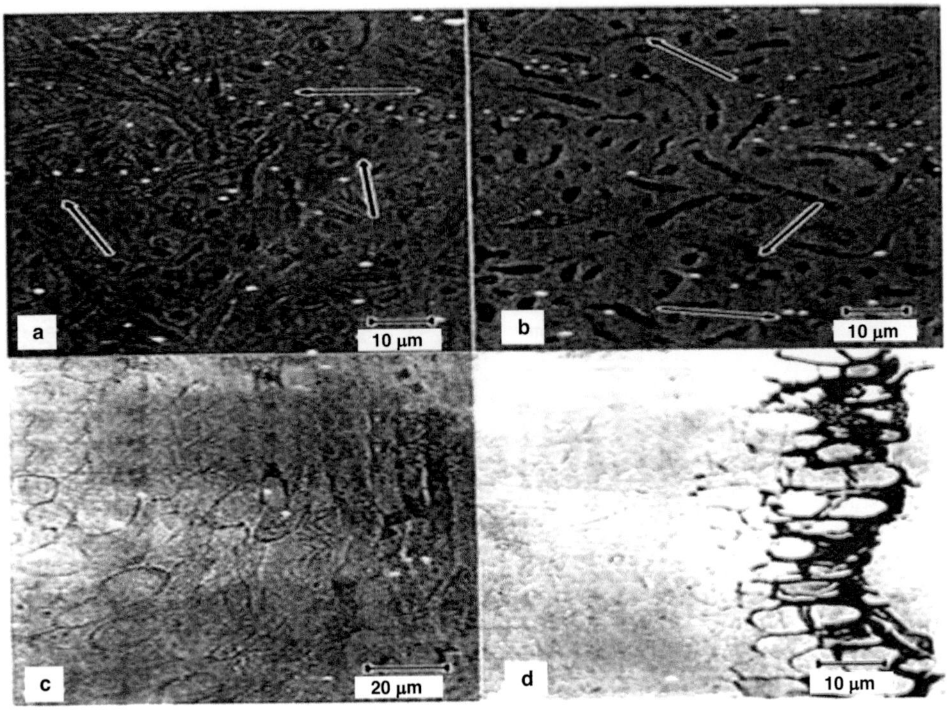

Fig. 5.25. (a) The medullary tissue section of sprouting sclerotia contains an intercellular matrix that is basophilic, as shown by the arrows in the pyronin staining. (b) Toluidine blue staining reveals basophilic characteristics in the intercellular matrix (arrows). (c) A longitudinal section of the early stipe shows that the inner cells on the right side exhibit noticeably higher succinate dehydrogenase activity compared to those on the left. (d) In a growing sclerotium, the nonprimordial region of the medulla shows a succinate dehydrogenase reaction comparable to that shown in section (c). (Adapted from publication of Saito, 1977, with permission.)

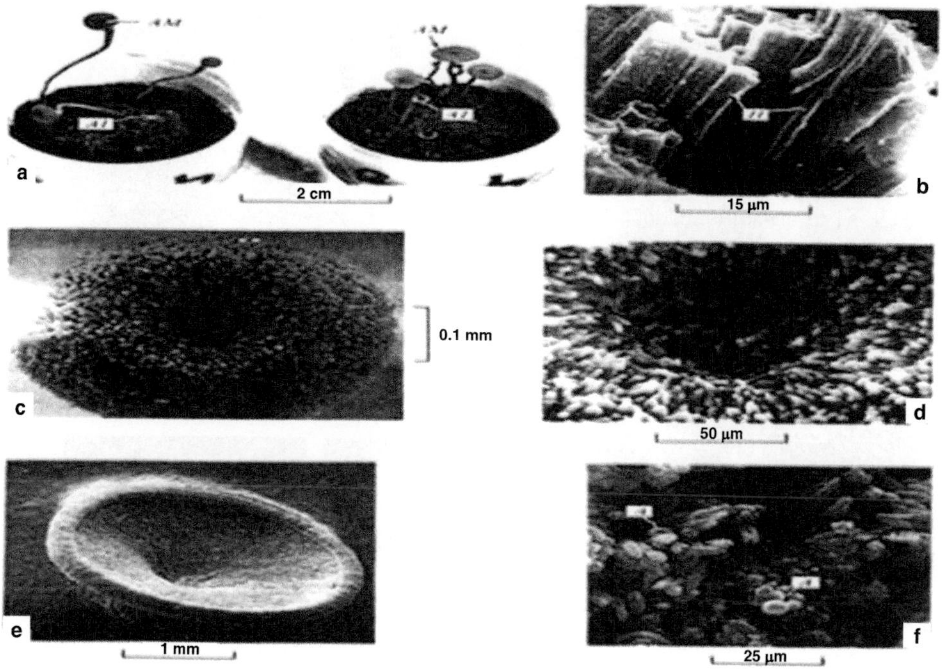

Fig. 5.26. (a) Sclerotia are seen to be thriving in the pots and exhibiting apothecia and stipes at different growth stages. In the context of detached stipes, scanning electron micrographs (b) through (f) (b), an immature apothecium (c and d), and a fully formed apothecium (e and f) are shown. AI: immature apothecium; AM: mature apothecium; H: hyphal threads; A: asci. (Adapted from publication of Jones, 1974, with permission.)

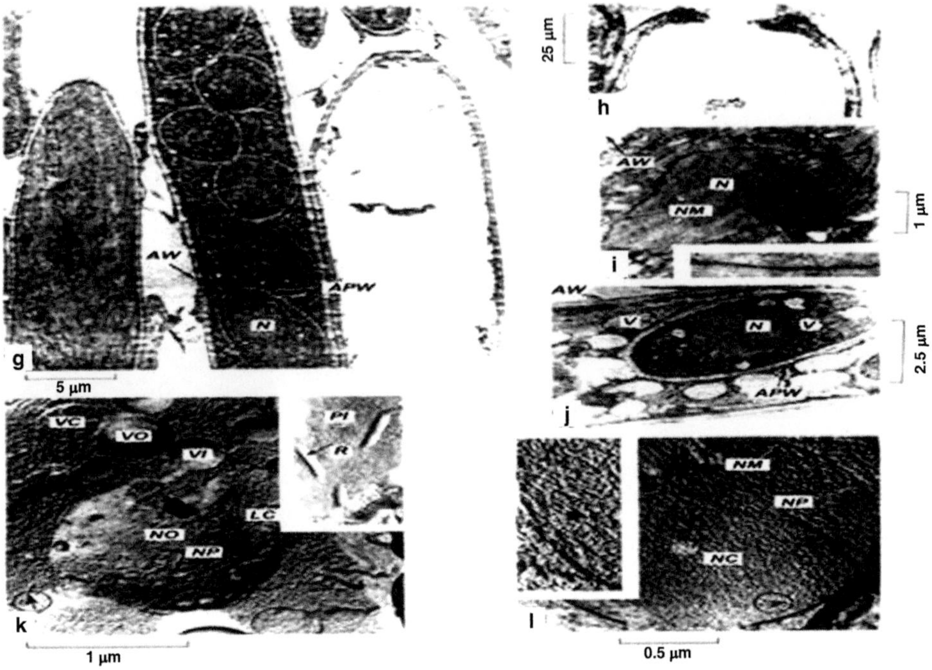

Fig. 5.27. (g) A longitudinal section of the apothecium displays both mature asci containing ascospores and asci without ascospores. (h) The longitudinal segment of an empty ascus reveals the opening near the tip. (i) The longitudinal section of an immature ascus indicates that the ascospore wall is still in the early stages of development. Inset: A closer look at the nuclear membrane. (j) A transverse segment through a fully formed ascus. (k) A portion of the ascospore showcasing multiple organelles. Inset: Ridges can be observed on the inner surface of the plasmalemma. (l) The nucleus inside an ascospore is cross fractured. Inset: A thorough depiction of the dual-layered membrane surrounding the nucleus. (Adapted from publication of Jones, 1974, with permission.)

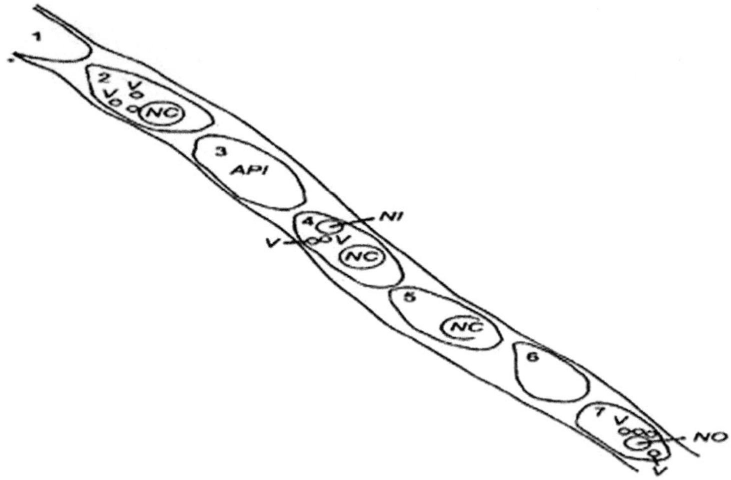

Fig. 5.28. A schematic image of a freeze-fractured structure shows an ascus harboring ascospores. (Adapted from publication of Jones, 1974, with permission.)

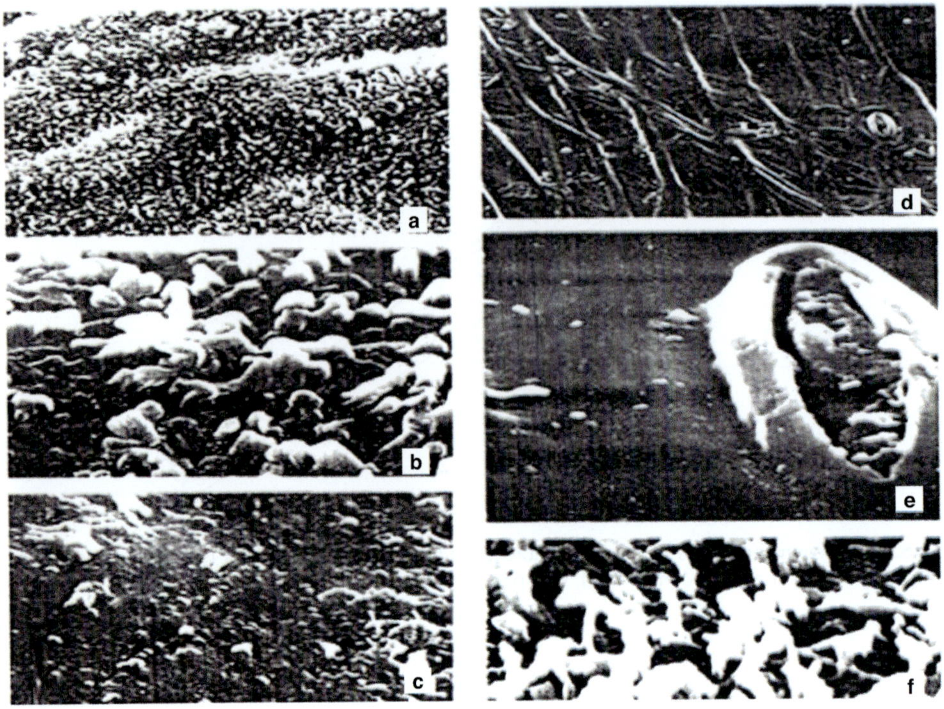

Fig. 5.29. The scanning electron micrographs illustrate the surface of pea leaf and that have encountered an infection caused by *S. sclerotiorum*. (a and b) The stem lesions exhibit a parchment-like tissue with an abundance of rod-like wax structures. Panel (c) shows the white area containing a few protuberances of cuticular wax surrounded by the healthy green sections of the stem. After being submerged in petroleum ether, the stem lesions are shown in panels (d) and (e), emphasizing the existence of stomatal apertures. Finally, panel (f) shows a cuticular wax on a leaf's lower surface that resembles a ribbon. (Adapted from publication of Jones, 1974, with permission.)

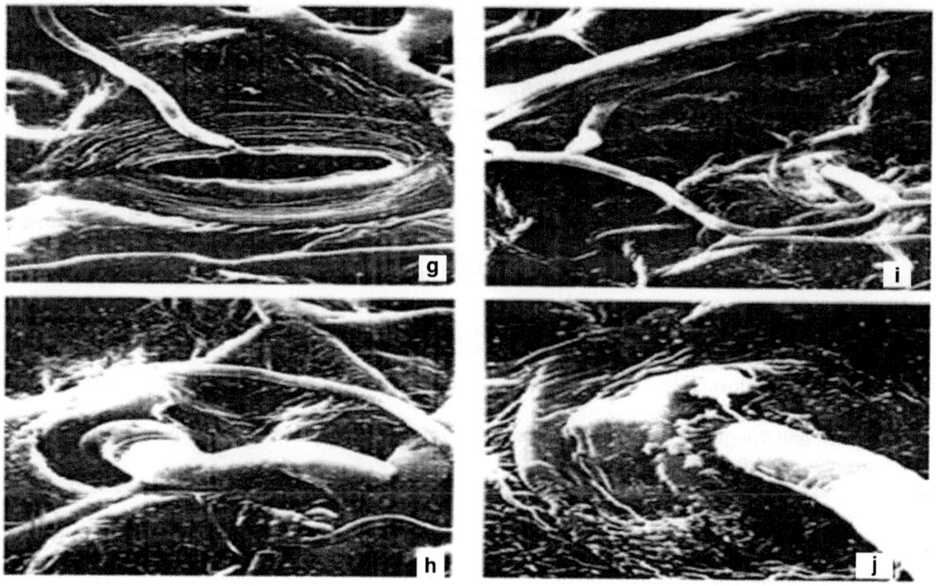

Fig. 5.30. *S.* sclerotiorum hyphae infiltration into stomatal holes on upper surface of potato leaf is depicted in the scanning electron micrographs in panel (g)–(j). (Adapted from publication of Jones, 1974, with permission.)

References

Abawi, G.S., Polach, F.J. and Molin, W.T. (1975) Infection of bean by ascospores of *Whetzelinia sclerotiorum*. *Phytopathology* 65(673), 678.

Boyle, C. (1921) Studies in the physiology of parasitism. VI. Infection by *Sclerotinia libertiana*. *Annals of Botany* 35(3), 337–347.

Bullock, S. and Willetts, H.J. (1996) Ultrastructural and histochemical studies on mycelial germination of sclerotia of *Sclerotinia minor*. *Mycological Research* 100(5), 561–570.

Bullock, S., Ashford, A.E. and Willetts, H.J. (1980a) The structure and histochemistry of sclerotia of *Sclerotinia minor* jagger II. histochemistry of extra cellular substances and cytoplasmic reserves. *Protoplasma* 104, 333–351.

Bullock, S., Willetts, H.J. and Ashford, A.E. (1980b) The structure and histochemistry of sclerotia of *Sclerotinia minor* jagger. *Protoplasma* 104(3–4), 315–331.

Butler, G.M. (1966) Vegetative structures. In: Ainsworth, G.C. and Sussman, A.S. (eds) *The Fungi—An Advanced Treatise*, Vol. 2. Academic, New York, pp. 83–112.

Calonge, F.D. (1970) Notes on the ultrastructure of the microconidium and stroma in *Sclerotinia sclerotiorum*. *Archives of Microbiology* 71(2), 191–195.

Codron, P.D. (1974) Ultra-structural study of some aspects of the ascus development of Sclerotinia sclerotiorum (lib.) *de bary*. In: *Annals Des Sciences Naturelles, Botanique, Paris*, Vol. 15. pp. 255–276.

Coley-Smith, J.R. and Cooke, R.C. (1971) Survival and germination of fungal sclerotia. *Annual Review of Phytopathology* 9(1), 65–92.

Colotelo, M. (1974) A scanning electron microscope study of developing sclerotia of *Sclerotinia sclerotiorum*. *Canadian Journal of Botany* 52(5), 1127–1130.

Huang, H.C. (1982) Morphologically abnormal sclerotia of *Sclerotinia sclerotiorum*. *Canadian Journal of Microbiology* 28(1), 87–91.

Huang, H.C. (1983) Histology, amino acid leakage, and chemical composition of normal and abnormal sclerotia of *Sclerotinia sclerotiorum*. *Canadian Journal of Botany* 61(5), 1443–1447.
Jones, D. (1970) Ultrastructure and composition of the cell walls of *Sclerotinia sclerotiorum*. *Transactions of the British Mycological Society* 54(3), 351–360.
Jones, D. (1974) Ultrastructure of the stipe and apothecium of *Sclerotinia sclerotiorum*. *Transactions of the British Mycological Society* 63(2), 386–389.
Jones, D. (1976) Infection of plant tissue by *Sclerotinia sclerotiorum*: A scanning electron microscope study. *Micron* 7(4), 275–279.
Kolattukudy, P.E. (1985) Enzymatic penetration of the plant cuticle by fungal pathogens. *Annual Review of Phytopathology* 23(1), 223–250.
Kosasih, B.D. and WIlletts, H.J. (1975) Types of abnormal apothecia produced by *Sclerotinia sclerotiorum*. *Mycologia* 67(1), 89–97.
Le Tourneau, D. (1966) Trehalose and acyclic polyols in sclerotia of *Sclerotinia sclerotiorum*. *Mycologia* 58(6), 934–942.
Lumsden, R.D. and Dow, R.L. (1973) Histopathology of *Sclerotinia sclerotiorum* infection of bean. *Phytopathology* 63(6), 708.
Lumsden, R.D. and Wergin, W.P. (1980) Scanning-electron microscopy of infection of bean by species of *Sclerotinia*. *Mycologia* 72(6), 1200–1209.
Prior, G.D. and Owen, J.H. (1964) Pathological anatomy of *Sclerotinia trifoliorum* on clover and alfalfa. *Phytopathology* 54, 784–787.
Purdy, L.H. (1958) Some factors affecting penetration and infection by *Sclerotinia sclerotiorum*. *Phytopathology* 48, 605–609.
Saito, I. (1973) Initiation and development of apothecial stipe primordia in sclerotia of *Sclerotinia sclerotiorum*. *Transactions of Mycological Society, Japan* 14, 343–351.
Saito, I. (1974) Ultra-structural aspects of the maturation of sclerotia of *Sclerotinia sclerotiorum* (lib.) de Bary. *Transaction of Mycological Society of Japan* 15, 384–400.
Saito, I. (1977) *Sclerotinia sclerotiorum* (lib.) de Bary. Studies on the maturation and germination of sclerotia of *Sclerotinia sclerotiorum* (lib.) de Bary, a causal fungus of bean stem rot. In: *Report Hokkaido Prefectural Agricultural Experiment Station*. NARO, Hitsujigaoka, Sapporo, Japan, p. 106.
Tariq, V.N. (1984) Sclerotinia disease of plants: Characterization of isolates and host pathogen interaction. PhD. thesis, University of Kent, Canterbury, UK.
Tariq, V.N. and Jeffries, P. (1984) Appressorium formation by *Sclerotinia sclerotiorum*: Scanning electron microscopy. *Transactions of the British Mycological Society* 82(4), 645–651.
Tariq, V.-N. and Jeffries, P. (1986) Cytochemical localization of lipolytic enzyme activity during penetration of host tissues by *Sclerotinia sclerotiorum*. *Physiological and Molecular Plant Pathology* 30(1), 77–91.
Townsend, B.D. and Willetts, H.J. (1954) The development of sclerotia of certain fungi. *Transactions of the British Mycological Society* 37(3), 213–221.
Weete, J.D., Weber, D.J. and Tourneau, D. (1970) Hydrocarbons, free fatty acids, and amino acids of sclerotia of *Sclerotinia sclerotiorum*. *Archives of Mikrobiology* 75(1), 59–66.

6 The Disease, Symptomatology, Assessment, and Economic Importance

Abstract

Depending on the distinctive symptoms that *Sclerotinia* causes in different agricultural plants, the disease is known by more than 60 different names. Although it can infect entire plant components, it usually targets the stem as its primary site of infection in natural settings. Elongated, water-soaked lesions are among the symptoms that show up on the stem. These lesions are followed by the emergence of cottony-white mycelium. The plant starts to droop and eventually dries out when the lesions completely encircle the stem. On the surface or inside the pith of the infected stem, there are numerous spherical, grayish-white to black sclerotia, which indicates degeneration. Early on in the infection, the plant may show signs of damping down, root rot, and eventually complete death. On the leaves, there are irregular, whitish lesions that occasionally have petals adhering to them. Symptoms of twig blight with entire cottony mycelia are highly prevalent in humid environments. *Sclerotinia* can lead to significant yield losses in vulnerable crops, with the extent of these losses varying widely and potentially surpassing 100% in some cases.

6.1 Introduction

Many biotic and abiotic stresses can affect crucifers. While nematodes, bacteria, viruses, phytoplasma, and fungi can cause biotic stress, fungi account for the majority of pathogens infecting crucifers. The range of hosts that crucifer pathogens infect, their geographic distribution, the amount of yield losses they cause, and the importance that scientists attach to genomic research in raising the productivity and quality of *Brassica* crops are all taken into account in the global evaluation of crucifer pathogens. The third most devastating disease worldwide among these infections is caused by *Sclerotinia sclerotiorum*, which causes *Sclerotinia* stem rot (SSR) in crucifers (Saharan *et al.*, 2021; Table 6.1).

6.2 The Disease and Its Nomenclature

Diseases caused by Sclerotinia are called by various names and affect a wide range of host plants. Some of the most commonly affected plants are artichoke, asparagus, beans, beet, broad bean, cabbage, chrysanthemum, camellia, carrot, celery, citrus, clover, cucumber, delphinium, dahlia, eggplant, hollyhock, lettuce, peony, peanut, potato, narcissus, red pepper, rapeseed–mustard, sweet potato, soybean, sunflower, safflower, columbine, snapdragon, stock, squash, tomato, and tobacco. These diseases can take on various forms, such as watery soft rot, stem rot, timber rot, dollar spot in turf grass, cottony rot, crown rot, damping-off, head rot, lettuce drop, pink joint, root and crown rot, stem blight, stem canker, stalk break, white rot, white mold, and white blight. It is important to note that the causative fungus, which was formerly thought to be *Sclerotinia homoeocarpa*, is now categorized as a species of *Moellerodiscus* and *Lenzia* (Saharan and Mehta, 2008). However, because of its distinctive symptoms on the stem, which farmers can easily identify, the disease is usually known as *Sclerotinia* stem rot (SSR) on crucifers. The geographic distribution of primary illnesses affecting cruciferous crops, the range of hosts they can infect, the degree of production decline they cause, and the amount of genomic research done on them are factors that influence how these diseases are classified globally.

A noteworthy finding was made by Li *et al.* (1998) when they identified a novel *Sclerotinia* sp. variation in lettuce samples that were gathered from central China. This newly identified variant

DOI: 10.1079/9781800629592.0006

Table 6.1. Global ranking of primary diseases affecting crucifers throughout the world. Adapted from publication of Saharan *et al*. (2021).

No.	Casual organism	Disease name	GD in No. of different countries	Number host range	Average/ Range) % yield loss	GR No. of published papers	Ranking				Average (Ranking)	Citations
							GD	HR	YL	GR		
1.	*Albugo candida*	White rust	100	300	34	122	6	79	4	33	30.5 (VI)	Saharan *et al*. (2014)
2.	*Alternaria brassicae*	Alternaria blight	96	4000	45.1	183	12	114	11	46	45.75 (II)	Saharan *et al*. (2016)
3.	*Colletotrichum higginsianum*	Anthracnose	100	100	50	9	2	2	1	4	2.25(–)	Cannon *et al*. (2012)
4.	*Erysiphe cruciferarum*	Powdery mildew	44	100	27.3	134	18	31	5	80	33.5(V)	Saharan *et al*. (2019)
5.	*Fusarium oxysporum*	Wilt	100	100	15.9	20	1	0	1	18	5.0(–)	Garibaldi *et al*. (2006), Matić *et al*. (2018), and Lange *et al*. (2007)
6.	*Hyaloperonospora parasitica*	Downy mildew	102	99	37.3	15	3	4	1	7	3.75(–)	Saharan *et al*. (2017)
7.	*Leptosphaeria maculans*	Blackleg	102	35	42	142	14	37	3	88	35.5(IV)	Saharan *et al*. (2021)
8.	*Plasmodiophora brassicae*	Clubroot	144	3700	28.5	273	21	127	17	108	68.25(I)	Saharan *et al*. (2021)
9.	*Neopseudocercosporella capsellae*	White leaf spot	29	22	15	4	1	3	0	0	1.0(–)	Gunasinghe *et al*. (2016); Gunasinghe *et al*. (2020)
10.	*Pyrenopeziza brassicae*	Light leaf spot	100	33	22	9	5	2	1	1	2.25(–)	Fitt *et al*. (1997)
11.	*Rhizoctonia solani*	Seedling rot	70	200	30	5	0	1	0	4	1.25(–)	Saharan *et al*. (2021)

Continued

Table 6.1. Continued

No.	Casual organism	Disease name	GD in No. of different countries	Number host range	Average/ Range) % yield loss	GR No. of published papers	Ranking				Average (Ranking)	Citations
							GD	HR	YL	GR		
12.	*Sclerotinia sclerotiorum*	Stem rot	102	900	40	166	22	70	3	71	41.5(III)	Saharan and Mehta (2008)
13.	*Turnip mosaic virus* (TuMV)	Crucifers Mosaic	106	8		77	5	18	2	52	19.25(VII)	Saharan *et al.* (2021)
14.	*Verticillium longisporum*	Wilt	10	9		27	5	8	0	14	6.75(–)	Tyvaert *et al.* (2014), Novakazi *et al.* (2015), and Depotter *et al.* (2017)
15.	*Xanthomonas campestris* pv. *Campestris*	Black rot	157	10		73	5	39	1	28	18.25(VIII)	Saharan *et al.* (2021)
16.	*Heterodera schachtii*	Nematodes Beet cyst eelworm	91			1				1		Saharan *et al.* (2021)
						3135[a]						

– denotes nonavailability of data.
[a]Total publications; the ranking of the disease is indicated within the parenthesis.
GD = geographical distribution, GR = genomics research, HR = host range, YL = yield losses.

was named *Let-19*. It was discovered through pathogenicity assays that the hyphae arising from *Let-19*'s myceliogenically sprouted sclerotia effectively infected healthy lettuce leaves. These hyphae, however, only infected damaged leaves of other plant species, such as rapeseed. *Let-19* does not exhibit pathogenicity against carrots, radish, or Chinese cabbage. Let-19 develops at a slower rate than *S. trifoliorum* and is best grown at a temperature of 20°C. *Let-19* sclerotia readily proliferate to form stipes, while apothecia seldom grow in low light. According to preliminary electrophoretic investigations of soluble proteins and enzymes, *Let-19* belongs to the *Sclerotinia* genus and is not to be confused with *S. sclerotiorum* and *S. minor*, two known pathogens associated with lettuce drop.

6.3 Symptomatology

6.3.1 General symptoms on different host plants

The host or parts of the host that are infected, as well as the ecological environment, can all influence the symptoms that *Sclerotinia* produces. One of the most noticeable early signs of SSR disease is the existence of a white, fluffy mycelial growth on the infected plant. Eventually, this development matures into sclerotia, which are big, compact resting bodies. The sclerotia are initially white, but gradually turn black and solidify on the outside. The diameter of these objects ranges from 2 to 10 mm, demonstrating a variety in size. Nonetheless, they typically have an extended, flattened form that differs from a spherical one.

The base of infected succulent or herbaceous plant stems may become pale or dark brown in color. Plants that are infected may go unnoticed at first since the leaves may not show obvious symptoms of the infection. However, the stem gradually decays as the fungus spreads throughout it. Moreover, the development of large oval or circular cavities on the leaves may result from ascosporic infection. The leaves that are above the impacted region then quickly wither and die. Sometimes, the illness begins on a leaf, then spreads to the stalk. The fungal sclerotia can grow externally on the stem, where they are easily visible, or internally within the pith of the stem, where they are not visible from the outside.

When the fungus infiltrates the lower leaves and stem at the base of crops such as lettuce, celery, and beet, the foliage and stalks quickly wilt and eventually die. The disease quickly penetrates and disperses throughout the entire stalk, causing the plant to completely collapse and die. Until they settle on the leaf below, the leaves progressively droop and bow. Sclerotia and mycelium are frequently seen on the underside of the outer leaves. When the weather is damp, the fungus colonizes the entire plant, causing its deterioration and producing a fluffy, white mycelial growth that covers the entire plant. However, the fungus will produce cankers in the stem if the infection is followed by dry weather, which will finally kill the plant without causing soft rot. In celery, a conspicuous pink or reddish-brown, damp region at the base of the affected petioles is typically hidden by the white mycelium. The plant may finally die as a result of the rot spreading throughout the stems.

Whether in the field or in storage, carrots and other juicy storage organs that are infected with SSR show a white, fluffy growth on their exterior. Black sclerotia, which occur as the disease penetrates the outer tissues, give the damaged areas a darker appearance than the healthy ones. The tissues that are infected become pliable and swollen. Should the illness surface in the storage facility after harvest, it could spread to nearby roots or any other exposed storage organs. As a result, pockets of decomposed organs may form, or the organ crate may get infected and collapse as a whole. The subsequent degradation is covered in fungal growth and has a delicate, watery appearance.

Sclerotinia is a fungus that can infect a variety of delicious fruits, including squash, cucumber, eggplant, and bean seed pods. The closest point to the ground, where it makes contact with the earth, or the desiccated parts of the flower is potential sites of infection. This fungus produces a moist decay that completely breaks down the fruit or pod by starting at the tip and working its way throughout the entire structure. Black sclerotia and white fungal mycelium are typically visible on the outside as well as the inside of the infected pods and fruits.

Narcissus and camellias (flowering plants) are particularly vulnerable to floral infection. Numerous tiny, watery, light-brown lesions appear on the petals. These lesions can spread, coalesce, and eventually cover the entire petal. The entire flower then undergoes a transformation, turning dark brown before eventually dropping off the plant. However, the flowers typically wilt once they have already dropped, especially during rainy conditions. The fungus produces a significant amount of mycelium and sclerotia during this time (Saharan and Mehta, 2008).

6.3.2 Symptoms on cruciferous crops

Sclerotinia causes disease symptoms on three major crops of the *Brassicaceae* family, which are discussed in the following section.

6.3.2.1 *Cabbage*

The Sclerotinia diseases that infect cabbage seed plants go by several names; some of them are cottony rot, drop, white blight, and watery soft rot. The disease initially manifests as little white spots on the stem or near the base of the leaf stalks in seed cabbage. These dots quickly enlarge to encircle the entire shoot and extend in an oval shape for about 4–6 inches along its central axis. The term 'white blight' refers to the grayish-white coloration of the surface of the affected area. The fungus produces a white, cottony mycelium. Once the stem is entirely surrounded, the transportation of nutrients stops, leading to the death of the plant. Frequently, the shoot fractures at the encircled area, and the underlying pith tissue is deteriorated by the fungus, leading to the development of solid, black sclerotia of varying sizes and shapes within the damaged tissue. These solid, dark, dormant bodies are typically oval-shaped and range in length from 1/8 to 1/2 inch, with a diameter that usually falls between 1/12 and 1/4 inch. While infections on the main stem are most commonly found in the lower regions, secondary branches can also be affected. When the main stem is affected, the decrease in seed production is less noticeable than when lesions appear in the leaf axils. According to a theory, the large, fleshy petioles that collect water in the leaf axils facilitate the spread of the spore-borne infection (Pound, 1946). During the transportation and storage of infected cabbage heads, a dense mycelial mass resembling white cotton develops, containing noticeable black sclerotia. This fungus possesses the capability to transmit from one head to another when they are placed together in a container or package. *Sclerotinia* infections can have a detrimental effect on the production of cauliflower, cabbage, rutabaga, and cauliflower crops for seed in the following year. White blight is the term for elongated, gray-white stem lesions that can encircle plants and possibly penetrate the pith, killing the plant before seeds have an opportunity to mature (Sherf and Macnab, 1986).

6.3.2.2 *Cauliflower*

The first signs of damage on cauliflower are marked by a decline in leaf firmness during the day, followed by a recovery at night. The infected plants develop a hazy white to light yellow coloration. The process of yellowing begins at the tips of mature leaves and slowly progresses downwards, eventually covering the whole leaf. These leaves have a tendency to fall prematurely. The lower foliage, particularly that touching the ground, exhibits a soft, brown to black decay on the central vein and leaf stalks, along with a powdery fungal development in cool and damp climatic conditions. Dark brown to black lesions result from tissue deterioration that progresses from the petioles to the shoot. The lesions enlarge and encircle the base of the stem. The decomposition of pith in the stem results in a hollow interior surrounded by sclerotia, fluffy mycelium, and a caustic material. Damaged cauliflower heads show several phases of degradation, from light brown to dark brown, beginning in the center. The fungus primarily targets the flower cluster in seed plants, which causes mycelium and sclerotia to grow (Singh, 1987).

6.3.2.3 *Rapeseed–mustard*

Different names have been assigned to the disease, including stem rot, rape canker, stem blight, stem canker, stalk break, white blight, and white rot, depending on the symptoms that are identified. Initially, symptoms such as damping-off, root rot, and complete plant collapse may become evident. The extent of the disease's impact is influenced by favorable weather conditions for the pathogen throughout the flowering period of the crop. Furthermore, the abundance of sclerotia of the pathogen in consistently mono-cultivated mustard fields is also a factor in the quick dissemination of the disease. SSR is recognized as a prevalent and consequential disease that greatly impacts the global production of oilseed rape.

There are two categories of infection in the field: myceliogenic infection and carpogenic infection. Initially, myceliogenic infection occurs in the field when sclerotia germinate. These sclerotia are buried in the soil during the previous crop season's harvest. The germinating seed or seedlings become infected a few days after sowing due to the germinated sclerotia. This infection is characterized by the development of white mycelium on the stem or cotyledon (Fig. 6.1a,b). The virus causes damping-off, wilting, and decay in the seedlings. The pathogen's white mycelium has the ability to infect the neighboring seedlings. The stem of the plant is usually the most infected in its natural habitat, but the *Sclerotinia* rot disease can infect any portion of the host plant that is above the ground. Longer water-soaked patches show up as lesions on the stem, which are later covered in a cottony-white mycelial growth of the fungus, especially near the soil surface (Fig. 6.1a,b). The white mycelial growth appears when these lesions completely encircle the stem. Plants that are impacted eventually wilt and die (Fig. 6.1c,d). Sometimes, the infection only affects a smaller portion of the pith, which results in delayed ripening and slows down the plant growth instead of causing a dramatic collapse (Fig. 6.1c,d). These plants can be easily identified in the field by their early maturation, browning of the plant in the end, and wilting of branches (Fig. 6.1 e, f, g, h).

Favorable environmental circumstances can lead to carpogenic germination. Under field conditions, in North India, the infection usually starts in the last week of December and lasts for the first 3 weeks of January. In the event of an outbreak, the average daily maximum and minimum air temperatures were recorded as 19.4 and 5.1°C, respectively, for the morning and afternoon hours. There was 1.4 mm of rain, 4.9 hour of bright sunshine, with 95% relative humidity in the morning and 62% in the afternoon. Sharma *et al.* (2015) established the ideal conditions for carpogenic infection and disease progression. It is impossible for *S. sclerotiorum* ascospores to directly infect healthy green mustard tissues. However, the blooming of flowers is mostly responsible for the infection and disease progression in rapeseed–mustard. Ascospores from the air land on mustard blossoms that are still attached to the upper portion of the plant. Infected flower petals fall and become trapped on stems, usually at the junction where leaves meet, or they land on the soil. Subsequently, fungal mycelia rapidly invade the blossoms during periods of elevated humidity within the upper portion of the plant (Fig. 6.1 e, f, g). Ascospores landing on leaves can lead to symptoms including grayish discoloration, irregular shapes, and frequently adhering petals. Ascosporic infection leads to the formation of large oval to round-shaped openings on leaves (Fig. 6.1i). The affected stem initially turns yellowish-brown, indicating the pathogen's growth along the stem (Fig. 6.1j). Sometimes, the affected stem may exhibit white mycelial growth on its surface (Fig. 6.1 k, l). The damaged stems take on a fragile appearance and, when they split, display a large number of spherical sclerotia, which can be grayish-white to black in color and can be found either outside or inside the compromised stem. Brown spots appear in the field when the cultivation moves closer to the stage of seed maturation (Fig. 6.1 l, m, n). This is caused by the plants bending and the seed pods touching the ground. These plants show signs of fungal development and siliquae decay even if they are not impacted by stem or aerial diseases (Fig. 6.1 o, p). Sclerotial bodies are also seen slightly above the soil surface. At times, stem decay is linked to gray mold (*Botrytis cinerea*), which covers the infected regions with mycelium that is brown to gray in color and fluffy. In humid conditions, twig blight symptoms with white cottony mycelial growth can be observed. Ascospores are found on decaying and expired plant matter on the soil, resulting in the sprouting and formation of mycelium and sclerotia. The infection takes place shortly after tainted flowers get stuck on branches within the plant's uppermost part or when branches come into contact with tainted fallen flowers

Fig. 6.1. Stem rot of *Brassica* crops. (a, b) The primary symptoms are expressed on the underside of the stem, surrounding the entire stem. ; (c, d) infection on the axis of the stem and white mycelial growth of the fungus on the infected site; (e–h) advanced stage of the infection on stem showing growth of pathogen and drying of the branch and stem; (i) ascospore infection on the leaves showing grayish-brown, shot hole-like lesions on leaves; (j) advancement of infection shown by yellowish color of the stem; (k) white mycelial growth spreads to more areas on the stem; (l, m) stem infection triggered dehydration of the shoots and formation of sclerotia over stem as well as inside it; (n) black rigid sclerotia inside the stem pith; (o) drying and lodging of diseased hosts visible under field conditions; (p) drying process of the crop is visibly apparent in the field.

Fig. 6.1. *Continued*

or decomposing plant tissues on the ground. The severely infected field appears dry from a distance, with the infected plants becoming dry and the field exhibiting dry/yellowish patches where seeds are usually not formed, or few seeds are formed which are light in weight and may not germinate (Fig. 6.1 o, p). If seeds do form, they are often shriveled or smaller in size (Saharan and Mehta, 2008).

6.3.3 Symptoms during storage period

Plants at different growth stages are susceptible to pathogen attacks. Occasionally, the pathogen invades the primary stem close to the soil level, swiftly infiltrating the cortical tissue without any immediate observable consequences. However, the plant eventually collapses suddenly. Superficial, white, cottony extrametrical mycelium is typically found in or on the diseased tissues, along with young sclerotia. The sclerotia begin as compact white clusters of hyphae, their dimensions changing based on the host plant and surrounding factors. Initially, they emerge either on the surface of the substrate or just beneath it; however, as time progresses, the external layers transform into a deep black shade. The fungus can infiltrate the central region of growing plants without producing a significant quantity of external mycelium in certain cases. Instead, it multiplies extensively and forms sclerotia inside the central cavity. This particular phase is commonly seen in rutabaga, cabbage, and turnip crops. The initial sign of the disease in cruciferous seed crops usually appears as the development of linear lesions with a grayish color on the main stalk, which originate from the nodes. A pith infection causes a plant to gradually slow down its growth and mature earlier. In contrast, fungal accumulation in the cortex at the base of the plant results in an abrupt collapse. The adult storage organs grow sclerotially and mycelially in comparable ways. The host tissue gets brittle and wet as the illness worsens. In the end, dehydration results from the rotting portion of the organ losing moisture, sometimes quickly. The impact on the plant is similar to that caused by bacterial soft rot, exhibiting a watery soft rot with visible mycelium in the initial stages and sclerotia in the advanced phases (Walker, 1969; Saharan and Mehta, 2008).

6.4 Disease Assessment

Different people have used different disease assessment scales to examine host resistance, control strategies, and the effect of the disease on crucifer crop production loss.

6.4.1 Rapeseed and mustard crops

A scale ranging from 0 to 4 is utilized to quantify the extent of stem rot in rapeseed–mustard (Sansford, 1995):

0 = no apparent signs of injury
1 = lesion is 0.1–2 cm long on stem;
2 = lesions are 2.1–4 cm long on stem;
3 = lesions are 4.1–6 cm long on stem; and
4 = lesions are more than 6 cm on stem and cause plant death.

6.4.2 Comparative analysis of resistance in rapeseed cultivars and lines using greenhouse and field evaluation

The evaluation of rapeseed–mustard lines under field and greenhouse conditions against SSR was proposed by Dalili (2013) as under:

6.4.2.1 Evaluation of rapeseed genotypes in a greenhouse

In order to carry out the experiment, a total of 11 rapeseed genotypes obtained from the Seed and Plant Improvement Institute in Iran were cultivated in plastic pots filled with sterilized sandy loam soil. The research was conducted in three replications using a randomized complete design. The inoculation procedure was performed using the technique described by Lewartowska *et al.* (1994). During the flowering phase, stems positioned 20 cm above the ground were exposed to wheat grains that had been colonized by *S. sclerotiorum* mycelium. The infected grains were enveloped in moist cotton and fastened to the stalk using parafilm. Individual row or variety consisted of nine plants that were inoculated and nine plants that served as non-inoculated controls, with three replications in total. Following artificial infection, disease incidence and severity were assessed within 1–2 weeks. A five-point scale was used, with 0 denoting no stem discoloration and 5 showing discoloration, sclerotia, early plant ripening, and decreased seed output. The 1–5 rating was associated with standard values of 0, 1.25, 2.5, 3.75, and 5, respectively. Equation 6.1, which was devised by Dueck *et al.* (1983), was utilized to calculate the disease index (DI). The data can be calculated utilizing MS Excel and the MSTAT-C program.

$$\mathrm{DI} = \frac{(1.255 \times Y2) + (2.5 \times Y3) + (3.75 \times Y4) + (5 \times Y5)}{\mathit{Total\ of\ plants}} \times \frac{1}{0.05} \quad (6.1)$$

where DI, disease index; Y_2, Y_3, Y_4, and Y_5, total plants with score 2–5, respectively.

Equation 6.1 is used for calculating the DI (Dueck *et al.*, 1983).

6.4.2.2 *Field evaluation of rapeseed–mustard genotypes*

In a study conducted in 2011–2012 at the Biekol Agriculture Research Station in Neka, Iran, 11 genotypes of *Brassica napus* were evaluated using a randomized complete block design, with three replications. The four rows that made up the experimental plot were 5 m long. To guarantee that the crops grew as best they could, a number of steps were taken: weed control, fertilizer application, crop rotation, and land preparation. To achieve the best outcomes, these procedures were carried out in compliance with the suggested guidelines for the particular area. Equation 6.2, developed by Krüger (1983), was utilized to calculate the disease occurrence and severity percentage, hence establishing the average infection level. The data analysis was carried out using MS Excel and MSTAT-C.

$$\%Q = \frac{(n \times 0) + (n \times 10) + (n \times 20) + \ldots\ldots.(n \times 100)}{n} \quad (6.2)$$

where Q, degree of infection; n, total number of plants assessed; 0–100, percentage of plant affected.

Equation 6.2 is used for calculating average infection level (Krüger, 1983).

6.4.3 Disease assessment scale and germplasm evaluation techniques in India

The disease assessment scales and germplasm evaluation techniques have been proposed by the ICAR–All India Coordinated Research Project (AICRP) on Rapeseed–Mustard group. The AICRP on Rapeseed–Mustard standardized the *Sclerotinia* rot scale for assessing the incidence and intensity, as mentioned in Table 6.2.

6.4.3.1 *Method of artificial inoculation for* Sclerotinia *rot*

For artificial inoculation of plants, following methods are used. Obtain 5 mm pieces of fungal mycelium, as well as the surrounding medium from a pure culture of *S. sclerotiorum*. This culture should have been cultivated on a thick layer of potato dextrose agar (PDA) medium for a duration of 7 days at a temperature of 20°C. Position one of these slices on the third internode of a flowering plant. Secure the stem and the fungal agar slice by wrapping them with a cotton swab soaked in sterile distilled water. After 3 weeks of inoculation, document the observations regarding the size of the lesions and the percentage of disease incidence.

Table 6.2. Genotype reaction to Sclerotinia rot can be rated on a scale of 0–4.

Disease reaction[a]	Disease rating	Lesion length (cm)
Resistant (no disease)	0	<3
Moderately resistant	1	3–5
Moderately susceptible	2	5–10
Susceptible	3	10–15
Highly susceptible	4	>15

[a]Stem diameter and percentage of disease incidence must also be noted down.

6.5 Economic Importance

6.5.1 Introduction

SSR infections can cause yield losses in susceptible crops that vary greatly and can even approach 100% (Purdy, 1979). Infections that cause rotting in storage settings as well as in the field prior to harvest cause losses in horticultural crops (Walker, 1969; Willetts and Wong, 1980). The combination of smaller seeds and diseased plants developing earlier is mostly responsible for the drop in seed yield (Kruger, 1973, 1975; Morrall *et al.*, 1976). Damage to seeds during crop harvesting is also a contributing factor. Sclerotia replace seeds in crops such as beans and sunflower, causing seed loss. Furthermore, instances of early seed pod rupture occurring before harvest have been observed, resulting in a decrease in seed quality. In crops such as sunflower and rapeseed, this results in the growth of smaller, withered, and husk-like seeds (Kruger, 1973; Morrall *et al.*, 1976; Kruger *et al.*, 1981). Moreover, beans, rapeseed, and sunflower seeds may become infected by fungal sclerotia (Hoes and Huang, 1976; Kruger *et al.*, 1981). Each year, *S. sclerotiorum* and other *Sclerotinia* species cause diseases in a large number of crops, causing millions of dollars' worth of financial damages.

Yield losses caused by Sclerotinia rot diseases in vulnerable crops can differ significantly, reaching up to 100% (Purdy, 1979). Horticultural crop losses occur due to infections that lead to rotting, both in the field before harvest and in storage conditions (Walker, 1969; Willetts and Wong, 1980). The decline in seed yield is mainly attributed to the combination of reduced size of seed and the early maturing of diseased plants (Kruger, 1973, 1975; Morrall *et al.*, 1976), along with the damage to seeds at the time of crop harvest. Crops such as sunflower and beans experience seed loss due to the replacement of seeds by sclerotia. In addition, there have been reports of premature seed pod rupture occurring prior to harvest, resulting in a decrease in seed quality. This leads to the development of smaller, withered, and husk-like seeds in crops such as rapeseed and sunflower (Kruger, 1973; Morrall *et al.*, 1976; Kruger *et al.*, 1981). Moreover, beans, rapeseed, and sunflower seeds may become infected by fungal sclerotia (Hoes and Huang, 1976; Kruger *et al.*, 1981).

Many crops experience diseases brought on by *S. sclerotiorum* and other *Sclerotinia* species each year, which causes enormous financial losses of millions of dollars. These losses can be linked to a drop in grade as a result of both a lower yield and impaired quality. Sadly, information about the financial effects of disease control strategies that target *S. sclerotiorum* regardless of their efficacy is lacking. This pathogen also causes farmers to forgo fields meant for high-value crops in favor of less lucrative substitutes or nonhost plants such as weeds or uncultivated land, which results in yet another type of loss (Purdy, 1979).

6.5.2 Rapeseed–mustard crops

The estimated damages due to Sclerotinia disease in oilseed rape fields in Alberta, Canada, have climbed to 28% (Morrall *et al.*, 1976). Sclerotinia rot disease resulted in seed yield reduction from 11.10 to 14.90% in Saskatchewan (Morrall *et al.*, 1976). Germany experienced the lowest productivity during periods of disease outbreaks (Kruger, 1975). The disease resulted in significant losses in the central and eastern regions of Finland, resulting in successful crop growth only in the southern and western regions of the country (Jamalainen, 1954). Additionally, premature ripening and shattering of siliquae caused by the disease could result in further decreases in yield (Morrall and Dueck, 1983). The fungus' sclerotia become mixed with the seeds during harvesting and threshing. This presence of the fungus' sclerotia in the seeds is considered an undesirable form of seed contamination, especially when exporting seeds from one country to another. Consequently, it significantly impacts the marketability of the crop. Additionally, partially infected plants also experience a negative impact on the quality of the seed (Kruger *et al.*, 1981). In Nepal, Chaudhary (1993) reported a substantial 75% decrease in seed yield, stunted plant growth, fewer siliquae per plant, and reduced 1000-grain weight, resulting in significant yield loss. Similar to this, SSR causes canola (*B. napus*) production in the southern part of New South Wales, Australia, to drop from 0.39 to 1.54 t/ha (Kirkegaard *et al.*, 2006). In plants afflicted with the disease, the percentage oil content and seed quality decrease (Aggarwal *et al.*, 1997). There has been a decrease in the productivity of the mustard harvest in India, with seed yield losses ranging from 50 to 75% (Kang and Chahal, 2000; Shivpuri *et al.*, 2000; Sharma *et al.*, 2001). In cases where the infection strikes the plants before or during flowering, there could be 100% loss of yield (Shukla, 2005). On the other hand, the yield loss is greater than 50% if the infection develops after the flowering stage. The SSR caused a 5–13% drop in seed yield in North Dakota and an 11.02–13.20% drop in seed yield in Minnesota, USA, between 1991 and 1997 (Lamey, 1999).

6.5.2.1 Yield loss in India

The estimation of the yield impact caused by SSR on Indian mustard was carried out at Bharatpur (Chattopadhyay *et al.*, 2003). An experiment was conducted to determine the relationship between disease incidence, plant height (PH), and seed yield (*Y*) for SSR disease. The analysis of the data uncovered a significant ($P < 0.01$) and opposite correlation between the occurrence of SSR incidence and the mustard yield. In order to predict *Y*, the DI was subjected to a linear equation (R^2: 0.89) utilizing the SSR-vulnerable cultivar Rohini (Fig. 6.2), resulting in the equation $Y = 310.25 - 2.04 \times$ disease incidence. The disease led to a reduction of 39.9% in the highest recorded incidence of SSR among mustard farmers in the field. Reduction in diseased plant over healthy plant for plant height (PH) was 12.3%, 36.1% for number of primary and 68.8% for number of secondary branches, 25.4% for main shoot length, 39.5% for number of siliquae per main shoot, 17% for siliquae length on main shoot, and 22.8% for seed per siliquae on main shoot. The disease caused a reduction of 6.4% in test weight of seed with a 2% reduction in their oil content. Reduction in seed yield could be due to the effect of the disease on different yield attributes. Such a quantitative relationship between the disease and different agronomic parameters is useful for quantification of yield losses.

The AICRP Rapeseed–Mustard, Bharatpur, India, developed a correlation matrix to estimate the yield losses in seed yield of rapeseed–mustard with regard to disease intensity/DI. Fig. 6.3 indicates that with an increase in

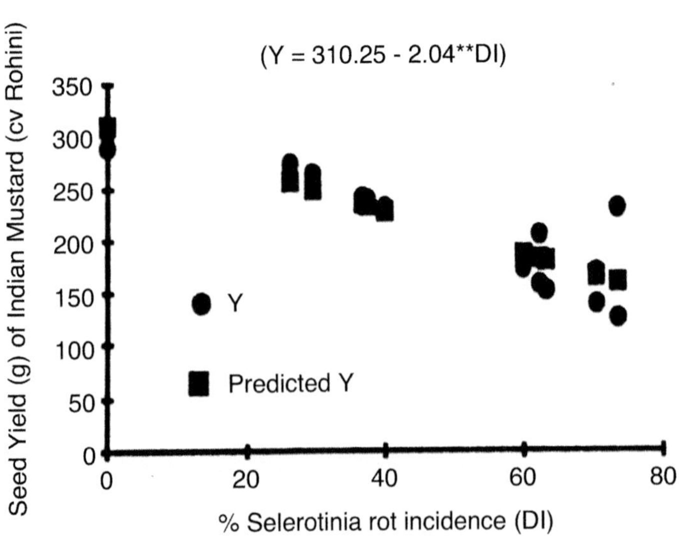

Fig. 6.2. The relationship between Indian mustard seed output (*Y*) and the prevalence of SSR (DI). (Adapted from publication of Chattopadhyay *et al.*, 2003, with permission.)

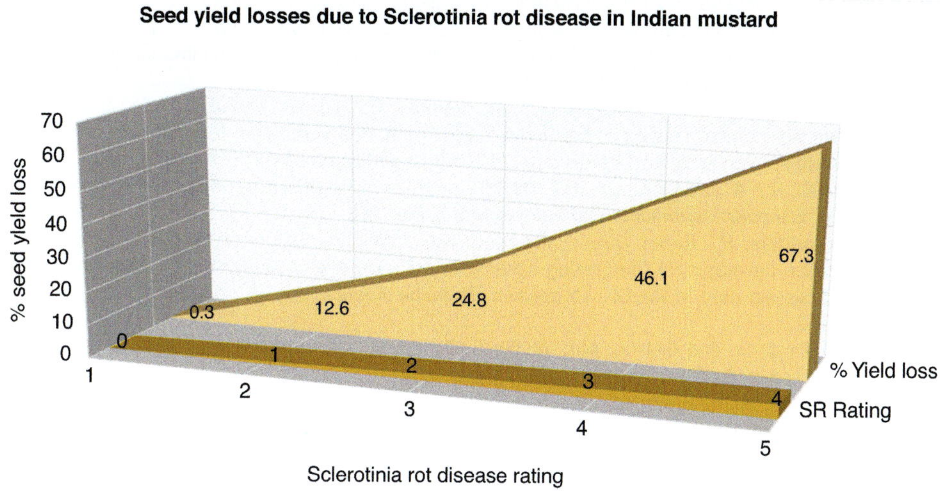

Fig. 6.3. This figure depicts a strong correlation between the severity of SSR of mustard and the estimation of seed yield loss caused by the disease. (AICRPR&M, Bharatpur, Rajasthan, with permission, 2024).

disease intensity among crucifers, there is a corresponding rise in yield losses, which may amount to as much as 67.3% at the highest level of disease intensity.

6.5.2.2 The loss of yield at Canada

Del Río *et al.* (2007) evaluated the effect of SSR on canola crop seed output in North Dakota. Research assessing the impact of SSR on oilseed rape seed output was conducted at various locations in North Dakota and Minnesota between 2000 and 2004. Either naturally occurring infections in commercial fields or ascospores generated in a lab were applied to the experimental plots. Due to varying fungicide treatment timing and concentrations, the incidence of SSR disease varied during the flowering stage. Before harvest, a disease incidence evaluation was carried out on 50–100 plants in each plot. The results of the study showed that for every unit percentage increase in the prevalence of SSR (which varied from 0.18 to 0.96%), there was a 0.5% loss in potential yield, or 12.75 kg/ha. A 17% incidence rate for SSR might result in losses equal to the costs of applying fungicides, using the canola market value and the present costs of fungicide applications. To strengthen canola plants' resistance to this specific disease, further measures are needed.

References

Aggarwal, R.A.K., Kumar, A. and Thakur, H.L. (1997) Effect of Sclerotinia rot on oil quality in low erucic acid cultivars of rapeseed. *Cruciferae Newsletter* 19, 103–104.

Cannon, P.F., Damm, U., Johnston, P.R. and Weir, B.S. (2012) *Colletotrichum*—Current status and future directions. *Studies in Mycology* 73(1), 181–213.

Chattopadhyay, C., Meena, P.D., Sastry, R.K. and Meena, R.L. (2003) Relationship among pathological and agronomic attributes for soil borne disease of three oilseed crops. *Indian Journal of Plant Protection* 31(2), 127–128.

Chaudhary, B.N. (1993) Yield loss estimation by *Sclerotinia sclerotiorum* (Lib) de Bary. *Journal of the Institute of Agriculture and Animal Science* 14, 113.

Dalili, A. (2013) Greenhouse and field evaluation of rapeseed cultivars and lines for resistance against *Sclerotinia sclerotiorum* (Lib) de Bary. *Journal of Oilseed Brassica* 4(2), 61–65.

del Río, L.E., Bradley, C.A., Henson, R.A., Endres, G.J., Hanson, B.K. *et al.* (2007) Impact of Sclerotinia stem rot on yield of canola. *Plant Disease* 91(2), 191–194.

Depotter, J.R.L., Seidl, M.F., van den Berg, G.C.M., Thomma, B.P.H.J. and Wood, T.A. (2017) A distinct and genetically diverse lineage of the hybrid fungal pathogen *Verticillium longisporum* population causes stem striping in British oilseed rape. *Environmental Microbiology* 19(10), 3997–4009.

Dueck, J., Morrall, R.A.A. and McKenzie, D.L. (1983) Control of *Sclerotinia sclerotiorum* in rapeseed with fungicides. *Canadian Journal of Plant Pathology* 5(4), 289–293.

Fitt, B.D.L., Gladders, P., Turner, J.A., Sutherland, K.G., Welham, S.J. and Davies, J.M.L. (1997) Prospects for developing a forecasting scheme to optimize use of fungicides for disease control on winter oilseed rape in the UK. *Optimising Pesticide Applications. Aspects of Applied Biology* 48, 135–142.

Garibaldi, A., Gilardi, G. and Gullino, M.L. (2006) Evidence for an expanded host range of *Fusarium oxysporum* f.sp.*raphani*. *Phytoparasitica* 34(2), 115–121.

Gunasinghe, N., You, M.P. and Barbetti, M.J. (2016) Phenotypic and phylogenetic studies associated with the crucifer white leaf spot pathogen, *Pseudocercosporella capsellae* in Western Australia. *Plant Pathology* 65(2), 205–217.

Gunasinghe, N., Barbetti, M.J., You, M.P., Burrell, D. and Neate, S. (2020) White leaf spot caused by *Neopseudocercosporella capsellae*: A Re-emerging disease of Brassicaceae. *Frontiers in Cellular and Infection Microbiology* 10, 588090. DOI: 10.3389/fcimb.2020.588090.

Hoes, J.A. and Huang, H.C. (1976) Importance of disease to sunflower in Manitoba in 1975. *Canadian Plant Disease Survey* 56(3), 75–76.

Jamalainen, E.A. (1954) Over-wintering of cultivated plants under snow. *FAO Plant Protection Bulletin* 2, 12.

Kang, I.S. and Chahal, S.S. (2000) Prevalence and incidence of white rot of rapeseed and mustard incited by *Sclerotinia sclerotiorum* in Punjab. *Plant Disease Research* 15, 232–233.

Kirkegaard, J.A., Robertson, M.J., Hamblin, P. and Sprague, S.J. (2006) Effect of blackleg and Sclerotinia stem rot on canola yield in the high rainfall zone of southern New South Wales, Australia. *Australian Journal of Agricultural Research* 57(2), 201–212.

Kruger, W. (1973) Control measures for *Sclerotinia sclerotiorum* in rape. *Phytopathology Z* 77, 125–137.

Kruger, W. (1975) Influence of weather on the attack of rape by *Sclerotinia sclerotiorum* (Lib) de Bary. *Nachrichtenble Devet Pflanzenschutzd (Braunschweig)* 27, 1–6.

Krüger, W. (1983) Oilseed rape. *Pests and Diseases*. Semundo, Hamburg, Germany, 119 pp.

Kruger, W., Marguard, W.R. and Schosser, E. (1981) Plant disease product quality. II. Influence of stem canker *Sclerotinia sclerotiorum* (Lib) de Bary on the quality of rapeseed. *Review of Plant Pathology* 60, 2949.

Lamey, H.A. (1999) Incidence of *Sclerotinia* stem rot on canola in North Dakota and Minnesota, 1991–1997. In: Nelson, B.D. and Gulya, T.J. (eds) *Proceedings of the 1998 International Sclerotinia Workshop*, Department of Plant Pathology, North Dakota State University, September 9–12, 1988, pp. 7–9.

Lange, R.M., Gossmann, M. and Büttner, C. (2007) Yield loss in susceptible cultivars of spring rapeseed due to Fusarium wilt caused by *Fusarium oxysporum*. *Communications in Agricultural and Applied Biological Sciences* 72(4), 723–734.

Lewartowska, E., Jedryczka, U. and Frencel, L. (1994) The methods of winter oilseed rape (*Brassica napus* L.) resistance evaluation against *Sclerotinia sclerotiorum* (Lib) de Bary. IV. Symposium on plant immunity to diseases and pests. Dobrich Bulgaria. *Plant Science* (Sofia) 252–254.

Li, G.Q., Wang, D.B., Jiang, D.H., Huang, H.Z. and Zhou, Q. (1998) A new kind of lettuce drop caused by *Sclerotinia* sp. in Central China. *Acta Phytopathologica Sinica* 28, 249–255.

Matić, S., Gilardi, G., Gullino, M.L. and Garibaldi, A. (2018) Evidence for an expanded host range of *Fusarium oxysporum* f. sp. *chrysanthemi*. *Journal of Plant Pathology* 100(1), 97–104.

Morrall, R.A.A. and Dueck, J. (1983) Sclerotinia stem rot of spring rapeseed in Western Canada. In: *Proceedings of the 6th International Rapeseed Conference*, GCIRC, Paris, pp. 957–962.

Morrall, R.A.A., Duczek, I.J., Mc Kenzie, D.L. and McGee, D.C. (1976) Some aspects of *Sclerotinia sclerotiorum* in Saskatchewan: 1970–1975. *Canadian Plant Disease Survey* 56:56.

Novakazi, F., Inderbitzin, P., Sandoya, G., Hayes, R.J., von Tiedemann, A. *et al.* (2015) The three lineages of the diploid hybrid *Verticillium longisporum* differ in virulence and pathogenicity. *Phytopathology* 105(5), 662–673.

Pound, G.S. (1946) In: *Diseases of Cabbage Plants Grown for Seeds in Western Washington*. International Agricultural Sciences, Agriculture Experiment Station Pullman, Washington, DC, Bulletin, p. 27.

Purdy, L.H. (1979) *Sclerotinia sclerotiorum*: History, diseases and symptomatology, host range, geographic distribution and impact. *Phytopathology* 69(8): 875.

Saharan, G.S. and Mehta, N. (2008) *Sclerotinia Diseases of Crop Plants: Biology, Ecology and Disease Management*. Springer Science, Dordrecht, The Netherlands, p. 485.

Saharan, G.S., Mehta, N. and Meena, P.D. (2016) *Alternaria Diseases of Crucifers: Biology, Ecology and Disease Management*. Springers Nature, Singapore, p. 297.

Saharan, G.S., Mehta, N. and Meena, P.D. (2017) *Downy Mildew Disease of Crucifers: Biology, Ecology and Disease Management*. Springers Nature, Singapore, p. 357.

Saharan, G.S., Mehta, N. and Meena, P.D. (2019) *Powdery Mildew Disease of Crucifers: Biology, Ecology and Disease Management*. Springers Nature, Singapore, p. 362.

Saharan, G.S., Mehta, N. and Meena, P.D. (2021) *Genomics of Crucifer's Host-Resistance*. Springer Nature, Singapore, 790 pp.

Saharan, G.S., Verma, P.R., Meena, P.D. and Kumar, A. (2014) *White Rust of Crucifers: Biology, Ecology and Management*. Springers Nature, New Delhi, p. 244.

Sansford, C. (1995) Oilseed rape: Development of stem rot (*Sclerotinia sclerotiorum*) and its effect on yield. In: *Proceedings of IX International Rapeseed Congress*, Rapeseed, Today and Tomorrow, pp. 634–636 (Vol. 2).

Sharma, P., Meena, P.D., Verma, P.R., Saharan, G.S., Mehta, N. *et al.* (2015) *Sclerotinia sclerotiorum* (Lib) de Bary causing Sclerotinia rot in oilseed brassicas: A review. *Journal of Oilseed Brassica* 6(Special), 1–44.

Sharma, S., Yadav, J.L. and Sharma, G.R. (2001) Effect of various agronomic practices on the incidence of white rot of Indian mustard caused by *Sclerotinia sclerotiorum*. *Journal of Mycology and Plant Pathology* 31, 83–84.

Sherf, A.F. and Macnab, A.A. (1986) *Vegetable Diseases and Their Control*, Vol. II. Wiley, New York, p. 728.

Shivpuri, A., Sharma, K.B. and Chhipa, H.P. (2000) Some studies on the stem rot (*Sclerotinia sclerotiorum*) disease of rapeseed-mustard in Rajasthan. *Journal of Mycology and Plant Pathology* 30:268.

Shukla, A.K. (2005) Estimation of yield losses to Indian mustard (*Brassica juncea*) due to Sclerotinia stem rot. *Journal of Phytopathological Research* 18, 267–268.

Singh, R.S. (1987) *Diseases of Vegetable Crops*. Oxford/IBH Publishing, New Delhi, 362 pp.

Tyvaert, L., França, S.C., Debode, J. and Höfte, M. (2014) The endophyte *Verticillium* Vt305 protects cauliflower against *Verticillium wilt*. *Journal of Applied Microbiology* 116(6), 1563–1571.

Walker, J.C. (1969) *Plant Pathology*, 3rd edn. McGraw-Hill, New York, p. 819.

Willetts, H.J. and Wong, J.A.L. (1980) The biology of *Sclerotinia sclerotiorum, S. trifoliorum* and *S. minor* with emphasis on specific nomenclature. *The Botanical Review* 46(2), 101–165.

7 Infection and Pathogenesis

Abstract

Sclerotinia penetration and subsequent infection are dependent on a number of variables, including the type of inoculum, the nutritional state of fungus, the susceptibility of host, and the existence of favorable conditions. Mycelium from sclerotia and ascospores, as well as germinated ascospores, is the two forms of inocula that can cause the infection. Hypha can penetrate the host cuticles directly through the straightforward, solitary appressorium that is produced by germinated ascospores. Following penetration, an enlarged, granular-looking vesicle forms between the epidermis and cuticle, giving rise to infection hyphae from infection cushions that invade the host tissues. Multiple enzymes break down cell walls and tissues before hyphae invade, allowing for host tissue colonization. *Brassica* has variable expression of its several metabolic enzymes during infection and disease. Numerous genes are expressed to control the development of ascospores, apothecia, sclerotial formation, pathogen pathogenicity, mycelial expansion, and pathogenic processes. The *SsPG1* and *SsCUTA* genes control the synthesis of enzymes. *Sclerotinia* induces various proteinaceous effector genes (*Ss-caf1*, *SsSSVP1*, *SsCP1*, and *SsNEs*) that result in cell death.

7.1 Introduction

Sclerotinia stem rot (SSR) disease of crucifers, caused by *Sclerotinia sclerotiorum*, can infect plants successfully and remain for extended periods of time, making it a highly destructive and common disease. The pathogenic aspect of this disease is what makes it unique. Sclerotia of fungi, which can germinate by carpogenic or myceliogenic means, are the main source of infection. When myceliogenic methods germinate sclerotia, hyphae are produced, which can pierce the cuticle and enter the host's epidermis. On the other hand, ascospores are released by apothecia, which are formed when sclerotia go through carpogenic germination. Following germination, these ascospores create an appressorium and pierce the host's cuticle through an infection peg. The pathogen releases enzymes that break down the oxalic acid (OA) and cell wall during the infection phase. These compounds are essential for the pathogen's colonization because they aid in the breakdown of the host's tissues. The pathogen's ability to proliferate and develop further depends on its mycelium's intracellular and intercellular invasion of host cells.

7.2 Penetration of Host Surface Cuticle

A number of factors influence the ability of *Sclerotinia* spp. to invade and penetrate host tissues, including the type of inoculum, the nutritional state of fungus, the features of the host, and the surrounding atmosphere. The commencement of infection can be triggered by two types of inoculums: mycelium from sclerotia and ascospores, as well as germinated ascospores. In several *Sclerotinia*-caused illnesses, ascospores are thought to be the primary source of inoculum. Developed ascospores can form a simple and unique appressorium if nutrients are available, which enables them to enter the host (Purdy, 1958; Abawi *et al.*, 1975a). In order for ascospores to infiltrate their host organism, they require external nutrition. According to de Bary (1886, 1887), *Sclerotinia* hyphae don't initiate an attack until they've had enough food and time to mature. When the hyphae are held in a nutritional solution drop on the host, penetration occurs. Ascospores will germinate in the absence of nutrients, but the hyphae won't be able to pierce the host surface and form appressoria there. In order for infection to

DOI: 10.1079/9781800629592.0007

occur, nutrients must be available to ascospores from all isolates. Furthermore, the pathogen usually enters through the cuticle. When the ascospores germinate, they release a material that gets into the stomatal opening and damages the cellular components. According to earlier studies, *Sclerotinia trifoliorum* can enter a certain genotype of clover (Prior and Owen, 1964) and *S. sclerotiorum* potato crop leaves (Jones, 1976) through exposed stomata to form germ tubes. However, rather than passing through stomata, the host is usually pierced by the cuticle directly (Boyle, 1921; Purdy, 1958; Prior and Owen, 1964; Lumsden and Dow, 1973; Abawi *et al.*, 1975a; b). One of the main ways that *S. sclerotiorum* infects cabbage is by freezing and bruising injuries (Hudyncia *et al.*, 2000).

Unless there is direct stomatal penetration, the development of appressoria is frequently observed in some host plants (Prior and Owen, 1964; Jones, 1976). Appressoria are complex structures composed of many cells. They are referred to by different names, such as appressorial clusters (Purdy, 1958), large appressoria (Boyle, 1921), cushion-shaped appressoria (Abawi *et al.*, 1975a), and infection cushions (Prior and Owen, 1964; Lumsden and Dow, 1973). When in touch, a stimulus is necessary for the creation of these ordered patterns (de Bary, 1886; Purdy, 1958; Abawi *et al.*, 1975a). The hyphal strands split into finger-shaped structures once they come into contact with the host, and they eventually expand into infection cushions that resemble domes (Boyle, 1921; Purdy, 1958; Lumsden and Dow, 1973; Abawi *et al.*, 1975a). Examining cross sections of infection cushions reveals three distinct kinds of hyphae. The top of the cushion is home to the first type of hyphae, which are thin in diameter and heavily stained with safranin. In the middle are inflated, granular, somewhat safranin-staining hyphae that identify the second type. Dichotomously branching penetration hyphae are the last form of hyphae; they resemble those around the cushion's center in texture (Lumsden and Dow, 1973). Strong attachment to the host's surface is shown by the infection cushions, which also appear to be linked by a mucilaginous material (Boyle, 1921). This compound has been shown to stain (Prior and Owen, 1964; Lumsden and Dow, 1973), giving it a darker appearance than the surrounding material. A significant amount of pressure can be applied to the outer layer of tissue due to the composition of this substance and the curved shape of the cushion. This pressure helps the tiny injection pegs that develop at the appressoria's terminals mechanically penetrate the host tissue (Boyle, 1921; Purdy, 1958; Lumsden and Dow, 1973; Abawi *et al.*, 1975a). Since there is no cuticle weakening, breakage, or alteration before penetration, the evidence implies that support is most likely provided mechanically (Boyle, 1921). According to Lumsden and Dow (1973), infection cushions frequently cause the surrounding vulnerable tissue to form a slightly convex mound. At the penetration sites, the cuticle is forced inward (Boyle, 1921; Lumsden and Dow, 1973). Even in the later stages of pathogenesis, the cuticle can continue to be impermeable to the fungus (Lumsden and Dow, 1973), as evidenced by the staining reaction being unchanged. Prior and Owen (1964) observed that the mechanical tension generated by the infected cushions allows for the penetration of lower leaves. On the other hand, there is also the disintegration of the epidermal cells and cuticle, which is believed to be caused by the enzyme activity of the infection cushion.

7.3 Early Phase of Host Surface Infection Process

A bigger granular structure called a "vesicle" forms between the cuticle and epidermis after the infection penetrates the host cuticle (Boyle, 1921; Purdy, 1958; Lumsden and Dow, 1973). The ensuing "infection hyphae" swiftly multiply from the infection cushions and exclusively use intercellular pathways to attack the host tissue (Lumsden and Dow, 1973). A variety of mechanisms has been used to characterize hyphae's invasion into host tissue, including intracellular or intercellular pathways (Purdy, 1958; Abawi *et al.*, 1975a) and growth without cell-to-cell distinction (de Bary, 1887). At the expanding infection border, however, things are different. According to Lumsden and Dow (1973), the infection process is painstakingly planned, following a set of procedures that lead to the complete infiltration and destruction of the host tissue. Hyphae emerge from the vesicles in bean tissue afflicted by *S. sclerotiorum* or *Sclerotinia minor*, causing an expanding granular "infection." In the cortex and between the cuticle and the epidermal cell layer, they proliferate intercellularly (Lumsden and Dow, 1973). The mycelium exhibits distinct variations in terms of size, shape, and response to safranin staining when compared to those observed in conventional cultures or on the host's surface. Their inflated, granular expression and light safranin-staining response are characteristics in common with the hyphae located beneath the cuticle and within infection cushions. Under the cuticle, the subcuticular hyphae extend and form a well-organized, fan-like infection front by arranging themselves in a parallel fashion (Lumsden and Dow, 1973). Hyphae in the intercellular spaces move into cortex region for further growth. The radial hyphal front splits into clusters of 18–20 hyphae following the initial 12–24 h infection period. These clusters align parallel to the bean hypocotyl axis and exhibit faster upward development than transverse or downward growth. In comparison to those in the cortex, the subcuticular hyphae migrate more quickly and develop more prominently on the side of the hypocotyl where the infection is located. There might be less impediment to hyphae's development in the subcuticular portion, enabling a quick and forceful ascent up the hypocotyl. This is probably essential to the illness developing successfully. The procedure is then finished when the hypocotyl is surrounded by the cortical infection hyphae. All hyphae up to this time are big and swollen, with an average size of 19.10 μM and a range of 8.5–34.0 μM. The intercellular tissue can only be penetrated by the infected hyphae. They are essential in breaking down the host's defenses and starting the colonization process inside the host tissue. As the infection progresses, the hyphae of the infection can be observed at the edges of apparent lesions on hosts. Abawi *et al.* (1975a) describe them as slightly leading the margin, while Boyle (1921) and Lumsden and Dow (1973) define them as slightly following. The hyphae most likely induce modifications in the infected host tissue. These include histologically documented alterations in pectic components in the cell walls of two or three cells prior to the hyphae (Lumsden and Dow, 1973). Moreover, early cell death is observed (Boyle, 1921; Thatcher, 1942; Hancock, 1972); fluid accumulation and water-soaking are observed at the edges that progress; permeability changes in cells occur at the forefront (Thatcher, 1942; Hancock, 1972; Newton, 1972); and substances associated with disease-causing qualities are released (Dow and Lumsden, 1975; Lumsden, 1976, Lumsden, 1979). Rapeseed is first invaded by the primary ascospore inoculum of *S. sclerotiorum* through the petals. Ascospores cause adhesion, germination, penetration, and eventual collapse of epidermal cells, which is the pathogenesis of immature petals. The mycelium then spreads to the tissues of leaves, continuing the infection process (Jamaux *et al.*, 1995).

7.4 Progression of Tissue Colonization and Host Infection

Hyphal branches of tiny diameter emerge roughly 55 μM behind the advancing hyphal tips within 12–24 h after the *Sclerotinia* infection hyphae invading the host tissue (Prior and Owen, 1964; Lumsden and Dow, 1973). These "ramifying hyphae," a form of branching hyphae, are easily able to pass through cell walls and show a broad infiltration of failing or disintegrating host tissue, both inside and between cells. It is significant to observe that the diameter of these hyphae is significantly less than that of the infectious hyphae from which they originate (Prior and Owen, 1964; Lumsden and Dow, 1973). With an average diameter of 8.5 μM, these hyphae are comparable in size and staining intensity with safranin to those found in culture or on the host surface (Lumsden and Dow, 1973). The significant changes in diameter, cell length, and shape are caused by the vast range of hyphal sizes, which include minuscule diameter branching hyphae and exaggerated infection hyphae (Prior and Owen, 1964; Abawi *et al.*, 1975a). Hyphae's extensive branching allows them to target the cells and intercellular spaces in the cortex effectively. Moreover, by disturbing the crystalline structure of the host cell walls, these hyphae might cause damage to them (Calonge *et al.*, 1969; Lumsden, 1979). The bean and clover's branching hyphae target the vascular tissue; however, there may be some challenges involved (Prior and Owen, 1964; Lumsden and Dow,

1973). *S. sclerotiorum* causes vascular blockage and wilting in sunflower plants by penetrating the vessels and interfascicular areas (Pawlowski and Hawn, 1964). Minerals found in xylem channels may also be a factor in the development of clogging and wilting.

After a protracted colonization period, branched hyphae primarily emerge from the host tissue through stomata or openings in the cuticle (Lumsden and Dow, 1973). Hyphae have not been shown to protrude from stomata on potato leaves (Jones, 1976); however, noticeable strands of projecting hyphal strands are visible on the undersides of bean leaves (Abawi *et al.*, 1975a). Over time, these threads will develop into mycelial wefts and become a Darvill growth on the surface of adult lesions. Sclerotic beginnings are clusters of small, barrel-shaped cells that give rise to complete sclerotia in 3–7 days (Lumsden and Dow, 1973; Abawi *et al.*, 1975a). The appearance of sclerotia signifies the end of the disease cycle since they might develop under rotting plant parts on the surface of the soil or on the host's surface inside the pit (Lumsden and Dow, 1973; Lumsden, 1979).

After a protracted colonization phase, branched hyphae primarily emerge from the host tissue through stomata or openings in the cuticle (Lumsden and Dow, 1973). On potato leaves, hyphae have not been observed to protrude from stomata (Jones, 1976), on the undersides of bean leaves, conspicuous strands of protruding hyphal strands are discernible (Abawi *et al.*, 1975a). These threads will eventually become mycelial wefts and evolve into a cottony growth on the surface of adult lesions over time. Within 3–7 days, sclerotia initial clusters of briefs, barrel-shaped cells will give rise to mature sclerotia (Lumsden and Dow, 1973; Abawi *et al.*, 1975a). Since, the sclerotia can develop under decomposing plant parts on the soil's surface as well on the host surface within the pit, which lead to completion of their life cycle.

Since sclerotia can occur under decomposing plant parts on the soil's surface or on the host's surface within the pit, their formation marks the end of the disease cycle. It should be underlined that sufficient nutrition from the infected tissue is necessary for the formation and maturity of sclerotia (Lumsden and Dow, 1973; Lumsden, 1979).

The development of a disease in a host plant involves a complex and ever-changing process known as pathogenesis. This process is predisposed by various aspects, comprising the inherent capabilities of the pathogen and mechanisms that govern its ability to penetrate and infect the host plant. However, before a disease can fully develop, the pathogen must overcome the host plant's defense mechanisms. These defense mechanisms act as barriers that the pathogen must breach, inactivate, or nullify. In addition to their innate capacities, the host plant and the pathogen interact with each other as a function of time and the surrounding environment. The pathogen employs a range of attacking mechanisms to facilitate its invasion. These mechanisms include the discharge of enzymes that degrade the cell wall and middle lamella, the production of toxins, the release of enzymes that break down host tissue and defense elements, and the speed at which the infection occurs. The pathogen *S. sclerotiorum* discharges manifold pectinolytic enzymes that aid in its penetration, establishment, and breakdown of the tissues of host plant. These facts regarding the biochemistry and physiology of the host–pathogen interaction can be found in Chapter 9 of this volume.

7.5 *Sclerotinia* Infection Process in the Cotyledon Tissue of Susceptible and Resistant *Brassica* Genotypes

The elements of *S. sclerotiorum* infection processes that lead to resistance in *Brassica napus* genotype (Charlton) have been recognized by Garg *et al.* (2010). The results indicate that there were no discernible variations in ascospore germination on glass slides, resistant varieties "Charlton," or susceptible varieties "RQ001-02M2" after using either water or Pi-glucose medium for spore suspensions at 4 h postinoculation (hpi). Ascospore germination occurs as a result of the presence of enough moisture, which has been previously reported by Purdy (1958), Sutton and Deverall (1983), and Jamaux *et al.* (1995). Moreover, the data suggests that the susceptible strain "RQ001-02M2" and the resistant strain "Charlton" are equally prone to infection at this early stage. Nevertheless, the growth of ascospore germ tubes was only observed when the spore suspension was prepared using Pi-glucose medium. The findings provide further evidence for earlier research demonstrating the essential role of nutrients in the growth of hyphae, their penetration, and the successful invasion of a susceptible host by the pathogen (Abawi and Grogan, 1975; Jamaux *et al.*, 1995). Additionally, the frequency of stomatal and cuticular penetration was similar on both the resistant "Charlton" and the susceptible "RQ001-02M2" genotypes.

As early as 1 dpi, an interaction was observed between *S. sclerotiorum* and cotyledons of the vulnerable "RQ001-02M2" and the resistant "Charlton." The resistant 'Charlton's' cotyledons exhibited a significant decrease in hyphae growth at 1, 2, and 3 dpi. These results corroborate an earlier study by Dow and Lumsden (1975), who found that disease-resistant *Phaseolus coccineus* tissues against *S. sclerotiorum* included tiny, deformed infection hyphae. Furthermore, it appears that indications from the vulnerable "RQ001-02M2" encourage hyphal movement on the host surface, as seen by the higher hyphal activity on susceptible tissue in comparison to the resistant "Charlton" and glass slides. Blakeman and Sztejnberg (1973) proposed in their study on beetroot (*Beta vulgaris*) against *Botrytis cinerea*, and Kowalska and Niks (1999) found evidence in a resistant flax (*Linum usitatissimum*) genotype against *Melampsora lini*, that the "Charlton" genotype inhibits superficial mycelial growth, suggesting the production of specific antifungal/fungistatic compounds. According to the study, the susceptible genotype displayed infection cushions and/or complex appressoria on the surface of glass slides, whereas the resistant genotype did not. The findings of Dow and Lumsden (1975) differ from the current study, as they observed secondary infection cushions in a resistant genotype of *P. coccineus*, located adjacent to the epidermal layer and beneath the cuticle.

Furthermore, it has been proposed that the inoculum's nutritional state and the fungus's physical resistance to the surface it encounters determine how complex the appressoria become (Abawi *et al.*, 1975a; b; Tariq and Jeffries, 1984). The absence of infection cushions on the resistant "Charlton" in the research conducted by Garg *et al.* (2010) indicates that, despite sufficient nutrients for fungal growth, the prevention of infection cushions is due to the incompatible relationship between the genotype and the pathogen. Previous findings on *S. sclerotiorum* and related pathogens, including *Sclerotium rolfsii* and *S. trifoliorum*, have also suggested that the infection cushion contributes to the breakdown of the plant surface by enzymatic dissolution (Prior and Owen, 1964; Smith *et al.*, 1986; Tariq and Jeffries, 1986) or mechanical pressure (Boyle, 1921; Purdy, 1958; Lumsden and Dow, 1973; Lumsden and Wergin, 1980).

According to Garg *et al.* (2010), the most important defense mechanism against this pathogen's penetration and colonization is the resistant "Charlton" genotype's vital destruction of infection pads. Protoplasts released from hyphal cells were only seen on the resistant "Charlton's" cotyledonary surface. Previous findings have confirmed that the breakdown of cell walls of fungal hyphae by artificial extracellular enzyme activities, for example, chitinase and beta-glucanase can induce protoplast extrusion from filamentous fungi, including ascomycetes (Emerson and Emerson, 1958; De Vries and Wessels, 1972; Dhar and Kaur, 2009). Protoplast ejection from hyphal cells in the *Brassica–S. sclerotiorum* pathosystem was documented by Garg *et al.* (2010).

Tariq and Jeffries (1986) documented a temporary phase of wall-less protoplasm in *S. sclerotiorum* in *Phaseolus* spp. During this phase, the infected peg penetrated the cuticle of the host epidermis. Subsequently, the invading protoplast expanded quickly to create a subcuticular vesicle. Eventually, the fungal cell wall was formed and the organism returned to its usual appearance. Their study differs from that of Garg *et al.* (2010) primarily in that extrusion take place on the surface of "Charlton," whereas the protoplast phase in Tariq and Jeffries (1986) study was restricted to the host tissue. These results suggest that the materials that were extruding from hyphal cells as protoplasts and leaving the fungal cells were not appressoria.

Initially, within *S. sclerotiorum*, the bifurcating division of the last hyphae consistently leads to the formation of appressoria on the surface of vulnerable host genotype (Lumsden, 1979). In contrast, the development of appressoria from the intercalary hyphal cells was seen. Furthermore, there is no previous documentation indicating a notable decline in the cytoplasmic content of the hyphal cells (Fig. 7.1b), the presence of hyphal cells devoid of cytoplasmic content subsequent to protoplast ejection (Fig. 7.1a), or the disintegration of the hyphal cells (Fig. 7.2n). Under a light microscope or scanning electron microscopy (SEM), it was observed that different forms of protoplast extrusion occurred. The forms encompassed the protoplast emerging from either the end or center of the fungal hyphae, taking the shape of a bud (Figs 7.1a and 7.2m), the protoplast extruding after the intercalary hyphal cells swelled (Fig. 7.2l), and the protoplast extruding after the hyphal cells evidently lacking cytoplasmic content (Fig. 7.1a). These findings are strikingly similar to earlier studies (Bachmann and Bonner, 1959; Talburt and Johnson, 1965; Bartnicki-Garcia and Lippman, 1966; Fenice *et al.*, 1999) that experimentally induced protoplast release of different filamentous fungus. This provides additional proof that the substances being released from the fungal cells are indeed protrusions of protoplasts. Moreover, it is possible that the strong "Charlton" is generating one or multiple enzymes that degrade cell walls, similar to the ones discovered in *B. napus* cultivar Zhong You 821. This is supported by the fact that hyphal cells extrude protoplasts even without being treated with any external enzyme arrangements (Zhao *et al.*, 2009). These enzymes may break down the hyphal cell wall on a pore-sized or bigger scale. Only the resistant genotype "Charlton" exhibits protoplast extrusion, which suggests the existence of an as-yet-unidentified defense mechanism that impedes pathogen growth.

The vulnerable variety "RQ001-02M2" showed signs of fungal invasion that reached the spongy mesophyll cells. On the other hand, the resistant variety "Charlton" showed that the pathogen was mainly confined to the upper epidermis. The occurrence of localized necrosis in the palisade mesophyll cells at the infection site at 4 and 6 dpi indicates that the hypersensitive reaction (HR) is the key factor behind "Charlton's" ability to resist fungal invasion. This result corresponds with the earlier studies carried out by Garg *et al.* (2008), who often noted hypersensitive resistance in "Charlton." Additionally, at 4 dpi, susceptible "RQ001-02M2" showed signs of cell death in the palisade mesophyll cells. The necrotrophic infection in "RQ001-02M2" may have grown more easily.

As a result of this cell death, it gets its nutrients from the necrotic host tissues. This finding is consistent with the results (Govrin and Levine, 2000; Dickman *et al.*, 2001; Kim *et al.*, 2008) in *Arabidopsis thaliana*. This is also shown in the investigation conducted by Garg *et al.* (2010). Following the death of these cells at 4 dpi, it was observed that the palisade mesophyll cells of the vulnerable "RQ001-02M2"6 were extensively colonized at 6 dpi. However, the resistant "Charlton" plant exhibited the production of phytoalexins, which are antifungal compounds that potentially hindered the infiltration of mycelia into the plant tissues beyond the HR lesion. Histochemical analysis of the HR lesions on "Charlton" revealed the presence of these substances; toluidine blue staining revealed darkly stained areas surrounding the dead cells of the palisade mesophyll layer. After these cells died at 4 dpi, they saw that the palisade mesophyll cells of the susceptible "RQ001-02M2" were heavily colonized at 6 dpi. Nonetheless, the resistant "Charlton" plant produced phytoalexins, an antifungal compound that might have stopped mycelia from invading the plant tissues outside of the HR lesion. Histochemical analysis of the HR lesions on "Charlton" revealed the presence of these substances; toluidine blue staining revealed darkly stained areas surrounding the dead cells of the palisade mesophyll layer.

In the resistant "Charlton" genotype, calcium oxalate crystals were only found in the upper epidermis at 8 dpi; in the susceptible genotype, they were present at 6 dpi and had spread throughout the cotyledon tissue by 8 dpi. The OA produced by *S. sclerotiorum* has been suggested to do more than just work in concert with pectolytic enzymes (Marciano *et al.*, 1983; Godoy *et al.*, 1990). It is also thought to extract calcium from plant cell walls and convert it to calcium oxalate prior to the middle lamella's enzymatic degradation (Lumsden, 1979; Punja *et al.*, 1985; Thompson *et al.*, 1995). In a previous work, Qiu *et al.* (2001) verified the presence of calcium oxalate crystals in sunflowers infected with *S. sclerotiorum* by SEM analysis. Furthermore, earlier studies by Lumsden and Dow (1973) and Smith *et al.* (1986) have shown that crystals form in *S. sclerotiorum*-infected bean and sunflower plants prior to hyphae

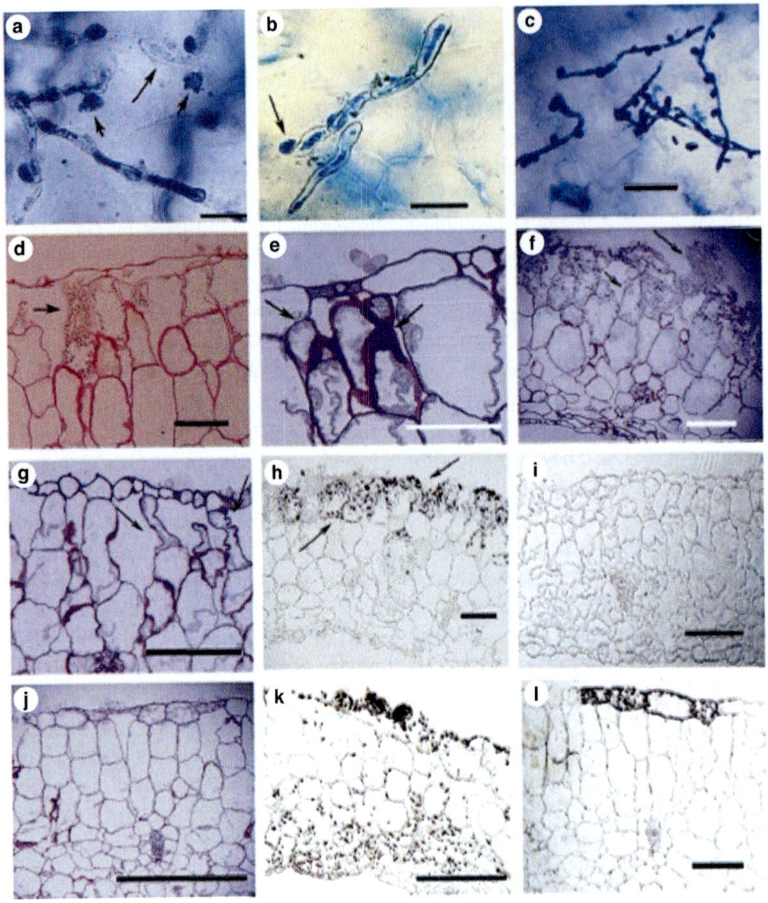

Fig. 7.1. Light micrographs demonstrating the histopathology of susceptible *Brassica napus* "RQ001-02M2" and resistant *B. napus* "Charlton" cotyledons after they were injected with *Sclerotinia sclerotiorum* isolate WW-3. Using a Zeiss Axioplan 2 microscope photographic system, the samples in (a–c) were cleaned in a solution of acetic acid, ethanol, and water (2:2:1), stained with 1% cotton blue, and then captured on camera. The same photographic apparatus used for (a–c) was used to take 2-mm-thick sections in (d–i), which were then stained as follows: (d) with periodic acid/Schiff's reagent; (e–g) and (j) with 0.5% toluidine blue; and (h), (i), (k), and (l) using the technique for calcium oxalate crystal detection as outlined by Yasue (1969). (a) After the protoplast is released on the resistant "Charlton" at 6 dpi, hyphal cells are free of cytoplasm (bigger arrow) and protoplast extrusion in an amorphous state (smaller arrows) occurs. (b) Bud-like protoplast extrusion (arrow) on resistant "Charlton" at the apex of the fungal hyphae. (c) Synchronous extrusion of protoplasts on resilient "Charlton." (d) Starch deposits (arrow) at 4 dpi in susceptible "RQ001-02M2" palisade mesophyll cells. (e) The resistant 'Charlton' with darkly stained regions (arrows) surrounding the dead cells of the palisade mesophyll layer at 4 dpi. (f) At 6 days postinfection, susceptible "RQ001-02M2" exhibits severe upper epidermal damage and fungal invasion (smaller arrow), with a noticeable presence of an infection cushion near the epidermis (bigger arrow). (g) Cytoplasmic disarray and necrotic cells (arrows) of palisade mesophyll cells beneath the intact epidermal layer at 6 dpi in the resistant "Charlton." (h) At 6 dpi, susceptible "RQ001-02M2" cells showed signs of calcium oxalate crystals (arrows) in the top epidermis and palisade mesophyll cells. (i)"Charlton" resistant at 6 dpi without calcium oxalate crystals visible. (j) In resistant "Charlton," fungal invasion is primarily limited to the top epidermis at eight dpi. (k) In susceptible "RQ001-02M2," calcium oxalate crystals are distributed throughout the leaf tissue at eight dpi. (l) In resistant "Charlton," calcium oxalate crystals are primarily restricted to the top epidermis at 8 dpi. The scale bars are as follows: (a, b) ¼ 20 mm; (c) ¼ 50 mm; (d, e) ¼ 100 mm; (f–i, l) ¼ 200 mm; (j, k) ¼ 500 mm. (Adapted from Garg *et al.*, 2010, with permission.)

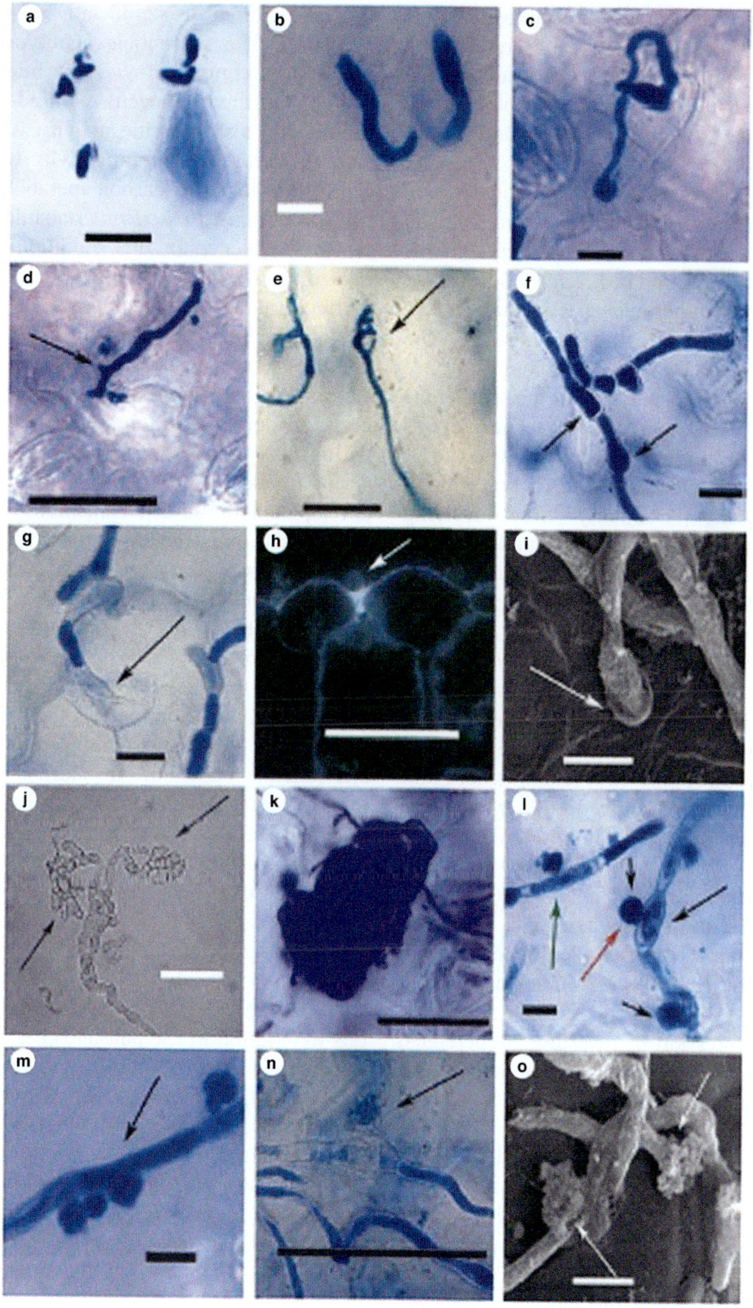

Fig. 7.2. Light and scanning electron micrographs taken after *Sclerotinia sclerotiorum* isolate WW-3 was injected onto the glass slide surface and the cotyledons of susceptible (RQ001-02M2) and resistant (Charlton) *Brassica napus*. For samples (a–g) and (k–n), acetic acid, ethanol, and water (2:2:1) were used for clearing. Lactophenol was stained with 1% cotton blue, and a Zeiss Axioplan 2 microscope photograph system was used for taking pictures. An AxioCam digital image system equipped with an exciter filter and a barrier filter was used to capture a 2-mm-thick section in (h) under a mercury incident light beam. (Adapted from Garg *et al*., 2010, with permission.)

invading. The calcium oxalate crystal levels of resistant "Charlton" variety are decreased due to the fungi's stunted growth. Prior studies have demonstrated a clear relationship between *S. rolfsii* mycelial growth and OA emission (Punja *et al.*, 1985).

The activation of an oxalate oxidase gene in the resistant "Charlton" cultivar may be the cause of the decrease in oxalate content. This gene can be found in *B. napus* (Dong *et al.*, 2008) and *Helianthus annuus* (Hu *et al.*, 2003). It helps convert oxalate into CO_2 and H_2O_2. The resistant variety "Charlton" was found to have 4 dpi fewer starch deposits than the susceptible variety "RQ001-02M2." Starch accumulations were detected in close proximity to the site of infection in various interactions between hosts and pathogens, particularly those involving biotrophic pathogens like rusts and powdery mildews (Long *et al.*, 1975; Minarcic and Janitor, 1994), as well as necrotrophic fungi such as *Kabatiella caulivora* (Bayliss *et al.*, 2001) or *Leptosphaeria maculans* (Li *et al.*, 2007). Furthermore, a number of studies—particularly those involving biotrophic pathogens—have shown how pathogens can interfere with the host genotype's source–sink relationship. Plant metabolism is reprogrammed in the pathogen's favor as a result of this interference, which causes a buildup of photosynthetic products at or close to the infection site, including sucrose. Other sections of the host plant suffer as a result of this buildup (Wright *et al.*, 1995; Berger *et al.*, 2007; Kocal *et al.*, 2008). Depending on the kind and stage of the host–pathogen interaction, increased expression of invertase is correlated with elevated levels of photosynthetic products at the infection site (Long *et al.*, 1975; Herbers *et al.*, 2000; Roitsch *et al.*, 2003).

Acid invertase hydrolyzes sucrose to generate glucose and fructose (Chou *et al.*, 2000). After going through glycolysis, these molecules produce triose phosphate. Triose phosphate has the ability to reach chloroplasts, where it is converted into starch by metabolism (Long *et al.*, 1975).

In the susceptible genotype "RQ001-02M2," it is possible that the pathogen's greater biomass led to an accumulation of photosynthase within the host tissues. According to Joosten *et al.* (1990), this buildup may have led to higher signals for invertase activity, which in turn caused a significant accumulation of starch. The susceptible genotypes having high starch content provide the more carbon sources to the pathogen required for both saprophytic and parasitic activity. Zhao *et al.* (2009) discovered alterations in the gene's expression linked to carbon metabolism in *B. napus* in response to *S. sclerotiorum* infection and provided evidence for this. Additionally, genes related to invertase activity were induced, suggesting that the pathogen accesses and utilizes the plant's carbon storage reserves through the photorespiration pathway. There are several factors that contribute to the slowing down of pathogen development on the plant surface and within host tissues. These factors include hindering fungal growth, actively suppressing the development of infection cushions, extruding protoplasts, and triggering the hypersensitive response, which is associated with resistance to *S. sclerotiorum* in *B. napus*. The development of disease-resistant genotypes and screening resistance markers against this pathogen can benefit greatly from these novel resistance mechanisms.

At 1, 2, and 3 days after inoculation (dpi), the growth of fungus was inhibited by the resistant cultivar "Charlton." In addition, it caused the protoplast to be ejected from hyphal cells, inhibited the development of appressoria and infection cushions, and induced a hypersensitive response. The top epidermis of "Charlton" was the main site of pathogen infiltration by 8 dpi. However, colonization reached the spongy mesophyll cells of the susceptible cultivar "RQ001-02M2." At 6 dpi, calcium oxalate crystals were found in the palisade cells and upper epidermis of "RQ001-02M2," and at 8 dpi, they were found throughout the leaf tissues. Crystals were absent in "Charlton" at 6 dpi, but by 8 dpi, they were mostly located in the top epidermis. Moreover, "RQ001-02M2" showed more starch deposits in comparison to "Charlton" (Garg *et al.*, 2010).

7.6 Infection of *Sclerotinia* in Apetalous and Fully Petalled *Brassica*

Near-isogenic lines (NILs) of oilseed rape with apetalous (AP) and fully petalled (FP) traits were utilized in the study of *S. sclerotiorum* infection. The primary mode of infection in FP oilseed rape is through infected petals that adhere to the leaves. In field experiments, the AP1 flowers displayed an average of 1.4 petals per flower, while in poly tunnel experiments, the average was 0.8 petals per flower. There were no significant differences in the counts of FP1 petals, FP1 stamens, and AP1 stamens adhered to leaves during flowering in the field experiments. At a specific sample time, a higher percentage of stamens tested positive for *S. sclerotiorum* on AP1 compared to FP1 NILs in 2004, with 37.5 and 24.2% of stamens testing positive on AP1 and FP1 NILs, respectively, during early flowering. In the poly tunnel experiments, there was a notable disparity in the number of *Sclerotinia* lesions per plant between the FP1 and AP1 NIL. Although the AP1 NIL did not entirely avoid infection, it still exhibited some resistance due to the production of petals. Lesions were observed to initiate from both adhered stamens and petals. Interestingly, while 8.5 and 16.3% of lesions were started from petals in the FP1 and AP1 NILs, respectively, only 2.5 and 1.0% of lesions were started from stamens in the FP1 and AP1 NILs, respectively. Stamens may exhibit reduced infectivity in comparison to petals, as indicated by the results. During field trials carried out in 2004, the AP1 NIL displayed a notably lower rate of SSR occurrence in contrast to the FP1 NIL (4.9 and 7.0%, respectively). Nevertheless, there was no important variance in SSR occurrence between AP and FP varieties in 2005 (3.6 and 4.3%, respectively) or 2006 (5.5 and 3.9%, respectively) (Young and Werner, 2012).

7.7 Impact of Wetness and Temperature on Ascosporic Infection

An experiment was conducted to examine the influence of moisture duration and incubation temperatures on the germination of *S. sclerotiorum* ascospores and the efficiency of ascosporic infection. Ascospore germination was most effective when incubated in constant moisture (free water) at 21°C, with significantly lower germination rates observed at 10 or 30°C. Disrupting the wet incubation of ascospores had adverse effects on germination. Dry ascospores were used to inoculate dry bean and canola flowers in infection efficiency studies, which were then placed on the leaves of dry bean and canola plants, respectively. The dry bean plants were subjected to an incubation period of 196 h at temperatures ranging from 18 to 20°C. During this time, the plants experienced alternating periods of 8–16 h of wetness followed by 12–24 h of dryness. On the other hand, the canola plants underwent an incubation period of 240 h at temperatures of 10, 15, 20, 25, and 30°C. These plants also experienced alternating periods of 6–18 h of wetness followed by 18–6 h of dryness. The occurrence of wet incubation interruptions resulted in a delay in the manifestation of symptoms and impeded the progression of epidemics on both types of plants (Cotton *et al.*, 2003). By employing logistic regression models, it was determined that there was a 50% probability of disease development on dry bean and canola plants when subjected to 68 and 48 h of wet incubation, respectively, at a temperature of 20°C over a span of 6 days. The validation of the canola model was conducted utilizing data obtained from field trials. The findings from these research endeavors will aid in the enhancement of precise warning models for diseases induced by *S. sclerotiorum* (Shahoveisi and del Río Mendoza, 2020).

7.8 Pathogenic and Saprophytic Stages of *Sclerotinia*

Due to its unusual dual life cycle, *Sclerotinia* is a very effective pathogen that becomes more challenging to control during the early phases of infection after the biotrophic phase and then the necrotrophic phase. *S. sclerotiorum* is unique among necrotrophic pathogens in that it requires dead tissues to start an infection and complete its life cycle. To explain the process, a model that splits the infection progression into pathogenic and saprophytic phases has been proposed. The primary difference is that the pathogen creates the dead tissues on its own at the later stage. Before spreading to infect healthy plant tissue, the pathogen receives the nourishment and resources it needs to establish itself in dead tissue. Peptides and oligogalacturonides are released by these processes, and this sets off the ensuing wave of degradative enzyme synthesis. When these enzymes work together, the plant tissue completely disintegrates. During the pathogenic stage, the host's defenses are weakened by OA and other metabolites and enzymes, and necrotic tissue is formed as a result of host cell reactions to other components. A secondary saprophytic phase occurs after the pathogenic stage. The body enters this phase in response to a fall in cAMP levels, which further inhibits the manufacture of hydrolytic enzymes and makes glucose available. Lower levels of cAMP combine with the acidic environment created by OA synthesis to promote the growth of sclerotia and the completion of the life cycle (Hegedus and Rimmer, 2005).

7.9 *Sclerotinia*-Induced Seed Infection

The agricultural crop production system's main input is seed. If tainted or diseased seed is sown, the pathogen may become epidemic while the crops are still in the early stages of growth. As the main source of inoculum, contaminated or diseased seeds may be able to spread the virus over great distances. *Sclerotinia*-infected or—infested seeds have been found in a wide range of crops, including sunflower (Young and Morris, 1927), safflower (Chaudhary and Puttoo, 1991), cabbage, cauliflower, kale (Neergaard, 1958), clover (Scott, 1981; Scott and Evans, 1984), beans (Starr *et al.*, 1953; Akai, 1981), lupine (Chamberlain, 1932), peanut (Porter and Beute, 1974; Akem and Melouk, 1990), rape, barley (Adams and Ayers, 1979), wheat, sorghum, flax (Miclaus *et al.*, 1988), and soybean (Anderson, 1985). Nevertheless, a comprehensive analysis of the part these seeds play in the epidemiology of *S. sclerotiorum* and its dissemination has not been done. Examining the seed-borne transmission of *S. sclerotiorum* is essential, especially in regions where disease-free seed production is the main goal. White mold of beans is a seed-borne disease that has been recognized (Nobel and Richardson, 1968). Various types of seeds have been found to carry the fungus (Blodgett, 1946; Baker and Davis, 1951; Hungerford and Pitts, 1953; Nicholson *et al.*, 1972). After a 3-year investigation, Steadman (1975) reported that there was no proof that sclerotia propagated by seeds in Nebraska bean fields. *S. sclerotiorum* was found in 48% of seed batches of the Pinto and Great Northern bean varieties that were taken from white mold-affected plants. On the other hand, it was found in only 6% of seed lots that came from Western Nebraska plants that seemed healthy. Less than 0.5% of normal seeds had the fungus identified in them, compared to roughly 12% of chalky, discolored, and shriveled seeds. Nevertheless, no white mold-infected bean plants emerged when these contaminated seed batches were planted in sterile soil inside a greenhouse humidity chamber. Additionally, there was no *in vitro* germination of *S. sclerotiorum*-infected seeds. Although *S. sclerotiorum* can spread by seeds, this knowledge has little relevance to epidemiology. When Hungerford and Pitts (1953) planted Pinto bean fields infected with *S. sclerotiorum* seeds in a greenhouse, they discovered that very few plants were impacted by white mold. According to Tu (1988), *S. sclerotiorum* can live in infected white bean seeds as dormant mycelium in the cotyledons and testa. Over the course of 3 years, the survival rate stays constant at 85–89%. Eighty-seven to 100% of these tainted bean seeds do not sprout and decompose when sown in sand or dirt. Three or six sclerotia are formed in place of a single seed. A minor percentage (2.5 and 11.5%, respectively) of these sclerotia grow carpogenically, either with or without preconditioning. According to Hartman *et al.* (1998), the frequency of *S. sclerotiorum* infection in soybeans that is transmitted through discolored, shriveled seeds ranges from 0 to 70%. Whether preconditioned or not, the germination of sclerotia by myceliogenic methods yields a germination rate of 35.5 and 70.5% on water agar and 81–93% on glucose agar, respectively. Both nonpreconditioned and preconditioned sclerotia have the capacity to go through myceliogenic germination when scattered on the soil's surface and infect bean leaves when they come into contact. This suggests that the latent mycelia present in contaminated seeds have a major impact on the disease's epidemiology in addition to aiding in the fungus's dissemination. The findings of this investigation contradict those of previous studies (Abawi and Grogan, 1975, 1979) which claimed that sclerotia could not infect bean tissues even after prolonged incubation under optimal conditions. The capacity of the pathogen to propagate in the soil may be aided by the presence of contaminated seeds. Sclerotia can develop on contaminated seeds under the right circumstances, and they may eventually give rise to apothecia later in the growing season (Hungerford and Pitts, 1953).

7.10 Role of Enzymes in Infection and Pathogenesis

During the process of infection, *Sclerotinia* produces several kinds of enzymes to penetrate the epidermal cuticle layer of the host surface. These enzymes help in the dissolution of the cell wall to colonize the host tissues.

7.10.1 Infection mechanism

Ascospore-bearing petals are usually the route via which *Sclerotinia* infection in oilseed rape crops takes place. The contaminated ascospores adhere to the leaves after the petals fall off, causing infections that may result in SSR lesions (McCartney *et al.*, 2001a, b). One important primary source of infection is the ascospores, which are expelled from the apothecia in the soil and found near the base of the plants. The existence of mycelium in the soil or mycelium produced from sclerotia is a less relevant main infection source since the fungus has a low ability for saprophytism (Newton and Sequeira, 1972). When ascospores are exposed to temperatures between 5 and 30°C, they can begin to grow in as little as 24 h. As long as there is some water available, the ideal temperature range for their germination is between 5 and 10°C. As the ascospores germinate, they create infection hyphae, which enter preexisting wounded or injured tissue or press against the cuticle to directly penetrate the host tissue. The mycelium can proliferate intracellularly or intercellularly once it has entered the host. Its primary mode of action is the enzymatic breakdown of the cell wall, which leads to tissue colonization and, as the hyphae proliferate, eventual cell death. Pectolytic enzymes cause indirect damage to the cell membrane, which leads to cellular death (Morrall *et al.*, 1972). It has been discovered that the rapeseed–mustard plant–infecting fungus *Sclerotinia* produces the enzymes cellulase (CX) and polymethyl galacturonase transeliminase (PGTE) (Rai and Dhawan, 1976; Dhawan and Srivastava, 1987). These enzymes are believed to contribute to the pathogenicity of different strains. Moreover, it has been demonstrated that protease activity contributes to mustard plant infection (Dhawan, 1980). According to Rai and Dhawan (1976), the infiltration of mustard tissues seems to be associated with the production of the toxin OA, which is produced in unhealthy plants. When the host plant is exposed to the fungus's culture filtrate, the poisonous material becomes heat resistant and transportable within the plant and causes infection.

7.10.2 Pathogenesis by *Sclerotinia*

S. sclerotiorum's etiology is still complex and poorly understood (Dickman and Mitra, 1992). The fungus produces a variety of degradative enzymes that are believed to help with colonization and breakdown of the host cell wall, including cellulases, hemicelluloses, endo- and exo-pectinases, and proteases (Marciano *et al.*, 1983; Riou *et al.*, 1991). Canola pollen contaminated with *S. sclerotiorum* enters the cell walls directly by the creation of appressoria, or infection cushions. The main pathway for hyphal penetration is through the germinative pores, not other parts of the pollen wall. The disintegration of the pollen cytoplasm and plasmolysis are caused by the widespread branching of hyphae within the infected pollen grains (Huang *et al.*, 1998).

Disease development is associated with the production of OA and enzymes that break down cell membranes. The ability to infiltrate membranes (cuticle), which are clearly disintegrated at the place of entry, is a sign of the capability to produce infection (de Bary, 1887). This ability is most likely dependent on the presence of a material, specifically an enzyme, that has the ability to degrade membranes. Moreover, enzymatic activity in conjunction with infection cushions may occur in the cuticle of clover leaves (Prior and Owen, 1964). However, the research did not find any changes in the host tissue brought about by the chemicals diffusing before they penetrated (Boyle, 1921; Purdy, 1958). Until there is experimental proof of possible enzymatic activity on the host tissue prior to penetration, the entry into the host should be considered purely histologically based and a mechanical process. Pectolytic enzymes have long been linked to Sclerotinia diseases (Held, 1955; Echandi and Walker, 1957; Hancock, 1966; Calonge *et al.*, 1969; Vanden Berg and Yang, 1969; Maxwell and Lumsden, 1970; Morrall *et al.*, 1972; Newton, 1972; Barkai-Golan, 1974; Lumsden, 1976Lumsden, 1979; Sharma and Sharma, 1984; Riou *et al.*, 1991). There is not much information on these enzymes, both in terms of number and quality. None the less, they have been connected to modifications in the histochemical and structural composition of the middle lamellae of the host cell (Calonge *et al.*, 1969; Lumsden, 1976) as well as a reduction in the pectic material content of the diseased tissue (Hancock, 1966). Furthermore, it has been explicitly found that there is pectolytic enzyme activity in the infected tissue.

7.11 Regulation of *Sclerotinia* Enzymes at Various Infection Phases

Sclerotinia produces different enzymes in a very systemic sequence to cause infection and spread into the host tissues at different stages of infection and pathogenesis. Regulatory genes are expressed at each step to complete the events. *S. sclerotiorum* releases hydrolytic enzymes that sequentially degrade the plant cuticle, middle lamellae, and primary and secondary cell walls. The cuticle was found to serve as a barrier against *S. sclerotiorum* infection, as leaves without epicuticular wax were more rapidly colonized. As a result, the variables influencing the gene regulation encoding a recently discovered cutinase (SsCUTA) and polygalacturonase1 (SsPG1) were noted. SsCutA transcripts were found *in vitro* within 1 h of leaf inoculation, and expression was mostly controlled by mycelia's interaction with solid surfaces. The interaction with solid surfaces, such as the leaf, moderately induced the expression of SsPg1, which was confined to the advancing edge of the lesion as the infection progressed. Galactose suppressed the expression of SsPg1, whereas carbohydrate shortage stimulated it. When glucose was given to the mycelia used to inoculate leaves, it enhanced the expression of SsPg1 at a basal level. These findings went against previous publications that suggested SsPg1 expression was promoted by galacturonic acid and suppressed by glucose. Pharmacological research unveiled that increased cAMP levels led to a decrease in virulence without affecting gene expression, whereas disruption of calcium signaling modified the expression of SsCutA and SsPg1, resulting in reduced virulence of *S. sclerotiorum*. These processes play a role in regulating the expression of hydrolytic enzymes produced by *S. sclerotiorum* for the period of diverse infection phases (Bashi *et al.*, 2012).

7.12 Oxalic Acid Production and Diversity of *Sclerotinia* Isolates

The development of an acidic environment in the plant is facilitated by OA, leading to the activation of hydrolytic enzymes. The mycotoxin, OA, is the main factor behind the aggressive nature and pathogenicity of this fungus. The formation of calcium oxalate crystals aids in the destabilization of pectin compounds present in the cell wall. Furthermore, OA acts as a suppressor of reactive oxygen species (ROS) (H_2O_2). The *S. sclerotiorum* isolates were obtained from three distinct locations. These isolates were referred to as populations. A total of 228 isolates of *S. sclerotiorum* were analysed to determine the number of acids produced by the pathogen. Additionally, molecular analysis using random amplified polymorphic DNA (RAPD) was conducted. The evaluation revealed significant variations in acid production among the specific *S. sclerotiorum* isolates collected. Furthermore, a correlation was observed between acid production and the rate of linear mycelium growth. RAPD analysis using 12 primers demonstrated a high similarity between pathogen isolates from the same locality, while small genetic differences were observed depending on the place of origin (Starzycka-Korbas *et al.*, 2021).

7.13 Oxalic Acid: A Plant Programmed Cell Death Inducer During the Development of Sclerotinia Disease

Necrotrophic plant infections may interact with their hosts by controlling cell death, according to mounting data. It has been established that OA is essential to this fungus's pathogenicity. The induction of a programmed cell death (PCD) response in plant tissue, which is necessary for the disease's advancement, is one way that oxalate contributes to fungal growth. This response highlights characteristics like DNA fragmentation and the existence of terminal deoxynucleotidyl transferase dUTP nick-end labeling (TUNEL) positive cells that are associated with mammalian apoptosis. When fungi infect plants, mutants lacking in OA formation do not harm plants or exhibit any signs of apoptosis. The pH-reducing qualities of this organic acid, which are required for the development of sclerotia, have no effect on the beginning of PCD by OA. Moreover, it has been discovered that oxalate causes a rise in the amounts of ROS in the plant, which results in PCD. Oxalate causes apoptotic-like cell death, which is avoided when ROS generation is stopped. *Sclerotinia* spp. release OA, which is responsible for initiating apoptotic-like traits in plants as the disease progresses and acts as an elicitor of PCD in plants. This PCD, which involves ROS, is essential to the pathogenicity of fungi. Thus, it would appear that OA works by triggering the plant processes that cause PCD. Furthermore, *Sclerotinia* spp. secrete OA, which appears to function as a signaling molecule rather than being directly damaging (Kim *et al.*, 2008).

7.13.1 Functional characterization of OA during pathogenesis

One important mechanism contributing to *S. sclerotiorum's* broad necrotrophic pathogenicity has been found as the deposition of large amounts of OA during the pathogenesis of the organism (Andrew *et al.*, 2012). Since "A" mutants, or oxalate-deficient mutants generated by ultraviolet (UV) radiation, are nonpathogenic on *Phaseolus vulgaris* (Godoy *et al.*, 1990) and *A. thaliana* (Dickman and Mitra, 1992), they are significant. This event provides strong evidence that oxalate plays a critical role in pathogenesis. It has been observed that these mutant strains tend to accumulate less OA and cause only mild lesions in a variety of hosts (Cessna *et al.*, 2000; Kim *et al.*, 2008; Williams *et al.*, 2011). Apart from its role in pathogenicity, the buildup of OA may also have an impact on *S. sclerotiorum's* growth and developmental processes. The "A" mutants have abnormal development and stunted growth, and they also lack the production of sclerotium and have fluffy hyphae. Conversely, a low ambient pH promotes both vegetative growth and sclerotium formation in the wild-type (WT) strain (Rollins and Dickman, 2001; Rollins, 2003). Because of its impact on the pH of the environment and its metabolic connections to important metabolism, OA may have an effect. Among the pathogenic and nonpathogenic species that produce the potent dicarboxylic organic acid known as OA are bacteria, fungus, plants, and certain mammals. Numerous processes, such as hydrolysis of oxaloacetate, oxidation of glyoxylate, and oxidation of glycolaldehyde, have been shown to be involved in the synthesis of oxaloacetate by filamentous fungi (Dutton and Evans, 1996). Previous studies in biochemistry have shown that the primary process by which *Ascomycota* fungi, including *S. sclerotiorum*, produce OA is the exchange of oxaloacetate to OA and CO_2, which is facilitated by oxaloacetate acetyl hydrolase (OAH, EC 3.7.1.1) (Maxwell, 1973; Kubicek *et al.*, 1988; Ruijter *et al.*, 1999). In *S. sclerotiorum*, the presence of a neutral pH significantly increases OA synthesis and oxaloacetate hydrolase activity (Maxwell, 1973; Rollins and Dickman, 2001). The zinc-finger transcriptional regulator Ss-Pac1, whose activity is reliant on the pH levels in the environment, is in charge of controlling this induction at neutral pH (Rollins and Dickman, 2001; Rollins, 2003). Depending on the carbon growth substrate, different carbon sources can promote the biogenesis of OA in *S. sclerotiorum*. In this process, intermediaries for the glyoxylate or tricarboxylic acid (TCA) cycles are essential (Liberti *et al.*, 2013).

The role of *S. sclerotiorum's* OAH in regard to OA metabolism and its importance in producing disease has been recognized (Liang *et al.*, 2015a). The researchers carried out gene deletion analysis and identified a gene known as *Ss-oah1*, which is thought to encode OAH. The results showed that *Ss-oah1* affects the pH-responsive proliferation, morphogenesis, and virulence of *S. sclerotiorum* on several plant hosts and is necessary for the accumulation of OA. It was observed that although *Ss-oah1* knockout mutants (Δss-oah1 mutants) might produce primary lesions, other parameters were important in defining the initial host compatibility, even though OA was not the only driver. This implies that several factors have an impact on primary lesions. According to the infection's histological features, *S. sclerotiorum* may be able to partially colonize its hosts at first without producing OA. The colonization and infection processes require dynamic control over the buildup of OA.

7.13.2 Cloning and characterization of OAH gene

Effective cloning and functional characterization of *S. sclerotiorum's* gene *Ss-oah1*, which codes for OAH (OAH, EC 3.7.1.1), have been achieved. The relationship between the increase in OA and Ss-oah1 transcripts was particularly evident in neutral pH settings. The pH-dependent stimulation of Ss-oah1 transcription was found to be regulated by the pH-responsive transcriptional regulator Ss-Pac1. The absence of OA accumulation was observed in both the culture and plant infection of the gene deletion mutants Δss-oah1, highlighting their distinctiveness. This is in contrast to the previously studied mutants (referred to as "A" mutants) that lacked oxalate formation due to UV radiation. However, when the WT *Ss-oah1* gene was reintroduced, this impairment in OA buildup was totally reversed. The Δss-oah1 mutants also showed a notable growth deficiency on media buffered at neutral pH, as well as deficiencies in the production of both compound appressorium and sclerotium. The Δss-oah1 mutants only produced extremely small lesions when infecting different plant hosts, and the infectious hyphae eventually became less viable. When the WT and Δss-oah1 strains were utilized to produce infections, cytological examination showed that the WT strain's subcuticular hyphae had low levels of OA accumulation, while the Δss-oah1 strain had no OA accumulation at all. At the infection front, living host epidermal cells briefly interacted with both the WT and mutant strains. These findings highlight the significance of OAH activity for the pathophysiology and biosynthesis of OA in *S. sclerotiorum*. However, they also suggest that factors other than OA might have an impact on the development of primary lesions (Liang *et al.*, 2015b).

7.14 Expression of Genes During Pathogenesis

Genes that aid in *Sclerotinia's* successful pathogenesis are expressed during infection and host–pathogen interactions. These genes have a critical role in both host infection and signal transduction. An investigation was carried out by Li RuGang *et al.* (2004a,b) to look at the genetic components of fungal development and disease. They used two cDNA libraries to produce 2232 expressed sequence tags (ESTs). Mycelia cultivated in pectin medium or tissues from *B. napus* infected stems were used to produce these libraries. They uncovered 774 distinct fungal genes through their study, 39 of which were only present in the EST collection of infected plants. They were able to annotate 534 uni-genes using the universal Gene Ontology approach and the categories utilized for *Saccharomyces cerevisiae*. Many putative pathogenicity factors, including four endo-polygalacturonases, two exo-polygalacturonases, and numerous metabolite transporters, are encoded by the annotated genes found within. When considering these genes' roles in the infection process, combined with those of signal transduction factors, it is impossible to overestimate how important they are.

7.14.1 Genes identified during the interaction between *Brassica Napus* and *Sclerotinia*

S. sclerotiorum isolate samples from *B. napus* were analysed using pectin gel overlays and isoelectric focusing, leading to the identification of five major and several minor PG isoenzymes. Four endo-polygalacturonase (PG) genes (*sspg1d*, *sspg3*, *sspg5*, and *sspg6*) and two exo-PG genes (*ssxpg1* and *ssxpg2*) were found using a combination of degenerate PCR and ESTs. In 1995, Fraissinet-Tachet *et al.* (1995) identified the *SSPG1d* gene as a member of the PG gene family. Mature *SSPG1d* acts as a neutral PG, but *SSPG3*, *SSPG5*, and *SSPG6* are acidic enzymes after they are completely digested. Under saprophytic growth conditions, pectin and galacturonic acid increase the production of sspg1d, sspg3, sspg5, and ssxpg1, but glucose represses this expression. It was unable to witness either sspg6 or manifesting themselves in any of the discovered circumstances. When mycelia are transferred from liquid media to solid substrates, sspg1d is upregulated, indicating that thigmotrophic interactions are involved in its regulation. Under pathological circumstances, *Sspg1d* has high expression levels, while *Sspg3* shows lower expression. Conversely, *Sspg5*, *Sspg6*, and *Ssxpg1* are characterized by low levels of expression (Li RuGang *et al.*, 2004a,b).

7.15 Understanding Molecular Biology of *Sclerotinia* to Control Pathogenicity, Virulence, and Growth

The many contradictions in the *S. sclerotiorum* literature have led to the widespread adoption of the gene, mutation, and protein catalog utilized by *Sclerotinia* researchers. The mutant gene is represented as *abc1*, while the WT (wild type) gene producing the Abc1 protein is italicized as *Abc1*. Any exceptions to this naming scheme are made clear in explicit terms. The notable variations in field isolates and local hosts utilized in these studies (Table 7.1) must be considered as they are most likely the cause of the variations in mutant phenotypes for the same genes (Xia *et al.*, 2020).

7.15.1 Gene expression and the control of virulence and mycelial growth

The capacity to get whole genome sequences and advancements in transformation, knockdown, and knockout methods have been crucial in facilitating the molecular analysis of *S. sclerotiorum* throughout the past 20 years. Involved in sclerotial and apothecial development, for example, the majority of the genes that have been investigated thus far are conserved genes that influence both growth and differentiation (Figs 7.3 and 7.4). These genes do not directly influence differentiation throughout development, but they do influence virulence and the pace at which hyphal growth occurs. It is crucial to keep in mind that the processes of differentiation and pathogenicity occur after mycelial expansion.

The mutants in this study that show significant growth limitations often show a decrease in pathogenicity. There are, however, some outliers, such as mat1-2-1, which grows more slowly and exhibits aberrant apothecial morphogenesis while maintaining its typical virulence (Doughan *et al.*, 2016). The rate of mycelial growth and pathogenicity decreases when the highly conserved transcription factors Ams2 or Mads are silenced with RNA interference (RNAi), suggesting their importance as early growth regulators. Similar to its function in *Schizosaccharomyces pombe*, Ams2 is a transcription factor with a GATA-box domain that is necessary for the correct expression of genes related to the cell cycle and histones. However, Mads, the yeast Mcm1 equivalent, has a MADS-box domain and is essential for stress tolerance, recombination, mating, viability, cell cycle, and mini chromosome maintenance. The ability to produce sclerotia implies that later stages of fungal differentiation pathways, such as sclerotial formation and apothecial development, may be less dependent on these initial growth factors, despite having a significant impact of mutations on these early growth factors (Qu *et al.*, 2014; Liu *et al.*, 2018).

The highly conserved protein known as Bax inhibitor-1 is present in both prokaryotes and eukaryotes (Bi1). It is capable of preventing apoptosis. The Bi1 knockdown mutants in *S. sclerotiorum* showed reduced virulence and enhanced aerial hyphae development. Due to varying pressures eliciting distinct gene expression patterns, these mutants exhibited reduced resilience to chemical stress. Although Bi1 plays relatively minor functions in growth and development, it seems to be a regulator that reacts specifically to biotic and abiotic stressors. Conversely, PemG1 has the opposite effect on pathogenicity and growth. PemG1's

Table 7.1. Genes with functional features from *Sclerotinia sclerotiorum* (The genes are arranged based on their chromosomal locations, Fig. 7.2). Adapted from Xia *et al*., 2020, with permission.

						Mutant phenotypes									
Gene code (new)	Gene code (old)	Strains	Mutant name	Gene full name	Mutant type	Hyphal growth	Sclerotial formation	Oxalate production	Virulence	Appressoria formation	Induce host HR/ resistance	Secretion signal	Tested hosts	Other functions	References
sscle_01g003850SS	*1G_02068*	Ep-1PNA367, Heilongjiang, China	ssvp1	Small secreted virulence-related protein	RNAi	–	NA	NA	+	NA	+	Yes	Canola (Zhongyou 821)	HR induction	Lyu *et al.* (2016)
sscle_01g004990SS	*1G_01919*	1980, Nebraska, USA	svf1	Survival factor 1 homologue	RNAi	–	NA	–	+	+	NA	No	*Arabidopsis* and canola (Zhongyou 821)	Cell wall integrity, ROS production	Yu *et al.* (2019)
sscle_01g006030SS	*1G_01788*	1980, Nebraska, USA	cna1	Catalytic subunit calcineurin encoding gene	RNAi	+	+	–	+	–	NA	No	*Arabidopsis* and tomato (Bonny Best)	Hyphal elongation	Harel *et al.* (2006)
sscle_01g007450SS	*1G_01614*	Ep-1PNA367, Heilongjiang, China	sop1	Microbial opsin homologue gene	RNAi	+	+	NA	+	NA	NA	No	*Arabidopsis*	Stress responses	Lyu *et al.* (2016)
sscle_01g011030SS	*1G_01151*	1980, Nebraska, USA	sfh1	GATA-box and SNF5 domains containing transcription factor	RNAi	+	+	NA	+	+	NA	No	Soybean, common bean, and tomato	ROS accumulation	Liu *et al.* (2018)
sscle_02g013550SS	*1G_12694*	HA61, Jiangsu, China	shk1	Histidine kinases	Deletion	+	+	NA	–	NA	NA	No	Rapeseed, strawberry, tomato, and cucumber	Oxidative stresses, glycerol accumulation	Duan *et al.* (2013)
sscle_02g020240SS	*1G_04003*	1980, Nebraska, USA	mat1-1-5	Mating-type gene	Deletion	–	–	NA	–	NA	NA	No	Tomato	Apothecial development	Doughan *et al.* (2016)
sscle_02g020250SS	*1G_04004*	1980, Nebraska, USA	mat1-1-1	Mating-type gene	Deletion	–	–	NA	–	NA	NA	No	Tomato	Apothecial development	Doughan *et al.* (2016)
1G_04005	*1G_04005*	1980, Nebraska, USA	mat1-2-4	Mating-type gene	Deletion	–	–	NA	–	NA	NA	No	Tomato	Apothecial development, ascospore production	Doughan *et al.* (2016)

sscle_02g020270SS	*1G_04006*	1980, Nebraska, USA	mat1-2-1	Mating-type gene	Deletion	–	–	NA	–	NA	NA	No	Tomato	Apothecial development	Doughan *et al.* (2016)
sscle_03g025030SS	*1G_00699*	WMA1, Washington USA	sod1	Cu/Zn superoxide dismutase	T-DNA	-	–	–	+	NA	NA	No	Pea (Guido)	Detoxification of host ROS	Xu and Chen (2013)
sscle_03g025030SS	*1G_00699*	1980, Nebraska, USA	sod1	Cu/Zn superoxide dismutase	Deletion	–	+	+	+	NA	NA	No	Tomato (Garden Peach) and tobacco	Oxidative stress tolerance, repression of host ROS	Veluchamy *et al.* (2012)
sscle_03g028510SS	*1G_00263*	Isolated from an infected canola stem Alberta, Canada	ssv263	Hypothetical secreted protein	Deletion	-	NA	NA	+	NA	NA	Yes	Canola	–	Liang *et al.* (2013)
sscle_03g031470SS	*1G_13314*	Isolated from diseased rapeseed, Alberta, Canada	scd1	Scytalone dehydratase	Deletion	+	+	NA	–	NA	NA	No	Rapeseed (Westar)	Hyphal branching	Liang *et al.* (2017)
sscle_03g031480SS	*1G_13315*	Isolated from diseased rapeseed Alberta, Canada	thr1	Tri-hydroxynaphthalene reductase	Deletion	+	+	NA	–	NA	NA	No	Rapeseed (Westar)	Hyphal branching	Liang *et al.* (2017)
sscle_03g031520SS	*1G_13322*	1980, Nebraska, USA.; UF1 Florida; WMA, Washington USA	pks13	Polyketide synthase	Deletion by CRISPR	–	–	NA	–	+	NA	No	Soybean, canola, tomato (Better Boy), faba bean (Windsor), and pea (Sugar Daddy)	Pigmentation of compound appressoria	Li *et al.* (2018)

Continued

Table 7.1. Continued

Gene code (new)	Gene code (old)	Strains	Mutant name	Gene full name	Mutant type	Mutant phenotypes: Hyphal growth	Sclerotial formation	Oxalate production	Virulence	Appressoria formation	Induce host HR/ resistance	Secretion signal	Tested hosts	Other functions	References
sscle_03g031670SS	*1G_13339*	274, Nebraska, USA	pth2	Peroxysomal carnitine acetyl transferase	Deletion	+	+	+	+	+	NA	No	Soybean	Oxalic acid accumulation	Liberti *et al.* (2013)
sscle_04g034810SS	*1G_02462*	Isolated from an infected canola stem, Alberta, Canada	axp	Arabinofuranosidase/ xylosidase	Deletion	NA	NA	NA	+	NA	NA	Yes	Canola	–	Yajima *et al.* (2009)
sscle_04g034960SS	*1G_02486*	Sunf-M, Inner Mongolia, China	caf1	Protein secreted that has an apparent EF-hand motif that binds Ca2+	RNAi, T-DNA	+	+	–	+	+	+	Yes	*Arabidopsis*, rapeseed, pak choi cabbage hot pepper, tomato, cucumber, and soybean	–	Xiao *et al.* (2014)
sscle_04g037170SS	*1G_02784*	1980, Nebraska, USA	scat1	Type A catalase	Deletion	+	+	NA	+	NA	NA	No	Tomato (Garden Peach)	Modulation of ROS	Yarden *et al.* (2014)
sscle_04g038020SS	*1G_02904*	Ep-1PNA367, Heilongjiang, China	cvnh	Secreted protein	RNAi	+	+	NA	+	NA	NA	Yes	*Arabidopsis*	–	Lyu *et al.* (2015)
sscle_05g046240SS	*1G_05917*	SUN-F-M, Inner Mongolia, China	sl2	Cell wall protein	RNAi	–	+	NA	–	NA	NA	Yes	Rapeseed and *Arabidopsis*	Cellular integrity	Yu *et al.* (2012)
sscle_05g046390SS	*1G_05899*	1980, Nebraska, USA	trr1	Thioredoxin reductase	RNAi	NA	+	NA	+	NA	NA	No	*Arabidopsis* and tobacco	Oxidative stress tolerance	Zhang *et al.* (2019)
sscle_05g046790SS	*1G_05839*	1980, Nebraska, USA	bi1	Bax inhibitor-1 protein	RNAi	+	–	–	+	NA	NA	No	*Arabidopsis* and tomato	Stress responses, hyphal tip branching	Yu *et al.* (2015)

sscle_05g046830SS	*1G_05834*	JRUF1, Florida, USA	foxe2	Fork head-box transcription factor family gene	Deletion, T-DNA	–	–	NA	–	–	NA	No	Tomato	Apothecial development	Wang *et al.* (2016)
sscle_05g048220SS	*1G_05661*	1980, Nebraska, USA	nox1	NADPH oxidase	RNAi	–	+	+	+	NA	NA	No	Tomato (Rutger)	ROS regulation	Kim *et al.* (2011)
sscle_06g049430SS	*1G_07404*	1980, Nebraska, USA	rhs1	Rearrangement hot spot repeat-containing protein	RNAi	+	+	–	+	+	NA	Yes	*Arabidopsis* and canola (Zhongshuang 9)	–	Yu *et al.* (2017)
sscle_06g049780SS	*1G_07360*	1980, Nebraska, USA	fkh1	Atypical forkhead (FKH)-box-containing protein	RNAi	–	+	NA	+	NA	NA	No	Tomato	Cellular integrity	Fan *et al.* (2017)
sscle_06g049830SS	*1G_07355*	1980, Nebraska, USA	pac1	pH-Responsive transcription factor	Deletion	–	+	+	+	NA	NA	No	*Arabidopsis* and tomato (Bonnie Best)	–	Rollins *et al.* (2003)
sscle_06g049890SS	*1G_07345*	NGA4, Anhui, China	pemg1	Elicitor-homologous protein	RNAi	+	NA	–	+	+	NA	No	Oilseed rape and soybean	Negative regulator of growth and virulence	Rollins *et al.* (2003)
sscle_06g051560SS	*1G_07136*	UF1, Florida, USA	ste12	Downstream transcription factor of Mitogen-Activated Protein Kinase (MAPK pathway	RNAi	+	+	NA	+	+	NA	No	Bush bean and tomato	–	Xu *et al.* (2018)
sscle_07g055970SS	*1G_03527*	SUN-F-M, Inner Mongolia, China	hex1	Woronin body major protein	RNAi	–	+	NA	NA	NA	NA	No	–	Cellular integrity	Yu *et al.* (2012)
sscle_07g058030SS	*1G_03252*	UF1, Florida, USA	ams2	Cell cycle-regulated GATA transcription factor	RNAi	+	–	NA	+	+	NA	No	Common bean	Chromosome segregation, number and distribution of sclerotia	Liu *et al.* (2018)
sscle_07g058620SS	*1G_03171*	1980, Nebraska, USA	pka1	Protein kinase A	Deletion	–	–	–	NA	NA	NA	No	–	PKA activity	Jurick *et al.* (2004)

Continued

Table 7.1. Continued

Gene code (new)	Gene code (old)	Strains	Mutant name	Gene full name	Mutant type	Mutant phenotypes: Hyphal growth	Sclerotial formation	Oxalate production	Virulence	Appressoria formation	Induce host HR/ resistance	Secretion signal	Tested hosts	Other functions	References
sscle_07g059700	*MK992913*	1980, Nebraska, USA	qdo	Quercetin dioxygenase gene	Deletion	–	NA	NA	+	NA	NA	Yes	*Arabidopsis*	Flavonol degradation	Chen *et al.* (2019)
sscle_08g064670SS	*1G_05163*	1980, Nebraska, USA	cry1	Cryptochrome family CRY-DASH ortholog	Deletion	–	–	NA	–	NA	NA	No	*Arabidopsis* and tomato	Sclerotial mass, response to UV	Veluchamy and Rollins (2008)
sscle_08g065550SS	*1G_05284*	NGA4, Anhui, China	nac	Nascent polypeptide-associated complex subunit	RNAi	–	–	–	+	NA	NA	No	Oilseed rape and tobacco	–	Li *et al.* (2015)
sscle_08g066770SS	*1G_05445*	1980, Nebraska, USA	smk3	Slt2 ortholog	Deletion	+	+	NA	+	+	NA	No	Canola (Westar)	Cuticle penetration, cell wall integrity	Bashi *et al.* (2016)
sscle_08g067830SS	*1G_05588*	1980, Nebraska, USA	mads	MADS-box proteins	RNAi	+	–	NA	+	NA	NA	No	Tomato	–	Qu *et al.* (2014)
sscle_08g068500SS	*1G_14133*	Ep-1 PNA 367, Heilongjiang, China	itl	Integrin-like protein	RNAi	+	+	NA	+	NA	+	Yes	*Arabidopsis* and canola	Hyphal branching, suppression of host defense	Zhu *et al.* (2013)
sscle_08g068530SS	*1G_14127*	1980, Nebraska, USA	ggt1	Glutamyl transpeptidase	Deletion	–	+	NA	–	+	NA	Yes	Tomato	–	Li *et al.* (2012)
sscle_09g069850SS	*1G_10796*	1980, Nebraska, USA	odc2	Oxalate decarboxylases	Deletion	–	–	+	+	+	NA	Yes	Common bean (Bush Blue Lake 47), soybean (Harosoy), tomato (Bonnie Best), and celery	–	Liang *et al.* (2015a)

sscle_10g075560SS	*1G_08218*	WMA1, Washington, USA	oah	Oxaloacetate acetylhydrolase	T-DNA	–	–	+	+	NA	NA	No	Faba bean (Broad Windsor), pea (Guido), green bean (Great North Tara), and soybean (Skylla)	–	Xu *et al.* (2015)
sscle_10g075560SS	*1G_08218*	1980, Nebraska, USA	oah1	Oxaloacetate acetylhydrolase	Deletion	–	+	+	+	+	NA	No	Tomato (Bonnie Best), common bean (Bush Blue Lake 47), soybean (Harosoy), canola, *Arabidopsis*, and sunflower	–	Liang *et al.* (2015b)
sscle_10g075560SS	*1G_08218*	1980, Nebraska, USA.; UF1, Florida; WMA, Washington, USA	oah1	Oxaloacetate acetylhydrolase	Deletion by CRISPR	–	+	+	+	+	NA	No	Soybean, canola, tomato (Better Boy), faba bean (Windsor), and pea (Sugar Daddy)	–	Li *et al.* (2018)
sscle_10g077630SS	*1G_08489*	1980, Nebraska, USA	pph1	Type 2A Ser/ Thr phosphatase catalytic subunit PP2Ac	RNAi	+	+	NA	NA	NA	NA	No	–	–	Erental *et al.* (2007)
sscle_10g079320SS	*1G_14065*	1980, Nebraska, USA	ssp1	Development-specific protein	Deletion	–	–	NA	NA	NA	NA	No	–	Resistance to glycoside-containing antibiotics	Li *et al.* (2010)
sscle_11g082700SS	*1G_07871*	1980, Nebraska, USA	rgb1	Type 2A Ser/Thr phosphatase B subunit	RNAi	–	+	NA	+	+	NA	No	*Arabidopsis* and tomato (Bonny Best)	MAPK pathway	Erental *et al.* (2007)

Continued

Table 7.1. Continued

Gene code (new)	Gene code (old)	Strains	Mutant name	Gene full name	Mutant type	Mutant phenotypes: Hyphal growth	Sclerotial formation	Oxalate production	Virulence	Appressoria formation	Induce host HR/ resistance	Secretion signal	Tested hosts	Other functions	References
sscle_11g083230SS	*1G_07798*	SUN-F-M, Inner Mongolia, China	gpd	Glyceraldehyde-3-phosphate dehydrogenase	RNAi	NA	+	NA	NA	NA	NA	No	–	–	Yu *et al.* (2012)
sscle_11g083680SS	*1G_07749*	1980, Nebraska, USA	xyl1	Endo-1, 4-xylanase	Deletion	+	+	NA	+	NA	NA	Yes	Canola and *Arabidopsis*	Apothecia formation	Yu *et al.* (2016)
sscle_11g083950SS	*1G_07715*	1980, Nebraska, USA	sac1	Adenylate cyclase	Deletion	+	+	–	+	NA	NA	No	Tomato (Bonnie Best)	Apothecia production cAMP signaling	Li and Rollins (2007)
sscle_12g087830SS	*1G_11172*	1980, Nebraska, USA	nox2	NADPH oxidase	RNAi	–	+	–	–	NA	NA	No	Tomato (Rutger)	ROS regulation	Kim *et al.* (2011)
sscle_12g090900SS	*1G_11866*	1980, Nebraska, USA	smk1	ERK-type MAP kinase	RNAi	+	+	NA	NA	NA	NA	No	–	pH-dependent regulation	Chen *et al.* (2004)
sscle_14g099710SS	*1G_08814*	1980, Nebraska, USA	odc1	Oxalate decarboxylases	Deletion	–	NA	NA	–	NA	NA	Yes	Common bean (Bush Blue Lake 47), soybean (Harosoy), and tomato (Bonnie Best)	–	Liang *et al.* (2015a)

sscle_16g107670SS	*1G_10096*	1980, Nebraska, USA	cp1	Cerato-platanin protein	Deletion	–	–	NA	+	NA	+	Yes	*Arabidopsis* and tobacco	Interaction with PR1	Yang *et al.* (2018)
sscle_16g107670SS	*1G_10096*	FXGD2, Anhui, China	sm1	Cerato-platanin protein	RNAi	+	+	–	+	NA	NA	Yes	Tobacco, oilseed-rape, and soybean	–	Pan *et al.* (2018)
sscle_16g107930SS	*1G_10135*	1980, Nebraska, USA	fdh1	Formaldehyde dehydrogenase	Deletion	–	+	NA	+	+	NA	No	Common bean and tomato	Osmotic oxidative stress resistance	Zhu *et al.* (2019)
sscle_16g109570	*ANQ80447*	1980, Nebraska, USA	nsd1	GATA-type Ivb zinc-finger transcription factor	Deletion	–	–	NA	+	+	NA	No	Tomato (Bonnie Best), celery, and tomato	Ascogonia formation, Apothecium development	Pan and Rollins (2017)

The assembly genome of ASM185786v1 was published in 2017 for the current version of the gene code, whereas ASM14694v1 was published in 2011 for the old version. A "+" in the table denotes a change in the associated phenotypic of the mutant strain in relation to the wild-type strain, while a "–" indicates no change in phenotype. NA denotes the absence of phenotypic assessment. "SignalP-5.0" was used to identify the secretion signals. All deletions, unless specified otherwise, were made via homologous recombination. *Apium graveolens* (celery), *Arabidopsis thaliana* (*Arabidopsis*), *Brassica napus* (canola, oilseed rape, rapeseed), *Brassica rapa* subsp. *chinensis* (pak choi cabbage), *Capsicum frutescens* (hot pepper), *Cucumis sativus* (cucumber), *Fragaria ananassa* (strawberry), *Glycine max* (soybean), *Helianthus annuus* (sunflower), *Lycopersicon solanum* (tomato), *Nicotiana benthamiana* (tobacco), *Phaseolus vulgaris* (bush bean, common bean, green bean), *Pisum sativum* (pea), and *Vicia faba* (faba bean) are the Latin names of the host species used for pathogenicity testing of mutants. Columbia-0 is the *A. thaliana* ecotype that is included in the table. The host species' particular local cultivars are listed in brackets. *ROS*, reactive oxygen species; WMA, World Medical Association.

RNAi mutants exhibit enhanced virulence and faster mycelial growth, indicating that it works as a negative regulator in these processes. In addition, these mutants showed elevated infection cushions, decreased tolerance to salt stress, and enhanced cell wall-degrading enzyme (CWDE) activity. Therefore, the precise mechanism by which this fungal-specific protein negatively regulates proliferation and pathogenicity remains unknown (Xia *et al.*, 2020).

7.15.2 Regulation mechanism of pathogenesis

The pathogenesis mechanism during host–parasite interaction follows a very systematic sequence of steps, starting with the expression of signaling genes to identify the host and ending with the production of genes for infection and pathogenesis. *S. sclerotiorum* increases cell death and breaks down the host cell wall during fungal colonization by causing tissue maceration and necrosis after infection. Following a brief biotrophic period, this happens. Toxins and CWDEs (cell wall degrading enzymes) are assumed to be necessary for these processes to take place. By analyzing mutants axp, which lack a precursor of secreted arabinofuranosidase/beta-xylosidase, the importance of CWDEs in *S. sclerotiorum* pathogenicity is revealed. The pathogenicity of canola plants infected with these mutants was reduced. Similarly, research employing mutations of sod1 that alter Cu/Zn superoxide dismutase demonstrates that ROS (reactive oxygen species) are necessary for pathogenicity. Moreover, it has long been postulated that OA, a major contributing factor, increases the pathogenesis of this particular disease.

During infection, *S. sclerotiorum* produces a significant amount of OA, as has been shown in earlier physiology and pharmacology research. Furthermore, despite the precise mutations were not identified, UV mutants that were unable to produce OA or form sclerotia showed a possible connection between OA and development. However, the significance of OA in *Sclerotinia* ecology did not become evident until the introduction of naturally different OA-lacking mutants. The odc2 RNAi strain exhibited decreased virulence due to reduced production of appressoria and non-functionality, which can be attributed to the silencing of the oxalate decarboxylase (ODD) gene, which is important for breaking down OA. The findings of this investigation provide support for the theory that OA accumulation, which is regulated during growth, is critical for penetration-based infections. The production of appressoria was hindered in tests involving the deletion or CRISPR (clustered regularly interspaced short palindromic repeats) induced mutation of OAH, the enzyme in charge of OA biosynthesis. This resulted in hypovirulence and delayed sclerotial development. However, a different study on OAH deletion or T-DNA (transfer DNA) mutants showed that virulence changed in a pH-dependent manner, but growth and development were unaffected. Additional research on other mutants deficient in OA biosynthesis may be able to resolve these discrepancies (Xia *et al.*, 2020). Pathogens try to decrease the host's immune response and/or increase their virulence by secreting different effectors into the host. Since secretome and RNA-seq studies make it easy to identify these possibilities, effector biology has primarily concentrated on studying secreted proteins. Different roles for several secreted proteins in *S. sclerotiorum's* growth and pathogenicity have been observed (Table 7.1). Multiple effectors that each have a unique effect on the virulence of *S. sclerotiorum* are present. Ssvp1, V263, and Qdo (quercetin dioxygenase) are some of these effectors. Although they are tiny secreted proteins that enhance virulence, Ssvp1 and V263 lack significant conservation. Conversely, Qdo is a more conserved effector that catalyzes the cleavage of the flavanol carbon skeleton, which is a critical step in the degradation of the host's antimicrobial flavanol. The process of degradation encourages the pathogenicity of fungi.

S. sclerotiorum has nonsecreted regulators in addition to secreted proteins that are only involved in pathogen penetration and virulence regulation. A few instances of proteins possessing distinct roles include Svf1, Pks13, and Nac α. It appears that Svf1 is involved in

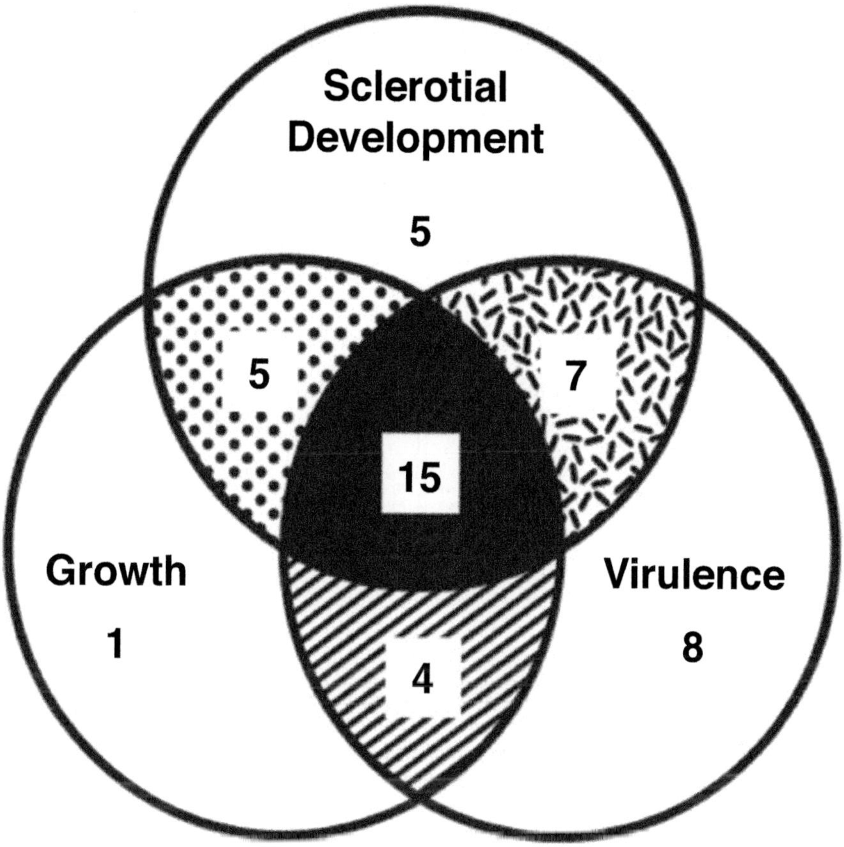

Fig. 7.3. An overview of the genes that have been the subject of mutant investigation, as reported in Table 7.1, using a Venn diagram (detailed in Table 7.1). (Adapted from Xia *et al*., 2020, with permission.)

managing ROS during Sclerotinia infection since it helps yeast survive in oxidative stress situations. Pks13 is most likely in control of melanin formation in appressoria because its CRISPR mutant produces albino compound appressoria without impairing virulence. Conversely, Nac α is a highly conserved protein that functions as a co-activator of transcription. It's interesting to note that it has a negative effect on the expression of genes producing polygalacturonase, which may help explain the mutant's increased virulence traits (Li *et al*., 2015; Chen *et al*., 2019; Yu *et al*., 2019; Xia *et al*., 2020).

7.16 Exhibition of Differential Expression of Multiple Metabolic Enzymes During Infection and Pathogenesis

Sclerotinia's wide host range is controlled by the variable expression of several metabolic enzymes, which allows it to influence the host's defense mechanisms. Numerous pathogenicity factors are used by necrotrophic plant diseases to aid in the invasion of their plant hosts. One of the contributing reasons is the secretion of carbohydrate-active enzymes (CAZymes) by *S. sclerotiorum*. These enzymes, which are often

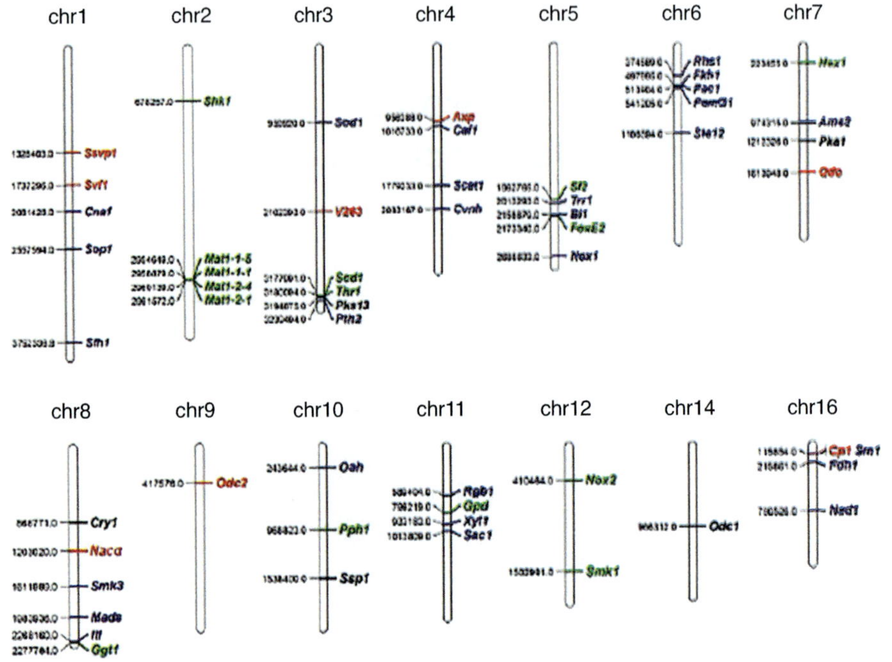

Fig. 7.4. Perusal of data in Table 7.1 displays the gene locations on a map. The positions of each gene are shown by the numbers on the left side of each chromosome. Genes associated with Sclerotinia sclerotiorum development, such as those involved in mycelial growth, sclerotia formation, and apothecia development, are indicated in green. The majority of the genes highlighted in red are involved in *S. sclerotiorum* virulence. The genes indicated in black are linked to other biological functions, whereas the genes highlighted in blue are involved in both processes. "Map Chart" software was used to construct the chromosomal map. (Adapted from Xia *et al.*, 2020, with permission.)

referred to as CWDEs and include xylanases, pectinases, and cellulases, are essential in breaking down plant cell walls. These enzymes facilitate the efficient penetration of plant tissues by fungal appressoria by the breakdown of cell walls. *S. sclerotiorum* possesses a set of genes that encode CAZymes; specific genes have been the subject of in-depth investigation. Enzymes that are essential for breaking down various components found in the cell wall structure, such as cellulose, hemicelluloses, and pectin, are produced by the genetic material found in these particular genes.

Besides releasing CWDEs, numerous necrotrophic fungi also synthesize phytotoxic secondary metabolites with broad-spectrum effects that trigger necrosis in the tissues of their host plants. These necrotrophs have evolved strategies to evade and influence the defense mechanisms of their host plants. During infection, *S. sclerotiorum* and other necrotrophs have the ability to metabolize phytotoxic secondary metabolites that are produced by the plant, like phytoalexins. This metabolic process helps them reduce the impact of these compounds on their growth and survival. As an alternative, fungal cells may use transporters to export phytotoxins generated from plants. But fungal mechanisms for controlling the plant immune response might involve more than just detoxification. It is believed that the early stages of *S. sclerotiorum* infection involve a biotrophic phase during which the pathogen secretes OA in an effort to weaken the host's defenses by using it as a "compatibility factor." The traditional classification of *S. sclerotiorum* as a necrotroph is called into question by this early biotrophic phase, which implies that a more accurate word would be hemibiotroph. The shift in understanding aligns with the expanding notion of a spectrum of trophic strategies among phytopathogens, indicating that necrotrophs and biotrophs may be more similar than previously thought (Liang and Rollins, 2018).

The assembly genome of ASM185786v1 was published in 2017 for the current version of the gene code, whereas ASM14694v1 was published in 2011 for the old version. A "+" in the table denotes a change in the associated phenotypic of the mutant strain in relation to the wild-type strain, while a "–" indicates no change in phenotype. NA denotes the absence of phenotypic assessment. "SignalP-5.0" was used to identify the secretion signals. All deletions, unless specified otherwise, were made via homologous recombination. *Apium graveolens* (celery), *Arabidopsis thaliana* (*Arabidopsis*), *Brassica napus* (canola, oilseed rape, rapeseed), *Brassica rapa* subsp. *chinensis* (pak choi cabbage), *Capsicum frutescens* (hot pepper), *Cucumis sativus* (cucumber), *Fragaria ananassa* (strawberry), *Glycine max* (soybean), *Helianthus annuus* (sunflower), *Lycopersicon solanum* (tomato), *Nicotiana benthamiana* (tobacco), *Phaseolus vulgaris* (bush bean, common bean, green bean), *Pisum sativum* (pea), and *Vicia faba* (faba bean) are the Latin names of the host species used for pathogenicity testing of mutants. Columbia-0 is the *A. thaliana* ecotype that is included in the table. The host species' particular local cultivars are listed in brackets. *ROS*, reactive oxygen species; WMA, World Medical Association.

Alternatively, fungal cells may export plant-derived phytotoxins using transporters. However, the strategies employed by fungi to manage the plant immune response may go beyond simple detoxification. The initial stage of *S. sclerotiorum* infection is thought to encompass a biotrophic phase, where the pathogen releases OA as a "compatibility factor" to subdue the host's defense mechanisms. This early biotrophic phase challenges the traditional classification of *S. sclerotiorum* as a necrotroph and suggests that it may be more accurately termed as a hemibiotroph. The change in perception coincides with the growing idea of a spectrum of trophic tactics within phytopathogens, suggesting that necrotrophs might have more commonalities with biotrophs than believed earlier (Liang and Rollins, 2018).

7.16.1 Role of broad-spectrum pathogenicity genes

The available evidence indicates that necrotrophic pathogens with a wide host range may utilize a mix of general and specific components in their approach to infecting hosts. An example of a general pathogenicity factor is the protein effector SsSSVP1, which is generated by *S. sclerotiorum*. Studies have demonstrated that *S. sclerotiorum* strains with silenced mutant SsSSVP1 are notably less aggressive in infecting both *B. napus* and *A. thaliana* compared to the control, highlighting the crucial function of this effector in the infection mechanism across various plant species. OA is a metabolite that *S. sclerotiorum* uses as another pathogenicity factor. It has broad-spectrum effects as well. The lower level of virulence seen in infection tests with *S. sclerotiorum* strains lacking in OA on various hosts highlights the important role that OA plays in the infection mechanism of *S. sclerotiorum* on a variety of host species. This broad range of pathogenicity probably helps *S. sclerotiorum* infect a wide variety of host plants. Our knowledge of host-specific infection pathways in host-generalist necrotrophs is still somewhat restricted, in contrast to these broad-spectrum determinants. According to studies on gene expression, some disease-causing substances might be directed toward particular hosts by changing the expression of certain genes. However, our understanding of the exact mechanisms underpinning the interactions between these variably expressed genes is still lacking. The investigation of *B. cinerea's* differential expression of CAZyme-associated genes on a variety of plant host tissues, such as lettuce leaves, tomato fruit, and grape berries has been observed (Blanco-Ulate *et al.*, 2014). The study concludes that, in contrast to lettuce leaves, which have lower quantities of pectin, tomato fruit and grape berries, which are high in pectin, have more abundant expression of pectin-modifying enzymes. This finding raises the possibility that variations in the expression of genes linked to CAZyme may have a role in *B. cinerea's* adaptability to various hosts, hence augmenting its capacity to infect several varieties of hosts.

Differential gene expression specific to the host has been observed in *S. sclerotiorum*. The expression patterns of 16 effector candidates varied significantly across diverse hosts, including *Nicotiana benthamiana*, *S. lycospersicum*, and resistant and susceptible *A. thaliana* accessions, according to a study done in 2014. Moreover, RNA-seq has been used to look into how *S. sclerotiorum* gene expression varies over time. Allan *et al.* (2019) have noted that a thorough examination of host-specific differential expression throughout *S. sclerotiorum's* whole genome is still lacking. An RNA-sequencing analysis was carried out to investigate the differential gene expression of *S. sclerotiorum* during its infection of *B. napus* and *Lupinus angustifolius*, the principal broad-leaf crops in the wheat belt of Western Australia. This study was unique from others because it examined the complete genome as opposed to certain subsets of potential pathogenicity variables. At first, it was thought that suspected pathogenicity-related genes, especially CAZymes, would be primarily responsible for the host-specific differential gene expression in *S. sclerotiorum*. All the host-specific differentially expressed genes, however, were not made up entirely of these genes. The vast array of genes exhibiting distinct expression by *S. sclerotiorum* in planta implies that this variation in expression could be involved in adjusting to defense responses in plants (Allan *et al.*, 2019).

The chemical processes driving broad host range necrotrophy remain mostly unknown. Researchers used RNA sequencing to examine the differential expression of *S. sclerotiorum* genes after infecting two different host crops, lupin (*L. angustifolius*) and canola (*B. napus*), in order to gain a deeper understanding of this phenomenon. A total of 10,864 of the 11,130 genes found in the *S. sclerotiorum* genome were found to be expressed. It was discovered that 68 of these expressed genes were upregulated in planta as opposed to *in vitro*, suggesting a potential role in the infection process as a whole. Among the prominent participants in this category were CAZymes and secondary metabolites, both of which are expected to have significant functions. On the other hand, a smaller subset of 53 genes showed differential expression between the two host plants. Among these, only six genes were identified as potential effectors, secondary metabolites, or CAZymes specific to the host plants. The other genes belonged to different functional groups, some of which were involved in the metabolism and excretion of foreign substances. The aforementioned categories encompassed cytochrome P450s, metal-beta-lactamases, tannases, and major facilitator superfamily transporters. These findings suggest that *S. sclerotiorum* might adjust the expression of detoxification-related genes in response to phytotoxins produced by different host species. This research presents a thorough examination of the entire transcriptome of *S. sclerotiorum* throughout the infection process across various host organisms (Allan *et al.*, 2019).

7.17 Investigation of Gene Expression in Different *Brassica* Genotypes During Infection

During infection by *Sclerotinia*, the resistant and susceptible *Brassica* genotypes have shown differential expression of genes. The evaluation of gene expression changes linked to *S. sclerotiorum* infection was conducted in two distinct genotypes of *B. napus*, with one genotype showing partial resistance and the other being susceptible. In order to analyze these changes, a comprehensive genome microarray sourced from *Arabidopsis* was employed. Total 686 genes were found to have differential expression in the resistant genotype, while 1547 genes exhibited differential expression in the susceptible genotype after infection. Remarkably, the number of genes that were differently expressed grew with time, with most of them being elevated in both genotypes. The diverse putative functions of these differentially expressed genes were found to include proteins related to pathogenesis, proteins involved in the oxidative burst, protein kinases, molecule transporters, proteins involved in cell maintenance and development, as well as proteins linked to abiotic stress and unknown functions (Fig. 7.5). The patterns of gene regulation indicated that the two genotypes differed in the timing and amount of their

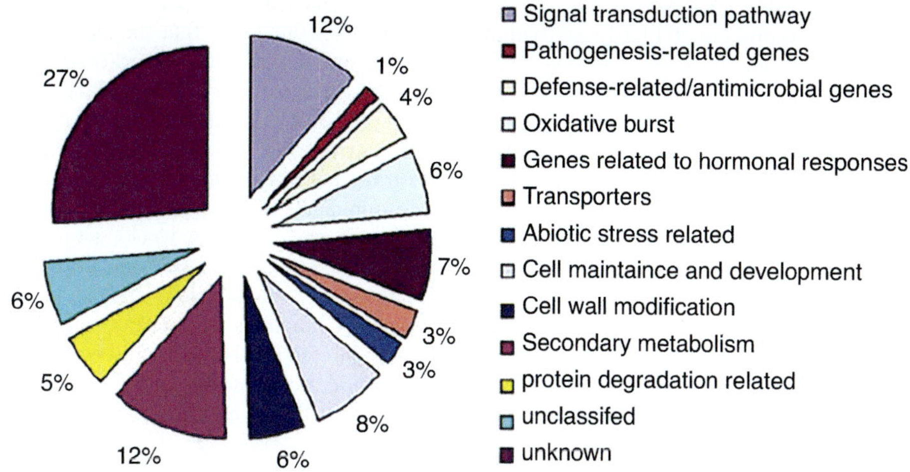

Fig. 7.5. The data in figure provides significant understanding of the particular genes implicated in RV289's response to *Sclerotinia sclerotiorum* infection at the 24-h time point. (Adapted from Zhao *et al.*, 2007, with permission.)

defense response. Genes linked to the ethylene and jasmonic acid (JA) signal transduction pathways were discovered to be activated; however, no genes affected by salicylic acid (SA) were detected. Potential defense genes were found by integrating the early response genes in the partially resistant line with known quantitative trait loci. Northern blot analyses were used to further confirm the expression levels of these genes. In summary, our results imply that a variety of proteins, including plant cell wall-associated proteins and WRKY transcription factors, may play a major part in the defense response against *S. sclerotiorum* illness (Zhao *et al.*, 2007). Many dicotyledonous plant species are known to be infected by the pathogenic fungus *S. sclerotiorum*, which causes stem rot in *B. napus*. Using a *B. napus* oligonucleotide microarray, researchers examined gene expression patterns in the partially resistant cultivar Zhong You 821 (ZY821) and the susceptible cultivar Westar to learn more about how plants fight off this pathogen. The cultivar ZY821 showed increased transcript levels at earlier stages of infection (6–12 hours postinoculation [hpi]) for a number of genes, including those encoding defense-related proteins like chitinases, glucanases, osmotins, and lectins, as well as genes encoding transcription factors from various classes like zinc finger, WRKY, APETALA2 (AP2), and MYB. It's interesting to note that the highest levels of differential gene expression were observed 48 h after inoculation in both cultivars. There was an increase in the genes that code for the enzymes involved in the synthesis of auxin, ethylene, and JA as well as the genes that break down gibberellin. Additionally, there were discernible modifications in the expression of genes related to the metabolism of carbohydrates and energy, suggesting a change in the direction of carbon stores to the TCA cycle and producing ROS. Secondary metabolites may also be involved in *B. napus* response to *S. sclerotiorum*, as suggested by the increased transcript levels from genes linked to phenylpropanoid and glucosinolate production in both cultivars (Zhao *et al.*, 2009).

7.18 Variation in the Expression of Multiple Metabolic Enzymes of *Sclerotinia* Depending on the Host Plant

The aim of the study was to examine variations in *S. sclerotiorum* gene expression after infecting two different host crops, *B. napus* and lupin (*L. angustifolius*), using RNA sequencing. The data showed that 10,864 out of 11,130 genes in the *S. sclerotiorum* genome were expressed. Sixty-eight of these expressed genes expressed more in planta than *in vitro* on at least one host, suggesting that they are involved in the wider infection process. Interestingly, CAZymes and secondary metabolites were among the increased genes. A smaller selection of 53 genes also showed different patterns of expression in the two plant hosts. It was discovered that just six of these host-specific genes could function as CAZymes, secondary metabolites, or effectors. The remaining genes belonged to various functional groupings, some of which had to do with how foreign chemicals were metabolized and eliminated. These include a number of categories, including major facilitator superfamily transporters, tannases, metal-beta-lactamases, and cytochrome P450s. The findings suggest that *S. sclerotiorum* might regulate the activation of genes associated with detoxification in response to phytotoxins produced by different host species. In this study, the researchers compared the comprehensive transcriptome analysis of *S. sclerotiorum* while infecting multiple hosts (Allan *et al.*, 2019).

7.19 Relationship Between Autophagy and Apoptosis During *Sclerotinia* Pathogenesis

In the realm of cellular biology, regulated cell death is a complex series of events that leads to the intentional demise of a cell. Autophagy and apoptosis are recognized as the primary mechanisms in multicellular organisms that execute genetically predetermined cell death. In the context of interactions between plants and microbes, pathways leading to cell death play a critical role in determining susceptibility or resistance. The necrotrophic phytopathogen *S. sclerotiorum* induces cell death in host plant tissue by influencing host pathways and utilizing OA. Characteristic apoptotic characteristics are displayed by this process, which depends on the interaction's dosage and timing. Mutants lacking in OA generate a certain kind of cell death in the host but do not cause disease. Interestingly, this cell death exhibits characteristics like callose deposition and a strong oxidative burst that are usually linked to the plant's hypersensitive response. This indicates that the plant has the capability to recognize this scenario and respond in a defensive manner. Kabbage *et al.* (2013) employed reverse genetics, chemical effectors, and electron and fluorescence microscopy to illustrate that this specific type of cellular death is autophagic. The nonpathogenic mutant phenotype was restored by blocking autophagy. These results indicate a new function for OA in autophagy suppression and show that autophagy functions as a defense mechanism in this interaction between the necrotrophic fungus and the plant. The data presented indicate that cell deaths are not uniform, and while all scenarios involve PCD, the outcome differs depending on the entity controlling the cell death process. According to the data, the way in which a particular plant–microbe interaction turns out is more important than the fact that cell death occurs.

7.20 Reduction of Host Resistance by *Sclerotinia* Secretory Protein in the First Phase of Disease

The host resistance in *Brassica* is suppressed by the expression of genes in the form of protein secretions at early stage of *Sclerotinia* infection. A signal peptide is present in the 302 amino acid residue protein SSITL (SS1G_14133) from *S. sclerotiorum*, also referred to as the *S. sclerotiorum* integrin like. Immune-fluorescence and immune-localization methods have verified its secretory characteristic. The three-dimensional structure of a fungal integrin-like protein and the human integrin a 4-subunit is remarkably similar. Additionally, SSITL belongs to the integrin alpha N-terminal domain superfamily. As *S. sclerotiorum* was injected into its host during the early infection phases (1.5–3.0 hpi), a high expression of SSITL was seen. Particularly, SSITL silencing led to a significant drop in pathogenicity. On the contrary, *Arabidopsis* exhibited a robust and rapid defense response when exposed to SSITL silenced transformant A10. The defense genes *PDF1.2* and *PR-1* showed peak expression levels at 3 hpi, which was 9 h earlier compared to plants inoculated with the WT *S. sclerotiorum* strain. Analysis of *PDF1.2* and *PR-1* expression indicated the systemic resistance induced by A10, which was also observed after inoculation with *B. cinerea*. A comparison between A10 and WT plants showed slightly larger lesions on mutants pad4 and NahG, and larger lesions on *Arabidopsis* mutants ein2 and jar1. Moreover, in *Arabidopsis*, the expression of SSITL, whether it is constant or temporary, hinders the expression of *PDF1.2*, thereby enhancing the vulnerability of the plants to *B. cinerea* and A10, the WT strain of *S. sclerotiorum*. These findings suggest that during the initial stages of infection, SSITL might function as an effector and play a vital role in suppressing the jasmonic/ethylene (JA/ET) signal pathway-mediated defense mechanism (Zhu *et al.*, 2013).

7.21 Role of Copper in Infection and Pathogenesis

In addition to acting as an important nutrient source for microbial pathogens, copper is a prominent cofactor in the scavenging of enzymes that produce ROS. In a study by Ding *et al.* (2020), the transcriptome alterations of *S. sclerotiorum* and *Brassica oleracea* during inoculation were examined concurrently. The findings demonstrated that *S. sclerotiorum* activates genes linked to copper ion absorption and transport, which facilitates ROS detoxification and increases fungal pathogenicity. However, *S. sclerotiorum* resistance and antioxidant activity are enhanced because the host's resistance mechanism preserves genes linked to copper ion homeostasis. Additionally, compared to the margin and healthy area, the necrotic area has a larger concentration of copper, suggesting that during infection, copper was transferred from the healthy area to the necrotic area. This demonstrates copper's intricate and varied role in the host–pathogen relationship.

Necrotrophic plant pathogens cause necrosis by causing the host to produce ROS. The infection can now obtain nutrients from the dead tissues because of this necrosis. More than 900 species experience SSR due to the widespread necrotrophic pathogen *S. sclerotiorum*, which causes significant financial losses. During the infection of *B. oleracea*, the upregulation of three genes in *S. sclerotiorum*, namely *SsCTR1*, *SsCCS*, and *SsATX1*, was observed. The importance and movement of copper ions are mediated by these genes. According to functional study, these genes are essential for the detoxification of ROS from fungi and their pathogenicity. On the host side, it was discovered that throughout infection, the expression of four genes linked to copper ion homeostasis—*BolCCS*, *BolCCH*, *BolMT2A*, and *BolDRT112*—was stable in resistant *B. oleracea* but downregulated in susceptible *B. oleracea*. It has been discovered that these genes improve *A. thaliana*'s resistance to *S. sclerotiorum* and boost antioxidant activity. Additionally, the copper concentration study has shown that during infection, copper moves from healthy to necrotic tissues. These important results lead to the proposal of a model (Fig. 7.6) which implies that *S. sclerotiorum* uses host copper to detoxify ROS inside its cells. None the less, by preserving copper ion homeostasis during infection, resistant hosts may restrict the availability of vital copper nutrients to *S. sclerotiorum* (Ding *et al.*, 2020).

7.22 Host Penetration and Development of Sclerotia Influenced by the *Ss-Caf1* Gene

The gene *Ss-Caf1* plays a crucial role during early phase of infection by *Sclerotinia* to destroy the host cell and absorb nutrients for the development of fungal mycelium for development of sclerotia. The necrotrophic pathogen is required to invade and eliminate its host in order to inhabit deceased cells and extract nutrients from lifeless tissues. *S. sclerotiorum* has managed to develop effective strategies to overcome the barriers posed by the plant's cuticle and cell wall, which serve as obstacles against pathogen invasion. Despite these challenges, the pathogen has found ways to penetrate and infect the plant. The cuticle and cell wall of the plant are known to offer protection against pathogen invasion; however, *S. sclerotiorum* has proven its capability to adapt and circumvent these defense mechanisms. Despite the presence of these initial barriers, the pathogen has managed to infect the plant by employing innovative strategies. Mechanistically, these include secretion of OA (Magro *et al.*, 1984; Godoy *et al.*, 1990; Cessna *et al.*, 2000; Favaron

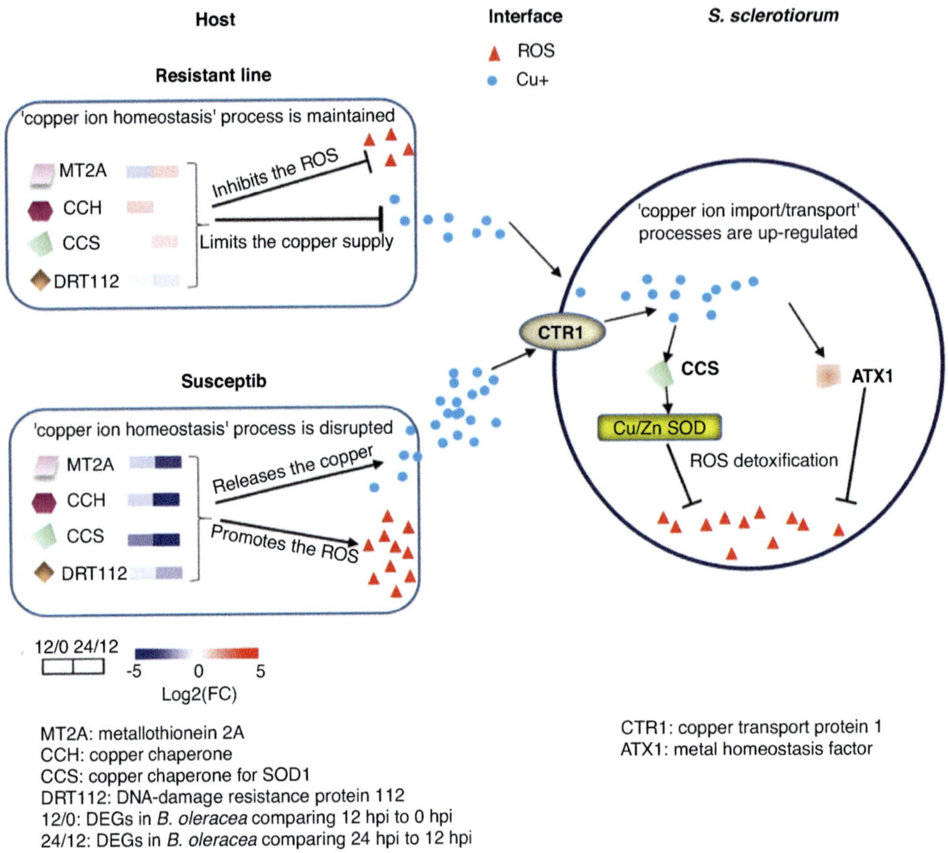

Fig. 7.6. The figure reveals gene expression levels at two distinct stages during infection are examined: 12 h postinfection against 0 hpi in the host, and 24 vs 12 hpi in the host. The goal of this gene expression analysis is to shed light on the molecular mechanisms underlying the host–Sclerotinia sclerotiorum competition for copper acquisition. This will help to understand how the host's defense mechanisms adapt over time to counter the pathogen's tactics for copper uptake and survival. (Adapted from Ding *et al.*, 2020, with permission.)

et al., 2004; Guimarães and Stotz, 2004; Kim *et al.*, 2008; Williams *et al.*, 2011; Kabbage *et al.*, 2013) and CWDE (Lumsden, 1969; Abawi *et al.*, 1975b; Riou *et al.*, 1991; Martel *et al.*, 1996; Poussereau *et al.*, 2001a; Zuppini *et al.*, 2005; Yajima *et al.*, 2009). Moreover, a multitude of other elements involved in the development and pathogenic properties of *S. sclerotiorum* have been identified, with many of them thoroughly examined on a molecular scale (Rollins, 2003; Jurick *et al.*, 2004; Harel *et al.*, 2006; Erental *et al.*, 2007; Jurick and Rollins, 2007; Kim *et al.*, 2011; Li *et al.*, 2012; Liberti *et al.*, 2013; Xu and Chen, 2013).

Zhu *et al.* (2013) discovered that an integrin-like protein in *S. sclerotiorum* is essential during the initial infection stages, acting as an effector to suppress the host's defense mechanisms regulated by the jasmonic/ethylene (JA/ET) signal pathway. Fungal phytopathogens, such as *S. sclerotiorum*, employ a unique strategy known as appressoria formation to invade their hosts. Detailed ultrastructural and cytochemical examinations revealed that the tips of *S. sclerotiorum* hyphae undergo significant swelling and branching before entering the host. These specialized multicellular structures, termed compound appressoria or infection cushions, are characterized by their high melanin content (Lumsden and Dow, 1973; Tariq and Jeffries, 1984; Jamaux *et al.*, 1995; Huang *et al.*, 2008). In addition, the presence of compound appressoria can be observed when *S. sclerotiorum* is grown on inanimate surfaces like aluminum foil, glass cover slips, or sides of plastic Petri dishes (Tariq and Jeffries, 1984). The sticky mucilage that covers these compound appressoria may contribute to the adhesion of *S. sclerotiorum* to the plant surface, aiding in its successful penetration into the host tissues (Tariq and Jeffries, 1984).

Furthermore, the plant's protective cuticle may be pierced by thin penetration pegs originating from flattened hyphae of the compound appressorium, resulting in visible infection holes on the host's surface once the compound appressorium is removed (Huang *et al.*, 2008). Additionally, compound appressoria are believed to exert a mechanical force and are thought to secrete enzymes and toxins (Jamaux *et al.*, 1995; Huang *et al.*, 2008). However, the exact molecular mechanism behind the formation of compound appressoria and their specific functions in *S. sclerotiorum* remains largely unknown.

A mutant library was generated by inserting T-DNA using the *Agrobacterium tumefaciens*-mediated transformation technique. The library was derived from the WT strain Sunf-M of *S. sclerotiorum*. All of the T-DNA insertional mutants were assessed for their ability to cause disease in *A. thaliana*, and mutants that showed a loss of pathogenicity were identified by Xiao *et al.* (2014). Among these mutants, one particularly intriguing mutant, Sunf-MT6, exhibited defects in both pathogenicity and sclerotial development. The mutant Sunf-MT6, which was obtained from the T-DNA insertional library, was found to have a disrupted gene called *S. sclerotiorum* compound appressorium formation related gene 1 (*Ss-caf1*). This gene disruption was identified as the cause of the mutant phenotype. Despite producing higher levels of OA compared to the WT strain, Sunf-MT6 was unable to form compound appressorium and cause lesions on rapeseed leaves. However, it could still penetrate host tissues through wounds and result in necrotic lesions. Additionally, Sunf-MT6 exhibited larger sclerotia at lower concentrations compared to the WT strain Sunf-M. These phenotypic changes were attributed to the disruption of the *Ss-caf1* gene caused by T-DNA insertion in Sunf-MT6. Through gene complementation and knockdown assays, it was determined that Ss-caf1 encodes a secretory protein containing a potential Ca^{2+} binding EF-hand domain. The expression of Ss-caf1 was found to be increased in immature sclerotia and during the initial stages of compound appressorium development. Interestingly, when *Ss-caf1* was produced without signal peptides, it induced cell death in *N. benthamiana*. These findings indicate that *Ss-caf1* plays a crucial role in the formation of sclerotial appressorium and compound appressorium in *S. sclerotiorum*. Additionally, it is possible that *Ss-Caf1* interacts with specific host proteins or unknown compounds within host cells, leading to the demise of the host cell (Xiao *et al.*, 2014).

7.23 SsERP1 Effector Protein Hinders Plant Ethylene Signaling to Promote Infection and Pathogenesis

Plant–microbe interactions heavily rely on the involvement of effectors, which are essential for the successful infection or transmission of various plant pathogens. The intricate workings of these effectors have been extensively researched in both biotrophic and host-specific necrotrophic plant pathogens (Giraldo and Valent, 2013; Lo Presti *et al.*, 2015; Toruno *et al.*, 2016). The effectors have implemented a range of tactics to evade plant immune defenses or influence plant–microbe–vector interactions. Nevertheless, the roles of effectors in necrotrophic plant pathogens with a wide host range remain largely unclear when contrasted with those in biotrophic pathogens that are specific to certain hosts (Giraldo and Valent, 2013; Toruno *et al.*, 2016). Growing amounts of evidence suggest that *S. sclerotiorum* and related necrotrophic plant pathogens with broad host ranges use protein or RNA effectors to suppress the host immune response in addition to releasing toxins like OA and enzymes that break down cell walls in order to successfully infect the plant (Wang *et al.*, 2014; Wang *et al.*, 2016; Cai *et al.*, 2018; Xu *et al.*, 2018). The *S. sclerotiorum* genome sequencing data offers unique prospects for a detailed and in-depth analysis of its effectors (Guyon *et al.*, 2014; Derbyshire *et al.*, 2017). The genome of *S. sclerotiorum* contains over 100 potential effector-encoding genes, according to bioinformatic study (Guyon *et al.*, 2014; Heard *et al.*, 2015; Derbyshire *et al.*, 2017). However, the majority of these suspected effectors aside from a small number have not yet undergone experimental characterization (Xu *et al.*, 2018; Seifbarghi *et al.*, 2020).

7.23.1 Effector gene cloning

The cloned effector gene *SsERP1* inhibits *Brassica* ethylene signaling pathways, hence promoting *S. sclerotiorum* infection. Fan *et al.* (2021) reported the finding of a new effector protein in *S. sclerotiorum* named SsERP1 (ethylene pathway repressor protein 1). It has been established by a number of studies, including ectopic overexpression, knockout mutant analysis, and spray-induced RNA silencing, that SsERP1 is crucial to *S. sclerotiorum's* pathogenicity. According to transcriptome data, SsERP1 preferentially blocks plant ethylene signaling to intensify *S. sclerotiorum* infections. The *S. sclerotiorum* genome contains a significant number of potential effectors, but nothing is known about them or how they work. The new effector SsERP1 was successfully cloned and described by Fan *et al.* (2021), providing insight into its function in *S. sclerotiorum*. Early in an *S. sclerotiorum* infection, there is an extensive production of the secretory protein SsERP1. Ectopic overexpression of SsERP1 in plant leaves promoted the infection process, whereas SsERP1 knockout mutants showed decreased pathogenicity without influencing mycelial growth and sclerotium formation. This suggests that SsERP1 is essential to the pathophysiology of *S. sclerotiorum*. According to transcriptome research, SsERP1 blocks the plant's ethylene signaling pathway, which in turn encourages infection. Furthermore, research by Fan *et al.* (2021) showed that inhibiting SsERP1 with *in vitro* synthesized double-strand RNAs successfully prevented *S. sclerotiorum* infection, confirming the functional significance of SsERP1 in the pathogenesis of *S. sclerotiorum* and pointing to a possible treatment approach for SSR disease (Fan *et al.*, 2021).

7.24 Effector Gene Characterization During Infection and Pathogenesis

Gene clustering was performed in order to accelerate the analysis of genes expressed during *S. sclerotiorum* infection. To help us better understand the infection process, these genes were grouped based on their functional annotation and predicted role in pathogenicity or virulence. Many enzymes have been linked to the infection and development of different *Brassica* species' pathogenic processes by *Sclerotinia*. These include hydrolytic, cutinase, lipase, well wall-destroying protease, synthesizing secondary metabolites, detoxicating, and OA-generating expressed genes.

7.24.1 Hydrolytic enzymes

The analysis shown that the gene's sizable fraction encodes hydrolytically active enzymes that are activated during infection (Table 7.2). The majority of these genes are responsible for encoding enzymes that exhibit activity on carbohydrates, commonly known as CAZymes. As expected in the *S. sclerotiorum* genome by Amselem *et al.* (2011), the majority of CAZyme genes demonstrated increased expression levels after infection (Lyu *et al.*, 2015; Seifbarghi *et al.*, 2017), influencing different stages of *S. sclerotiorum* growth. These findings consistently highlight the critical role these enzymes play in infection. The majority of the genes encoding CAZymes belonged to the glycoside hydrolase (GH) and carbohydrate esterase (CE) families. The pathogen's capacity to infect a broad range of plant hosts is indicated by the emergence of many genes that produce the GH and CE enzymes (Orshinsky *et al.*, 2012).

7.24.2 Lipases and cutinases

The main defense against pathogen invasion is the plant cuticle, which is made up of a mixture of C:16 and C:18 fatty acids as well as their derivatives that produce surface waxes and cutin (Yeats and Rose, 2013). A gene in the CE family showed increased expression during the early

Table 7.2. An explanation of the upregulated genes that encode hydrolytic enzymes and their expression. (Adapted from Seifbarghi *et al*., 2017, with permission.)

Gene ID	Description[a]	Expression level (hpi)[b]					
		1	3	6	12	24	48
1. Lipid Degradation							
SS1G_07661	Cutinase (*SsCuta*)	3	3.9	4.3	4.7	3.4	-
SS1G_13386	Cutinase	-	-	-	-	14.9	29.6
SS1G_12907	Cutinase	-	-	-	-	29.7	236
SS1G_09557	Lipase/esterase	3.4	5.5	6.5	6.1	2.6	-
SS1G_11473	Lipase/esterase	2.2	-	-	-	-	3.2
SS1G_05990	Lipase/esterase	-	-	-	-	2.2	-
SS1G_00767	Lipase/esterase	-	-	-	-	-	34.4
SS1G_03597	Lipase/esterase	-	-	-	-	-	5.1
SS1G_01849	Lipase/esterase	-	-	-	-	-	2.7
SS1G_08869	Lipase class 3	-	-	-	-	-	2.8
SS1G_14146	Extracellular lipase	-	-	-	-	31.5	38.7
SS1G_04490	Extracellular lipase	-	-	-	-	2.5	3
SS1G_00877	Extracellular lipase	-	-	-	-	-	4.3
SS1G_14441	Triacylglycerol lipase	-	-	-	-	5	6.9
SS1G_13982	Triacylglycerol lipase	-	-	-	-	4.8	27.5
SS1G_01472	Triacylglycerol lipase	-	-	-	-	3.6	13.1
SS1G_03007	Alpha/beta-hydrolase (esterase)	-	2.6	2.6	-	-	-
SS1G_09718	Alpha/beta-hydrolase (esterase/lipase)	-	2.1	2.9	-	-	-
SS1G_01703	Alpha/beta-hydrolase (esterase/lipase)	-	3.9	3.5	2.2	3.8	6.5
SS1G_13361	Alpha/beta-hydrolase (esterase/lipase)	-	-	-	2.3	3.2	2.2
SS1G_11402	Alpha/beta-hydrolase (esterase/lipase)	-	-	-	3.6	4.8	3.7
SS1G_13263	Alpha/beta-hydrolase (esterase/lipase)	-	-	-	-	3.8	3.8
SS1G_08133	Alpha/beta-hydrolase (esterase/lipase)	-	-	-	-	-	2.5
SS1G_02163	Alpha/beta-hydrolase (esterase/lipase)	-	-	-	-	-	12.5
SS1G_01953	GDSL lipase acylhydrolase family protein	3.6	4	5	4.9	4.8	5.6
SS1G_13560	GDSL lipase acylhydrolase family protein	-	-	-	-	2.7	2.7
SS1G_06389	GDSL lipase acylhydrolase family protein	-	-	-	-	-	2.7
SS1G_02708	Cellulose-binding GDSL lipase	-	-	-	-	3.3	7.8
SS1G_14289	Cellulose-binding GDSL lipase	-	-	-	-	5.3	6.6
SS1G_03610	Cellulose-binding GDSL lipase	-	-	-	-	5.1	22.4
SS1G_04592	Cellulose-binding GDSL lipase	-	-	-	-	11.8	9.4
SS1G_11930	Carboxylesterase/lipase (cholinesterase)	4	3.9	8.9	5.2	-	-

Continued

Table 7.2. Continued

Gene ID	Description[a]	Expression level (hpi)[b]					
		1	3	6	12	24	48
SS1G_00376	Carboxylesterase/lipase (cholinesterase)	-	-	-	-	4.4	9.2
SS1G_09613	Carboxylesterase/lipase (cholinesterase)	-	-	-	-	-	4.7
SS1G_04422	Carboxylesterase/lipase (cholinesterase)	-	-	-	-	-	3.2
SS1G_11853	Carboxylesterase/lipase (cholinesterase)	-	-	-	-	-	3
SS1G_04030	Lysophospholipase	-	-	3.6	-	2.3	-
SS1G_10482	Lysophospholipase	-	-	-	-	-	4.1
2. Polysaccharide degradation							
2.1. Cellulose							
SS1G_08493	Beta-1,4-endo-glucanase	2.6	-	-	-	-	-
SS1G_00891	Beta-1,4-endo-glucanase (cellulase)	-	4.5	5.8	3.5	-	3.8
SS1G_01485	Beta-glucanase	-	-	-	-	-	2.1
SS1G_09365	Endo-glucanase	-	-	-	-	2.7	-
SS1G_03387	Endo-glucanase	-	-	-	-	5.3	41.4
SS1G_08837	Endo-glucanase	-	-	-	-	10.3	36.4
SS1G_04945	Endo-glucanase	-	-	-	-	-	2.3
SS1G_01828	Endo-glucanase	-	-	-	-	-	2.6
SS1G_09821	Endo-glucanase	-	-	-	-	-	144
SS1G_03041	Endo-glucanase	-	-	-	-	-	292
SS1G_00321	Endo-glucanase	-	-	-	-	-	4.2
SS1G_00471	Endo-glucanase	-	-	-	-	-	2.1
SS1G_06037	Exo-glucanase	-	3	3.2	-	9.9	12.3
SS1G_02334	Exo-glucanase (cellobiohydrolyase)	-	-	-	-	3.8	-
SS1G_09020	Exo-glucanase	-	-	-	-	15.1	79
SS1G_00892	Exo-glucanase	-	-	-	-	5.4	22.2
SS1G_09118	Exo-glucanase	-	-	-	-	-	2.5
SS1G_02245	Exo-glucanase	-	-	-	-	-	3.3
SS1G_13872	Exo-glucanase	-	-	-	-	-	7.6
SS1G_02501	Concanavalin A-like lectin/glucanase	5.1	8.7	6.1	3.1	-	-
SS1G_08907	Glycoside hydrolase (beta-glucanase)	2.6	-	-	-	-	-
SS1G_07863	Cellobiose dehydrogenase	3.7	-	-	-	-	22.1
SS1G_05151	Cellobiose dehydrogenase	-	-	-	-	5.9	6.6
SS1G_05118	Beta-glucosidase	2.4	-	-	-	-	-
SS1G_01662	Beta-glucosidase	-	-	-	-	2.8	9.5
SS1G_09366	Beta-glucosidase	-	-	-	-	2.8	8

Continued

Table 7.2. Continued

Gene ID	Description[a]	Expression level (hpi)[b]					
		1	3	6	12	24	48
SS1G_06304	Beta-glucosidase	-	-	-	-	2.6	6.8
SS1G_07847	Beta-glucosidase	-	-	-	-	3.1	4.7
SS1G_07146	Beta-glucosidase	-	-	-	-	7.4	18.7
SS1G_13255	Beta-glucosidase	-	-	-	-	40.4	54.8
SS1G_05368	Beta-glucosidase	-	-	-	-	3.2	4.6
SS1G_12622	Beta-glucosidase	-	-	-	-	-	2.1
SS1G_07162	Beta-1,4-glucosidase	-	-	-	-	-	2.3
SS1G_09129	Beta-glucosidase	-	-	-	-	-	13.1
SS1G_04264	Beta-glucosidase	-	-	-	-	-	2.2
SS1G_01021	Beta-glucosidase	-	-	-	-	-	5.2
2.2. Pectin							
SS1G_10167	Endo-polygalacturonase (SSPG1)	-	-	-	-	3.7	64.6
SS1G_10698	Endo-polygalacturonase (SSPG3)	-	-	-	-	-	5.5
SS1G_05832	Exo-polygalacturonase	4.1	6.5	-	-	4.2	32.4
SS1G_02553	Exo-polygalacturonase (exoPG2)	-	-	-	-	-	22.6
SS1G_04207	Exo-polygalacturonase (exoPG1)	-	-	-	-	14.6	168
SS1G_12057	Exo-polygalacturonase	-	-	-	-	-	4.2
SS1G_03540	Pectin lyase	-	-			4.5	3.3
SS1G_10071	Pectin lyase	-	-	-	-	-	6.2
SS1G_14449	Pectin lyase	-	-	-	-	-	2.4
SS1G_04551	Pectin methylesterase	-	-	-	-	11.6	35.8
SS1G_00332	Pectin methylesterase	-	-	-	-	33.9	178
SS1G_03286	Pectin methylesterase	-	-	-	-	4.8	16.5
SS1G_00468	Pectin methylesterase	-	-	-	-	-	21.4
SS1G_04095	Rhamnogalacturananacetylhydrolase	2.8	6.2	7.7	5.3	2.7	-
SS1G_12048	Rhamnogalacturananacetylhydrolase	-	-	-	-	-	33.5
SS1G_11992	Rhamnogalacturonan acetylesterase	-	-	-	-	-	37.5
SS1G_09857	Rhamnogalacturonyl hydrolase	-	-	-	-	-	8.6
SS1G_12964	Alpha-l-rhamnosidase	-	-	17.5	-	-	-
SS1G_13501	Alpha-l-rhamnosidase	-	-	-	-	3.1	21.2
SS1G_04541	Alpha-l-rhamnosidase	-	-	-	-	2	8.3
SS1G_08229	Rhamnogalacturonase	2.1	4.5	4.5	2.6	-	-
SS1G_07039	Rhamnogalacturonase	-	-	-	-	-	6
SS1G_04552	Endo-xylogalacturonan hydrolase	-	-	-	-	7.1	15.9
2.3. Hemicellulose							
SS1G_12191	Endo-1,4-beta-xylanase	-	5.3	7.9	4.6	6.4	25
SS1G_10092	Endo-beta-xylanase	-	-	-	-	-	155

Continued

Table 7.2. Continued

Gene ID	Description[a]	Expression level (hpi)[b]					
		1	3	6	12	24	48
SS1G_03618	Endo-beta-xylanase	-	-	-	-	-	451
SS1G_07749	Endo-beta-xylanase	-	-	-	-	-	56.7
SS1G_05140	Xylanase	3.6	3.5	4.2	4.3	3.4	8.1
SS1G_08104	Acetylxylan esterase	23	14.2	58.3	42.1	26	107
SS1G_05434	Acetylxylan esterase	-	-	-	-	4	11.8
SS1G_00746	Beta-mannosidase	-	-	-	-	6.6	42.3
SS1G_05977	Beta-mannosidase	-	-	-	-	3.6	6.4
SS1G_08208	Endo-1,4-beta-mannosidase	-	-	-	-	3.6	8.9
SS1G_08118	Alpha-xylosidase	-	-	-	-	4.5	3.8
SS1G_09367	Alpha-xylosidase	-	-	-	-	2.2	-
SS1G_11535	Alpha-fucosidase	-	-	-	-	-	2.5
SS1G_04662	Alpha-galactosidase	-	-	-	-	4.5	3.9
SS1G_03386	Alpha-galactosidase	-	-	-	-	-	3.7
SS1G_07904	Feruloyl esterase	-	-	-	-	3.6	12
SS1G_02462	Alpha-l-arabinofuranosidase	-	-	-	-	5.3	21
SS1G_03602	Alpha-l-arabinofuranosidase	-	-	-	-	6.3	17.6
2.4. Arabinogalactans							
SS1G_01216	Arabinogalactan endo-beta-galactosidase	4.1	9.6	11.9	7.2	4.1	-
SS1G_11585	Arabinogalactan endo-beta-galactosidase	-	-	-	-	-	19.3
SS1G_02618	Galactan 1,3-beta-galactosidase	-	-	-	-	5.7	11
SS1G_10842	Beta-galactosidase	4.1	2.5	2.4	-	2.2	5.8
SS1G_11763	Beta-galactosidase	-	-	-	-	4.7	-
SS1G_01572	Beta-galactosidase	-	-	-	-	6.8	18.3
SS1G_03647	Beta-galactosidase	-	-	-	-	3.5	6.9
SS1G_02781	Beta-galactosidase	-	-	-	-	-	2.7
SS1G_09866	1,6-beta-galactanase	-	4.8	4.5	3.1	3.0	12.4
SS1G_11922	Arabinan endo-1,5-alpha-l-arabinosidase	-	-	-	-	144.3	245
SS1G_01238	Beta-d-glucuronidase	7.6	-	4.7	-	-	-
SS1G_02620	Beta-glucuronidase	-	-	-	-	3.5	8.2
2.5. Lignin							
SS1G_04196	Dihydrogeodin oxidase/laccase	-	3.1	2.6	4.6	-	-
SS1G_06365	Dihydrogeodin oxidase/laccase	-	3.1	5.7	4.1	-	3.4
SS1G_05112	Dihydrogeodin oxidase/laccase	-	-	-	-	-	6
2.6. Starch							
SS1G_01776	Alpha-amylase	-	-	2.1	-	-	-

Continued

Table 7.2. Continued

Gene ID	Description[a]	Expression level (hpi)[b]					
		1	3	6	12	24	48
SS1G_11100	Alpha-amylase	-	-	-	-	6.6	5.2
SS1G_01083	Alpha-glucosidase	2.4	4.7	3	-	4.6	5.3
SS1G_01005	Alpha-glucosidase	-	-	-	-	9.2	7.2
2.7. Mannans							
SS1G_10867	Endo-1,6-alpha-mannosidase	-	-	-	-	6.4	-
SS1G_04468	Endo-1,6-alpha-mannosidase	-	-	-	-	7.5	25.3
SS1G_12937	Endo-1,6-alpha-mannosidase	-	-	-	-	-	10.4
SS1G_05110	Endo-1,6-alpha-mannosidase	-	-	-	-	-	3.5
SS1G_11579	Endo-1,6-alpha-mannosidase	-	-	-	-	-	3.5
SS1G_09229	Alpha 1,2 mannosidase	-	-	-	-	55	40.2
SS1G_00505	Alpha-1,2-mannosidase	-	-	-	-	9.8	4.7
SS1G_04148	Alpha-mannosidase	-	-	-	-	4.1	4
SS1G_02022	Alpha-mannosidase	-	-	-	-	12.5	10.9
SS1G_04200	Alpha-mannosidase	-	-	-	-	6.0	56.3
SS1G_01334	Alpha-mannosidase	-	-	-	-	-	11.5
2.8. Callose							
SS1G_01422	1,3 (4)-beta-d-glucanase	2.2	-		-	-	-
SS1G_10048	1,3 (4)-beta-d-glucanase	-	2.1	-	-	2.1	-
3. Protein degradation							
3.1. Proteases							
SS1G_10992	Caspase domain-containing protease	2.1	-	-	-	-	7.1
SS1G_00862	Cysteine protease (calpain family)	6.1	5.1	6.3	3.4	3.2	-
SS1G_09978	Peptidase (family 41 protein)	3.2	-	3.2	2.9	-	6.4
SS1G_07836	(acid) Nonaspartyl protease (ACP1)	-	-	-	-	-	40.4
SS1G_05329	Aspartyl protease	12.2	-	-	-	-	4.6
SS1G_02870	Aspartyl protease	2	2.6	3	3	2.7	3.3
SS1G_03181	Aspartyl protease	-	-	-	-	3.9	16.6
SS1G_06534	Serine protease (trypsin-like)	2.8	3	-	-	-	3.9
SS1G_12419	Serine protease (subtilisin-like)	-	-	3	-	-	-
SS1G_07655	Serine protease (subtilisin-like)	-	-	-	-	8.5	19.8
SS1G_02423	Serine protease (subtilisin-like)	-	-	-	-	-	2.2
SS1G_03282	Serine protease (subtilisin-like)	-	-	-	-	-	2.2
SS1G_07168	Serine protease (subtilisin-like)	-	-	-	-	-	2.5
SS1G_12210	Serine protease (subtilisin-like)	-	-	-	-	-	7.3
SS1G_05348	Metalloprotease	-	-	-	-	-	2.6
SS1G_05349	Metalloprotease	-	-	-	-	-	3.1
3.2. Peptidases							

Continued

Table 7.2. Continued

Gene ID	Description[a]	Expression level (hpi)[b]					
		1	3	6	12	24	48
SS1G_04565	Cytosolic nonspecific dipeptidase	-	-	-	-	26	9.2
SS1G_10529	Cytosolic nonspecific dipeptidase	-	-	-	-	-	2.2
SS1G_04140	Dipeptidyl-peptidase	2.5	-	-	-	-	-
SS1G_03087	Membrane dipeptidase	-	-	-	-	-	2.6
SS1G_03392	Proline dipeptidase	-	-	4.5	-	-	-
SS1G_08920	Proline dipeptidase	-	-	2.4	-	-	2
SS1G_04958	Tripeptidyl-peptidase	-	-	5	-	-	22.9
SS1G_07268	Tripeptidyl-peptidase	-	-	-	-	3.5	2.1
SS1G_13922	Tripeptidyl-peptidase	-	-	-	-	8.9	50.4
SS1G_09225	Tripeptidyl-peptidase	-	-	-	-	6.9	16.9
SS1G_09268	Tripeptidyl-peptidase	-	-	-	-	5.9	3.4
SS1G_02857	Tripeptidyl-peptidase	-	-	-	-	6.4	2.5
SS1G_03518	Tripeptidyl-peptidase	-	-	-	-	-	6.1
SS1G_01236	Tripeptidyl-peptidase	-	-	-	-	-	9.7
SS1G_08558	Prolyl aminopeptidase	2.5	4.3	4.4	3.4	2.4	2.2
SS1G_00617	Prolyl aminopeptidase	-	-	-	9.5	4.9	5.2
SS1G_12775	Prolyl aminopeptidase	-	-	-	-	-	3
SS1G_05449	Carboxypeptidase	-	2.5	-	-	3.3	13.3
SS1G_12413	Carboxypeptidase	-	-	-	-	3.5	4.2
SS1G_08855	Carboxypeptidase	-	-	-	-	3.7	4.1
SS1G_03361	Carboxypeptidase	-	-	-	-	4.2	21.1
SS1G_09475	Carboxypeptidase	-	-	-	-	-	13.1
SS1G_13633	Carboxypeptidase	-	-	-	-	-	7.6
SS1G_12499	Carboxypeptidase	-	-	-	-	-	11.5
SS1G_04819	Carboxypeptidase	-	-	-	-	-	5.5
4. Other hydrolytic enzymes							
SS1G_01113	Metallo-dependent amidohydrolase	-	-	-	-	3.2	-
SS1G_09143	Metallo-dependent amido-hydrolase	-	-	-	-	-	6.1
SS1G_02141	Alpha/beta-hydrolase	4.6	-	-	-	-	-
SS1G_11096	Alpha/beta-hydrolase	-	-	-	2.1	5.7	2.6
SS1G_08093	Alpha/beta-hydrolase (epoxide hydrolase)	-	-	-	-	4.4	9.2
SS1G_04475	Endo-alpha-1,4-poly-galactosaminidase	3	-	-	-	-	3.8
SS1G_11842	Sialidase	3.1	-	3.6	-	-	9.8
SS1G_01389	Polysaccharide lyase family 7 protein	-	4.9	4.9	5.2	-	-
SS1G_01493	Glycoside hydrolase family 3 protein	-	-	-	-	4.2	17

Continued

Table 7.2. Continued

Gene ID	Descriptionᵃ	Expression level (hpi)ᵇ					
		1	3	6	12	24	48
SS1G_09000	Glycoside hydrolase family 5 protein	2.7	-	-	-	2.7	-
SS1G_02369	Glycoside hydrolase family 12 protein	-	-	-	-	-	3.4
SS1G_04497	Glycoside hydrolase family 16 protein	-	-	-	-	-	2
SS1G_09789	Glycoside hydrolase family 16 protein	-	-	-	-	-	2.8
SS1G_06426	Glycoside hydrolase family 43 protein	-	-	-	-	2.4	5.9
SS1G_07515	Glycoside hydrolase family 43 protein	-	-	-	-	138	518
SS1G_07656	Glycoside hydrolase family 61 protein	-	-	-	-	-	93.6
SS1G_09251	Glycoside hydrolase family 61 protein	-	-	-	-	-	9.1
SS1G_12106	Glycoside hydrolase family 76 protein	-	-	-	-	3.0	13
SS1G_12083	Glycoside hydrolase family 115 protein	-	-	-	-	7.8	17
SS1G_04152	Glycoside hydrolase family 125 protein	-	2.2	-	-	6.0	12.5
SS1G_12917	Glycoside hydrolase family 128 protein	-	-	-	-	2.3	17.3

–(–) No significant change in expression. GDSL, Glycine, Asparagine, Serine, Leucine.
[a]Annotation based on Basic Local Alignment Search Tool (BLAST) results and the existence of conserved Protein Families Database (PFAM) domains.
[b]Fold change relative to 0 hpi.

stages of infection, between 1 and 24 h after the initial infection. More specifically, there was increased expression of the *SsCuta* gene (*SS1G 07661*), which codes for the cutinase. The fact that *SsCuta* was activated shortly after coming into contact with the leaf surface lends credence to its predicted role in breaking down the plant cuticle. This suggests that *SsCuta* is activated upon mycelia's interface with stiff sides (Bashi *et al.*, 2012). Considering that host penetration normally occurs within 24 h postinfection, it is expected that the expression of this gene will decrease after that time. In addition to *SsCuta*, it was discovered that three additional lipid degradation group genes (*SS1G_09557*, *SS1G_01953*, and *SS1G_11930*) were also elevated during the early phases of infection (Table 7.2). Since the expression patterns of these genes resemble those of *SsCuta*, it is likely that the lipolytic enzymatic activity of these genes also plays a role in host penetration. Research has demonstrated that lipase contributes to the pathogenicity of *B. cinerea* (Lip1), a fungal phytopathogen (Reis *et al.*, 2005). Another virulence factor that promotes the infection of cereals is the lipase released by *Fusarium graminearum*, which is encoded by FgL1. (Voigt *et al.*, 2005). Both 24 and 48 h after infection, there were notable increases in the expression of genes that produce more lipases and members of the CE family, such as cutinases (SS1G_13386 and SS1G_12907) and extracellular lipase (SS1G_14146) (Table 7.2).

7.24.3 Plant-derived enzymes that degrade cell wall

After penetrating the cuticle, the infection must establish itself inside the host and spread throughout the host tissues. Enzyme production facilitates this process by providing nutrients to fuel the infection process and physically dissolving components of plant cell walls (Zhao *et al.*, 2013). The primary constituents of the main plant cell wall are cellulose, hemicellulose, and pectin. Additionally, phenolic esters such as ferulic and coumaric acid as well as structural glycoproteins such as hydroxyproline-rich extensins are also present. On the other hand, lignin, a highly interconnected phenolic macromolecule, makes up the majority of the secondary cell wall. The enzymes known as polygalacturonases (PGs), which break down pectin in cell walls, are members of the GH28 subfamily. Prior research has demonstrated that when *B. napus* is infected with *S. sclerotiorum*, four endo-PG genes (*SsPG1*, *SsPG3*, *SsPG5*, and *SsPG6*) and two exo-PG genes (*SsXPG1* and *SsXPG2*) are expressed (Li *et al.*, 2004). Between the time points of 24–48 hpi and at 48 hpi, it was shown that the *SsPG1* (SS1G_10167) and *SsPG3* (SS1G_10698) genes were upregulated, with expression levels ranging from 3.7 to 64.6 times higher than the original inoculum. Similarly, between 24 and 48 h postinfection and 48 h postifection, two exo-PG genes, *SsXPG1* (SS1G_04207) and *SsXPG2* (SS1G_02553), demonstrated an increase in expression. In comparison to the initial inoculum, these genes were found to express at levels 14.6–168.3 and 22.6 times higher, respectively. The homopolymer consisting of α-(1$\rightarrow$4) bonds connecting D-galacturonic acid residues is the primary structural component of the pectin backbone. This strength is altered, therefore, by side chains that can be both branched and unbranched and contain different kinds of sugar. Together with the endo-PG genes *SsPG1* and *SsPG3*, the majority of the genes encoding these additional pectin-degrading enzymes demonstrated high expression levels at 24 and 48 h postinfection. Notably, SS1G_05832 (exo-PG), SS1G_04095 (rhamno-galacturananacetyl-hydrolase), and SS1G_08229 (rhamno-galacturonase) all displayed upregulation at 1–3 hpi, 1–24 hpi, and 1–12 hpi, respectively. In contact with hard surfaces, *SsPG1* expression might be increased (Li *et al.*, 2004). Conversely, interaction with *B. napus* leaves considerably raises *SsPG1*, but not *SsPG2* (Bashi *et al.*, 2012). Furthermore, it was discovered that the expression of *SsPG1* is limited to the lesion's front edge. According to the researchers' hypothesis, galacturonic acid activates *SsPG1* via regulating and being connected with carbon deficiency, suggesting that it may be involved in lesion development as well as the early stages of infection. Additionally, whereas *SsPG6* exhibits a twofold expression pattern with peaks at 6 and 48 hpi, *SsPG3* exhibits elevated expression at 12 hpi. *SsPG3* and *SsPG6* are both potent inducers of light-induced necrotic reactions, according to Bashi *et al.* (2013). In contrast, Kars *et al.* (2005) discovered that *B. cinerea BcPG1* and *BcPG2* have a considerable capacity to induce necrosis, and that the pathogenicity of *B. cinerea* is decreased in the absence of both genes (ten Have *et al.*, 1998; Kars *et al.*, 2005). The enzymes' capacity to induce necrosis in tissue and their ability to postpone activating *SsPG* or related genes after host penetration imply that their primary function is to facilitate lesion enlargement and the spread of the pathogen throughout the host tissues.

This work has demonstrated the overexpression of numerous members of the GH (Glycoside Hydrolase) and CE (Carbohydrate Esterase) families, which suggests that they play a role in the breakdown of cellulose and hemicellulose. Cellulose, a homopolymer composed of beta-(1, 4)-linked D-glucose, is broken down successively into glucose by enzymes known as cellulases, cellobiosidases, and beta-glucosidases. During the infection, especially in its later stages, the expression of many genes that could encode cellulases was increased. Interestingly, at 48 h postinfection, there was a notable upregulation of *SS1G_09821* and *SS1G_03041*, with fold increases of 144 and 292, respectively (Table 7.2). Genes encoding putative exo-glucanases and beta-glucosidases also showed similar tendencies of elevated expression, especially in the later stages of the infection. Hemicellulose, which comprises a range of sugars such as xylose, mannose, galactose, rhamnose, and arabinose, is more complex than cellulose and accounts for around 20% of the biomass in higher plants.

This means that the breakdown of hemicellulose requires a more intricate set of enzymatic procedures. The genes responsible for breaking down hemicellulose were also active later in the infection, in contrast to the genes responsible for breaking down cellulose. There were a few outliers, though. In actuality, the genes responsible for releasing xylose from xylan were activated early in the infection. *SS1G_12191*, *SS1G_05140*, and *SS1G_08104* are among the first genes that were triggered. Furthermore, in this study, the two additional genes that encode xylanases, *SS1G_10092* and *SS1G_03618*, showed the highest levels of activation. After 48 h of infection, it was discovered that their expression levels were 155 and 451 times higher than those of the original inoculum. These results suggest that plant cell walls contain a significant amount of xylose, which may be essential for *S. sclerotiorum's* feeding. Prior to the formation of obvious necrotic signs on the stem, the gene *SS1G_01493* (beta-xylosidase) expressed more when *S. sclerotiorum* and *P. vulgaris* first started interacting. When apparent stem lesions started to appear later in the infection, the genes encoding the cellulose-degrading enzymes, *SS1G_07146* (cellobio hydrolase) and SS1G_13255 (beta-1, 4-glucanase), remained active. Depending on the host plant, the expression patterns of the *B. cinerea* genes that generate xyloglucan-degrading enzymes differ significantly.

Within 24–48 h postinfection, two genes were discovered to be upregulated: *SS1G_02462* and *SS1G_03602*. These genes encode hemicellulose-degrading enzymes that produce alpha-L-arabinofuranosidase. They specifically target the side chains of pectin that are present in the L-arabinofuranose residues of the hemicellulose. This discovery in *S. sclerotiorum* was first reported by Yajima and Kav (2006). Arabinogalactans are branching galactans with side chains of arabinose; their structural makeup is complex. Beta-1, 4-galactans are found in two different forms: those linked to proteins and those associated with rhamnogalacturonan I in pectin (Sakamoto and Ishimaru, 2013). In line with other enzymes that break down the cell wall, the majority of the genes encoding arabinogalactan-degrading enzymes were active in the latter phases of infection (24–48 hpi), as Table 7.2 illustrates. Still, there were two anomalies.

Two genes were already activated at 1 hpi: *SS1G_10842*, which encodes a beta-galactosidase, and *SS1G_01216*, which encodes an arabinogalactan endo-beta-galactosidase. Manganese polymers, or storage polysaccharides in plants, are one type of polymer. Manganese are abundant in branches that are linked by α (1-3) and α (1-2) and have α (1-6)-linked backbones. These branches are often linked to glycoproteins. As was previously mentioned, the construction of the cell wall depends on these glycoproteins. Multiple genes encoding mannosidases were expressed more regularly in the later stages of the infection. These mannosidases have the ability to hydrolyze and degrade these chemical bonds (Table 7.2).

Lignin is essential for giving plant cell walls their rigidity and structure. Many times, biotic and abiotic stressors lead to increased lignification. During the mid-to-late stages of infection, it was observed that three genes were upregulated, resulting in the production of extracellular dihydrogeodin oxidases (Barros *et al.*, 2015). Three multi-copper oxidase domains, which are frequently linked to laccases and other enzymes that oxidize phenolic substances, are present in these genes. Laccases are commonly recognized for their role in breaking down lignin, and some laccases may also assist in neutralizing phenolic secondary metabolites. *Sclerotinia* species possess a wide range of CAZymes which enable them to effectively degrade most host polysaccharides and obtain nutrients from diverse hosts. Further evidence that these CAZyme genes predominantly contribute to tissue maceration comes from the expression patterns of these genes, which are primarily detected during the latter phases of infection (24–48 hpi). The expression of several CAZyme genes is also affected by other factors, such as the kind of tissues being colonized, the species of host plant, and the surrounding conditions. There were discernible variations in the expression of PG genes, particularly *BcPG1* and *BcPG2*, when *B. cinerea* was infected on distinct hosts or on the same host under various temperature conditions (Blanco-Ulate *et al.*, 2014). Similar patterns were also observed in the genes encoding several pectin lyases, pectate lyases, and pectin methyl esterases.

The significance of cell wall degradation activity was underscored by the observation that the genes responsible for this activity were not only highly upregulated but also contributed to numerous significantly enriched upregulated GO- Gene Ontology categories. Around 29 out of 90 highly upregulated genes (top 50 genes upregulated by 25-fold or more with a log2 fold change of 4.64) in *S. sclerotiorum* were found to be associated with enzymatic activities related to cell wall degradation. The CWDEGO categories that showed significant enrichment were as follows: hydrolases (GO:0016787) that act on O-glucosyl bonds (GO:0004553), which include cellulase (GO:0008810), glucosidase (GO:0015926), galactosidase (GO:0015925), polygalacturonase (GO:0004650), alpha-L-arabinofuranosidase (GO:0046556), and

Table 7.3. An explanation of the genes upregulated and responsible for the expression of secondary chemicals. (Adapted from Seifbarghi *et al*., 2017, with permission.)

Gene ID	Description[a]	Expression level (hpi)[b]					
		1	3	6	12	24	48
1.Polyketides							
SS1G_02356	Polyketide synthase (PKS1)[c]	–	–	–	5.1	7.5	5
SS1G_05681	PKS4[c]	–	–	3.4	2.5	–	–
SS1G_09237	PKS6[c]	–	–	–	–	–	34.3
SS1G_01997	PKS8[c]	3.1	4.9	6.4	4.5	–	–
SS1G_11789	PKS12[c]	–	2.3	–	–	–	2.4
SS1G_11404	PKS14[c]	2.8	2.1	2.9	4.7	4.7	9.2
SS1G_04125	PKS16[c]	2.7	3	–	–	3.5	–
SS1G_05787	PKS18[c]	–	–	–	–	3.5	3
SS1G_03591	PKS21[c]	–	–	2.5	–	2.2	–
SS1G_04127	PKS	2.1	–	–	–	2.9	–
SS1G_10474	PKS	4	–	–	–	5.2	–
SS1G_05788	PKS	–	–	–	2.5	4.9	3.5
SS1G_11352	DBSA oxidoreductase (FmE)	4.4	2.8	3.3	–	–	–
SS1G_09239	FAD-dependent monooxygenase (similar to BcBOA8)	–	–	–	–	10.9	17.6
SS1G_02592	Zinc-binding oxidoreductase (ToxD)	4	–	–	4.2	5.8	17.8
SS1G_02338	Chalcone and stilbene synthase (CHS1)	–	–	–	–	2.1	–
2. Nonribosomal peptides							
SS1G_08561	Nonribosomal peptide synthase (NRPS1)[c]	–	–	–	4.7	8.9	32.5
SS1G_03693	NRPS2[c]	–	2.2	2.3	–	4	3.1
SS1G_06185	NRPS3[c]	3.6	8.2	8.8	9.3	4.7	–
SS1G_10563	NRPS5[c]	–	–	–	–	2.3	–
SS1G_12986	NRPS-like enzyme	–	6.3	5.2	4.3	5.8	7
SS1G_09846	NRPS-like enzyme	2.9	2.2	2.6	–	2.1	–
SS1G_03440	NRPS-like enzyme	–	–	–	3.5	4	–
SS1G_00726	NRPS-like enzyme	–	–	–	18.8	–	–
SS1G_01217	NRPS-like enzyme	–	–	–	5.4	10.5	7.7
3. Other							
SS1G_09177	Siroheme synthase	10.9	8.7	8.3	5.7	–	–

[a]Annotation based on BLAST results and the existence of preserved PFAM domains.
[b]Fold change relative to 0 hpi. (–) No significant change in expression.
[c]Gene abbreviations (Amselem *et al*., 2011).

beta-mannosidase (GO:0004567); ester bonds (GO:0016788), which include aspartyl esterase (GO:0045330), lipase (GO:0016298), cutinase (GO:0050525), and pectinesterase (GO:0030599); and peptidase (GO:0008233), which include serine-type endopeptidase (GO:0004252), tripeptidyl-peptidase (GO:0008240), and serine-type carboxypeptidase (GO:0004185) activities. The genome of *S. sclerotiorum* harbors a substantial collection of active enzymes that target both plant cell walls and fungal cell walls (Lyu *et al.*, 2015). A considerable number of these enzymes exhibited upregulation (Table 7.3). Comparative transcriptome analyses revealed a significant upregulation of various CWDEs during pathogenesis (Li *et al.*, 2004; Sexton *et al.*, 2006; Amselem *et al.*, 2011; Lyu *et al.*, 2015; Peng *et al.*, 2017; Seifbarghi *et al.*, 2017; Westrick *et al.*, 2019). In the meantime, a functional study demonstrated that the absence of the CWDE arabinofuranosidase/xylosidase (*SS1G_02462*) resulted in reduced virulence on *B. napus* (Yajima *et al.*, 2009).

7.24.4 Proteases

Despite the predominant focus on CAZymes as the primary group of hydrolytic enzymes associated with pathogenesis, it is important to acknowledge the significant roles played by other groups of hydrolytic enzymes. In the context of infection, it has been observed that several genes responsible for encoding mono-, di-, and tripeptidyl peptidases, along with endoproteases, are upregulated. This finding is supported by the data presented in Table 7.2. The substantial upregulation of Acp1, which was discovered to be 40 times greater than the original inoculum at 48 h postinfection, is of special relevance. This emphasizes the significance of Acp1 in the necrotic phase. These results are in line with previous research that indicates a variety of environmental factors, including deprivation of glucose and nitrogen as well as acidification, precisely control the expression of acp1. Additionally, it has been noted that in acidic conditions, the Pac C transcription factor controls the expression of acp1 (Poussereau *et al.*, 2001a).

According to previous studies, *S. sclerotiorum* induces the expression of the aspartyl protease gene (*aspS*, *SS1G_03629*) at the onset of infection, which contributes to the degradation of host defense proteins (Poussereau *et al.*, 2001b). Furthermore, it has been demonstrated that aspartyl protease, which is secreted by both *B. cinerea* and *S. sclerotiorum*, is one of the elements accountable for cell death. Other genes encoding asp-like proteases (*SS1G_05329* and *SS1G_02870*) were shown to be upregulated in the early stages of infection, which is in line with previous studies. Surprisingly, the aspS gene itself did not exhibit any upregulation. Furthermore, *SS1G_03181*, another aspartyl protease-encoding gene, was found in the interaction between *S. sclerotiorum* and *P. vulgaris*. Throughout the spreading necrosis phase, this gene's expression levels were increased (Oliveira *et al.*, 2015). The results of previous studies are consistent with the upregulation of *SS1G_03181* at 24–48 hpi.

The expression of many genes (*SS1G_07655*, *SS1G_02423*, *SS1G_03282*, *SS1G_12210*, and *SS1G_07168*) that encode subtilisin-like serine proteases was shown to be upregulated throughout the later phases of the infection (24–48 hpi). Olivieri *et al.* (2002) state that these subtilisins are thought to contribute to the penetration and colonization processes by hydrolyzing disease-associated proteins or glyco proteins found in plant cell walls. Surprisingly, patterns of expression for these genes suggest that they are also implicated in necrotic stage events. Additionally, at 48 hpi, there was an upregulation of the genes *SSG_05348* and *SSG_05349*. These genes encode metalloprotease-active enzymes, and their role in dissolving plant components for nitrogen usage was indicated. Similar to the genes that encode serine and metalloproteases, the majority of the exo-peptidase genes were triggered during the later stages of the infection. Endo-protease-produced peptides are broken down into individual amino acids by exo-peptidases (Table 7.2).

The investigation revealed one gene, designated as *SS1G_00862*, from the calpain family of cysteine proteases. It was shown to be initiated at 1 h postinfection and to be active throughout the duration of the infection's intermediate phases, lasting up to 24 h. The *SS1G_10992* gene, which codes for a unique cysteine protease with a caspase domain, showed elevated expression levels at 48 hpi, when it increased by 7.1 times over the initial inoculum. This peak occurred at 1 hpi. Studies have indicated that these endo-peptidases are involved in pathways that are connected to PCD (Vardar and Unal, 2008). Specifically, it is known that cysteine proteases like caspases have a major role in PCD, which is also known as apoptosis (Vardar and Unal, 2008). Moreover, the mechanisms involved in PCD are facilitated by cysteine proteases, along with a variety of other proteolytic systems such as calpain, 26S proteasome, granzyme B, cathepsin D, and matrix metalloproteinases (Beers *et al.*, 2000). A paper states that the toxin victorin, produced by *Cochliobolus victoriae*, activates the host cysteine protease. This activation then causes the large subunit of Rubisco to be cleaved by a protease. In light of these earlier findings, additional research is necessary to completely comprehend the function of these genes in the pathogenesis of *S. sclerotiorum*. They might be engaged in procedures that have to do with phase shifts that occur during an infection (Hegedus *et al.*, 2016). As approximately 10% of the protoplasm in host cells is composed of protein, proteases are necessary for feeding *S. sclerotiorum*. The transcriptome of *Sclerotinia homoeocarpa* also contained a large number of proteases, primarily serine endo-proteases (Orshinsky *et al.*, 2012). A larger variety of catalytic pathways were demonstrated by the upregulated endoprotease genes. Early on in the infection process, it was found that many genes encoding putative aspartyl proteases, which function well in acidic pH conditions, and cysteine proteases, which are active in acidic to neutral pH environments, were elevated. Subsequently, however, two genes encoding metalloproteases and the majority of serine proteases active at high pH levels increased. The pathogen's nutritional requirements, its interactions with the host's defense systems at different stages of infection, and the particular circumstances inside the lesion environment may all be reflected in the deliberate and methodical use of proteolytic enzymes using a variety of catalytic activities.

7.24.5 Production of secondary compounds through biosynthesis

During the current analysis, the expression levels of many genes involved in the manufacture of secondary metabolites were elevated (Table 7.3). These genes included key enzymes connected to the processes of hazardous compounds, such as polyketide synthase (PKS), chalcone synthase (CHS), nonribosomal peptide synthase (NRPS), and hybrid PKS/NRPS. According to Choquer *et al.* (2007), fungi can disrupt the normal functions of host cells, which might decrease defensive mechanisms in plants and possibly hasten the onset of disease symptoms. Among the several phytotoxic metabolites that *B. cinerea* is known to produce, botrydial is one of the most well researched (Colmenares *et al.*, 2002). According to Siewers *et al.* (2005), the *BcBOT1* gene encodes a cytochrome P450 monooxygenase that is necessary for the production of botrydial. Moreover, aflatoxin's manufacture involves cytochrome P450 enzymes (Kelkar *et al.*, 1997). Aflatoxin biosynthesis in *Aspergillus nidulans* is linked to the *StcL* gene, which is comparable to the genes *SS1G_09638* and *SS1G_11247*. Although these genes were expressed, their levels did not increase in comparison to the inoculum. Conversely, from 1 to 6 h after immunization, *SS1G_13923* was upregulated (Table 7.4). It is important to stress that, in some fungal species, cytochrome P450 enzymes assist in the breakdown of host compounds in addition to their contribution to the synthesis of secondary metabolites and mycotoxins (Cresnar and Petric, 2011). Malonyl-CoA decarboxylates to form a class of secondary metabolites called polyketides, which includes a range of mycotoxins. The genes *BcPKS6* and *BcPKS9*, which encode PKS enzymes that cooperate to produce the phytotoxin botcinic acid, are upregulated in tomato plants infected with *B. cinerea* (Dalmais *et al.*, 2011). Forty-eight hours following the experiment, transcripts from the *S. sclerotiorum* ortholog of PKS6 (*SS1G_09237*) were discovered, indicating a dynamic regulatory response in the botcinic acid production. In contrast, PKS9 (*SS1G_09240*) exhibited expression without upregulation in this study. This result is consistent with that of Pedras and Ahiahonu's (2004) study, which

Table 7.4. Identification and expression of upregulated genes encoding cytochrome p450 enzymes. (Adapted from Seifbarghi *et al.*, 2017, with permission.)

Gene ID	Description[a]	Expression level (hpi)[b]					
		1	3	6	12	24	48
SS1G_02340	Cytochrome p450 (pistatin demethylase)	3.0	5.4	6.7	5.1	–	–
SS1G_04805	CYP51 (eburicol 14 alpha-demethylase)	3.4	3.6	2.4	2.2	–	–
SS1G_00406	CYP55a3 (nitric oxide reductase)	2.1	2.5	2.1	–	–	–
SS1G_00623	CYp450	2.3	–	–	–	–	2.5
SS1G_13470	CYP450 (alkane hydroxylase)	2.2	–	–	–	–	–
SS1G_05491	CYp450	–	86	271	495	1067	5372
SS1G_13909	CYP450	–	2.1	2.2	–	–	–
SS1G_05384	CYp450	–	2.6	–	2.4	–	–
SS1G_02363	CYp450 (trichodiene oxygenase)	–	–	–	4.2	3.3	3.9
SS1G_04088	CYp450 (oxidoreductase)	–	–	–	3.5	7.5	18.7
SS1G_11553	CYp450	–	–	–	2.0	2.3	2.9
SS1G_03436	CYp450 (benzoate 4-monooxygenase)	–	–	–	4.0	2.6	–
SS1G_05490	CYp450 (benzoate 4-monooxygenase)	–	–	–	–	4.6	8.6
SS1G_01843	CYp450 (benzoate 4-hydroxylase)	–	–	–	–	5.5	3.0
SS1G_13957	CYp450	–	–	–	–	14.9	20.4
SS1G_02157	CYp450	–	–	–	–	3.6	9.9
SS1G_04780	CYp450		–	–	–	2.2	2.8
SS1G_06101	CYp450	–	–	–	–	4.9	12.7
SS1G_10037	CYp450	–	–	–	–	8.6	9.0
SS1G_11430	CYp450 (alkane hydroxylase)	–	–	–	–	2.1	4.7
SS1G_11607	CYp450	–	–	–	–	2.4	2.5
SS1G_14168	CYp450	–	–	–	–	2.1	6.0
SS1G_11768	CYp450	–	–	–	–	5.3	–
SS1G_01006	CYp450	–	–	–	–	–	3.5
SS1G_08136	CYp450	–	–	–	–	–	7.0
SS1G_14163	CYp450	–	–	–	–	–	3.1
SS1G_08677	CYP450 (monooxygenase)	–	–	–	–	–	2.8

[a]Annotation established through the recognition of conserved PFAM domains and BLAST data.
[b]Fold change relative to 0 hpi. (–) No significant change in expression.

found that *S. sclerotiorum* was unable to produce botcinic acid since both enzymes were needed for its synthesis. The pathogenicity of *B. cinerea* is impacted by both botrydial and botcinic acid toxins (Dalmais *et al.*, 2011). Furthermore, the *SSHEG_02592* gene is linked to the production of the Tox D-like protein in *A. terreus* and the polyketide lovastatin (Kennedy *et al.*, 1999). The overexpression of the *SS1G_02592* gene in this investigation started at 1 hpi and peaked at 48 hpi. During infections induced by *B. cinerea* and *S. sclerotiorum*, genes involved in the synthesis of melanin (PKS13), coprogen (NRPS6), and intracellular siderophores (NRPS2, NRPS3) are frequently active. Ascomycetes all share these genes. Amselem *et al.* (2011) Elevated expression levels were observed at 3–48 and 1–24 hpi for the siderophore biosynthesis genes *NRPS2* and *NRPS3* homologs (*SS1G_03693T0* and *S1G_06185T0*, respectively). Conversely, NRPS6

Table 7.5. Description and expression of upregulated genes encoding transporters. (Adapted from Seifbarghi *et al*., 2017, with permission.)

Gene ID	Description[a]	Expression level (hpi)[b]					
		1	3	6	12	24	48
*1. ATP-binding cassette (**ABC**) transporter super family*							
1.1. Pleiotropic drug resistance							
SS1G_06715	Pleiotropic drug resistance (PDR)	2	–	–	–	–	–
SS1G_13659	PDR-like	–	–	–	–	3.1	13.5
SS1G_04483	PDR-like	–	–	–	–	–	4.6
1.2. Multidrug resistance							
SS1G_04756	ABC multidrug transporter	3.4	7.9	6.5	3.7	2.4	–
SS1G_07287	ABC multidrug transporter	3.5	3.2	2.6	–	–	2.2
SS1G_03324	ABC multidrug transporter	3.5	4.8	4.4	–	–	–
SS1G_13112	ABC multidrug transporter	–	4.1	6.2	3.9	–	–
SS1G_02000	ABC multidrug transporter	–	–	3.9	3.2	–	–
SS1G_06062	ABC multidrug transporter	–	–	–	–	2.4	2.8
1.3. Other families							
SS1G_04757	Mitochondrial ATP-binding cassette protein involved in iron homeostasis	–	6.5	5.1	3.4	–	–
SS1G_07963	ABC transporter with iron-containing redox enzyme with death domains	–	–	–	–	4	8.5
SS1G_12070	ABC lipid transporter	–	–	–	–	–	2.6
SS1G_10747	ABC peroxysomal fatty acyl-CoA transporter	–	–	–	–	–	3.2
SS1G_05904	ABC peroxysomal fatty acyl-CoA transporter	–	–	–	–	–	2
2. MFS transporter superfamily							
2.1. Efflux pumps and drug resistance							
SS1G_10155	MFS drug efflux and drug resistance	10.1	5.5	3.7	–	–	–
SS1G_11948	MFS drug efflux and drug resistance	2.8	6.7	5.5	–	–	–
SS1G_00919	MFS gliotoxin efflux transporter	–	2.7	3	–	–	–
SS1G_09759	MFS aflatoxin efflux transporter	–	8	–	–	–	–
SS1G_05556	MFS drug efflux and drug resistance	–	–	2.4	–	–	–
SS1G_06662	MFS toxin efflux pump	–	–	3	–	–	–
SS1G_05572	MFS drug resistance transporter	11.4	2.9	–	–	–	–
SS1G_02358	MFS drug resistance transporter	–	–	–	4.1	3.7	4.3
SS1G_12101	MFS drug resistance transporter	–	–	–	–	2.4	–
SS1G_02394	MFS drug resistance transporter	–	–	–	–	–	4.6

Continued

Table 7.5. Continued

Gene ID	Description[a]	Expression level (hpi)[b]					
		1	3	6	12	24	48
SS1G_05756	MFS drug resistance transporter	–	–	–	–	–	15.2
SS1G_10566	MFS drug resistance transporter	–	–	–	–	–	5.5
SS1G_05556	MFS drug resistance transporter	–	–	–	–	–	2
SS1G_05095	MFS fungal trichothecene efflux pump (TRI12)	–	–	–	–	–	7.1
SS1G_05145	MFS fungal trichothecene efflux pump (TRI12)	–	–	–	–	5	–
SS1G_13024	MFS drug resistance transporter	–	–	–	–	–	3.6
SS1G_11919	MFS multidrug transporter	–	2.7	2.3	–	–	–
SS1G_02623	MFS multidrug transporter	–	–	–	–	–	4.7
SS1G_08554	MFS multidrug transporter	–	–	–	–	–	2.9
SS1G_10279	MFS multidrug transporter	–	–	–	–	–	2.1
SS1G_09892	MFS multidrug transporter	–	–	–	–	–	10.2
SS1G_02931	MFS multidrug transporter	–	–	–	–	–	2.8
SS1G_02048	Spermidine drug resistance transporter	2.1	–	–	–	–	3.2
2.2. Sugar transporters							
SS1G_04841	Hexose transporter	4.1	2.9	–	–	–	–
SS1G_13734	Hexose transporter HXT13	–	2.1	–	–	–	–
SS1G_04273	Hexose transporter	–	–	–	–	6.1	15.8
SS1G_12412	Hexose transporter	–	–	2.1	–	2.5	2.4
SS1G_09710	MFS sugar transporter	–	–	2.7	2.6	4.8	6.5
SS1G_06023	MFS monosaccharide transporter	6.9	6.7	7	4.9	–	–
SS1G_10125	MFS monosaccharide transporter	2.1	2.4	–	–	–	–
SS1G_06620	MFS monosaccharide transporter	–		–	–	2.4	–
SS1G_07618	MFS monosaccharide transporter	–	–	–	–	–	2.1
SS1G_08794	MFS sucrose transporter	–	–	2.6	–	–	–
SS1G_04208	MFS sugar transporter	–	–	–	4	36.7	49.2
SS1G_10413	MFS sugar transporter	–	–	–	2.3	4.6	4.8
SS1G_06751	MFS sugar transporter	–	–	–	–	2	–
SS1G_08117	MFS sugar transporter	–	–	–	–	2.8	–
SS1G_11591	MFS sugar transporter	–	–	–	–	10.7	9.4
SS1G_01523	MFS sugar transporter	–	–	–	–	13.1	12.5
SS1G_00928	MFS sugar transporter	–	–	–	–	7.2	10.6
SS1G_08467	MFS sugar transporter	–	–	–	–	2.9	2.5
SS1G_07132	MFS sugar transporter	–	–	–	–	8.3	6.3
SS1G_01759	MFS sugar transporter	–	–	–	–	–	2.8
SS1G_14316	MFS sugar transporter	–	–	–	–	–	6.1

Continued

Table 7.5. Continued

Gene ID	Description[a]	Expression level (hpi)[b]					
		1	3	6	12	24	48
SS1G_06402	MFS sugar transporter	–	–	–	–	–	16.2
SS1G_01302	MFS sugar transporter	–	–	–	–	–	47.6
SS1G_03579	MFS sugar transporter	–	–	–	–	–	4.1
SS1G_05006	MFS sugar transporter	–	–	–	–	–	11.8
SS1G_08982	MFS quinate transporter	–	–	–	–	18.9	26.1
SS1G_09293	MFS quinate transporter	–	–	–	–	3.5	2.9
SS1G_08981	MFS quinate transporter	–	–	–	–	49.3	38.4
SS1G_00139	MFS carbohydrate transporter	–	3.1	2.2	2	–	–
SS1G_04149	MFS carbohydrate transporter	–	3.2	2.5	–	–	8.2
SS1G_07210	MFS carbohydrate transporter	–	–	–	–	2.5	2.3
SS1G_08328	MFS carbohydrate transporter	–	–	–	–	5.8	4.1
SS1G_13671	MFS carbohydrate transporter	–	–	–	–	2.4	–
SS1G_14028	MFS carbohydrate transporter	–	–	–	–	2.1	2
SS1G_13819	MFS carbohydrate transporter	–	–	–	–	–	8.1
SS1G_11142	MFS monocarboxylate transporter	–	–	–	–	3.6	–
SS1G_01656	MFS monocarboxylate transporter	–	–	–	–	–	2.2
SS1G_07861	MFS galactonate transporter	2.8	–	–	–	–	–
SS1G_09368	MFS galactonate transporter	–	–	3.6	5.8	8.3	9
SS1G_09803	MFS galactonate transporter	–	–	–	4.2	14.9	13.1
SS1G_01055	MFS galactonate transporter	–	–	–	–	3.8	–
SS1G_10065	MFS galactonate transporter	–	–	–	–	73.2	487.6
SS1G_11224	MFS galactonate transporter	–	–	–	–	3.3	2.2
SS1G_11609	MFS galactonate transporter	–	–	–	–	3	13
SS1G_12649	MFS galactonate transporter	–	–	–	–	6.9	2.9
SS1G_12698	MFS galactonate transporter	–	–	–	–	4	5.7
SS1G_05090	MFS galactonate transporter	–	–	–	–	7.2	5.7
SS1G_10896	MFS alpha glucoside:h + symporter	–	–	–	–	4.4	2.3
SS1G_13094	MFS alpha glucoside: h + symporter	–	–	–	–	5	18.6
SS1G_03673	4-hydroxyphenylacetate permease and sugar transporter	–	–	–	–	–	5.3
SS1G_04849	MFS-fucose permease	–	–	–	–	–	3.1
SS1G_02117	Lactose permease	–	–	–	2.8	23	31.3
SS1G_08398	Maltose permease	–	4.2	–	–	–	3.8
SS1G_07922	Maltose permease	–	–	–	–	2.2	–
SS1G_12954	Maltose permease	–	–	–	–	–	8.2
2.3. Oligopeptides							
SS1G_02676	Oligopeptide transporter	2	–	–	–	4	5.2

Continued

Table 7.5. Continued

Gene ID	Description[a]	Expression level (hpi)[b]					
		1	3	6	12	24	48
SS1G_09296	Oligopeptide transporter	–	–	–	–	3.3	3.1
SS1G_09297	Oligopeptide transporter	–	–	–	–	–	2.8
SS1G_10121	MFS peptide transporter	5.9	7.4	5.7	3	–	–
SS1G_07522	MFS peptide transporter	–	–	2.1	–	–	4.1
2.4. Others							
SS1G_08628	MFS transporter	5.9	6.6	5.6	3	2.4	–
SS1G_04769	MFS transporter	–	–	–	–	–	7.3
SS1G_13941	MFS general substrate transporter	–	3	–	–	–	–
SS1G_13194	MFS general substrate transporter	–	–	–	–	2.2	–
SS1G_07517	MFS vacuole effluxer Atg22 transporter	–	3.5	3.5	–	2.7	2.4
SS1G_04978	MFS sulfate transporter	–	–	–	–	2.2	–
SS1G_12690	MFS nitrite transporter	–	7	12.2	10.7	–	–
SS1G_02906	MFS nicotinic acid transporter	–	–	–	–	3.8	2.1
3. Amino acid transporters							
SS1G_08387	Amino acid transporter	–	–	–	3	10.4	9.5
SS1G_11233	Amino acid transporter	–	–	–	–	2.7	–
SS1G_01802	Amino acid transporter	–	–	–	–	–	3.9
SS1G_09839	Amino acid transporter	–	–	–	–	–	13.3
SS1G_14102	Choline, amino acid permease transport	–	–	–	2.3	–	–
SS1G_05203	Choline, amino acid permease transport	–	–	–	–	–	5
SS1G_06535	Choline, amino acid permease transport	–	–	–	–	–	5.1
SS1G_06536	Choline, amino acid permease transport	–	–	–	–	–	3.1
SS1G_04884	Ammonium transporter	–	–	–	–	6.8	–
SS1G_11781	Arginine permease	2.2	–	–	–	–	–
SS1G_11563	Amino acid permease	3.2	4.7	4.9	5.5	–	–
SS1G_02549	Amino acid permease	3.1	4.2	3.9	2.2	–	–
SS1G_14381	Amino acid permease	2.6	4.6	3.8	3.3	–	–
SS1G_13916	Amino acid permease	9.1	11.5	8.6	6.9	2.3	–
SS1G_10633	Amino acid permease	–	2.1	2.7	2.1	2.4	4.9
SS1G_03403	Amino acid permease	–	–	–	–	3	–
SS1G_11780	Amino acid permease	–	–	–	–	–	2.9
SS1G_06841	Amino acid permease	–	–	–	–	–	2.7
Others							
SS1G_03654	Formate nitrate transporter	3	3.2	3.1	3.4	4	3.9

Continued

Table 7.5. Continued

Gene ID	Description[a]	Expression level (hpi)[b]					
		1	3	6	12	24	48
SS1G_09621	Mitochondrial phosphate carrier protein	3.8	5	4.4	3.3	–	3
SS1G_01720	C4-dicarboxylate transporter	20.2	19	18.8	8.5	–	–
SS1G_02619	C4-dicarboxylate transporter	2.2	2.7	2.6	–	–	–
SS1G_12337	Calcium proton exchanger	2.5	3	2.2	–	–	2.6
SS1G_02802	Putative Mg2 +transporter	2.6	2.8	2.4	–	–	3.8
SS1G_02548	UDP-N-acetylglucosamine transporter	2.4	2.8	2.8	–	–	–
SS1G_11564	Lysosomal cystine transporter	2.1	3	2.5	–	–	–
SS1G_06298	Cation efflux family transporter	3.1	4	3.5	–	–	–
SS1G_09822	Monocarboxylate permease-like protein	2.1	2.5	–	–	–	2.7
SS1G_06910	Vacuolar iron transporter	2.7	–	2.3	–	3.1	4
SS1G_03019	Carnitine mitochondrial carrier protein	2.4	–	2.6	–	–	3.4
SS1G_08146	AGZA family xanthine/uracil permease	3	–	–	–	–	–
SS1G_10280	Polyamine transport protein	–	2.7	2.1	–	–	–
SS1G_04686	Phosphate/sulfate permease	–	4.2	–	–	3.3	–
SS1G_12991	Cation, potassium transporter	–	–	2.4	–	–	–
SS1G_08795	Succinate fumarate, mitochondrial transporter	–	–	5.4	–	–	8
SS1G_06006	Purine-cytosine permease	–	–	–	6.4	3.7	–
SS1G_03305	Solute carrier protein	–	–	–	3	–	–
SS1G_11712	Cation, magnesium transporter	–	–	–	–	2.4	–
SS1G_07645	Cation, potassium transporter	–	–	–	–	2.2	2.5
SS1G_02434	Mitochondrial 2-oxoglutarate malate carrier protein	–	–	–	–	45.3	44.7
SS1G_06806	3-oxoacyl-[acyl-carrier protein] reductase	–	–	–	–	3.7	5.3
SS1G_10890	Acetyl-CoA acetyltransferases	–	–	–	–	3	3.4
SS1G_04606	Arsenite efflux transporter	–	–	–	–	–	2.8
SS1G_01231	Sulfate permease	–	–	–	–	–	5.7
SS1G_01111	Cytosine permease	–	–	–	–	–	3.6
SS1G_10204	Mitochondrial carrier protein	–	–	–	–	–	2.2
SS1G_06998	Peroxisomal, mitochondrial carrier protein	—	–	–	–	–	2
SS1G_04019	Mitochondrial carrier protein	–	–	–	–	–	2.5
SS1G_08663	Solute carrier protein	–	–	–	–	–	2.5

Continued

Table 7.5. Continued

Gene ID	Descriptionᵃ	Expression level (hpi)ᵇ					
		1	3	6	12	24	48
SS1G_05281	Fatty acid transporter involves in acyl-CoA synthetase	–	–	–	–	–	4.3
SS1G_06664	Nucleoside transporter	3.7	5.7	5.7	4.3	2.8	–
SS1G_06212	Nucleoside transporter	3.2	–	–	–	–	–
SS1G_04537	Nucleoside transporter	–	–	–	–	2.6	–
SS1G_09667	Nucleoside transporter	–	–	–	–	–	2.3

[a]Annotation based on the presence of conserved PFAM domains and BLAST reports.
[b]Fold change relative to 0 hpi. (–) No significant change in expression.

(*SS1G_04250*) and PKS13 (*SS1G_13322*) were expressed but did not show any signs of upregulation. Siderophores play a major role in the pathogenicity of some diseases and are essential for scavenging iron. Heme and siroheme, two iron-containing cofactors, are essential for the catalytic activity of enzymes involved in sulfur and nitrogen absorption as well as xenobiotic detoxification (Tripathy *et al.*, 2010).

A significant increase in the expression of the siroheme synthase gene (*SS1G_09177*), also known as *S*-adenosyl methionine: urophototransferase, was observed early in the infection phase (1–12 hpi). This finding emphasizes the significance of iron metabolism and sequestration. Transporters are required for the release or secretion of secondary metabolites in addition to being necessary enzymes for their production (Table 7.5). The *HTS1* gene encodes a non-ribosomal peptide synthetase (NRPS) responsible for the production of HC-toxin, a key virulence factor of *Cochliobolus carbonum* in maize (Walton, 2006). This toxin is then removed from the cell by the two major facilitator superfamily ([MFS]) transporters, TOXA and TOXB (Pitkin *et al.*, 1996). The results of the analysis showed that the MFS toxin efflux pump protein of *B. cinerea*, TOXA from *A. lentulus*, and TOXA from *Glarea lozoyensis* are possible HC-toxin efflux carriers. They exhibited 51, 61, and 84% of similarities with *SS1G_09759*, *SS1G_00919*, and *SS1G_06662*. These findings imply that there was an increase in the three transporter genes in the research that resemble TOXA proteins.

The information suggests that when *S. sclerotiorum* is infected, these transporters play a role in the release of toxins into the host plant. Further research will be necessary to determine their specific substrate preferences. According to the research, during *S. sclerotiorum* infection, these transporters may contribute to the release of toxins into the host plant. To ascertain their particular substrate preferences, more investigation is required. As of right now, the only phytotoxin known to exist in *S. sclerotiorum* is sclerin. Plants like *B. napus*, *Brassica juncea*, and *Sinapsis alba* develop chlorotic and necrotic tissues when they come into contact with sclerin. It is thought that these plants are weak. *Erucastrum gallicum* and other resistant species are unaffected by it, according to Pedras and Ahiahonu (2004). The genes that cause the production of sclerin are still unknown. Nonetheless, the high expression levels of these genes during infection and their role in the enzymatic production of identified secondary metabolites from *S. sclerotiorum* suggest that this fungus, like *B. cinerea*, is able to secrete a wide range of secondary metabolites.

7.24.6 Expression of detoxification of genes

During the infection process, plant pathogens face numerous biochemical defense mechanisms from the host. They can overcome these mechanisms through different strategies. One strategy is avoidance, where pathogens grow inside the host cells to evade extracellular phytoalexins. Another strategy is resistance, which involves mutations that change the sensitivity of the pathogens to antimicrobial compounds. The third strategy is detoxification, where the pathogens modify or degrade the host phytoalexins. The expulsion of harmful phytochemicals through energy-dependent mechanisms by transporters associated with membranes is a common detoxification process found in diseases that affect a wide range of hosts (van Etten *et al.*, 2001). In the genome of *S. sclerotiorum*, researchers have found 218 MFS transporters and 33 genes encoding ATP-binding cassette (ABC) transporters (Amselem *et al.*, 2011). These transporters, ABC and MFS (major facilitator super-family), have a wide range of specificities, including those for medicines, sugars, polysaccharides, amino acids, peptides, heavy metals, and inorganic ions. Perlin *et al.* (2014) have linked a few of these transporters to the discharge of toxins originating from fungus as well as the outflow of phytoalexins from the host.

According to findings shown in Table 7.5, there were indications of upregulation in 91 MFS transporters and 14 ABC transporters. The function ABC transporters play in an infection's resistance to phytoalexins is widely recognized. It has been observed that during the early stages of the infection and in the presence of camalexin, the *B. cinerea* gene that codes for the ABC transporter BcAtrB becomes active. When the *BcAtrB* gene is inactive, the pathogen becomes more vulnerable to camalexin, which reduces its pathogenicity (Stefanato *et al.*, 2009). Furthermore, a variety of compounds, including the phytoalexin resveratrol, the plant component phenyl-propanoideugenol, and certain antibiotics and fungicides, have been linked to the BcAtrB transporter (Schoonbeek *et al.*, 2001Schoonbeek *et al.*, 2003). The *S. sclerotiorum* homolog of BacAtrB (SS1G_13659) was also elevated between 24 and 48 h postinfection, and it most likely serves a similar purpose in avoiding phytoalexins. Even though BcAtrA has no influence on *B. cinerea's* pathogenicity on bean plants, it is thought to function as a multidrug transporter since it can reduce yeast susceptibility to cycloheximide and catechol (Del Sorbo *et al.*, 2008). An hour following immunization, there was a slight rise in the expression of the BcAtrA homolog from *S. sclerotiorum* (*SS1G 06715*). Both *B. cinerea* (Hayashi *et al.*, 2002a) and *S. homoeocarpa* (Hulvey *et al.*, 2012) have demonstrated resistance to demethylation inhibitor (DMI) fungicides in the presence of the ABC transporter AtrD. On the other hand, there was no upregulation of the *S. sclerotiorum* homolog (*SS1G_02407*) in the setting of *B. napus* infection. On the other hand, the *S. sclerotiorum* counterpart of BMR- Brown Midrib (*SS1G_04483*) rose in expression 48 h after immunization. This is associated with *B. cinerea's* ability to withstand toxicants such polyoxin and iprobenfos (Makizumi *et al.*, 2002).

Although BcAtrA has no effect on *B. cinerea's* pathogenicity on bean plants, it can reduce yeast sensitivity to cycloheximide and catechol, suggesting that it functions as a multidrug transporter (Del Sorbo *et al.*, 2008). The BcAtrA homolog for *S. sclerotiorum* (SS1G 06715) showed a slight increase in expression 1 h after vaccination. Both *B. cinerea* (Hayashi *et al.*, 2002a) and *S. homoeocarpa* (Hulvey *et al.*, 2012) have demonstrated resistance to DMI- Demethylation Inhibitors. These fungicides inhibiting the ergosterol biosynthesis pathway in fungi a crucial process for forming functional cell membranes fungicides in the presence of the ABC transporter AtrD. On the other hand, there was no upregulation of the *S. sclerotiorum* equivalent (SS1G_02407) in the setting of *B. napus* infection. Conversely, 48 h after BMR1, the *S. sclerotiorum* analogs (SS1G_04483) showed increased expression. Camptothecin and cercosporin are examples of naturally occurring toxic chemicals that can be detoxified with the help of the MFS transporter, which is encoded by the *BcMFS1* gene in *B. cinerea*. Moreover, Hayashi *et al.* (2002b) stated that it offers resistance to specific fungicides such DMIs. Conversely, the orthologous gene *SSOEG_12842*, which codes for *BcMFS1*, is expressed but not elevated. If other upregulated MFS transporters are involved in detoxification, more research is required to find out. In order to neutralize the effects of the host phytoalexins, phytoalexins from the pathogen can be transferred by enzymes, which can also permanently deactivate or change these molecules into a less harmful form. One such gene is *SsBGT1*, which codes for brassinin glucosyltransferase1 (*SS1G_09997*). It was upregulated upon infection with *S. sclerotiorum* isolate UQ1280 in *B. napus* cultivar Surpass 400 leaves. Plant phytoalexins such as brassinin induce SsBGT1, which is implicated in the glucosylation-mediated detoxification of plant defense chemicals

(Sexton *et al.*, 2009). Additionally, this gene was expressed at levels that were 223 and 28 times higher than those of the inoculum in *S. sclerotiorum* at 24 and 48 h after infection, respectively. Another gene that codes for the cytochrome P450 enzyme, SS1G_02340, was upregulated between 1 and 12 h after infection (Table 7.4). Given their structural similarities, *B. napus* phytoalexins may benefit from detoxification (Darvill and Albersheim, 1984).

Derivatives of benzoic acid from the beta-ketoadipate pathway are aromatic chemicals produced by plant defense systems that are targeted by plant diseases (Harwood and Parales, 1996). One significant function of the cytochrome P450 enzymes in the CYP53 family is the transformation of benzoic acid into 4-hydroxybenzoate. The gene *SS1G_01843*, which codes for the enzyme benzoate 4-hydroxylase, is involved in detoxification and expressed more when it came to 24 and 48 h following infection. This gene may contribute to the plant's defense mechanism against infections by aiding in the breakdown of benzoic acid derivatives. Conversely, propiconazole is a sterol DMI fungicide that acts on CYP51, also known as eburicol 14 alpha-demethylase, to suppress ergosterol production (Yoshida, 1993). CYP51 is required for the synthesis of fungal sterol, which is required to maintain membrane integrity (Parks and Casey, 1995). A possible defense against DMI has been associated in certain studies with increased expression of CYP51 or its paralogues in *S. homoeocarpa* and *Monilinia fructicola* (Luo and Schnabel, 2008; Hulvey *et al.*, 2012). Between 1 and 12 h after infection, the paralogous gene of *S. homoeocarpa* CYP51, also referred to as the *SS1G_04805* gene, exhibited increased levels of expression. It most likely has a similar function in phytoalexin clearance. Glucosinolates, which are mostly found in the *Brassicaceae* family, and hydroxyl-nitrile glycosides, which are more widely distributed, are significant compounds implicated in plant defense. Damage to the molecules causes enzymatic modifications that produce toxic metabolites, such as hydrogen cyanide, which is known to severely impair cell respiration (Frisch *et al.*, 2015). Studies have indicated that cyanide hydratases have a role in the detoxification of hydrogen cyanide in *L. maculans* (Sexton and Howlett, 2000) and *B. cinerea* (González-Fernández *et al.*, 2014). Four genes that encode cyanide hydratases or cyanate hydrolases *SS1G_13754*, *SS1G_10174*, *SS1G_01652*, and *SS1G_11485* (Table 7.6) were shown to be considerably upregulated at various phases of infection, from 1 to 48 h postinfection. These genes may be necessary for detoxifying the glucosinolate-derived compounds produced by *B. napus* infections. It is commonly recognized that the glutathione *S*-transferases, or GSTs, can effectively detoxify xenobiotic substances by joining glutathione with them (Dowd *et al.*, 1997). It was shown that several genes producing Glutathione S-transferases (GSTs) were significantly elevated in various infections; the majority of the stimulation was observed 24 h postinfection (Table 7.6). One particular gene, *SS1G_01918*, exhibited significant induction levels from 1 to 48 hpi and peaked at 189 times the initial inoculum's expression level. It has recently been determined that GSTs also assist in the detoxification of xenobiotics in *Phanerochaete chrysosporium*. Originally discovered in plants, GSTs were known for their capacity to detoxify herbicides (Riechers *et al.*, 1997).

The first GST discovered in a filamentous fungus was Bcgst1 from *B. cinerea*; nevertheless, subsequent investigation revealed that altering the Bcgst1 gene had no effect on virulence (Prins *et al.*, 2000). The putative virulence factor Bcgst1 is hypothesized to play a part in plant defense chemical tolerance; further research is necessary to pinpoint the precise mechanism of action. It was discovered that the ortholog of Bcgst1 in *S. sclerotiorum*, *SS1G_07195*, was upregulated 24 and 48 h after infection, indicating that it might be involved in the infection process. Nitroalkanes like 2-nitropropane can be converted into carbonyl compounds and nitrite by the enzyme 2-nitropropane dioxygenase. Mijatovic (2008) states that 2-nitropropane dioxygenase helps detoxify nitroalkanes, which are known to be harmful molecules, when *Trichoderma harzianum* and tomatoes interact (Samolski *et al.*, 2009). Furthermore, *Pseudomonas jessenii* uses this enzyme to break down xenobiotics (Garrido-Sanz *et al.*, 2016). Four genes that encode 2-nitropropane dioxygenases (*SS1G_10881*, *SS1G_14466*, *SS1G_00355*, and *SS1G_11235*) were shown to exhibit significant upregulation during a

Table 7.6. Description and expression of upregulated genes involved in detoxification. (Adapted from Seifbarghi *et al.*, 2017, with permission.)

Gene ID	Description[a]	Expression level (hpi)[b]					
		1	3	6	12	24	48
SS1G_01918	Glutathione *S*-transferase	19.1	24.8	34.6	45.0	189.0	19.1
SS1G_10108	Glutathione *S*-transferase	–	–	–	3.5	4.1	2.3
SS1G_08210	Glutathione *S*-transferase	–	–	–	–	2.6	–
SS1G_04914	Glutathione *S*-transferase	–	–	–	–	2.1	–
SS1G_07195	Glutathione *S*-transferase	–	–	–	–	2.7	2.8
SS1G_14440	Glutathione *S*-transferase	–	–	–	–	4.5	3.7
SS1G_09479	Glutathione *S*-transferase	–	–	–	–	–	40.1
SS1G_08258	Glutathione *S*-transferase with glutathionyl-hydroquinone reductase, ECM4 domain	–	–	–	–	3.1	2.4
SS1G_09997	UDP-glucuronosyl and udp-glucosyltransferase (brassinin)	–	–	–	–	28.0	223.2
SS1G_03517	UDP-glucoronosyl and udp-glucosyltransferase family protein	4.5	3.7	–	5.0	8.2	10.8
SS1G_13524	Glucosyltransferase family	6.3	4.9	4.2	2.5	–	2.9
SS1G_13754	Nitrilase-cyanide hydratase	3.2	3.5	3.1	2.2	4.6	4.4
SS1G_10174	Nitrilase-cyanide hydratase	–	11.6	16.3	19.7	29.8	58.6
SS1G_01652	Nitrilase-cyanide hydratase	–	–	–	–	3.4	2.3
SS1G_11485	Cyanate lyase, cyanase superfamily	–	–	–	–	2.4	–
SS1G_10881	2-Nitropropane dioxygenase	2.8		–	–	–	–
SS1G_14466	2-Nitropropane dioxygenase	2.8	4.0	2.5	–	–	3.0
SS1G_00355	2-Nitropropane dioxygenase	–	–	–	–	3.4	5.4
SS1G_11235	2-Nitropropane dioxygenase	–	–	–	–	2.7	3.5

[a]Annotation based on BLAST results and the existence of conserved PFAM domains.
[b]Fold change relative to 0 hpi. (–) No significant change in expression.

range of infection durations, from 1 to 48 h postinfection (Table 7.6).

These genes might be required for the detoxification of harmful nitroalkanes produced during a *B. napus* infection.

7.24.7 Suppression of host defense by the synthesis of OA

OA plays a critical role in the *S. sclerotiorum* infection process because it suppresses host defenses (Williams *et al.*, 2011), controls the synthesis and activity of hydrolytic enzymes (Favaron *et al.*, 2004), and starts the processes that result in tissue necrosis (Kabbage *et al.*, 2013) and host colonization (Davidson *et al.*, 2016). Oxaloacetate is hydrolyzed by fungi, usually with the aid of an enzyme called OAH, which is present in *S. sclerotiorum*, to create OA (Maxwell, 1973). It has been demonstrated that disruption of the OAH gene reduces OA synthesis and pathogenicity in *Aspergillus niger*, *B. cinerea* (Han *et al.*, 2007), and *S. sclerotiorum* (Liang *et al.*, 2015a). These results demonstrate how important

OA is to *S. sclerotiorum's* pathogenicity. The *SS1G_08218* gene, commonly referred to as OAH, was expressed in *S. sclerotiorum* as early as 1-hpi. However, a notable upregulation did not occur until 48 hpi, at which point levels were five times higher than those of the original inoculum. It's interesting to note that the ODD (oxalate decarboxylase) gene (*SS1G_10796*) was stimulated at 24 and 48 hpi, with maximal expression levels occurring at 7 and 10 times greater than the inoculum, respectively (Magro *et al.*, 1988). The opposing functions of ODD and OA genes are expressed simultaneously, indicating that the complex biological roles of OA in the infection process include a careful balance between its production and breakdown (Kabbage *et al.*, 2013). The results confirm the ODD gene and OAH's synchronized expression, which is consistent with the findings of other studies (Amselem *et al.*, 2011). Ozone (OA) produces an acidic environment that has a significant impact on *S. sclerotiorum's* pathogenesis and virulence. Throughout the infection process, the release of OA causes the pH to drop, which in turn promotes the synthesis of hydrolytic enzymes. In addition, OA contributes to sclerotiogenesis in the advanced phases of infection. It is important to remember that OA is not the only factor that determines tissue acidity; other factors may also be involved. It is interesting to note that early host–pathogen interactions depend on the accurate regulation of OA levels by OA biogenic and degradative enzyme activity. This involves the growth of lesions and the development of compound appressoria.

7.25 Role of *GRF* Genes in the Pathogenesis of *Sclerotinia* to *Brassica*

The G-box regulatory factors (GRFs) are essential components in numerous signal transduction pathways and are vital for the physiological, developmental, and stress responses in plants. Previous studies have identified GRF proteins in various plant species, such as *B. napus*. Sun *et al.* (2022) conducted an extensive genome-wide analysis of GRFs in *B. napus* utilizing the existing genome sequence data. They looked at how these genes expressed themselves in different tissues after being infected with *S. sclerotiorum* and after receiving various hormone therapies. In rapeseed, the study found 46 putative *BnaGRF* genes spread over 18 chromosomes. The 46 *BnaGRFs* were categorized into two main evolutionary lineages: the epsilon (+) group and the nonepsilon (non-+) group, resembling 14-3-3 proteins found in other plant species. According to evolutionary study, both populations experienced an expansion of the *BnaGRF* gene family before speciation, which is similar to the recent gene duplication event that was noticed in the 14-3-3 gene family. Collinearity analysis revealed that syntenic linkages were present in the majority of *BnaGRF* genes. A total of 41.3% (19/46) of the *BnaGRF* genes responded to *S. sclerotiorum* infection, according to RNA-seq data. The signaling pathways for SA and JA may have been the source of this response. The results provide insight into how 14-3-3s respond to biotic stress and advance our knowledge of the variety of roles they play in *B. napus* (Sun *et al.*, 2022).

References

Abawi, G.S. and Grogan, R.G. (1975) Source of primary inoculum and effects of temperature and moisture on infection of beans by *Whetzelinia sclerotiorum*. *Phytopathology* 65(3), 300. DOI: 10.1094/Phyto-65-300.

Abawi, G.S. and Grogan, R.G. (1979) Epidemiology of diseases caused by *Sclerotinia species*. *Phytopathology* 69, 899–904. DOI: 10.1094/Phyto-69-899.

Abawi, G.S., Polach, F.J. and Molin, W.T. (1975a) Infection of bean by ascospores of *Whetzelinia sclerotiorum*. *Phytopathology* 65(6), 673. DOI: 10.1094/Phyto-65-673.

Abawi, G.S., Provvidenti, R., Grogan, R.G. and Hunter, J.E. (1975b) Predisposition of beans to infection by ascospores of *Whetzelinia sclerotiorum* prior to blossoming. *Proceedings of the American Phytopathological Society* 2, 61.

Adams, P.B. and Ayers, W.A. (1979) Ecology of *Sclerotinia species*. *Phytopathology* 69, 896–899. DOI: 10.1094/Phyto-69-896.

Akai, J. (1981) Studies on the epidemiology and control of *Sclerotinia* disease of beans caused by *Sclerotinia sclerotiorum* (Llib) de Bary. *Hokkaido Prefectural Agricultural Experimentation Station Reporter* 36, 77–79.

Akem, C.N. and Melouk, H.A. (1990) Transmission of *Sclerotinia minor* in peanut from infected seed. *Plant Disease* 74(3), 216. DOI: 10.1094/PD-74-0216.

Allan, J., Regmi, R., Denton-Giles, M., Kamphuis, L.G. and Derbyshire, M.C. (2019) The host generalist phytopathogenic fungus *Sclerotinia sclerotiorum* differentially expresses multiple metabolic enzymes on two different plant hosts. *Scientific Reports* 9(1), 19966. DOI: 10.1038/s41598-019-56396-w.

Amselem, J., Cuomo, C.A., Kan, J.A.L., Viaud, M., Benito, E.P *et al.* (2011) Genomic analysis of the necrotrophic fungal pathogens *Sclerotinia sclerotiorum* and *Botrytis cinerea*. *PLoS Genetics* 7(8), e1002230. DOI: 10.1371/journal.pgen.1002230.

Anderson, T.R. (1985) Seed molds of soybean in ontario and the influence of production area on the incidence of *Diaporthe phaseolarum var. caulivara* and *Phomopsis* sp.. *Journal of Plant Pathology* 7, 74–78.

Andrew, M., Barua, R., Short, S.M. and Kohn, L.M. (2012) Evidence for a common toolbox based on necrotrophy in a fungal lineage spanning necrotrophs, biotrophs, endophytes, host generalists and specialists. *PLoS ONE* 7(1), e29943. DOI: 10.1371/journal.pone.0029943.

Bachmann, B.J. and Bonner, D.M. (1959) Protoplasts from *Neurospora crassa*. *Journal of Bacteriology* 78(4), 550–556. DOI: 10.1128/jb.78.4.550-556.1959.

Baker, K.F. and Davis, L.H. (1951) An unusual occurrence of sclerotia of *Sclerotinia species* with seed of *Centaurea cygnus*. *Plant Disease Reporter* 35, 39–41.

Barkai-Golan, R. (1974) Production of cellulase and polygalacturonase by *Sclerotinia minor*. *Mycopathologia et Mycologia Applicata* 54(3), 297–302. DOI: 10.1007/BF02050164.

Barros, J., Serk, H., Granlund, I. and Pesquet, E. (2015) The cell biology of lignification in higher plants. *Annals of Botany* 115(7), 1053–1074. DOI: 10.1093/aob/mcv046.

Bartnicki-Garcia, S. and Lippman, E. (1966) Liberation of protoplasts from the mycelium of *Phytophthora*. *Journal of General Microbiology* 42, 411–416.

Bashi, Z.D., Rimmer, S.R., Khachatourians, G.G. and Hegedus, D.D. (2012) Factors governing the regulation of *Sclerotinia sclerotiorum* cutinase A and polygalacturonase 1 during different stages of infection. *Canadian Journal of Microbiology* 58, 605–616. DOI: 10.1139/w2012-031.

Bashi, Z.D., Gyawali, S., Bekkaoui, D., Coutu, C., Lee, L *et al.* (2016) The *Sclerotinia sclerotiorum* Slt2 mitogen-activated protein kinase ortholog, *SMK3*, is required for infection initiation but not lesion expansion. *Canadian Journal of Microbiology* 62(10), 836–850. DOI: 10.1139/cjm-2016-0091.

Bashi, Z.D., Rimmer, S.R., Khachatourians, G.G. and Hegedus, D.D. (2013) *Brassica napus polygalacturonase* inhibitor proteins inhibit *Sclerotinia sclerotiorum* polygalacturonase enzymatic and necrotizing activities and delay symptoms in transgenic plants. *Canadian Journal of Microbiology* 59, 79–86. DOI: 10.1139/cjm-2012-0352.

Bayliss, K.L., Kuo, J., Sivasithamparam, K., Barbetti, M.J. and Lagudah, M.J. (2001) Infection of subterranean clover (*Trifolium subterraneum*) by *Kabatiella caulivora*. *Journal of Phytopathology* 149, 699–705.

Beers, E.P., Woffenden, B.J. and Zhao, C. (2000) Plant proteolytic enzymes: Possible roles during programmed cell death. *Plant Molecular Biology* 44(3), 399–415. DOI: 10.1023/a:1026556928624.

Berger, S., Sinha, A.K. and Roitsch, T. (2007) Plant physiology meets phytopathology: Plant primary metabolism and plant-pathogen interactions. *Journal of Experimental Botany* 58(15–16), 4019–4026. DOI: 10.1093/jxb/erm298.

Blakeman, J.P. and Sztejnberg, A. (1973) Effect of surface wax on inhibition of germination of *Botrytis cinerea* spores on beetroot leaves. *Physiological Plant Pathology* 3(2), 269–278. DOI: 10.1016/0048-4059(73)90089-1.

Blanco-Ulate, B., Morales-Cruz, A., Amrine, K.C.H., Labavitch, J.M., Powell, A.L.T. *et al.* (2014) Genome-wide transcriptional profiling of *Botrytis cinerea* genes targeting plant cell walls during infections of different hosts. *Frontiers in Plant Science* 5, 435. DOI: 10.3389/fpls.2014.00435.

Blodgett, E.C. (1946) The *Sclerotinia* rot disease of beans in Idaho. *Plant Disease Reporter* 30, 137–144.

Boyle, C. (1921) Studies in the physiology of parasitism. VI. infection by *Sclerotinia libertiana*. *Annals of Botany* 35(3), 337–347. DOI: 10.1093/oxfordjournals.aob.a089761.

Cai, Q., Qiao, L., Wang, M., He, B., Lin, F.-M *et al.* (2018) Plants send small rnas in extracellular vesicles to fungal pathogen to silence virulence genes. *Science* 360(6393), 1126–1129. DOI: 10.1126/science.aar4142.

Calonge, F.D., Fielding, A.H., Byrde, R.J.W. and Akinrefon, O.A. (1969) Changes in ulrastructure following fungal invasion and the possible relevance of extracellular enzymes. *Journal of Experimental Botany* 20(2), 350–357. DOI: 10.1093/jxb/20.2.350.

Cessna, S.G., Sears, V.E., Dickman, M.B. and Low, P.S. (2000) Oxalic acid, a pathogenicity factor for *Sclerotinia sclerotiorum*, suppresses the oxidative burst of the host plant. *Plant Cell* 12(11), 2191–2200. DOI: 10.1105/tpc.12.11.2191.

Chamberlain, E.E. (1932) Sclerotinia disease (*Sclerotinia sclerotiorum*) of tomatoes. *New Zealand Journal of Agriculture* 45, 260–268.

Chaudhary, K.C.B. and Puttoo, B.L. (1991) Seed borne nature of *Sclerotinia sclerotiorum* in safflower in india. *Indian Journal of Mycology and Plant Pathology* 21, 196.

Chen, C., Harel, A., Gorovoits, R., Yarden, O. and Dickman, M.B. (2004) MPAK regulation of sclerotial development in *Sclerotinia sclerotiorum* is linked with pH and cAPM sensing. *Molecular Plant Microbe Interactions* 17, 404–413.

Chen, J., Ullah, C., Reichelt, M., Gershenzon, J. and Hammerbacher, A. (2019) *Sclerotinia sclerotiorum* circumvents flavonoid defenses by catabolizing flavonol glycosides and aglycones. *Plant Physiology* 180(4), 1975–1987. DOI: 10.1104/pp.19.00461.

Choquer, M., Fournier, E., Kunz, C., Levis, C., Pradier, J.-M. *et al.* (2007) *Botrytis cinerea* virulence factors: New insights into a necrotrophic and polyphageous pathogen. *FEMS Microbiology Letters* 277(1), 1–10. DOI: 10.1111/j.1574-6968.2007.00930.x.

Chou, H.M., Bundock, N., Rolfe, S.A. and Scholes, J.D. (2000) Infection of *Arabidopsis thaliana* leaves with *Albugo candida* causes a reprogramming of host metabolism. *Molecular Plant Pathology* 1, 99–113.

Colmenares, A.J., Aleu, J., Durán-Patrón, R., Collado, I.G. and Hernández-Galán, R. (2002) The putative role of botrydial and related metabolites in the infection mechanism of *Botrytis cinerea*. *Journal of Chemical Ecology* 28, 997–1005.

Cotton, P., Kasza, Z., Bruel, C., Rascle, C. and Fevre, M. (2003) Ambient pH controls the expression of endogalacturonase genes in the necrotrophic fungus *Sclerotinia sclerotiorum*. *FEMS Microbiology Letters* 277, 163–169.

Cresnar, B. and Petric, S. (2011) Cytochrome P450 enzymes in the fungal kingdom. *Biochimica et Biophysica Acta* 1814, 29–35.

Dalmais, B., Schumacher, J., Moraga, J., Pecheur, P.L.E., Tudzynski, B. *et al.* (2011) The *Botrytis cinerea* phytotoxin botcinic acid requires two polyketide synthases for production and has a redundant role in virulence with botrydial. *Molecular Plant Pathology* 12, 564–579.

Darvill, A.G. and Albersheim, P. (1984) Phytoalexins and their elicitors: A defense against microbial infection in plants. *Annual Review of Plant Physiology* 35, 243–275.

Davidson, A.L., Blahut-Beatty, L., Itaya, A., Zhang, Y., Zheng, S. *et al.* (2016) Histopathology of *Sclerotinia sclerotiorum* infection and oxalic acid function in susceptible and resistant soybean. *Plant Pathology* 65(6), 878–887. DOI: 10.1111/ppa.12514.

de Bary, A. (1886) *Ueber einige Sclerotinien and Sclerotienkrankheiten*. *Botanische Zeitung* 44, 377–474.

de Bary, A. (1887) *Comparative Morphology and Biology of Fungi*, Vol. 44. Clarendon Press, Oxford, pp. 377–474. DOI: 10.5962/bhl.title.1274.

De Vries, O.M.H. and Wessels, J.G.H. (1972) Release of protoplasts from *Schizophyllum* commune by a lytic enzyme preparation from *Trichoderma viride*. *Journal of General Microbiology* 73(1), 13–22. DOI: 10.1099/00221287-73-1-13.

Del Sorbo, G., Ruocco, M., Schoonbeek, H.J., Scala, F., Pane, C. *et al.* (2008) Cloning and functional characterization of *BcatrA*, a gene encoding an ABC transporter of the plant pathogenic fungus *Botryotinia fuckeliana* (*Botrytis cinerea*). *Mycological Research* 112, 737–746.

Derbyshire, M., Denton-Giles, M., Hegedus, D., Seifbarghy, S., Rollins, J. *et al.* (2017) The complete genome sequence of the phytopathogenic fungus *Sclerotinia sclerotiorum* reveals insights into the genome architecture of broad host range pathogens. *Genome Biology and Evolution* 9(3), 593–618. DOI: 10.1093/gbe/evx030.

Dhar, P. and Kaur, G. (2009) Optimization of different factors for efficient protoplast release from entomopathogenic fungus *Metarhizium anisopliae*. *Annals of Microbiology* 59(1), 183–186. DOI: 10.1007/BF03175619.

Dhawan, S. (1980) Production of protease by *Sclerotinia sclerotiorum* cause of white rot of *Brassica juncea*. *Current Science* 49, 361–362.

Dhawan, S. and Srivastava, S.K. (1987) Trans-eliminative (lytic) pectic enzymes in *Sclerotinia* infection of *Brassica juncea*. *Indian Journal of Mycology and Plant Pathology* 17, 325–327.

Dickman, M.B. and Mitra, A. (1992) *Arabidopsis thaliana* as a model for studying *Sclerotinia sclerotiorum* pathogenesis. *Physiological and Molecular Plant Pathology* 41(4), 255–263. DOI: 10.1016/0885-5765(92)90025-Q.

Dickman, M.B., Park, Y.K., Oltersdorf, T., Li, W., Clemente, T. *et al.* (2001) Abrogation of disease development in plants expressing animal antiapoptotic genes. *Proceedings of the National Academy of Sciences* 98(12), 6957–6962. DOI: 10.1073/pnas.091108998.

Ding, L.-N., Li, M., Guo, X.-J., Tang, M.-Q., Cao, J. *et al.* (2020) *Arabidopsis* GDSL1 overexpression enhances rapeseed *Sclerotinia sclerotiorum* resistance and the functional identification of its homolog in *Brassica napus*. *Plant Biotechnology Journal* 18(5), 1255–1270. DOI: 10.1111/pbi.13289.

Dong, X., Ji, R., Guo, X., Foster, S.J., Chen, H *et al.* (2008) Expressing a gene encoding wheat oxalate oxidase enhances resistance to *Sclerotinia sclerotiorum* in oilseed rape (*Brassica napus*). *Planta* 228(2), 331–340. DOI: 10.1007/s00425-008-0740-2.

Doughan, B., Rollins, J.A. and Avery, S.V. (2016) Characterization of MAT gene functions in the life cycle of *Sclerotinia sclerotiorum* reveals a lineage-specific MAT gene functioning in apothecium morphogenesis. *Fungal Biology* 120(9), 1105–1117. DOI: 10.1016/j.funbio.2016.06.007.

Dow, R.L. and Lumsden, R.D. (1975) Histology of resistance of *Phaseolus coccineus* (Scarlet runner bean) to *Sclerotinia sclerotiorum*. *Proceedings of the American Phytopathological Society* 2, 122.

Dowd, C.A., Buckley, C.M. and Sheehan, D. (1997) Glutathione S-transferases from the white-rot fungus, *Phanerochaete chrysosporium*. *The Biochemical Journal* 324(Pt 1), 243–248. DOI: 10.1042/bj3240243.

Duan, Y., Ge, C., Liu, S., Wang, J. and Zhou, M. (2013) A two-component histidine kinase shk1 controls stress response, sclerotial formation and fungicide resistance in *Sclerotinia sclerotiorum*. *Molecular Plant Pathology* 14(7), 708–718. DOI: 10.1111/mpp.12041.

Dutton, M.V. and Evans, C.S. (1996) Oxalate production by fungi: Its role in pathogenicity and ecology in the soil environment. *Canadian Journal of Microbiology* 42(9), 881–895. DOI: 10.1139/m96-114.

Echandi, E. and Walker, J.C. (1957) Pectolytic enzymes produced by *Sclerotinia sclerotiorum*. *Phytopathology* 47, 303–306.

Emerson, S. and Emerson, M.R. (1958) Production, reproduction, and reversion of protoplast-like structures in the osmotic strain of *Neurospora crassa*. *Proceedings of the National Academy of Sciences of the United States of America* 44(7), 668–671. DOI: 10.1073/pnas.44.7.668.

Erental, A., Harel, A. and Yarden, O. (2007) Type 2A phosphoprotein phosphatase is required for asexual development and pathogenesis of *Sclerotinia sclerotiorum*. *Molecular Plant-Microbe Interactions* 20(8), 944–954. DOI: 10.1094/MPMI-20-8-0944.

Fan, H., Yu, G., Liu, Y., Zhang, X., Liu, J *et al.* (2017) An atypical fork head-containing transcription factor ssfkh1 is involved in sclerotial formation and is essential for pathogenicity in *Sclerotinia sclerotiorum*. *Molecular Plant Pathology* 18(7), 963–975. DOI: 10.1111/mpp.12453.

Fan, H., Yang, W., Nie, J., Zhang, W., Wu, J. *et al.* (2021) A novel effector protein SsERP1 inhibits plant ethylene signaling to promote *Sclerotinia sclerotiorum* infection. *Journal of Fungi* 7(10), 825. DOI: 10.3390/jof7100825.

Favaron, F., Sella, L. and D'Ovidio, R. (2004) Relationships among endo-polygalacturonase, oxalate, pH, and plant polygalacturonase-inhibiting protein (PGIP) in the interaction between *Sclerotinia sclerotiorum* and soybean. *Molecular Plant-Microbe Interactions* 17(12), 1402–1409. DOI: 10.1094/MPMI.2004.17.12.1402.

Fenice, M., Giambattista, R.D., Leuba, J.-L. and Federici, F. (1999) Inactivation of *Mucor plumbeus* by the combined actions of chitinase and high hydrostatic pressure. *International Journal of Food Microbiology* 52(1–2), 109–113. DOI: 10.1016/S0168-1605(99)00114-2.

Fraissinet-Tachet, L., Reymond-Cotton, P. and Fevre, M. (1995) Characterization of a multi-gene family encoding an endopolygalacturonase in *Sclerotinia sclerotiorum*. *Current Microbiology* 29, 96–100.

Frisch, T., Motawia, M.S., Olsen, C.E., Agerbirk, N., Møller, B.L. *et al.* (2015) Diversified glucosinolate metabolism: Biosynthesis of hydrogen cyanide and of the hydroxynitrile glucoside alliarinoside in relation to sinigrin metabolism in *Alliaria petiolata*. *Frontiers in Plant Science* 6, 926. DOI: 10.3389/fpls.2015.00926.

Garg, H., Sivasithamparam, K., Banga, S.S. and Barbetti, M.J. (2008) Cotyledon assay as a rapid and reliable method of screening for resistance against *Sclerotinia sclerotiorum* in *Brassica napus* genotypes. *Australasian Plant Pathology* 37(2), 106. DOI: 10.1071/AP08002.

Garg, H., Sivasithamparam, K. and Barbetti, M.J. (2010) Scarification and environmental factors that enhance carpogenic germination of sclerotia of *Sclerotinia sclerotiorum*. *Plant Disease* 94(8), 1041–1047. DOI: 10.1094/PDIS-94-8-1041.

Garrido-Sanz, D., Meier-Kolthoff, J.P., Göker, M., Martín, M., Rivilla, R. *et al.* (2016) Genomic and genetic diversity within the *Pseudomonas fluorescens* complex. *PLoS ONE* 11(2), e0150183. DOI: 10.1371/journal.pone.0150183.

Giraldo, M.C. and Valent, B. (2013) Filamentous plant pathogen effectors in action. *Nature Reviews. Microbiology* 11(11), 800–814. DOI: 10.1038/nrmicro3119.

Godoy, G., Steadman, J.R., Dickman, M.B. and Dam, R. (1990) Use of mutants to demonstrate the role of oxalic acid in pathogenicity of *Sclerotinia sclerotiorum* on *Phaseolus vulgaris*. *Physiological and Molecular Plant Pathology* 37(3), 179–191. DOI: 10.1016/0885-5765(90)90010-U.

González-Fernández, R., Aloria, K., Valero-Galván, J., Redondo, I., Arizmendi, J.M. *et al.* (2014) Proteomic analysis of mycelium and secretome of different *Botrytis cinerea* wild-type strains. *Journal of Proteomics* 97, 195–221. DOI: 10.1016/j.jprot.2013.06.022.

Govrin, E.M. and Levine, A. (2000) The hypersensitive response facilitates plant infection by the necrotrophic pathogen *Botrytis cinerea*. *Current Biology* 10(13), 751–757. DOI: 10.1016/S0960-9822(00)00560-1.

Guimarães, R.L. and Stotz, H.U. (2004) Oxalate production by *Sclerotinia sclerotiorum* deregulates guard cells during infection. *Plant Physiology* 136(3), 3703–3711. DOI: 10.1104/pp.104.049650.

Guyon, K., Balagué, C., Roby, D. and Raffaele, S. (2014) Secretome analysis reveals effector candidates associated with broad host range necrotrophy in the fungal plant pathogen *Sclerotinia sclerotiorum*. *BMC Genomics* 15(1), 336. DOI: 10.1186/1471-2164-15-336.

Hancock, J.G. (1966) Degradation of pectic substances associated with pathogenesis by *Sclerotinia sclerotiorum* in sunflower and tomato stems. *Phytopathology* 56, 975–979.

Hancock, J.G. (1972) Changes in cell membrane permeability in sunflower hypocotyls infected with *Sclerotinia sclerotiorum*. *Plant Physiology* 49(3), 358–364. DOI: 10.1104/pp.49.3.358.

Han, Y., Joosten, H.-J., Niu, W., Zhao, Z., Mariano, P.S *et al.* (2007) Oxaloacetate hydrolase, the C-C bond lyase of oxalate secreting fungi. *The Journal of Biological Chemistry* 282(13), 9581–9590. DOI: 10.1074/jbc.M608961200.

Harel, A., Bercovich, S. and Yarden, O. (2006) Calcineurin is required for sclerotial development and pathogenicity of *Sclerotinia sclerotiorum* in an oxalic acid-independent manner. *Molecular Plant-Microbe Interactions* 19(6), 682–693. DOI: 10.1094/MPMI-19-0682.

Hartman, G.L., Kull, L. and Huang, Y.H. (1998) Occurrence of *Sclerotinia sclerotiorum* in soybean fields in east-central illinois and enumeration of inocula in soybean seed lots. *Plant Disease* 82(5), 560–564. DOI: 10.1094/PDIS.1998.82.5.560.

Harwood, C.S. and Parales, R.E. (1996) The beta-ketoadipate pathway and the biology of self-identity. *Annual Review of Microbiology* 50, 553–590. DOI: 10.1146/annurev.micro.50.1.553.

Hayashi, K., Schoonbeek, H. and Waard, M.A. (2002a) Expression of the ABC transporter BcatrD from *Botrytis cinerea* reduces sensitivity to sterol demethylation inhibitor fungicides. *Pesticide Biochemistry and Physiology* 73(2), 110–121. DOI: 10.1016/S0048-3575(02)00015-9.

Hayashi, K., Schoonbeek, H.J. and Waard, M.A. (2002b) Bcmfs1, a novel major facilitator superfamily transporter from *Botrytis cinerea*, provides tolerance towards the natural toxic compounds camptothecin and cercosporin and towards fungicides. *Applied and Environmental Microbiology* 68(10), 4996–5004. DOI: 10.1128/AEM.68.10.4996-5004.2002.

Heard, S., Brown, N.A. and Hammond-Kosack, K. (2015) An interspecies comparative analysis of the predicted secretomes of the necrotrophic plant pathogens *Sclerotinia sclerotiorum* and *Botrytis cinerea*. *PLoS ONE* 10(6), e0130534. DOI: 10.1371/journal.pone.0130534.

Hegedus, D.D. and Rimmer, S.R. (2005) *Sclerotinia sclerotiorum*: When "to be or not to be" a pathogen? *FEMS Microbiological Letters* 251, 177–184.

Hegedus, D.D., Gerbrandt, K. and Coutu, C. (2016) The eukaryotic protein kinase superfamily of the necrotrophic fungal plant pathogen. *Sclerotinia Sclerotiorum*. *Molecular Plant Pathology* 17(4), 634–647. DOI: 10.1111/mpp.12321.

Held, Y.M. (1955) Physiological differences between a normal and a degenerate strain of *Sclerotinia trifoliorum*. *Phytopathology* 45, 39–42.

Herbers, K., Takahata, Y., Melzer, M., Mock, H.P., Hajirezaei, M. *et al.* (2000) Regulation of carbohydrate partitioning during the interaction of potato virus Y with tobacco. *Molecular Plant Pathology* 1(1), 51–59. DOI: 10.1046/j.1364-3703.2000.00007.x.

Hu, X., Bidney, D.L., Yalpani, N., Duvick, J.P., Crasta, O. *et al*. (2003) Over expression of a gene encoding hydrogen peroxide-generating oxalate oxidase evokes defense responses in sunflower. *Plant Physiology* 133(1), 170–181. DOI: 10.1104/pp.103.024026.

Huang, H.C., Chang, C. and Kozub, G.C. (1998) Effect of temperature during sclerotial formation, sclerotial dryness, and relative humidity on myceliogenic germination of sclerotia of *Sclerotinia sclerotiorum*. *Canadian Journal of Botany* 76(3), 494–499. DOI: 10.1139/b98-016.

Huang, L., Buchenauer, H., Han, Q., Zhang, X. and Kang, Z. (2008) Ultrastructural and cytochemical studies on the infection process of *Sclerotinia sclerotiorum* in oilseed rape. *Journal of Plant Diseases and Protection* 115(1), 9–16. DOI: 10.1007/BF03356233.

Hudyncia, J., Shew, H.D., Cody, B.R. and Cubeta, M.A. (2000) Evaluation of wounds as a factor to infection of cabbage by ascospores of *Sclerotinia sclerotiorum*. *Plant Disease* 84(3), 316–320. DOI: 10.1094/PDIS.2000.84.3.316.

Hulvey, J., Popko, J.T.J., Sang, H., Berg, A. and Jung, G. (2012) Over expression of ShCYP51B and shatrD in *Sclerotinia homoeocarpa* isolates exhibiting practical field resistance to a demethylation inhibitor fungicide. *Applied and Environmental Microbiology* 78(18), 6674–6682. DOI: 10.1128/AEM.00417-12.

Hungerford, C.W. and Pitts, R. (1953) The Sclerotinia disease of beans in Idaho. *Phytopathology* 43, 519–521.

Jamaux, I., Gelie, B. and Lamarque, C. (1995) Early stages of infection of rapeseed petals and leaves by *Sclerotinia sclerotiorum* revealed by SEM. *Plant Pathology* 44(1), 22–30. DOI: 10.1111/j.1365-3059.1995.tb02712.x.

Jones, D. (1976) Infection of plant tissue by *Sclerotinia sclerotiorum*: A scanning electron microscope study. *Micron* 7(4), 275–279. DOI: 10.1016/0047-7206(76)90029-7.

Joosten, M.H.A.J., Hendrickx, L.J.M. and DeWit, P.J.G.M. (1990) Carbohydrate composition of apoplastic fluids isolated from tomato leaves inoculated with virulent or avirulent races of *Cladosporium fulvum* (syn. *Fulvia Fulva*). *Netherlands Journal of Plant Pathology* 96(2), 103–112. DOI: 10.1007/BF02005134.

Jurick, W.M. and Rollins, J.A. (2007) Deletion of the adenylate cyclase (*sac1*) gene affects multiple developmental pathways and pathogenicity in *Sclerotinia sclerotiorum*. *Fungal Genetics and Biology* 44(6), 521–530. DOI: 10.1016/j.fgb.2006.11.005.

Jurick, W.M., II, Dickman, M.B. and Rollins, J.A. (2004) Characterization and functional analysis of a camp dependent protein kinase a catalytic subunit gene (*pka1*) in *Sclerotinia sclerotiorum*. *Physiological and Molecular Plant Pathology* 64(3), 155–163. DOI: 10.1016/j.pmpp.2004.07.004.

Kabbage, M., Williams, B. and Dickman, M.B. (2013) Cell death control: The interplay of apoptosis and autophagy in the pathogenicity of *Sclerotinia sclerotiorum*. *PLoS Pathogens* 9(4), e1003287. DOI: 10.1371/journal.ppat.1003287.

Kars, I., Krooshof, G.H., Wagemakers, L., Joosten, R., Benen, J.A.E. *et al*. (2005) Necrotizing activity of five *Botrytis cinerea* endopolygalacturonases produced in *Pichia pastoris*. *The Plant Journal* 43(2), 213–225. DOI: 10.1111/j.1365-313X.2005.02436.x.

Kelkar, H.S., Skloss, T.W., Haw, J.F., Keller, N.P. and Adams, T.H. (1997) *Aspergillus nidulans* stcL encodes a putative cytochrome P-450 monooxygenase required for bisfuran desaturation during aflatoxin/sterigmatocystin biosynthesis. *The Journal of Biological Chemistry* 272, 1589–1594.

Kennedy, J., Auclair, K., Kendrew, S.G., Park, C., Vederas, J.C. *et al*. (1999) Modulation of polyketide synthase activity by accessory proteins during lovastatin biosynthesis. *Science* 284(5418), 1368–1372. DOI: 10.1126/science.284.5418.1368.

Kim, H., Chen, C., Kabbage, M. and Dickman, M.B. (2011) Identification and characterization of *Sclerotinia sclerotiorum* NADPH oxidases. *Applied and Environmental Microbiology* 77(21), 7721–7729. DOI: 10.1128/AEM.05472-11.

Kim, K.S., Min, J.Y. and Dickman, M.B. (2008) Oxalic acid is an elicitor of plant programmed cell death during *Sclerotinia sclerotiorum* disease development. *Molecular Plant Microbe Interactions* 21, 605–612.

Kleman -Leyer, K.M., Siika-Aho, M., Teeri, T.T. and Kirk, T.K. (1996) The cellulases endoglucanase i and cellobiohydrolase II of *Trichoderma reesei* act synergistically to solubilize native cotton cellulose but not to decrease its molecular size. *Applied and Environmental Microbiology* 62, 2883–2887.

Kocal, N., Sonnewald, U. and Sonnewald, S. (2008) Cell wall-bound invertase limits sucrose export and is involved in symptom development and inhibition of photosynthesis during compatible interaction between tomato and *Xanthomonas campestris* pv. *vesicatoria*. *Plant Physiology* 148, 1523–1536.

Kowalska, A. and Niks, R.E. (1999) Histology of quantitative resistance in flax to the flax rust fungus (*Melampsora lini*). *Canadian Journal of Plant Pathology* 21(4), 354–360. DOI: 10.1080/07060669909501171.

Kubicek, C.P., Kunar, G.S. and Rohr, W.W.M. (1988) Evidence for a cytoplasmic pathway of oxalate biosynthesis in *Aspergillus niger*. *Applied Environment Microbiology* 54(3), 633–637.

Li, W.M.J. and Rollins, J.A. (2007) Deletion of the adenylate cyclase (sac1) gene affects multiple developmental pathways and pathogenicity in *Sclerotinia sclerotiorum*. *Fungal Genetics and Biology* 44(6), 521–530. DOI: 10.1016/j.fgb.2006.11.005.

Li, H., Stone, V., Dean, N., Sivasithamparam, K. and Barbetti, M.J. (2007) Breaching by a new strain of *Leptosphaeria maculans* of anatomical barriers in cotyledons of *Brassica napus* cultivar surpass 400 with resistance based on a single dominant gene. *Journal of General Plant Pathology* 73(5), 297–303. DOI: 10.1007/s10327-007-0011-2.

Li, J.T., Zhang, Y.H., Zhang, Y.C., Yu, P.-L., Pan, H. *et al.* (2018) Introduction of large sequence inserts by CRISPR-Cas9 to create pathogenicity mutants in the multinucleate filamentous pathogen *Sclerotinia sclerotiorum*. *MBio* 9(3), e00567–18. DOI: 10.1128/mBio.00567-18.

Li, M., Liang, X. and Rollins, J.A. (2012) *Sclerotinia sclerotiorum* γ-glutamyl transpeptidase (Ss-Ggt1) is required for regulating glutathione accumulation and development of sclerotia and compound appressoria. *Molecular Plant-Microbe Interactions* 25(3), 412–420. DOI: 10.1094/MPMI-06-11-0159.

Li, R., Rimmer, R., Buchwaldt, L., Sharpe, A.G., Séguin-Swartz, G. *et al.* (2004) Interaction of *Sclerotinia sclerotiorum* with *Brassica napus*: Cloning and characterization of endo- and exo-polygalacturonases expressed during saprophytic and parasitic modes. *Fungal Genetics and Biology* 41(8), 754–765. DOI: 10.1016/j.fgb.2004.03.002.

Li, X., Ximenes, E., Kim, Y., Slininger, M., Meilan, R. *et al.* (2010) Lignin monomer composition affects *Arabidopsis* cell-wall degradability after liquid hot water pretreatment. *Biotechnology for Biofuels* 3, 27. DOI: 10.1186/1754-6834-3-27.

Li, X., Guo, M., Xu, D., Chen, F., Zhang, H. *et al.* (2015) The nascent-polypeptide-associated complex alpha subunit regulates the polygalacturonases expression negatively and influences the pathogenicity of *Sclerotinia sclerotiorum*. *Mycologia* 107(6), 1130–1137. DOI: 10.3852/14-250.

Li RuGang, L.R., Rimmer, R., Buchwaldt, L., Sharpe, A.G., Séguin-Swartz, G. and Hegedus, D.D. (2004a) Interaction of *Sclerotinia sclerotiorum* with *Brassica napus*: Cloning and characterization of endo- and exo-polygalacturonases expressed during saprophytic and parasitic modes. *Fungal Genetics and Biology* 41(8), 754–765. DOI: 10.1016/j.fgb.2004.03.002.

Li RuGang, L.R., Rimmer, R., Buchwaldt, L., Sharpe, A.G., Séguin-Swartz, G., Coutu, C. *et al.* (2004b) Interaction of *Sclerotinia sclerotiorum* with a resistant *Brassica napus* cultivar: Expressed sequence tag analysis identifies genes associated with fungal pathogenesis. *Fungal Genetics and Biology* 41(8), 735–753. DOI: 10.1016/j.fgb.2004.03.001.

Liang, X. and Rollins, J.A. (2018) Mechanisms of broad host range necrotrophic pathogenesis in *Sclerotinia sclerotiorum*. *Phytopathology* 108(10), 1128–1140. DOI: 10.1094/PHYTO-06-18-0197-RVW.

Liang, X., Moomaw, E.W. and Rollins, J.A. (2015a) Fungal oxalate decarboxylase activity contributes to *Sclerotinia sclerotiorum* early infection by affecting both compound appressoria development and function. *Molecular Plant Pathology* 16(8), 825–836. DOI: 10.1111/mpp.12239.

Liang, X., Liberti, D., Li, M., Kim, Y.T., Hutchens, A. *et al.* (2015b) Oxaloacetate acetylhydrolase gene mutants of *Sclerotinia sclerotiorum* do not accumulate oxalic acid, but do produce limited lesions on host plants. *Molecular Plant Pathology* 16(6), 559–571. DOI: 10.1111/mpp.12211.

Liang, Y., Yajima, W., Davis, M.R., Kav, N.N.V. and Stephen, E. (2013) Disruption of a gene encoding a hypothetical secreted protein from *Sclerotinia sclerotiorum* reduces its virulence on canola (*Brassica napus*) disruption of a gene encoding a hypothetical secreted protein from *Sclerotinia sclerotiorum* reduces its virulence on canola (*Brassica napus*). *Canadian Journal of Plant Pathology* 35, 46–55. DOI: 10.1080/07060661.2012.745904.

Liang, Y.U.E., Xiong, W.E.I., Steinkellner, S. and Feng, J.I.E. (2017) Deficiency of the melanin biosynthesis genes *SCD1* and *THR1* affects sclerotial development and vegetative growth, but not pathogenicity, in *Sclerotinia sclerotiorum*. *Molecular Plant Pathology* 19(6), 1444–1453. DOI: 10.1111/mpp.12627.

Liberti, D., Rollins, J.A. and Dobinson, K.F. (2013) Peroxysomal carnitine acetyl transferase influences host colonization capacity in *Sclerotinia sclerotiorum*. *Molecular Plant Microbe Interactions* 26, 768–780. DOI: 10.1094/MPMI-03-13-0075-R.

Liu, L., Wang, Q., Zhang, X., Liu, J., Zhang, Y. *et al.* (2018) *Ssams2*, a gene encoding GATA transcription factor, is required for appressoria formation and chromosome segregation in *Sclerotinia sclerotiorum*. *Frontiers in Microbiology* 9, 1–12. DOI: 10.3389/fmicb.2018.03031.

Long, D.E., Fung, A.K., McGEE, E.E.M., Cooke, R.C. and Lewis, D.H. (1975) The activity of invertase and its relevance to the accumulation of storage polysaccharides in leaves infected by biotrophic fungi. *New Phytologist* 74(2), 173–182. DOI: 10.1111/j.1469-8137.1975.tb02603.x.

Lo Presti, L., Lanver, D., Schweizer, G., Tanaka, S., Liang, L. *et al.* (2015) Fungal effectors and plant susceptibility. *Annual Review of Plant Biology* 66, 513–545. DOI: 10.1146/annurev-arplant-043014-114623.

Lumsden, R.D. (1969) *Sclerotinia sclerotiorum* infection of bean and the production of cellulose. *Phytopathology* 59, 653–657.

Lumsden, R.D. (1979) Histology and physiology of pathogenesis in plant diseases caused by *Sclerotinia* species. *Phytopathology* 69(8): 890. DOI: 10.1094/Phyto-69-890.

Lumsden, R.D. (1976) Pectolytic enzymes of *Sclerotinia sclerotiorum* and their localization in infected bean. *Canadian Journal of Botany* 54(23), 2630–2641. DOI: 10.1139/b76-283.

Lumsden, R.D. and Dow, R.L. (1973) Histopathology of *Sclerotinia sclerotiorum* infection of bean. *Phytopathology* 63(6): 708. DOI: 10.1094/Phyto-63-708.

Lumsden, R.D. and Wergin, W.P. (1980) Scanning-electron microscopy of infection of bean by species of *Sclerotinia*. *Mycologia* 72(6), 1200–1209. DOI: 10.1080/00275514.1980.12021302.

Luo, C.X. and Schnabel, G. (2008) The cytochrome P450 Lanosterol 14alpha-demethylase gene is a demethylation inhibitor fungicide resistance determinant in *Monilinia fructicola* field isolates from Georgia. Applied and Environmental Microbiology 74(2), 359–366. DOI: 10.1128/AEM.02159-07.

Lyu, X., Shen, C., Fu, Y., Xie, J., Jiang, D. *et al.* (2015) Comparative genomic and transcriptional analyses of the carbohydrate-active enzymes and secretomes of phytopathogenic fungi reveal their significant roles during infection and development. *Scientific Reports* 5, 15565. DOI: 10.1038/srep15565.

Lyu, X., Shen, C., Fu, Y., Xie, J., Jiang, D. *et al.* (2016) A small secreted virulence-related protein is essential for the necrotrophic interactions of *Sclerotinia sclerotiorum* with its host plants. *PLoS Pathogens* 12(2), e1005435. DOI: 10.1371/journal.ppat.1005435.

Magro, P., Marciano, P. and Lenna, P. (1984) Oxalic acid production and its role in pathogenesis of *Sclerotinia sclerotiorum*. *FEMS Microbiology Letters* 24(1), 9–12. DOI: 10.1111/j.1574-6968.1984.tb01234.x.

Magro, P., Marciano, P. and Lenna, P. (1988) Enzymatic oxalate decarboxylation in isolates of *Sclerotinia sclerotiorum*. *FEMS Microbiological Letters* 49, 49–52.

Makizumi, Y., Takeda, S., Matsuzaki, Y., Nakaune, R., Hamamoto, H. *et al.* (2002) Cloning and selective toxicant-induced expression of BMR1 and BMR3, novel ABC transporter genes in *Botrytis cinerea*. *Journal of General Plant Pathology* 68(4), 338–341. DOI: 10.1007/PL00013100.

Marciano, P., Lenna, P.D. and Magro, P. (1983) Oxalic acid, cell wall-degrading enzymes and ph in pathogenesis and their significance in the virulence of two *Sclerotinia sclerotiorum* isolates on sunflower. *Physiological Plant Pathology* 22(3), 339–345. DOI: 10.1016/S0048-4059(83)81021-2.

Martel, M.B., Letoublon, R. and Fevre, M. (1996) Purification of endopolygalacturonases from *Sclerotinia sclerotiorum*: Multiplicity of the complex enzyme system. *Current Microbiology* 33, 243–248.

Maxwell, D.P. (1973) Oxalate formation in *Whetzelinia sclerotiorum* by oxaloacetate acetylhydrolase. *Physiological Plant Pathology* 3(2), 279–288. DOI: 10.1016/0048-4059(73)90090-8.

Maxwell, D.P. and Lumsden, R.D. (1970) Oxalic acid production by *Sclerotinia sclerotiorum* in infected bean and in culture. *Phytopathology* 60(9), 1395. DOI: 10.1094/Phyto-60-1395.

McCartney, A., Heran, A. and Li, Q. (2001a) Infection of oilseed rape (*Brassica napus*) by petals containing ascospores of *Sclerotinia sclerotiorum*. In: Young, C.S. and Hughes, K.J.D. (eds) *Proceedings of Sclerotinia 2001, the XI International Sclerotinia Workshop*, Central Science Laboratory, pp. 183–184.

McCartney, A., Heran, A., Li, Q. and Freeman, J. (2001b) Petal fall, petal retention and petal duration in oilseed rape crops. In: Young, C.S. and Hughes, K.J.D. (eds) *Proceedings of Sclerotinia 2001, the XI International Sclerotinia Workshop*, Central Science Laboratory, pp. 185–186.

Miclaus, D., Sin, G., Damian, V. and Guram, M. (1988) Presence and spread of *Sclerotinia sclerotiorum* (Lib) de Bary. Seeds of various cereals and industrial crops. In: *I Analele Institutului de Cercetari Pentru Cereale Si Plante Technice*, Vol. Fundulea 56. pp. 169–188.

Mijatovic, S. (2008) Biochemical characterization of 2-nitropropane dioxygenase from hansenula MRAKII. MSc. thesis. Atlanta: Georgia State University. Available at: https://scholarworks.gsu.edu/chemistry_theses/8

Minarcic, P. and Janitor, A. (1994) Changes of starch content in above-ground parts of barley plants infected by powdery mildew (*Erysiphe graminis*). *Bulletin Slovenskej Botanickej Spolocnosti* 49, 77–82.

Morrall, R.A.A., Duczek, L.J. and Sheard, J.W. (1972) Variations and correlations within and between morphology, pathogenicity, and pectolytic enzyme activity in *Sclerotinia* from Saskatchewan. *Canadian Journal of Botany* 50(4), 767–786. DOI: 10.1139/b72-095.

Neergaard, P. (1958) Mycelial seed infection of certain crucifers by *Sclerotinia sclerotiorum* (Lib) de bary. *Plant Disease Reporter* 42, 1105–1106.

Newton, H.K. (1972) *Sclerotinia sclerotiorum* an incitant of lettuce drop: Sources of inoculum, host resistance and mechanism of pathogenesis. PhD. thesis, Murofolm 72-V.292, University of Murofilm, Ann Arborr, MI.

Newton, H.K. and Sequeira, L. (1972) Ascospores as the primary infective propagules of *Sclerotinia sclerotiorum* in Wisconsin. *Plant Disease Reporter* 56, 798–802.

Nicholson, J.F., Dhingra, O.D. and Sinclair, J.B. (1972) Internal seed-borne nature of *Sclerotinia sclerotiorum* and *Phomopsis* spp. and their effects on soybean seed quality. *Phytopathology* 62(11), 1261. DOI: 10.1094/Phyto-62-1261.

Nobel, M. and Richardson, M.J. (1968) *An Annotated List of Seed Borne Diseases*. Common Wealth Mycological Institute, Kew, Surrey, UK, p. 191.

Oliveira, M.B., Andrade, R.V., Grossi-de-Sá, M.F. and Petrofeza, S. (2015) Analysis of genes that are differentially expressed during the *Sclerotinia sclerotiorum-Phaseolus vulgaris* interaction. *Frontiers in Microbiology* 6, 1162. DOI: 10.3389/fmicb.2015.01162.

Olivieri, F., Eugenia Zanetti, M., Oliva, C.R., Covarrubias, A.A. and Casalongué, C.A. (2002) Characterization of an extracellular serine protease of *Fusarium eumartii* and its action on pathogenesis related proteins. *European Journal of Plant Pathology* 108(1), 63–72. DOI: 10.1023/A:1013920929965.

Orshinsky, A.M., Hu, J., Opiyo, S.O., Reddyvari-Channarayappa, V., Mitchell, T.K. *et al.* (2012) RNA-seq analysis of the *Sclerotinia homoeocarpa*—Creeping bentgrass pathosystem. *PLoS ONE* 7(8), e41150. DOI: 10.1371/journal.pone.0041150.

Pan, H. and Rollins, J.A. (2017) The GATA-Type IVb zinc-finger transcription factor SsNsd1 regulates asexual–sexual development and appressoria formation in *Sclerotinia sclerotiorum*. *Molecular Plant Pathology* 19(7), 1679–1689. DOI: 10.1111/mpp.12651.

Pan, Y., Wei, J., Yao, C., Reng, H. and Gao, Z. (2018) Plant science SsSm1, a cerato-platanin family protein, is involved in the hyphal development and pathogenic process of *Sclerotinia sclerotiorum*. *Plant Science* 270, 37–46. DOI: 10.1016/j.plantsci.2018.02.001.

Parks, L.W. and Casey, W.M. (1995) Physiological implications of sterol biosynthesis in yeast. *Annual Review of Microbiology* 49, 95–116. DOI: 10.1146/annurev.mi.49.100195.000523.

Pawlowski, S.H. and Hawn, E.J. (1964) Host–parasite relationship in sunflower wilt incited by *Sclerotinia sclerotiorum* as determined by the twin technique. *Phytopathology* 54, 33–35.

Pedras, M.S.C. and Ahlahonu, P.W.K. (2004) Phytotoxin production and phytoalexin elicitation by the phytopathogenic fungus *Sclerotinia sclerotiorum*. *Journal of Chemical Ecology* 30(11), 2163–2179. DOI: 10.1023/B:JOEC.0000048781.72203.6c.

Peng, Q., Xie, Q., Chen, F., Zhou, X., Zhang, W. *et al.* (2017) Transcriptome analysis of *Sclerotinia sclerotiorum* at different infection stages on *Brassica napus*. *Current Microbiology* 74(10), 1237–1245. DOI: 10.1007/s00284-017-1309-8.

Perlin, M.H., Andrews, J. and Toh, S.S. (2014) Essential letters in the fungal alphabet: ABC and MFS transporters and their roles in survival and pathogenicity. *Advances in Genetics* 85, 201–253. DOI: 10.1016/B978-0-12-800271-1.00004-4.

Pitkin, J.W., Panaccione, D.G. and Walton, J.D. (1996) A putative cyclic peptide efflux pump encoded by the *TOXA* gene of the plant-pathogenic fungus *Cochliobolus carbonum*. *Microbiology* 142, 1557–1565. DOI: 10.1099/13500872-142-6-1557.

Porter, D.M. and Beute, M.K. (1974) Sclerotinia blight of peanuts. *Phytopathology* 64(2), 263. DOI: 10.1094/Phyto-64-263.

Poussereau, N., Creton, S., Billon-Grand, G., Rascle, C. and Fevre, M. (2001a) Regulation of acp1 encoding a non-aspatyl acid protease expressed during pathogenesis of *Sclerotinia sclerotiorum*. *Microbiology* 147, 717–726.

Poussereau, N., Gente, S., Rascle, C., Billon-Grand, G. and Fever, M. (2001b) AspS encoding an unusual aspartyl protease from *Sclerotinia sclerotiorum* is expressed during phytopathogenesis. *FEMS Microbiology Letters* 194(1), 27–32. DOI: 10.1111/j.1574-6968.2001.tb09441.x.

Prins, T.W., Wagemakers, L., Schouten, A. and Kan, J.A.L. (2000) Cloning and characterization of a glutathione S-transferase homologue from the plant pathogenic fungus *Botrytis cinerea*. *Molecular Plant Pathology* 1, 169–178. DOI: 10.1046/j.1364-3703.2000.00021.

Prior, G.D. and Owen, J.H. (1964) Pathological anatomy of *Sclerotinia trifoliorum* on clover and alfalfa. *Phytopathology* 54, 784–787.

Punja, Z.K., Huang, J.S. and Jenkins, S.F. (1985) Relationship of mycelial growth and production of oxalic acid and cell wall degrading enzymes to virulence in *Sclerotium rolfsii*. *Canadian Journal of Plant Pathology* 7(2), 109–117. DOI: 10.1080/07060668509501485.

Purdy, L.H. (1958) Some factors affecting penetration and infection by *Sclerotinia sclerotiorum*. *Phytopathology* 48, 605–609.

Qiu, L., Jicheng, Y. and Dechun, F. (2001) Studies on the toxin from *Sclerotinia sclerotiorum*. *Journal of Shenyang Agricultural University (China); Shenyang Nongye Daxue Xuebao* 32, 422–425.

Qu, X., Yu, B., Liu, J., Zhang, X., Li, G *et al.* (2014) MADS-box transcription factor SsMADS is involved in regulating growth and virulence in *Sclerotinia sclerotiorum*. *International Journal of Molecular Sciences* 15(5), 8049–8062. DOI: 10.3390/ijms15058049.

Rai, J.N. and Dhawan, S. (1976) Production of polymethyl-galacturonase and cellulase and its relationship with virulence in isolates of *Sclerotinia sclerotiorum* (Lib.) de Bary. *Indian Journal of Experimental Biology* 14(2), 196–197.

Reis, H., Pfiffi, S. and Hahn, M. (2005) Molecular and functional characterization of a secreted lipase from *Botrytis cinerea*. *Molecular Plant Pathology* 6(3), 257–267. DOI: 10.1111/j.1364-3703.2005.00280.x.

Riechers, D.E., Irzyk, G.P., Jones, S.S. and Fuerst, E.P. (1997) Partial characterization of glutathione S-transferases from wheat (*Triticum* spp.) and purification of a safener-induced glutathione S-transferase from *Triticum tauschii*. *Plant Physiology* 114(4), 1461–1470. DOI: 10.1104/pp.114.4.1461.

Riou, C., Freyssinet, G. and Fevre, M. (1991) Production of cell wall-degrading enzymes by the phytopathogenic fungus *Sclerotinia sclerotiorum*. *Applied and Environmental Microbiology* 57(5), 1478–1484. DOI: 10.1128/aem.57.5.1478-1484.1991.

Roitsch, T., Balibrea, M.E., Hofmann, M., Proels, R. and Sinha, A.K. (2003) Extracellular invertase: Key metabolic enzyme and PR protein. *Journal of Experimental Botany* 54(382), 513–524. DOI: 10.1093/jxb/erg050.

Rollins, J.A. (2003) The *Sclerotinia sclerotiorum pac1* gene is required for sclerotial development and virulence. *Molecular Plant-Microbe Interactions* 16(9), 785–795. DOI: 10.1094/MPMI.2003.16.9.785.

Rollins, J.A. and Dickman, M.B. (2001) PH signaling in *Sclerotinia sclerotiorum*: Identification of a *pacc/rim1* homolog. *Applied and Environmental Microbiology* 67(1), 75–81. DOI: 10.1128/AEM.67.1.75-81.2001.

Rollins, J.A., Pathology, P. and Hall, F. (2003) The *Sclerotinia sclerotiorum pac1* gene is required for sclerotial development and virulence. *Molecular Plant Microbe Interactions* 16, 785–795. DOI: 10.1094/MPMI.2003.16.9.785.

Ruijter, G.J.G., Vondervoort, P.J.I. and Visser, J. (1999) Oxalic acid production by *Aspergillus niger*: An oxalate-non-producing mutant produces citric acid at pH 5 and in the presence of manganese. *Microbiology* 145, 2569–2576.

Sakamoto, T. and Ishimaru, M. (2013) Peculiarities and applications of galactanolytic enzymes that act on type I and II arabinogalactans. *Applied Microbiology and Biotechnology* 97(12), 5201–5213. DOI: 10.1007/s00253-013-4946-2.

Samolski, I., de Luis, A., Vizcaíno, J.A., Monte, E. and Suárez, M.B. (2009) Gene expression analysis of the biocontrol fungus *Trichoderma harzianum* in the presence of tomato plants, chitin, or glucose using a high-density oligonucleotide microarray. *BMC Microbiology* 9, 217. DOI: 10.1186/1471-2180-9-217.

Schoonbeek, H., Del Sorbo, G. and Waard, M.A. (2001) The ABC transporter bcatrB affects the sensitivity of *Botrytis cinerea* to the phytoalexin resveratrol and the fungicide fenpiclonil. *Molecular Plant-Microbe Interactions* 14, 562–571. DOI: 10.1094/MPMI.2001.

Schoonbeek, H., van Nistelrooy, J.G.M. and de Waard, M.A. (2003) Functional analysis of ABC transporter genes from *Botrytis cinerea* identifies bcatr B as a transporter of Eugenol. *European Journal of Plant Pathology* 109, 1003–1011.

Scott, S.W. (1981) Contamination of seed of white clover (*Trifolium repens*) with mycelium of a Sclerotinia species. *Transactions of the British Mycological Society* 77(3), 671–674. DOI: 10.1016/S0007-1536(81)80128-3.

Scott, S.W. and Evans, D.R. (1984) Sclerotia of *Sclerotinia trifoliorum* in red clover seed. *Transactions of the British Mycological Society* 82(3), 567–569. DOI: 10.1016/S0007-1536(84)80030-3.
Seifbarghi, S., Borhan, M.H., Wei, Y., Coutu, C., Robinson, S.J. *et al.* (2017) Changes in the *Sclerotinia sclerotiorum* transcriptome during infection of *Brassica napus*. *BMC Genomics* 18(1), 266. DOI: 10.1186/s12864-017-3642-5.
Seifbarghi, S., Borhan, M.H., Wei, Y., Ma, L., Coutu, C. *et al.* (2020) Receptor-like kinases BAK1 and SOBIR1 are required for necrotizing activity of a novel group of *Sclerotinia sclerotiorum* necrosis-inducing effectors. *Frontiers in Plant Science* 11, 1021. DOI: 10.3389/fpls.2020.01021.
Sexton, A.C. and Howlett, B.J. (2000) Characterisation of a cyanide hydratase gene in the phytopathogenic fungus *Leptosphaeria maculans*. *Molecular and General Genetics MGG* 263(3), 463–470. DOI: 10.1007/s004380051190.
Sexton, A.C., Cozijnsen, A.J., Keniry, A., Jewell, E., Love, C.G. *et al.* (2006) Comparison of transcription of multiple genes at three developmental stages of the plant pathogen *Sclerotinia sclerotiorum*. *FEMS Microbiology Letters* 258(1), 150–160. DOI: 10.1111/j.1574-6968.2006.00212.x.
Sexton, A.C., Minic, Z., Cozijnsen, A.J., Pedras, M.S.C. and Howlett, B.J. (2009) Cloning, purification and characterisation of brassinin glucosyltransferase, a phytoalexin-detoxifying enzyme from the plant pathogen *Sclerotinia sclerotiorum*. *Fungal Genetics and Biology* 46(2), 201–209. DOI: 10.1016/j.fgb.2008.10.014.
Shahoveisi, F. and del Río Mendoza, L.E. (2020) Effect of wetness duration and incubation temperature on development of ascosporic infections by *Sclerotinia sclerotiorum* plant disease 104(6), 1817–1823. DOI: 10.1094/PDIS-06-19-1304-RE.
Sharma, R.C. and Sharma, S.L. (1984) Studies on pectolytic enzymes associated with stalk rot of cauliflower caused by *Sclerotinia sclerotiorum* (Lib) de Bary. *Indian Journal of Plant Pathology* 2, 69–72.
Siewers, V., Viaud, M., Jimenez-Teja, D., Collado, I.G., Gronover, C.S. *et al.* (2005) Functional analysis of the cytochrome P450 monooxygenase gene *bcbot1* of *Botrytis cinerea* indicates that botrydial is a strain-specific virulence factor. *Molecular Plant-Microbe Interactions®* 18(6), 602–612. DOI: 10.1094/MPMI-18-0602.
Smith, V.L., Punja, Z.K. and Jenkins, S.F. (1986) A histological study of infection of host tissue by *Sclerotinia rolfsii*. *Phytopathology* 76, 755–759. DOI: 10.1094/Phyto-76-755.
Starr, G.H., Walters, H.J. and Bridgmon, G.H. (1953) White mould (*Sclerotinia*) of beans. *Wyoming Agricultural Experimental Station Technical Bulletin* 322, 11.
Starzycka-Korbas, E., Weber, Z., Matuszczak, M., Bocianowski, J., Budzianowski, G. *et al.* (2021) The diversity of *Sclerotinia sclerotiorum* (Lib.) de Bary isolates from western Poland. *Journal of Plant Pathology* 103(1), 185–195. DOI: 10.1007/s42161-020-00705-0.
Steadman, J.R. (1975) Nature and epidemiological significance of infection of bean seed by *Whetzelinia sclerotiorum*. *Phytopathology* 65(11), 1323. DOI: 10.1094/Phyto-65-1323.
Stefanato, F.L., Abou-Mansour, E., Buchala, A., Kretschmer, M., Mosbach, A. *et al.* (2009) The ABC transporter bcatrB from *Botrytis cinerea* exports camalexin and is a virulence factor on *Arabidopsis thaliana*. *The Plant Journal* 58(3), 499–510. DOI: 10.1111/j.1365-313X.2009.03794.x.
Sun, Q., Xi, Y., Lu, P., Lu, Y., Wang, Y. *et al.* (2022) Genome-wide analysis of the G-box regulating factors protein family reveals its roles in response to *Sclerotinia sclerotiorum* infection in rapeseed (*Brassica napus* L.). *Frontiers in Plant Science* 13, 986635. DOI: 10.3389/fpls.2022.986635.
Sutton, D.C. and Deverall, B.J. (1983) Studies on infection of bean (*Phaseolus vulgaris*) and soybean (*Glycine max*) by ascospores of *Sclerotinia sclerotiarum*. *Plant Pathology* 32(3), 251–261. DOI: 10.1111/j.1365-3059.1983.tb02832.x.
Talburt, D.E. and Johnson, G.T. (1965) Plasmoptysis of filamentous fungi. *Mycologia* 57(4), 660. DOI: 10.2307/3756743.
Tariq, V.N. and Jeffries, P. (1984) Appressorium formation by *Sclerotinia sclerotiorum*: Scanning electron microscopy. *Transactions of the British Mycological Society* 82(4), 645–651. DOI: 10.1016/S0007-1536(84)80105-9.
Tariq, V.N. and Jeffries, P. (1986) Ultra-structure penetration of *Phaseolus* spp. by *Sclerotinia sclerotiorum*. *Canadian Journal of Botany* 64, 2909–2915.
ten Have, A., Mulder, W., Visser, J. and Kan, J.A. (1998) The endopolygalacturonase gene *Bcpg*-1 is required for full virulence of *Botrytis cinerea*. *Molecular Plant Microbe Interactions* 11, 1009–1016.
Thatcher, F.S. (1942) Further studies of osmotic and permeability relations in parasitism. *Canadian Journal of Research* 283–311. DOI: 10.1139/cjr42c-027.

Thompson, C., Dunwell, J.M., Johnstone, C.E., Lay, V., Ray, J. *et al.* (1995) Degradation of oxalic acid by transgenic oilseed rape plants expressing oxalate oxidase. *Euphytica* 85(1–3), 169–172. DOI: 10.1007/BF00023945.

Toruno, T.Y., Stergiopoulos, I. and Coaker, G. (2016) Plant-pathogen effectors: Cellular probes interfering with plant defenses in spatial and temporal manners. *Annual Review of Phytopathology* 54, 419–441.

Tripathy, B.C., Sherameti, I. and Oelmüller, R. (2010) Siroheme: An essential component for life on earth. *Plant Signaling and Behavior* 5(1), 14–20. DOI: 10.4161/psb.5.1.10173.

Tu, J.C. (1988) The role of white mold infected white bean (*Phaseolus vulgaris* L.) seeds in the dissemination of *Sclerotinia sclerotiorum* (Lib) de bary. *Journal of Phytopathology* 121, 40–50. DOI: 10.1111/j.1439-0434.1988.tb00951.x.

van Etten, H., Temporini, E. and Wasmann, C. (2001) Phytoalexin (and phytoanticipin) tolerance as a virulence trait: Why is it not required by all pathogens. *Physiological and Molecular Plant Pathology* 59(2), 83–93. DOI: 10.1006/pmpp.2001.0350.

Vanden Berg, L. and Yang, S.M. (1969) Effect of relative humidity on production of extracellular pectolytic enzymes by *Botrytis cinerea* and *Sclerotinia sclerotiorum*. *Canadian Journal of Botany* 47(6), 1007–1010. DOI: 10.1139/b69-141.

Vardar, F. and Unal, M. (2008) Proteolytic enzymes in plant programmed cell death. *Türk Bilim Derlemeler Derg* 1, 65–78.

Veluchamy, S. and Rollins, J.A. (2008) A CRY-DASH-type photolyase/cryptochrome from *Sclerotinia sclerotiorum* mediates minor UV-a-specific effects on development. *Fungal Genetics and Biology* 45, 1265–1276. DOI: 10.1016/j.fgb.2008.06.004.

Veluchamy, S., Williams, B., Kim, K. and Dickman, M.B. (2012) The CuZn superoxide dismutase from *Sclerotinia sclerotiorum* is involved with oxidative stress tolerance, virulence, and oxalate production. *Physiological and Molecular Plant Pathology* 78, 14–23. DOI: 10.1016/j.pmpp.2011.12.005.

Voigt, C.A., Schäfer, W. and Salomon, S. (2005) A secreted lipase of *Fusarium graminearum* is a virulence factor required for infection of cereals. *Plant Journal* 42, 364–375.

Walton, J.D. (2006) HC-toxin. *Phytochemistry* 67, 1406–1413. DOI: 10.1016/j.phytochem.2006.05.

Wang, L., Liu, Y., Liu, J., Zhang, Y., Zhang, X. *et al.* (2016) The *Sclerotinia sclerotiorum FoxE2* gene is required for apothecial development. *Phytopathology* 106(5), 484–490. DOI: 10.1094/PHYTO-08-15-0181-R.

Wang, X., Jiang, N., Liu, J., Liu, W. and Wang, G.L. (2014) The role of effectors and host immunity in plant-necrotrophic fungal interactions. *Virulence* 5(7), 722–732. DOI: 10.4161/viru.29798.

Westrick, N.M., Ranjan, A., Jain, S., Grau, C.R., Smith, D.L. *et al.* (2019) Gene regulation of *Sclerotinia sclerotiorum* during infection of glycine max: On the road to pathogenesis. *BMC Genomics* 20(1), 157. DOI: 10.1186/s12864-019-5517-4.

Williams, B., Kabbage, M., Kim, H.J., Britt, R. and Dickman, M.B. (2011) Tipping the balance: *Sclerotinia sclerotiorum* secreted oxalic acid suppresses host defenses by manipulating the host redox environment. *PLoS Pathogens* 7(6), e1002107. DOI: 10.1371/journal.ppat.1002107.

Wright, D.P., Baldwin, B.C., Shephard, M.C. and Scholes, J.D. (1995) Source sink relationships in wheat leaves infected with powdery mildew.I. Alterations in Carbohydrate Metabolism. *Physiological and Molecular Plant Pathology* 47, 237–253.

Xiao, X., Xie, J., Cheng, J., Li, G., Yi, X., Jiang, D. and Fu, Y. (2014) Novel secretory protein Ss-Caf1 of the plant-pathogenic fungus *Sclerotinia sclerotiorum* is required for host penetration and normal sclerotial development. *Molecular Plant-Microbe Interactions* 27(1), 40–55. DOI: 10.1094/MPMI-05-13-0145-R.

Xia, S., Xu, Y., Hoy, R., Zhang, J., Qin, L. and Li X. (2020) The notorious soilborne pathogenic fungus *Sclerotinia sclerotiorum*: An update on genes studied with mutant analysis. *Pathogens* 9(1): 27. DOI: 10.3390/pathogens9010027.

Xu, L. and Chen, W. (2013) Random T-DNA mutagenesis identifies a cu/Zn superoxide dismutase gene as a virulence factor of *Sclerotinia sclerotiorum*. *Molecular Plant-Microbe Interactions* 26(4), 431–441. DOI: 10.1094/MPMI-07-12-0177-R.

Xu, L., Xiang, M., White, D. and Chen, W. (2015) PH dependency of sclerotial development and pathogenicity revealed by using genetically defined oxalate-minus mutants of *Sclerotinia sclerotiorum*. *Environmental Microbiology* 17(8), 2896–2909. DOI: 10.1111/1462-2920.12818.

Xu, T., Li, J., Yu, B., Liu, L., Zhang, X. *et al.* (2018) Transcription factor SsSte12 was involved in mycelium growth and development in *Sclerotinia sclerotiorum*. *Frontiers in Microbiology* 9, 2476. DOI: 10.3389/fmicb.2018.02476.

Yajima, W. and Kav, N.N.V. (2006) The proteome of the phytopathogenic fungus *Sclerotinia sclerotiorum*. *Proteomics* 6(22), 5995–6007. DOI: 10.1002/pmic.200600424.

Yajima, W., Liang, Y. and Kav, N.N.V. (2009) Gene disruption of an arabinofuranosidase/beta-xylosidase precursor decreases *Sclerotinia sclerotiorum* virulence on canola tissue. *Molecular Plant-Microbe Interactions* 22(7), 783–789. DOI: 10.1094/MPMI-22-7-0783.

Yang, G., Tang, L., Gong, Y., Xie, J., Fu, Y. *et al*. (2018) A cerato-platanin protein SsCp1 targets plant PR1 and contributes to virulence of *Sclerotinia sclerotiorum*. *The New Phytologist* 217(2), 739–755. DOI: 10.1111/nph.14842.

Yarden, O., Veluchamy, S., Dickman, M.B. and Kabbage, M. (2014) *Sclerotinia sclerotiorum* catalase SCAT1 affects oxidative stress tolerance, regulates ergosterol levels and controls pathogenic development. *Physiological and Molecular Plant Pathology* 85, 34–41. DOI: 10.1016/j.pmpp.2013.12.001.

Yasue, T. (1969) Histochemical identification of calcium oxalate. *Acta Histochemica et Cytochemica* 2, 83–95.

Yeats, T.H. and Rose, J.K.C. (2013) The formation and function of plant cuticles. *Plant Physiology* 163(1), 5–20. DOI: 10.1104/pp.113.222737.

Yoshida, Y. (1993) Lanosterol 14α-demethylase (cytochrome P45014DM). In: Schenkman, J.B. and Greim, H. (eds) *Cytochrome P450*. Springer, Berlin, Heidelberg, Germany, pp. 627–639. DOI: 10.1007/978-3-642-77763-9_40.

Young, C.S. and Werner, C.P. (2012) Infection routes for *Sclerotinia sclerotiorum* in apetalous and fully petalled winter oilseed rape. *Plant Pathology* 61(4), 730–738. DOI: 10.1111/j.1365-3059.2011.02556.x.

Young, P. and Morris, H.E. (1927) Sclerotinia wilt of sunflower. *University of Montana, Agricultural Experiment Station Bulletin* 208, 32.

Yu, Y., Jiang, D., Xie, J., Cheng, J., Li, G. *et al*. (2012) Ss-sl2, a novel cell wall protein with PAN modules, is essential for sclerotial development and cellular integrity of *Sclerotinia sclerotiorum*. *PLoS ONE* 7(4), e34962. DOI: 10.1371/journal.pone.0034962.

Yu, Y., Xiao, J., Yang, Y., Bi, C., Qing, L. *et al*. (2015) Physiological and molecular plant pathology Ss-Bi1 encodes a putative BAX inhibitor-1 protein that is required for full virulence of *Sclerotinia sclerotiorum*. *Physiological and Molecular Plant Pathology* 90, 115–122. DOI: 10.1016/j.pmpp.2015.04.005.

Yu, Y., Xiao, J., Du, J., Yang, Y., Bi, C. *et al*. (2016) Disruption of the gene encoding endo-β-1, 4-xylanase affects the growth and virulence of *Sclerotinia sclerotiorum*. *Frontiers in Microbiology* 7, 1787. DOI: 10.3389/fmicb.2016.01787.

Yu, Y., Xiao, J., Zhu, W., Yang, Y., Mei, J. *et al*. (2017) Ss-Rhs1, a secretory rhs repeat-containing protein, is required for the virulence of *Sclerotinia sclerotiorum*. *Molecular Plant Pathology* 18(8), 1052–1061. DOI: 10.1111/mpp.12459.

Yu, Y., Du, J., Wang, Y., Zhang, M., Huang, Z. *et al*. (2019) Survival factor 1 contributes to the oxidative stress response and is required for full virulence of *Sclerotinia sclerotiorum*. *Molecular Plant Pathology* 20(7), 895–906. DOI: 10.1111/mpp.12801.

Zhang, J., Wang, Y., Du, J., Huang, Z., Fang, A. *et al*. (2019) *Sclerotinia sclerotiorum* thioredoxin reductase is required for oxidative stress tolerance, virulence, and sclerotial development. *Frontiers in Microbiology* 10, 233. DOI: 10.3389/fmicb.2019.00233.

Zhao, J., Wang, J., An, L., Doerge, R.W., Chen, Z.J *et al*. (2007) Analysis of gene expression profiles in response to *Sclerotinia sclerotiorum* in *Brassica napus*. *Planta* 227(1), 13–24. DOI: 10.1007/s00425-007-0586-z.

Zhao, J., Buchwaldt, L., Rimmer, S.R., Sharpe, A., Mcgregor, L. *et al*. (2009) Patterns of differential gene expression in *Brassica napus* cultivars infected with *Sclerotinia sclerotiorum*. *Molecular Plant Pathology* 10, 635–649. DOI: 10.1111/j.1364-3703.00558.

Zhao, Z., Liu, H., Wang, C. and Xu, J.R. (2013) Comparative analysis of fungal genomes reveals different plant cell wall degrading capacity in fungi. *BMC Genomics* 14(1), 274. DOI: 10.1186/1471-2164-14-274.

Zhu, G., Yu, G., Zhang, X., Liu, J., Zhang, Y. *et al*. (2019) The formaldehyde dehydrogenase SsFdh1 is regulated by and functionally cooperates with the GATA transcription. *MSystems* 4(5), 1–19. DOI: 10.1128/mSystems.00397-19.

Zhu, W., Wei, W., Fu, Y., Cheng, J., Xie, J. *et al*. (2013) A secretory protein of necrotrophic fungus *Sclerotinia sclerotiorum* that suppresses host resistance. *PLoS ONE* 8(1), e53901. DOI: 10.1371/journal.pone.0053901.

Zuppini, A., Navazio, L., Sella, L., Castiglioni, C., Favaron, F. *et al*. (2005) An endopoly-galacturonase from *Sclerotinia sclerotiorum* induces calcium mediated signaling and programmed cell death in soybean cells. *Molecular Plant-Microbe Interactions* 18(8), 849–855. DOI: 10.1094/MPMI-18-0849.

8 Molecular Mechanisms of Infection and Pathogenesis

Abstract

The molecular mechanism in *Brassica–Sclerotinia* host pathosystem is another important event in disease development. In this system, small RNAs are regulated, and long noncoding RNAs regulate the expression of plant genes to respond to *Sclerotinia*. Various metabolic enzymes show the differential expression in host plants on infection by *Sclerotinia*, impacting pathogenesis. Several genes are differentially expressed for biological, biometabolotical, and pathological functions in response to *Sclerotinia*. The sequencing of mRNA, sRNA, and degradome in *Brassica* demonstrates a multifaceted response to *Sclerotinia*. Several effector genes, viz. *SsNEP1, Ss NEP2, Sscaf1, SsSSVP1, SsCP1, SsSNE's, SsITL, SsCM1, SsCVNH*, and *ssv263*, are secreted by *Sclerotinia* for pathogenesis. *SSITL*, the effector gene, plays a significant role in suppressing the jasmonate (JA)/ethylene (ET) signal pathways in *Sclerotinia* infection. The *Sclerotinia* oxaloacetate acetylhydrolase gene (*Ss-Oah1*) also plays a crucial role in oxaloacetate biogenesis and pathogenesis. The interaction reveals the expression of genes, viz. *BNa NPR1, SSITL, SsCTR1, SsCCS, SsATX1, SsTrx1, SsCaf1, SsERP1, YML079*, and *SsYCP1*, which contributes to pathogenesis. Effector genes during pathogenesis have been functionally characterized as *CAZymes, Ss-cuta, PGs, Acp1, SSIG, MFS, GST, OAH, NoX, TFs, MAPK, NEP, SsCUNH, PaC1, Ss-RHs1, SsCM1, SsNEP1, SsCP1*, and *SsSSVP1*, which regulate the growth and development of pathogens.

8.1 Introduction

Stem rot of crucifers caused by hemibiotrophic pathogen *Sclerotinia sclerotiorum* is one of the widest spread and destructive diseases throughout the world. The sequencing of host and pathogen genomes has allowed understanding molecular biology of the pathogen and molecular mechanisms of infection and pathogenesis. The analysis of *Sclerotinia* transcriptomes has revealed the expression of effector gene for infection, regulation of reproductive structures, biosynthetic pathways, production of metabolites, signaling molecules, and differential expression of genes to regulate pathogenesis. The molecular mechanisms of host–pathogen interaction have revealed that the dual lifestyle of *Sclerotinia* acts as biotrophs for short period postinfection and later as necrotrophs to colonize the host tissues. This has strengthened the notion of hemibiotrophic pathogens over necrotrophic pathogens believed earlier for *Sclerotinia*.

8.2 Transcriptome Functional Mechanism in *Brassica* During *Sclerotinia* Pathogenesis

The molecular processes behind Sclerotinia disease have been uncovered by the *Brassica* transcriptome's integrated mRNA, sRNA, and degradome sequencing. The functional mechanisms of the *Sclerotinia* effector genes have been uncovered by transcriptome analysis conducted throughout the various stages of infection. To look at what happened during the early (1 hour) and middle (48 hour) stages of infection, RNA-seq analysis was done. The examination of transcripts showed the timing and degree of gene activation associated with pathogenicity or virulence as *S. sclerotiorum* infected *Brassica napus*. The genes are classified into eight functional groups: secreted effectors, signaling pathways, development, detoxification, hydrolytic enzymes, secondary metabolites, oxalic acid (OA), and reactive oxygen species (ROS) production. Most of these genes' expression

 Sclerotinia Stem Rot of Crucifers Volume 1: Biology and Ecology (G.S. Saharan *et al.*)
DOI: 10.1079/9781800629592.0008

arrays matched the responsibilities that were anticipated for them. Using principal component analysis, distinct patterns of gene expression were found in *S. sclerotiorum* at various stages of pathogenicity, such as necrotic phases and host penetration ramification. Moreover, after host penetration, a transient biotrophic stage was seen. The fungus employs a range of techniques to successfully infiltrate the host plant and get past its defenses. Comprehensive investigation of the temporal expression of genes linked to virulence during *S. sclerotiorum* infection of *B. napus* yields important information on the interactions between the pathogen and its host (Seifbarghi *et al.*, 2017). Joshi *et al.* (2016a) examined the transcript level alterations in canola postinfection with *S. sclerotiorum* in a time period of a compatible interaction utilizing strand-specific whole transcriptome sequencing analysis in order to comprehend the specific molecular pathways underpinning host response to *Sclerotinia* infection.

Joshi *et al.* (2016a) used strand-specific whole transcriptome sequencing to analyze the transcript level changes in canola postinfection with *S. sclerotiorum* in a time course of a compatible interaction in order to comprehend the specific molecular mechanisms underlying host response to *Sclerotinia* infection. After infection, 24 and 23 genes were repressed at 24 and 48 hours post inoculation (hpi), respectively, while 161 and 52 genes ($P<0.001$) were activated. This shows that only a smaller number of host-specific genes were expressed later in the infection due to a steady rise in host cell lyses and an increase in the pathogen's pathogenicity. Key pathogen-responsive genes, such as glucanases, chitinases, peroxidases, and WRKY transcription factors (TFs), were rapidly induced within 24 hour postinfection, suggesting the early detection of the pathogen by the host. Merely 16 genes exhibited significant induction during both time intervals, indicating a pathogen-induced coordinated suppression of host responses. Together with genes related to plant–pathogen interactions, a large number of unique genes responsive to disease were found, such as different TFs and those connected to ET and JA signaling. This implies that canola employs a variety of tactics to mediate plant reactions to the pathogen invasion. The validation of a specific collection of genes using quantitative real-time polymerase chain reaction (qRT-PCR) revealed a pattern that was consistent with the findings of RNA-seq analysis, suggesting that the host may utilize these genes to protect itself against pathogen invasion. A thorough examination is provided showing the relationship between pathogen virulence and host susceptibility in the agriculturally significant *B. napus*–*S. sclerotiorum* pathosystem (Joshi *et al.*, 2016a).

8.3 Molecular Components of *Sclerotinia* for Pathogenesis

Many complex molecular processes are involved in the pathogenesis of *S. sclerotiorum*, and a number of essential molecular components are involved in the pathogen's capacity to infect plants. Following the establishment of gene knockout techniques for *S. sclerotiorum*, numerous researchers investigated the roles of pleiotropic genes in growth, development, and infection. The primary effector genes that manipulate the host defense system during infection and disease have been identified from the enzymatic and proteinaceous categories.

8.3.1 Oxalic acid

Research has demonstrated that the pathogenicity of *S. sclerotiorum* depends on OA, an organic acid found in many different organisms (Dutton and Evans, 1996). The fungus's capacity to infect host species is largely dependent on OA, as was shown through early research on ultraviolet (UV)-induced mutants that produced less of the acid than the wild type (WT). Additional research (Williams *et al.*, 2011) showed that OA not only reduces host defenses in the early stages of infection but also, by changing the host cell's redox environment later on, causes programmed cell death. The finding that UV-induced mutants were unable to prevent the oxidative burst at the start of the infection process served as inspiration for part of this theory. Utilizing particular null mutants for the OA biosynthetic gene, oxaloacetate acetylhydrolase (oah), produced different results. According to preliminary studies, oah null mutants did not produce OA and instead showed aberrant sclerotia and compound appressoria formation defects (Liang *et al.*, 2015a; Liang *et al.*, 2015b). These problems persisted even when complementation strains restored the production of OA, indicating that the lack of OA was not the only cause. Infected damaged plants demonstrated the effect of OA on pathogenicity, allowing the penetration of host tissues without the need for appressoria. The results showed that limited damages were caused in planta by both the UV-induced mutant strains and the oah null mutants. Although their lesions were slightly larger, which may have been caused by their faster growth rate, the oah null mutants showed similar characteristics to the UV-induced mutants in this aspect (Liang *et al.*, 2015a,Liang *et al.*, 2015b).

An alternative investigation looked into independent strains known as oah null strains, which were unable to manufacture OA. These strains may nevertheless make fumaric acid in spite of this shortage, but less so than OA. The pH of the medium decreased slightly when fumaric acid was present, although not as much as when OA was present. The growth of these strains was inhibited, and proper sclerotia formation did not occur when exposed to alkaline environments. Reduced pathogenicity was the outcome of the oah null mutants' incapacity to efficiently lower the medium's pH in comparison to WT strains. Lesions of the disease like those caused by WT strains were seen on faba beans; however, pathogenicity was essentially nonexistent on soybeans, severely restricted on green beans, and diminished by 30–40% on peas. Artificially reducing the pH could somewhat correct the flaws on green beans, but potassium oxalate cannot. Depending on the host tissue's pH and buffering ability, the oah null mutants' lesions varied in size. The disruption of oah was identified as the main genetic aberration by complementation tests, which showed that all deficiencies in these strains could be restored (Xu *et al.*, 2015). Similar symptoms in oah null mutants across different strains were independently validated by another work using CRISPR/Cas9 insertional mutagenesis (Li *et al.*, 2018).

Based on the available data, it appears that OA mostly increases the virulence of *S. sclerotiorum* by lowering the pH of host tissues, as opposed to having more focused effects on host physiology. Furthermore, studies indicate that these strains' OA mutations caused by UV light produce unexplained anomalies that prevent normal growth. It is interesting to observe that these mutants can still create some OA in response to changes in the ambient pH, but even in artificially lowered medium pH circumstances, their growth rates are unaffected and they are unable to develop sclerotia. This suggests that, contrary to what would normally happen in response to a low pH, they are unable to control the expression of genes linked to growth and development (Godoy *et al.*, 1990; Liang *et al.*, 2015a).

8.3.2 Proteinaceous effectors

Plant pathogens secrete tiny proteins called effectors during infection, and these proteins contribute to the progression of the disease (Toruño *et al.*, 2016). A considerable number of *S. sclerotiorum* effectors that may be active have been identified in recent years. It has been demonstrated that the majority of these proteinaceous effectors from *S. sclerotiorum* cause host tissue cell death. Since *S. sclerotiorum* is a necrotrophic pathogen, it needs to induce host cell death in order to obtain nutrients. *Agrobacterium tumefaciens*-mediated expression in *Nicotiana benthamiana* leaves, which resulted in cell death, allowed for the initial identification of SsNEP1 and SsNEP2 as effectors of *S. sclerotiorum* (Bashi *et al.*, 2010). Later, further *S. sclerotiorum* necrosis-inducing effectors such as Ss-Caf1, SsSSVP1, and SsCP1, and six 'SsNEs' have also been described using this method (Xiao *et al.*, 2014; Lyu *et al.*, 2016; Yang *et al.*, 2018; Seifbarghi *et al.*, 2020).

Transfer DNA insertion mutagenesis was used to find the gene *Ss-Caf1*. The mutant with a transfer DNA insertion in this gene showed no compound appressoria, uneven sclerotia formation, and decreased pathogenicity. Furthermore, *N. benthamiana* cells died when the mature Ss-Caf1 protein was expressed without its original signal peptide, demonstrating the protein's role as an intracellular necrosis-inducing effector (Xiao *et al.*, 2014). Likewise, SsSSVP1 is an additional effector that induces intracellular necrosis. Its localization to mitochondria is disrupted by its physical interaction with QCR8, a component of the cytochrome b–c1 complex in the host's mitochondrial respiratory chain. According to Lyu *et al.* (2016), knocking down SsSSVP1 reduced growth *in vitro* little but significantly increased pathogenicity. Therefore, it is possible that *Ss-Caf1* and *SsSSVP1* function as pathogenicity and developmental factors simultaneously. Yang *et al.* (2018) investigated SsCP1, a protein belonging to the *cerato-platanin* family in *S. sclerotiorum*. This protein demonstrated a variety of roles in the pathogenicity of fungi. It has been established that SsCP1 induces cell death in *N. benthamiana* and has a critical role in virulence, both intracellularly and extracellularly. However, the induction of cell death was only detected when SsCP1 was overexpressed via a tobacco rattling virus vector; this result was not repeated in a follow-up study (Yang *et al.*, 2018; Seifbarghi *et al.*, 2020). Moreover, SsCP1 may hinder PR1's antifungal effects by interacting with it in the apoplast. Notably, *Arabidopsis* with overexpressed SsCP1 showed increased salicylic acid (SA) levels and enhanced resistance to a range of fungal infections. Considering the part SsCP1 plays in *S. sclerotiorum* virulence and the possibility of its interaction with PR1, this finding is noteworthy.

In addition to its ability to induce necrosis, the effector SsITL contributes to the pathogenicity of the plant by inhibiting its immune system. Following the infection, SsITL has elevated expression levels in *Arabidopsis* plants 1.5 and 3 hour after inoculation, indicating its involvement in the early stages of infection. SsITL has the ability to enter host cells and prevent the activation of genes that react to SA and JA/ET, hence reducing the host's immune response (Zhu *et al.*, 2013). Tang *et al.* (2020) found that SsITL inhibits the buildup of SA by physically interacting with a calcium-sensing receptor located in the chloroplast. It is hypothesized that SsITL may interfere with the JA/ET signaling pathway by interacting with other elements in the host. In addition to acting as an effector, SsITL is essential for the healthy development of sclerotia, hyphae, and carpogenic germination (Zhu *et al.*, 2013).

SA synthesis may be hampered by the putative effector SsCM1 from *S. sclerotiorum* (Kabbage *et al.*, 2013). The *Ustilago maydis* effector Cmu1, which is essential for causing illness in maize by lowering SA levels, is structurally similar to the predicted chorismate mutase SsCM1 (Djamei et al., 2011).

SsCM1's function in *S. sclerotiorum* virulence is still thought to be a possible effector. Similar to other proteins like Ss-Caf1, SsSSVP1, and SsITL, the effector SsCVNH is upregulated during infection. According to Lyu *et al.* (2015), SsCVNH and other proteins are essential for *B. napus* proliferation, sclerotia development, and pathogenicity. Moreover, SsCVNH might be able to bind to chitin oligomers, which would help it evade recognition by host extracellular receptors that are involved in pathogen awareness. This method is comparable to other fungal effectors with Lys M domains that have been reported (Sánchez-Vallet *et al.*, 2020). The virulence of *B. napus* depends on a putative secreted effector called ssv263, which resembles a secreted *Botrytis cinerea* protein (Liang *et al.*, 2013). However, not much is known about this effector. It has been discovered through recent studies on proteinaceous effectors that *S. sclerotiorum* interacts with its hosts in a more complex way than previously believed. It may be possible to improve genetic resistance to *S. sclerotiorum* in crops by developing novel tactics based on an understanding of the mechanisms of action of *S. sclerotiorum* proteinaceous effectors. CRISPR/Cas9 was utilized by Zhang *et al.* (2021) to alter one or more copies of the SsSSVP1 target QCR8 in *B. napus*. The resultant mutants showed enhanced resistance to *B. cinerea* and *S. sclerotiorum*, but they had no appreciable effect on important agronomic characteristics. A viable method to find host resistance or susceptibility genes is to determine the host targets of other *S. sclerotiorum* effectors. In order to improve *S. sclerotiorum* resistance in commercially significant crops, these discovered genes or their equivalents would make great candidates for gene editing (Derbyshire *et al.*, 2022).

8.4 Functional Characterization of Resistance Genes for Pathogenesis

In general speaking, plants fight off infections by expressing quantitative genes that have little effect and qualitative, dormant genes. Resistance to key genes (*R*-genes) in *Brassica* species is unknown. The nucleotide-binding site leucine-rich repeat (NLR) gene family contains genes that encode complete resistance to disease; these genes are also known as resistance (*R*) genes. Numerous studies have demonstrated the diverse biological functions exhibited by the genes linked to quantitative disease resistance (QDR) (Poland *et al.*, 2011; Roux *et al.*, 2014; Corwin and Kliebenstein, 2017). The molecular functions involving transporters (Krattinger *et al.*, 2009), kinases (Fu *et al.*, 2009; Huard-Chauveau *et al.*, 2013), peptidases (Poland *et al.*, 2011; Badet *et al.*, 2017), and actin-related proteins (Moscou *et al.*, 2011) are among the many molecular functions linked to your genes. The potential of NLRs highlights the close link between QDR and elements of *R*-mediated immunity to improve QDR, as well as signaling variables commonly linked to *R*-mediated resistance genes (Debieu *et al.*, 2015; Iakovidis *et al.*, 2016; Lee *et al.*, 2016).

Quantitative disease resistance (QDR) genes present a challenge for functional characterization due to their intricate connections and limited phenotypic influence. The model plant *Arabidopsis thaliana* has a large number of mutant lines, which makes it possible to quickly analyze the function of specific putative QDR genes (Huard-Chauveau *et al.*, 2013; Corwin *et al.*, 2016; Rajarammohan *et al.*, 2018). Comprehensive quantitative analyses, such as the identification of ET and ROS (Perchepied *et al.*, 2010; Zhang *et al.*, 2013), measurement of lesion size, and estimation of fungal biomass (Zhang *et al.*, 2013; Badet *et al.*, 2017), are frequently necessary to clarify the role of specific *A. thaliana* genes in QDR against *S. sclerotiorum*. For assessing disease severity and understanding the effect of stress on plant health, quantitative image analysis is a useful technique (Chen *et al.*, 2014; Baranowski *et al.*, 2015; Mutka *et al.*, 2016; Badet *et al.*, 2017; Karisto *et al.*, 2018; Czedik-Eysenberg *et al.*, 2018; Nelson *et al.*, 2018). In order to improve quantitative phenotyping's accuracy and resilience, additional measurements are frequently needed. Scientific studies and agricultural breeding have demonstrated the benefits of applying high-throughput image-based plant phenotyping (Araus and Cairns, 2014; Fahlgren *et al.*, 2015; Coppens *et al.*, 2017). Improving the ability to generate large-scale, low-cost, quantitative plant disease measurement datasets with less space needed and human intervention is crucial (Czedik-Eysenberg *et al.*, 2018). Because image-based disease measurement is primarily noninvasive, it makes it possible to track the dynamics of disease progression. This has made it easier to study the temporal and spatial transmission of pathogens in plant tissues (Mutka *et al.*, 2016) and distinguish between plants that are infected and those that are not, even before symptoms become apparent (Czedik-Eysenberg *et al.*, 2018). Even though methods and technology have advanced significantly, automated plant phenotyping still has limited potential to progress plant pathology. Currently, there is no low-cost, low-footprint, high-throughput, high-resolution technique for image-based phenotyping of plant disease signs. The creation of a system like this is crucial to understanding this intricate trait better and to the identification of novel QDR genes. The Navautron system was introduced by Barbacci *et al.* (2020). It is a hardware and software solution that is integrated and can measure the area of disease lesions on detached leaves that have necrotic lesions caused by pathogens automatically over time. A total of 800 *A. thaliana* leaves may be analyzed per square meter by the Navautron system. With very minor variations in lesion doubling times of up to 12 min, its automated leaf and lesion tracking features enable the distinction of disease symptom progression with minimal human intervention.

The rate at which lesions spread was significantly influenced by the host plant's genetic makeup, which served as a reliable indicator of the plant's level of QDR. The presence of specific polymorphisms with significant effects, seen in the two most resistant *A. thaliana* accessions but absent from the other four, was thought to indicate the presence of genes associated with disease susceptibility. The resistant accessions Rubezhnoe and Lip-0 have different LAZ5 alleles, but the other *A. thaliana* accessions under study had the same alleles. Plants with mutations in the closely related *CSA1* gene responded similarly to the WT, but two LAZ5-deficient mutant lines in the Col-0 genetic background showed enhanced QDR against *S. sclerotiorum*. The results indicate that *S. sclerotiorum*, a widely distributed necrotrophic fungus, uses the aberrant activation of LAZ5 that causes cell death to take advantage of the *R*-gene-driven plant cell death (Barbacci *et al.*, 2020).

Plant genes that offer complete protection against *S. sclerotiorum* are not known to exist. Instead, plant populations show varying degrees of partial resistance—also referred to as QDR in response to this pathogen. The complexity of QDR genes and the small effects that are seen have made a complete understanding of these genes difficult. The majority of the mechanisms via which necrotrophic fungi that are extensively spread affect plant-induced programmed cell death remain unclear. However, Barbacci *et al.* (2020) have presented a comprehensive and economical method of evaluating disease lesions with their time-resolved automated disease phenotyping pipeline. It has been effectively demonstrated that distinct disease responses in various pathosystems can be identified using this technique. By employing this phenotyping technique, the scientists assessed the emergence of disease symptoms caused by seven *S. sclerotiorum* isolates on six *A. thaliana* wild accessions with an unprecedented level of clarity. The leucine-rich repeat gene LAZ 5, which is a nucleotide-binding site, exhibits a notable degree of genetic diversity, particularly in the highly resistant *A. thaliana* accessions Rubezhnoe and Lip-0. Reduced LAZ 5 expression was discovered in *A. thaliana's* Rub natural accession and LAZ 5.1 mutant lines, and this was connected to increased QDR to *S. sclerotiorum*. These results underscore the significance of time-resolved image-based phenotyping in elucidating the genetic foundation of intricate traits such as QDR. The findings further imply that in order to cause disease and promote programmed cell death, *S. sclerotiorum* modifies plant sphingolipid pathways, which are controlled by LAZ 5 (Barbacci *et al.*, 2020).

8.5 Regulation of *Brassica* Small RNAs During Infection and Pathogenesis

Brassica species express small RNAs genes during infection by *Sclerotinia*, which allows the understanding of molecular mechanisms of host pathosystem. Small RNAs play a crucial role as gene regulators in eukaryotes, controlling various biological processes. In response to pathogen attacks, plants can regulate specific sets of sRNAs. However, there is a lack of research on how sRNA regulation happens in the *S. sclerotiorum* and *B. napus* pathosystem. Regmi *et al.* (2021) utilized high-throughput sequencing to identify different classes of sRNAs from *B. napus*, with 3999 sRNA loci being highly expressed, including 730 that were significantly upregulated during infection. These upregulated sRNAs targeted 64 genes, such as disease resistance proteins and transcriptional regulators. Additionally, 73 conserved miRNA families were found in the dataset, with degradome sequencing revealing 2124 cleaved mRNA products from these microRNAs (miRNAs). Among them, 50 genes were specific to infection. Furthermore, 20 conserved miRNAs showed differential expression, with eight transcripts being cleaved by differentially expressed miRNAs like miR159, miR5139, and miR390, indicating a potential role in the *S. sclerotiorum* response. A miR1885-triggered disease resistance gene-derived secondary sRNA locus was also confirmed through degradome sequencing. Moreover, evidence of silencing a plant immunity-related ET response factor gene by a novel RNA was observed using 5′-RACE and RT-qPCR. These findings contribute to a better understanding of the molecular mechanisms of the *S. sclerotiorum* and *B. napus* pathosystems at the sRNA level (Regmi *et al.*, 2021).

8.5.1 The contribution of small RNAs to the progression of *Sclerotinia* infection and pathogenesis

The initial findings indicate that cross-kingdom RNAi plays a crucial role in *S. sclerotiorum* pathogenesis. *S. sclerotiorum* generates numerous small RNAs during its infection of both *Phaseolus vulgaris* and *Arabidopsis* (Derbyshire *et al.*, 2019). It was discovered that the small RNAs targeted genes in *Arabidopsis* that displayed decreased expression during infection, indicating a possible fungal small RNA-mediated silencing mechanism. Moreover, changes linked to quantitative resistance to *S. sclerotiorum* in a wild population were found in these genes. When compared with the WT, the deletion of three *Arabidopsis* genes targeted by *S. sclerotiorum* small RNAs increased susceptibility to infection, indicating an evolutionary basis for the interaction between pathogen small RNAs and specific mRNAs. This work did not, however, provide concrete evidence of the interaction between plant mRNAs and short RNAs from *Sclerotiorum*. Though research on short RNAs' functions in *S. sclerotiorum* pathogenesis is still in its early stages, it has the potential to shed light on a number of important aspects of their functions.

Small RNAs have the ability to traverse species boundaries and are also responsible for modulating internal gene expression. Within *S. sclerotiorum*, internal small RNAs are crucial for the regulation of genes that govern sclerotial development (Xia *et al.*, 2020). Earlier studies on *S. sclerotiorum* have explored the influence of small RNAs on pathogen response (Neupane *et al.*, 2019). The existence of a functional RNAi pathway in *S. sclerotiorum* indicates that it may serve as a promising target for novel disease management techniques, such as host-induced gene silencing (HIGS) and spray-induced gene silencing.

Glycine max showed a decrease in disease susceptibility when double-stranded RNAs targeting the *S. sclerotiorum* OA biosynthetic gene *oah* were used (McCaghey *et al.*, 2021). These double-stranded RNAs were found to be able to be ingested by fungal hyphae in laboratory experiments, suggesting a possible transfer from host to pathogen. These are likely to enter the endogenous RNAi pathway, which produces short RNAs that regulate the expression of target genes, given the understanding of RNAi in fungi (Dang *et al.*, 2011). *S. sclerotiorum* is very good at absorbing double-stranded RNAs, which makes it a good candidate for topical application of RNAs targeting virulence genes (Qiao *et al.*, 2021; Derbyshire *et al.*, 2022).

The use of double-stranded RNAs targeting the *S. sclerotiorum* OA biosynthesis gene *oah* led to a reduction in disease susceptibility in *G. max* (McCaghey *et al.*, 2021). In laboratory studies, it was observed that these double-stranded RNAs could be taken in by fungal hyphae, indicating a potential transfer from the host to the pathogen. Considering understanding of RNAi in fungi, it is probable that these double-stranded RNAs would enter the endogenous RNAi pathway, producing small RNAs that control the expression of target genes (Dang *et al.*, 2011). The absorption of double-stranded RNAs is particularly viable in *S. sclerotiorum*, making it suitable for the topical application of double-stranded RNAs targeting virulence genes (Qiao *et al.*, 2021; Derbyshire *et al.*, 2022).

8.6 Functions of Long Noncoding RNAs in Pathogenesis

According to recent research, long noncoding RNAs (lncRNAs) have a major role in controlling how genes are expressed in plants under different stress scenarios. Despite the fact that lncRNAs have mostly been researched in a small number of plant species, little is known about how precisely they work in *B. napus* in response to biotic stress. Joshi *et al.* (2016a) employed specialized lncRNA detection methods in conjunction with sophisticated RNA-sequencing techniques to identify novel lncRNAs sensitive to *S. sclerotiorum* infection in *B. napus* at two distinct time points (24 and 48 hpi). A total of 2821 intergenic lncRNA candidates, 111 naturally occurring antisense transcripts, 76 exhibiting exonic overlap with known coding transcripts, and the remaining 173 unique lnc-isoforms were among the 3181 lncRNA candidates. It has been found that 41 lncRNAs are the progenitors of miRNAs, including miR156, miR169, and miR394, and that these miRNAs are essential for plant defense against fungal phytopathogens. A total of 12 of the 931 lncRNAs that were identified as having differential expression in response to *S. sclerotiorum* infection were validated by qRT-PCR. TCONS_00000966, a *B. napus* antisense lncRNA that has 90% sequence overlap with a plant defensin gene, was found to be significantly upregulated during infection stages, suggesting that it plays a function in controlling the expression of genes associated with defense during *S. sclerotiorum* infection. Furthermore, it was shown that nine lncRNAs in *B. napus* overlapped with the cis-regulatory areas of genes that were differentially expressed. A set of *S. sclerotiorum* sensitive sense/antisense transcript pairs were validated using qRT-PCR, which showed divergent expression patterns. These results lend credence to the theory that conflicts between transcriptional machinery may cause sense promoters to become inactive. This emphasizes how lncRNAs may affect how plant genes that respond to biotic stress are expressed (Joshi *et al.*, 2016b).

8.7 Gene Encoding Survival Factor 1 Crucial for Pathogenesis of *Sclerotinia*

During an attempt to find mutations that might be complemented by the expression of the human antiapoptotic gene *Bcl-xL*, the gene survival factor 1 (SVF1) was originally discovered in *Saccharomyces cerevisiae* (Vander Heiden *et al.*, 2002; Brace *et al.*, 2005). It is significant to remember that *SVF1* and *Bcl-xL* play distinct functions in regulating cell survival. An increased sensitivity to chemical precursors of ROS in *S. cerevisiae* cells lacking the Svf1 protein revealed the need of *SVF1* for survival under oxidative stress conditions (Brace *et al.*, 2005). Additional investigation revealed that *S. cerevisiae's* sphingolipid metabolism is changed to facilitate Svf1-mediated cell survival under oxidative stress (Brace *et al.*, 2007). It was anticipated that the gene *SsSvf1* (*S. sclerotiorum SVF1*) in *S. sclerotiorum* would encode a protein resembling yeast Svf1. Reverse genetic analysis was used to determine the function of *SsSvf1*, and its relationship to pathogenicity and the oxidative stress response was investigated. The knowledge of *Svf1's* function in fungal plant diseases and *S. sclerotiorum* pathogenicity has improved as a result of present work (Yu *et al.*, 2019).

The development of *S. sclerotiorum* is significantly influenced by ROS. During infection, the fungus uses antioxidants that are both enzymatic and nonenzymatic to fight oxidative stress. In *S. cerevisiae*, a survival factor known as SVF1 (*SsSvf1*) has been discovered and described. It enhances survival in the face of oxidative stress. Interestingly, a gene called *SsSvf1* has been found in *S. sclerotiorum*, and it is predicted to encode a homologue of SVF1. The *SsSvf1* transcripts exhibited notably greater expression levels in hyphae under oxidative stress. Silencing *SsSvf1* in the fungal culture raised intracellular ROS levels and made the cells more susceptible to oxidative stress. Additionally, there was a noticeable increase in *SsSvf1* transcripts at the beginning of the infection, and strains with silenced genes exhibited less virulence against *A. thaliana* and oilseed rape. It is interesting to note that strains with the *SsSvf1* gene silenced partially recovered their virulence when plant ROS generation was inhibited. Notably, these strains continued to produce oxalate normally, but they had issues with cell wall integrity and compound appressorium development. Based on these findings, it appears that *SsSvf1* regulates ROS levels during the fungal–host interaction, which affects *S. sclerotiorum's* pathogenic potential.

8.8 Detecting the Pathogenic Process in *Brassica* via Integrated mRNA, sRNA, and Degradome Sequencing Approaches

Genes encoding receptor-like kinases, proteins containing nucleotide-binding site leucine-rich repeats, proteins involved in mitogen-activated protein kinase cascades, G proteins, calcium-binding proteins, hormones, TFs, pathogenesis-related (PR) proteins, proteins involved in phenylpropanoid and glucosinolate metabolism, and transport proteins are among those in *B. napus* that are responsible for responding to pathogen invasion, according to transcriptome analyses (Li *et al.*, 2012). Endogenous noncoding RNAs known as miRNAs adversely regulate gene expression, which is why they are important for a number of biological processes at both the transcriptional and posttranscriptional levels. These miRNAs play crucial roles in gene regulation circuits linked to developmental, maturation, stress response, germination, and signal transduction activities.

Furthermore, a plethora of research has demonstrated the involvement of miRNAs in the immune responses triggered by pathogen invasions. *A. thaliana's* defense mechanism against *Pseudomonas syringae* pv. *tomato* is significantly aided by miR393, which inhibits certain target genes, including as auxin receptors TIR1, AFB2, and AFB3, which negatively impact auxin signaling. Furthermore, miR393 is necessary for the glucosinolate pathway, which is essential for plant defense against infections. MiR158 and miR1885 levels in *B. rapa* increased after the *Turnip Mosaic Virus* infection. Consequently, this stopped the emergence of genes that encode disease resistance proteins, specifically the leucine-rich repeat nucleotide-binding site. The *A. thaliana* miR396 functions as a positive regulator during a cyst nematode infection. According to Li *et al.* (2012), miR6019 and miR6020 direct the cleavage of N genes in *Nicotiana tabacum* (tobacco), increasing resistance to *tobacco mosaic virus*. Numerous investigations have been conducted to detect miRNAs in oilseed rape. The first study to use computational methods to find 21 putative *B. napus* miRNAs was Xie *et al.* (2007). Next-generation sequencing technologies have since been used to find miRNAs in various tissues or under various stress scenarios. Only 92 mature *B. napus* miRNA sequences are currently available in the miRBase database. Many *B. napus* miRNAs may still be unknown given the wide variety of miRNAs discovered in other species, such as *A. thaliana*, *Arabidopsis lyrata*, and *Oryza sativa*.

The lack of resistance lines makes it difficult to properly understand the interaction between *S. sclerotiorum* and *B. napus*. Five lines with some resistances were found in a previous analysis (Jian *et al.*, 2016). They performed RNA-seq, sRNA-seq, and degradome deep sequencing on mock-inoculated and *S. sclerotiorum*-inoculated stems at 48 hpi to examine the oilseed rape response to the infection at the mRNA and miRNA levels. Genes related to protein kinases, hormones, TFs, secondary metabolism, transport proteins, signal transduction (including CDPKs, G proteins, and MAPKs), and secondary metabolism were among those with differential expression. At 48 hour postinfection with *S. sclerotiorum*, the sRNA-Seq analysis revealed differential expression in ten recognized and 41 new miRNAs. Moreover, degradome sequencing identified 64 miRNAs and 80 cleavage sites. These findings offer a thorough knowledge of the modifications in gene expression that occur after oilseed rape infection with *S. sclerotiorum* (Jian *et al.*, 2016).

The molecular mechanisms governing oilseed rape's transcriptional and post-transcriptional responses to *S. sclerotiorum* infection are currently poorly understood. The researchers studied this response using a complete omics approach in order to address this. Oilseed rape plants that were either infected with *S. sclerotiorum* or not were examined for RNA expression and posttranscriptional profiles using transcriptome, sRNAome, and degradome sequencing methods on the Illumina platform.

A total of 7065 differentially expressed genes (DEGs) were identified 48 hour after inoculation in comparison to the mock-inoculated control. Numerous biological activities, such as protein kinases, signal transduction, TFs, hormones, proteins relevant to pathogenesis, secondary metabolism, and transport, have been associated with these DEGs. Furthermore, it was discovered that the expression of just ten known and 41 novel miRNAs varied between samples with and without *S. sclerotiorum* infection. There were 176 new and 77 recognized miRNAs found by the sRNA-Seq analysis. A total of 64 miRNAs are expected to be present at 80 cleavage sites by degradome sequencing. Overall, the detailed analysis of degradome sequencing, sRNA, and mRNA offered insightful knowledge regarding the intricate responses of oilseed rape to *S. sclerotiorum* infection. Jian *et al.* (2016) provided clarification on the miRNA and mRNA expression profiles after *S. sclerotiorum* infection in oilseed rape.

8.9 Mechanisms and Functions of Gene Oxaloacetate Acetylhydrolase for Pathogenesis

The ability of *S. sclerotiorum* to initiate necrotrophic pathogenicity in a variety of hosts has been found to be critically dependent on the accumulation of significant quantities of OA during the pathogenesis process (Andrew *et al.*, 2012). 'A' mutants, also known as UV-induced oxalate-deficient mutants, have been found and demonstrated to be nonpathogenic on *A. thaliana* (Dickman and Mitra, 1992) and *P. vulgaris* (Godoy *et al.*, 1990). This is the most convincing proof of the role that OA plays in pathogenesis to date. Research has shown that these genetic variations result in reduced levels of OA deposition and only marginally affect various hosts (Cessna *et al.*, 2000; Kim *et al.*, 2008; Williams *et al.*, 2011). Moreover, the development, growth, and ability to cause disease of *S. sclerotiorum* can all be impacted by OA. The mutants containing the 'A' gene show stunted growth and aberrant development, with fluffy hyphae and no formation of sclerotium. Conversely, in low pH conditions, the normal strain grows more vegetatively and forms its sclerotium more effectively. These effects of OA could be related to how it affects pH levels and how it interacts metabolically with vital body functions. Strong dicarboxylic OA is produced by a wide range of pathogenic and nonpathogenic species, including fungus, plants, bacteria, and some animals. Earlier biochemical research has suggested that the transformation of oxaloacetate into OA and carbon dioxide, facilitated by oxaloacetate acetylhydrolase (OAH, EC 3.7.1.1), represents the main function linked to OA biogenesis in *S. sclerotiorum* (Maxwell, 1973; Kubicek *et al.*, 1988; Ruijter *et al.*, 1999). This procedure has been validated by a substantial amount of subsequent biochemical research. In *S. sclerotiorum*, a neutral pH significantly increases the accumulation of OA and OAH activity (Maxwell, 1973; Rollins and Dickman, 2001). The zinc finger ambient pH-dependent transcriptional regulator Ss-Pac1 is essential for its induction at neutral pH (Rollins and Dickman, 2001; Rollins, 2003). *S. sclerotiorum* can utilize a range of carbon sources to create OA, depending on the carbon growth substrate. According to Liberti *et al.* (2013), intermediates can originate from either the glyoxylate or the tricarboxylic acid (TCA) cycle.

Liang *et al.* (2015a) investigated the metabolism of OA and its role in the pathogenic process in order to investigate the effect of OAH on *S. sclerotiorum*. The researchers carried out gene deletion analysis and identified a gene known as Ss-oah1, which is thought to encode OAH. The findings demonstrated that, in addition to being required for the buildup of OA, *Ss-oah1* affects the pathogenicity, morphogenesis, and pH-responsive proliferation of *S. sclerotiorum* on a variety of plant hosts. Interestingly, the *Ss-oah1* deletion mutants (Δss oah1 mutants) may still cause primary lesions, indicating that early host compatibility is defined by variables other than OA. According to the histological characteristics of infection, *S. sclerotiorum* can partially colonize its hosts at first without accumulating OA. It appears that a dynamic regulation of OA buildup is involved in the infection and colonization process.

From *S. sclerotiorum*, the *OAH, EC 3.7.1.1* gene, or *Ss-oah1*, was successfully cloned and given a functional description. OA buildup and the increase in *Ss-oah1* transcripts were found to be significantly correlated, particularly at neutral pH values. It was found that *Ss-oah1*'s transcription is pH-dependent, and that *Ss-Pac1*, a pH-responsive transcriptional regulator, regulates it. Compared with previously characterized mutants weak in oxalate production due to UV radiation (referred to as 'A' mutants), the Δss-oah1 mutants had the *Ss-oah1* gene deleted and were unable to accumulate OA in culture or during plant infection. However, when the WT *Ss-oah1* gene was reintroduced, this impairment in OA buildup was totally reversed. The Δss-oah1 mutants also showed deficits in the development of sclerotia and compound appressoria, as well as a significant growth impairment on medium buffered to a neutral pH. The Δss-oah1 mutants exhibited modest lesions upon infection of various plant hosts, and the infected hyphae eventually lost their viability. Analysis of infections generated by the wild type (WT) and Δssoah1 strains showed that the WT strain's subcuticular hyphae had very little accumulation of osteoarthritis (OA), while the Δss-oah1 strain had no accumulation of OA at all. At the infection site, both the WT and mutant strains showed a transient attachment to living host epidermal cells. These findings highlight the crucial part that OAH activity plays in the development and pathophysiology of OA in *S. sclerotiorum*. However, they also suggest that the development of primary lesions is not only due to OA (Liang *et al.*, 2015a).

8.10 Functional Characterization of *BNaNPR1* Gene for Pathogenesis

In *Brassica* plants, several NPR1-like genes have been identified but *NBaNPR1* gene in *B. napus* reduces resistance to *Sclerotinia*. The NPR1, also known as nonexpressor of PR1, is a crucial gene involved in SA perception. According to Cao *et al.* (1994), it encodes a protein that functions as a transcription cofactor. Several investigations on *A. thaliana* have demonstrated the important role NPR1 plays in regulating defense responses. Mutations in AtNPR1 prevent SA from activating a variety of defense genes, suggesting that AtNPR1 is essential for controlling the SA-mediated defensive response (Cao *et al.*, 1994; Roetschi *et al.*, 2001). Furthermore, Mou *et al.* (2003) found that AtNPR1 controls the SA-mediated defensive response in a critical manner.

Furthermore, *NPR1* suppresses the defense reaction that is triggered by the well-known signaling chemical JA. As a result, it controls the communication between JA and SA (Spoel *et al.*, 2003).

Numerous plant species have been shown to possess two to six NPR1-like genes. There are five paralogs in *Arabidopsis*, excluding AtNPR1: *AtNPR2*, *AtNPR3*, *AtNPR4*, *AtNPR5* (also known as *AtBOP1*, Blade on Petiole 1), and *AtNPR6 (BOP2)*. Research has since altered our understanding of *AtNPR2*'s role in SA perception from the time it was thought to be a redundant or nonfunctional paralog of *AtNPR1* (Canet *et al.*, 2010; Backer *et al.*, 2019). The suppressive effect of *AtNPR3* and *AtNPR4 (AtNPR3/4)* on SAR has been shown, as mentioned by Zhang *et al.* (2006) and Liu *et al.* (2005). However, *At*BOP1 and *AtBOP2* are involved in the lateral organ development (Hepworth *et al.*, 2005). Numerous domains, including an N-terminal BTB/POZ domain, a central ankyrin repeat domain, and a C-terminal trans-activation domain, are present in these proteins (Spoel *et al.*, 2009). Despite the publishing of the crop's genome in 2014 (Chalhoub *et al.*, 2014), a thorough examination of the NPR1-like genes in *B. napus* has not yet been published, despite the identification of these genes in a number of plant species.

AtNPR1 is the member of the NPR1-like gene family that has been studied the most, mostly because of its importance to the plant immune system. Plants having mutations in the protein encoded by *AtNPR1* express PR1 less and are more susceptible to virulent diseases than plants with the *WT NPR1* (Cao *et al.*, 1994; Glazebrook *et al.*, 1996). Reduced PR1 activation and enhanced susceptibility to *S. sclerotiorum* were seen in the npr1 mutant of *Arabidopsis* (Guo and Stotz, 2007). However, in contrast to the WT, the npr1 mutant showed no appreciable difference in immunity against *S. sclerotiorum*; therefore, another trial was unable to identify this increased vulnerability in it (Perchepied *et al.*, 2010). There is still conflicting evidence regarding *AtNPR1*'s role in *S. sclerotiorum* resistance. Investigating the roles played by *B. napus* NPR1 (*BnaNPR1*), a critical regulator of defensive mechanisms, in the host's resistance to *S. sclerotiorum* may provide important insights into how *B. napus* and *S. sclerotiorum* interact. Wang *et al.* (2020) performed a comprehensive investigation of the NPR1-like genes in *B. napus*, encompassing phylogenetic analysis, gene structure, conserved motifs, gene chromosome position, and expression profiling. Additionally, they looked into how *BnaNPR1* influences resistance to *S. sclerotiorum* by performing a loss-of-function investigation of the protein using RNA interference (RNAi).

B. napus was found to have genes similar to *BnaNPR1*, and it was discovered that BnaNPR1 knockdown dramatically decreased resistance to *S. sclerotiorum*. Furthermore, it increases the generation of ROS and modifies the expression of genes linked to signaling pathways involved in the pathogen response in SA and (JA)/ET pathways. Controlling plant defense mechanisms requires the NPR1 gene and its NPR1-like equivalents. Nineteen BnaNPR1-like genes have been found in the genome of *B. napus*, and they have been called after the closest match they have in *A. thaliana* (At). This makes it possible to determine which *AtNPR1*-like gene in *B. napus* is related to another. Analyses of significant protein domains and functional motifs revealed that these counterparts were both conserved and varied. Phylogenetic analysis of these BnaNPR1-like proteins and their *Arabidopsis* counterparts revealed six distinct subclades, suggesting that the protein naming method accurately reflects the evolutionary relationships between the proteins. Subsequent examination of the transcriptome data of *B. napus* revealed that the principal fungal disease impacting the crop, *Sclerotiorum*, drastically lowered the expression of three BnaNPR1s. However, there were no overall notable alterations in the expression of BnaNPR 2/3/4/5/6 s. *B. napus* BnaNPR1-RNAi lines were generated by Wang *et al.* (2020) in order to investigate the impact of BnaNPR1s' downregulated expression on resistance to *S. sclerotiorum*. The findings of the study demonstrated that employing BnaNPR1-RNAi greatly decreased resistance to *S. sclerotiorum*. Additionally, it was noted that BnaNPR1-RNAi altered defensive mechanisms and elevated ROS production during the plant–pathogen interaction. These findings suggest that BnaNPR1 may be used by *S. sclerotiorum* to regulate particular physiological processes in *B. napus*, such as the generation of ROS and the SA defense response, in order to promote infection.

8.11 Effector-Like Proteins (Genes) Regulate Pathogenesis

S. sclerotiorum has been shown to contain a large number of proteins having properties and/or functions similar to those of effectors in plants. These proteins include SSITL, an integrin-like protein that inhibits defense responses in *A. thaliana*. Nevertheless, deletion mutants for the corresponding gene showed abnormal hyphal tip branching, slower growth *in vitro*, and abnormally small sclerotia in addition to SSITL's activity in planta (Zhu *et al.*, 2013). Two proteins that cause necrosis and ET in *N. benthamiana* when expressed heterologously are SsNep1 and SsNep2 (Bashi *et al.*, 2010); sspg1d is a polygalacturonase enzyme that works in conjunction with a protein that is part of the *B. napus* C2 domain (Wang *et al.*, 2009). Cutinase, also referred to as SsCut, is the enzyme that kills cells in a variety of plant species, including nonhost wheat (*Triticum aestivum*) (Zhang *et al.*, 2014); polygalacturonase, also referred to as SsSVP1, is a hypothetical secreted protein that decreases virulence in *S. sclerotiorum* when disrupted (Liang *et al.*, 2013); a small, cysteine-rich secreted protein with a CVNH domain that decreases virulence when deleted (Lyu *et al.*, 2015); and a secreted protein with no known functional domains, SsSSVP1, appears to be exclusive to *S. sclerotiorum* and the closely related species *B. cinerea*, which causes necrosis in *N. benthamiana*. It has been shown that this protein interacts with *N. benthamiana* QCR8, a subunit of the mitochondrial respiratory chain's cytochrome b–c1 complex. This interaction is assumed to be the reason for the necrosis observed in the host tissue, as stated by Lyu *et al.* (2016). First, the fungal genome must be carefully examined in order to understand the formation and molecular function of fungal effector genes. Numerous studies, focusing on *S. sclerotiorum*, have looked for proteins that resemble effectors or are secreted and may be involved in infection (Amaral *et al.*, 2012; Guyon *et al.*, 2014; Nemri *et al.*, 2014; Heard *et al.*, 2015). This research has employed a number of techniques to distinguish effector-like from non-effector-like genes. Small, cysteine-rich proteins with an expected secretion signal have drawn special attention due to the commonalities among the effectors found through testing (Sperschneider *et al.*, 2015). Recently, effector prediction has also made use of a machine learning technique that incorporates many criteria derived from known effectors. When compared with the conventional way of manually choosing criteria, this methodology showed higher sensitivity and specificity (Sperschneider *et al.*, 2016; Derbyshire *et al.*, 2017).

8.12 The *Thioredoxin1 (SsTrx1)* gene for pathogenesis and oxidative stress tolerance

S. sclerotiorum infects host plant tissues by inducing necrosis to source nutrients needed for its establishment. Tissue necrosis results from an enhanced generation of ROS at the site of infection and apoptosis. Pathogens have evolved ROS scavenging mechanisms to withstand host-induced oxidative damage. However, the genes associated with ROS scavenging pathways are yet to be fully investigated in *S. sclerotiorum*. Rana *et al.* (2021) selected the *S. sclerotiorum Thioredoxin1* gene (*SsTrx1*) for its expression since it is significantly induced during *S. sclerotiorum* infection. RNAi-induced silencing of *SsTrx1* in *S. sclerotiorum* affected the hyphal growth rate, mycelial morphology, and sclerotial development under *in vitro* conditions. These outcomes confirmed the involvement of *SsTrx1* in promoting pathogenicity and oxidative stress tolerance of *S. sclerotiorum*. An *SsTrx1*-based *HIGS* vector is then constructed and mobilized it into *A. thaliana* (HIGS-*A*) and *N. benthamiana* (*HIGS-N*). The disease resistance analysis revealed significantly reduced pathogenicity and disease progression in the transformed genotypes compared with the nontransformed and empty vector controls. The relative gene expression of *SsTrx1* increased under oxidative stress. The results show that the normal expression of *SsTrx1* is crucial for pathogenicity and oxidative stress tolerance of *S. sclerotiorum* (Rana *et al.*, 2021).

8.13 Effector Genes and Secretory YML079-Like Cupin Protein for Pathogenesis

Recent studies suggest that, in addition to secreting cell wall-degrading enzymes (CWDEs) and toxins, *S. sclerotiorum* employs effector proteins to facilitate its infection. Cysteine-rich effector SsSSVP1 can interact with QCR8 (a subunit of the mitochondrial respiratory chain cytochrome b–c1 complex) and disturb its subcellular localization to disrupt the function of QCR8 and induce plant cell death (Lyu *et al.*, 2016). During the initial phase of infection, the integrin-like effector SsITL hinders the accumulation of SA and suppresses the host's immune response by interacting with a calcium sensor that is localized in the chloroplast (Zhu *et al.*, 2013; Tang *et al.*, 2020). Cerato-platanin effector SsCP1 interacts with plant pathogenesis-related protein (PR1) in the apoplast to facilitate *S. sclerotiorum* infection (Yang *et al.*, 2018). Despite this groundbreaking study, most of the potential effectors encoded by *S. sclerotiorum*—which, based on prior bioinformatic projections, may number in the hundreds—have not been the subject of experimental investigation (Xia *et al.*, 2019; Seifbarghi *et al.*, 2020). In 2021, Fan *et al.* (2021) discovered and described SsYCP1, a new secretory protein derived from *Sclerotiorum*. YML079-like cupin protein SsYCP1 is extensively produced and released after infection with *S. sclerotiorum*. While silencing has the opposite impact of ectopic overexpression of SsYCP1, it enhances *S. sclerotiorum* infection, indicating that SsYCP1 may be an effector protein that contributes to the pathogenicity of *S. sclerotiorum*. It is interesting to note that the secretory signal peptide lacking from YML079-like cupin proteins encoded by numerous other ascomycetes suggests that SsYCP1 may be a recently developed secretory protein to facilitate *S. sclerotiorum* infection.

A YML079-like cupin protein (SsYCP1) from *S. sclerotiorum* has been characterized which showed that SsYCP1 is strongly expressed and secreted during *S. sclerotiorum* infection. The *S. sclerotiorum* infection was promoted by SsYCP1 over expression and inhibited by silencing this gene with synthetic double-stranded RNA. These results collectively indicate SsYCP1 as a putative effector protein that contributes to *S. sclerotiorum* pathogenicity. These findings extend the understanding of effector-mediated *S. sclerotiorum* pathogenesis and suggest a novel role for YML079-like cupin proteins in plant–pathogen interactions (Fan *et al.*, 2021).

8.14 The Mechanism of Pathogenic Disruption via *A. Tumefaciens*-Mediated Gene Mediation

Several candidate genes linked to *S. sclerotiorum* have been discovered through the examination of expressed sequence tags derived from mycelial-infected *B. napus* and the formation of infection cushions (Li *et al.*, 2004; Sexton *et al.*, 2006). In their study, Sexton *et al.* (2006) employed real-time reverse transcription polymerase chain reaction (RT-PCR) to investigate the transcriptional patterns of a number of these genes, validating their possible role in the development of infection cushions and pathogenesis. Gene expression assays are valuable for assigning a putative function to a specific gene, but gene disruption or silencing experiments are essential for obtaining definitive information about the role(s) of the gene in pathogenesis. Despite the well-documented pathogenic lifestyle of *S. sclerotiorum*, genetic transformation of the fungus has only recently been achieved. Targeted gene inactivation and antisense RNA procedures have been used to study specific genes, and *A. tumefaciens*-mediated transformation (ATMT) has proven to be a useful tool for functional genetic analysis of many fungi. The challenges of T-DNA transformation in *S. sclerotiorum* have been described anecdotally, but successful transformation using ATMT has been reported.

An effective method for disrupting genes in *S. sclerotiorum* using *A. tumefaciens* was developed. To demonstrate this method, a transformation vector with the full-length sequence of the malate synthase gene (mlsl) from the glyoxylate cycle disrupted by a gene cassette encoding resistance to hygromycin B was utilized. Gene disruption mutants were created by introducing the disruption construct into ascospores, selecting transformants on hygromycin-containing medium, and purifying transformants through rigorous hyphal tip purification. This process resulted in stable, hygromycin-resistant transformants, with DNA blot hybridization analysis showing a high rate of gene disruption through homologous recombination. The knockout mutants exhibited impaired glyoxylate cycle function and loss of pathogenicity (Liberti *et al.*, 2007).

8.15 Analysis of Genome-Wide Alternative Splicing Patterns in *Sclerotinia* During Pathogenesis

A well-known plant disease called Sclerotinia stem rot is dangerous to more than 900 host plants, including both domesticated and untamed varieties. It has been determined which molecular pathways enable *S. sclerotiorum* to grow in a wide variety of hosts. One ubiquitous and regulated biological mechanism that increases the diversity of proteins and RNA in cells is called alternative splicing (AS). *S. sclerotiorum's* genome contains spliceosome genes, whose expression patterns have been investigated both *in vitro* and during the colonization of six different host species (Ibrahim *et al.*, 2021). Differential expression of numerous spliceosome genes was detected in planta, suggesting alterations in AS during infection. Strict criteria were used to identify 1487 *S. sclerotiorum* genes that showed alternative transcripts and were differentially expressed in planta. During the colonization of all host plants, retained introns and alternate 3′ receiver sites were the most frequent AS occurrences. Furthermore, Ibrahim *et al.* (2021) found that *S. sclerotiorum* genes were expressed in planta, where the accumulation of alternative transcripts varied according to the host being colonized and these transcripts had distinct protein domains. Notably, study includes 42 genes encoding putative secreted proteins with high-confidence AS events.

The plant pathogenic fungus *S. sclerotiorum* is known to undergo AS events during the colonization of host plants, which may result in functional variety in the repertoire of proteins generated by the organism during infection, according to the conclusions drawn from Figs 8.1 and 8.2.

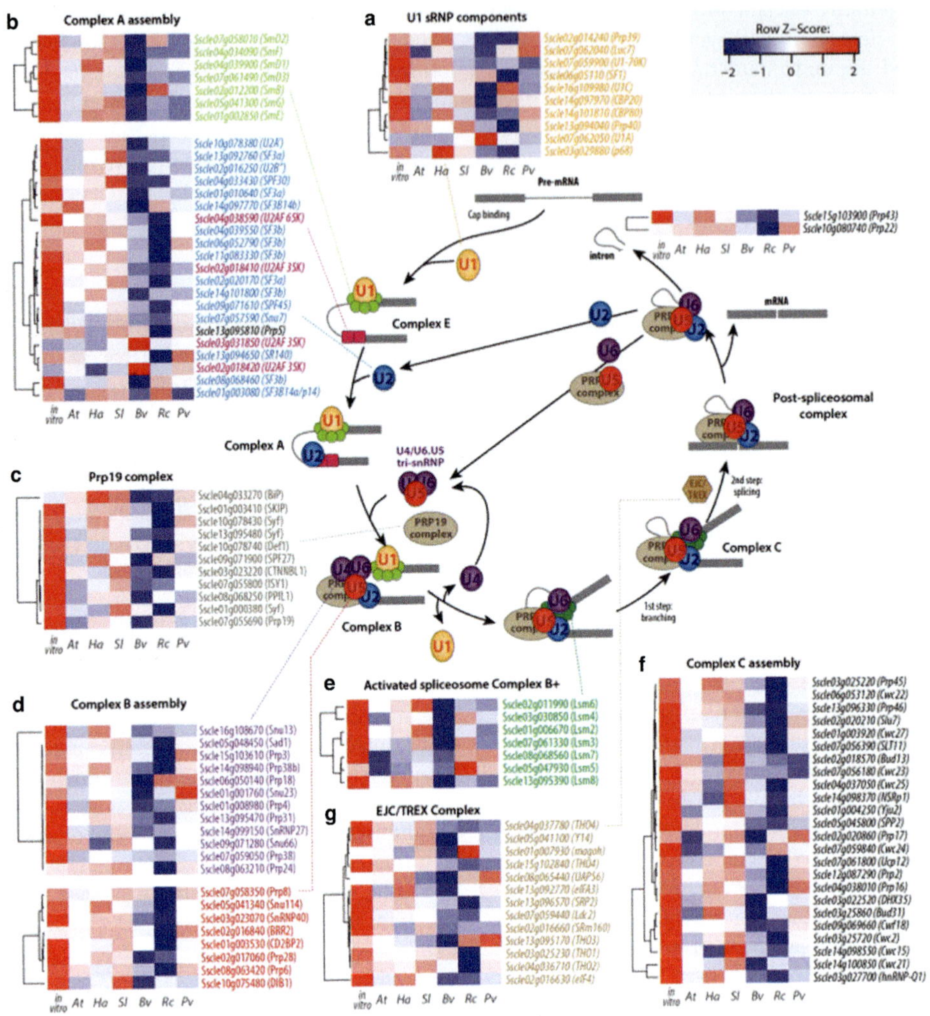

Fig. 8.1. Spliceosome components and their control during plant infection have been found in six different botanical groups of *S. sclerotiorum*. Each step of the mRNA splicing process is depicted in a schematic with the *S. sclerotiorum* genes involved. An intron is shown as a dark gray line, symbolizing a potential pre-mRNA molecule, and two exons are shown as dark gray boxes. Protein complexes are shown as circles, rounded rectangles, and hexagons. The heat maps illustrate the differential expression of 116 spliceosome genes in the periphery of *S. sclerotiorum* mycelium both *in vitro* and during the infection of plants from six distinct species. It is evident that several spliceosomal complexes are involved in pre-mRNA splicing. Initially, U1 snRNP (a) to small nuclear ribonucleo-protein-associated proteins (Sm), U2 associated factors (U2AF), and splicing factors (SF), leading to the recruitment of U2 and the formation of complex A (b). The PRP19C/Prp19 complex/NTC/19 complex (c) stabilizes the U4/U5/U6 tri-snRNP spliceosomal complex leading to complex B assembly (d). The U1/U4 snRNPs are released to form the activated spliceosome complex B+ (e) triggering branching, intron excision, conformational rearrangements into complex C (f) and ligation of the proximal and distal exons. EJC/TREX is recruited to spliced mRNAs to mediate export to the cytoplasm (g). *At, Arabidopsis thaliana; Bv, Beta vulgaris; Ha, Helianthus annuus; Pv, Phaseolus vulgaris; Rc, Ricinus communis; Sl, Solanum lycospersicum*. (Adapted from publication of Ibrahim *et al.*, 2021 with permission.)

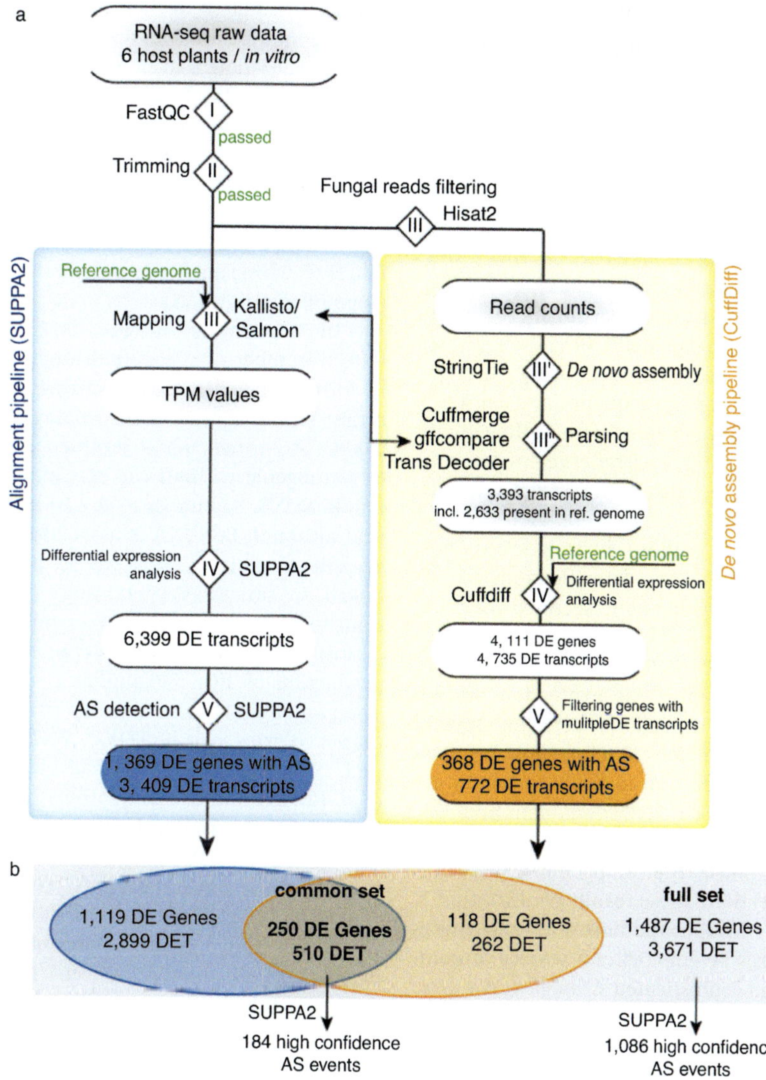

Fig. 8.2. Pipeline for the genome-wide detection of alternative splicing (AS) in Sclerotinia sclerotiorum. The raw RNA-seq data were examined using Fast QC (I) and then trimmed using (a) Trimmomatic (II) found and examined novel transcripts using the transcriptome alignment process (blue box) and the de novo assembly workflow (yellow box). Readings were mapped to the *S. sclerotiorum* reference genome using HISAT (III) in order to find novel transcripts; this information was then utilized in a modified String Tie de novo assembly. (III′). Novel transcripts were found and compared with the reference gene annotation (III″) using Cuff merge, gff compare, and Trans decoder. This resulted in the creation of a new reference annotation. Differential expression analysis (IV) was undertaken utilizing the new annotation and the reference genome. A selection was made of the differentially expressed (DE) genes that encoded two or more DE transcripts (DET; V). Maps were created using Salmon (III), DE analysis, and SUPPA2 (IV and V) as part of the transcriptome alignment route. (b) A Venn diagram that summarizes the findings for both pipelines from the DE analysis in (a); only genes encoding multiple transcripts are given numbers. Transcripts per million (TPM), alternative splicing (AS), differentially expressed (DE), and differentially expressed transcript (DET). (Adapted from publication of Ibrahim *et al.*, 2021 with permission.)

8.16 Analysis of *Sclerotinia* Transcriptome During Pathogenesis

Significant progress has been made in identifying the precise components that contribute to *S. sclerotiorum*'s pathogenicity. Nevertheless, little is now known about the modifications in gene expression that take place while an *S. sclerotiorum* infection spreads across a host plant. In a study by Seifbarghi *et al.* (2017), the processes that take place during the initial (1 hour) and intermediate (48 hour) phases of infection were examined using RNA-seq analysis. The temporal sequence and degree of gene activation associated with pathogenicity or virulence during *S. sclerotiorum* infection on *B. napus* were revealed by transcript analysis. Eight functional groups were identified by these genes: OA synthesis, hydrolytic enzymes, secondary metabolites, detoxification, signaling pathways, development, secreted effectors, and production of ROS. Most of these genes' expression profiles matched the roles that were predicted for them. Gene expression profiles that indicated transitions between several stages of pathogenicity, such as host penetration, ramification, and necrotic phases, were identified using principal component analysis. Additionally, it provided evidence of a transient biotrophic phase that happened right after host penetration. These results confirm that *S. sclerotiorum* invades the host and undermines host defenses by employing a variety of components and sophisticated strategies. The gene expression sequences linked to pathogenicity and virulence during *S. sclerotiorum* infection on *B. napus* are thoroughly reviewed in this chapter. Furthermore, as shown in Fig. 8.3, it offers crucial information for interpreting the genes involved in the communication between *S. sclerotiorum* and its host plant.

8.17 Generation of Reactive Oxygen Species for Pathogenesis

All aerobic organisms create ROS, which include hydroxyl radicals, superoxide, and hydrogen peroxide (H_2O_2) (Apel and Hirt, 2004). Plants release ROS as a defense against pathogen attacks, in addition to their roles in ion transport, signal transmission, and cell development and specialization (Baker and Orlandi, 1995; Foreman *et al.*, 2003). The oxidative burst that generates ROS is initiated by the enzymes known as nicotinamide adenine dinucleotide phosphate (NADPH) oxidases (NOX) (Segmüller *et al.*, 2008). Superoxide dismutase (SOD) converts superoxide, a critical precursor of many ROS, into H_2O_2 when superoxide is generated by NOXs (Segmüller *et al.*, 2008).

In filamentous fungus, NOX enzymes have a number of specialized functions, such as forming penetration structures and facilitating sexual reproduction (Malagnac *et al.*, 2004). *B. cinerea* cannot produce sclerotia or be pathogenic without the enzymes BcNOXA and BcNOXB. Segmüller *et al.* (2008) asserted that although BcNOXA is associated with the growth of lesions, BcNOXB is particularly linked to processes pertaining to penetration. It is important to keep in mind that *B. cinerea's* NOX enzymes do not contribute to the production of ROS (Segmüller *et al.*, 2008). The two NADPH oxidase genes that *S. sclerotiorum* carries are *SsNOX1 (SS1G_05661)* and *SsNOX2 (SS1G_11172)*, which Kim *et al.* (2011) found. The synthesis of oxalate, fungal growth, and pathogenicity all depend on *SsNOX1*. *SsNOX1* and *SsNOX2* expression was detectable but not elevated. Notably, Kim *et al.* (2011) suggested that the *SS1G_10796*-encoded bicupin domain enzyme might exhibit oxalate oxidase function. Given that oxalate oxidase is the enzyme that transforms oxalate into H_2O_2, it is possible that oxalate mediates the ROS that *S. sclerotiorum* produces during pathogenesis. The necrotic stage started at 24 and 48 hpi, when *SS1G_10796* was upregulated. This finding is consistent with a recent study (Kim *et al.*, 2008) that showed oxalate-generated H_2O_2 in the host is involved in programmed cell death.

8.18 Analysis of Transcription Factors for Development and Pathogenesis

Controlling developmental processes and triggering pathogenicity factors during infection

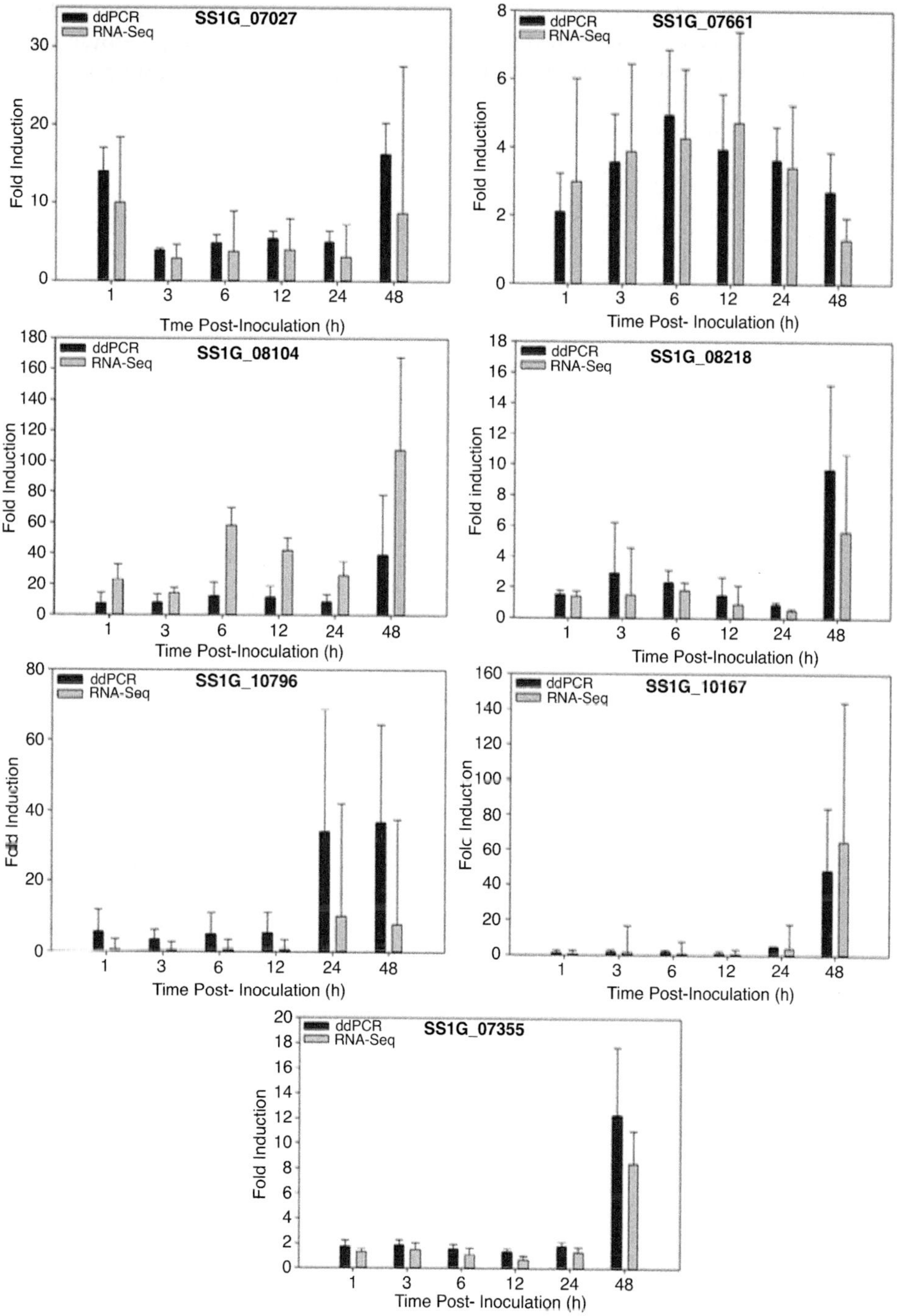

Fig. 8.3. The expression of several *Sclerotinia sclerotiorum* genes during *Brassica napus* infection was ascertained by the use of droplet digital PCR (ddPCR) analysis and RNA-Seq. Histograms are revealed as averages and standard errors of three biological replicates for 1 hpi, and they display the relative expression level (fold change). (Adapted from publication of Seifbarghi *et al.*, 2017 with permission.)

depend on TFs. Several genes that encode potential TFs were found to be upregulated from 1 to 48 hpi. Zinc-binding TFs from the Cys2 His2, Cys4, and Zn2 Cys6 families—which are commonly seen in fungi—were among those found.

More precisely, the gene encoding the Pac1 zinc finger domain TF (SS1G 07355) was eight times more highly expressed at 48 hpi. Pac1, the ortholog of *Aspergillus nidulans* PacC, is recognized to be crucial for regulating the expression of genes sensitive to pH. Pac1 is necessary for *S. sclerotiorum* to regulate a wide range of physiological and PR activities (Rollins, 2003). When the ambient pH level is high, Pac1 triggers the creation of OA. According to Rollins and Dickman (2001), this pH reduction then increases SsPG1 expression and promotes sclerotial development. Pac1 regulates ambient pH levels, which is expected to play a major impact in the accumulation of OA and PG, making it essential to the pathophysiology of *S. sclerotiorum*. Together with the upstream signaling molecules phospholipase C (BcPLC1) and calcineurin, the Cys2 His2 zinc finger TF BcCRZ1 is responsible for controlling the generation of the *B. cinerea* phytotoxin botrydial, according to (Schumacher *et al.*, 2008a, b). The expression of four phospholipase C-like enzyme genes (*SS1G_03548*, *SS1G_05073*, *SS1G_08747*, and *SS1G_13589*) was observed to be upregulated at 48 hour postinfection, indicating a possible role for them in signaling pathways for the generation of secondary metabolites or events connected to later stages of infection. A Zn2Cys6 TF that controls growth, conidiation, and conidial germination is encoded by *Penicillium roqueforti* Pcz1 (Gil-Durán *et al.*, 2015). A Cys2 His2 TF that controls the synthesis of secondary metabolites is encoded by the *BcYOH1* gene in *B. cinerea* (Simon *et al.*, 2013). The expression of other genes, such as *SS1G_10532*, *SS1G_01109*, and *SS1G_02054* (which encode zinc cluster TFs comparable to *BcYOH1* and *Pcz1*), altered at different stages of infection (Table 8.1). Additionally, a large number of upregulated genes encoding different TF types involved in fungal development will be discussed here.

8.19 Expression of Phosphorylation-Dependent Signaling Pathways Genes for Pathogenesis

The *Sclerotinia* genome investigation has revealed many kinases involved in signaling pathways for gene expression that regulate fungal growth, infection, and illness. Several kinases associated with signaling pathways, such as the G protein-coupled receptor, MAP kinase, heterotrimeric G protein, cAMP, and Ca^{2+}-related signaling pathways, are present in the genome of *S. sclerotiorum* (Amselem *et al.*, 2011). During the infection, the expression levels of many of the genes linked to these pathways increased (Table 8.2). It has been suggested that genes encoding the components of signaling pathways may play important roles or serve as regulatory checkpoints during the infection process, even though the activation of these processes is not certainly necessary. It is commonly known that two-component histidine kinases regulate how bacteria and fungi react to external stimuli (Tanaka and Izumitsu, 2010). Only at the onset of the infection (1 hpi) was the gene encoding the two-component sensor histidine protein kinase Shk1 (*SS1G_10091*) upregulated. Though previously thought to be unnecessary for pathogenicity on plant leaves, Shk1 has been connected to hyphal development and sclerotial formation in *S. sclerotiorum* (Duan *et al.*, 2013). According to Hegedus *et al.* (2016), eukaryotic protein kinases are encoded by the ++ gene group. There was evidence of upregulation of two MAPKKK genes (*SS1G_00606* and *SS1G_10983*) at 48 and 1–6 hpi, respectively. Pathogenicity was lost when the orthologs of STE7 and STE11 in *B. cinerea* (Schamber *et al.*, 2010) or MST7 and MST11 in *Magnaporthe grisea* (Zhao *et al.*, 2005) could not produce infection structures. In *S. cerevisiae*, *SS1G_10021* is an ortholog of the CMK1/CMK2 kinases, while in *S. sclerotiorum*, it is a member of the *CAMK1* family. It was shown to be upregulated at 48 hpi. Ding *et al.* (2014) state that CAMKs are known to play a part in regulating the integrity of cell walls and their response to oxidative stress. Another gene (*SS1G_03455*) that is similar to *S. cerevisiae's* SAT4/HAL 4 showed

Table 8.1. Characterization and designation of overexpressed genes involved in encoding transcription-related variables (TF). (Adapted from publication of Seifbarghi *et al.*, 2017 with permission.)

Gene ID	Description[a]	Expression level (hpi)[b]					
		1	3	6	12	24	48
1. Zn2Cys6 (C6)							
SS1G_06255	Zn2 Cys6 TF	7.0	4.6	3.9	2.4	–	–
SS1G_14383	Zn2 Cys6 TF	3.5	4.4	3.5	2.1	–	–
SS1G_08819	Zn2 Cys6 TF	3.6	2.5	2.1	3.0	–	–
SS1G_06907	Zn2 Cys6 TF	4.1	4.1	3.3	–	–	–
SS1G_02793	Zn2 Cys6 TF	6.0	3.0	–	–	–	–
SS1G_10324	Zn2 Cys6 TF	2.0	–	–	–	2.0	–
SS1G_02851	Zn2 Cys6 TF	2.2	–	–	–	–	–
SS1G_12532	Zn2 Cys6 TF	2.2	–	–	–	–	–
SS1G_12799	Zn2 Cys6 TF	–	2.5	2.5	–	–	–
SS1G_11949	Zn2 Cys6 TF	–	2.7	2.5	–	–	–
SS1G_00732	Zn2 Cys6 TF	–	2.4	–	5.0	–	–
SS1G_00398	Zn2 Cys6 TF	–	4.1	–	–	–	4.8
SS1G_01905	Zn2 Cys6 TF	–	–	–	–	2.7	–
SS1G_06876	Zn2 Cys6 TF	–	–	–	–	2.3	–
SS1G_00170	Zn2 Cys6 TF	–	–	–	–	2.3	–
SS1G_02339	Zn2 Cys6 TF	–	–	–	–	2.0	–
SS1G_07143	Zn2 Cys6 TF	–	–	–	–	2.7	–
SS1G_02791	Zn2 Cys6 TF	–	–	–	–	2.9	3.0
SS1G_00392	Zn2 Cys6 TF	–	–	–	–	2.2	2.2
SS1G_13144	Zn2 Cys6 TF	–	–	–	–	2.1	3.0
SS1G_02054	Zn2 Cys6 TF	–		–		0.0	36.9
SS1G_05109	Zn2 Cys6 TF	–	–	–	–	6.7	7.2
SS1G_05755	Zn2 Cys6 TF	–	–	–	–	2.6	3.8
SS1G_07003	Zn2 Cys6 TF	–	–	–	–	–	4.6
SS1G_10447	Zn2 Cys6 TF	–	–	–	–	–	2.5
SS1G_01353	Zn2 Cys6 TF	–	–	–	–	–	4.1
SS1G_11367	Zn2 Cys6 TF	–	–	–	–	–	3.3
SS1G_08351	Zn2 Cys6 TF	–	–	–	–	–	2.1
SS1G_03775	Zn2 Cys6 TF	–	–	–	–	–	6.8
SS1G_00787	Zn2 Cys6 TF	–	–	–	–	–	2.8
SS1G_01733	Zn2 Cys6 TF	–	–	–	–	–	2.6
SS1G_04056	Zn2 Cys6 TF	–	–	–	–	–	2.7
SS1G_11081	Zn2 Cys6 TF	–	–	–	–	–	2.8
SS1G_09741	Fungal-specific TF domain	2.2	3.7	3.9	2.7	–	–
SS1G_04333	Fungal-specific TF domain	2.0	2.3	2.0	–	–	–
SS1G_12561	Fungal-specific TF domain	2.3	2.1	–	–	–	–
SS1G_09823	Fungal-specific TF domain	–	–	2.1	–	2.2	2.4
SS1G_04846	Fungal-specific TF domain	–	–	–	–	2.6	4.0
SS1G_13882	Fungal-specific TF domain	–	–	–	–	2.0	–

Continued

Table 8.1. Continued

Gene ID	Description[a]	Expression level (hpi)[b]					
		1	3	6	12	24	48
SS1G_13729	Fungal-specific TF domain	–	–	–	–	2.4	–
SS1G_02758	Fungal-specific TF domain	–	–	–	–	2.5	–
SS1G_08406	Fungal-specific TF domain	–	–	–	–	2.3	–
SS1G_05809	Fungal-specific TF domain	–	–	–	–	3.9	–
SS1G_04057	Fungal-specific TF domain	–	–	–	–	3.9	2.4
SS1G_11395	Fungal-specific TF domain	–	–	–	–	–	2.3
SS1G_06361	Fungal-specific TF domain	–	–	–	–	–	2.1
2. Cys2His2							
SS1G_01109	C2H2 TF	19.8	26.1	26.6	15.6	–	–
SS1G_01684	C2H2 TF	2.6	3.1	2.7	2.0	–	–
SS1G_04676	C2H2 TF	2.0	–	–	–	–	7.3
SS1G_07355	C2H2 TF (Pac1)	–	–	–	–	–	8.4
SS1G_10532	C2H2 TF	–	–	–	–	–	20.5
SS1G_00104	C2H2 binding site	–	–	2.5	3.1	–	5.4
SS1G_09588	C2H2 binding site	–	–	–	–	2.9	4.4
SS1G_06370	C2H2 binding site	–	–	–	–	–	2.1
SS1G_09499	C2H2 binding site	–	–	–	–	–	3.1
SS1G_07425	C2H2 binding site	–	–	–	–	–	4.3
SS1G_06044	C2H2 binding site	–	–	–	–	–	2.3
3. Basic leucine zipper (bZIP)							
SS1G_10244	bZIP TF	4.4	5.2	–	5.6	–	–
SS1G_01411	bZIP TF	–	–	–	–	2.0	–
SS1G_00816	bZIP TF	–	–	–	–	–	5.4
SS1G_06075	bZIP TF	–	–	–	–	–	7.7
4. Others							
SS1G_14328	Transcription regulator BDF1	2.7	6.5	5.5	3.6	–	–
SS1G_11030	CP2 TF	4.9	3.2	2.5	–	–	–
SS1G_04107	MYB family TF	2.5	2.9	–	–	–	–
SS1G_06257	pex2/pex12 super family, zinc finger of C3HC4-type	3.0	2.4	–	–	–	–
SS1G_07430	Zinc finger domain	3.0	–	–	–	–	2.8
SS1G_03881	Regulator of G protein signaling domain protein (RGS) (GTPase-activating proteins (GAPs))	2.0	–	–	–	–	–
SS1G_14385	CBF/Mak21 TF	2.4	–	–	–	–	–
SS1G_03280	Helix-loop-helix DNA-binding protein (HIH)	2.4	–	–	–	–	–
SS1G_10206	CHY and ring, zinc finger protein	–	2.1	–	–	–	–
SS1G_04050	NF-X1 zinc finger TF	–	3.0	–	–	–	–

Continued

Table 8.1. Continued

Gene ID	Description[a]	Expression level (hpi)[b]					
		1	3	6	12	24	48
SS1G_11663	Large tegument protein UL36 and similar to TFIIIC transcription initiation factor complex subunits Tfc3	–	–	2.1	2.0	–	–
SS1G_03992	Transcription mediator subunit Med12	–	–	2.2	–	–	–
SS1G_13930	Bro1-Alix-like domain and pH-response regulator protein	–	–	–	–	2.3	–
SS1G_07542	RING-H2 zinc finger protein	–	–	–	–	2.3	–
SS1G_06124	mads-box mef2 type TF (SRF type)	–	–	–	–	–	3.5
SS1G_03098	Homeobox TF	–	–	–	–	–	2.2
SS1G_03835	Homeobox C2H2 TF	–	–	–	–	–	3.1
SS1G_06987	Yippee zinc-binding protein	–	–	–	–	–	2.4
SS1G_01859	MYB DNA-binding containing domain protein	–	–	–	–	–	6.2
SS1G_08831	(vWA) TF (Von Willebrand factor type A)	–	–	–	–	–	2.3
SS1G_09890	RAP TF	–	–	–	–	–	2.6
SS1G_02756	TF protein	–	–	–	–	–	8.5
SS1G_10304	Tetra tricopeptide (TPR) repeat transcriptional corepressor	–	–	–	–	–	2.3
SS1G_10511	Lipopoly-saccharide (LPS)-induced TF- and LITAF-like zinc ribbon domain	–	–	–	–	–	2.7

(–) Expression shows no notable difference.
[a]Annotation is performed by detecting conserved PFAM domains and examining BLAST reports.
[b]Relative fold change compared with time 0 hpi.

induction at 24–48 hpi. The *HAL* family kinases control the membrane permeases that carry glucose and amino acids (Pérez-Valle *et al.*, 2010). Forty-8 hours, after infection, two more protein kinases, *SS1G_09511* and *SS1G_06542*, showed notable upregulation. A cyclin-dependent kinase called PHO85 (*SS1G_07226*) controls cell division in response to outside stimuli. It is comparable to the kinase that *SS1G_09511* encodes (Nishizawa, 2015). Conversely, an ortholog of *SHA3/SKS1*, *SS1G_06542*, is associated with the integration of the hyphal growth response to glucose (Johnson *et al.*, 2014). Despite being specific to multicellular fungi, the FunK1 protein kinases have characteristics in common with eukaryotic protein kinases (Stajich *et al.*, 2010). Within the *S. sclerotiorum* genome, three *FunK1* family members (*SS1G_09355*, *SS1G_12423*, and *SS1G_14212*) were all upregulated during the early phases of infection. Furthermore, GTP-binding proteins are encoded by *SS1G_06571*, *SS1G_03234*, and *SS1G_10333*, which

Table 8.2. Identification of overexpressed genes in gene regulation networks or signaling cascades. (Adapted from publication of Seifbarghi *et al*. (2017) with permission.)

Gene ID	Description[a]	Expression level (hpi)[b]					
		1	3	6	12	24	48
1. *Protein Kinases*							
SS1G_10983	MAPKK kinase (SsBCK1)	4.7	5.4	4.4	2.4	–	–
SS1G_11525	Protein kinase-like protein	2.5	3.6	6.9	6.1	–	–
SS1G_00606	MAPKK kinase (SsSTE11)	2.1	2.2	–	–	–	2.1
SS1G_08085	AGC protein kinase (SsRIM15)	–	2.1	2	–	–	–
SS1G_03455	HAL family protein kinase	–	–	–	–	3.3	3.7
SS1G_06203	CAMK protein kinase (SsRCK2)	–	–	–	–	–	2.3
SS1G_10021	CAMK protein kinase (SsCMK2)	–	–	–	–	–	4.1
SS1G_06542	Ran1-like protein kinase (SsSKS1)	–	–	–	–	–	2.3
SS1G_09511	Protein kinase (Ss-Other-03)	–	–	–	–	–	5.8
SS1G_14212	Funk1 serine threonine-protein kinase	2.1	–	–	–	–	–
SS1G_12423	Funk1 serine threonine-protein kinase	2.3	2.4	–	–	–	–
SS1G_09355	Funk1 serine threonine-protein kinase	2.1	2.3	–	–	–	–
SS1G_10091	Two-component histidine protein kinase (SHK1)	2.6	–	–	–	–	–
2. *Phosphatases*							
SS1G_11340	Tyrosine phosphatase	2.6	2.1	2.2	–	–	2
SS1G_01711	Protein phosphatase type one complex subunit hex2 reg1	3.3	3.2	2.2	–	–	–
SS1G_04320	Tyrosine-protein phosphatase non-receptor-type partial	2.1	3.5	2.5	–	–	–
SS1G_06382	Protein phosphatase regulator	2.1	–	–	–	–	–
SS1G_12383	Histidine acid phosphatase	–	2.8	–	–	–	–
SS1G_08513	Serine threonine-protein phosphatase PP2A	–	–	–	–	–	2.8
SS1G_10466	Diketo-5-methylthio-1-phosphopentane phosphatase	–	–	–	–	–	2.5
3. *GTPase/GTP-binding*							
SS1G_03234	rho GTPase activator	3.1	5	4.6	4.8	–	–
SS1G_06571	GTP-binding protein rho2	3.2	3	2.1	–	–	–
SS1G_01564	Nuclear GTP-binding protein NUG1	2.9	–	–	–	–	–
SS1G_04075	ARF GTPase activator	–	–	–	–	–	2.5
SS1G_10333	GTP-binding protein	–	–	–	–	–	2.3
SS1G_08371	CLP1 GTPase	–	–	–	–	–	2.9
SS1G_13589	Phosphatidyl inositol phospholipase C (PL-PLC)	–	–	–	–	–	2.1
SS1G_03548	Phosphatidyl inositol phospholipase C (PL-PLC)	–	–	–	–	–	2.1

Continued

Table 8.2. Continued

Gene ID	Description[a]	Expression level (hpi)[b]					
		1	3	6	12	24	48
SS1G_05073	Phosphatidyl inositol phospholipase C (PL-PLC)	–	–	–	–	–	5
SS1G_08747	Phosphatidyl inositol phospholipase C (PL-PLC)	–	–	–	–	–	9
4. *Other*							
SS1G_06667	sir2 chromatin regulatory protein	2.1	3	2.6	2.1	–	–
SS1G_06180	Pal1 morphogenesis-related protein	9.5	8.7	7.9	4.5	–	–
SS1G_04402	Arrestin (chitin synthesis regulation)	–	–	–	–	2.1	–
SS1G_03525	Ankaryin repeat protein	–	–	–	–	2.7	–
SS1G_04325	SUR7/Pal1 family (pH-response regulator)	–	–	–	–	–	2.2
SS1G_10456	SUR7/Pal1 family (pH-response regulator)	–	–	–	–	–	12.6
SS1G_09665	Inositol-pentakisphosphate 2-kinase	–	–	–	–	–	2.8
SS1G_00378	SRP19 signal recognition particle protein	–	–	–	–	–	3
SS1G_08048	Sok1 cAMP-mediated signaling protein	–	–	–	–	–	2.7

(–) No significant change in expression.
[a]Annotation based on the presence of conserved PFAM domains and BLAST reports.
[b]Fold change relative to 0 hpi.

were initially elevated at 1, 1, and 48 hpi, respectively. One of the functions of the small GTPase Rap-1 is to modulate the inhibitory effects of cAMP on the SMK1 MAPK signaling cascade and the mechanisms leading to sclerotial development in *S. sclerotiorum*. As downstream effectors of cAMP, GTP-binding proteins of the *Ras* superfamily aid in the suppression of MAPK (Chen and Dickman, 2005). Studies have also demonstrated the significance of other small GTPases, such as Rap-1 and Rho/Rac/Cdc42, in addition to *Ras*, in signal transmission through the activation of MAPK cascades (Zimmermann and Moelling, 1999). Kinase-dependent signaling pathways can be activated or inhibited by protein dephosphorylation. Harel *et al.* (2006) state that calcineurin, a Type 2B serine/threonine-phosphatase, is required for proper sclerotial development and hyphal cell wall formation. The synthesis of melanin and other secondary metabolites, sclerotia development, infection cushion formation, hyphal growth, and other aspects of *S. sclerotiorum* pathogenesis have all been linked to the serine/threonine-phosphatase Type 2A (PP2A), which is encoded by SS1G_08489 (PPH1) (Erental *et al.*, 2007). The serine/threonine-phosphatase gene (*SS1G_08513*), which encodes a catalytic subunit of PP2A at 18 hpi, did show modest upregulation even though the PPH1 and calcineurin genes did not. Though more investigation is needed to determine its precise functions, SSG_08513 may be able to support PPH1 during the *S. sclerotiorum* infection process. Additionally, the SMK1 MAPK favorably controls PPH1 activity through a nitrous oxide-dependent mechanism.

8.20 Expression of Genes for Development of Pathogenic Structures and Virulence

Like the majority of other multicellular pathogens, *S. sclerotiorum* goes through numerous stages of infection, during which it undergoes noticeable morphological and biochemical changes. The genome of *S. sclerotiorum* contains homologues of *Magnaporthe oryzae* genes, which are essential for the growth and

Table 8.3. Description and characterization of genes that are upregulated during fungal development. (Adapted from publication of Seifbarghi *et al*. (2017) with permission.)

Gene ID	Description[a]	Expression level (hpi)[b]					
		1	3	6	12	24	48
1. Reserve mobilization							
SS1G_05192	Acid trehalase	–	–	–	–	6.7	5.8
SS1G_01494	1,3-alpha-glucanase/mutanase	–	–	–	–	–	30.7
SS1G_09861	1,3-alpha-glucanase/mutanase	–	–	–	–	–	15.0
1.1 Cell wall turnover							
SS1G_05454	Chitinase	–	–	2.4	–	3.5	2.7
SS1G_11700	Chitinase	–	–	–	–	3.7	69.1
SS1G_05897	Chitinase	–	–	–	–	3	2.6
SS1G_11304	Chitinase	–	–	–	–	–	4.8
SS1G_08695	Class III chitinase	–	–	–	5.8	8.2	11.7
SS1G_11212	Class III chitinase	–	–	–	–	–	9.3
SS1G_12510	Class V chitinase	–	–	–	–	11.2	66.9
SS1G_09403	Alpha-N-acetylglucosaminidase	–	–	–	–	2.9	–
SS1G_12837	Beta-N-acetylglucosaminidase	–	–	–	–	2.3	–
SS1G_10038	Beta-N-acetylglucosaminidase	–	–	–	–	–	3
SS1G_04898	Polysaccharide (chitin) deacetylase	–	–	15.2	–	–	12.7
SS1G_01131	Polysaccharide (chitin) deacetylase	–	–	–	–	3	3.6
SS1G_00642	Polysaccharide (chitin) deacetylase	–	–	–	–	–	8.3
SS1G_12836	N-acetylglucosamine-6-phosphate deacetylase	–	–	–	–	–	6.7
SS1G_01229	Exo-beta 1,3 glucanase	–	–	–	–	3.4	22.8
SS1G_09858	Exo-beta 1,3 glucanase	–	–	–	–	–	2.8
SS1G_12930	GPI-anchored cell wall beta-1,3-endoglucanase	8.5	6.4	3.9	–	–	2.2
SS1G_04852	GPI-anchored cell wall beta-endoglucanase	2.1	–	–	–	–	–
2.2 Biosynthesis							
SS1G_04969	Glycosyltransferase (cell wall synthesis)	2.2	–	–	–	–	–
SS1G_04062	Glycosyltransferase (cell wall synthesis)	–	3.6	3.8	–	–	–
SS1G_07313	Lipopolysaccharide biosynthesis protein	–	–	–	–	–	2.4
3. Other							
SS1G_02742	Heterokaryon incompatibility protein	3.8	3.6	3.1	2.8	–	–
SS1G_02744	Heterokaryon incompatibility protein (SEC1)	3.7	5.5	–	5	3	3
SS1G_02602	Heterokaryon incompatibility protein	2.3	–	–	–	–	–
SS1G_03889	Heterokaryon incompatibility protein	–	3.2	–	–	–	–
SS1G_11165	Heterokaryon incompatibility protein	–	3.3	2.2	–	–	
SS1G_06800	Heterokaryon incompatibility protein	–	3.6	–	–	–	

Continued

Table 8.3. Continued

Gene ID	Description[a]	Expression level (hpi)[b]					
		1	3	6	12	24	48
SS1G_08974	Heterokaryon incompatibility protein (WD40 repeat)	–	–	–	–	2.7	–
SS1G_06855	Heterokaryon incompatibility protein	–	–	–	–	5.2	2.4
SS1G_12973	Heterokaryon incompatibility protein	–	–	–	–	–	2
SS1G_09167	Heterokaryon incompatibility protein	–	–	–	–	–	2.5
SS1G_11315	Heterokaryon incompatibility protein	–	–	–	–	–	4
SS1G_07526	Ferritin-like sexual development protein	3	2.5	2.5	–	–	23.9
SS1G_04316	Acyltransferase (hard surface induced)	2.5	–	–	–	–	–
SS1G_10311	DUF cell surface protein (MAS2 ortholog)	–	2.1	–	2.5	2.5	
SS1G_11468	CAS1 appressorium-specific protein (MAS3 ortholog)	–	–	2.9	7.3	12.9	10.5
SS1G_14127	Gamma-glutamyltranspeptidase (SsGGT1)	–	–	3.4	6.5	4.8	2.7
SS1G_05330	Gamma-glutamyltranspeptidase	–	–	–	–	2.1	–
SS1G_12877	Conidiation-specific expression protein	–	–	–	–	2.3	2.3
SS1G_12133	Predicted protein (SSP2)	–	–	–	–	4.7	9.6
SS1G_07404	Predicted protein (Ss-Rh1)	–	–	–	–	4.9	3.8
SS1G_01614	G protein-coupled receptor (SOP1)	–	–	–	–	17.7	17.1
SS1G_07626	Velvet family	–	–	–	–	–	5.2
SS1G_13339	Choline carnitine O-acyltransferase	–	–	–	–	–	3.4
SS1G_02422	UDP-galactopyranose mutase (GLF)	–	–	–	–	–	3.0
SS1G_10940	Gamma-glutamyltranspeptidase	–	–	–	–	–	4
SS1G_14065	Predicted protein (SSP1)	–	–	–	–	–	47

(–) No significant change in expression.
[a]Annotation based on the presence of conserved PFAM domains and BLAST reports.
[b]Fold change relative to 0 hpi.

infiltration of infection structures (Amselem *et al.*, 2011). Notably, Table 8.3 demonstrates the upregulation of genes *SS1G_13339*, *SS1G_10311*, and *SS1G_11468*. *SS1G_10311* and *SS1G_11468* are involved in the development of appressoria in the early stages of infection, much like mas2 and mas3 in *M. oryzae* (Xue *et al.*, 2002). But *SS1G_10311* was upregulated at 3 and 12 hpi, *SS1G_11468* was upregulated at 6 and 48 hpi, and *SS1G_13339* was the only one that was upregulated at 48 hpi, indicating that this necrotrophic pathogen may play many roles. Environmental cues can also cause morphological alterations. The enzyme known as γ-glutamyl transpeptidase regulates the amounts of glutathione and the redox potential of cells. While it is not required for host colonization and symptom expression, the γ-glutamyl transpeptidase encoded by *SS1G_14127 (SsGgt1)* aids in the production of compound appressoria during host penetration and sclerotia growth in *S. sclerotiorum*. The genes encoding two additional γ-glutamyl transpeptidases (*SS1G_05530* and *SS1G_10940*) were upregulated at later stages, specifically at 24 and 48 hpi, whereas the expression of the *SS1G_14127* gene rose from 6 to 48 hpi. Forty-8 hours after the experiment, the SSP1 gene (*SS1G_14065*) had 47-fold increased expression. In addition, SSP1-related gene *SSP2* (*SS1G_12133*) expressed more when the infection progressed. *SSP1*, a distinct sclerotia protein only present during sclerotial formation, is linked to apothecial and sclerotial development (Li and Rollins, 2009). Sclerotia development may have started based on the increase of SSP1 and SSP2 expression between 24 and 48 hour postinjury. According to Lyu *et al.* (2015), SOP1 was stimulated twice: once in the early phases of *A. thaliana* infection and once more 3 days postinoculation, during the stage of sclerotial growth. Significantly upregulated (17-fold) at 24 and 48 hour postinfection, the SOP1-encoding gene (*SS1G_01614*) suggests its role in the pathosystem of *S. sclerotiorum–B. napus* sclerotial development program. According to Bayram and Braus (2012), members of the velvet protein family are in charge of organizing several fungal differentiation processes, such as the spore, sclerotia, and fruiting body development. The velvet protein-coding gene, *SS1G_07626*, was only active 48 hour after infection, indicating that it might be involved in *S. sclerotiorum* pathogenesis. *Ss1G_07404* (Ss-Rhs1, *S. sclerotiorum* rearrangement hotspot repeats 1) is another gene that has been connected to *S. sclerotiorum* virulence and sclerotial development; it was upregulated at 24 and 48 hpi (Yu *et al.*, 2016).

8.20.1 Expression of genes that control various processes during the growth of fungi

MADS-box proteins represent a category of TFs that play a crucial role in the regulation of various processes, such as cell identity, the cell cycle, and primary metabolism (Qu *et al.*, 2014). These TFs are also highly conserved. It was shown that the MADS-box TF-encoding gene (*SS1G_06124*) is crucial to *S. sclerotiorum* mating (Amselem *et al.*, 2011). Its expression increased threefold at 48 hpi. SsMADS (*SS1G_05588*) is another significant MADS-box TF that is essential for both virulence and proliferation (Qu *et al.*, 2014). On the other hand, Kim *et al.* (2009) found that homeobox genes are essential for regulating several aspects of fungal anatomical development, such as hyphal growth, appressorium formation, and conidia production. In *S. sclerotiorum*, the homeobox C2H2TF-encoding gene *SS1G_03835* was found to be active at 48 hpi. However, the exact purpose of this gene in the organism is yet to be totally known. Furthermore, *B. cinerea* has been shown to contain a homeobox TF called BcHOX8, which supports conidiogenesis and vegetative development (Antal *et al.*, 2012). The interactions of proteins containing heterokaryon incompatibility (HET) domains control the process of programmed cell death linked to vegetative incompatibility in fungus (Paoletti and Clavé, 2007). It was shown that during the infection, 11 genes that produce HET proteins were upregulated. Depending on when they are activated, these genes can be divided into two categories. Early on in the infection process, *SS1G_02742*, *SS1G_02602*, *SS1G_03889*, *SS1G_11165*, and *SS1G_06855*, among others, were activated. The second group (*SS1G_08974*, *SS1G_06855*, *SS1G_12973*, *SS1G_09167*, and *SS1G_11315*) was formed in the later phases of infection (Table 8.3). Many *S. sclerotiorum* genes encoding HET-domain proteins are paralogs of the *HET-E-1* family genes identified in *Podospora anserine* according to Paoletti and Saupe (2009). It is critical to keep in mind that there was no evidence of overexpression of these genes. Moreover, it has been proposed that HET domain-containing proteins contributed to the speciation of *B. cinerea* and *S. sclerotiorum* (Amselem *et al.*, 2011). Later, he reported that there was increased expression of two orthologs of *A. nidulans* genes related to mating and fruiting body development, *SS1G_09861* and *SS1G_07526*. Additionally, there was a notable increase in the *A. nidulans* orthologs *SS1G_06124* (a TF) and *SS1G_00606* (STE 11 kinase), which are implicated in mating process signaling pathways (Amselem *et al.*, 2011). Major morphological phase changes, such as the emergence of sclerotia, deplete storage reserves, especially those

Table 8.4. Expression study of genes linked to potential effector proteins that are upregulated. (Adapted from publication of Seifbarghi *et al.* 2017 with permission.)

Gene ID	Description[a]	Expression level (hpi)[b]					
		1	3	6	12	24	48
SS1G_08557	Salicylate hydroxylase	4.1	5.4	5.1	3.6	5.1	10.8
SS1G_00849	22kda glycoprotein (AltA-1 allergen)	7.4	3.2	–	–	–	7.1
SS1G_10096	Cerato-platanin (PF07249)	3.8	3.7	–	–	–	3.3
SS1G_07295	CFEM domain-containing protein	2	–	–	–	3.9	4.8
SS1G_12336	Chitin-binding protein	–	3.3	–	–	–	–
SS1G_12509	LysM domain protein	–	–	–	3.8	11.8	35.4
SS1G_03611	Cysteine-rich protein	–	–	–	–	85.1	247
SS1G_11912	npp1 domain protein (NEP2)	–	–	–	–	5.8	8.2
SS1G_03282	Serine protease inhibitor	–	–	–	–	–	2.2
SS1G_02904	CyanoVirin-N homology (SsCVNH)	–	–	–	–	–	3.9
SS1G_00263	Protein unique to *S. sclerotiorum and B. cinerea* (Ssv263)	–	–	–	–	–	49.2
SS1G_02068	Predicted protein (SsSSVP1)	–	–	–	–	–	21.5

(–) No significant change in expression.
[a]Annotation based on the presence of conserved PFAM domains and BLAST findings.
[b]Fold change relative to 0 hpi.

resulting from rebuilding and turnover of the cell wall. The gene encoding an acid trehalase (*SS1G_05192*) was activated in the later stages of the infection. In fungus, trehalose is a typical storage carbohydrate. Mutan, also known as alpha-1,3-glucan, is a crucial energy source and a part of the fungal cell wall (Wei *et al.*, 2001). The 1,3-alpha-glucanase/mutanase genes (*SS1G_01494* and *SS1G_09861*) were significantly activated 48 hour after infection (15- and 30-fold, respectively). The expression of these genes was highly associated with a broad range of genes that break down fungal cell walls, including chitinases, chitin deacetylases, *N*-acetyl-glycosaminidases, and many endoglucanases. A different study revealed that during infection, only 19% of all *S. sclerotiorum* genes that break down fungal cell walls were upregulated, indicating that these genes were involved in the reorganization of the cell wall as the pathogen moved through different developmental stages (Seifbarghi *et al.*, 2017).

8.21 Secreted Effectors (Genes) for Infection, Pathogenesis, and Virulence

Many effector proteins are produced by the fungus *S. sclerotiorum*, which may help explain why it can inflict injury or disease (Guyon *et al.*, 2014). It was discovered that the *B. napus* infection resulted in a rise in the quantity of certain of these proteins as well as others (Table 8.4). The ability of two necrosis and ethylene-inducing protein (NEP) proteins from infection, *S. sclerotiorum* (SsNEP1 and SsNEP2), to cause necrosis has been shown by studies on their ability to rapidly create necrotic tissue (Bashi *et al.*, 2010). In the intermediate to late phases of infection, both genes were increased, with SsNEP2 expressing themselves at notably higher levels than SsNEP1. According to Frías *et al.* (2013), these proteins have the ability to trigger plant defenses, which results in systemic acquired resistance. One of the proteins that *B.*

cinerea releases in unusually large quantities is cerato-platanin. According to Frías *et al.* (2011), it induces a marked hypersensitive reaction in the host plant and results in the development of localized necrotic lesions. It was found that *S. sclerotiorum* induces upregulation of the cerato-platanin coding gene (SS1G_10096) during both the early and late stages of infection. In contrast, SsSSVP1 (SS1G_02068) encodes a tiny cysteine-rich secreted protein that disrupts host energy metabolism and causes plant cell death. According to Lyu *et al.* (2016), this protein is essential to *S. sclerotiorum's* pathogenicity. Within 48 hour postinoculation, the expression level of SS1G_02068 increased remarkably by 21 times. SsSSVP1 showed a significant upregulation during the interaction between *S. sclerotiorum* and *A. thaliana*, starting at 3 hour postinfection and continuing steadily to increase from 6 to 12 hour postinfection (Lyu *et al.*, 2016). These results suggest that the host plant may have an effect on the gene's expression profile. It has been shown that a wide range of hydrolytic enzymes, including certain xylanases and polygalacturonases (Kars *et al.*, 2005; Bashi *et al.*, 2013), can induce host necrosis. In earlier studies using a variety of host plants, the gene encoding a cysteine-rich protein with a common fungal extracellular membrane (CFEM) domain (SS1G 07295) did not demonstrate a significant induction. It was demonstrated that during the infection phase, the induction level of effector genes in *S. sclerotiorum* varied between 2- and 4.8-fold, suggesting a potential host influence.

While common in fungal extracellular membrane (CFEM) proteins are connected to biofilm production and iron uptake in Candida species, the CFEM protein Pth11 in *M. grisea* is implicated in appressorium growth (DeZwaan *et al.*, 1999; Ding *et al.*, 2011). The CVNH domain is present in a tiny, secreted protein that is encoded by the gene *SS1G_02904* or *SsCVNH*. According to the earlier research by Guyon *et al.* (2014), *SsCVNH* may act as a possible effector of *S. sclerotiorum*.

The significant role of *SsCVNH* in the infection, the development of sclerotia, and overall growth of *S. sclerotiorum* were illustrated by Lyu *et al.* (2015). After 48 hpi, an upregulation of *SsCVNH* was observed. Furthermore, a remarkable increase in the expression of *SsCVNH* was recorded during the early phases of sclerotial development, specifically 3 days after immunization. Surprisingly, Pac1, a gene involved in sclerotial development through OA-mediated pH decrease, was upregulated in conjunction with the induction of *SsCVNH*, indicating a possible impact of pH levels on *SsCVNH* expression.

Surprisingly, one of the most strongly upregulated genes was found to be SS1G 03611, which codes for an alternative cysteine-rich protein. At 48 hpi, the gene's expression had increased by 247 times. In addition, 49 times throughout that same time frame, the gene *SS1G_00263*, which codes for a protein unique to *S. sclerotiorum* and *B. cinerea*, was elevated. The protein encoded by SS1G_00849 is a homologue of *Colletotrichum higginsianum* effector candidate 91 (CHEC91) (Kleemann *et al.*, 2012). It is similar to the *Alternaria alternata* AltA-1 allergen (Guyon *et al.*, 2014), although it lacks the domains associated with fungal effectors. At 1–3 and 48 hpi, *SS1G_00849* showed considerable overexpression, suggesting a possible function as an *S. sclerotiorum* effector. Certain effectors impede the host's capacity to mount the necessary defenses, which leads to infection. By attaching to and disguising chitin fragments from the fungal cell wall, Lysine motif (LysM) effectors impede the host from recognizing the pathogen, hence averting host defense reactions (Mentlak *et al.*, 2012). In the middle to later phases of the infection, one gene (*SS1G_12509*) encoding a LysM protein was upregulated, and at 3 hpi, another gene (*SS1G_12336*) producing a chitin-binding protein was upregulated. Plant defenses against a range of biotic and abiotic stresses are activated by the signaling molecule SA. Some fungal endophytes generate SA-breaking enzymes to prevent these defenses from activating (Ambrose *et al.*, 2015). The salicylate hydroxylase gene (*SS1G_08557*) was upregulated from the start of the infection, even in the lack of a signal peptide. Seifbarghi *et al.* (2017) suggest that this gene may be secreted by a distinct Golgi-independent protein secretion pathway or endoplasmic reticulum.

Effector proteins that are secreted are vital to the pathophysiological mechanism. The *S. sclerotiorum* genome's probable effector genes were located using bioinformatic analysis (Guyon *et al.*, 2014). Chittem *et al.* (2020) compared the expression of these putative effector genes in *S. sclerotiorum* in order to

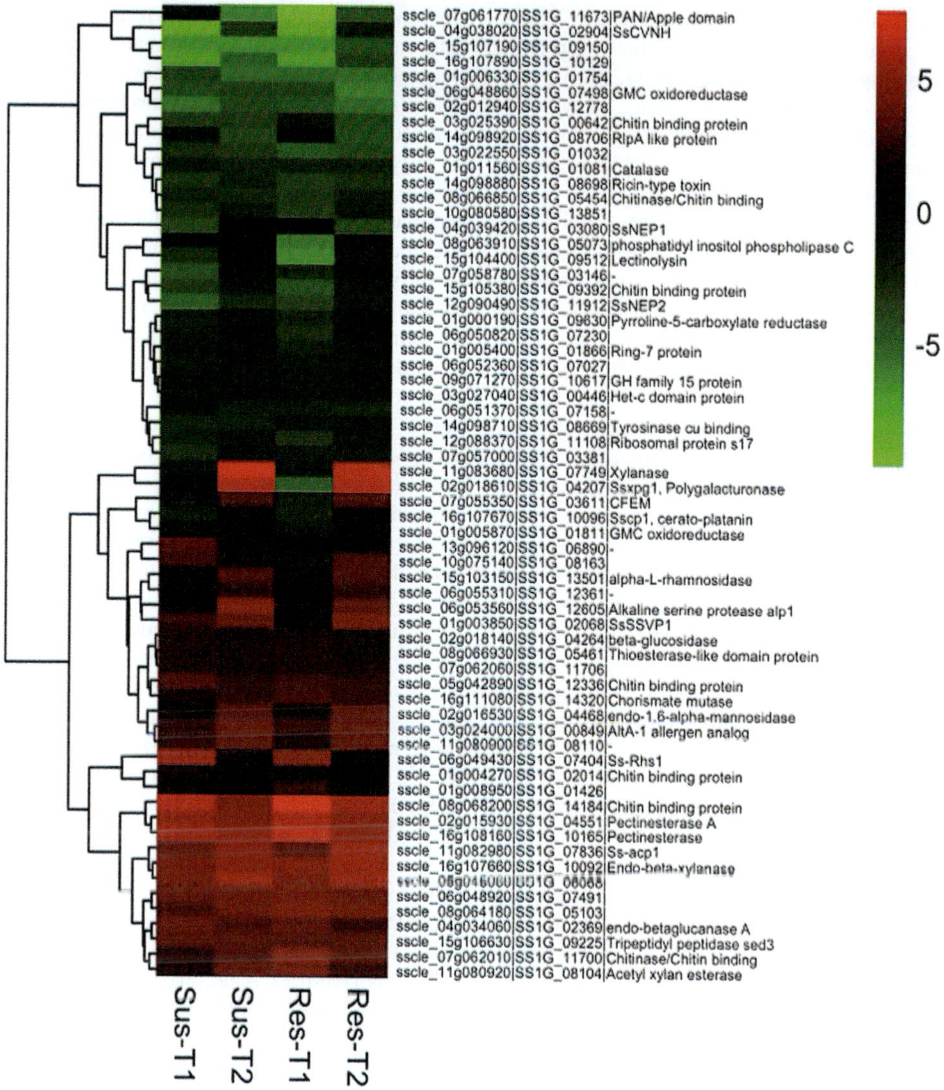

Fig. 8.4. Heat maps showing Sclerotinia *sclerotiorum* effector gene expression profiles. Using hierarchical clustering, genes are grouped according to how they express themselves. The interaction's early (8 and 16 hpi) and later (24 and 48 hpi) time points are denoted by the letters T1 and T2, respectively. The colour gradient displays the log2 fold change in gene expression compared with the *in vitro* control, with black denoting no change, green representing downregulation, and red representing upregulation. (Adapted from publication of Chittem *et al*., 2020 with permission.)

identify specific temporal shifts in their regulation. Throughout the duration of the study, 64 putative effector candidates exhibited differential regulation, with 37 genes displaying at least one instance of upregulation (Fig. 8.4). Most of these upregulated genes were consistently elevated at different times in susceptible and resistant lines. Among the upregulated effector genes that were primarily linked to CWDE activity were a non-aspartyl acid protease (acp1, sscle_11g082980), a protein containing an Rhs repeat (Ss-rhs1, sscle_06g049430), chorismate mutase (SsCm1, sscle_16g111080), and two chitin-binding domain proteins (sscle_08g068200, sscle_11g082980). For example, the synthesis of the gene *acp1*, which is controlled by pH, carbon, and nitrogen deficit, is induced by the presence of cell walls, as shown in Fig. 8.4 (Poussereau *et al.*, 2001).

The upregulation of this gene from 12.5 fold (l2fc 3.64) at T1 to 48 -fold (l2fc 5.58) at T2. The similar expression pattern was observed in *S. sclerotiorum* interaction with *P. vulgaris*, *B. napus*, and *G. max*, as reported by Oliveira *et al.* (2015), Seifbarghi *et al.* (2017), and Westrick *et al.* (2019). Ss-Rhs1 was shown to be important for sclerotial formation and early infection on *B. napus* and *Arabidopsis* (Yu *et al.*, 2017). During the early phases of sclerotial formation and hyphal infection, this gene was highly expressed. Additionally, Westrick *et al.* (2019) found that Ss-Rhs1 displayed a noteworthy upregulation on *G. max* during the early infection phase (24 hpi) as opposed to the later infection phase (48–96 hpi), suggesting a possible critical function in the early infection process. In a similar vein, both susceptible and resistant lines showed a 26–48-fold upregulation (l2fc 4.72–5.56) of Ss-Rhs1 exclusively at T1. On the other hand, the gene SsCm1 showed an upregulation of 4–8.5-fold (l2fc 2.02–3.08), which is in line with other research works that have demonstrated high expression levels in early infection stages contacts with *B. napus* (Nováková *et al.*, 2014) and *G. max* (Westrick *et al.*, 2019). It has been discovered that during *Zea mays* infection, the biotrophic pathogen *U. maydis* chorismate mutase gene cmu1 lowers SA levels in infected host tissue, increasing virulence (Djamei *et al.*, 2011).

It has been postulated that SsCm1 might play a role in controlling the host, analogous to cmu1 in the *U. maydis–Z. mays* interaction, via disrupting SA signaling in *B. napus*. Prior studies have focused on how LysM effectors and chitin-binding proteins contribute to avoiding host recognition. Some possible effectors, however, indicated a change in regulation, with T1 showing downregulation and T2 showing upregulation. Four of these genes *SsCP1*, *sscle_16g107670*, *sscle_07g055350*, *sscle_01g003850*, and *sscle_04g039420*, are known to act as necrotrophic effectors. Small secreted proteins rich in cysteine, known as chymotrypsins, are peculiar to fungi and can both activate plant defenses and increase virulence by causing localized necrosis in host tissue (Pazzagli *et al.*, 2014). It has been shown that SsCP1 increases virulence in *S. sclerotiorum* by directly interacting with the pathogenesis-related protein PR1 (Yang *et al.*, 2018). Sscp1 was downregulated at T1 by 4.9-fold (l2fc−2.39) but upregulated by 1.8-fold (l2fc 0.85) at T2. Conserved fungal extracellular membrane protein, or CFEM domain, is a fungal-specific domain thought to play a role in fungal diseases. It contains eight conserved cysteine residues (Kulkarni *et al.*, 2003; Zhang *et al.*, 2015). The CFEM protein pth11 is necessary for the pathogenesis and appressoria production in *M. oryzae* (DeZwaan *et al.*, 1999; Kou *et al.*, 2017). The ablation of pth11 affects the development of appressoria during pathogenesis by upsetting redox homeostasis, according to research by Kou *et al.* (2017).

The CFEM-containing gene *BcCFEM1*, which is likewise capable of acting as an elicitor, is essential for stress tolerance and virulence in *B. cinerea*, a closely related broad-host-range necrotroph (Zhu *et al.*, 2017). The gene *Sscle_07g055350*'s expression was upregulated by 9-fold (l2fc 3.18) at T2 and downregulated by 4.4-fold (l2fc−2.13) at T1

A similar expression pattern, characterized by upregulation solely during late infection, was noted by Seifbarghi *et al.* (2017) and Guyon *et al.* (2014) in the context of canola and soybean infections (Westrick *et al.*, 2019) caused by *S. sclerotiorum*. Additionally, this pattern was also observed by Thatcher *et al.* (2016) in the interaction between *Fusarium oxysporum* and *Medicago truncatula*. SsSSVP1, which is another identified virulence factor of *S. sclerotiorum*, is a small, cysteine-rich secreted protein that is crucial for achieving full virulence. Plant cell death is the result of SsSSVP1's interaction with the QCR8 member of the cytochrome b–c1 complex, which impairs the host mitochondrial respiratory pathway (Lyu *et al.*, 2016). According to Seifbarghi *et al.* (2017), the expression of SsSSVP1 was exclusively found in *B. napus* during the late infection stage (96 hpi).

On the other hand, SsNep1 and SsNep2 have been shown to cause necrosis in host plants (Bashi *et al.*, 2010). Until T2, when it showed a 2.5-fold increase (l2fc 1.35), the expression of SsNep1 was not significantly raised. This suggests that it rose gradually from the initial infection phase, when it observed a 15-fold reduction (l2fc−3.89). SsNep2 showed a similar expression pattern between T1 and T2, with a change ranging from −33 to −3 (l2fc−5.07 to −1.76). According to Westrick *et al.* (2019) and Seifbarghi *et al.* (2017), these genes were active in *B. napus* and *G. max* during the intermediate to later phases of infection. While *S. sclerotiorum* is mostly known for its necrotrophic behavior, a two-phase model is supported by both genetic and cytological data. This concept includes a necrotrophic phase that follows a first biotrophic, or compatibility, phase that is characterized by the downregulation of host defenses (Liang and Rollins, 2018). The reported temporal differential expression patterns of effector candidates (Kabbage *et al.*, 2013; Kabbage *et al.*, 2015; Liang and Rollins, 2018) corroborate this two-phase infection model for *S. sclerotiorum*. Early in an infection, there is an upregulation of biotrophic effectors. These effectors include LysM, chitin-binding proteins, chorismate mutase (SsCm1), and genes related to reactive nitrogen species (RNS) and ROS scavenging. However, necrosis-causing effectors, such as SsCp1, SsSSVp1, SsNep1, and SsNep2, are either not controlled until later in the infection process or downregulated in the early stages. Westrick *et al.* (2019) observed a comparable pattern in the expression of necrotrophic effectors in the interaction between *S. sclerotiorum* and *G. max*. Furthermore, at both the time points that were studied, there was a noticeable downregulation of two effector genes that had been identified: *SsITL* (Zhu *et al.*, 2013) and *SsCVNH* (Lyu *et al.*, 2015). On the other hand, Chittem *et al.* (2020) observed that *Ssggt2* (*sscle_09g068730*), an effector gene that encodes γ-glutamyl transpeptidase, consistently showed upregulation (2–3-fold, l2fc 1.12–1.65) at both time points.

8.22 Transcriptome Analysis for Mechanism of Pathogenesis and Genes' Differential Expression In Resistant and Susceptible *Brassica*

Two canola kinds with varying degrees of vulnerability have been observed to exhibit the distinct gene expression patterns of *S. sclerotiorum* over the course of the disease. About 164 million Illumina reads were obtained from the examination of mRNA libraries derived from infected petioles and mycelium cultivated in liquid culture. These readings were composed of 95 million single reads of 75 base pairs and 69 million paired-end reads of 50 base pairs. A little over 36% of the reads that met the quality requirements coincided with the reference genome of *Sclerotiorum*. The susceptible variety showed differential expression for *S. sclerotiorum* genes *1214* and *1301* in the early (8–16 hour postinoculation) and late (24–48 hour postinoculation) phases of infection, respectively.

Gene sets that are elevated in lines that are vulnerable and resistant revealed a noteworthy augmentation in Gene Ontology (GO) classifications pertaining to the breakdown of cell walls, detoxification of host metabolites, and peroxisome-associated procedures such as fatty acid beta-oxidation, glyoxylate cycle, and oxidoreductase activity. Six chosen DEGs were subjected to qRT-PCR in order to validate the RNA-seq study of differential gene expression. These findings point to a two-phase model of infection, with an initial short biotrophic phase characterized by effector gene control that helps the infection suppress or evade the host's defense mechanisms. The expression of effectors that aid in the demise of host cells in the latter stages of infection comes next. Functional class enrichment analysis and the expression profiles of significantly upregulated genes highlight the predominant roles played by peroxisome-related pathways in the pathogenesis of *S. sclerotiorum* in *B. napus*, along with mechanisms of cell wall degradation and detoxification of host metabolites (Chittem *et al.*, 2020).

8.23 Detoxification Mechanism of Xenobiotic Compounds for Pathogenesis

The collection of genes that exhibit a significant upregulation following the application of CWDEs (Table 8.5) comprises genes that are crucial for detoxification and evasion of host

Table 8.5. Variations in the expression levels of genes from *Sclerotinia sclerotiorum* associated with the detoxification of xenobiotic substances and peroxisome-related pathways observed during interactions with both susceptible (Sus) and resistant (Res) *Brassica* lines at early (T1, 8–16 hour postinoculation) and late (T2, 24–48 hour postinoculation) stages. (Adapted from publication of Chittem *et al*. (2020) with permission.)

Gene ID	Version 1 ID	Sus-T1		Sus-T2		Res-T1		Res-T2		Activity/pathway
		l2fc	fdr	l2fc	fdr	l2fc	fdr	l2fc	fdr	
sscle_01g007350	*SS1G_01627*	1.00	0.572	−1.74	0.267	−0.75	0.714	−1.71	0.269	Laccase
sscle_03g023030	*SS1G_00974*	5.13	0.007	5.34	0.005	5.15	0.007	5.17	0.006	
sscle_02g018680	*SS1G_04196*	−1.96	0.121	−4.09	0.003	−2.52	0.085	−4.53	0.001	
sscle_08g064260	*SS1G_05112*	−1.26	0.435	−3.28	0.029	−5.94	0.038	−3.48	0.017	
sscle_12g090390	*SS1G_11927*	−3.80	0.067	−4.05	0.018	−2.53	0.171	−3.18	0.030	
sscle_02g021570	*SS1G_13036*	3.80	0.034	0.67	0.744	6.15	0.003	0.19	0.939	
sscle_13g092370	*SS1G_06365*	0.75	0.711	−1.91	0.341	−1.74	0.461	−0.21	0.936	
sscle_01g003110	*SS1G_09997*	4.10	0.001	5.42	0.000	6.13	0.000	5.82	0.000	Glucosyltransferase
sscle_02g016980	*SS1G_04416*	0.30	0.709	1.22	0.057	1.34	0.060	1.28	0.043	
sscle_03g023640	*SS1G_00894*	0.90	0.292	−0.62	0.484	0.06	0.968	−0.34	0.727	
sscle_05g046370	*SS1G_05901*	−1.47	0.243	−1.26	0.249	−3.60	0.039	−1.30	0.215	
sscle_08g062730	*SS1G_04910*	−0.14	0.951	0.41	0.802	−0.19	0.937	0.52	0.733	
sscle_11g081870	*SS1G_07979*	−0.80	0.134	−0.31	0.584	−1.09	0.066	−0.42	0.423	
sscle_12g088170	*SS1G_11129*	0.80	0.210	0.69	0.279	0.34	0.662	0.64	0.316	
sscle_15g103340	*SS1G_13524*	−3.05	0.064	−4.36	0.012	−4.70	0.033	−4.34	0.010	
sscle_15g106380	*SS1G_09252*	−2.18	0.070	−2.86	0.008	−2.71	0.039	−2.54	0.010	

sscle_10g073920	*SS1G_13754*	−4.52	0.002	−0.28	0.838	−4.38	0.002	0.40	0.761	Cyanide hydratase
sscle_16g108230	*SS1G_10174*	6.55	0.000	3.41	0.000	7.17	0.000	6.51	0.000	
sscle_01g007130	*SS1G_01652*	−3.22	0.021	−0.84	0.402	−3.23	0.053	−1.03	0.286	
sscle_07g060330	*SS1G_11485*	−0.17	0.859	1.02	0.130	−0.68	0.391	0.98	0.142	
sscle_01g008520	*SS1G_01487*	−2.53	0.020	−2.66	0.006	−5.99	0.002	−3.52	0.001	
sscle_15g104110	*SS1G_09546*	−1.08	0.132	−0.49	0.455	−1.70	0.050	−0.69	0.256	
sscle_16g108980	*SS1G_10281*	−0.42	0.519	−0.54	0.339	−1.24	0.070	−0.17	0.818	
sscle_03g030680	*SS1G_13215*	−0.73	0.628	0.25	0.863	−0.49	0.772	0.06	0.973	
sscle_13g096730	*SS1G_14415*	0.43	0.823	0.49	0.743	−1.01	0.580	0.33	0.837	
sscle_01g005000	*SS1G_01918*	7.74	0.000	7.51	0.000	8.20	0.000	7.70	0.000	Glutathione-*S*-transferase
sscle_16g107740	*SS1G_10108*	−3.67	0.016	−4.00	0.004	−3.39	0.029	−4.99	0.001	
sscle_10g075490	*SS1G_08210*	−1.61	0.069	−0.59	0.471	−2.60	0.021	−0.72	0.358	
sscle_08g062750	*SS1G_04914*	−0.91	0.419	−0.99	0.311	−1.24	0.305	−1.39	0.125	
sscle_06g051110	*SS1G_07195*	−0.49	0.398	0.66	0.237	−0.81	0.167	0.79	0.142	
sscle_13g096550	*SS1G_14440*	0.25	0.892	0.72	0.588	1.29	0.297	1.17	0.317	
sscle_15g104750	*SS1G_09479*	−0.33	0.830	0.38	0.749	−0.60	0.661	0.73	0.480	
sscle_10g075830	*SS1G_08258*	−0.87	0.457	0.09	0.954	1.71	0.092	0.20	0.892	
sscle_06g053300	*SS1G_12640*	1.97	0.006	2.36	0.001	1.83	0.009	2.24	0.001	
sscle_08g067590	*SS1G_05554*	−1.46	0.102	−0.59	0.498	−1.69	0.082	−0.34	0.729	
sscle_01g008020	*SS1G_01545*	−2.18	0.038	−1.68	0.048	−1.99	0.068	−1.70	0.036	
sscle_13g094100	*SS1G_06623*	−0.19	0.873	−0.48	0.589	−0.66	0.485	−0.29	0.771	
sscle_01g004950	*SS1G_01922*	−0.06	0.962	−0.31	0.757	0.24	0.835	−0.01	0.990	
sscle_11g083650	*SS1G_07752*	3.97	0.001	3.54	0.001	4.93	0.000	3.56	0.001	

Table 8.5. Continued

Gene ID	Version 1 ID	Sus-T1		Sus-T2		Res-T1		Res-T2		Activity/pathway
		l2fc	fdr	l2fc	fdr	l2fc	fdr	l2fc	fdr	
s scle_01g004270	*SS1G_02014*	1.66	0.043	0.03	0.979	0.51	0.600	0.06	0.961	Chitin binding or LysM effector
s scle_02g014090	*SS1G_04786*	−0.48	0.735	−0.70	0.534	−0.70	0.604	−0.30	0.822	
sscle_03g024480	*SS1G_00772*	2.35	0.012	1.53	0.067	2.35	0.015	1.11	0.174	
sscle_03g025390	*SS1G_00642*	−3.63	0.037	−3.63	0.011	−2.16	0.223	−4.09	0.005	
sscle_05g042890	*SS1G_12336*	3.25	0.001	2.49	0.004	2.89	0.003	2.57	0.003	
sscle_06g054140	*SS1G_12513*	−8.68	0.010	−5.19	0.003	−4.41	0.030	−4.31	0.005	
sscle_06g054180	*SS1G_12509*	−10.20	0.032	−10.20	0.009	−10.20	0.033	−10.20	0.007	
sscle_08g068200	*SS1G_14184*	6.39	0.018	4.59	0.049	7.56	0.006	5.74	0.015	
sscle_15g105380	*SS1G_09392*	−3.52	0.028	−1.60	0.192	−4.45	0.028	−1.67	0.158	
sscle_15g107050	*SS1G_09169*	−1.39	0.605	−4.37	0.071	−0.52	0.902	−3.87	0.072	
sscle_07g062010	*SS1G_11700*	2.54	0.020	3.40	0.002	4.87	0.000	3.59	0.001	
sscle_08g066850	*SS1G_05454*	−2.69	0.077	−2.89	0.031	−3.45	0.056	−2.59	0.033	
sscle_08g066840	*SS1G_05453*	−2.36	0.282	−1.11	0.533	−7.05	0.067	−2.04	0.215	
sscle_05g041720	*SS1G_12200*	4.27	0.005	5.19	0.001	3.51	0.018	4.62	0.002	
sscle_03g027760	*SS1G_00355*	2.33	0.013	2.99	0.003	1.83	0.041	2.89	0.003	Nitronate monooxygenase
sscle_12g087340	*SS1G_11235*	−1.62	0.374	1.97	0.222	−3.25	0.157	2.04	0.198	
sscle_11g085500	*SS1G_14466*	1.85	0.064	2.00	0.049	1.66	0.095	2.00	0.045	
sscle_09g069190	*SS1G_10881*	0.57	0.702	0.17	0.914	−1.42	0.384	0.20	0.897	

sscle_04g038190	*SS1G_02923*	0.03	0.976	2.06	0.007	−1.13	0.189	2.19	0.005	Fatty acid beta-oxidation
sscle_16g108660	*SS1G_10238*	4.00	0.004	3.30	0.008	3.58	0.006	3.21	0.009	
sscle_03g028710	*SS1G_00237*	−0.31	0.727	1.42	0.045	0.05	0.970	1.65	0.020	
sscle_07g055730	*SS1G_11414*	0.81	0.238	2.21	0.004	1.24	0.092	2.26	0.003	
sscle_12g090420	*SS1G_11923*	−0.37	0.773	1.59	0.101	0.13	0.943	1.56	0.101	
sscle_09g069100	*SS1G_10890*	−0.38	0.793	1.35	0.231	−1.93	0.198	1.36	0.217	
sscle_10g076540	*SS1G_08354*	0.62	0.573	1.61	0.098	−0.38	0.785	1.97	0.042	
sscle_14g099760	*SS1G_08821*	0.64	0.479	0.83	0.363	0.23	0.835	1.02	0.246	
sscle_02g018210	*SS1G_04249*	1.89	0.276	1.27	0.406	1.29	0.459	2.42	0.069	
sscle_05g041330	*SS1G_12152*	−0.16	0.925	−0.53	0.690	0.29	0.840	−0.54	0.690	
sscle_11g086520	*SS1G_14001*	0.69	0.554	0.48	0.707	0.32	0.820	0.41	0.764	
sscle_09g069850	*SS1G_10796*	4.43	0.042	2.48	0.220	4.78	0.035	2.54	0.198	Oxalic acid
sscle_14g099710	*SS1G_08814*	0.20	0.862	0.91	0.231	1.01	0.239	1.13	0.121	
sscle_10g075560	*SS1G_08218*	2.18	0.010	1.87	0.020	2.04	0.013	2.12	0.010	
sscle_08g062640	*SS1G_04900*	0.29	0.601	1.20	0.017	0.40	0.476	1.35	0.008	Glyoxylate cycle
sscle_08g063200	*SS1G_04975*	3.50	0.001	4.20	0.000	3.72	0.001	4.39	0.000	
sscle_08g067810	*SS1G_05583*	1.29	0.047	2.78	0.001	2.06	0.005	2.92	0.000	
sscle_04g038190	*SS1G_02923*	0.03	0.976	2.06	0.007	−1.13	0.189	2.19	0.005	Acetyl-CoA acetyl transferase
sscle_09g069100	*SS1G_10890*	−0.38	0.793	1.35	0.231	−1.93	0.198	1.36	0.217	
sscle_10g075450	*SS1G_08207*	3.01	0.001	3.07	0.001	2.47	0.003	3.13	0.000	
sscle_05g041330	*SS1G_12152*	−0.16	0.925	−0.53	0.690	0.29	0.840	−0.54	0.690	

defenses. This collection features two laccases (*sscle_03g023030*, *sscle_02g021570*), a nitrilase (*sscle_16g108230*), a brassinin glycosyltransferase (*sscle_01g003110*, *SsBGT1*), two glutathione-*S*-transferases (GST), two cytochrome P450 monooxygenases, and three transporter genes, which include two major facilitator superfamily transporters and three *S*-adenosyl methionine (SAM)-dependent methyltransferases, one of which is a thiolmethyl transferase. Laccases function as enzymes that detoxify phenolic compounds through oxidation (Shraddha *et al.*, 2011). Of the seven laccase genes predicted in the *S. sclerotiorum* genome, only *sscle_02g021570* (*sslacc6*) showed consistent upregulation at both time points (fold change of 35–40, l2fc 5.13–5.34), whereas sslacc2 (*sscle_03g023030*) only demonstrated consistent upregulation at T1 (fold change of 15–70, l2fc 3.80–6.15). In the interaction between soybean and *S. sclerotiorum*, a similar expression pattern was observed (Westrick *et al.*, 2019). The laccase gene *BcLCC2* and the ABC transporter BcAtrB were necessary for *B. cinerea* to detoxify the antifungal phenolic antibiotic 2, 4-diacetylphloroglucinol (Schouten *et al.*, 2008). The host's detoxification of phenolic compounds was aided by the catechol 1, 2-dioxygenase gene, *sscle_04g037100*, which was included in the significantly enriched GO category for xenobiotic metabolic activities (GO: 0006805). In reaction to phenolic chemicals generated by host plants, catechol-dioxygenases are increased (Shalaby and Horwitz, 2015). According to Shanmugam *et al.* (2010), phenolics from maize activated the catechol-dioxygenase-encoding gene *CCHD1*, which has been identified as an avirulence factor in the spruce pathogen *Endoconidiophora polonica* (Wadke *et al.*, 2016). *Brassica* species are known to synthesize a variety of phytoalexins, which are secondary metabolites induced by biotic and abiotic stress, in addition to phenolic substances (Pedras *et al.*, 2011). Additionally, these compounds are synthesized in response to fungal pathogen attacks (Dahiya and Rimmer, 1988; Storck and Sacristan, 1995; Pedras *et al.*, 2008). In certain host–pathogen interactions, phytoalexins are essential for mediating resistance (Ferrari *et al.*, 2003).

Many studies have been conducted on the conversion of these potent antifungal compounds by pathogenic fungi, both in controlled settings and in plant systems (Pedras *et al.*, 2008). *S. sclerotiorum* uses a unique process called glucosylation to detoxify cruciferous phytoalexins (Pedras and Hossain, 2006). Plant pathogens use this mechanism to their advantage. Sexton *et al.* (2009) discovered many glucosyltransferases, including a brassinin glucosyltransferase (*SsBGT1*, *sscle_01g003110*), by genome analysis and transcriptional profiling. Table 8.5 shows that the *SsBGT1* gene showed a considerable upregulation spanning from 17 to 70 times (l2fc 4.10–6.13), which is consistent with the findings of Sexton *et al.* (2009). Further research has also shown that this gene becomes more active during canola and soybean infections (Seifbarghi *et al.*, 2017; Westrick *et al.*, 2019).

The ability of *Brassicas* to manufacture phytoanticipins, such as cyanogenic glucosides and glucosinolates, is widely recognized (Osbourn, 1996; Pedras *et al.*, 2008). These substances, which are naturally produced by plant metabolism, may have potent antibacterial qualities. Highly toxic isothiocyanates (ITC) and nitriles are produced when nontoxic glucosinolates undergo cellular destruction (Osbourn, 1996). Enzymes like nitrilases or cyanide hydratases are important because they help phytopathogenic fungi transform hydrogen cyanide into the less toxic formamide (Fry and Evans, 1977; Wang and Etten, 1992). Certain fungi can use this formamide as a source of nitrogen (Barclay *et al.*, 1998; Sexton and Howlett, 2000). At each of the observed time points, there was a significant overexpression of the cyanide hydratase gene, *sscle_16g108230*, by 85–145 times (l2fc 6.41–7.17). When *Alternaria brassicicola*, *Leptosphaeria maculans*, and *S. sclerotiorum* infected *Brassica* hosts, cyanide hydratases and nitrilases were likewise upregulated in the secretomes of *B. cinerea* (Sexton and Howlett, 2000; Cramer and Lawrence, 2004; González-Fernández *et al.*, 2014; Seifbarghi *et al.*, 2017). According to Sexton and Howlett (2000), *L. maculans's* cyanide hydratase gene expression rose in response to potassium cyanide or *Brassica* glucosinolate derivatives. Moreover, another cyanate hydratase gene, *sscle_07g060330*, was only increased later (>2 fold in susceptible lines (l2fc 1.02), 2-fold (l2fc 0.98 in resistant lines)). Seifbarghi *et al.* (2017) also observed this delayed gene activation. Glutathione S-transferases (GSTs) are known to have a role in the detoxification of xenobiotics and endogenous toxic chemicals in fungi by conjugation with glutathione (Gullner *et al.*, 2018). Data in Table 8.5 show that the GST gene, *sscle_01g005000*, showed a significant upregulation spanning from 182 to 294 times (l2fc 7.51–8.20). When infections occurred in soybean (Westrick *et al.*, 2019) and canola (Sexton and Howlett, 2000), similar amounts of upregulation for this gene were seen. *Sscle_11g083650*, a membrane-associated protein implicated in glutathione and eicosanoid metabolism (MAPEG), is another GST that showed a 12–30-fold increase (l2fc 3.54–4.93). Furthermore, as shown in Table 8.5, the GST gene *sscle_06g053300* consistently increased by four to five times (l2fc 1.83–2.36). Similarly, in the case of *A. brassicicola* to reach maximum virulence on *Brassica oleracea*, the ITC-inducible GST from the MAPEG class, *AbMAPEG1*, was necessary (Calmes *et al.*, 2015). *A. brassicicola* became less pathogenic and more susceptible to ITC when two more ITC-inducible GSTs, *AbGSOT1* and *AbUre2pB1*, were deleted (Calmes *et al.*, 2015). The significance of detoxifying cyanogenic chemicals during the pathogenesis process is highlighted by these findings. SAM-dependent methyltransferases can enable the transfer of methyl groups from SAM to a variety of substrates (Struck *et al.*, 2012). The upregulated SAM-dependent methyltransferases, specifically *sscle_09g073010* (with a fold change ranging from 68 to 192, log2 fold change of 6.08–7.59), *sscle_15g106060* (with a fold change between 8.5 and 27, log2 fold change of 3.09–4.74), and *sscle_08g065070* (with a fold change of 19–30, log2 fold change of 4.24–4.92), demonstrated marked increases in their expression levels. In the wood decomposing fungus *Phanerochaete chrysosporium*, SAM transferases contributed to the detoxification of phenolic compounds (Jeffers *et al.*, 1997). Additionally, thiolmethyl-transferases found in *Brassica* and *Arabidopsis* are recognized for their role in detoxifying glucosinolates (Attieh *et al.*, 2000).

Proteins known as secreted effectors are included in the lysine motif, or LysM (Jonge and Thomma, 2009; Kombrink and Thomma, 2013). By attaching to chitin present in the fungal cell walls, these proteins serve a critical role in hiding the presence of plant pathogenic fungi within plant tissues (Jonge *et al.*, 2010; Mentlak *et al.*, 2012; Takahara *et al.*, 2016). The plant's ability to identify the fungus is interfered with by this binding activity (Shibuya and Minami, 2001; Kaku *et al.*, 2006; Fesel and Zuccaro, 2016). Two genes, *sscle_08g068200* (fold changes between 24 and 189, with a log2 fold change ranging from 4.60 to 7.56) and *sscle_05g041720* (fold changes between 11 and 36, with a log2 fold change of 3.51–5.19), that encode proteins with chitin-binding domains were found to be significantly upregulated. Furthermore, two additional chitin-binding domain-containing genes (*sscle_01g004270* and *sscle_07g062010*) and a LysM effector gene (*sscle_03g024480*) displayed upregulation (Table 8.5). The regulatory patterns of these six genes exhibited a consistent tendency, with the most pronounced upregulation occurring at T1 and a slight decline at T2. Remarkably, prior research on the interactions between *S. sclerotiorum* and *B. napus* has shown the upregulation of LysM and chitin-binding domain genes (Seifbarghi *et al.*, 2017) as well as *G. max* (Westrick *et al.*, 2019). According to Wojtaszek (1997), the term 'oxidative burst' describes a sudden and brief rise in ROS, which is an essential plant defensive mechanism against pathogen invasions. Because of the unfavorable oxidative stress circumstances this event creates, the invasive pathogens must be able to control ROS in order to survive and successfully infect the host. Fungal pathogens have developed a number of defense mechanisms against ROS, such as the employment of peroxidases, catalases, NADPH oxidases, and SOD (NOX). Three peroxidases were found to be considerably upregulated upon examination of the data in Table 8.5: *sscle_01g000730* (10–26-fold, l2fc 3.35–4.73), *sscle_08g065740* (3–24-fold, l2fc 1.43–4.59), and *sscle_04g035020* (6–29-fold, l2fc 2.58–4.86). Additionally, it is acknowledged that flavin mononucleotide (FMN) enzymes called nitronate monooxygenases (NMOs) play a significant part in the oxidative detoxification of nitroalkanes (Francis *et al.*, 2012).

According to Marroquin-Guzman *et al.* (2017), NMOs are necessary to maintain cellular redox equilibrium, which in turn regulates the host's immunological responses and improves tolerance to RNS stress. Throughout the duration of infection, it was consistently observed that two specific NMO genes, *sscle_03g027760* and *sscle_11g085500*, were upregulated in both susceptible and resistant plant lines. Conversely, the third NMO gene, *sscle_12g087340*, was only upregulated at time point T2, having been downregulated at time point T1. Further evidence points to *S. sclerotiorum* going through a brief biotrophic phase throughout the infection process. Therefore, in addition to their role in mitigating nitro-oxidative stress, it is hypothesized that *S. sclerotiorum's* NMO may impair host defense systems during their interaction with canola plants. The literature has extensively documented the impact of *S. sclerotiorum* on the host's redox environment, which is aided by the generation of OA (Williams *et al.*, 2011). Numerous studies have been conducted on OA as a virulence factor connected to *S. sclerotiorum*. Apart from its function of modifying the redox state of the host, OA amplifies virulence by other pathways, such as triggering programmed cell death (Kim *et al.*, 2008), calcium chelation, and pH signaling mediation (Rollins, 2003; Liang *et al.*, 2015a). The gene *sscle_10g075560*, which encodes the enzyme OAH (Ssoah1) responsible for the production and accumulation of OA (Liang *et al.*, 2015b), was consistently upregulated during the infection process, according to a study by Chittem *et al.* (2020), which looked into the expression of genes related to OA metabolism (Table 8.5).

Together with Ssoah1, the oxalate decarboxylase (OxDC) gene *sscle_09g069850 (Ssodc2)* was shown to be upregulated. According to Heard *et al.* (2015), these oxalate decarboxylase genes (OxDC) genes are critical for preventing the accumulation of OA in fungal cells, protecting the pathogen from its harmful effects. *Ssoah1* and *Ss-odc2* showed similar expression patterns when infected with *B. napus* and *P. vulgaris*, respectively (Oliveira *et al.*, 2015; Seifbarghi *et al.*, 2017); however, *G. max* did not exhibit these patterns (Westrick *et al.*, 2019). Conversely, another OxDC gene, *Ss-odc1* (*sscle_14g099710*), did not reveal significant differential expression (1–2-fold change, l2fc 0.20–1.13). Additionally, there was a high enrichment of GO concepts associated with the oxidation–reduction process and the generation of OA precursors (GO: 0055114) in the sets of genes with differential expression.

8.24 Pathways Linked to the Pathogenesis of Peroxisome Biogenesis Genes

The genes that were upregulated exhibited a consistent and significant enrichment in those linked to peroxisomal pathways. Among this extensive category, notable GO classifications that were significantly enriched or overrepresented include peroxisome organization (GO: 0007031), which involves processes, such as protein targeting (GO: 0006625) and protein import (GO: 0016558) into the peroxisome matrix, in addition to fatty acid beta-oxidation (GO: 0006635) and the glyoxylate cycle (GO: 0006097). According to research, peroxisome-related metabolic processes are essential for the pathogenic growth of a variety of plant pathogenic fungus (Kubo, 2012). Proteins called peroxins or PEX genes, which are involved in peroxisome biogenesis, are necessary for the production of peroxisomes. Fungal pathogens like *M. oryzae* (Wang *et al.*, 2007; Goh *et al.*, 2011; Wang *et al.*, 2015), *Colletotrichum orbiculare* (Kimura *et al.*, 2001; Fujihara *et al.*, 2010; Kubo, 2012), and *A. alternata* (Imazaki *et al.*, 2010) have been found to require these PEX genes for their pathogenicity and virulence. The significant upregulation of numerous PEX genes indicates their potential importance in the pathogenicity and virulence of *S. sclerotiorum*.

The breakdown of long-chain fatty acids, which enables the synthesis of nutrients and energy, is dependent on the fatty acid beta-oxidation metabolic pathway (Poirier *et al.*, 2006; Chen *et al.*, 2016). There are four separate phases in this enzyme-mediated process, which results in the synthesis of acetyl-CoA. The produced acetyl-CoA can then go through the citric acid cycle to the mitochondria where it can be used to produce energy, or it can go through the glyoxylate cycle. The *pth2* gene is essential for *M. oryzae* to infect rice because it encodes the enzyme carnitine acetyl transferase, which aids in the transport of acetyl-CoA (Ramos-Pamplona and Naqvi, 2006). The similar gene, known as Ss-pth2, in *S. sclerotiorum* has been identified as essential for host colonization (Liberti *et al.*, 2013). This suggests that the processes of host colonization and disease progression involve peroxisomal pathways. Interestingly, in interactions with both susceptible and resistant hosts, the gene (*sscle_03g031670*) showed higher expression levels in both the early and late stages of infection (Chittem *et al.*, 2020). The glyoxylate cycle plays a crucial role in gluconeogenesis, a process that uses acetyl-CoA from fatty acid beta-oxidation to help produce glucose when nutrients are scarce. Isocitratelyase (ICL1) and malate synthase (MSL1) are two important enzymes in this cycle (Chen *et al.*, 2016). It has been established that ICL1 is essential to *L. maculans'* pathogenicity, a pathogen that affects canola (Idnurm and Howlett, 2002). Moreover, studies have demonstrated that it is necessary for *M. oryzae* (Wang *et al.*, 2003) and *C. orbiculare* (Asakura *et al.*, 2006) to be fully virulent. Furthermore, it has been determined that *S. sclerotiorum's* malate synthase gene (mls1) is conditionally essential for fatty acid metabolism and its pathogenicity in tomato plants (Liberti *et al.*, 2007).

8.25 Reactive Oxygen Species Modulation by *SCAT1* Gene for *Sclerotinia* Pathogenic Development

The development of *S. sclerotiorum* as a pathogen is significantly influenced by ROS. A key question for *S. sclerotiorum* and many other pathogens concerns how fungi tolerate/dampen the oxidative environment during growth and pathogenesis. Regulatory components of oxidative stress include both enzymatic and nonenzymatic antioxidants. Catalases are a ubiquitous family of enzymes that play an important role in the enzymatic detoxification of ROS by converting H_2O_2 to water and molecular oxygen. The genome of *S. sclerotiorum* contains even predicted catalase genes. Yarden *et al.* (2014) evaluated and functionally characterized the type Acatalase (*Scat1*) in *S. sclerotiorum*, whose expression is highly induced during host infection. Insertional inactivation of *Scat1* (Δ*Scat1*) resulted in hyper-branching of hyphae accompanied by slower growth and smaller sclerotia. The Δ*Scat1* strains were attenuated in pathogenicity and rendered the fungus hypersensitive to sodium dodecyl sulfate, as well as to osmotic and salt stresses. Unexpectedly, Δ*Scat1* exhibited increased tolerance to H_2O_2, suggesting that although a member of the catalase family, generally associated with amelioration of oxidative stress, *Scat1* is probably not required for detoxification of this oxygen species and presumably has different function(s). The Δ*Scat1* strains had a twofold decrease in ergosterol content, and overall lower sterol levels compared with the WT strain. These observations are consistent with increased resistance to the polyene drugs amphotericin B and nystatin. The findings indicate that Scat1 plays a role in the modulation of ROS in a way that diverges from the detoxification of H_2O_2, affects membrane integrity, and aids in the pathogenic effectiveness of *S. sclerotiorum* (Yarden *et al.*, 2014).

References

Amaral, A.M.D., Antoniw, J., Rudd, J.J. and Hammond-Kosack, K.E. (2012) Defining the predicted protein secretome of the fungal wheat leaf pathogen *Mycosphaerella graminicola*. *PLoS ONE* 7(12), e49904. DOI: 10.1371/journal.pone.0049904.

Ambrose, K.V., Tian, Z., Wang, Y., Smith, J., Zylstra, G. *et al.* (2015) Functional characterization of salicylate hydroxylase from the fungal endophyte *Epichloë festucae*. *Nature* 5(1), 10939. DOI: 10.1038/srep10939.

Amselem, J., Cuomo, C.A., Kan, J.A.L., Viaud, M., Benito, E.P. *et al.* (2011) Genomic analysis of the necrotrophic fungal pathogens *Sclerotinia sclerotiorum* and *Botrytis cinerea*. *PLoS Genetics* 7(8), e1002230. DOI: 10.1371/journal.pgen.1002230.

Andrew, M., Barua, R., Short, S.M. and Kohn, L.M. (2012) Evidence for a common toolbox based on necrotrophy in a fungal lineage spanning necrotrophs, biotrophs, endophytes, host generalists and specialists. *PLoS ONE* 7(1), e29943. DOI: 10.1371/journal.pone.0029943.

Antal, Z., Rascle, C., Cimerman, A., Viaud, M., Billon-Grand, G. *et al.* (2012) The homeobox *BcHOX8* gene in *Botrytis cinerea* regulates vegetative growth and morphology. *PLoS ONE* 7(44), 48. DOI: 10.1371/journal.pone.0048134.

Apel, K. and Hirt, H. (2004) Reactive oxygen species: Metabolism, oxidative stress, and signal transduction. *Annual Review of Plant Biology* 55, 373–399. DOI: 10.1146/annurev.arplant.55.031903.141701.

Araus, J.L. and Cairns, J.E. (2014) Field high-throughput phenotyping: The new crop breeding frontier. *Trends in Plant Science* 19(1), 52–61. DOI: 10.1016/j.tplants.2013.09.008.

Asakura, M., Okuno, T. and Takano, Y. (2006) Multiple contributions of peroxisomal metabolic function to fungal pathogenicity in *Colletotrichum lagenarium*. *Applied and Environmental Microbiology* 72(9), 6345–6354. DOI: 10.1128/AEM.00988-06.

Attieh, J., Kleppinger-Sparace, K.F., Nunes, C. and Sparace, S.A. (2000) Evidence implicating a novel thiol methyltransferase in the detoxification of glucosinolate hydrolysis products in *Brassica oleracea* L. *Plant, Cell and Environment* 23(2), 165–174. DOI: 10.1046/j.1365-3040.2000.00541.x.

Backer, R., Naidoo, S. and Berg, N. (2019) The nonexpressor of pathogenesis-related genes 1 (*NPR1*) and related family: Mechanistic insights in plant disease resistance. *Frontiers in Plant Science* 10, 102. DOI: 10.3389/fpls.2019.00102.

Badet, T., Peyraud, R., Mbengue, M., Navaud, O., Derbyshire, M. *et al.* (2017) Codon optimization underpins generalist parasitism in fungi. *eLife* 6, e22472. DOI: 10.7554/eLife.22472.

Baker, C.J. and Orlandi, E.W. (1995) Active oxygen in plant pathogenesis. *Annual Review of Phytopathology* 33, 299–321. DOI: 10.1146/annurev.py.33.090195.001503.

Baranowski, P., Jedryczka, M., Mazurek, W., Babula-Skowronska, D., Siedliska, A. *et al.* (2015) Hyperspectral and thermal imaging of oilseed rape (*Brassica napus*) response to fungal species of the genus *Alternaria*. *PLoS ONE* 10(3), e0122913. DOI: 10.1371/journal.pone.0122913.

Barbacci, A., Navaud, O., Mbengue, M., Barascud, M., Godiard, L. *et al.* (2020) Rapid identification of an *Arabidopsis NLR* gene as a candidate conferring susceptibility to *Sclerotinia sclerotiorum* using time-resolved automated phenotyping. *The Plant Journal* 103(2), 903–917. DOI: 10.1111/tpj.14747.

Barclay, M., Tett, V.A. and Knowles, C.J. (1998) Metabolism and enzymology of cyanide/metallocyanide biodegradation by *Fusarium solani* under neutral and acidic conditions. *Enzyme and Microbial Technology* 23(5), 321–330. DOI: 10.1016/S0141-0229(98)00055-6.

Bashi, Z.D., Hegedus, D.D., Buchwaldt, L., Rimmer, S.R. and Borhan, M.H. (2010) Expression and regulation of *Sclerotinia sclerotiorum* necrosis and ethylene-inducing peptides (NEPs). *Molecular Plant Pathology* 11(1), 43–53. DOI: 10.1111/j.1364-3703.2009.00571.x.

Bashi, Z.D., Rimmer, S.R., Khachatourians, G.G. and Hegedus, D.D. (2013) *Brassica napus* polygalacturonase inhibitor proteins inhibit *Sclerotinia sclerotiorum* polygalacturonase enzymatic and necrotizing activities and delay symptoms in transgenic plants. *Canadian Journal of Microbiology* 59(2), 79–86. DOI: 10.1139/cjm-2012-0352.

Bayram, O. and Braus, G. (2012) Coordination of secondary metabolism and development in fungi: The velvet family of regulatory proteins. *FEMS Microbiology Reviews* 36(1), 1–24. DOI: 10.1111/j.1574-6976.2011.00285.x.

Brace, J.L., Vanderweele, D.J. and Rudin, C.M. (2005) SVF1 inhibits reactive oxygen species generation and promotes survival under conditions of oxidative stress in *Saccharomyces cerevisiae*. *Yeast* 22, 641–652. DOI: 10.1002/yea.1235.

Brace, J.L., Lester, R.L., Dickson, R.C. and Rudin, C.M. (2007) SVF1 regulates cell survival by affecting sphingolipid metabolism in *Saccharomyces cerevisiae*. *Genetics* 175(1), 65–76. DOI: 10.1534/genetics.106.064527.

Calmes, B., Morel-Rouhier, M., Bataillé-Simoneau, N., Gelhaye, E., Guillemette, T. *et al.* (2015) Characterization of glutathione transferases involved in the pathogenicity of *Alternaria brassicicola*. *BMC Microbiology* 15(1), 123. DOI: 10.1186/s12866-015-0462-0.

Canet, J.V., Dobón, A., Roig, A. and Tornero, P. (2010) Structure-function analysis of npr1 alleles in *Arabidopsis* reveals a role for its paralogs in the perception of salicylic acid. *Plant, Cell and Environment* 33(11), 1911–1922. DOI: 10.1111/j.1365-3040.2010.02194.x.

Cao, H., Bowling, S.A., Gordon, A.S. and Dong, X. (1994) Characterization of an *Arabidopsis* mutant that is nonresponsive to inducers of systemic acquired resistance. *The Plant Cell* 6, 1583–1592. DOI: 10.1105/tpc.6.11.1583.

Cessna, S.G., Sears, V.E., Dickman, M.B. and Low, P.S. (2000) Oxalic acid, a pathogenicity factor for *Sclerotinia sclerotiorum*, suppresses the oxidative burst of the host plant. *The Plant Cell* 12(11), 2191–2200. DOI: 10.1105/tpc.12.11.2191.

Chalhoub, B., Denoeud, F., Liu, S., Parkin, I.A.P., Tang, H. *et al.* (2014) Plant genetics. Early allopolyploid evolution in the post-Neolithic *Brassica napus* oilseed genome. *Science* 345(6199), 950–953. DOI: 10.1126/science.1253435.

Chen, C. and Dickman, M.B. (2005) CAMP blocks MAPK activation and sclerotial development via rap-1 in a PKA-independent manner in *Sclerotinia sclerotiorum*. *Molecular Microbiology* 55(1), 299–311. DOI: 10.1111/j.1365-2958.2004.04390.x.

Chen, D., Neumann, K., Friedel, S., Kilian, B., Chen, M. *et al.* (2014) Dissecting the phenotypic components of crop plant growth and drought responses based on high-throughput image analysis. *The Plant Cell* 26(12), 4636–4655. DOI: 10.1105/tpc.114.129601.

Chen, X.L., Wang, Z. and Liu, C. (2016) Roles of peroxisomes in the rice blast fungus. *BioMed Research International* 2016, 9343417. DOI: 10.1155/2016/9343417.

Chittem, K., Yajima, W.R., Goswami, R.S. and Río Mendoza, L.E. (2020) Transcriptome analysis of the plant pathogen *Sclerotinia sclerotiorum* interaction with resistant and susceptible canola (*Brassica napus*) lines. *PLoS ONE* 15(3), e0229844. DOI: 10.1371/journal.pone.0229844.

Coppens, F., Wuyts, N., Inzé, D. and Dhondt, S. (2017) Unlocking the potential of plant phenotyping data through integration and data-driven approaches. *Current Opinion in Systems Biology* 4, 58–63. DOI: 10.1016/j.coisb.2017.07.002.

Corwin, J.A. and Kliebenstein, D.J. (2017) Quantitative resistance: More than just perception of a pathogen. *The Plant Cell* 29(4), 655–665. DOI: 10.1105/tpc.16.00915.

Corwin, J.A., Copeland, D., Feusier, J., Subedy, A., Eshbaugh, R. *et al.* (2016) The quantitative basis of the *Arabidopsis* innate immune system to endemic pathogens depends on pathogen genetics. *PLoS Genetics* 12(1), 29. DOI: 10.1371/journal.pgen.1005789.

Cramer, R.A. and Lawrence, C.B. (2004) Identification of *Alternaria brassicicola* genes expressed in planta during pathogenesis of *Arabidopsis thaliana*. *Fungal Genetics and Biology* 41(2), 115–128. DOI: 10.1016/j.fgb.2003.10.009.

Czedik-Eysenberg, A., Seitner, S., Güldener, U., Koemeda, S., Jez, J. *et al.* (2018) The "Phenobox", a flexible, automated, open-source plant phenotyping solution. *The New Phytologist* 219(2), 808–823. DOI: 10.1111/nph.15129.

Dahiya, J.S. and Rimmer, S.R. (1988) Phytoalexin accumulation in tissues of *Brassica napus* inoculated with *Leptosphaeria maculans*. *Phytochemistry* 27(10), 3105–3107. DOI: 10.1016/0031-9422(88)80009-8.

Dang, Y., Yang, Q., Xue, Z. and Liu, Y. (2011) RNA interference in fungi: Pathways, functions, and applications. *Eukaryotic Cell* 10(9), 1148–1155. DOI: 10.1128/EC.05109-11.

Debieu, M., Huard-Chauveau, C., Genissel, A., Roux, F. and Roby, D. (2015) Quantitative disease resistance to the bacterial pathogen *Xanthomonas campestris* involves an *Arabidopsis* immune receptor pair and a gene of unknown function. *Molecular Plant Pathology* 17, 510–520.

Derbyshire, M., Denton-Giles, M., Hegedus, D., Seifbarghy, S., Rollins, J. *et al.* (2017) The complete genome sequence of the phytopathogenic fungus *Sclerotinia sclerotiorum* reveals insights into the genome architecture of broad host range pathogens. *Genome Biology and Evolution* 9(3), 593–618. DOI: 10.1093/gbe/evx030.

Derbyshire, M., Mbengue, M., Barascud, M., Navaud, O. and Raffaele, S. (2019) Small RNAs from the plant pathogenic fungus *Sclerotinia sclerotiorum* highlight host candidate genes associated with quantitative disease resistance. *Molecular Plant Pathology* 20(9), 1279–1297. DOI: 10.1111/mpp.12841.

Derbyshire, M.C., Newman, T.E., Khentry, Y. and Taiwo, A.O. (2022) Pathogen profile: The evolutionary and molecular features of the broad host range plant pathogen *Sclerotinia sclerotiorum*. *Molecular Plant Pathology* 23(8), 1–16.

DeZwaan, T.M., Carroll, A.M., Valent, B. and Sweigard, J.A. (1999) *Magnaporthe grisea* pth11p is a novel plasma membrane protein that mediates appressorium differentiation in response to inductive substrate cues. *The Plant Cell* 11, 2013–2030.

Dickman, M.B. and Mitra, A. (1992) *Arabidopsis thaliana* as a model for studying *Sclerotinia sclerotiorum* pathogenesis. *Physiological and Molecular Plant Pathology* 41(4), 255–263. DOI: 10.1016/0885-5765(92)90025-Q.

Ding, C., Vidanes, G.M., Maguire, S.L., Guida, A., Synnott, J.M. *et al.* (2011) Conserved and divergent roles of *Bcr1* and *CFEM* proteins in *Candida parapsilosis* and *Candida albicans*. *PLoS ONE* 6(12), e28151. DOI: 10.1371/journal.pone.0028151.

Ding, X., Yu, Q., Zhang, B., Xu, N., Jia, C. *et al.* (2014) The type II Ca2+/calmodulin-dependent protein kinases are involved in the regulation of cell wall integrity and oxidative stress response in *Candida albicans*. *Biochemical and Biophysical Research Communications* 446(4), 1073–1078. DOI: 10.1016/j.bbrc.2014.03.059.

Djamei, A., Schipper, K., Rabe, F., Ghosh, A., Vincon, V. *et al.* (2011) Metabolic priming by a secreted fungal effector. *Nature* 478(7369), 395–398. DOI: 10.1038/nature10454.

Duan, Y., Ge, C., Liu, S., Wang, J. and Zhou, M. (2013) A two-component histidine kinase *Shk1* controls stress response, sclerotial formation and fungicide resistance in *Sclerotinia sclerotiorum*. *Molecular Plant Pathology* 14(7), 708–718. DOI: 10.1111/mpp.12041.

Dutton, M.V. and Evans, C.S. (1996) Oxalate production by fungi: Its role in pathogenicity and ecology in the soil environment. *Canadian Journal of Microbiology* 42(9), 881–895. DOI: 10.1139/m96-114.

Erental, A., Harel, A. and Yarden, O. (2007) Type 2A phosphoprotein phosphatase is required for asexual development and pathogenesis of *Sclerotinia sclerotiorum*. *Molecular Plant-Microbe Interactions* 20, 944–954. DOI: 10.1094/MPMI-20-8-0944.

Fahlgren, N., Gehan, M.A. and Baxter, I. (2015) Lights, camera, action: High-throughput plant phenotyping is ready for a close-up. *Current Opinion in Plant Biology* 24, 93–99.

Fan, H., Yang, W., Nie, J., Lin, C., Wu, J. *et al.* (2021) Characterization of a secretory YML079-like cupin protein that contributes to *Sclerotinia sclerotiorum* pathogenicity. *Microorganisms* 9, 2519. DOI: 10.3390/microorganisms9122519.

Ferrari, S., Plotnikova, J.M., Lorenzo, G. and Ausubel, F.M. (2003) *Arabidopsis* local resistance to *Botrytis cinerea* involves salicylic acid and camalexin and requires *Eds4* and *Pad2*, but not SID2, EDS5 or PAD4. *The Plant Journal* 35(2), 193–205. DOI: 10.1046/j.1365-313x.

Fesel, P.H. and Zuccaro, A. (2016) β-glucan: Crucial component of the fungal cell wall and elusive MAMP in plants. *Fungal Genetics and Biology* 90, 53–60. DOI: 10.1016/j.fgb.2015.12.004.

Foreman, J., Demidchik, V., Bothwell, J.H.F., Mylona, P., Miedema, H. *et al.* (2003) Reactive oxygen species produced by NADPH oxidase regulate plant cell growth. *Nature* 422(6930), 442–446. DOI: 10.1038/nature01485.

Francis, K., Nishino, S.F., Spain, J.C. and Gadda, G. (2012) A novel activity for fungal nitronate monooxygenase: Detoxification of the metabolic inhibitor propionate-3-nitronate. *Archives of Biochemistry and Biophysics* 521(1–2), 84–89. DOI: 10.1016/j.abb.2012.03.015.

Frías, M., Brito, N. and González, C. (2013) The *Botrytis cinerea* cerato-platanin BcSpl1 is a potent inducer of systemic acquired resistance (SAR) in tobacco and generates a wave of salicylic acid expanding from the site of application. *Molecular Plant Pathology* 14(2), 191–196. DOI: 10.1111/j.1364-3703.2012.00842.x.

Fry, W.E. and Evans, P.H. (1977) Association of formamide hydro-lyase with fungal pathogenicity to cyanogenic plants. *Phytopathology* 77(8), 1001. DOI: 10.1094/Phyto-67-1001.

Fu, D., Uauy, C., Distelfeld, A., Blechl, A., Epstein, L. *et al.* (2009) A *kinase-START* gene confers temperature-dependent resistance to wheat stripe rust. *Science* 323(5919), 1357–1360. DOI: 10.1126/science.1166289.

Frías, M., González, C. and Brito, N. (2011) BcSpl1, a cerato-platanin family protein, contributes to *Botrytis cinerea* virulence and elicits the hypersensitive response in the host. *The New Phytologist* 192(2), 483–495. DOI: 10.1111/j.1469-8137.2011.03802.x.

Fujihara, N., Sakaguchi, A., Tanaka, S., Fujii, S., Tsuji, G. *et al.* (2010) Peroxisome biogenesis factor *pex13* is required for appressorium-mediated plant infection by the anthracnose fungus *Colletotrichum orbiculare*. *Molecular Plant-Microbe Interactions* 23(4), 436–445. DOI: 10.1094/MPMI-23-4-0436.

Gil-Durán, C., Rojas-Aedo, J.F., Medina, E., Vaca, I., García-Rico, R.O. *et al.* (2015) The *pcz1* gene, which encodes a Zn (II) 2Cys6 protein, is involved in the control of growth, conidiation, and conidial germination in the filamentous fungus *Penicillium roqueforti*. *PLoS ONE* 10(3), e0120740. DOI: 10.1371/journal.pone.0120740.

Glazebrook, J., Rogers, E.E. and Ausubel, F.M. (1996) Isolation of *Arabidopsis* mutants with enhanced disease susceptibility by direct screening. *Genetics* 143(2), 973–982. DOI: 10.1093/genetics/143.2.973.

Godoy, G., Steadman, J.R., Dickman, M.B. and Dam, R. (1990) Use of mutants to demonstrate the role of oxalic acid in pathogenicity of *Sclerotinia sclerotiorum* on *Phaseolus vulgaris*. *Physiological and Molecular Plant Pathology* 37(3), 179–191. DOI: 10.1016/0885-5765(90)90010-U.

Goh, J., Jeon, J., Kim, K.S., Park, J., Park, S.Y. *et al.* (2011) The PEX7-mediated peroxisomal import system is required for fungal development and pathogenicity in *Magnaporthe oryzae*. *PLoS ONE* 6(12), e28220. DOI: 10.1371/journal.pone.0028220.

González-Fernández, R., Aloria, K., Valero-Galván, J., Redondo, I., Arizmendi, J.M. *et al.* (2014) Proteomic analysis of mycelium and secretome of different *Botrytis cinerea* wild-type strains. *Journal of Proteomics* 97, 195–221. DOI: 10.1016/j.jprot.2013.06.022.

Gullner, G., Komives, T., Kiraly, L. and Schroder, P. (2018) Glutathione S-transferase enzymes in plant-pathogen interactions. *Frontiers in Plant Science* 9, 1836. DOI: 10.3389/fpls.2018.01836.

Guo, X. and Stotz, H.U. (2007) Defense against *Sclerotinia sclerotiorum* in *Arabidopsis* is dependent on jasmonic acid, salicylic acid, and ethylene signaling. *Molecular Plant-Microbe Interactions* 20(11), 1384–1395. DOI: 10.1094/MPMI-20-11-1384.

Guyon, K., Balagué, C., Roby, D. and Raffaele, S. (2014) Secretome analysis reveals effector candidates associated with broad host range necrotrophy in the fungal plant pathogen *Sclerotinia sclerotiorum*. *BMC Genomics* 15(1), 336. DOI: 10.1186/1471-2164-15-336.

Harel, A., Bercovich, S. and Yarden, O. (2006) Calcineurin is required for sclerotial development and pathogenicity of *Sclerotinia sclerotiorum* in an oxalic acid independent manner. *Molecular Plant-Microbe Interactions* 19, 682–693.

Heard, S., Brown, N.A. and Hammond-Kosack, K. (2015) An interspecies comparative analysis of the predicted secretomes of the necrotrophic plant pathogens *Sclerotinia sclerotiorum* and *Botrytis cinerea*. *PLoS ONE* 10(6), e0130534. DOI: 10.1371/journal.pone.0130534.

Hegedus, D.D., Gerbrandt, K. and Coutu, C. (2016) The eukaryotic protein kinase super family of the necrotrophic fungal plant pathogen *Sclerotinia sclerotiorum*. *Molecular Plant Pathology* 17, 634–647.

Hepworth, S.R., Zhang, Y., McKim, S., Li, X. and Haughn, G.W. (2005) BLADE-ON-PETIOLE-dependent signaling controls leaf and floral patterning in *Arabidopsis*. *Plant Cell* 17, 1434–1448. DOI: 10.1105/tpc.104.030536.

Huard-Chauveau, C., Perchepied, L., Debieu, M., Rivas, S., Kroj, T. *et al.* (2013) An atypical kinase under balancing selection confers broad-spectrum disease resistance in *Arabidopsis*. *PLoS Genetics* 9(9), e1003766. DOI: 10.1371/journal.pgen.1003766.

Iakovidis, M., Teixeira, P.J.P.L., Exposito-Alonso, M., Cowper, M.G., Law, T.F. *et al.* (2016) Effector-triggered immune response in *Arabidopsis thaliana* is a quantitative trait. *Genetics* 204(1), 337–353. DOI: 10.1534/genetics.116.190678.

Ibrahim, H.M.M., Kusch, S., Didelon, M. and Raffaele, S. (2021) Genome-wide alternative splicing profiling in the fungal plant pathogen *Sclerotinia sclerotiorum* during the colonization of diverse host families. *Molecular Plant Pathology* 22(1), 31–47. DOI: 10.1111/mpp.13006.

Idnurm, A. and Howlett, B.J. (2002) Isocitrate lyase is essential for pathogenicity of the fungus *Leptosphaeria maculans* to canola (*Brassica napus*). *Eukaryotic Cell* 1(5), 719–724. DOI: 10.1128/EC.1.5.719-724.2002.

Imazaki, A., Tanaka, A., Harimoto, Y., Yamamoto, M., Akimitsu, K. *et al.* (2010) Contribution of peroxisomes to secondary metabolism and pathogenicity in the fungal plant pathogen *Alternaria alternata*. *Eukaryotic Cell* 9(5), 682–694. DOI: 10.1128/EC.00369-09.

Jeffers, M., McRoberts, W. and Harper, D. (1997) Identification of a phenolic 3-O-methyltransferase in the lignin-degrading fungus *Phanerochaete chrysosporium*. *Microbiology* 143(6), 1975–1981.

Jian, H., Wang, J., Wang, T., Wei, L., Li, J. *et al.* (2016) Identification of rapeseed microRNAs involved in early stage seed germination under salt and drought stresses. *Frontiers in Plant Science* 7, 658. DOI: 10.3389/fpls.2016.00658.

Johnson, C., Kweon, H.K., Sheidy, D., Shively, C.A., Mellacheruvu, D. *et al.* (2014) The yeast Sks1p kinase signaling network regulates pseudohyphal growth and glucose response. *PLoS Genetics* 10(3), e1004183. DOI: 10.1371/journal.pgen.1004183.

Jonge, R. and Thomma, B.P.H.J. (2009) Fungal lysm effectors: Extinguishers of host immunity. *Trends in Microbiology* 17(4), 151–157. DOI: 10.1016/j.tim.2009.01.002.

Jonge, R., Esse, H., Kombrink, A., Shinya, T., Desaki, Y. *et al.* (2010) Conserved fungal LysM effector Ecp6 prevents chitin-triggered immunity in plants. *Science* 329(5994), 953–955. DOI: 10.1126/science.1190859.

Joshi, R.K., Megha, S., Rahman, M.H., Basu, U. and Kav, N.N.V. (2016a) A global study of transcriptome dynamics in canola (*Brassica napus* L.) responsive to *Sclerotinia sclerotiorum* infection using RNA-seq. *Gene* 590(1), 57–67. DOI: 10.1016/j.gene.2016.06.003.

Joshi, R.K., Megha, S., Basu, U., Rahman, M.H. and Kav, N.N.V. (2016b) Genome wide identification and functional prediction of long non-coding RNAs responsive to *Sclerotinia sclerotiorum* infection in *Brassica napus*. *PLoS ONE* 11(7), e0158784. DOI: 10.1371/journal.pone.0158784.

Kabbage, M., Williams, B. and Dickman, M.B. (2013) Cell death control: The interplay of apoptosis and autophagy in the pathogenicity of *Sclerotinia sclerotiorum*. *PLoS Pathogens* 9(4), e1003287. DOI: 10.1371/journal.ppat.1003287.

Kabbage, M., Yarden, O. and Dickman, M.B. (2015) Pathogenic attributes of *Sclerotinia sclerotiorum*: Switching from a biotrophic to necrotrophic lifestyle. *Plant Science* 233, 53–60. DOI: 10.1016/j.plantsci.2014.12.018.

Kaku, H., Nishizawa, Y., Ishii-Minami, N., Akimoto-Tomiyama, C., Dohmae, N. *et al.* (2006) Plant cells recognize chitin fragments for defense signaling through a plasma membrane receptor. *Proceeding of National Academy of Sciences, USA* 103(29), 11086–11091. DOI: 10.1073/pnas.0508882103.

Karisto, P., Hund, A., Yu, K., Anderegg, J., Walter, A. *et al.* (2018) Ranking quantitative resistance to *Septoria tritici* blotch in elite wheat cultivars using automated image analysis. *Phytopathology* 108(5), 568–581. DOI: 10.1094/PHYTO-04-17-0163-R.

Kars, I., Krooshof, G.H., Wagemakers, L., Joosten, R., Benen, J.A.E. *et al.* (2005) Necrotizing activity of five *Botrytis cinerea* endopolygalacturonases produced in *Pichia pastoris*. *Plant Journal* 43, 213–225.

Kim, H., Chen, C., Kabbage, M. and Dickman, M.B. (2011) Identification and characterization of *Sclerotinia sclerotiorum* NADPH oxidases. *Applied and Environmental Microbiology* 77(21), 7721–7729. DOI: 10.1128/AEM.05472-11.

Kim, K.S., Min, J.Y. and Dickman, M.B. (2008) Oxalic acid is an elicitor of plant programmed cell death during *Sclerotinia sclerotiorum* disease development. *Molecular Plant-Microbe Interactions* 21, 605–612.

Kim, S., Park, S.-Y., Kim, K.S., Rho, H.-S., Chi, M.-H. *et al.* (2009) Homeobox transcription factors are required for conidiation and appressorium development in the rice blast fungus *Magnaporthe oryzae*. *PLoS Genetics* 5(12), e1000757. DOI: 10.1371/journal.pgen.1000757.

Kimura, A., Takano, Y., Furusawa, I. and Okuno, T. (2001) Peroxisomal metabolic function is required for appressorium-mediated plant infection by *Colletotrichum lagenarium*. *The Plant Cell* 13(8), 1945–1957. DOI: 10.1105/TPC.010084.

Kleemann, J., Rincon-Rivera, L.J., Takahara, H., Neumann, U., Themaat, E. *et al.* (2012) Sequential delivery of host-induced virulence effectors by appressoria and intracellular hyphae of the phytopathogen *Colletotrichum higginsianum*. *PLoS Pathogens* 8(4), e1002643. DOI: 10.1371/journal.ppat.1002643.

Kombrink, A. and Thomma, B.P.H.J. (2013) LysM effectors: Secreted proteins supporting fungal life. *PLOS Pathogens* 9(12), e1003769. DOI: 10.1371/journal.ppat.1003769.

Kou, Y., Tan, Y.H., Ramanujam, R. and Naqvi, N.I. (2017) Structure–function analyses of the Pth11 receptor reveal an important role for CFEM motif and redox regulation in rice blast. *New Phytologist* 214(1), 330–342. DOI: 10.1111/nph.14347.

Krattinger, S.G., Lagudah, E.S., Spielmeyer, W., Singh, R.P., Huerta-Espino, J. *et al.* (2009) A putative ABC transporter confers durable resistance to multiple fungal pathogens in wheat. *Science* 323(5919), 1360–1363. DOI: 10.1126/science.1166453.

Kubicek, C.P., Kunar, G.S. and Rohr, W.W.M. (1988) Evidence for a cytoplasmic pathway of oxalate biosynthesis in *Aspergillus niger*. *Applied Environment Microbiology* 54(3), 633–637. DOI: 10.1128/AEM.54.3.633-637.1988.

Kubo, Y. (2012) Appressorium function in *Colletotrichum orbiculare* and prospect for genome-based analysis. In: Perez-Martın, J. and Pietro, A. (eds) *Morphogenesis and Pathogenicity in Fungi*. Springer, Berlin, Heidelberg, Germany, pp. 115–131.

Kulkarni, R.D., Kelkar, H.S. and Dean, R.A. (2003) An eight-cysteine-containing CFEM domain unique to a group of fungal membrane proteins. *Trends in Biochemical Sciences* 28(3), 118–121. DOI: 10.1016/S0968-0004(03)00025-2.

Lee, T.G., Diers, B.W. and Hudson, M.E. (2016) An efficient method for measuring copy number variation applied to improvement of nematode resistance in soybean. *The Plant Journal* 88(1), 143–153. DOI: 10.1111/tpj.13240.

Li, M. and Rollins, J.A. (2009) The development-specific protein (Ssp1) from *Sclerotinia sclerotiorum* is encoded by a novel gene expressed exclusively in sclerotium tissues. *Mycologia* 101(1), 34–43. DOI: 10.3852/08-114.

Li, J., Zhang, Y., Zhang, Y., Yu, P.-L., Pan, H. *et al.* (2018) Introduction of large sequence inserts by CRISPR-cas9 to create pathogenicity mutants in the multinucleate filamentous pathogen *Sclerotinia sclerotiorum*. *mBio* 9(3), e00567–18. DOI: 10.1128/mBio.00567-18.

Li, M., Liang, X. and Rollins, J.A. (2012) *Sclerotinia sclerotiorum*-glutamyl transpeptidase (Ss-Ggt1) is required for regulating glutathione accumulation and development of sclerotia and compound appressoria. *Molecular Plant-Microbe Interactions* 25(3), 412–420. DOI: 10.1094/MPMI-06-11-0159.

Li, R., Rimmer, R., Buchwaldt, L., Sharpe, A.G., Séguin-Swartz, G. *et al.* (2004) Interaction of *Sclerotinia sclerotiorum* with *Brassica napus*: Cloning and characterization of endo- and exo-polygalacturonases expressed during saprophytic and parasitic modes. *Fungal Genetics and Biology* 41(8), 754–765. DOI: 10.1016/j.fgb.2004.03.002.

Liang, X. and Rollins, J.A. (2018) Mechanisms of broad host range necrotrophic pathogenesis in *Sclerotinia sclerotiorum*. *Phytopathology* 108(10), 1128–1140. DOI: 10.1094/PHYTO-06-18-0197-RVW.

Liang, Y., Yajima, W., Davis, M.R., Kav, N.N.V. and Stephen, E. (2013) Disruption of a gene encoding a hypothetical secreted protein from *Sclerotinia Sclerotiorum* reduces its virulence on canola (*Brassica napus*) disruption of a gene encoding a hypothetical secreted protein from *Sclerotinia Sclerotiorum* reduces its virulence on canola (*Brassica napus*). *Canadian Journal of Plant Pathology* 35, 46–55. DOI: 10.1080/07060661.2012.745904.

Liang, X., Liberti, D., Li, M., Kim, Y.T., Hutchens, A. *et al.* (2015a) Oxaloacetate acetylhydrolase gene mutants of sclerotinia sclerotiorum do not accumulate oxalic acid, but do produce limited lesions on host plants. *Molecular Plant Pathology* 16(6), 559–571. DOI: 10.1111/mpp.12211.

Liang, X., Moomaw, E.W. and Rollins, J.A. (2015b) Fungal oxalate decarboxylase activity contributes to *Sclerotinia Sclerotiorum* early infection by affecting both compound appressoria development and function. *Molecular Plant Pathology* 16, 825–836. DOI: 10.1111/mpp.

Liberti, D., Grant, S.J., Benny, U., Rollins, J.A. and Dobinson, K.F. (2007) Development of an *Agrobacterium tumefaciens* mediated gene disruption method for *Sclerotinia Sclerotiorum*. *Canadian Journal of Plant Pathology* 29(4), 394–400.

Liberti, D., Rollins, J.A. and Dobinson, K.F. (2013) Peroxysomal carnitine acetyl transferase influences host colonization capacity in *Sclerotinia Sclerotiorum*. *Molecular Plant-Microbe Interactions* 26(7), 768–780. DOI: 10.1094/MPMI-03-13-0075-R.

Liu, G., Holub, E.B., Alonso, J.M., Ecker, J.R. and Fobert, P.R. (2005) An Arabidopsis NPR1-like gene, NPR4, is required for disease resistance. *Plant Journal of Cell and Molecular Biology* 41, 304–318.

Lyu, X., Shen, C., Fu, Y., Xie, J., Jiang, D. *et al.* (2015) Comparative genomic and transcriptional analyses of the carbohydrate-active enzymes and secretomes of phytopathogenic fungi reveal their significant roles during infection and development. *Scientific Reports* 5, 15565. DOI: 10.1038/srep15565.

Lyu, X., Shen, C., Fu, Y., Xie, J., Jiang, D. *et al.* (2016) A small secreted virulence-related protein is essential for the necrotrophic interactions of *Sclerotinia sclerotiorum* with its host plants. *PLoS Pathogens* 12(2), e1005435. DOI: 10.1371/journal.ppat.1005435.

Malagnac, F., Lalucque, H., Lepère, G. and Silar, P. (2004) Two NADPH oxidase isoforms are required for sexual reproduction and ascospore germination in the filamentous fungus *Podospora anserina*. *Fungal Genetics and Biology* 41(11), 982–997. DOI: 10.1016/j.fgb.2004.07.008.

Marroquin-Guzman, M., Hartline, D., Wright, J.D., Elowsky, C., Bourret, T.J. *et al.* (2017) The *Magnaporthe oryzae* nitrooxidative stress response suppresses rice innate immunity during blast disease. *Nature Microbiology* 2, 17054. DOI: 10.1038/nmicrobiol.2017.54.

Maxwell, D.P. (1973) Oxalate formation in *Whetzelinia sclerotiorum* by oxalo-acetate acetylhydrolase. *Physiological Plant Pathology* 3(2), 279–288. DOI: 10.1016/0048-4059(73)90090-8.

McCaghey, M., Shao, D., Kurcezewski, J., Lindstrom, A., Ranjan, A. *et al.* (2021) Host-induced gene silencing of a *Sclerotinia sclerotiorum oxaloacetate acetylhydrolase* using bean pod mottle virus as a vehicle reduces disease on soybean. *Frontiers in Plant Science* 12, 677631. DOI: 10.3389/fpls.2021.677631.

Mentlak, T.A., Kombrink, A., Shinya, T., Ryder, L.S., Otomo, I. *et al.* (2012) Effector-mediated suppression of chitin-triggered immunity by *Magnaporthe oryzae* is necessary for rice blast disease. *The Plant Cell* 24(1), 322–335. DOI: 10.1105/tpc.111.092957.

Moscou, M.J., Lauter, N., Steffenson, B. and Wise, R.P. (2011) Quantitative and qualitative stem rust resistance factors in barley are associated with transcriptional suppression of defense regulons. *PLoS Genetics* 7(7), e1002208. DOI: 10.1371/journal.pgen.1002208.

Mou, Z., Fan, W. and Dong, X. (2003) Inducers of plant systemic acquired resistance regulate *NPR1* function through redox changes. *Cell* 113(7), 935–944. DOI: 10.1016/S0092-8674(03)00429-X.

Mutka, A.M., Fentress, S.J., Sher, J.W., Berry, J.C., Pretz, C. *et al.* (2016) Quantitative, image-based phenotyping methods provide insight into spatial and temporal dimensions of plant disease. *Plant Physiology* 172(2), 650–660. DOI: 10.1104/pp.16.00984.

Nelson, J.M., Hauser, D.A., Hinson, R. and Shaw, A.J. (2018) A novel experimental system using the liverwort *Marchantia polymorpha* and its fungal endophytes reveals diverse and context-dependent effects. *The New Phytologist* 218(3), 1217–1232. DOI: 10.1111/nph.15012.

Nemri, A., Saunders, D.G.O., Anderson, C., Upadhyaya, N.M., Win, J. *et al.* (2014) The genome sequence and effector complement of the flax rust pathogen *Melampsora lini*. *Frontiers in Plant Science* 5, 98. DOI: 10.3389/fpls.2014.00098.

Neupane, A., Feng, C., Mochama, P.K., Saleem, H. and Lee Marzano, S.Y. (2019) Roles of argonautes and dicers on *Sclerotinia sclerotiorum* antiviral RNA silencing. *Frontiers in Plant Science* 10, 976. DOI: 10.3389/fpls.2019.00976.

Nishizawa, M. (2015) The regulators of yeast PHO system participate in the transcriptional regulation of G1 cyclin under alkaline stress conditions. *Yeast* 32(3), 367–378. DOI: 10.1002/yea.3064.

Nováková, M., Šašek, V., Dobrev, P.I., Valentová, O. and Burketová, L. (2014) Plant hormones in defense response of *Brassica napus* to *Sclerotinia sclerotiorum*—reassessing the role of salicylic acid in the interaction with a necrotroph. *Plant Physiology and Biochemistry* 80, 308–317. DOI: 10.1016/j.plaphy.2014.04.019.

Oliveira, M.B., Andrade, R.V., Grossi-de-Sá, M.F. and Petrofeza, S. (2015) Analysis of genes that are differentially expressed during the *Sclerotinia sclerotiorum–Phaseolus vulgaris* interaction. *Frontiers in Microbiology* 6, 1162. DOI: 10.3389/fmicb.2015.01162.

Osbourn, A.E. (1996) Preformed antimicrobial compounds and plant defense against fungal attack. *The Plant Cell* 8(10), 1821. DOI: 10.2307/3870232.

Paoletti, M. and Clavé, C. (2007) The fungus-specific HET domain mediates programmed cell death in *Podospora anserina*. *Eukaryotic Cell* 6(11), 2001–2008. DOI: 10.1128/EC.00129-07.

Paoletti, M. and Saupe, S.J. (2009) Fungal incompatibility: Evolutionary origin in pathogen defense? *BioEssays* 31(11), 1201–1210. DOI: 10.1002/bies.200900085.

Pazzagli, L., Seidl-Seiboth, V., Barsottini, M., Vargas, W.A., Scala, A. *et al.* (2014) Cerato-platanins: Elicitors and effectors. *Plant Science* 228, 79–87. DOI: 10.1016/j.plantsci.2014.02.009.

Pedras, M.S.C. and Hossain, M. (2006) Metabolism of crucifer phytoalexins in *Sclerotinia sclerotiorum*: Detoxification of strongly antifungal compounds involves glucosylation. *Organic and Biomolecular Chemistry* 4(13), 2581. DOI: 10.1039/b604400j.

Pedras, M.S.C., Zheng, Q.A., Gadagi, R.S. and Rimmer, S.R. (2008) Phytoalexins and polar metabolites from the oilseed's canola and rapeseed: Differential metabolic responses to the biotroph *Albugo candida* and to abiotic stress. *Phytochemistry* 69(4), 894–910. DOI: 10.1016/j.phytochem.2007.10.019.

Pedras, M.S.C., Yaya, E.E. and Glawischnig, E. (2011) The phytoalexins from cultivated and wild crucifers: Chemistry and biology. *Natural Product Reports* 28(8), 1381–1405. DOI: 10.1039/c1np00020a.

Perchepied, L., Balagué, C., Riou, C., Claudel-Renard, C., Rivière, N. *et al.* (2010) Nitric oxide participates in the complex interplay of defense-related signaling pathways controlling disease resistance to

Sclerotinia sclerotiorum in arabidopsis thaliana. *Molecular Plant-Microbe Interactions* 23(7), 846–860. DOI: 10.1094/MPMI-23-7-0846.

Pérez-Valle, J., Primo, C., Martínez Pastor, M. and Ariño, J. (2010) *Hal4* and *Hal5* protein kinases are required for general control of carbon and nitrogen uptake and metabolism. *Eukaryotic Cell* 9(12), 1881–1890. DOI: 10.1128/EC.00184-10.

Poirier, Y., Antonenkov, V.D., Glumoff, T. and Hiltunen, J.K. (2006) Peroxisomal β-oxidation—A metabolic pathway with multiple functions. *Biochimica et Biophysica Acta (BBA)—Molecular Cell Research* 1763(12), 1413–1426. DOI: 10.1016/j.bbamcr.2006.08.034.

Poland, J.A., Bradbury, P.J., Buckler, E.S. and Nelson, R.J. (2011) Genome-wide nested association mapping of quantitative resistance to northern leaf blight in maize. *Proceedings of the National Academy of Sciences* 108(17), 6893–6898. DOI: 10.1073/pnas.1010894108.

Poussereau, N., Creton, S., Billon-Grand, G., Rascle, C. and Fevre, M. (2001) Regulation of acp1, encoding a non-aspartyl acid protease expressed during pathogenesis of *Sclerotinia sclerotiorum*. *Microbiology* 147(Pt 3), 717–726. DOI: 10.1099/00221287-147-3-717.

Qiao, L., Lan, C., Capriotti, L., Ah-Fong, A., Nino Sanchez, J. *et al.* (2021) Spray-induced gene silencing for disease control is dependent on the efficiency of pathogen RNA uptake. *Plant Biotechnology Journal* 19(9), 1756–1768. DOI: 10.1111/pbi.13589.

Qu, X., Yu, B., Liu, J., Zhang, X., Li, G. *et al.* (2014) MADS-box transcription factor SsMADs is involved in regulating growth and virulence in *Sclerotinia sclerotiorum*. *International Journal of Molecular Sciences* 15(5), 8049–8062. DOI: 10.3390/ijms15058049.

Rajarammohan, S., Pradhan, A.K., Pental, D. and Kaur, J. (2018) Genome-wide association mapping in *Arabidopsis* identifies novel genes underlying quantitative disease resistance to *Alternaria brassicae*. *Molecular Plant Pathology* 19(7), 1719–1732. DOI: 10.1111/mpp.12654.

Ramos-Pamplona, M. and Naqvi, N.I. (2006) Host invasion during rice-blast disease requires carnitine-dependent transport of peroxisomal acetyl-CoA. *Molecular Microbiology* 61(1), 61–75. DOI: 10.1111/j.1365-2958.2006.05194.x.

Rana, K., Ding, Y., Banga, S.S., Liao, H., Zhao, S. *et al.* (2021) *Sclerotinia sclerotiorum Thioredoxin1 (SsTrx1)* is required for pathogenicity and oxidative stress tolerance. *Molecular Plant Pathology* 22(11), 1413–1426. DOI: 10.1111/mpp.13127.

Regmi, R., Newman, T.E., Kamphuis, L.G. and Derbyshire, M.C. (2021) Identification of *Brassica napus* small RNAs responsive to infection by a necrotrophic pathogen. *BMC Plant Biology* 21(1), 366. DOI: 10.1186/s12870-021-03148-6.

Roetschi, A., Si-Ammour, A., Belbahri, L., Mauch, F. and Mauch-Mani, B. (2001) Characterization of an *Arabidopsis-phytophthora* pathosystem: Resistance requires a functional *PAD2* gene and is independent of salicylic acid, ethylene and jasmonic acid signalling. *The Plant Journal* 28(3), 293–305. DOI: 10.1046/j.1365-313x.2001.01148.x.

Rollins, J.A. and Dickman, M.B. (2001) PH signaling in *Sclerotinia sclerotiorum*: Identification of a pac C/RIM 1 homolog. *Applied and Environmental Microbiology* 67(1), 75–81. DOI: 10.1128/AEM.67.1.75-81.2001.

Rollins, J.A. (2003) The *Sclerotinia sclerotiorum pac1* gene is required for sclerotial development and virulence. *Molecular Plant-Microbe Interactions* 16(9), 785–795. DOI: 10.1094/MPMI.2003.16.9.785.

Roux, F., Voisin, D., Badet, T., Balagué, C., Barlet, X. *et al.* (2014) Resistance to phytopathogens e tutti quanti: Placing plant quantitative disease resistance on the map. *Molecular Plant Pathology* 15(5), 427–432. DOI: 10.1111/mpp.12138.

Ruijter, G.J.G., Vondervoort, P.J.I. and Visser, J. (1999) Oxalic acid production by *Aspergillus niger*: An oxalate-non-producing mutant produces citric acid at pH 5 and in the presence of manganese. *Microbiology* 145(Pt 9), 2569–2576. DOI: 10.1099/00221287-145-9-2569.

Sánchez-Vallet, A., Tian, H., Rodriguez-Moreno, L., Valkenburg, D.-J., Saleem-Batcha, R. *et al.* (2020) A secreted LysM effector protects fungal hyphae through chitin-dependent homodimer polymerization. *PLoS Pathogens* 16(6), e1008652. DOI: 10.1371/journal.ppat.1008652.

Schamber, A., Leroch, M., Diwo, J., Mendgen, K. and Hahn, M. (2010) The role of mitogen-activated protein (MAP) kinase signalling components and the Ste12 transcription factor in germination and pathogenicity of *Botrytis cinerea*. *Molecular Plant Pathology* 11(1), 105–119. DOI: 10.1111/j.1364-3703.2009.00579.x.

Schouten, A., Baarlen, P. and Kan, J.A.L. (2008) Phytotoxic *Nep1*-like proteins from the necrotrophic fungus *Botrytis cinerea* associate with membranes and the nucleus of plant cells. *The New Phytologist* 177(2), 493–505. DOI: 10.1111/j.1469-8137.2007.02274.x.

Schumacher, J., Larrinoa, I.F. and Tudzynski, B. (2008a) Calcineurin-responsive zinc finger transcription factor *CRZ1* of *Botrytis cinerea* is required for growth, development, and full virulence on bean plants. *Eukaryotic Cell* 7(4), 584–601. DOI: 10.1128/EC.00426-07.

Schumacher, J., Viaud, M., Simon, A. and Tudzynski, B. (2008b) The Gα subunit BCG1, the phospholipase c (*BcPLC1*) and the calcineurin phosphatase co-ordinately regulate gene expression in the grey mould fungus *Botrytis cinerea*. *Molecular Microbiology* 67(5), 1027–1050. DOI: 10.1111/j.1365-2958.2008.06105.x.

Segmüller, N., Kokkelink, L., Giesbert, S., Odinius, D., Kan, J. *et al.* (2008) NADPH oxidases are involved in differentiation and pathogenicity in *Botrytis cinerea*. *Molecular Plant-Microbe Interactions* 21(6), 808–819. DOI: 10.1094/MPMI-21-6-0808.

Seifbarghi, S., Borhan, M.H., Wei, Y., Coutu, C., Robinson, S.J. *et al.* (2017) Changes in the *Sclerotinia sclerotiorum* transcriptome during infection of *Brassica napus*. *BMC Genomics* 18, 266. DOI: 10.1186/s12864-017-3642-5.

Seifbarghi, S., Borhan, M.H., Wei, Y., Ma, L., Coutu, C. *et al.* (2020) Receptor-like kinases *BAK1* and *SOBIR1* are required for necrotizing activity of a novel group of *Sclerotinia sclerotiorum* necrosis-inducing effectors. *Frontiers in Plant Science* 11, 1021. DOI: 10.3389/fpls.2020.01021.

Sexton, A.C. and Howlett, B.J. (2000) Characterisation of a cyanide hydratase gene in the phytopathogenic fungus *Leptosphaeria maculans*. *Molecular and General Genetics MGG* 263(3), 463–470. DOI: 10.1007/s004380051190.

Sexton, A.C., Cozijnsen, A.J., Keniry, A., Jewell, E., Love, C.G. *et al.* (2006) Comparison of transcription of multiple genes at three developmental stages of the plant pathogen *Sclerotinia sclerotiorum*. *FEMS Microbiology Letters* 258(1), 150–160. DOI: 10.1111/j.1574-6968.2006.00212.x.

Sexton, A.C., Minic, Z., Cozijnsen, A.J., Pedras, M.S.C. and Howlett, B.J. (2009) Cloning, purification and characterisation of brassinin glucosyltransferase, a phytoalexin-detoxifying enzyme from the plant pathogen *Sclerotinia sclerotiorum*. *Fungal Genetics and Biology* 46(2), 201–209. DOI: 10.1016/j.fgb.2008.10.014.

Shalaby, S. and Horwitz, B.A. (2015) Plant phenolic compounds and oxidative stress: Integrated signals in fungal–plant interactions. *Current Genetics* 61(3), 347–357. DOI: 10.1007/s00294-014-0458-6.

Shanmugam, V., Ronen, M., Shalaby, S., Larkov, O., Rachamim, Y. *et al.* (2010) The fungal pathogen *Cochliobolus heterostrophus* responds to maize phenolics: Novel small molecule signals in a plant–fungal interaction. *Cellular Microbiology* 12(10), 1421–1434. DOI: 10.1111/j.1462-5822.2010.01479.x.

Shibuya, N. and Minami, E. (2001) Oligosaccharide signalling for defence responses in plant. *Physiological and Molecular Plant Pathology* 59(5), 223–233. DOI: 10.1006/pmpp.2001.0364.

Shraddha, R., Shekher, S., Sehgal, M., Kamthania, A. and Kumar, A. (2011) Laccase: Microbial sources, production, purification, and potential biotechnological applications. *Enzyme Research* 2011(1), 1–11. DOI: 10.4061/2011/217861.

Simon, A., Dalmais, B., Morgant, G. and Viaud, M. (2013) Screening of a *Botrytis cinerea* one-hybrid library reveals a *Cys2His2* transcription factor involved in the regulation of secondary metabolism gene clusters. *Fungal Genetics and Biology* 52, 9–19. DOI: 10.1016/j.fgb.2013.01.006.

Sperschneider, J., Dodds, P.N., Gardiner, D.M., Manners, J.M., Singh, K.B. *et al.* (2015) Advances and challenges in computational prediction of effectors from plant pathogenic fungi. *PLoS Pathogens* 11(5), e1004806. DOI: 10.1371/journal.ppat.1004806.

Sperschneider, J., Gardiner, D.M., Dodds, P.N., Tini, F., Covarelli, L. *et al.* (2016) Effector P: Predicting fungal effector proteins from secretomes using machine learning. *The New Phytologist* 210(2), 743–761. DOI: 10.1111/nph.13794.

Spoel, S.H., Koornneef, A., Claessens, S.M.C., Korzelius, J.P., Pelt, J.A. *et al.* (2003) *NPR1* modulates cross-talk between salicylate- and jasmonate-dependent defense pathways through a novel function in the cytosol. *The Plant Cell* 15(3), 760–770. DOI: 10.1105/tpc.009159.

Spoel, S.H., Mou, Z., Tada, Y., Spivey, N.W., Genschik, P. *et al.* (2009) Proteasome-mediated turnover of the transcription coactivator *NPR1* plays dual roles in regulating plant immunity. *Cell* 137(5), 860–872. DOI: 10.1016/j.cell.2009.03.038.

Stajich, J.E., Wilke, S.K., Ahrén, D., Au, C.H., Birren, B.W. *et al.* (2010) Insights into evolution of multicellular fungi from the assembled chromosomes of the mushroom *Coprinopsis cinerea* (*Coprinus cinereus*). *Proceedings of the National Academy of Sciences* 107(26), 11889–11894. DOI: 10.1073/pnas.1003391107.

Storck, M. and Sacristan, M.D. (1995) The role of phytoalexins in the seedling resistance to *Leptosphaeria maculans* in some crucifers. *Zeitschrift Für Naturforschung C* 50(1–2), 15–20. DOI: 10.1515/znc-1995-1-204.

Struck, A.W., Thompson, M.L., Wong, L.S. and Micklefield, J. (2012) S-adenosyl-methionine-dependent methyltransferases: Highly versatile enzymes in biocatalysis, biosynthesis and other biotechnological applications. *Chem Biochem* 13(18), 2642–2655. DOI: 10.1002/cbic.201200556.

Takahara, H., Hacquard, S., Kombrink, A., Hughes, H.B., Halder, V, *et al.* (2016) *Colletotrichum higginsianum* extracellular lysm proteins play dual roles in appressorial function and suppression of chitin triggered plant immunity. *New Phytologist* 211(4), 1323–1337. DOI: 10.1111/nph.13994.

Tanaka, C. and Izumitsu, K. (2010) Two-component signaling system in filamentous fungi and the mode of action of dicarboximide and phenylpyrrole fungicides. In: Carisse, O. (ed.) *Fungicides. IntechOpen*. Rijeka, Croatia, pp. 523–538. DOI: 10.5772/555pp.

Tang, L., Yang, G., Ma, M., Liu, X., Li, B. *et al.* (2020) An effector of a necrotrophic fungal pathogen targets the calcium-sensing receptor in chloroplasts to inhibit host resistance. *Molecular Plant Pathology* 21(5), 686–701. DOI: 10.1111/mpp.12922.

Thatcher, L.F., Williams, A.H., Garg, G., Buck, S.A.G. and Singh, K.B. (2016) Transcriptome analysis of the fungal pathogen *Fusarium oxysporum* f. sp. *medicaginis* during colonisation of resistant and susceptible *Medicago truncatula* hosts identifies differential pathogenicity profiles and novel candidate effectors. *BMC Genomics* 17(1), 860. DOI: 10.1186/s12864-016-3192-2.

Toruño, T.Y., Stergiopoulos, I. and Coaker, G. (2016) Plant-pathogen effectors: Cellular probes interfering with plant defenses in spatial and temporal manners. *Annual Review of Phytopathology* 54, 419–441. DOI: 10.1146/annurev-phyto-080615-100204.

Vander Heiden, M.G., Choy, J.S., VanderWeele, D.J., Brace, J.L., Harris, M.H. *et al.* (2002) Bcl-x L complements *Saccharomyces cerevisiae* genes that facilitate the switch from glycolytic to oxidative metabolism. *The Journal of Biological Chemistry* 277(47), 44870–44876. DOI: 10.1074/jbc.M204888200.

Wadke, N., Kandasamy, D., Vogel, H., Lah, L., Wingfield, B.D. *et al.* (2016) The bark-beetle-associated fungus, *Endoconidiophora polonica*, utilizes the phenolic defense compounds of its host as a carbon source. *Plant Physiology* 171(2), 914. DOI: 10.1104/pp.15.01916.

Wang, P. and Etten, H.D. (1992) Cloning and properties of a cyanide hydratase gene from the phytopathogenic fungus *Gloeocercospora sorghi*. *Biochemical and Biophysical Research Communications* 187(2), 1048–1054. DOI: 10.1016/0006-291x(92)91303-8.

Wang, J., Li, L., Zhang, Z., Qiu, H., Li, D. *et al.* (2015) One of three pex11 family members is required for peroxisomal proliferation and full virulence of the rice blast fungus *Magnaporthe oryzae*. *PLoS ONE* 10(7), e0134249. DOI: 10.1371/journal.pone.0134249.

Wang, X., Li, Q., Niu, X., Chen, H., Xu, L. *et al.* (2009) Characterization of a canola C2 domain gene that interacts with PG, an effector of the necrotrophic fungus *Sclerotinia sclerotiorum*. *Journal of Experimental Botany* 60(9), 2613–2620. DOI: 10.1093/jxb/erp104.

Wang, Z., Ma, L.Y., Li, X., Zhao, F.Y., Sarwar, R. *et al.* (2020) Genome-wide identification of the NPR1-like gene family in *Brassica napus* and functional characterization of *BnaNPR1* in resistance to *Sclerotinia sclerotiorum*. *Plant Cell Reports*. DOI: 10.1007/s00299-020-02525-z.

Wang, Z.Y., Thornton, C.R., Kershaw, M.J., Debao, L. and Talbot, N.J. (2003) The glyoxylate cycle is required for temporal regulation of virulence by the plant pathogenic fungus *Magnaporthe grisea*. *Molecular Microbiology* 47(6), 1601–1612. DOI: 10.1046/j.1365-2958.2003.

Wang, Z.Y., Soanes, D.M., Kershaw, M.J. and Talbot, N.J. (2007) Functional analysis of lipid metabolism in *Magnaporthe grisea* reveals a requirement for peroxisomal fatty acid β-oxidation during appressorium-mediated plant infection. *Molecular Plant-Microbe Interactions* 20(5), 475–491. DOI: 10.1094/MPMI-20-5-0475.

Wei, H., Scherer, M., Singh, A., Liese, R. and Fischer, R. (2001) *Aspergillus nidulans* alpha-1,3 glucanase (mutanase), mutA, is expressed during sexual development and mobilizes mutan. *Fungal Genetics and Biology* 34(3), 217–227. DOI: 10.1006/fgbi.2001.1303.

Westrick, N.M., Ranjan, A., Jain, S., Grau, C.R., Smith, D.L. *et al.* (2019) Gene regulation of *Sclerotinia sclerotiorum* during infection of *Glycine max*: On the road to pathogenesis. *BMC Genomics* 20(1), 157. DOI: 10.1186/s12864-019-5517-4.

Williams, B., Kabbage, M., Kim, H.J., Britt, R. and Dickman, M.B. (2011) Tipping the balance: *Sclerotinia sclerotiorum* secreted oxalic acid suppresses host defenses by manipulating the host redox environment. *PLoS Pathogens* 7(6), e1002107. DOI: 10.1371/journal.ppat.1002107.

Wojtaszek, P. (1997) Oxidative burst: An early plant response to pathogen infection. *Biochemical Journal* 322(3), 681–692. DOI: 10.1042/bj3220681.

Xia, S., Xu, Y., Hoy, R., Zhang, J., Qin, L. *et al.* (2019) *Sclerotinia sclerotiorum* the notorious soilborne pathogenic fungus: An update on genes studied with mutant analysis. *Pathogens* 9(1), 27. DOI: 10.3390/pathogens9010027.

Xia, Z., Wang, Z., Kav, N.N.V., Ding, C. and Liang, Y. (2020) Characterization of microRNA-like RNAs associated with sclerotial development in *Sclerotinia sclerotiorum*. *Fungal Genetics and Biology* 144, 103471. DOI: 10.1016/j.fgb.2020.103471.

Xiao, X., Xie, J., Cheng, J., Li, G., Yi, X. *et al.* (2014) Novel secretory protein ss-caf1 of the plant-pathogenic fungus *Sclerotinia sclerotiorum* is required for host penetration and normal sclerotial development. *Molecular Plant-Microbe Interactions* 27(1), 40–55. DOI: 10.1094/MPMI-05-13-0145-R.

Xie, F.L., Huang, S.Q., Guo, K., Xiang, A.L., Zhu, Y.Y. *et al.* (2007) Computational identification of novel micrornas and targets in *Brassica napus*. *FEBS Letters* 581(7), 1464–1474. DOI: 10.1016/j.febslet.2007.02.074.

Xu, L., Xiang, M., White, D. and Chen, W. (2015) PH dependency of sclerotial development and pathogenicity revealed by using genetically defined oxalate-minus mutants of *Sclerotinia sclerotiorum*. *Environmental Microbiology* 17(8), 2896–2909. DOI: 10.1111/1462-2920.12818.

Xue, C., Park, G., Choi, W., Zheng, L., Dean, R.A. *et al.* (2002) Two novel fungal virulence genes specifically expressed in appressoria of the rice blast fungus. *The Plant Cell* 14(9), 2107–2119. DOI: 10.1105/tpc.003426.

Yang, G., Tang, L., Gong, Y., Xie, J., Fu, Y. *et al.* (2018) A cerato-platanin protein SsCP1 targets plant PR1 and contributes to virulence of *Sclerotinia sclerotiorum*. *The New Phytologist* 217(2), 739–755. DOI: 10.1111/nph.14842.

Yarden, O., Veluchamy, S., Dickman, M.B. and Kabbage, M. (2014) Physiological and molecular plant pathology *Sclerotinia sclerotiorum* catalase SCAT1 affects oxidative stress tolerance, regulates ergosterol levels and controls pathogenic development. *Physiological and Molecular Plant Pathology* 85, 34–41. DOI: 10.1016/j.pmpp.2013.12.001.

Yu, Y., Xiao, J., Du, J., Yang, Y., Bi, C. *et al.* (2016) Disruption of the gene encoding endo-β-1, 4-xylanase affects the growth and virulence of *Sclerotinia sclerotiorum*. *Frontiers in Microbiology* 7, 1–9. DOI: 10.3389/fmicb.2016.01787.

Yu, Y., Xiao, J., Zhu, W., Yang, Y., Mei, J. *et al.* (2017) Ss-Rhs1, a secretory Rhs repeat-containing protein, is required for the virulence of *Sclerotinia sclerotiorum*. *Molecular Plant Pathology* 18(8), 1052–1061. DOI: 10.1111/mpp.12459.

Yu, Y., Du, J., Wang, Y., Zhang, M., Huang, Z. *et al.* (2019) Survival factor 1 contributes to the oxidative stress response and is required for full virulence of *Sclerotinia sclerotiorum*. *Molecular Plant Pathology* 20(7), 895–906. DOI: 10.1111/mpp.12801.

Zhang, H., Wu, Q., Cao, S., Zhao, T., Chen, L. *et al.* (2014) A novel protein elicitor (SsCut) from *Sclerotinia sclerotiorum* induces multiple defense responses in plants. *Plant Molecular Biology* 86(4–5), 495–511. DOI: 10.1007/s11103-014-0244-3.

Zhang, W., Fraiture, M., Kolb, D., Löffelhardt, B., Desaki, Y. *et al.* (2013) *Arabidopsis* receptor-like protein30 and receptor-like kinase suppressor of *BIR1-1/EVERSHED* mediate innate immunity to necrotrophic fungi. *The Plant Cell* 25(10), 4227–4241. DOI: 10.1105/tpc.113.117010.

Zhang, X., Cheng, J., Lin, Y., Fu, Y., Xie, J. *et al.* (2021) Editing homologous copies of an essential gene affords crop resistance against two cosmopolitan necrotrophic pathogens. *Plant Biotechnology Journal* 19(11), 2349–2361. DOI: 10.1111/pbi.13667.

Zhang, Y., Cheng, Y.T., Qu, N., Zhao, Q., Bi, D. *et al.* (2006) Negative regulation of defense responses in *Arabidopsis* by two *NPR1* paralogs. *The Plant Journal* 48(5), 647–656. DOI: 10.1111/j.1365-313X.2006.02903.x.

Zhang, Z.N., Wu, Q.Y., Zhang, G.Z., Zhu, Y.Y., Murphy, R.W. *et al.* (2015) Systematic analyses reveal uniqueness and origin of the CFEM domain in fungi. *Scientific Reports* 5, 13032. DOI: 10.1038/srep13032.

Zhao, X., Kim, Y., Park, G. and Xu, J.R. (2005) A mitogen-activated protein kinase cascade regulating infection-related morphogenesis in *Magnaporthe grisea*. *The Plant Cell* 17(4), 1317–1329. DOI: 10.1105/tpc.104.029116.

Zhu, W., Wei, W., Fu, Y., Cheng, J., Xie, J. *et al*. (2013) A secretory protein of necrotrophic fungus *Sclerotinia sclerotiorum* that suppresses host resistance. *PLoS ONE* 8(1), e53901. DOI: 10.1371/journal.pone.0053901.

Zhu, W., Ronen, M., Gur, Y., Minz-Dub, A., Masrati, G. *et al*. (2017) *BcXYG1*, a secreted xyloglucanase from *Botrytis cinerea*, triggers both cell death and plant immune responses. *Plant Physiology* 175(1), 438–456. DOI: 10.1104/pp.17.00375.

Zimmermann, S. and Moelling, K. (1999) Phosphorylation and regulation of Raf by Akt (protein kinase B). *Science* 286(5445), 1741–1744. DOI: 10.1126/science.286.5445.1741.

9 Biochemistry and Physiology of Host–Pathogen Interaction

Abstract

Sclerotinia produces a number of cell wall-degrading enzymes during infection to kill host cell components. Numerous genes have been discovered that control the functions and synthesis of enzymes that break down cell walls to aid in the colonization of the host. In addition to other isozymes that are critical to pathogenesis, these genes include pectinases, β-1,3-glucanases, glycosidases, cellulases, xylases, cutinases, and phosphatidase β. The endo-polygalacturonase gene (*pg1*) has been successfully cloned and characterized. The gene sequence is described in detail. By modifying the host redox environment and causing programmed cell death, detoxifying calcium, mediating pH signaling, lowering host pH, chelating Ca^{2+}, suppressing oxidative burst, modifying guard cell functions, blocking polyphenol oxidase, and influencing transcriptional regulation, the oxalic acid-producing gene performs a variety of functions. The *BnaNPR1* gene is used by *Sclerotinia* to control particular physiological processes, including the generation of reactive oxygen species and the salicylic acid defense mechanism against infection. Numerous metabolic substances are produced, including phytotoxins and siderophore pigments, which increase pathogenicity. Eighty biosynthetic gene clusters have been found in the *Sclerotinia* genome using genomic and transcriptome analyses. Many techniques, including transcriptional induction, genome mining, measurement of phytoalexins, protein expression, and enzyme purification, have been found to be effective in cloning, purifying, and characterizing genes that detoxify phytoalexins, such as brassinin glusyltransferase 1 (*SsBGT1*).

9.1 Introduction

The stem rot pathogen of crucifers induces biochemical and physiological changes through the synthesis and production of biochemical compounds to metabolize host defence compounds after infection and colonization of host tissues, releasing nutrients to feed itself for growth and ramification. Plants produce a variety of toxic compounds, both constitutive and inducible, including glucosinolates, which are essential for their defence strategies. Nevertheless, *Sclerotinia* can circumvent this component of the plant's inherent immune response by releasing oxalic acid (OA) and several enzymes. Phytoalexins produced by the host are detoxified by the enzymes produced by *Sclerotinia* to subvert host defence mechanisms.

9.2 Synthesis of Biochemical Compounds During Host–Pathogen Interaction

Sclerotinia produces several kinds of enzymes to cause infection in host tissues. These enzymes are capable of destroying cellular components of host tissues to penetrate, rectify, and develop so that the pathogen goes deeper into the host organs. This process is achieved in a sequence of events as follows.

9.2.1 Production of cell wall-degrading enzymes

The breakdown of cell walls by enzymes that are also capable of destroying cellular components

DOI: 10.1079/9781800629592.0009

and producing OA is linked to the progression of a number of diseases. According to de Bary (1887), "The ability to penetrate the membranes (cuticle), which are clearly dissolved at the point of penetration, demonstrates the power of infecting." Hence, it is very probable that this power depends on the presence of a substance which can dissolve a membrane, ferment (enzyme), in fact. Furthermore, Purdy (1958) and Boyle (1921) did not find any changes in host tissue as a result of the substances' diffusion prior to penetration; in the lack of experimental evidence demonstrating enzymatic activity on host tissue prior to penetration, the process of entry into the host should be considered mechanical, supported by histological observations. Prior and Owen (1964) also investigated the potential enzymatic functions occurring on the cuticle regarding the infection cushions present on clover leaves.

9.2.2 Production of pectolytic enzymes

The different pectolytic enzymes are produced in order to reduce the amount of pectin substances in host tissues. Several studies have demonstrated a relationship between pectolytic enzymes and *Sclerotinia*-related diseases (Held, 1955; Echandi and Walker, 1957; Hancock, 1966; Calonge *et al.*, 1969; Vanden Berg and Yang, 1969; Maxwell and Lumsden, 1970; Morrall *et al.*, 1972; Newton, 1972; Barkai-Golan, 1974; Lumsden, 1976, 1979; Sharma and Sharma, 1984; Riou *et al.*, 1991). These enzymes have been linked to modifications in the histochemical and structural characteristics of the middle lamellae in host cells (Calonge *et al.*, 1969; Lumsden, 1976) as well as a decrease in the amount of pectic substance content within infected tissues (Hancock, 1966). This is despite the paucity of qualitative and quantitative data that are currently available. Moreover, pectolytic enzymes have been identified in infected tissues; distinct pectolytic enzymes are generated in infected bean tissue, and *Sclerotinia sclerotiorum* cultures appear to produce similar enzymes (Lumsden, 1976). Polygalacturonase (PG) was discovered by Lumsden (1976) to decrease viscosity as early as 12 hour after injection into bean plants; the enzyme's activity peaks 24 hour after inoculation, coinciding with the infection's irreversible onset. Following this peak, the activity progressively declines over the course of the following 48 hour, at which point growing lesions exhibit a second peak of viscosity-reducing activity. Because it hydrolyses the pectin polymer substrate at random, the initial peak of the enzyme activity has been recognized as an endo-polygalacturonase (endo-PG) (Hancock, 1966; Lumsden and Dow, 1973; Lumsden, 1976). In the early stages of infection (up to 48 hour), endo-PG is mostly detected near the advancing edges of infected bean hypocotyls; however, its abundance decreases in later stages and it is not associated with mycelium in culture (Lumsden, 1976). Because the introduction of glucose into the growth media inhibits its synthesis, the environment has an impact on the production of PG (Lumsden and Dow, 1973; Lumsden, 1976). Since the disease's initial 24 hour incubation period is followed by a discernible decline in enzyme activity, the hydrolysis products found in affected tissues are recognized for their suppressive qualities (Lumsden, 1976). An ideal pH range for the enzyme's activity is between 4.5 and 5.5. Furthermore, compared to pectin, it has greater activity on pectate. To determine the enzyme activity, thin sections of sick tissue have been cultivated on a pectate medium; regions with considerable enzyme activity are encircled by opaque halos. Pectinase activity in earlier lesions produces a more diffuse response when tissue slices are placed on the pectate medium, in contrast to the conspicuous reaction of endo-PG. This enzyme breaks down the pectin substrate entirely into galacturonic acid and facilitates the rapid release of reducing endo-groups (Lumsden, 1976). This study's primary target is exo-polygalacturonase (exo-PG), while it may potentially target endo-polygalacturonase (endo-PG). This activity is most noticeable in the older areas of lesions and mycelium in aged cultures, as well as at the borders of new lesions that are just beginning. The enzyme activity is in line with findings from Hancock (1966) regarding various host tissues (Morrall *et al.*, 1972; Newton, 1972; Barkai-Golan, 1974) and infected sunflower plants that were tested for enzyme activity following 2–4 days of incubation. It's interesting to note that glucose does not inhibit the synthesis of this enzyme (Lumsden, 1976). According to observations, the optimal pH range for enzyme activity is between 4.5

and 5.5 (Echandi and Walker, 1957; Hancock, 1966; Lumsden, 1976). Galactosterone in plant cell walls stimulates *S. sclerotiorum* to synthesize polygalacturonases and pectinases (Fraissinet-Tachet and Fèvre, 1996; Fraissinet-Tachet *et al.*, 1996).

9.2.3 The synthesis of hemicelluloses and cellulases enzymes to affect cell wall for pathogenesis

Early on in the disease's development, pectin methylesterase (PME) was identified in bean tissue, and its presence has been linked to the development of lesions as the disease progresses (Lumsden, 1976). The fungal PME can be easily separated from the host PME due to its independent activation without the need for salt and its much lower optimal pH of 5.0 compared to the host PME's pH 8.0. Further studies (Hancock, 1966; Morrall *et al.*, 1972; Barkai-Golan, 1974) have confirmed the presence of PME in impacted tissues, and it is known to have a role in the demethylation of pectin in host plants (Hancock, 1966; Lumsden, 1976). The absence of pectin transeliminase production is a characteristic of *Sclerotinia* species (Hancock, 1966; Morrall *et al.*, 1972; Barkai-Golan, 1974; Lumsden, 1976). Furthermore, pectinolytic enzymes found outside of cells have been isolated and described (Riou *et al.*, 1992).

Sclerotinia spp. and the beginning of a number of diseases have frequently been linked to the enzymes cellulase and hemicellulose (Hancock, 1967; Calonge *et al.*, 1969; Lumsden, 1969; Newton, 1972; Barkai-Golan, 1974; Bauer *et al.*, 1977; Sharma *et al.*, 1983; Riou *et al.*, 1991). However, it is unclear exactly how these enzymes contribute to the development of various diseases. It is believed that *S. sclerotiorum* breaks down native, insoluble cellulose in a sequential fashion using the C1 enzyme, whereas the Connexin (Cx) enzyme is responsible for breaking down soluble cellulose. Moreover, the β−1,3-glucosidase enzyme facilitates the conversion of cellobiose into glucose (Lumsden, 1969). A pH of 3.0 is ideal for the activity of the Cx enzyme. If this set of enzymes is in fact in charge of breaking down natural cellulose, it implies that *S. sclerotiorum* has an extensive system that can use natural cellulose as a source of energy. Infected tissues produce a significant amount of cellulase as a kind of adaptive response (Lumsden, 1969; Barkai-Golan, 1974). Moreover, when the infection worsens, the amount of cellulase in the impacted tissue decreases noticeably (Lumsden, 1969), and changes in the cellulose structure of the infected tissue are also visible (Boyle, 1921; Calonge *et al.*, 1969). Likewise, there is notable deterioration of the araban and galactan components of infected sunflower tissues (Hancock, 1967). Galactanase's actions are not measured, despite arabanase being linked to the contaminated tissue. On the other hand, xylanase appears to be limited in its ability to significantly break down native xylan, even if it is present in infected tissues at levels that can do so. This suggests that arabans and galactans are more susceptible to enzymatic degradation than xylans. According to Bauer *et al.* (1977), partially pure galactanase efficiently breaks down carbohydrates, including the galactan portion present in potato and sycamore cell walls. Despite the release of galactose, it does not, however, soften the tissue of potato tubers. A number of additional enzymes have been investigated for their function in pathogenesis in addition to those that break down cell walls. Early in the evolution of the disease in beans, phosphatidase β may be discovered in abundance in culture and is responsible for hydrolysing the phosphatide components of cell membranes (Lumsden, 1970; Newton, 1972). This enzyme is extracellular, inducible, calcium-activated, and functions best at a pH of 4.0.

9.2.4 Production of proteolytic enzymes to degrade host protoplasm

Infected tissues, as well as *S. sclerotiorum* and *S. minor* cultures, have been shown to contain active proteolytic enzymes (Khare and Bompeix, 1976). These enzymes have the ability to degrade host protoplasm and potentially components of the cell wall. Extracts from infected carrot, cucumber, and celery show significant protease activity 2 days after inoculation, peaking 10 days later (Khare and Bompeix, 1976). The optimal pH for protease activity is found to be 3.0.

9.2.5 Characterization of genes regulating cell wall-damaging enzymes

Sclerotinia causes host tissues to produce genes during infection and pathogenesis in order to aid in host colonization by producing a variety of enzymes. Fungal pathogens on plants have the capacity to facilitate host colonization by generating a range of enzymes known as cell wall-degrading enzymes. These include cellulases, xylanases, glycosidases, pectinases, and β−1,3-glucanases (Annis and Goodwin, 1997). Numerous cell wall degrading enzymes (CWDEs) exist in different forms referred to as isozymes, which differ in processing, transcriptional regulation, molecular weight, and isoelectric point (Keon *et al.*, 1987). Because of this diversity, the pathogen may enter and colonize the host with great flexibility and can produce an assimilation-ready reservoir of nutrients. According to Alghisi and Favaron (1995), the majority of CWDEs have strictly regulated transcriptional levels in culture, which are reliant on the availability of carbon and/or nitrogen supply. The fungus uses a regulatory mechanism called carbon catabolite suppression to modify its metabolic processes when glucose or other metabolisable carbon sources are present. Moreover, transcriptionally mediated enzyme level modulation is impacted by the pH of the surrounding environment (Rollins and Dickman, 2001; Cotton *et al.*, 2003).

9.2.6 Functional characterization of genes encoding enzymes

During its contact with the host, the pathogen *S. sclerotiorum* secretes a wide spectrum of cell wall-degrading enzymes (CWDEs) that help with penetration, tissue breakdown, and the breakdown of plant cell wall elements (Table 9.1; Hancock, 1966; Lumsden, 1969; Riou *et al.*, 1991). Among these, pectinases preferentially attack and break down pectin, an essential component of plant cell walls. Pectin's hydrolysis provides a carbon supply that is necessary for the fungus to thrive while also weakening the cell wall and facilitating the fungus's increased penetration and colonization. *S. sclerotiorum* synthesizes a number of pectinolytic enzymes that are essential to the organism's pathogenicity, as evidenced by their capacity to independently cause plant cell death and macerate plant tissues (Alghisi and Favaron, 1995). The structural polysaccharides known as unesterified pectate polymers, which are essential to the middle lamella and primary cell wall of higher plants, are broken down by PGs, which are pectinases. The presence of simple sugars inhibits the action of PG in *S. sclerotiorum*, whereas pectin or pectin monomers like galacturonic acid promote this activity (Riou *et al.*, 1992; Fraissinet-Tachet and Fevre, 1996). PGs may function as virulence factors in specific pathosystems, according to targeted gene disruption studies (Shieh *et al.*, 1997; Ten Have *et al.*, 1998; Wagner *et al.*, 2000; García-Maceira *et al.*, 2001; Kars *et al.*, 2005). But as of right now, there isn't enough solid evidence to back up the idea that PGs play a general function in virulence (Scott-Craig *et al.*, 1990; Gao *et al.*, 1996). Enzymes known as endo-PGs aid in the degradation of homogalacturonan within the cellular environment. Conversely, exo-PGs are responsible for the removal of glycosyl groups from pectic polysaccharides within the cell wall, leading to the breakdown and subsequent release of nutrients from the substrate (Kars and van Kan, 2004). Numerous endo- and exo-PGs in *S. sclerotiorum* have been found and studied (Table 9.1). Studies on molecular genetics have shown that fungi can produce several types (isozymes) of endo-PGs by modifications that occurs either posttranslation (glycosylation) or postsecretion (proteolysis). Alternatively, these forms might come from a multigene family, as suggested by Bussink *et al.* (1992), Caprari *et al.* (1993), Fraissinet-Tachet and Fevre (1996), and Fraissinet-Tachet *et al.* (1996). Five endo-PGs from the similarly related necrotrophic fungus *Botrytis cinerea* were identified to have distinct metabolic features and necrotizing effects on several host plants (Kars *et al.*, 2005). Similar to this, *S. sclerotiorum* generates a variety of molecular pectinases with comparable enzymatic characteristics (Riou *et al.*, 1991).

Several pure acidic endo-PGs derived from *S. sclerotiorum* show changes in their isoelectric values, despite maintaining a consistent molecular weight. This result

Table 9.1. Genes responsible for the production of CWDEs in *Sclerotinia sclerotiorum*. (Adapted from publication of Bolton *et al*., 2006, with permission.)

Gene or protein designation	Accession no.	Reference
Acidic endo-PGs		
PG2	S13661	Waksman *et al*. (1991)
PG3	B60155	Waksman *et al*. (1991)
sspg3	AY312510	Li *et al*. (2004)
pg5	Y13669	Kasza *et al*. (2004)
sspg5	AY496277	Li *et al*. (2004)
pg6	AJ539086	Kasza *et al*. (2004)
pg7	AJ539087	Cotton *et al*. (2003)
PGa	CAF05669	Favaron *et al*. (2004)
Neutral or basic endo-PGs		
pg1	L12023	Reymond *et al*. (1994)
Pg2	L29040	Fraissinet-Tachet and Fevre (1996)
Pg3	L29041	Fraissinet-Tachet *et al*. (1996)
sspg1d	AF501307	Li *et al*. (2004)
Sspg6	AF501307	Li *et al*. (2004)
PGb	CAF05670	Favaron *et al*. (2004)
Exo-PGs		
ssxpg1(partial cDNA)	AY312511	Li *et al*. (2004)
Ssxpg2(partial cDNA)	AY312512	Li *et al*. (2004)
Nonaspartyl acid protease		
acp1	AF221843	Poussereau *et al*. (2001)
Nonaspartyl acid protease		
acpS	AF221387	Poussereau *et al*. (2001)

implies that differential glycosylation, which changes molecular mass without changing the overall charge of proteins, may have some effect on the number of PGs in *S. sclerotiorum* (Waksman *et al*., 1991). A seven-membered multigene family organized into two subfamilies controls the diversity of pectinolytic enzymes and PG isozymes in *S. sclerotiorum*, according to research by Echandi and Walker (1957); Fraissinet-Tachet and Fevre (1996) and Fraissinet-Tachet *et al*. (1996). It is conceivable that the presence of many copies of functionally related genes provides flexibility and adaptability to a disease with a wide host range, even if the benefits of having multiple copies of the pectinase gene have not been empirically validated for this fungus. Different expression levels during the pathogenesis process have been discovered by further research on the endo-PGs of *S. sclerotiorum* (Cotton *et al*., 2002; Kasza *et al*., 2004; Li *et al*., 2004). The expression levels of endo-PGs pg1, pg2, and pg3 were investigated by Kasza *et al*. (2004) in their study. The expression levels peaked 36 hours after inoculation (hpi), during the colonization phase of healthy plant tissues. At 96 hpi, these endo-PGs were not discernible, nevertheless. On the other hand, pg6 and pg7 were detected by RT-PCR analysis starting at 24 hpi and continuing until the experiment's conclusion at 96 hpi. Interestingly, pg5 transcripts were only found between 48 and 72 hpi, which corresponds to the maceration's latter stage. According to the study's findings, a number of factors may be reduced that either encourage or inhibit the transcription of particular endo-PG genes. These factors include polygalacturonic acid, catabolite repression brought on by the build-up of end products, and an acidic pH of the surrounding environment (Kasza *et al.*, 2004).

Plants possess the capability to synthesize glycoproteins linked to the cell wall, which are recognized for their role in inhibiting fungal endo-PGs, referred to as polygalacturonase-inhibiting proteins (PGIPs) (De Lorenzo *et al.*, 2001). Research has shown that PGIP derived from soybean displays varying degrees of inhibitory effectiveness against the endo-PGs of *S. sclerotiorum* (Favaron *et al.*, 1994, 2004). Plant cell walls are the site of endo-PGs' enzymatic activity, which releases oligogalacturonides. According to Davis *et al.* (1986), these substances are endogenous elicitors of the hypersensitive reaction. Furthermore, it has been shown that the *Phaseolus vulgaris* PGIP may efficiently block the HR caused by the endo-PGs of *S. sclerotiorum* (Zuppini *et al.*, 2005).

One of the first defence mechanisms in plants is the oxidative burst, which produces ROS (reactive oxygen species) such as H_2O_2 and O_2. According to Wojtaszek (1997), this procedure is connected to the hypersensitive response (HR), which attempts to lessen pathogen invasion. Although it is generally believed that the HR inhibits the colonization of biotrophic pathogens, the virulence of necrotrophic fungi such as *S. sclerotiorum* and *Botrytis* species that derive their nutrients from dead host tissues may actually be enhanced by the HR (Govrin and Levine, 2000; Thomma *et al.*, 2001). Research indicates that plants unable to initiate the HR have a higher resistance to *S. sclerotiorum* in both tobacco and *Arabidopsis thaliana* (Govrin and Levine, 2000; Dickman *et al.*, 2001).

Data in Table 9.1 show that pectinolytic CWDEs have been the main focus of study on *S. sclerotiorum*. However, this fungus has also been investigated for producing other enzymes, such as glucoamylases, proteases, and cellulases (Lumsden, 1969; Riou *et al.*, 1991; Poussereau *et al.*, 2001; Martel *et al.*, 2002). Since proteins make up about 10% of the plant cell wall (Carpita and Gibeaut, 1993), proteases could be essential to *S. sclerotiorum's* pathogenicity.

9.2.7 Cloning and sequencing analysis of *Sclerotinia* gene producing PG

Many extracellular pectin-degrading enzymes are produced by *S. sclerotiorum*. The whole sequence of a cloned gene (*pg1*) encoding endo-polygalacturonase (PG1) has been determined. The coding area consisted of a single 1143 bp open reading frame. The *S. sclerotiorum pg1* gene was compared to other PG-encoding genes in fungi. It was discovered that the five noncoding regions had essential transcription regulatory sequences. Seven fungal PG sequences showed a significant degree of similarity (41.5–59.8%) when compared to the deduced amino acid (aa) sequence (380 aa) of the enzyme. A comparison of the predicted secondary structures showed a comparable protein arrangement, most likely in antiparallel beta sheets. By means of hybridization analysis, a *pg1* 0.65 kb Bam HI fragment was employed as a probe to identify seven different recombinant phages from a genomic library. A study of the hybridizing restriction segments suggests that the PG-encoding genes are organized as a family (Reymond *et al.*, 1994).

9.2.8 Differential expression patterns of enzymatic genes for virulence by *Sclerotinia*

Many broad host-range infections share the function of hydrolytic CWDEs, which have been the focus of much research into the molecular mechanisms behind *S. sclerotiorum's* pathogenicity. It is known that *S. sclerotiorum* synthesizes a range of pectinases, including exo- (Ssxpg1 and Ssxpg2) and endo- (Sspg1, Sspg3, Sspg5, and Sspg6) PGs. Moreover, the silencing of genes encoding *endo-β–1,4-xylanase (SsXyl1)* and arabinofuranosidase/β-xylosidase precursor (Ssaxp) in mutant strains resulted in reduced or eliminated pathogenicity, highlighting their significance as virulence factors (Yang *et al.*, 2016). To date, OA has been the most extensively studied virulence factor of *S. sclerotiorum*. OA contributes to *S. sclerotiorum's* virulence in a number of ways, including by changing the redox environment of the host, causing programmed cell death, detoxifying calcium, and promoting pH signalling. Although OA has long been thought to be crucial to *S. sclerotiorum's* pathogenicity, studies using targeted mutations of the oxaloacetate acetylhydrolase gene (*Ssoah1*) have shown that, while OA is necessary for virulence, it is not necessary for pathogenicity in all host species. Some have proposed that the basic prerequisite for the colonization phase of pathogenesis is not the synthesis of OA alone, but rather the formation of an acidic pH environment during the host's colonization (Xu *et al.*, 2015).

9.2.8.1 *Expression of effector genes for pathogenesis*

Many investigations have demonstrated the vital role secreted effector genes play in this particular pathogen's pathogenicity and virulence. Two essential genes are required for *S. sclerotiorum* to be fully virulent: Ssv263, which codes for a tiny putative secreted protein, and SsCVNH, which codes for a short cysteine-rich secreted protein with a cyanovirin-N-homology domain. Furthermore, the presence of Ss-Rhs1, which encodes a repeat-containing protein known as Rhs, and *SsCaf1*, a gene linked to appressorium formation, is necessary for the creation of compound appressoria and the first hyphal invasion of the host. Furthermore, it is known that secreted effectors such as the integrin-like protein SsITL and the chorismate mutase SsCM1 can reduce host resistance by interfering with the salicylic acid (SA) and jasmonic acid/ethylene signalling pathways, respectively. On the other hand, it is known that certain secreted effectors cause host cell death and necrosis. These include the peptides SsNep1 and SsNep2, which cause necrosis and ethylene, as well as SsSSVP1, a tiny secreted protein connected to virulence, and SsCP1, a cerato-platanin (Yang *et al.*, 2018). A study was conducted to examine the global changes in gene expression that *S. sclerotiorum* causes in canola plants with varying degrees of disease susceptibility (Chittem *et al.*, 2020). The researchers have presented a two-phase infection model that differs from the widely held belief that the pathogen functions as a necrotroph. According to this hypothesis, before entering a necrotrophic phase, the pathogen first goes through a brief biotrophic phase that is characterized by the suppression of host resistance and the manipulation of host defences. Genomic, molecular, and cytological findings support this theory. In the first stage, *S. sclerotiorum* uses OA and secreted effectors to block host defence signalling pathways and avoid being detected by the host. To further our comprehension of the molecular pathways underlying the interaction between *S. sclerotiorum* and canola, a transcriptome sequencing study was carried out. It produced important new information about the temporal dynamics of the vital routes and processes that *S. sclerotiorum* uses to successfully infect canola plants.

9.2.8.2 *Differential expression of genes on R and S genotypes of the host*

S. sclerotiorum gene expression profiles were examined in connection with the development of the illness in two canola cultivars that exhibited varying degrees of susceptibility to the pathogen. A total of roughly 164 million Illumina reads were produced from mRNA libraries collected from infected petioles and mycelium cultivated in liquid medium. There were 69 million paired-end reads of 50 base pairs and 95 million single reads of 75 base pairs in this sample. Thirty-six percent of the reads that cleared quality filtering were successfully aligned to the *S. sclerotiorum* reference genome. The *S. sclerotiorum* genes *1301*, *1214*, and *816* showed differential expression in the susceptible canola line during the early (8–16 hour after inoculation) and late (24–48 hour after inoculation) stages of infection, respectively. On the other hand, *1311* and *1335* genes showed differentiating expression in the resistant line throughout these relevant phases. The gene sets that showed significant enrichment in Gene Ontology (GO) categories associated with multiple functions, such as cell wall degradation, host metabolite detoxification, and peroxisome-related processes like fatty acid β-oxidation, the glyoxylate cycle, and oxido-reductase activity, were upregulated in both the susceptible and resistant lines. Six differentially expressed genes (DEGs) were chosen, and quantitative RT-PCR was done on each to further validate the RNA-seq differential gene expression study. The findings also provide credence to a two-phase infection paradigm in which effector genes involved in evading or reducing host defence mechanisms are expressed during the late stages of infection, while those related to host cell death are controlled during the early stages of infection. In addition to cell wall disintegration and host metabolite detoxification, peroxisome-related pathways are the main processes underpinning the pathogenesis of *S. sclerotiorum* in *B. napus* (Chittem *et al.*, 2020).

9.2.9 The *BnaNPR1* gene family's role to regulate pathogenesis of *Sclerotinia*

A gene family corresponding to *BnaNPR1* has been discovered in *B. napus*. This indicates that *BnaNPR1* downregulation causes changes in the expression of genes linked to SA and jasmonic acid/ethylene (JA/ET) signalling pathways in response to pathogens, as well as decreased resistance to *S. sclerotiorum* and an increase in ROS accumulation. Plant defence systems are regulated by the *NPR1* gene and other *NPR1*-like genes, which do not activate genes involved to disease. The *B. napus* genome had 19 *BnaNPR1*-like genes in total. These genes were named by their closest *Arabidopsis thaliana* homologs, which made it easier to find the equivalent *B. napus* genes for each *AtNPR1*-like gene. The examination of essential protein domains and functional motifs demonstrated that these homologs exhibit both conservation and divergence. Six separate sub-clades were found by a phylogenetic examination of the BnaNPR1-like proteins and their *Arabidopsis* counterparts, confirming that the naming approach correctly reflects their evolutionary connections. Moreover, transcriptome investigation of *B. napus* revealed that the main pathogen affecting this crop, *S. sclerotiorum*, significantly reduced the expression of three *BnaNPR1* genes after infection. However, for *BnaNPR2*, *BnaNPR3*, *BnaNPR4*, *BnaNPR5*, and *BnaNPR6* in general, no substantial expression changes were found. Wang *et al.* (2020) created *B. napus* BnaNPR1-RNAi lines to investigate the impact of downregulated BnaNPR1 expression on resistance to *S. sclerotiorum* in more detail. The plant's resistance to this disease was found to be severely weakened by the lowered expression of BnaNPR1, according to the results. Furthermore, it was discovered that BnaNPR1-RNAi altered the defence responses during the contact between the pathogen and the plants and increased the formation of ROS. These results suggest that BnaNPR1 may be used by *S. sclerotiorum* to manipulate the production of ROS and the SA-mediated defensive response in *B. napus*, therefore facilitating infection (Wang *et al.*, 2020).

9.2.10 Metabolites associated to *Sclerotinia*-mediated sclerotia formation

Along with significant cellular alterations, the development of sclerotia involves the movement and accumulation of numerous chemicals. The initial phase's contributing components have not been fully examined. Vegetative hyphae typically develop in opposite directions, according to Willetts (1978). In order to facilitate hyphal fusion and the creation of early structures, an attraction is needed. Since the earliest structures usually appear after the substrate has been fully colonized, their formation could be a reaction to variations in the availability of nutrients. The finding that some fungi, such as *S. sclerotiorum*, generate sclerotia when their mycelial mats are placed in nutrient-poor media lends credence to this notion (Christias and Lockwood, 1973). The earliest structures form only once certain metabolic products are produced in the surrounding media or the mycelium. It has been shown that some *Sclerotinia* species produce organic acids, which may have a role in the development of sclerotia (Wang and Le Tourneau, 1971; Humpherson-Jones and Cooke, 1977). *S. sclerotiorum* produces a variety of organic acids in varying amounts and types depending on the isolate and medium. These acids typically include succinic, fumaric, malic, and OA (Corsini and Le Tourneau, 1973; Humpherson-Jones and Cooke, 1977). Moreover, additional acids have been identified, including citric, glycolic, glyoxylic, and α-ketoglutaric (Colotelo, 1973; Humpherson-Jones and Cooke, 1977). The enzymes required for the formation of OA and Krebs cycle acids are present in *S. sclerotiorum* (Corsini and Le Tourneau, 1973; Maxwell, 1973).

9.2.10.1 *Sclerotia development under in vitro conditions*

The cultivation of *S. sclerotiorum* in a liquid medium results in an initial reduced pH level of culture filtrate between pH 3 and 4, followed by a slight increase (Wang and Le Tourneau, 1971). This reduction in pH aligns with the peak production of acids and the heightened specific activity of enzymes associated with the Krebs cycle. When the fungus is grown in conditions that favour optimal growth, a significant number of sclerotia are produced, which corresponds with the observed decrease in pH. In contrast, when the fungus is cultivated in environments that limit growth or sclerotium formation, the pH of the culture filtrate tends to rise (Wang and Le Tourneau, 1971). The introduction of oxalic and Krebs cycle acids into the medium does not enhance sclerotium formation (Wang and Le Tourneau, 1971; Humpherson-Jones and Cooke, 1977). After sclerotia first develop, resources must be made available for the synthesis of reserves and other components that are suggestive of mature sclerotia. Droplets around sclerotia in development are attached to the inside of the sclerotium (Colotelo, 1974). Numerous organic and inorganic substances, including numerous enzymes, are packed inside these membrane-encased droplets; the concentrations of these substances change as the sclerotium grows (Colotelo, 1973). The exudate's precise role is still unknown; however, it might facilitate the sclerotium's water release during maturation or serve as an osmotic agent during translocation (Willetts, 1978).

9.2.10.2 *Production of enzymes during sclerotial morphogenesis*

A variety of enzymes have been discovered in sclerotia; however, the mechanisms governing their regulation and coordination, as well as the specific enzymes essential for morphogenesis, remain ambiguous. Based on results from polyacrylamide gel electrophoresis, Wong and Willetts (1974) proposed that the compaction and maturation of sclerotia are influenced by the suppression of glycolysis and the Krebs cycle, as well as the stimulation of the pentose phosphate pathway. Research has identified a large number of enzymes associated with phenol oxidation (Jones, 1970; Colotelo, 1973; Wong and Willetts, 1974). Tyrosinase may be the enzyme that initiates the formation of sclerotia, and other phenol oxidases may be involved in the development of the darker rind, according to Wong and Willetts (1974).

9.2.10.3 *Regulation of metabolites at transcription level*

To aid in their colonization of host species, plant pathogenic fungi have the ability to create a wide variety of CWDEs, including pectinases, β−1,3-glucanases, glycosidases, cellulases, xylanases, and cutinases (Annis and Goodwin, 1997). According to Keon *et al.* (1987), there are numerous unique isozymes of CWDEs that vary in processing, transcriptional regulation, molecular weight, and isoelectric point. The pathogen's variety allows it to invade and colonize hosts with remarkable adaptability, producing a nutritional store that is ready for absorption. Depending on the availability of supplies of carbon and/or nitrogen, the majority of CWDEs in culture have well-regulated transcriptional expression (Alghisi and Favaron, 1995). The pathogen can alter its metabolic pathways in response to the presence of glucose or other usable carbon sources, thanks to this mechanism of carbon catabolite suppression. Furthermore, transcriptional enzyme levels can be impacted by the pH of the surrounding environment (Rollins and Dickman, 2001; Cotton *et al.*, 2003).

9.2.11 Metabolites secreted by *Sclerotinia* under axenic condition

The metabolites released by the pathogenic strains of *S. sclerotiorum* under axenic culture have been identified using untargeted high-resolution UPLC-QTOF-ESI-MS/MS. The analysis of the ethyl acetate extracts of the *S. sclerotiorum* culture revealed ten major secondary metabolites, namely, sclerin, sclerotinin-B, sclerone, melanin, bostrycoidin, botcinin-D, botcinin-A, gliovirin, scleramide, and botcinic acid. In the phylogenetic analysis based on the overlapping and unique informative peaks in the chromatograms, the six

S. sclerotiorum strains were grouped into three major clades. The clustering based on metabolic profiles does not substantiate the diversity based on morphology or virulence differences on the host. The findings of the study signified that the metabolites secreted under axenic conditions vary based on their growth and developmental stages and may not necessarily be the determining factors for their differential aggressiveness and virulence over the host (Gupta *et al.*, 2022).

9.2.12 Biosynthesis of secondary metabolites and their functional analysis within the *Sclerotinia* genome

Secondary metabolites are produced by pathogenic fungi for a variety of purposes, such as signalling, defence mechanisms, nutrient acquisition, and the compromise of host cell integrity and activity. It has been established that some secondary metabolites, including phytotoxins, siderophores, and pigments, increase the pathogenicity of plant-dwelling fungi (Chooi and Solomon, 2014; Scharf *et al.*, 2014). In the context of gene expression, pigments and siderophores are categorized as secondary metabolites, despite the possibility that they function as primary metabolites because of their vital roles in survival. Usually, these genes are strictly associated with the manufacture of secondary metabolites. In particular, fungi generate microscopic compounds known as siderophores, which are capable of absorbing iron from the surrounding milieu and facilitating its internal absorption. Iron is necessary for fungi to respire, run the tricarboxylic acid cycle, and produce amino acids, lipids, sterols, and deoxyribonucleotides, among other metabolic processes (Philpott, 2006). Because iron is highly reactive in its reduced form and has low solubility in anaerobic, nonacidic environments, it is difficult to absorb and needs to be carefully managed within cells. For this reason, extracellular and intracellular siderophores are critical to the pathogenicity of different types of plant-disease-causing fungi. Furthermore, the pigment melanin is essential for strengthening cell walls and shielding cells from outside stresses like UV rays. Melanin plays a major role in the virulence of many plant pathogenic fungi by helping to generate sufficient turgor pressure in appressoria, which facilitates the invasion of host tissues (Collemare *et al.*, 2008).

9.2.12.1 Expression and function of secondary metabolites genes

The genetic material that governs the synthesis of secondary metabolites in fungi is often organized in a consolidated manner within the genome, allowing for simultaneous regulation (Keller, 1997; Keller and Hohn, 1997). Biosynthetic gene clusters (BGCs) typically encompass one or more essential "backbone" enzymes, such as polyketide synthases (PKSs), nonribosomal peptide synthases (NRPSs), hybrid PKS/NRPSs, terpene synthases, or dimethylallyl tryptophan synthases (DMATS). "Decorating" enzymes that alter the backbone chemical via oxidation, reduction, methylation, or glycosylation are also included in these clusters. According to Fedorova *et al.* (2012), additional genes within a cluster may encode transcriptional regulators unique to the route, transporters that aid in the export of the finished product from the cell, or enzymes involved in precursor biosynthesis. It is now possible to find gene clusters in fungal genomes by using bioinformatics techniques that search for protein domains related to key enzymes and auxiliary genes. Notable tools include FunGeneClusterS (Vesth *et al.*, 2016) for evaluating gene co-expression, MultiGeneBlast (Medema *et al.*, 2013) for comparative genomic research, and antiSMASH (Weber *et al.*, 2015) and SMURF (secondary metabolite unique region finder) (Khaldi *et al.*, 2010) for homology searches. The significance of BCGs (biosynthetic gene clusters) extends beyond their role in virulence; they also play a crucial part in evolutionary processes. BGCs are frequently located near the ends of chromosomes within sub-telomeric regions rich in transposable elements (TEs) (McDonagh *et al.*, 2008; Palmer and Keller, 2010; Cairns and Meyer, 2017), which are characterized by elevated rates of recombination and mutation relative to other genomic areas (Brown *et al.*, 2010; Palmer and Keller, 2010). This close proximity of BGCs to TE-rich sub-telomeric regions is thought to be influenced by the selective pressure for enhanced adaptability of fungal metabolite production in response to fluctuating environmental conditions (Raffaele and Kamoun, 2012).

9.2.12.2 *Identification of secondary metabolites clusters in Sclerotinia genome*

The secondary metabolites that have been identified from *S. sclerotiorum* to date include β-carotene, dihydroxynaphthalene (DHN) melanin, and six aromatic phytotoxins extracted from liquid culture (Pedras and Ahiahonu, 2004), although their specific functions in the infection process remain unclear. While certain genes associated with the synthesis of β-carotene and DHN (dihydroxynaphthalene) melanin in *S. sclerotiorum* have been characterized (Butler *et al.*, 2009; Amselem *et al.*, 2011), the BGCs for these and other metabolites have yet to be fully defined. Using already available genomic and transcriptome data, Graham-Taylor *et al.* (2020) predicted and characterized BGCs within the *S. sclerotiorum* genome, resulting in the identification of 80 possible BGCs. The majority of the genes discovered in these clusters were situated in sub-telomeric areas. These sub-telomeric clusters were discovered to be richer in paralogous genes and showed a strong correlation with genomic sequences rich in repetitive elements, indicating that BGCs had evolved through processes of duplication and neo-functionalization in sites of recombination. Furthermore, it was demonstrated that genes contained in BGCs were more likely to exhibit presence/absence polymorphisms and had higher average sequence diversity than non-BGC genes. It is noteworthy that during the advanced phases of infection, there was a notable increase in the quantity of secondary metabolites produced in planta as opposed to *in vitro*. This suggests that secondary metabolite production is crucial for the necrotrophic growth of *S. sclerotiorum*.

9.2.12.3 *Production of secondary metabolites clusters in Sclerotinia genome*

Among the 80 identified putative clusters of secondary metabolites, over half are characterized by the presence of at least three genes that are co-regulated at the transcriptional level. Comparative genomic analyses have revealed that these clusters share similarities with those found in *Botrytis cinerea*, a closely related plant pathogen. The manufacture of botcinic acid, hydroxamate siderophores, DHN melanin, and carotenoids is associated with these clusters. Potential phytotoxin clusters that can produce the polyketide sclerin and an epipolythiodioxopiperazine were also found by Graham-Taylor *et al.* (2020). Secondary metabolite clusters have been observed to be primarily found in sub-telomeric genomic areas; those that contain paralogues exhibit an especially significant association with repetitive sequences. Intraspecific comparisons, which show that potential secondary metabolite genes have a higher frequency of presence/absence polymorphisms and much larger sequence diversity than other genes, have been used to support this positional bias. *S. sclerotiorum* produces a wide variety of secondary metabolites during plant infection, and it has been noted that the gene clusters in charge of these compounds undergo increased rates of duplication, recombination, and mutation in sub-telomeric areas. It is still unclear exactly which microevolutionary processes underlie the diversity of secondary metabolites in *S. sclerotiorum*. However, the discovery of a number of putative phytotoxins establishes the framework for upcoming functional studies (Table 9.2; Fig. 9.1).

9.2.13 Detoxification of plant glucosinolate hydrolysis products through isothiocyanate hydrolase by *Sclerotinia* pathogenesis

The *Brassica* plants produce glucosinolates and myrosinases that generate toxic isothiocyanates, conferring broad resistance against pathogens and herbivorous insects. Nevertheless, some cosmopolitan fungal pathogens, such as the necrotrophic white mold *S. sclerotiorum*, are able to infect many plant hosts, including glucosinolate producers. Chen *et al.* (2020) showed that *S. sclerotiorum* infection activates the glucosinolate–myrosinase system, and isothiocyanates contribute to resistance against this fungus. *S. sclerotiorum* metabolizes isothiocyanates via two independent pathways: conjugation to glutathione and, more effectively, hydrolysis to amines. The latter pathway features an isothiocyanate hydrolase that is homologous to a previously

Table 9.2. The genome analysis of *Sclerotinia sclerotiorum* has revealed the presence of 80 predicted clusters of secondary metabolites. (Adapted from publication of Graham-Taylor *et al*., 2020, with permission.)

Name	Backbone gene type	Number of genes	Upregulated *in Planta*	Co-expressed	Probable produce
01_1	Isoprenoid	25	Yes	Yes	-
01_2	NRPS-like protein, PKS/NRPS-like protein	14	Yes	No	-
01_3	PKS/NRPS-like protein	22	Yes	Yes	-
01_4	Type I PKS (HR), NRPS-like protein	34	Yes	Yes	-
01_5	PKS/NRPS-like protein	19	Yes	No	-
01_6	Type I PKS (HR), Type I PKS (HR), PKS/NRPS-like Protein	29	Yes	Yes	-
01_7	Hybrid PKS-NRPS, NRPS	12	No	No	-
01_8	NRPS-like protein, PKS/NRPS-like protein	15	Yes	Yes	-
02_1	PKS/NRPS-like protein	18	Yes	No	-
02_2	PKS/NRPS-like protein	20	Yes	Yes	-
02_3	PKS/NRPS-like protein	59	Yes	Yes	Carotenoid
02_4	PKS/NRPS-like protein, NRPS	25	No	No	Coprogen/fusarine (extracellular siderophore)
02_5	PKS/NRPS-like protein, Hybrid PKS-NRPS	17	Yes	No	-
02_6	NRPS-like protein	19	Yes	No	-
03_1	PKS/NRPS-like protein	9	No	Yes	-
03_2	Isoprenoid	82	No	Yes	-
03_3	NRPS-like protein, PKS/NRPS-like protein	22	Yes	Yes	-
03_4	PKS/NRPS-like protein	34	No	No	-
03_5	PKS/NRPS-like protein	9	No	No	-
03_6	Isoprenoid	17	No	No	-
03_7	Type I PKS (NR), PKS/NRPS-like Protein	13	No	No	Dihydroxynaphthalene (DHN) melanin in appressoria
04_1	PKS/NRPS-like protein	7	No	No	-
04_2	Type I PKS (HR)	19	No	No	-
04_3	Type III PKS, Type I PKS (HR)	28	Yes	Yes	-
04_4	PKS/NRPS-like protein, Isoprenoid	17	Yes	Yes	-

Continued

Table 9.2. Continued

Name	Backbone gene type	Number of genes	Upregulated *in Planta*	Co-expressed	Probable produce
04_5	PKS/NRPS-like protein, NRPS-like protein	65	Yes	No	-
04_6	Isoprenoid	31	No	Yes	-
05_1	NRPS-like protein, PKS/NRPS-like protein	18	No	Yes	-
05_2	PKS/NRPS-like protein	13	No	Yes	Botcinic acid
05_3	NRPS	27	No	Yes	-
05_4	2 x PKS/NRPS-like protein	30	Yes	No	-
05_5	Isoprenoid	36	No	Yes	-
05_6	4 x PKS/NRPS-like protein	43	Yes	Yes	-
05_7	Type I PKS (NR)	15	Yes	Yes	Aromatic polyketide/ Sclerotinin
05_8	PKS/NRPS-like protein, Type I PKS (NR)	23	Yes	Yes	Aromatic polyketide/ Sclerotinin
06_1	PKS/NRPS-like protein	14	Yes	Yes	-
06_2	NRPS-like protein	32	No	Yes	-
06_3	2 x PKS/NRPS-like protein	33	Yes	Yes	-
06_4	Type I PKS (HR)	21	No	Yes	-
06_5	Other	40	No	Yes	-
06_6	Other	39	No	Yes	-
07_1	Type I PKS (HR)	29	No	No	-
07_2	NRPS-like protein, PKS/NRPS-like protein	28	Yes	Yes	-
07_3	PKS/NRPS-like protein	12	No	No	-
07_4	NRPS-like protein	24	Yes	Yes	-
07_5	Type I PKS (HR), PKS/ NRPS-like protein	24	Yes	Yes	-
07_6	PKS/NRPS-like protein	44	Yes	Yes	-
08_1	NRPS-like protein	37	No	Yes	-
08_2	PKS/NRPS-like protein	32	Yes	No	-
08_3	PKS/NRPS-like protein	25	No	No	-
09_1	Other	14	No	Yes	-
09_2	PKS/NRPS-like protein	21	Yes	No	-
09_3	PKS/NRPS-like protein, NRPS-like protein	17	Yes	No	-
09_4	PKS/NRPS-like protein	30	Yes	Yes	-

Continued

Table 9.2. Continued

Name	Backbone gene type	Number of genes	Upregulated *in Planta*	Co-expressed	Probable produce
09_5	NRPS	18	Yes	No	Ferricrocin (intracellular siderophore)
10_1	PKS/NRPS-like protein	12	Yes	No	-
10_2	PKS/NRPS-like protein	25	Yes	Yes	-
10_3	PKS/NRPS-like protein	26	No	Yes	-
10_4	PKS/NRPS-like protein	22	Yes	No	-
10_5	NRPS-like protein, NRPS	26	Yes	Yes	Epipolythiodioxopiperazine
10_6	PKS/NRPS-like protein	27	No	Yes	-
11_1	PKS/NRPS-like protein	16	No	No	-
11_2	NRPS-like protein, PKS/NRPS-like protein	38	No	Yes	-
11_3	Other	40	No	No	-
11_4	NRPS-like protein	71	No	Yes	-
12_1	Type I PKS (NR)	20	Yes	No	Sclerotial melanin
13_1	PKS/NRPS-like protein	20	Yes	Yes	-
13_2	Type I PKS (HR)	14	No	No	-
14_1	Hybrid PKS-NRPS(NR)	19	No	Yes	-
14_2	PKS/NRPS-like protein	14	No	Yes	-
15_1	PKS/NRPS-like protein	30	Yes	No	-
15_2	PKS/NRPS-like protein, Type I PKS (HR), NRPS-like protein	29	Yes	No	-
15_3	PKS/NRPS-like protein, Type I PKS (PR), Type I PKS (HR)	24	Yes	Yes	Botcinic acid
15_4	PKS/NRPS-like protein	53	No	Yes	-
16_1	NRPS-like protein	18	Yes	Yes	-
16_2	Isoprenoid	38	No	Yes	-
16_3	PKS/NRPS-like protein	15	No	No	-
16_4	DMAT	5	No	No	-
16_5	Other	20	No	No	-
16_6	PKS/NRPS-like protein	21	No	No	-

characterized bacterial enzyme and converts isothiocyanate into products that are not toxic to the fungus. The isothiocyanate hydrolase promotes fungal growth in the presence of the toxins and contributes to the virulence of *S. sclerotiorum* on glucosinolate-producing plants.

9.2.14 Expression of genes for metabolic process of compounds involved in host defence during pathogenesis

Plants have evolved a variety of strategies to defend themselves against pathogens and herbivores. A prominent strategy involves the

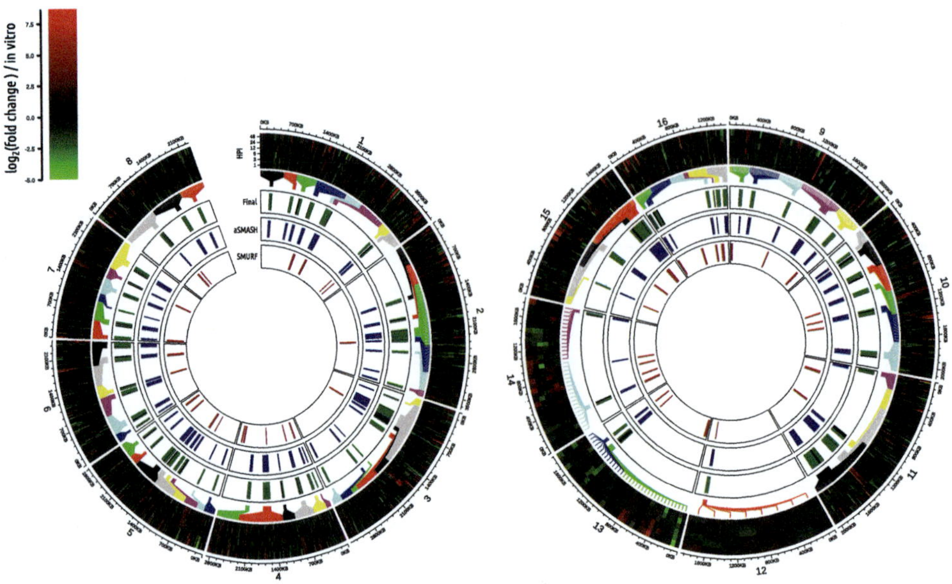

Fig. 9.1. The goal of the *Sclerotinia sclerotiorum* genome research was to pinpoint the 80 clusters in charge of secondary metabolite production. Chromosomes 1 through 8 are shown on the left side of the circular representation, while chromosomes 9 through 16 are shown on the right. The margins of the depiction show the genomic coordinates in kilobases (kb). The outermost track displays expression data from time-course research carried out by Seifbarghi *et al.* (2017) using detached *B. napus* leaves that were collected at 1, 3, 6, 12, 24, and 48 hour postinoculation (hpi). The log (fold change) expression data is visually displayed using a color gradient that goes from green, which represents low expression, to black, which indicates no change, to red, which indicates high expression. The following track, "Final," displays the genomic locations of the 80 BGCs for secondary metabolites that have been projected to be present in the genome. The lines connecting the genes in these clusters on the heat map demonstrate the evolution of these genes throughout time, with black lines denoting high co-expression and green lines denoting low co-expression. The locations of SMURF predictions for secondary metabolite BGCs are shown by the "SMURF" track, which is displayed in dark red, whereas the "a SMASH" track, which is displayed in blue, shows the locations of antiSMASH predictions for secondary metabolite clusters. By manually integrating and curating these two prediction outputs, the final gene clusters in the second track were created. (Adapted from publication of Graham-Taylor *et al.*, 2020, with permission.)

synthesis of toxic compounds, which can be either constitutive or inducible, functioning as a defensive barrier (Tiku, 2020). Glucosinolates, a significant category of these compounds, are essential for plant defence mechanisms against herbivores and pathogens alike (Rask *et al.*, 2000; Buxdorf *et al.*, 2013). These compounds are inherently produced by plants belonging to the *Brassicas*. Upon tissue damage, enzymes known as myrosinases activate glucosinolates, resulting in the formation of toxic isothiocyanates and nitriles (Rask *et al.*, 2000). Nevertheless, pathogens that target *Brassicas* plants must devise methods to circumvent this inherent immune response. For example, *S. sclerotiorum* employs the conserved enzyme SsSaxA to degrade isothiocyanates. *In vitro* experiments have demonstrated that purified SsSaxA can hydrolyse various isothiocyanates, and knockout strains of *S. sclerotiorum* that lack the gene for *SsSaxA* exhibit increased vulnerability to isothiocyanates. Furthermore, these knockout strains displayed diminished pathogenicity on wild-type plants, underscoring the importance of isothiocyanate metabolism in virulence (Chen *et al.*, 2020). Studies on transcriptomes have demonstrated the importance of the host's metabolism of several antimicrobial substances in the pathogenicity of *S. sclerotiorum*.

According to Allan *et al.* (2019), 53 genes were identified that showed differential expression during infections of *Lupinus angustifolius* (lupin) and *B. napus*. Several of these genes were found to possess domains related to the metabolism and efflux of xenobiotic substances, including cytochrome P450, major facilitator superfamily, and tannase domains. The colonization of different host species led to a unique induction signature of cytochrome P450 (Kusch *et al.*, 2022). This observation is consistent with the general finding that the infection of various host species elicits distinct transcriptional responses in *S. sclerotiorum*. Importantly, a group of 70 genes was found to be upregulated in *S. sclerotiorum* during infections of *Arabidopsis* and in response to the Brassicales-specific antimicrobial camalexin. The genes present in *S. trifoliorum* remained conserved, despite their lack of activation upon exposure to camalexin, a compound that inhibits infection in *Brassicas* species. The identification of a particular promoter motif in the genes of *S. sclerotiorum*, absent in those of *S. trifoliorum*, suggests that regulatory modifications may have facilitated the capacity to infect Brassicales hosts, resulting in a host-specific transcriptional response. As noted by Andrew *et al.* (2012), the evolution of a core set of genes has been instrumental in the adaptation to new host environments, significantly contributing to the broadened host range of *S. sclerotiorum* (Derbyshire *et al.*, 2022).

9.3 Physiological Process Involved in the Interaction between Hosts and Pathogens

The insights gained from histopathological analysis, ultrastructural alterations, and the interactions between host and pathogen contribute to an indirect comprehension of the physiological processes that underlie pathogenesis. Conversely, a direct assessment of these physiological processes yields the critical evidence required to elucidate the mechanisms of pathogenesis.

9.3.1 Alterations in the host's physiology during the process of host tissue colonization

Sclerotinia species' infectious hyphae are capable of penetrating host tissues only. Enzymes that break down the host cell's middle lamella promote this invasion. To do this, the pathogen uses three different pectolytic enzymes, the most important of which is endo-PG in the early stages of pathogenesis. It is hypothesized that endo-PG is essential for hydrolysing the middle lamella after PME demethylates the pectin, which allows the fungus to move quickly between the tissue's cells. Furthermore, there is oblique evidence supporting the significance of endo-PG. It is known that *S. sclerotiorum* isolates that exhibit high virulence on bean plants generate sizable amounts of this enzyme *in vitro* (Lumsden, 1976). Moreover, endo-PG effectively degrades the tissues of sensitive beans, cucumbers, and tubers, as demonstrated by Echandi and Walker (1957) and corroborated in 1955; on the other hand, it has a different effect on resistant potato tuber tissue (Lumsden, 1976). The early emergence and subsequent deactivation of this enzyme in infected tissue may account for the lack of correlation observed between PG activity and virulence (Held, 1955; Morrall *et al.*, 1972; Newton, 1972). The role of PME (pectin methylesterase) appears to be essential to endo-PG's quick response. These enzymes work together to degrade highly methylated pectin. PME aids in the removal of methyl groups from pectin within the middle lamella, forming pectate, the preferred substrate for both *Sclerotinia* exo- and endo-PG (Hancock, 1966; Lumsden, 1976). PME continues to function during the later phases of pathogenesis when exo-PG is largely active, despite the fact that the exact relationship between PME and the virulence of *Sclerotinia* isolates is still unknown (Morrall *et al.*, 1972; Lumsden, 1976). Exo-PG is more efficient at breaking down pectate than pectin, suggesting that exo-PG and PME worked together to break down middle lamellar pectin. According to Lumsden (1976), the production of exo-PG is linked to the expansion of *Sclerotinia* and may also be important for the pathogen's growth and nourishment in the afflicted tissue.

9.3.2 Nutritional requirements during the process of disease development

Sclerotinia dietary requirements play a crucial role in the disease's progression and have a significant impact on how the infection affects the host plant. Access to a suitable food source prior to infection is usually necessary for the disease to successfully establish itself (de Bary, 1887; Purdy, 1958; Lumsden and Dow, 1973; Abawi *et al.*, 1975). During the infection phase, the fungus develops specialized infection hyphae, which require a substantial amount of energy as well as a rich and readily accessible nutrient source. The susceptibility of a host to the disease can be influenced by the nutritional profile of the substrate. *S. sclerotiorum* infection hyphae can be grown on cellophane atop an appropriate agar medium to produce the fungus in a laboratory setting (Lumsden, 1975). The nutritional composition of the media and direct contact with the cellophane surface are prerequisites for the production of the larger, parallel infection hyphae. Comparative investigation reveals that the medium with bean stem extract beneath cellophane had more inflated hyphae and a higher degree of parallel hyphal alignment than the medium with cornmeal agar covered in cellophane. When compared to nonhost media, media formed from host plant tissues typically encourage a more significant production of infection hyphae (Lumsden, 1979). In the course of pathogenesis, enzymes such as phosphatidase, exo-PG, cellulase, hemicellulose, and proteolytic enzymes may have a nutritional role. The action of these enzymes on cell walls and their contents can generate a rich source of carbon and nitrogen when the infection hyphae swiftly penetrate host tissues. This is crucial for the vigorous metabolic processes of *Sclerotinia* spp. (Lumsden, 1979). According to reports, *Sclerotinia* spp.'s enormous hyphae (Lumsden and Dow, 1973) contain enzymes that break down cell walls, which can cause significant cell wall disintegration and the subsequent release of a large amount of carbohydrates (Hancock, 1967; Lumsden, 1969, 1976; Bauer *et al.*, 1977). The lengthy hyphae appear capable of invading host cells intracellularly, as they extend much beyond the terminals of the infected hyphae. This emphasizes how crucial it is to produce cellulolytic enzymes that can break down cell walls and facilitate hyphal growth. Further research indicates that the breakdown of cell walls by enzymes could potentially have a nutritional or secondary function throughout the pathogenic process. Two days after inoculation, observations show a small decrease in the amount of cellulose in infected tissues, indicating the start of the illness. This decline increases with the progression of the pathophysiology (Lumsden, 1969). Similarly, as the illness worsens, the hemicellulose breakdown likewise gets worse (Hancock, 1967). When the early phases of pathogenesis or the borders of lesions are examined, no changes in the birefringence of infected host tissues are found. But as the disease worsens, the birefringence becomes less stable. This suggests that after the hyphae of *Sclerotinia* infection colonize the tissues, the crystalline structure of the cell walls is altered (Lumsden, 1979). Additionally, compared to their less resistant counterparts, resistant cultivars have a higher total soluble protein content (Kamara *et al.*, 1991).

Nitrogen is required for the growth and synthesis of extracellular enzymes in living things, and it comes in a variety of forms. Studies conducted by Lumsden (1970), Newton (1972), Khare and Bompeix (1976) have revealed that proteases and phosphatidase are putative sources of nitrogen. Nevertheless, until more research is done, it is unknown exactly how these enzymes contribute to the provision of nutrients. Furthermore, the invasion of cells contributes to the provision of nitrogen. Held (1955) observed, for instance, that *S. sclerotiorum* appears to prefer organic or ammoniacal nitrogen forms over nitrate. Most likely, the primary source of nutrients in the impacted tissues is the hydrolysed plant material. Nonetheless, Thatcher (1942) suggested that modifications in cell permeability prior to hyphal invasion might meet the nutritional requirements in the early stages of infection, prior to the disintegration of cell wall components and the demise of protoplasts.

9.3.3 Regulation of alterations in permeability and water relationship during pathogenesis

The concept of increased permeability in host cells following infection was initially introduced by Thatcher in 1942. A significant fourfold increase in permeability was noted in infected celery petioles, even in areas distant from necrotic regions. These changes in permeability are thought to contribute to the symptoms of water soaking and subsequent cell death. In 1972, Hancock presented a different viewpoint on the changes in permeability in tissues affected by *Sclerotinia* infections. His research revealed that the permeability of sunflower hypocotyl sections, as determined by the measurement of water and urea influx and efflux as well as electrolyte leakage, is reduced in sections located above lesions caused by *S. sclerotiorum* when compared to similar sections from healthy specimens. It is believed that the decreased permeability of host cells above the lesions is connected to alterations in the nonlipid components of the plasmalemma. Furthermore, Newton's (1972) study was unable to conclusively link the electrolyte leakage seen in healthy tissues to any one of the hydrolytic enzymes, such as peroxidase, cellulase, phosphatidase, and protease. Hancock's (1972) study provides an explanation for the differences seen in Thatcher's (1942) research. Hancock claims that Thatcher's use of aged and detached celery stalks exhibits increased electrolyte loss and permeability. On the other hand, intact sunflower or celery plants show a reduction in permeability rather than these alterations. This decrease in permeability may not have a major impact on *Sclerotinia* species, but it may be harmful to the host tissue. It inhibits growth, makes the tissue more prone to injury, and has a negative effect on the tissue's ability to fend off infections. As a result, these results call into question the previously accepted theories of fluid build-up surrounding infection hyphae and water soaking of tissues (Lumsden and Dow, 1973; Newton *et al.*, 1973). The seeping from host cells is a substitute explanation (Thatcher, 1942; Newton *et al.*, 1973). The build-up of liquid may be due to a rise in osmotic pressure within the invading tissue. Water is said to be moved by the fungus from the lower parts of the plant to its own place (Thatcher, 1942). The osmotic flow of water from surrounding regions can be accelerated by the solutes in the fluid surrounding the pathogen hyphae and their elevated osmotic pressure. This fluid may serve as a medium for enzyme activity, facilitate the long-distance movement of nutrients, and allow OA to diffuse into uninvaded tissues that are separated by multiple cells.

9.3.4 OA generation in relation to host–pathogen interactions

Numerous research studies have demonstrated the function of OA in the process of infection. This assertion is corroborated by the discovery of OA in infected host tissues and the capacity of OA and culture filtrates to cause symptoms of disease (Marciano *et al.*, 1983; Riou *et al.*, 1991). Using *Arabidopsis thaliana* as a model organism, Dickman and Mitra (1992) investigated the pathogenesis of *S. sclerotiorum*, emphasizing the role of OA in the infection mechanism. The initial connection between OA and *Sclerotinia* infections was made by de Bary in 1886. The observed pH shift was attributed to nonvolatile acids due to the damaged carrot tissue exhibiting a strong acidic reaction. Subsequent analysis showed that approximately 0.319% of the carrot tissue was composed of oxalate, mostly as calcium salt. A study by Maxwell and Lumsden (1970) found that bean tissue inoculated with *S. sclerotiorum* had 1.1, 31.4, and 48.3 mg of oxalate per gram of dry weight at 0, 2, and 4 days postinoculation. Moreover, Rai and Dhawan (1976) discovered that *S. sclerotiorum* produces OA, a poisonous chemical that causes white rot in cruciferous plants. All kinds of mushrooms produce OA, also known as ethanedioic acid, demonstrating the great diversity found in fungal ecology and physiology. Regarding the potential roles of OA in various ecological contexts, numerous theories have been put out. These include its involvement in pathogenicity, interspecies competition among fungi, and the control of ambient nutrients and toxins (Dutton and Evans, 1996).

9.3.5 OA production and function

There are various suggested mechanisms of action that are crucial to the ways by which OA contributes to pathogenicity.

1. Early in the pathogenesis, OA builds up in the infected tissues, and as the host gets more colonized, the concentration of OA increases (Bateman, 1964; Bateman and Beer, 1965). Since many CWDEs have an optimum pH of less than 5.0, OA lowers the extracellular pH to about 4–5. According to various research (Bateman and Beer, 1965; Maxwell and Lumsden, 1970; Marciano *et al.*, 1983; Magro *et al.*, 1984), the pH reduction greatly boosts their activity.
2. OA may be able to lessen the inhibitory effects of plant defense PGIPs connected to acidity by adjusting the pH to optimal values for the activity of CWDEs (Favaron *et al.*, 2004).
3. OA is released in tandem with the chelation of Ca^{2+} and the extraction of pectic chemicals from the cell walls, three to five cell layers ahead of the moving fungal hyphae (Smith *et al.*, 1986). Studies have shown that hydrolysed Ca^{2+} pectate is able to produce in the middle lamellae on its own, and PG works in tandem with OA. According to Bateman and Beer (1965) and Kurian and Stelzig (1979), OA chelates Ca^{2+} in the cell wall, enabling PG to hydrolyze the pectate and undermine the integrity of the host cell wall. Moreover, defense mechanisms in plants that depend on Ca^{2+} could be compromised (Bateman and Beer, 1965).
4. Despite the fact that tobacco and soybean plants are unrelated, OA has the power to inhibit the oxidative burst, a crucial main defense mechanism (Cessna *et al.*, 2000).
5. OA inhibits the closure of stomata caused by abscisic acid and promotes their opening, which controls the activity of guard cells. Thus, when *S. sclerotiorum* infects the plant, this causes foliar wilting (Guimarães and Stotz, 2004).
6. OA inhibits the enzymatic activities of polyphenol oxidase (PPO), a plant-synthesized enzyme (Marciano *et al.*, 1983; Magro *et al.*, 1984).
7. A drop in the ambient environmental pH can have an impact on the transcriptional regulation of pH-responsive genes, which are critical for the pathogenesis and developmental life cycle of *S. sclerotiorum*. The putative transcription factor pacl, for example, in *S. sclerotiorum* has been found to be auto-regulated, with transcript accumulation observed in response to increasing ambient pH levels. Pacl is a pathogenicity factor that acts similarly to other known fungal pH-sensing transcription factors based on homology (Rollins and Dickman, 2001; Rollins, 2003). Similarly, in acidic pH conditions caused by OA accumulation, peak expression of smk1, a MAPK necessary for sclerotial development, was seen (Chen *et al.*, 2004).
8. Because of the acidic environment that its synthesis creates, OA can have a detrimental effect on plants (Noyes and Hancock, 1981). According to Rollins and Dickman (2001), the pH of the environment affects both the synthesis of OA and the action of endo-PG in *S. sclerotiorum*. This suggests that low pH can weaken the plants and make them more vulnerable to fungal infections.

Research has demonstrated that the generation of OA is highly influenced by the pH of the culture medium through a self-regulating mechanism (Maxwell and Lumsden, 1970; Rollins and Dickman, 2001). The pH falls as OA concentrations grow, which in turn limits the amount of OA that can be produced further. It has been noted that an alkaline environment encourages the production of OA, regardless of the carbon source used, even though the carbon source is essential to the synthesis of OA (Rollins and Dickman, 2001).

9.3.5.1 *Evidence of OA production during infection*

Researchers have found a relationship between the pathogenicity of some *Sclerotinia* species and the synthesis of OA (Higgins, 1927; Bateman and Beer, 1965; Maxwell and Lumsden, 1970; Kritzman *et al.*, 1977; Noyes and Hancock, 1981; Marciano *et al.*, 1983; Magro *et al.*, 1984; Stone and Armentrout,

1985; Godoy *et al.*, 1990; Pierson and Rhodes, 1992; Dutton and Evans, 1996). The discovery of OA in tissues damaged by infections (de Bary, 1886; Marciano *et al.*, 1983; Godoy *et al.*, 1990; Ferrar and Walker, 1993) lends credence to this link. Additionally, there is evidence linking the amount of OA present to the severity of the condition (Bateman and Beer, 1965; Maxwell and Lumsden, 1970; Noyes and Hancock, 1981; Magro *et al.*, 1984). It's interesting to note that injecting the toxin directly into plants has produced symptoms similar to Sclerotinia diseases (Bateman and Beer, 1965; Noyes and Hancock, 1981). Before fungal hyphae invade, lesions and tissues soaked in water form as a result of *S. sclerotiorum* releasing OA (Lumsden and Dow, 1973; Tu, 1985). It's interesting to note that mutants possessing all of the CWDEs but unable to produce OA do not cause illness in dry beans. But these mutants exhibit conventional virulence patterns once they are able to produce OA again (Godoy *et al.*, 1990). Similar results were obtained using *Arabidopsis thaliana* as the host plant (Dickman and Mitra, 1992). It is important to stress that the decreased activity of CWDEs, which rely on the OA-facilitated acidity of the plant environment, may be the cause of the lack of pathogenicity seen in OA-deficient mutants.

9.3.6 The multiple roles of OA in host infected tissues

OA can have a wide range of consequences on tissues that are impacted. OA may collaborate with pectolytic enzymes in the periphery of developing lesions and in the initial phases of the disease's development, as observed in other conditions of a similar nature (Maxwell and Lumsden, 1970). It is commonly known that OA has strong chelating capabilities with calcium and other cations. Tissue maceration is inhibited by its capacity to bind to mono- and divalent cations, which also inhibits the effects of endo- and exo-PG (Hancock, 1966). Experimental research indicates that OA facilitates the degradation of pectic compounds by endo-PG, exo-PG, and PME (Lumsden, 1979).

9.3.6.1 OA influence pH level in host

OA changes the host's pH levels. Research indicates that these tissues experience pH changes, particularly at the leading edges of lesions (Hancock, 1966; Maxwell and Lumsden, 1970; Morrall *et al.*, 1972; Lumsden, 1972, 1976). The pH at the advancing edges of lesions on bean hypocotyls drops from 5.0 to 4.0, according to observations made using microspectrophotometric techniques (Lumsden, 1972). The increasing acidity in the forming lesions activates various hydrolytic enzymes, including endo- and exo-PG (Hancock, 1966; Lumsden, 1976), cellulase (Lumsden, 1969), hemicellulose (Hancock, 1967), and several others (Lumsden, 1970). These enzymes prefer pH values that are significantly lower than those of healthy host tissues and cell sap. Significant pH fluctuations also adversely affect the ability of cells to resist pathogen invasion and remain viable. The increased acidity causes the fungus to grow more quickly (Maxwell and Lumsden, 1970).

9.3.6.2 Toxic effects of OA on host tissues

de Bary (1886), Overell (1952), Hancock (1972), and Thatcher (1942) all state that the toxicity of OA against host cells, which results in cell death, is related to either the cation-chelation characteristics of OA or large pH shifts. Maxwell and Lumsden's (1970) research, which Hancock (1972) supported, shows that oxalate present in lesions at specific pH values and concentrations is sufficient to cause sunflower cell death. According to de Bary (1886) and Newton (1972), a portion of the fatal component in lesions has been observed to be heat-sensitive. Hancock (1972) suggested that pectolytic enzymes might have a role in the mechanism of cell death, similar to other disorders. Overell (1952) refuted the notion that OA is a toxin released in plant tissues by *S. sclerotiorum*, arguing instead that OA is exclusively linked to aging cultures. However, as Maxwell and Lumsden (1970) pointed out, the pH, buffering capacity, and glucose levels in *Sclerotinia* spp. cultures are significant determinants affecting the quantity and production of oxalate produced.

9.3.6.3 OA causes wilting in host tissues

The wilting symptoms that are frequently linked to *Sclerotinia* spp.-related diseases could be produced by OA (Pawlowski and Hawn, 1964; Noyes and Hancock, 1981). Oxalate crystals have been seen to have the ability to clog xylem channels (Lumsden and Dow, 1973). The vascular obstruction may cause wilting, or oxalate may have a direct impact on the water equilibrium in the lamellar tissue (Pawlowski and Hawn, 1964; Noyes and Hancock, 1981). According to Noyes and Hancock (1981), OA appears to enter the vascular system and spread throughout the leaves as lesions form. When the amount of OA in the leaves reaches a "critical" level, wilting symptoms are triggered. One mobile toxin that exacerbates wilt syndrome is OA. Marciano *et al.* (1983) looked at the relationship between OA, CWDEs, and virulence in sunflower stems infected with two isolates of *S. sclerotiorum* that exhibit differing degrees of virulence. Although OA does not directly affect the activity of enzymes that break down cell walls, it does cause noticeable changes in pH. Furthermore, OA inhibits the polyphenol oxidase (PPO) enzyme, which restricts the synthesis of phenolic oxidation molecules. These molecules are vital to plant defense mechanisms and protect the pectic compounds present in cell walls.

9.3.6.4 OA regulates multiple functions of host cells

Research using *S. sclerotiorum* mutants devoid of the capacity to make OA has brought attention to the significance of OA (Godoy *et al.*, 1990). These OA-deficient mutants do not infect *Phaseolus vulgaris* L. like the wild-type and revertant mutants do. This conclusion is supported by the findings of Favaron *et al.* (1988), who found that OA stimulates the synthesis of the phytoalexin glyccollin I in soybean hypocotyls. However, the release of OA causes a pH drop in bean pods that renders oxiphenoloxidase ineffective. This strategy may increase the infection's success by reducing the host's resistance, according to some ideas (Ferrar and Walker, 1993). Alfalfa seedlings from four different cultivars were chosen for OA resistance, and the results showed a significant gain in resistance in all cultivars except for one that displayed greater resistance to *S. trifoliorum* (Rowe and Welty, 1984). Callahan and Rowe (1991) used a host–pathogen interaction system (HPIS) in their research to show how macromolecular components (greater than 3500 mol wt.) and OA found in *S. trifoliorum* exudates are essential in inhibiting seedling growth in a bioassay by preventing fungal mycelia from contacting potential hosts. Rowe (1993) asserts that OA can select for resistance to the fungus in alfalfa or to *S. sclerotiorum* in red clover. The production of oxalate by *S. sclerotiorum* is crucial for the regulation of guard cells during the infection phase. According to Guimarães and Stotz (2004), oxalate works by encouraging the build-up of osmotically active molecules, which causes stomatal opening and suppresses ABA-induced stomatal closure.

9.3.7 Differential behaviour of OA in hosts exhibiting tolerance and susceptibility

When compared to the susceptible cultivars Kentwood, Seafarer, and Fleetwood, the bean cultivar Ex Rico-23, which exhibits tolerance, shows less severe disease signs and a slower spread of the disease (Tu and Beversdorf, 1982). Tu (1985) found in a follow-up study that white beans' resistance to OA is correlated with their tolerance to *S. sclerotiorum*, and that differences in disease severity correspond to the pace at which OA solution diffuses through leaf tissues. Similar findings were made by Noyes and Hancock (1981), who found that cultivars of sunflower that are resistant to *S. sclerotiorum* have a greater tolerance to OA than those that are sensitive. Additionally, Tariq and Jeffries (1985) suggested that the OA secreted by the fungus during *Sclerotinia* infection is connected to the degeneration of chloroplasts. This may be the consequence of OA-induced rupture of chloroplast membranes. The susceptibility of chloroplasts to OA may be inversely connected with the plasma membrane's capacity to tolerate OA. This means that the less tolerant plasma membranes lose their stability more quickly than the more tolerant membranes when exposed to a given dosage of OA of varying tolerance levels. Ultimately, this instability breaks down the selective permeability, which permits OA to freely enter the cell. In plants that are vulnerable, brown-rot lesions enlarge more quickly during the *S. sclerotiorum* pathogenesis. The varying rates of membrane denaturation between tolerant and susceptible cultivars could account for this phenomenon. Tu (1989) used electron microscopy to study the effects of OA on the cytology of both sensitive and tolerant bean cultivars infected by white mold. Compared to the susceptible cultivars, the resistant cultivar's plasma membrane showed higher resistance to damage caused by OA. The sensitive cultivar's plasma membrane and chloroplasts suffered more severe damage and burst more quickly than those of the tolerant cultivars at equal OA concentrations. The ruptured organelles appeared partially or totally disordered in thin slices. Extended exposure to OA was found to cause an increase in protrusions, wrinkles, breaking, and ruptures in preparations of the plasma membrane that had been freeze-fractured. In comparison to the tolerant cultivar, the vulnerable cultivar's plasma membrane showed noticeably more damage. In addition, the vulnerable cultivar had more conductivity than the resistant one.

9.3.8 Pathways for the detoxification of phytoalexins in *Sclerotinia*

The study concentrated on the alterations of significant cruciferous phytoalexins by the stem rot pathogen *S. sclerotiorum*. A number of new metabolic products were discovered, and their chemical structures were elucidated using spectroscopic methods, subsequently confirmed through synthesis. Notably, these metabolic products did not demonstrate any measurable antifungal activity against *S. sclerotiorum*, implying that the transformations were primarily detoxification mechanisms. The results suggest that *S. sclerotiorum* possesses various enzymes that can detoxify cruciferous phytoalexins via multiple pathways. In particular, while potent and moderately effective antifungal phytoalexins such as brassilexin, sinalexin, and 1-methoxybrassinin were swiftly converted into glucosylated derivatives, the less effective antifungal phytoalexins brassicanal A, spirobrassinin, and 1-methoxyspirobrassinin experienced gradual transformations, leading to the formation of nonglucosylated compounds.

9.3.8.1 *Identification of phytoalexin inhibitors*

The aim of this research was to create potential selective inhibitors for the brassinin detoxification enzyme, BGT (β-glucosyltransferases). Two distinct groups of inhibitors were formulated:

1. the first group was based on the brassinin structure, incorporating modifications such as substituting the indole ring with benzofuran, thianaphthene, 7-azaindole, and pyrazolo [1,5-a]pyridine, as well as repositioning the side chain from C-3 to C-2; and
2. the second group was inspired by the camalexin structure, where the thiazole ring was replaced with a phenyl group.

The synthesis and chemical characterization of these potential inhibitors, along with their antifungal properties, were thoroughly examined. Screening was performed utilizing fungal cultures and cell-free extracts of *S. sclerotiorum*. The results from the screening indicated that compounds such as 3-phenylindoles, 3-phenylbenzofuran, 5-fluorocamalexin, methyl (indol-2-yl)methyl-dithiocarbamate, methyl (benzofuran-3-yl)methyl-dithiocarbamate, and methyl (benzo-furan-2-yl) methyl-dithiocarbamate exhibited the potential to inhibit brassinin detoxification in both fungal cultures and cell-free extracts of *S. sclerotiorum*. Notably, 3-phenylindole exhibited the most significant inhibitory effect in both experimental setups. Metabolic studies conducted with *S. sclerotiorum* fungal cultures revealed that all synthesized compounds were converted into glucosyl derivatives, albeit at reduced rates. This finding implies that inhibitors that can decelerate brassinin metabolism may serve as valuable templates for the development of more effective BGT inhibitors (Hossain, 2007).

Table 9.3. Synthesis and accumulation of phytoalexins in *Brassica napus* affected by *Sclerotinia sclerotiorum*. (Adapted from publication of Sexton *et al*. (2009), with permission.)

Phytoalexins[a]	Time postinoculation (h)	Quantity (nmol/g of fresh leaf tissue)[b]
Brassinin	48	11.5–14.5
	72	1.4–5.9
1-Methoxybrassinin	48	9.3–12.9
	72	7.0–21.3
1-Methoxyindole-3-acetonitrile	48	0.05–1.3
	72	1.1–2.4
1-b-D-Glucopyranosylbrassinin	48	4.0–16.4
	72	40.2–47.6
Cyclobrassinin	48	<0.1
	72	<0.1

[a]None of these metabolites were identified in uninoculated plants.
[b]Values from four independent experiments.

9.3.9 Cloning, purification, and characterization of brassinin glucosyltransferase that facilitates the detoxification of phytoalexins obtained from *Sclerotinia*

The fungus *Sclerotinia sclerotiorum* possesses the capability to detoxify cruciferous phytoalexins, such as brassinin, through the process of glucosylation. In their 2009 study, Sexton *et al.* outlined several methodologies, including genome mining, transcriptional induction, phytoalexin quantification, protein expression, and enzyme purification, which culminated in the identification of a glucosyltransferase from *S. sclerotiorum* responsible for the detoxification of brassinin. The gene encoding this enzyme, designated *SsBGT1*, exhibited significant transcriptional induction in response to various cruciferous phytoalexins, including camalexin, cyclobrassinin, brassilexin, brassinin, and 3-phenylindole, a camalexin analogue (Table 9.3). Additionally, the expression of this gene was notably upregulated during the infection of *Brassica napus* leaves. Observations indicated a marked reduction in brassinin levels between 48 and 72 hour postinoculation, coinciding with an increase in the concentration of 1-β-D-glucopyranosylbrassinin, the product formed through the action of *SsBGT1*. These findings strongly imply the role of this gene in the infection process of *B. napus*. The gene was successfully cloned and expressed in *Saccharomyces cerevisiae*, and the purified recombinant enzyme exhibited the capacity to glucosylate brassinin along with two additional phytoalexins, although with reduced efficiency.

References

Abawi, G.S., Provvidenti, R., Grogan, R.G. and Hunter, J.E. (1975) Predisposition of beans to infection by ascospores of *Whetzelinia sclerotiorum* prior to blossoming. *Proceedings of the American Phytopathological Society* 2, 61. DOI: 10.1094/Phyto-65-673.

Alghisi, P. and Favaron, F. (1995) Pectin-degrading enzymes and plant-parasite interactions. *European Journal of Plant Pathology* 101(4), 365–375. DOI: 10.1007/BF01874850.

Allan, J., Regmi, R., Denton-Giles, M., Kamphuis, L.G. and Derbyshire, M.C. (2019) The host generalist phytopathogenic fungus *Sclerotinia sclerotiorum* differentially expresses multiple metabolic enzymes on two different plant hosts. *Scientific Reports* 9(1), 19966. DOI: 10.1038/s41598-019-56396-w.

Amselem, J., Cuomo, C.A., van Kan, J.A.L., Viaud, M., Benito, E.P. *et al.* (2011) Genomic analysis of the necrotrophic fungal pathogens *Sclerotinia sclerotiorum* and *Botrytis cinerea*. *PLoS Genetics* 7(8), e1002230. DOI: 10.1371/journal.pgen.1002230.

Andrew, M., Barua, R., Short, S.M. and Kohn, L.M. (2012) Evidence for a common toolbox based on necrotrophy in a fungal lineage spanning necrotrophs, biotrophs, endophytes, host generalists and specialists. *PLoS ONE* 7(1), e29943. DOI: 10.1371/journal.pone.0029943.

Annis, S.L. and Goodwin, P.H. (1997) Recent advances in molecular genetics of plant cell wall degrading enzymes in plant pathogenic fungi. *European Journal of Plant Pathology* 103(1), 1–14. DOI: 10.1023/A:1008656013255.

Barkai-Golan, R. (1974) Production of cellulase and polygalacturonase by *Sclerotinia minor*. *Mycopathologia et Mycologia Applicata* 54(3), 297–302. DOI: 10.1007/BF02050164.

Bateman, D.F. (1964) An induced mechanism of tissue resistance to polygalacturonase in *Rhizoctonia* infected hypocotyls of bean. *Phytopathology* 54, 438–445.

Bateman, D.F. and Beer, S.V. (1965) Simultaneous production and synergistic action of oxalic acid and polygalacturonase during pathogenesis by *Sclerotium rolfsii*. *Phytopathology* 55, 204–211.

Bauer, W.D., Bateman, D.F. and Whalen, C.H. (1977) Purification of an endo-β-1,4 galactanase produced by *Sclerotinia sclerotiorum*: Effects on isolated plant cell walls and potato tissue. *Phytopathology* 77(7), 862. DOI: 10.1094/Phyto-67-862.

Bolton, M.D., Thomma, B.P.H.J. and Nelson, B.D. (2006) Pathogen profile: *Sclerotinia sclerotiorum* (Lib.) de Bary, biology and molecular traits of a cosmopolitan pathogen. *Molecular Plant Pathology* 7, 1–16.

Boyle, C. (1921) Studies in the physiology of parasitism. VI. infection by *Sclerotinia libertiana*. *Annals of Botany* 35(3), 337–347. DOI: 10.1093/oxfordjournals.aob.a089761.

Brown, C.A., Murray, A.W. and Verstrepen, K.J. (2010) Rapid expansion and functional divergence of subtelomeric gene families in yeasts. *Current Biology* 20(10), 895–903. DOI: 10.1016/j.cub.2010.04.027.

Bussink, H.J., Buxton, F.P., Fraaye, B.A., de Graaff, L.H. and Visser, J. (1992) The polygalacturonases of *Aspergillus niger* are encoded by a family of diverged genes. *European Journal of Biochemistry* 208(1), 83–90. DOI: 10.1111/j.1432-1033.1992.tb17161.x.

Butler, M.J., Gardiner, R.B. and Day, A.W. (2009) Melanin synthesis by *Sclerotinia sclerotiorum*. *Mycologia* 101(3), 296–304. DOI: 10.3852/08-120.

Buxdorf, K., Yaffe, H., Barda, O. and Levy, M. (2013) The effects of glucosinolates and their breakdown products on necrotrophic fungi. *PLoS ONE* 8(8), e70771. DOI: 10.1371/journal.pone.0070771.

Cairns, T. and Meyer, V. (2017) *In silico* prediction and characterization of secondary metabolite biosynthetic gene clusters in the wheat pathogen *Zymoseptoria tritici*. *BMC Genomics* 18, 631. DOI: 10.1186/s12864-017-3969-y.

Callahan, F.E. and Rowe, D.E. (1991) Use of a host-pathogen interaction system to test whether oxalic acid is the sole pathogenic determinant in the exudate of *Sclerotinia trifoliorum*. *Phytopathology* 81(12), 1546. DOI: 10.1094/Phyto-81-1546.

Calonge, F.D., Fielding, A.H., Byrde, R.J.W. and Akinrefon, O.A. (1969) Changes in ulrastructure following fungal invasion and the possible relevance of extracellular enzymes. *Journal of Experimental Botany* 20(2), 350–357. DOI: 10.1093/jxb/20.2.350.

Caprari, C., Bergmann, C., Migheli, Q., Salvi, G., Albersheim, P. *et al.* (1993) *Fusarium moniliforme* secretes four endopolygalacturonases derived from a single gene product. *Physiological and Molecular Plant Pathology* 43(6), 453–462. DOI: 10.1006/pmpp.1993.1073.

Carpita, N.C. and Gibeaut, D.M. (1993) Structural models of primary cell walls in flowering plants: Consistency of molecular structure with the physical properties of the walls during growth. *The Plant Journal* 3(1), 1–30. DOI: 10.1111/j.1365-313x.1993.tb00007.x.

Cessna, S.G., Sears, V.E., Dickman, M.B. and Low, P.S. (2000) Oxalic acid, a pathogenicity factor for *Sclerotinia sclerotiorum* suppresses the oxidative burst of the host plant. *The Plant Cell* 12, 2119–2199.

Chen, C., Harel, A., Gorovoits, R., Yarden, O. and Dickman, M.B. (2004) MPAK regulation of sclerotial development in *Sclerotinia sclerotiorum* is linked with pH and cAMP sensing. *Molecular Plant Microbe Interactions* 17, 404–413.

Chen, J., Ullah, C., Reichelt, M., Beran, F., Yang, Z.L. *et al.* (2020) The phytopathogenic fungus *Sclerotinia sclerotiorum* detoxifies plant glucosinolate hydrolysis products via an isothiocyanate hydrolase. *Nature Communications* 11, 3090. DOI: 10.1038/s41467-020-16921-2.

Chittem, K., Yajima, W.R., Goswami, R.S. and de Río Mendoza, L.E. (2020) Transcriptome analysis of the plant pathogen *Sclerotinia sclerotiorum* interaction with resistant and susceptible canola (*Brassica napus*) lines. *PLoS ONE* 15(3), e0229844. DOI: 10.1371/journal.pone.0229844.

Chooi, Y.H. and Solomon, P.S. (2014) A chemical ecogenomics approach to understand the roles of secondary metabolites in fungal cereal pathogens. *Frontiers in Microbiology* 5, 1–7. DOI: 10.3389/fmicb.2014.00640.

Christias, C. and Lockwood, J.L. (1973) Conservation of mycelial constituents in four *Sclerotium* forming fungi in nutrient-deprived conditions. *Phytopathology* 63(5), 602. DOI: 10.1094/Phyto-63-602.

Collemare, J., Billard, A., Böhnert, H.U. and Lebrun, M.H. (2008) Biosynthesis of secondary metabolites in the rice blast fungus *Magnaporthe grisea*: The role of hybrid PKS-NRPS in pathogenicity. *Mycological Research* 112(2), 207–215. DOI: 10.1016/j.mycres.2007.08.003.

Colotelo, N. (1973) Physiological and biochemical properties of the exudate associated with developing sclerotia of *Sclerotinia sclerotiorum* (Lib.) De Bary. *Canadian Journal of Microbiology* 19(1), 73–79. DOI: 10.1139/m73-011.

Colotelo, N. (1974) A scanning electron microscope study of developing sclerotia of *Sclerotinia sclerotiorum*. *Canadian Journal of Botany* 52(5), 1127–1130. DOI: 10.1139/b74-143.

Corsini, D.L. and Le Tourneau, D. (1973) Organic acid metabolism in *Sclerotinia sclerotiorum*. *Archiv of Mikrobiologie* 90(1), 59–64. DOI: 10.1007/BF00424824.

Cotton, P., Rascle, C. and Fevre, M. (2002) Characterization of PG2, an early endo PG produced by *Sclerotinia sclerotiorum* expressed in yeast. *FEMS Microbiology Letters* 213(2), 239–244. DOI: 10.1111/j.1574-6968.2002.tb11312.x.

Cotton, P., Kasza, Z., Bruel, C., Rascle, C. and Fèvre, M. (2003) Ambient pH controls the expression of endopolygalacturonase genes in the necrotrophic fungus *Sclerotinia sclerotiorum*. *FEMS Microbiology Letters* 227(2), 163–169. DOI: 10.1016/S0378-1097(03)00582-2.

Davis, K.R., Darvill, A.G., Albersheim, P. and Dell, A. (1986) Host-pathogen interactions XXIX oligogalacturonides released from sodium polypectate by endopolygalact-uronic acid lyase are elicitors of phytoalexins in soybean. *Plant Physiology* 80(2), 568–577. DOI: 10.1104/pp.80.2.568.

de Bary, A. (1886) Ueber einige *Sclerotinien* and *Sclerotienkrankheiten*. *Botanische Zeitimg* 44, 377–474.

de Bary, A. (1887) *Comparative Morphology and Biology of Fungi, Mycetozoa and Bacteria*. Clerenden, New York, p. 525. DOI: 10.5962/bhl.title.5372.

De Lorenzo, G., D'Ovidio, R. and Cervone, F. (2001) The role of polygalacturonase-inhibiting proteins (pgips) in defense against pathogenic fungi. *Annual Review of Phytopathology* 39, 313–335. DOI: 10.1146/annurev.phyto.39.1.313.

Derbyshire, M.C., Newman, T.E., Khentry, Y. and Taiwo, A.O. (2022) Pathogen profile: The evolutionary and molecular features of the broad host range plant pathogen *Sclerotinia sclerotiorum*. *Molecular Plant Pathology* 23(8), 1075–1090. DOI: 10.1111/mpp.13221.

Dickman, M.B. and Mitra, A. (1992) *Arabidopsis thaliana* as a model for studying *Sclerotinia sclerotiorum* pathogenesis. *Physiological and Molecular Plant Pathology* 41(4), 255–263. DOI: 10.1016/0885-5765(92)90025-Q.

Dickman, M.B., Park, Y.K., Oltersdorf, T., Li, W., Clemente, T. *et al.* (2001) Abrogation of disease development in plants expressing animal antiapoptotic genes. *Proceedings of the National Academy of Sciences* 98(12), 6957–6962. DOI: 10.1073/pnas.091108998.

Dutton, M.V. and Evans, C.S. (1996) Oxalate production by fungi: Its role in pathogenicity and ecology in the soil environment. *Canadian Journal of Microbiology* 42(9), 881–895. DOI: 10.1139/m96-114.

Echandi, E. and Walker, J.C. (1957) Pectolytic enzymes produced by *Sclerotinia sclerotiorum*. *Phytopathology* 47, 303–306.
Favaron, F., Alghisi, P., Marciano, P. and Magro, P. (1988) Polygalacturonase isoenzymes and oxalic acid produced by *Sclerotinia sclerotiorum* in soybean hypocotyls as elicitors of glyceollin. *Physiological and Molecular Plant Pathology* 33(3), 385–395. DOI: 10.1016/0885-5765(88)90005-7.
Favaron, F., D'Ovidio, R., Porceddu, E. and Alghisi, P. (1994) Purification and molecular characterization of a soybean polygalacturonase-inhibiting protein. *Planta* 195(1), 80–87. DOI: 10.1007/BF00206295.
Favaron, F., Sella, L. and D'Ovidio, R. (2004) Relationships among endo-polygalacturonase, oxalate, pH, and plant polygalacturonase-inhibiting protein (PGIP) in the interaction between *Sclerotinia sclerotiorum* and soybean. *Molecular Plant-Microbe Interactions* 17(12), 1402–1409. DOI: 10.1094/MPMI.2004.17.12.1402.
Fedorova, N.D., Moktali, V. and Medema, M.H. (2012) Bioinformatics approaches and software for detection of secondary metabolic gene clusters. *Methods in Molecular Biology* 944, 23–45. DOI: 10.1007/978-1-62703-122-6_2.
Ferrar, P.H. and Walker, J.R.L. (1993) O-diphenol oxidase inhibition—An additional role for oxalic acid in the phytopathogenic arsenal of *Sclerotinia sclerotiorum* and *Sclerotium rolfsii*. *Physiological and Molecular Plant Pathology* 43(6), 415–422. DOI: 10.1006/pmpp.1993.1069.
Fraissinet-Tachet, L. and Fevre, M. (1996) Regulation by galacturonic acid of pectinolytic enzyme production by *Sclerotinia sclerotiorum*. *Current Microbiology* 33(1), 49–53. DOI: 10.1007/s002849900073.
Fraissinet-Tachet, L., Reymond-Cotton, P. and Fevre, M. (1996) Characterization of a multigene family encoding an endopolygalacturonase in *Sclerotinia sclerotiorum*. *Current Genetics* 29(1), 96–99. DOI: 10.1007/BF00313199.
Gao, S., Choi, G.H., Shain, L. and Nuss, D.L. (1996) Cloning and targeted disruption of enpg-1, encoding the major *in vitro* extracellular endopolygalacturonase of the chestnut blight fungus *Cryphonectria parasitica*. *Applied and Environmental Microbiology* 62(6), 1984–1990. DOI: 10.1128/aem.62.6.1984-1990.1996.
García-Maceira, F.I., Pietro, A., Huertas-González, M.D., Ruiz-Roldán, M.C. and Roncero, M.I. (2001) Molecular characterization of an endopolygalacturonase from *Fusarium oxysporum* expressed during early stages of infection. *Applied and Environmental Microbiology* 67(5), 2191–2196. DOI: 10.1128/AEM.67.5.2191-2196.2001.
Godoy, G., Steadman, J.R., Dickman, M.B. and Dam, R. (1990) Use of mutants to demonstrate the role of oxalic acid in pathogenicity of *Sclerotinia sclerotiorum* on *Phaseolus vulgaris*. *Physiological and Molecular Plant Pathology* 37(3), 179–191. DOI: 10.1016/0885-5765(90)90010-U.
Govrin, E.M. and Levine, A. (2000) The hypersensitive response facilitates plant infection by the necrotrophic pathogen *Botrytis cinerea*. *Current Biology* 10(13), 751–757. DOI: 10.1016/s0960-9822(00)00560-1.
Graham-Taylor, C., Kamphuis, L.G. and Derbyshire, M.C. (2020) A detailed *in silico* analysis of secondary metabolite biosynthesis clusters in the genome of the broad host range plant pathogenic fungus *Sclerotinia sclerotiorum*. *BMC Genomics* 21(1), 7. DOI: 10.1186/s12864-019-6424-4.
Guimarães, R.L. and Stotz, H.U. (2004) Oxalate production by *Sclerotinia sclerotiorum* deregulates guard cells during infection. *Plant Physiology* 136(3), 3703–3711. DOI: 10.1104/pp.104.049650.
Gupta, N.C., Arora, S., Kundu, A., Sharma, P., Rao, M. *et al.* (2022) UPLC-Q-TOF-MS-based untargeted studies of the secondary metabolites secreted by *Sclerotinia sclerotiorum* under the axenic condition. *Journal of Plant Science and Phytopathology* 6, 173–182. DOI: 10.29328/journal.jpsp.1001095.
Hancock, J.G. (1966) Degradation of pectic substances associated with pathogenesis by *Sclerotinia sclerotiorum* in sunflower and tomato stems. *Phytopathology* 56, 975–979.
Hancock, J.G. (1967) Hemicellulose degradation in sunflower hypocotyls infected with *Sclerotinia sclerotiorum*. *Phytopathology* 57, 203–206.
Hancock, J.G. (1972) Changes in cell membrane permeability in sunflower hypocotyls infected with *Sclerotinia sclerotiorum*. *Plant Physiology* 49(3), 358–364. DOI: 10.1104/pp.49.3.358.
Held, Y.M. (1955) Physiological differences between a normal and a degenerate strain of *Sclerotinia trifoliorum*. *Phytopathology* 45, 39–42.
Higgins, B.B. (1927) Physiology and parasitism of *Sclerotium rolfsii* (Sacc.). *Phytopathology* 17, 417–448.
Hossain, M. (2007) Phytoalexins from crucifers: Probing detoxification pathways in *Sclerotinia sclerotiorum*. Doctor of Philosophy, Thesis, Department of Chemistry, University of Saskatchewan, Saskatoon, Canada.

Humpherson-Jones, F.M. and Cooke, R.C. (1977) Induction of sclerotium formation by acid staling compounds in *Sclerotinia sclerotiorum* and *Sclerotium rolfsii*. *Transactions of the British Mycological Society* 68(3), 413–420. DOI: 10.1016/S0007-1536(77)80195-2.

Jones, D. (1970) Ultrastructure and composition of the cell walls of *Sclerotinia sclerotiorum*. *Transactions of the British Mycological Society* 54(3), 351–360, IN5-IN6. DOI: 10.1016/S0007-1536(70)80148-6.

Kamara, A.M., Aboshosha, S.S. and Saeed, F.A. (1991) Effects of pathogenesis of *Sclerotinia sclerotiorum* on soluble protein in infected sunflower tissues. *Asscut Journal of Agriculture Science* 22, 65–80.

Kars, I. and van Kan, J.A.L. (2004) Extracellular enzymes and metabolites involved in pathogenesis of *Botrytis*. In: Elad, Y., Williamson, B., Tudzynski, P. and Delen, N. (eds) *Botrytis: Biology, Pathology and Control*. Kluwer, Dordrecht, The Netherlands, pp. 99–118. DOI: 10.1007/978-1-4020-2626-3_7.

Kars, I., Krooshof, G.H., Wagemakers, L., Joosten, R., Benen, J.A.E. *et al.* (2005) Necrotizing activity of five *Botrytis cinerea* endopolygalacturonases produced in *Pichia pastoris*. *The Plant Journal* 43(2), 213–225. DOI: 10.1111/j.1365-313X.2005.02436.x.

Kasza, Z., Vagvölgyi, C., Févre, M. and Cotton, P. (2004) Molecular characterization and in planta detection of *Sclerotinia sclerotiorum* endopolygalacturonase genes. *Current Microbiology* 48(3), 208–213. DOI: 10.1007/s00284-003-4166-6.

Keller, H. (1997) Metabolic pathway gene clusters in filamentous fungi. *Fungal Genet Biology* 21, 17–29.

Keller, N.P. and Hohn, T.M. (1997) Metabolic pathway gene clusters in filamentous fungi. *Fungal Genetics and Biology* 21(1), 17–29. DOI: 10.1006/fgbi.1997.0970.

Keon, J.P.R., Byrde, R.J.W. and Cooper, R.M. (1987) Some aspects of fungal enzymes that degrade plant cell walls. In: Pegg, G.F. and Ayres, P.G. (eds) *Fungal Infection of Plants*. Cambridge University Press, Cambridge, UK, pp. 133–157.

Khaldi, N., Seifuddin, F.T., Turner, G., Haft, D., Nierman, W.C. *et al.* (2010) SMURF: Genomic mapping of fungal secondary metabolite clusters. *Fungal Genetics and Biology* 47(9), 736–741. DOI: 10.1016/j.fgb.2010.06.003.

Khare, K.B. and Bompeix, G. (1976) Activities proteolytiques des *Sclerotinia sclerotiorum* et *S. minor*: Role possible loss de la pathogenese. *Review of Mycology* 40, 65–84.

Kritzman, G., Chet, I. and Henis, Y. (1977) The role of oxalic acid in the pathogenic behavior of *Sclerotium rolfsii* Sacc. *Experimental Mycology* 1(4), 280–285. DOI: 10.1016/S0147-5975(77)80003-0.

Kurian, P. and Stelzig, D.A. (1979) The synergistic role of oxalic acid and endopolygalacturonase in bean leaves infected by *Cristulariella pyramidalis*. *Phytopathology* 69(12), 1301. DOI: 10.1094/Phyto-69-1301.

Kusch, S., Larrouy, J., Ibrahim, H.M.M., Mounichetty, S., Gasset, N. *et al.* (2022) Transcriptional response to host chemical cues underpins the expansion of host range in a fungal plant pathogen lineage. *The ISME Journal* 16(1), 138–148. DOI: 10.1038/s41396-021-01058-x.

Li, R., Rimmer, R., Buchwaldt, L., Sharpe, A.G., Séguin-Swartz, G. *et al.* (2004) Interaction of *Sclerotinia sclerotiorum* with *Brassica napus*: Cloning and characterization of endo- and exo-polygalacturonases expressed during saprophytic and parasitic modes. *Fungal Genetics and Biology* 41(8), 754–765. DOI: 10.1016/j.fgb.2004.03.002.

Lumsden, R.D. (1969) *Sclerotinia sclerotiorum* infection of bean and the production of cellulose. *Phytopathology* 59, 653–657.

Lumsden, R.D. (1970) Phosphatidase of *Sclerotinia sclerotiorum* produced in culture and in infected bean. *Phytopathology* 60(7), 1106. DOI: 10.1094/Phyto-60-1106.

Lumsden, R.D. (1972) Localization of pH changes in beans hypocotyls infected with *Sclerotinia sclerotiorum*. *Phytopathology* 62, 774.

Lumsden, R.D. (1975) Production in culture of *Sclerotinia sclerotiorum* hyphae similar to infection hyphae in bean. *Proceedings of American Phytopathological Society* 2, 125.

Lumsden, R.D. (1976) Pectolytic enzymes of *Sclerotinia sclerotiorum* and their localization in infected bean. *Canadian Journal of Botany* 54(23), 2630–2641. DOI: 10.1139/b76-283.

Lumsden, R.D. (1979) Histology and physiology of pathogenesis in plant diseases caused by *Sclerotinia species*. *Phytopathology* 69(8), 890. DOI: 10.1094/Phyto-69-890.

Lumsden, R.D. and Dow, R.L. (1973) Histopathology of *Sclerotinia sclerotiorum* infection of bean. *Phytopathology* 63(6), 708. DOI: 10.1094/Phyto-63-708.

Magro, P., Marciano, P. and Lenna, P. (1984) Oxalic acid production and its role in pathogenesis of *Sclerotinia sclerotiorum*. *FEMS Microbiology Letters* 24(1), 9–12. DOI: 10.1111/j.1574-6968.1984.tb01234.x.

Marciano, P., Di Lenna, P. and Magro, P. (1983) Oxalic acid, cell wall-degrading enzymes and pH in pathogenesis and their significance in the virulence of two *Sclerotinia sclerotiorum* isolates on sunflower. *Physiological Plant Pathology* 22(3), 339–345. DOI: 10.1016/S0048-4059(83)81021-2.
Martel, M.B., Hervé du Penhoat, C., Létoublon, R. and Fèvre, M. (2002) Purification and characterization of a glucoamylase secreted by the plant pathogen *Scierotinia sclerotiorum*. *Canadian Journal of Microbiology* 48(3), 212–218. DOI: 10.1139/w02-011.
Maxwell, D.P. (1973) Oxalate formation in *Whetzelinia sclerotiorum* by oxalo-acetate acetyl hydrolase. *Physiological Plant Pathology* 3(2), 279–288. DOI: 10.1016/0048-4059(73)90090-8.
Maxwell, D.P. and Lumsden, R.D. (1970) Oxalic acid production by *Sclerotinia sclerotiorum* in infected bean and in culture. *Phytopathology* 60(9), 1395. DOI: 10.1094/Phyto-60-1395.
McDonagh, A., Fedorova, N.D., Crabtree, J., Yu, Y., Kim, S. *et al.* (2008) Sub-telomere directed gene expression during initiation of invasive aspergillosis. *PLoS Pathogens* 4(9), e1000154. DOI: 10.1371/journal.ppat.1000154.
Medema, M.H., Takano, E. and Breitling, R. (2013) Detecting sequence homology at the gene cluster level with multigeneblast. *Molecular Biology and Evolution* 30(5), 1218–1223. DOI: 10.1093/molbev/mst025.
Morrall, R.A.A., Duczek, L.J. and Sheard, J.W. (1972) Variations and correlations within and between morphology, pathogenicity, and pectolytic enzyme activity in *Sclerotinia* from Saskatchewan. *Canadian Journal of Botany* 50(4), 767–786. DOI: 10.1139/b72-095.
Newton, H.K. (1972) *Sclerotinia sclerotiorum* an incitant of lettuce drop: Sources of inoculum, host resistance and mechanism of pathogenesis. PhD. thesis, Murofolm 72-V.292, University of Murofilm, Ann Arbor, MI.
Newton, H.K., Maxwell, D.P. and Sequeira, L. (1973) A conductivity assay for measuring virulence of *Sclerotinia sclerotiorum*. *Phytopathology* 63(3), 424–428. DOI: 10.1094/Phyto-63-424.
Noyes, R.D. and Hancock, J.G. (1981) Role of oxalic acid in the Sclerotinia wilt of sunflower. *Physiological Plant Pathology* 18(2), 123–132. DOI: 10.1016/S0048-4059(81)80033-1.
Overell, B.T. (1952) A toxin in culture filtrates of *Sclerotinia sclerotiorum*. *Australian Journal of Science* 14, 197–198.
Palmer, J.M. and Keller, N.P. (2010) Secondary metabolism in fungi: Does chromosomal location matter? *Current Opinion in Microbiology* 13(4), 431–436. DOI: 10.1016/j.mib.2010.04.008.
Pawlowski, S.H. and Hawn, E.J. (1964) Host–parasite relationship in sunflower wilt incited by *Sclerotinia sclerotiorum* as determined by the twin technique. *Phytopathology* 54, 33–35.
Pedras, M.S.C. and Ahiahonu, P.W.K. (2004) Phytotoxin production and phytoalexin elicitation by the phytopathogenic fungus *Sclerotinia sclerotiorum*. *Journal of Chemical Ecology* 30(11), 2163–2179. DOI: 10.1023/b:joec.0000048781.72203.6c.
Philpott, C.C. (2006) Iron uptake in fungi: A system for every source. *Biochimica et Biophysica Acta (BBA)—Molecular Cell Research* 1763(7), 636–645. DOI: 10.1016/j.bbamcr.2006.05.008.
Pierson, P.E. and Rhodes, L.H. (1992) Effect of culture medium on the production of oxalic acid by *Sclerotinia trifoliorum*. *Mycologia* 84(3), 467–469. DOI: 10.1080/00275514.1992.12026161.
Poussereau, N., Creton, S., Billon-Grand, G., Rascle, C. and Fevre, M. (2001) Regulation of acp1, encoding a non-aspartyl acid protease expressed during pathogenesis of *Sclerotinia sclerotiorum*. *Microbiology* 147(Pt 3), 717–726. DOI: 10.1099/00221287-147-3-717.
Prior, G.D. and Owen, J.H. (1964) Pathological anatomy of *Sclerotinia trifoliorum* on clover and alfalfa. *Phytopathology* 54, 784–787.
Purdy, L.H. (1958) Some factors affecting penetration and infection by *Sclerotinia sclerotiorum*. *Phytopathology* 48, 605–609.
Raffaele, S. and Kamoun, S. (2012) Genome evolution in filamentous plant pathogens: Why bigger can be better. *Nature Reviews Microbiology* 10(6), 417–430. DOI: 10.1038/nrmicro2790.
Rai, J.N. and Dhawan, S. (1976) Studies on purification on identification of toxic metabolite produced by *Sclerotinia sclerotiorum* causing white rot disease of crucifers. *Indian Phytopathology* 29, 407–411.
Rask, L., Andréasson, E., Ekbom, B., Eriksson, S., Pontoppidan, B. *et al.* (2000) Myrosinase. Gene family evolution and herbivore defense in Brassicaceae. *Plant Molecular Biology* 42(1), 93–114. DOI: 10.1023/A:1006380021658.
Reymond, P., Deléage, G., Rascle, C. and Fèvre, M. (1994) Cloning and sequence analysis of a polygalacturonase-encoding gene from the phytopathogenic fungus *Sclerotinia sclerotiorum*. *Gene* 146(2), 233–237. DOI: 10.1016/0378-1119(94)90298-4.

Riou, C., Freyssinet, G. and Fevre, M. (1991) Production of cell wall-degrading enzymes by the phytopathogenic fungus *Sclerotinia sclerotiorum*. *Applied and Environmental Microbiology* 57(5), 1478–1484. DOI: 10.1128/aem.57.5.1478-1484.1991.

Riou, C., Freyssinet, G. and Fevre, M. (1992) Purification and characterization of extracellular pectinolytic enzymes produced by *Sclerotinia sclerotiorum*. *Applied and Environmental Microbiology* 58(2), 578–583. DOI: 10.1128/aem.58.2.578-583.1992.

Rollins, J.A. and Dickman, M.B. (2001) pH signaling in *Sclerotinia sclerotiorum*: Identification of a *pacC/RIM1* homolog. *Applied and Environmental Microbiology* 67(1), 75–81. DOI: 10.1128/AEM.67.1.75-81.2001.

Rowe, D.E. and Welty, R.E. (1984) Indirect selection for resistance to Sclerotinia crown and stem rot on alfalfa. *Canadian Journal of Plant Science* 64(1), 145–150. DOI: 10.4141/cjps84-018.

Rowe, D.E. (1993) Oxalic acid effects in exudates of *Sclerotinia trifoliorum* and *S. sclerotiorum* and potential use in selections. *Crop Science* 33, 1146–1149.

Rollins, J.A. (2003) The *Sclerotinia sclerotiorum pac1* gene is required for sclerotial development and virulence. *Molecular Plant-Microbe Interactions* 16(9), 785–795. DOI: 10.1094/MPMI.2003.16.9.785.

Scharf, D.H., Heinekamp, T. and Brakhage, A.A. (2014) Human and plant fungal pathogens: The role of secondary metabolites. *PLoS Pathogens* 10(1), e1003859. DOI: 10.1371/journal.ppat.1003859.

Scott-Craig, J.S., Panaccione, D.G., Cervone, F. and Walton, J.D. (1990) Endopoly-galacturonase is not required for pathogenicity of *Cochliobolus carbonum* on maize. *The Plant Cell* 2(12), 1191–1200. DOI: 10.1105/tpc.2.12.1191.

Seifbarghi, S., Borhan, M.H., Wei, Y., Coutu, C., Robinson, S.J. *et al.* (2017) Changes in the *Sclerotinia sclerotiorum* transcriptome during infection of *Brassica napus*. *BMC Genomics* 18(1), 266. DOI: 10.1186/s12864-017-3642-5.

Sexton, A.C., Minic, Z., Cozijnsen, A.J., Pedras, M.S.C. and Howlett, B.J. (2009) Cloning, purification and characterisation of brassinin glucosyltransferase, a phytoalexin-detoxifying enzyme from the plant pathogen *Sclerotinia sclerotiorum*. *Fungal Genetics and Biology* 46(2), 201–209. DOI: 10.1016/j.fgb.2008.10.014.

Sharma, R.C. and Sharma, S.L. (1984) Studies on pectolytic enzymes associated with stalk rot of cauliflower caused by *Sclerotinia sclerotiorum* (Lib) de Bary. *Indian Journal of Plant Pathology* 2, 69–72.

Sharma, S.L., Sharma, R.C. and Sharma, I. (1983) Cellulase activity of *Sclerotinia sclerotiorum* causing stalk rot of cauliflower. *Indian Journal of Mycology and Plant Pathology* 13, 286–289.

Shieh, M.T., Brown, R.L., Whitehead, M.P., Cary, J.W., Cotty, P.J. *et al.* (1997) Molecular genetic evidence for the involvement of a specific polygalacturonase, P2c, in the invasion and spread of *Aspergillus flavus* in cotton bolls. *Applied and Environmental Microbiology* 63(9), 3548–3552. DOI: 10.1128/aem.63.9.3548-3552.1997.

Smith, V.L., Punja, Z.K. and Jenkins, S.F. (1986) A histological study of infection of host tissue by *Sclerotium rolfsii*. *Phytopathology* 76(8), 755. DOI: 10.1094/Phyto-76-755.

Stone, H.E. and Armentrout, V.N. (1985) Production of oxalic acid by *Sclerotinia cepivorum* during infection of onion. *Mycologia* 77, 526–530.

Tariq, V.N. and Jeffries, P. (1985) Changes occurring in chloroplasts of *Phaseolus* following infection by *Sclerotinia*: A cytochemical study. *Journal of Cell Science* 75, 195–205. DOI: 10.1242/jcs.75.1.195.

Ten Have, A., Mulder, W., Visser, J. and Kan, J.A. (1998) The endopolygalacturonase gene *Bcpg1* is required for full virulence of *Botrytis cinerea*. *Molecular Plant-Microbe Interactions* 11(10), 1009–1016. DOI: 10.1094/MPMI.1998.11.10.1009.

Thatcher, F.S. (1942) Further studies of osmotic and permeability relations in parasitism. *Canadian Journal of Research* 20(5), 283–311. DOI: 10.1139/cjr42c-027.

Thomma, B.P., Penninckx, I.A., Cammue, B.P. and Broekaert, W.F. (2001) The complexity of disease signaling in *Arabidopsis*. *Current Opinion in Immunology* 13(1), 63–68. DOI: 10.1016/S0952-7915(00)00183-7.

Tiku, A.R. (2020) Antimicrobial compounds (phytoanticipins and phytoalexins) and their role in plant defense. In: Merillon, J.M. and Ramawat, K. (eds) *Co-Evolution of Secondary Metabolites*. Springer, New York, pp. 845–868.

Tu, J.C. (1985) Tolerance of white bean (*Phaseolus vulgaris*) to white mold (*Sclerotinia sclerotiorum*) associated with tolerance to oxalic acid. *Physiological Plant Pathology* 26(1), 111–117. DOI: 10.1016/0048-4059(85)90034-7.

Tu, J.C. (1989) Oxalic acid induced cytological alterations differ in beans tolerant or susceptible to white mould. *The New Phytologist* 112(4), 519–525. DOI: 10.1111/j.1469-8137.1989.tb00346.x.

Tu, J.C. and Beversdorf, W.D. (1982) Tolerance to white mold (*Sclerotinia sclerotiorum*) (Lib) de Bary in Ex. Rico. 23, a cultivar of white bean (*Phaseolus vulgaris* L). *Rico* 23, 65–69. DOI: 10.4141/cjps82-010.

Vanden Berg, L. and Yang, S.M. (1969) Effect of relative humidity on production of extra-cellular pectolytic enzymes by *Botrytis cinerea* and *Sclerotinia sclerotiorum*. *Canadian Journal of Botany* 47, 1007–1010.

Vesth, T.C., Brandl, J. and Andersen, M.R. (2016) Fun Gene Cluster S: Predicting fungal gene clusters from genome and transcriptome data. *Synthetic and Systems Biotechnology* 1(2), 122–129. DOI: 10.1016/j.synbio.2016.01.002.

Wagner, F., Kusserow, H. and Schäfer, W. (2000) Cloning and targeted disruption of two polygalacturonase genes in *Penicillium olsonii*. *FEMS Microbiology Letters* 186(2), 293–299. DOI: 10.1111/j.1574-6968.2000.tb09120.x.

Waksman, G., Keon, J.P. and Turner, G. (1991) Purification and characterization of two endopolygalacturonases from *Sclerotinia sclerotiorum*. *Biochimica et Biophysica Acta* 1073(1), 43–48. DOI: 10.1016/0304-4165(91)90180-o.

Wang, S.C. and Le Tourneau, D. (1971) Carbon sources, growth, sclerotium formation and carbohydrate composition of *Sclerotinia sclerotiorum*. *Archives of Microbiology* 80, 219–233.

Wang, Z., Ma, L.Y., Li, X., Zhao, F.Y., Sarwar, R. *et al.* (2020) Genome-wide identification of the NPR1-like gene family in *Brassica napus* and functional characterization of *bnanpr1* in resistance to *Sclerotinia sclerotiorum*. *Plant Cell Reports* 39(6), 709–722. DOI: 10.1007/s00299-020-02525-z.

Weber, T., Blin, K., Duddela, S., Krug, D., Kim, H.U. *et al.* (2015) Anti SMASH 3.0—A comprehensive resource for the genome mining of biosynthetic gene clusters. *Nucleic Acids Research* 43(W1), W237–W243. DOI: 10.1093/nar/gkv437.

Willetts, H.J. (1978) Sclerotium formation. In: Smith, J.E. and Berry, D.R. (eds) *The Filamentous Fungi*, Vol. 3. Wiley, New York, pp. 297–213.

Wojtaszek, P. (1997) Oxidative burst: An early plant response to pathogen infection. *Biochemistry Journal* 322(Pt 3), 681–692. DOI: 10.1042/bj3220681.

Wong, J.A.L. and Willetts, H.J. (1974) Polyacrylamide-gel electrophoresis of enzymes during morphogenesis of sclerotia of *Sclerotinia sclerotiorum*. *Journal of General Microbiology* 81(1), 101–109. DOI: 10.1099/00221287-81-1-101.

Xu, L., Xiang, M., White, D. and Chen, W. (2015) pH dependency of sclerotial development and pathogenicity revealed by using genetically defined oxalate-minus mutants of *Sclerotinia sclerotiorum*. *Environmental Microbiology* 17(8), 2896–2909. DOI: 10.1111/1462-2920.12818.

Yang, G., Tang, L., Gong, Y., Xie, J., Fu, Y. *et al.* (2018) A cerato-platanin protein SsCP1 targets plant PR1 and contributes to virulence of *Sclerotinia sclerotiorum*. *The New Phytologist* 217(2), 739–755. DOI: 10.1111/nph.14842.

Yang, Y., Jifon, X., Jiao, D., Yuheng, Y., Chaowei, B. *et al.* (2016) Disruption of the gene encoding endo-β-1, 4-xylanase affects the growth and virulence of *Sclerotinia sclerotiorum*. *Frontiers in Microbiology* 7, 1787.

Zuppini, A., Navazio, L., Sella, L., Castiglioni, C., Favaron, F. *et al.* (2005) An endopolygalacturonase from *Sclerotinia sclerotiorum* induces calcium-mediated signaling and programmed cell death in soybean cells. *Molecular Plant-Microbe Interactions* 18(8), 849–855. DOI: 10.1094/MPMI-18-0849.

10 Future Research Priorities of *Sclerotinia* Biology and Ecology

Abstract

Brassica production and productivity all over the globe need continuous fortification of research priorities with the latest approaches and multi-omics technology to manage diseases by hemibiotrophs such as *Sclerotinia*. The sequencing of *Brassica-Sclerotinia* host-pathosystem genomes with a next-generation sequencing approach will allow new discoveries and understanding of the genetic relationship with the host defence system and the virulence nature of the pathogen. Due to individual diversity, not all *R*-genes are present in the sequenced *Brassica* accessions; nonetheless, the application of pangenomics enables gene mining across a larger platform. To boost resistance, more *R*-genes should be found; hence, these efforts should be maintained. The creation of broad-spectrum disease-resistant crops may be aided by the engineering of known *R*-gene promoters to identify several effectors. The majority of effectors and avirulence genes are yet unidentified at the molecular level. A thorough understanding of the molecular mechanisms behind plant resistance in seedlings and adults, including epigenetics, is necessary. Strong cultivars of *Brassica* crops that are resistant to *Sclerotinia* can be developed with the use of the processes underpinning host defence *R*-genes against several diseases and their functional traits, such as signaling molecules.

10.1 Introduction

The stem rot of crucifers incited by *Sclerotinia sclerotiorum* was established as the major species of the *Sclerotinia* genus in 1945, which now has the widest host range, infecting over 900 species of plants all over the globe. Cruciferous oilseed crops are greatly affected by the third most destructive pathogen in the *Brassica* crop group, leading to significant losses in yield and quality. This hemibiotrophic pathogen employs both biotrophic and necrotrophic infection tactics to successfully attack and thrive within cruciferous plants. The dual lifestyle of this pathogen often dictates the host defence strategy and manipulates it for successful colonisation. There have been no reports of robust single-gene host resistance to this pathogen, which complicates the enhancement of resistance through traditional breeding techniques. Omics technologies, including genomics, proteomics, transcriptomics, effectomics, secretomics, interactomics, and metabolomics, provide valuable insights into the interactions between hosts and pathogens. By utilizing integrated data from multi-omics and systems biology, we can develop *Sclerotinia* disease-resistant *Brassica* crops with enhanced quality through a comprehensive, focused, and precise approach. During the 77 years since its establishment as a serious pathogen of crops, enormous data have been generated on all aspects of its effects on crops and the world economy, but there is still a lot of scope to strengthen research areas to manage this notorious soilborne pathogen at the global level. In the present century, global climate change has aggravated the severity of the disease, which fluctuates and needs special attention to search for climate-resilient resistant genes for breeding durable, high-yielding *Brassica* cultivars through a gene pyramiding approach.

10.2 Research Priorities

In biological science, especially in plant sciences, research is a continuous process to fortify the production system through improved tools and approaches to achieve quality crop commodities for a better economy and health of the human population. Strengthening the following areas of research will further improve the *Brassica*

DOI: 10.1079/9781800629592.0010

disease management system at large. Crucifers researchers may improve their research achievements by their efforts to get a clearer vision of future challenges of more demand of quality products and the threat of adverse weather fluctuations.

10.2.1

Establishing a global database for *Sclerotinia* through the creation of an international research network is essential to resolving the nomenclature problem with *R*-genes and pathotypes. Due to the possibility of changes in host cultivars and diseases, such as the decline of cultivar resistance or changes in pathogenicity and virulence, it is imperative to maintain an up-to-date global database.

10.2.2

The ability to study genes within the resistance-regulating network has improved due to developments in molecular and omics techniques, including transcriptomics, proteomics, and metabolomics. These approaches can be directly applied to resistance breeding or used in pre-breeding studies to enhance our comprehension of the interactions that occur between pathogens and *Brassica*. Due to individual variances, it is significant to highlight that not all *R*-genes are present in the sequenced *Brassica* accessions. On the other hand, by offering a larger platform, the development of pangenomes may broaden the use of gene mining.

10.2.3

Improved knowledge of the molecular interactions between infections and hosts may provide insightful information for managing disease. Extensive research has been conducted on the connections between proteins in the *Brassica*–pathogen system. According to recent research, TuRB01/01b/04's viral avirulence is largely determined by the cytoplasmic inclusion protein, while TuRB03/05's avirulence is determined by P3.

10.2.4

While many avirulence or interaction genes in the pathogens are still unknown, over ten *R*-genes have been found so far in oilseed *Brassica* crops. Thus, the focus now needs to be on creating international collections of pathogen isolates in order to pinpoint crucial avirulence or interaction genes. Furthermore, it would be advantageous for the host to generate a sequence of introgression lines (ILs) through successive backcrossing in order to 'Mendalize' the quantitative loci.

10.2.5

The importance of certain effector proteins has been highlighted by recent studies, although many of them are still poorly understood in terms of their molecular targets and functions. Molecular identification of most effector/avirulence genes remains necessary in the proposed gene-for-gene strategy. Additionally, it is expected that ongoing developments in molecular and biological research aimed at clarifying the functions of various isoforms of a single gene will make it easier to identify potential isoforms that shield cruciferous plants against infections. A comprehensive strategy for data production is required in order to gain a deeper understanding of the molecular processes underlying plant resistance in adults and seedlings.

10.2.6

The research on developments in sequencing technology is quite appreciable, and more sequences of various *Sclerotinia* isolates, including several isolates from each race and more sequences of isolates of closely related pathotypes, could be accessible shortly. Multiple sequence analysis will provide more information about the important genes, including effectors that are pathogenicity related. The function and evolutionary history of these genes will become clearer when certain genes are present or absent, as well as when isolates differ from one another. Further examination of these genes' functions can confirm their role in plant–pathogen interactions.

10.2.7

During the biotrophic stage of plants, the effector proteins ChELP1 and ChELP2 are essential for inhibiting chitin-triggered immune responses. To guarantee efficient substrate penetration, these proteins' basal expression levels must be maintained in plants and appressoria *in vitro*. To fully comprehend these proteins' unique roles in appressorial activity and the shift to invasive hyphal development, more investigation is required.

10.2.8

The absence of chromosome 11 results in the termination of infection during biotrophy, suggesting that several genes located on chromosome 11 play crucial roles in manipulating the responses of plant hosts. These genes can be specifically examined in future studies to assess their potential functions.

10.2.9

Because of *Sclerotinia's* brief biotrophic phase and subsequent transition to the necrotrophic stage, it is extremely difficult to detect and ascribe temporal, geographical, and tissue-specific roles to virulence factors. However, there is hope that the numerous mysteries surrounding *S. sclerotiorum* will be solved by the expected advancements in cell biology and the creation of biochemical instruments specifically designed for this fungus.

10.2.10

To tackle the difficult subject of how the host's signaling network reacts to various stresses, more research is needed. It is anticipated that promising developments in nanoscale imaging methods and proteomics strategies aimed at dissecting protein complexes will improve our comprehension of the underlying mechanisms. In particular, these developments will clarify the critical function that a solitary Ca^{2+} sensor plays in coordinating many signaling pathways, allowing plants to successfully respond to a wide range of environmental stressors.

10.2.11

In order to reduce the disparities in transgenic plant tolerance levels between laboratory and field conditions, a comprehensive understanding of the intricate plant stress-response system is required.

10.2.12

More investigation is needed to comprehend how calmodulin-binding transcription activators (*CAMTA*) and *WRKY* genes contribute to *Brassica* species' resistance against *Sclerotinia*.

10.2.13

More research work is required to learn about the fitness cost of multiple disease resistance in *Brassica*, especially with respect to linkage disequilibrium and confused pleiotropy. More investigations are required in respect to ecological and physiological research.

10.2.14

A national and worldwide collaborative network system comprising institutes and scientists is urgently needed. To make sure that the produced varieties are adequately assessed against all pathogen races and pathotypes, this system might be put into place gradually.

10.2.15

The generation of comprehensive data is crucial for understanding the intricate molecular mechanisms underlying seedling and adult plant resistance.

10.2.16

The *Brassica*–pathogen interaction can be investigated using the CRISPR-Cas9 system, which is typified by palindromic repeats that are clustered and consistently spaced. This allows for the functional investigation of candidate genes. Two particular genes, *BnWRKY11* and *BnWRKY70*, which encode WRKY transcription factors in the case of *B. napus*, were successfully broken using CRISPR-Cas9. This disruption indicates that the latter gene controls disease resistance to *S. sclerotiorum*. But additional data need to be generated in order to have a more comprehensive understanding. Since polyploid crops like *Brassica* have several copies of the same gene as homeologs in their genomes, making them suitable subjects for gene functional research, more information is needed for this complex gene knockout study.

10.2.17

In *Brassica rapa*, the early blooming trait's genome has been successfully modified by the application of CRISPR-Cas9 technology, and further improvement is required.

10.2.18

A thorough examination of the vast amounts of information gathered over time about the expression of the molecular mechanisms underlying the interaction between hosts and pathogens, as well as the characterisation of potential genes, is essential. The engineering of *Brassica* crops resistant to disease will be made possible by this analysis for development of highly yielding varieties.

10.2.19

The creation of robust, resistant cultivars of *Brassica* crops can be facilitated by the functional qualities of signaling molecules and the mechanisms of multipathogen and multipathotype host defence *R*-genes.

10.2.20

For the principal pathosystems in *Brassica*, the knowledge of controlling pathogen infections at multiple developmental phases of the host, such as seedlings and adults, as well as across different plant organs, such as leaves, stems, and pods, has become more accessible. This is important because plant productivity can be negatively impacted by pathogen infection and colonisation, especially when *S. sclerotiorum* affects stems. By expanding our knowledge of the mechanisms governing pathogen infection and colonisation life stages as well as the host's developmental stages and their interactions, we can open up new avenues for improving the comprehension and putting into practice practical controls to manage pathogen infections and lessen ensuing yield losses.

10.2.21

Different *Brassica* host–pathogen systems provide an opportunity to explore the documentation and inheritance of *R*-genes, which govern resistance in both seedling and adult plants. Additionally, the sensitivity of these genes to temperature, light, and soil conditions can also be investigated.

10.2.22

High-throughput sequencing data-based mapping techniques have advanced quickly, particularly in the field of SNP-based methods like Genome-wide association study (GWAS), Bulked segregant analysis with whole-genome sequencing (BSA seq), KASP (Kompetitive allele specific PCR) markers, and microarrays.

With GWAS, genes linked to target traits in natural populations can be quickly and precisely identified by analysing changes across the genome and their correlation with trait values. Additionally, KASP technology offers remarkable assay precision and dependability, saving a substantial amount of money and time. These approaches demand further research on diseases that are *Brassica* hemibiotrophs, like *Sclerotinia*.

Subject Index

Printed and bound by CPI Group (UK) Ltd, Croydon, CR0 4YY

13/09/2026

14961084-0006